Avian Medicine and Surgery in Practice

This book is an invaluable quick reference resource for clinicians and a useful study guide for veterinary students and residents.

This practical and beautifully illustrated book takes you on a walk through avian medicine, beginning with the basics, such as common species, husbandry, nutrition, and grooming, before moving on to the more clinical-focused material, such as how to perform a physical examination, collect samples and interpret diagnostic tests, take good quality radiographs and then interpret them, and how to support a sick bird while in hospital.

It then moves into the disorders affecting the different organs and body systems. Each chapter begins with a description of clinical anatomy and physiology before going on to explore the disorders associated with the effects of disease or trauma. The book also addresses behavioural problems, incubation of eggs, paediatrics, anaesthesia, and surgery.

For this edition, Bob Doneley is joined by avian, zoo/wildlife and exotic pet veterinarian Shangzhe Xie. Together they have thoroughly updated the book by incorporating the latest research and covering a larger range of avian species. Building on the practical aspects that make this book stand out from other resources, they have added step-by-step pictorial guides to common procedures and surgeries, as well as videos for longer, more complicated ones. *Avian Medicine and Surgery in Practice* will therefore allow the practitioner to immediately help the patient at hand, whether to prescribe a medication or perform a life-saving surgery.

Avian Medicine and Surgery in Practice

Companion and Aviary Birds

THIRD EDITION

BOB DONELEY AND SHANGZHE XIE

CRC Press
Taylor & Francis Group
Boca Raton London New York

CRC Press is an imprint of the
Taylor & Francis Group, an **informa** business

Cover Image Credit: David Tan, Mandai Wildlife Group, Singapore

Third edition published 2025
by CRC Press
2385 NW Executive Center Drive, Suite 320, Boca Raton FL 33431

and by CRC Press
4 Park Square, Milton Park, Abingdon, Oxon, OX14 4RN

CRC Press is an imprint of Taylor & Francis Group, LLC

ISBN: 9781032224220 (hbk)
ISBN: 9781032224237 (pbk)
ISBN: 9781003272519 (ebk)

DOI: 10.1201/9781003272519

Typeset in Janson
by Deanta Global Publishing Services, Chennai, India

Access the Support Material: www.routledge.com/9781032224220

MIX
Paper | Supporting responsible forestry
FSC® C013056
FSC www.fsc.org

Printed and bound in Great Britain by
TJ Books, Padstow, Cornwall

CONTENTS

ABOUT THE AUTHORS

Dr Bob Doneley graduated from the University of Queensland in 1982 and worked in veterinary practices in Bundaberg, Brisbane, Toowoomba, and the UK before opening his own practice, the West Toowoomba Veterinary Surgery, in Toowoomba in 1988.

He was awarded the Fellowship of the Australian and New Zealand College of Veterinary Scientists (FANZCVS) in 2003, becoming Queensland's first specialist in bird medicine, the third in Australia. In the same year, he was awarded the College Prize by the Australian College for outstanding contributions to veterinary science in Australia.

In 2010 he sold his practice after 22 years to take up a position at the University of Queensland's Gatton campus. He is now a professor and head of the Avian and Exotic Pet Service, a specialist bird practice, as well as treating reptiles, small mammals and wildlife. In 2015 he was awarded the Meritorious Service Award by the Australian and New Zealand College of Veterinary Scientists. In 2018 he was awarded the prestigious international T.J. Lafeber Avian Practitioner of the Year by the Association of Avian Veterinarians.

He lectures to both veterinary and veterinary technology students on avian and exotic pet medicine; and has published two textbooks on bird medicine – one of which, *Avian Medicine and Surgery in Practice*, is now in its third edition and has been translated into German, Turkish, and Chinese. He has also written and edited a text on reptile medicine and surgery, as well as chapters for many other textbooks, and has published numerous papers in veterinary journals.

His crowning achievement is his marriage to Maree for 40 years and their two wonderful children, Liz and Pat.

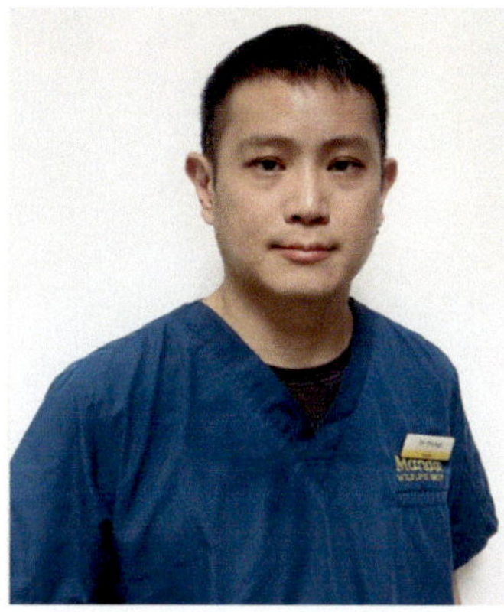

Dr Shangzhe Xie graduated from Murdoch University in 2008 and began a nomadic start to his veterinary career in small animal and exotic pet practices in Australia, New Zealand, and USA. He returned to Adelaide, South Australia, in 2013 to pursue a PhD, studying the physiology and pathology of heat stress in Australian desert birds, and founded the Adelaide Exotic Pets Referral Services.

In 2017, he moved back to Singapore to take up a position at the Jurong Bird Park, Mandai Wildlife Group. He is currently the vice president of Veterinary Healthcare at Mandai Wildlife Group, in charge of the two veterinary hospitals caring for the animals under the care of the organization. In 2020 he achieved board-certification status with the American Board of Veterinary Practitioners in avian practice and became recognized as a diplomate of the Asian College of Conservation Medicine the following year.

He continues to mentor and teach veterinary students, research students, and veterinarians through continuing education courses, veterinary student extramural rotations, internships and residencies. He has contributed to scientific publications and textbook chapters, culminating in *Avian Medicine and Surgery in Practice*, third edition.

All of these were made possible by the support of his wife, veterinary educator Dr Ji Zhen Low, and their rambunctious sons, Sean, Seth, and Jude.

PREFACE

THERE is a saying, attributed to Sir Isaac Newton (1642–1727), "If I have seen further, it is by standing on the shoulders of Giants". This is a metaphor for "discovering truth by building on previous discoveries" or "using the understanding gained by major thinkers who have gone before, in order to make intellectual progress".

I have been working with birds as a clinician now for a little over 40 years. In that time, I have seen veterinary interest in avian medicine as an almost new field, but mainly as mild background noise in the midst of veterinary practice at the time. It then really started to take off in the 1980s and 1990s, particularly with practice management consultants telling their readers that this was a service their clients wanted and would pay for. Then, as it became clear to many vets that this was a complex and demanding field, interest dropped off again – but never to the levels of the 1970s and 1980s. But now I see that interest growing with new generations of veterinarians graduating and wanting to learn.

But all through all those years of ups and downs, there was a group of veterinarians – the giants – building their knowledge and expertise around the world. They came together in the early 1980s when the Association of Avian Veterinarians was formed and passed on their knowledge to fellow enthusiasts. People like Bob Altman (USA), Walter Rosskopf (USA), Ross Perry (Australia), Susan Clubb (USA), Helga Gerlach (Germany), Margaret Petrak (USA), David Graham (USA), Greg Harrison (USA), and Alan Jones (UK) are just a few of those Giants on whose shoulders we stand. They didn't just keep the fire burning, they stoked and built it to become the mature body of science that we now call avian medicine and surgery. Knowledge in this field continues to grow, with contributions from both private practitioners and university academics. It is continually challenged by both new knowledge and a fresh view. And that is how it should be.

This book would not have been possible without the giants on whose shoulders I stand. It is written with a view to bringing together the knowledge that they gained. It is not written as a peer-reviewed textbook but rather as a text for veterinary students, new graduates, and busy clinicians to gain a solid understanding of the field and to 'get a feel' for birds and their problems. Hopefully, more experienced veterinarians will find it useful when studying for residency exams or when confronted with a difficult case.

I have been fortunate in this third edition to have Dr Shangzhe Xie, of Singapore, join me as a co-editor. Shangzhe, through his work as a senior vet, firstly at Jurong Bird Park and now at Mandai Wildlife Group, has been exposed to more bird species and conditions than many of us could imagine. Behind his laconic, almost dry manner lies a keen brain, a thirst for knowledge, and a willingness to share it. I am indeed lucky to count him as a friend and colleague.

Most importantly, no man is an island. I could not have done any of this without my wife, Maree, and our children, Liz and Pat. To them, I dedicate this book.

Bob Doneley BVSc FANZCVS (Avian Medicine)
Professor, Avian and Exotic Pet Medicine
School of Veterinary Science
The University of Queensland

I do not remember the species nor presenting signs of my first avian patient in practice. What I do remember was being stuck and not knowing what to do, simply because it had wings rather than legs and feathers rather than fur. I also remember looking to the internet for help, and by some stroke of luck, coming across Bob Doneley's profile and contact details. I proceeded to email Bob for help and he replied to me at an ungodly hour (Australian Eastern Standard Time) with an unexpected level of detail. That advice was a perfect balance between evidence-based theoretical information and practical steps to directly help the patient. This was the stage upon which my avian veterinary career was set, and the textbooks *Avian Medicine and Surgery in Practice*, first and second editions, became my go-to when I needed guidance.

Over the years, Bob grew to be a friend and mentor, and we got to spend time together in Singapore for a couple of weeks before COVID-19 changed the way everyone lived. It came as a surprise when Bob asked if I would like to work on the third edition together, and it was not something I could turn down, even if it seemed daunting at the time. Avian medicine and surgery have been advancing at an unprecedented rate, with new discoveries changing our understanding of birds every three to five years. We set out on the task of updating the book based on this new information, but more importantly, we wanted to build on the practical aspects of the book, which was what set it apart from other textbooks on the topic. The content has been reorganised so that the anatomical and physiological theory is placed at the start of each organ-based chapter, which then leads on to the disorders associated with disruption in the normal anatomy and physiology of each organ. Furthermore, we wanted anyone with the book to be able to use it to immediately help the patient at hand, whether it is to prescribe a medication that will provide immediate relief to the problem or a life-saving surgery. It is with this purpose that we added step-by-step pictorial guides to common procedures and surgeries, as well as videos for longer, more complicated ones.

Ultimately, Bob and I are just two veterinarians working to improve the lives of millions of birds, and we hope that more veterinarians become comfortable and competent at caring for avian patients, so that more bird lives can be saved and improved. Bob humbly omitted himself from the list of giants, but his shoulders are definitely one of those my generation of veterinarians and beyond will be standing on as we build on the work done.

My involvement with this comes on the back of the support provided by my colleagues holding the clinical fort, so I would like to thank Dr Daniel Carrasco, Dr Trent van Zanten, Dr Ellen Rasidi, Dr Gabrina Goh, Dr Chiharu Okumura, Marcus Tan, Zhi Ying Tang, Keiko Watanabe, Fauzi Awang, and Alan Tan for taking care of all the avian patients at Mandai Wildlife Group. The first two avian veterinary interns of the Mandai Wildlife Institute, Dr Serena Shum and Dr Joey Lo, also contributed to the materials found in the book. Pathologists Dr Chia-da Hsu and Dr Shin Min Chong also took many of the photos and videos added to the new edition. Last, but not least, my wife, Dr Ji Zhen Low, and children, Sean, Seth, and Jude, deserve all the credit for putting up with me and my dad jokes. My contributions to this book would not have been possible without them.

Shangzhe Xie BSc/BVMS, MVS, PhD, DABVP (Avian Practice), DACCM
Vice President, Veterinary Healthcare
Mandai Wildlife Group
Adjunct Lecturer, School of Animal and Veterinary Sciences
Adelaide University

CHAPTER 1

COMMONLY KEPT SPECIES

1

WHO KEEPS BIRDS, AND WHY?

There are many reasons for keeping birds.

- Companionship: Whether it be a senior citizen's canary or a young girl's cockatiel, birds can make great companions. Hand-reared birds are, rightly or wrongly, highly sought after for their pet qualities.
- Aviculture: The keeping and breeding of birds is not only a popular hobby around the world, it is also a high-dollar industry. The value of birds can range from under $100 to well over $100,000.
- Food: Birds have long been kept as a food source – poultry, meat pigeons, ostriches and emus, and even quail and pheasants.
- Sport and entertainment: Racing pigeons are the most obvious form of sport, but in some countries and cultures, falconry and cockfighting are other sports utilising birds (**Fig. 1.1**).

THE HISTORY OF BIRD KEEPING

Keeping wild birds is a deeply ingrained and widespread practice going back thousands of years. The Sumerians (c. 5000 BC) had a word, *subura*, for birdcage, while in the 5th century BC, Greek writers recorded that certain captive birds spoke in human voices. Alexander the Great's army returned from India in 327 BC, bringing home with them native parrots that are now known as the Alexandrine parrot (*Psittacula eupatria*).

Ancient mariners – Babylonians, Hindu merchants of the 5th century BC, Polynesians, and Vikings – often carried caged birds on long ocean journeys. When seeking land, they would release a bird and observe its flight. If the bird saw land in the distance, it would fly in that direction and not be seen again. If no land was detected, the bird would return to the ship.

Ancient Greeks kept parrots, and Romans used mockingbirds in the entryways of their homes to announce visitors. Ravens were a prized possession of Romans for their ability to talk. The Romans also had elaborate garden aviaries, and credit is given to them for bringing different varieties of birds to Great Britain and Europe.

Bird keeping was popular in China during the Qing dynasty (1636–1921), with the Manchu noblemen and eunuchs keeping pet birds as a hobby to pass the time. The earliest Japanese records on keeping small birds as pets are found from the 12th and 13th centuries, the Kamakura Era.

Falconry, the sport of hunting using raptors, began in the Middle East thousands of years ago and eventually spread to Europe. In the 13th century, the emperor Friedrich the Second (1194–1250) wrote a book on falconry, *About the Art to Hunt with Birds*.

After the fall of the Roman Empire, parrots were rare in Europe, but interest in them rose again after Christopher Columbus brought some back from his voyages to the Americas. In medieval Europe, bird keeping was mostly for the wealthy. Kings, queens, and the clergy would often keep parrots. Both King Henry VIII and Queen Marie Antoinette owned an African grey. Pope Martin V (1369–1431) appointed a Keeper of Parrots during his time in the Vatican. For hundreds of years, women's quarters in the royal courts of Europe were livened with cages housing local species of birds (chaffinches, greenfinches, siskins, and bullfinches), which could be trained to mimic a variety of melodies.

As early as 1341, Portuguese sailors visiting the Canary Islands and the neighbouring Azores and Madeira noted that the islands were inhabited by great numbers of small, plainly marked birds with extraordinary vocal abilities. These birds, now named canaries, would be popular in Europe for

DOI: 10.1201/9781003272519-1

Figure 1.1 **Bird shows are becoming popular in many zoos. Birds used in these shows have been trained using falconry techniques.**

many hundreds of years. In the 17th century, miners in the Austrian Alps valued canaries not only for their entertaining songs but also for their tendency to die more quickly than humans in the presence of toxic gases, thus serving as early warning systems – a practice which continued into the 20th century.

Following the Portuguese explorer Vasco de Gama's completion of a sea route to India in 1499, traders began transporting large numbers of parrots from Africa, India, and Java to the capitals of Europe, where they were purchased as pets by merchants and patricians. They were soon followed by American species such as macaws and Amazons.

In the 19th and 20th centuries, the popularity of pet birds crossed class lines as well as racial and ethnic barriers. Solitary caged birds were common in working-class households, whereas wealthy families often kept a variety of birds in aviaries. In the USA in the mid-1800s, decades before radio and recorded music, caged birds were among the most popular forms of home entertainment. The most desirable birds included wild European and North American species such as linnets, thrushes, nightingales, European robins, bullfinches, goldfinches, purple finches, mockingbirds, cardinals, grosbeaks, and buntings, but by far the most popular was the canary. Small neighbourhood bird stores were as common in 19-century cities and towns in America as barbershops. (Many barbershops, in fact, also sold birds.) Budgerigars imported from Australia in the mid-1800s to Europe and the USA started the widespread hobby of parrot keeping for millions of people.

COMMONLY KEPT SPECIES

Parrots

A parrot is any bird of the over 380 species within the order *Psittaciformes*. They have a short, blunt, rounded bill with a curved maxillary portion fitting over the thick mandibular portion ('hookbill'), a thick muscular tongue, and a fleshy cere at the top of the beak, around the external nares. The foot is four-toed (digits numbered medial to lateral), with digits 2 and 3 directed anteriorly and digits 1 and 4 posteriorly (zygodactyly). They are predominantly Southern Hemisphere birds, originating from Africa, India, Southeast Asia, Australia, New Zealand, the Southwest Pacific Islands, and South America. They have evolved in extremely diverse ecosystems, from jungle to desert, from the tropics to snowfields, and from rainforest to open grasslands.

Care needs to be taken with the interpretation of the following Table 1.1. Only the more commonly kept species are listed. Many of the common health problems listed can be seen in all species but may be more prevalent in a few.

Passerines

Passerines, also known as perching birds, songbirds, or softbills, consist of more than 110 families and about 6,409 known species and are found on every continent except Antarctica. They are terrestrial birds, and include canaries, finches, starlings, wrens, and even lyrebirds. Passerines can be differentiated from other orders by their toe arrangement, where three toes point forward while one toe points backward. They have a tendon on the caudal tarsometatarsus, connecting the muscle behind the tibiotarsus to the underside of the toe, which tightens and pulls automatically whenever the leg bends, therefore making the foot curl and become stiff whenever the birds perch on a branch. This characteristic allows them to sleep while they are perching without falling.

Passerines kept as pets belong to the sub-order Passeri. They have been kept as pets and cage birds for a very long time; it is known that by the 5th century BC, several varieties of songbirds such as starlings, magpies, thrushes, and nightingales were kept by the Greeks. Canaries were introduced to Europe in the 16th century from their native land in the Canary Islands and have since been bred into different varieties of domesticated birds.

Table 1.1 **Commonly kept bird species and some of their characteristics**

COMMON NAME(S)	LATIN NAME	CHARACTERISTICS	COMMON HEALTH PROBLEMS
AUSTRALIAN PARROTS			
Budgerigar (parakeet, budgie)	*Melopsittacus undulatus*	• Have been bred in captivity since the 1850s and are now one of the world's best known pet birds. • Very animated and can mimic human speech. • Has been bred from the native green bird with a yellow head into a huge range of colours and patterns from mauve, olive, and blue to pure white. • Sexually dimorphic.	• APV (French moult) • PBFD • Chlamydiosis • Cnemidocoptes • Intestinal worms • Trichomonas • Coccidiosis • Neoplasia • Polyfolliculosis • Macrorhabdus
Cockatoos			

Greater sulphur-crested cockatoo
Long-billed (slender-bill) corella
Western corella
Short-billed (little) corella
Cockatiel (quarrion, weero) (**Fig. 1.2**)
Galah (rose-breasted cockatoo)
Major Mitchell cockatoo (Leadbeater's)
Red-tailed black cockatoo
Yellow-tailed black cockatoo
White-tailed black cockatoos (2)
Glossy black cockatoo
Gang-gang cockatoo
	Cacatua galerita galerita
	Cacatua tenuirostris
	Cacatua pastinator
	Cacatua sanguinea
	Nymphicus hollandicus
	
	Eolophus roseicapilla
	Lophochroa leadbeateri
	
	Calyptorhynchus banksii
	C. funereus
	C. baudinii
	C. latirostris
	C. lathami
	Callocephalon fimbriatum
		• The crest, a tuft of feathers on the top of their heads of varying length, shape, and colour; the bird has conscious control over the position of these feathers, e.g., the crest is raised when the bird is excited, interactive, or alarmed.
		• Powder-down feathers over the thighs that produce a fine white powder that assists in keeping the bird clean.
		• A loud, often raucous call.
		• Social birds that do not do well in isolation.
		• Affectionate birds that can bond well to people.
		• Destructive.
		• Often good talkers.
		• Most are sexually dimorphic, but the corellas are monomorphic.
			• PBFD (except cockatiels)
			• Chlamydiosis
			• Sternal trauma (poor wing clips)
			• Obesity related problems:
			  • Hepatic lipidosis
			  • Atherosclerosis
			  • Lipomas
			  • Arthritis
			• Poor socialisation if hand-reared
			• Reproductive problems:
			  • Excessive egg laying
			  • Cloacal prolapse
			  • Egg binding
			  • Yolk peritonitis
			• Behavioural problems
			  • Screaming
			  • Aggression
			  • Feather-damaging behaviour
			• Xanthomas

COMMON NAME(S)	LATIN NAME	CHARACTERISTICS	COMMON HEALTH PROBLEMS
Long-tailed parrots			
Princess parrot Superb parrot (Barraband) Regent parrot	*Polytelis alexandrae* *Polytelis swainsonii* *Polytelis anthopeplus*	• Placid birds that can be kept as a companion bird, or as multiple pairs per aviary, or in a mixed collection with other placid species. • Sexually dimorphic.	• Intestinal worms • Chlamydiosis • PBFD
Lorikeets			

Rainbow lorikeet (**Fig. 1.3**)
Red-collared lorikeet
Scaly-breasted lorikeet (**Fig. 1.3**)
Musk lorikeet
Purple-crowned lorikeet
Little lorikeet
Varied lorikeet
	Trichoglossus moluccanus
	T. rubritorquis
	T. chlorolepidotus
	Glossopsitta concinna
	G. porphyrocephala
	G. pusilla
	Psitteuteles versicolor
		• Bright coloured plumage.
		• Elongated, sharp bill.
		• The tongue is tipped with elongated papillae that form a "brush-tip", allowing the extraction of nectar and pollen from flowers.
		• Sexually monomorphic.
		• Lorikeets can make difficult pets because they:
		  • Are constantly active and noisy.
		  • Require a specialised liquid diet.
		  • Produce projectile liquid droppings.
			• PBFD
			• Feather-damaging behaviour
			• Screaming
			• Haemorrhagic enteritis
			• Obesity
			• Hypovitaminosis A
			• Hypervitaminosis A (due to over-supplementation)

Neophemas			
Blue-winged parrot Elegant parrot Rock parrot Orange-bellied parrot Turquoise parrot Scarlet-chested parrot Bourke parrot (**Fig. 1.4**)	*Neophema chrysostoma* *N. elegans* *N. petrophila* *N. chrysogaster* *N. pulchella* *N. splendida* *Neopsephotus bourkii*	• Calm, quiet, placid birds. • Do well in mixed collections with other quiet species. • The most commonly kept species are the scarlet-chested parrot, the turquoise parrot, and the elegant parrot. • Several colour mutations have been selectively bred. • Usually sexually dimorphic	• Intestinal worms • Chlamydiosis • PBFD
Rosellas			
Eastern rosella Crimson rosella Yellow rosella Adelaide rosella Pale-headed rosella Green (Tasmanian) rosella Western rosella Northern rosella	*Platycercus eximius* *P. elegans* *P. elegans flaveolus* *P. elegans subadelaidae* *P. adscitus* *P. caledonicus* *P. icterotis* *P. venustus*	• The more colourful rosella species are popular as companion and aviary birds. • They can live for longer than 20 years, and they are relatively easy to breed. • All have a reputation for being aggressive in captivity and are hence recommended be kept separate from other birds. • Most are sexually monomorphic.	• Intestinal worms • Chlamydiosis • PBFD

COMMON NAME(S)	LATIN NAME	CHARACTERISTICS	COMMON HEALTH PROBLEMS
Psephotus and other similar grass parrots			
Red-rump parrot Hooded parrot Golden-shoulder parrots Blue-bonnet parrot Mulga parrot	*Psephotus haematonotus* *P. dissimilis* *P. chrysopterygius* *Northiella haematogaster* *Psephotellus varius*	• Small, usually very quiet birds. • Can be aggressive to other birds. • Several colour mutations have been developed. • Obvious sexual dimorphism in most species.	• Intestinal worms • Chlamydiosis • PBFD
Australian ringnecks			
Mallee ringneck Twenty-eight parrot Port Lincoln parrot Cloncurry parrot	*Barnardius zonarius barnadi* *B. z. semitorquatus* *B. z. zonarius* *B. z. macgillivrayi*	• Temperate, calm, and inquisitive pet. • Their energy levels are moderate, but they will still enjoy environmental enrichment. • More common as pets in the UK than in Australia and the USA. • Subtle sexual dimorphism.	• Intestinal worms • Chlamydiosis • PBFD
Other parrots			
Swift parrot Australian king parrot Red-capped parrot	*Lathamus discolor* *Alisterus scapularis* *Purpureicephalus spurius*		• Intestinal worms • Chlamydiosis • PBFD
African parrots			
Grey parrots			
Grey parrot (African grey, Congo African grey) (**Fig. 1.5**) Timneh (Timneh African grey)	*Psittacus erithacus* *P. timneh*	• Popular pets based on their intelligence and superior talking and mimicking ability. • Require socialisation and exposure to different environments to increase their adaptability to change. • Sexually monomorphic.	• Hypocalcaemic tetany • Axillary dermatitis • PBFD • Aspergillosis • Hypovitaminosis A
Poicephalus parrots			
Senegal (**Fig. 1.6**) Red-fronted parrot (Jardine's parrot) Red-bellied Rüppell's parrot Brown-necked parrot Cape parrot Meyer's parrot Yellow-fronted parrot Brown-headed parrot	*Poicephalus senegalus* *P. gulielmi* *P. rufiventris* *P. rueppellii* *P. fuscicollis* *P. robustus* *P. meyeri* *P. Flavifrons*	• Popular pets due to their small size, colourful feathering, independence, and relatively quiet nature. • They are known to make good apartment birds. • They can be territorial and unpredictable, especially during the breeding season. • Some are sexually dimorphic, others are monomorphic.	• PBFD • Adenovirus • Capillaria infections • Hypovitaminosis A

COMMON NAME(S)	LATIN NAME	CHARACTERISTICS	COMMON HEALTH PROBLEMS
Lovebirds			

Peach-faced lovebird
Masked lovebird (yellow-collared, black-masked)
Fischer's lovebird
Nyasa lovebird (Lilian's lovebird)
Black-cheeked lovebird
	Agapornis roseicollis
	A. personatus
	A. fischeri
	A. lilanae
	A. nigrigenis
		• Small (45–65 g), stocky birds with short, rounded tails.
		• There are two main groups of lovebirds:
		  • Peach-faced lovebirds.
		  • White eye-ring lovebirds (Fischer's, black-cheeked, and black-masked lovebirds).
		• All lovebirds have similar pet quality.
		• Can be aggressive and territorial with other birds, but with the right owner and environment, they can be interactive and social.
		• Sexually monomorphic.
			• PBFD
			• Polyfolliculosis
			• Patagial dermatitis

South American parrots (New World)			
Macaws			

• The giant macaws
  • Buffon macaw
  • Hyacinth macaw
• The large macaws
  • Blue and gold macaw (**Fig. 1.7**)
  • Scarlet macaw
  • Green-wing macaw
• The miniature (mini) macaws
  • Hahn's (Noble's) macaw
  • Severe (chestnut-breasted) macaw
  • Yellow-collared macaw (**Fig. 1.8**)
	
	Ara ambiguus
	
	Anodorhynchus hyacinthinus
	Ara arauna
	
	Ara macao
	
	Ara arachlorolepta
	Ara nobilis
	Ara severa
	Ara auricollis
		• Large beak and a long tail.
		• Featherless facial patch that can be completely bare or have widely spaced rows of short feathers.
		• Brightly coloured.
		• Sexually monomorphic.
			• Behavioural problems
			  • Screaming
			  • Feather-damaging behaviour
			  • Aggression
			• Scissor beak
			• Internal papillomatosis (herpesvirus)
			• Avian ganglioneuritis
			• Aspergillosis
			• Hypovitaminosis A

Amazon parrots			

Lilac-crowned Amazon
Blue-fronted Amazon
Orange-winged Amazon
Yellow-naped Amazon
Double yellow-headed Amazon
Red-lored Amazon (**Fig. 1.9**)
Hispaniolan Amazon
	Amazona finschi
	A. aestiva
	A. amazonica
	A. auropalliata
	A. oratrix
	A. autumnalis
	A. ventralis
		• Medium to large-sized, stocky birds with short, slightly rounded tails and beaks with the characteristically narrow.
		• Rhinotheca.
		• Common in the pet trade with many different species (>30) readily available.
		• Usually sexually monomorphic.
		• Independent, gregarious, and can be headstrong.
			• Behavioural problems
			  • Screaming
			  • Feather-damaging behaviour
			  • Aggression
			• Scissor beak
			• Internal papillomatosis (herpesvirus)
			• Avian ganglioneuritis
			• Hypovitaminosis A
			• Obesity issues:
			  • Hepatic lipidosis
			  • Diabetes mellitus
			  • Atherosclerosis
			  • Lipomas

COMMON NAME(S)	LATIN NAME	CHARACTERISTICS	COMMON HEALTH PROBLEMS
Pyrrhura conures			
Green-cheeked conure (**Fig. 1.10**) Maroon-bellied conure Pearly conure Black-capped conure Crimson-bellied conure	*Pyrrhura molinae* *P. frontalis* *P. lepida* *P. rupicola* *P. perlata*	• Usually small birds with scalloped edges to their chest feathers. • Often better suited for the first-time bird owner because they are sociable, adaptable, and quiet. • Sexually monomorphic.	• Egg binding • Feather-damaging behaviour • Hypovitaminosis A
Aratinga conures			

Sun conure
Jenday conure (**Fig. 1.11**)
Blue-crowned conure
Nanday conure
Gold-capped conure
	Aratinga solsitidialis
	A. jandaya
	A. acetoacetate
	A. nenday
	A. auricapillus
		• Larger than the Pyrrhura conures.
		• Playful, interactive, and can easily be taught tricks.
		• Very loud on occasion.
		• Sexually monomorphic.
			• Egg binding
			• Hypovitaminosis A
			• Behavioural issues
			  • Screaming
			  • Feather-damaging behaviour

Guaruba conures			

Queen of Bavaria conure (golden conure) (**Fig. 1.12**)	*Guaruba guarouba*	• Plumage is mostly bright yellow, but also has green remiges.
		• Uncommon in captivity.
		• An affectionate bird that can be loud and active.
		• Sexually monomorphic.
			• Hypovitaminosis A
			• Behavioural problems:
			  • Screaming
			  • Feather-damaging behaviour

Patagonian conures			
Patagonian conure (burrowing parrot)	*Cyanoliseus patagonus*	• A popular companion parrot. • Known for being playful, gentle, and affectionate with humans. • Can also learn to talk and mimic sounds from its environment. • Sexually monomorphic.	
Caiques (Pionites)			
Black-headed caique White-bellied caique	*Pionites melanocephala* *P. leucogaster*	• Small parrots with short, square tails and prominent colours like green, black, yellow, and orange, sharply delineated. • Affectionate and bond well with humans. • Can be aggressive to other species. • Very active. • Sexually monomorphic.	• Feather-damaging behaviour • Hypovitaminosis A
Pionus			
Blue-headed pionus White-crowned/capped pionus Maximillian's pionus Dusky pionus	*Pionus menstruus* *P. senilis* *P. maximillian* *P. fuscus*	• Medium-sized birds with short squarish tails and prominent naked ceres. • Not as common in the pet trade as the other South American parrots. Known for their gentle and inquisitive nature, as well as their adaptability and quiet nature. • Have a distinct musky odour that becomes more prominent when they are excited. • Make a wheezing noise they make when stressed. • Sexually monomorphic.	• Aspergillosis • Hypovitaminosis A

COMMON NAME(S)	LATIN NAME	CHARACTERISTICS	COMMON HEALTH PROBLEMS
***Brotogeris* species**			
Grey-cheek parrot White-winged parrot (Canary-winged parrot) Yellow-chevroned parrot Orange-chinned parrot (Tovi parrot) Golden-winged parrot	*Brotogeris pyrrhoptera* *B. versicolurus* *B. chiriri* *B. jugularis* *B. chrysoptera*	• Active, very energetic birds. • Like to sleep in tight spaces such as clothing, blankets, etc. • Can be aggressive to other birds and pets. • Sexually monomorphic.	• Mycobacterial infections
Lineolated parrot			
Lineolated parrot (Linnie, barred parrot)	*Bolborhynchus lineola*	• Sociable birds, like interaction with owner or other birds. • Calm. • Sexually monomorphic.	• Lipomas
***Forpus* species (Parrotlets)**			
Pacific parrotlet Green-rumped parrotlet	*Forpus coelestis* *F. passerinus*	• Friendly, active birds, although feisty at times. • Aggressive to other birds. • Sexually dimorphic.	• Feather-damaging behaviour • Chlamydiosis
Other South American species			
Quaker parrot (monk parrot)	*Myiopsitta monarchus*	• Loud, active, and affectionate bird. • Build stick nests in a tree or on a man-made structure, rather than using a hole in a tree. A gregarious species, they often breed colonially, building a single, large nest with separate entrances for each pair. In the wild, the colonies can become quite large (the size of a small car). • For this reason, they are banned in some areas.	• Intestinal worms • Quaker mutilation syndrome • Screaming • Feather-damaging behaviour • Hypovitaminosis A
Hawk-headed parrot (red-fan parrot)	*Deroptyus accipitrinus*	• Have a crest of feathers which they can raise above their head when excited or alarmed. • Not commonly kept. • Juveniles are usually docile, but adults can be aggressive, ill-tempered, stubborn, and unpredictable. • Sexually monomorphic.	• Hypovitaminosis A
Asiatic parrots			
Indian ringneck (rose-ringed parrot) Alexandrine parrot Slatey head parrot Plum-headed parrot Derbyan parrot	*Psittacula krameri krameri* *P. eupatria* *P. himalayana* *P. cyanocephala* *P. derbiana*	• Popular as aviary birds. • Less common as pets, as they tend not to bond closely with people. However, they can make good pets, given the right owner, as they require daily handling and interaction to remain tame. • They are sexually dimorphic, although this may not be apparent till 2–3 years of age.	• Feather-damaging behaviour • Zinc toxicoses

COMMON NAME(S)	LATIN NAME	CHARACTERISTICS	COMMON HEALTH PROBLEMS
Oceanic parrots			
Cockatoos			

White cockatoo (umbrella cockatoo)
Moluccan cockatoo (salmon-crested cockatoo) (**Fig. 1.13**)
Lesser sulphur-crested cockatoo
Triton cockatoo
Goffin cockatoo
Citron-crested cockatoo
	Cacatua alba
	Cacatua moluccensis
	Cacatua sulphurea sulphurea
	Cacatua galerita triton
	Cacatua goffini
	Cacatua sulphurea citrinocristata
		• Common pets outside Australia.
		• Females are said to have a reddish-brown iris, while males have a blackish-brown iris, but this does not seem to be consistent.
		• Loud, destructive, and messy.
		• Produce powder down (white powder) that keeps their feathers clean.
			• PBFD (except cockatiels)
			• Chlamydiosis
			• Sternal trauma (poor wing clips)
			• Obesity related problems:
			• Hepatic lipidosis
			• Atherosclerosis
			• Lipomas
			• Arthritis
			• Poor socialisation if hand-reared
			• Reproductive problems:
			  • Excessive egg laying
			  • Cloacal prolapse
			  • Egg binding
			  • Yolk peritonitis
			• Behavioural problems:
			  • Screaming
			  • Aggression
			  • Feather-damaging behaviour
			  • Mate aggression
			• Xanthomas

Eclectus			
New Guinea eclectus (red-sided eclectus) (**Fig. 1.14**) Vosmaeri eclectus Solomon's Island eclectus Grand eclectus Australian eclectus	*Eclectus roratus polychloros* *E. r. vosmaeri* *E. r. solomnensis* *E. r. roratus* *E. r. macgillivrayi*	• Large, stocky parrot (405–600 g) well known for its obvious sexual dimorphism. • Dimorphism is believed to be due to different roles they play in breeding. Unlike other parrots, Eclectus exhibit cooperative breeding: females occupy the nest up to 8 weeks before laying and will stay in the nest tree for the duration of the breeding period (8–9 months) while multiple males forage for and deliver food to her. The different colours reflect the need for concealment. • Intelligent and curious birds but more independent than other parrot species. • The different subspecies are thought to have similar pet quality.	• Feather-damaging behaviour • APV • PBFD • Toe tapping • Egg binding • Beak overgrowth (idiopathic) • Hypovitaminosis A

COMMON NAME(S)	LATIN NAME	CHARACTERISTICS	COMMON HEALTH PROBLEMS
Lories and lorikeets			
Lories		• Lories are large birds with a short, broad, squarish tail. • Lorikeets have a long, pointed tail and tend to be smaller.	• PBFD • Feather-damaging behaviour • Screaming • Haemorrhagic enteritis • Obesity • Iron storage disease • Hypovitaminosis A • Hypervitaminosis A (from excessive supplementation)
***Chalcopsitta* species**			
Black lory Duyvenbode's lory (brown lory)	*Chalcopsitta atra* *C. duivenbodei*		
***Eos* species**			
Blue-streaked lory Red lory	*Eos reticulata* *E. bornea*		
***Pseudos* species**			
Dusky lory	*Pseudos fuscata*		
***Lorius* species**			
Chattering lory Black-capped lory Yellow-bibbed lory	*Lorius garrulus* *L. lory* *L. chlorocercus*		
Lorikeets			
***Trichoglossus* species**			
Green-naped lorikeet (coconut lorikeet) Edward's lorikeet	*Trichoglossus haematodus* *T. h. capistrasus*		
***Saudareos* species**			
Ornate lorikeet Iris lorikeet	*Saudareos ornatus* *S. iris*		
Glossuptilus			
Goldies lorikeet	*Glossuptilus goldei*		
***Vini* species**			
Collared lory	*Vini solitaria*		

COMMON NAME(S)	LATIN NAME	CHARACTERISTICS	COMMON HEALTH PROBLEMS
Canaries			

Canary (**Fig. 1.15**)	*Serinus canaria*	• A small songbird in the finch family originating from the Macaronesian Islands (the Azores, Madeira, and the Canary Islands.
		• Many breeds arose through selective breeding.
		  • Colour-bred canaries: bred for their many colour mutations, e.g., red factor canaries.
		  • Type canaries: bred for their shape and conformation, e.g., Border, Gloster, and Yorkshire canaries.
		  • Song canaries: bred for their unique and specific song patterns, e.g., Spanish Timbrado, German Roller
		• Kept as companion birds but are also shown in many parts of the world.
			• Moulting problems
			• Cnemidocoptes
			• Air sac and tracheal mites
			• Feather cysts
			• Straw feathers
			• Macrorhabdus
			• Trichomonas

Finches			
Zebra finch Gouldian finch (**Fig. 1.16**) Society finch (Bengalese mannikin, Bengalese finch) Double-barred finch (owl finch, Bicheno finch) Spice finch (scaly-breasted munia, nutmeg finch, spotted mannikin, mascot finch) Java sparrow (Java finch) Common chaffinch Grosbeaks Goldfinch Siskin Bullfinch	*Taeniopygia guttata* *Erythrura gouldiae* *Lonchura striata domestica* *Taeniopygia bichenovii* *Lonchura punctulata* *Lonchura oryzivora* *Fringilla coelebs* *Mycerobas* spp. *Carduelis carduelis* *Spinus spinus* *Pyrrhula pyrrhula*	• Two families: • Fringillidae finches have stout conical bills adapted for eating seeds and nuts and often have colourful plumage. They have a worldwide distribution except for Australia and the polar regions. It includes siskins, canaries, redpolls, serins, grosbeaks, and euphonias. • Estrildidae finches of the Old World tropics and Australia. • Many are kept as aviary birds, but they are also common as pets.	• Cestodes • Coccidiosis • Mycobacterial infections • Amyloidosis • APV
Mynahs (Myna)			
Common mynah Hill mynah	*Acridotheres tristis* *Gracula religiosa*	• A group of birds in the starling family (Sturnidae), a group of passerine birds native to southern Asia, especially India, Pakistan, and Bangladesh. • Mynahs are not a natural group; instead, the term mynah is used for any starling in the Indian subcontinent. • Chatty, intelligent, and social birds, making for good companions. • Good talkers.	• Iron storage disease
Starlings			
European starling (common starling)	*Sturnus vulgaris*	• Active, social birds that seemingly enjoy spending time with their owners. Pet European starlings are known for bonding closely with their caretakers and seeking them out for companionship. • Excellent mimics and talkers. • May be illegal in some countries.	• Iron storage disease

Figure 1.2 Cockatiel (*Nymphicus hollandicus*).

Figure 1.3 A lutino scaly-breasted lorikeet (*Trichoglossus chlorolepidotus*) and a rainbow lorikeet (*T. moluccanus*).

Figure 1.4 Bourke parrot (*Neopsephotus bourkii*).

Figure 1.5 A juvenile grey parrot (*Psittacus erithacus*).

Figure 1.6 Senegal parrot (*Poicephalus senegalus*).

Figure 1.7 Blue and gold macaw (*Ara arauna*).

Figure 1.8 Yellow-collared macaw (*Ara auricollis*).

Figure 1.10 Green-cheek conure (*Pyrrhura molinae*).

Figure 1.9 Red-lored Amazon (*Amazona autumnalis*).

Figure 1.11 Jenday conure (*Aratinga jendaya*).

Figure 1.12 Queen of Bavaria conure (*Guaruba guarouba*).

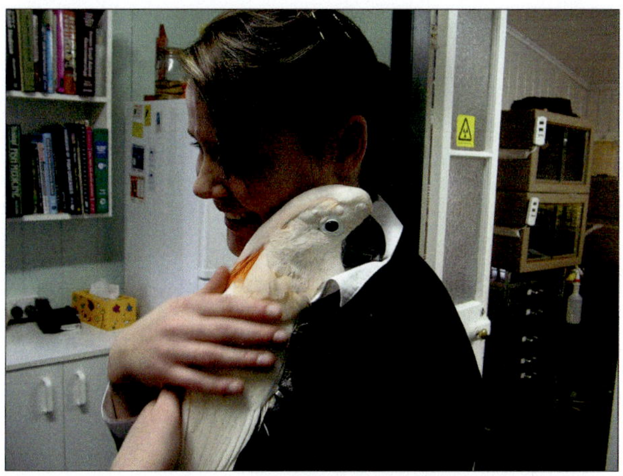

Figure 1.13 Moluccan cockatoo (*Cacatua moluccensis*).

Figure 1.14 Red-sided eclectus (*Eclectus roratus polychloros*).

Figure 1.16 Gouldian finch (*Erythrura gouldiae*).

Figure 1.15 Canary (*Serinus canaria*).

CHAPTER 2

HUSBANDRY AND NUTRITION

Although many veterinarians are comfortable and familiar with the husbandry of companion animals such as dogs and cats, many are much less comfortable or familiar with the world of bird keeping. But without an understanding of how birds are kept and fed, it is difficult to assess conditions that lead to disease or make recommendations on maintaining a bird's health and happiness. There are some simple and subtle differences between the husbandry and nutrition of pet birds and aviary birds, differences that the veterinarian needs to be aware of.

HUSBANDRY

Companion birds

Birds have been kept as pets for many hundreds of years, but the last 50 years have seen a surge in their popularity. A large number of species are kept as pets, with interactions ranging from a bird that becomes a part of the household to a bird that rarely leaves its cage, serving more as a decoration than a pet. They all share a common feature in that their living arrangements nearly always involve life in a cage for all or part of their day. These cages have a direct effect on the health of the occupant, and it is important that veterinarians include an assessment of the cage and the owner's husbandry as part of their initial workup of a patient.

The cage

While the size of the cage must obviously reflects the size of the bird it houses, it must also reflect the bird's lifestyle. A bird that is allowed free range of the household, spending only a small amount of time in the cage each day, can do well in a smaller cage than a similar-sized bird that spends most of its time in the cage (**Fig. 2.1**). As a general rule, the minimum size of the cage should be large enough so that the bird can stretch out its wings without touching the sides and sit on a perch without the tail touching the floor of the cage. If the bird has restricted out-of-cage time, the cage must be large enough for the bird to engage in normal behaviours such as grooming, playing, and moving. Most birds will do better in a cage that is wider, rather than high, as they fly or hop in a horizontal direction rather than vertical.

Figure 2.1 **A clear example of an inappropriately small cage for this size of bird if it never has the opportunity to leave it.**

The cage should be constructed from material that is strong enough to withstand the power of the bird's beak, easy to clean, and inon-toxic if ingested. Plastic-coated wire is commonly used for small bird cages while powder-coated bars, or even stainless steel, are used for larger cages. Small gauge wire, e.g., chicken wire netting, is easily chewed and ingested by large parrots, resulting in heavy metal intoxication, and should be avoided.

The floor of cage should be lined with an easily replaced, non-toxic substrate. Newspaper or paper towel are appropriate; sand, wood shavings, corn cob

DOI: 10.1201/9781003272519-2

bedding, and crushed nut shells are not. The latter style of cage liners can cause problems if ingested and may harbour fungal spores and bacteria. Many cages have a wire floor; this is appropriate so long as it is cleaned regularly to prevent spoilt food and dried faecal material adhering to the wire.

Unless very large, most cages are placed on a stand so that the bird is able to perch at a height level with a person's chest. Placing the cage on the floor, with human companions and predators such as dogs and cats looming over the top of the bird, may lead to the bird becoming fearful. The stand should be strong and stable enough to hold the weight of the cage and bird, and should ideally be mobile.

Cage furniture

Perches, while giving the birds somewhere to sit, should not be so numerous that the bird cannot move freely in the cage. Perches should be positioned so that they do not sit above food and water dishes as faecal contamination will otherwise occur (**Fig. 2.2**). The perches should be destructible, i.e., it is normal for parrots to chew on them. Using metal pipe to avoid replacing perches is not a sound practice. Cotton fibre perches have to be monitored carefully, as gastrointestinal obstructions following ingestion of these fibres are becoming a more common diagnosis. Perches should be irregular in shape and diameter, to exercise the birds' feet, and the diameter should be such that the birds' nails just meet when gripping the perch. A general guide to perch diameter is that the perch should not be so small that the bird's nails overlap excessively while perched, and not so large that the bird is defecating on the perch. The perches should be removed and replaced when they are soiled or badly damaged. Natural non-toxic branches usually make the best perches. The provision of a concrete 'grooming perch' can assist the bird in keeping its beak clean and blunting of the nails. The practice of covering perches with sandpaper should be discouraged as it can lead to abrasions of the plantar aspect of the foot and subsequent pododermatitis.

Figure 2.2 **A feed bowl, placed under a perch, contaminated with faeces.**

Feed and water bowls should be of a material suitable for the species; while plastic dishes are fine for a canary, a cockatoo or macaw will destroy them with ease. Stainless steel, glass, or glazed ceramic bowls are generally suitable for most species. Galvanised water dishes, coated in zinc and often soldered with lead, can be highly toxic to birds (**Fig. 2.3**). Terracotta dishes, although they keep water cool, are both absorbent and difficult to clean. The practice of placing water and feed dishes side by side should be avoided, as many parrots are fond of dunking their food in water, leading to contamination and rapid bacterial overgrowth in the water.

The provision of nest boxes and other sleeping/privacy shelters in a cage is a contentious subject. While some species, e.g., conures and small macaws, prefer a secure sleeping area, other birds will see it

Figure 2.3 **A water dish constructed using lead solder.**

Figure 2.4 Examples of commercially available cotton fibre toys.

as a stimulus for reproductive behaviour. Each situation may have to be weighed on its own merits, but if egg laying is not desired, it may be best to remove potential nesting sites.

Many owners provide their birds with toys to relieve boredom. Some of these toys are suitable for the inquisitive but destructive nature of parrots; others are not. Toys with cotton fibres should be avoided as ingestion of these fibres is a common cause of gastrointestinal obstruction (**Figs. 2.4, 2.5**). Cheaply made plastic toys or toys with metallic components are easily destroyed and ingested, presenting the possibility of toxicosis or obstruction. Alternatives include non-toxic seed pods (e.g., pine cones, eucalyptus nuts), natural browse, blocks of soft untreated wood, and small cardboard boxes.

Positioning

While pet birds can enjoy interacting with their human companions, they can also become stressed by constant noise and movement, lack of privacy and security, irregular diurnal cycles, and the possibility of the cage been knocked over. Indoor cages should be positioned with these stressors in mind. Some suggestions include: position the cage in a corner

Figure 2.5 An obstruction largely composed of cotton fibres which had to be removed from a parrot's proventriculus.

of a room, so that at least two approaches to it are secure; keep the cage off main traffic routes through the house, lessening close contact with people and other pets; and have a day cage and a night cage, so that birds can be 'put to bed' (rather than been covered) at nightfall.

Management

Feeding practices

Birds should be fed at least daily; it is a common practice to fill a feed bowl and leave it till empty. This practice, while labour saving, is misguided; empty seed husks can be mistaken for seed, leaving the bird to starve, and food left for long periods of time becomes contaminated with water and faeces, and attracts insect and mice. Fresh water should also be provided every day.

Cleaning

Cleaning of both the cage and the cage furniture needs to be done frequently and regularly. Cleaning should be aimed at removing organic material and then, where possible, disinfection should follow with a non-irritant chemical (**Fig. 2.6**).

Environmental enrichment

Most researchers agree that wild birds spend 80% of their day searching for, and consuming, food. The rest of the day is spent socialising, grooming, and sleeping. In captivity the 80:20 ratio is virtually flipped; with the food in the same place at the same time each day, the bird only needs to spend 10–20% of its time eating. This leaves 80% of the day to be filled with other activities. When a bird is housed by

Figure 2.6 **Example of poor cage hygiene which must be avoided.**

itself, with limited social opportunities available, it is no wonder that some captive birds develop behavioural problems such as overgrooming, stereotypic behaviours, and mental dullness (**Figs. 2.7, 2.8**).

Environmental enrichment is defined as "the stimulation of the brain by its physical and social surroundings in an attempt to reduce or overcome problems caused by containment". Its goals are to (1) alter the bird's captive behaviour so that it is similar to the normal behaviour of birds in the wild, as described earlier, and (2) to decrease the number of unnatural or unwanted behaviours a bird exhibits. Birds that are behaviourally well-adapted to their environment display "species-specific behaviour" as permitted by their social and physical environment and not abnormal behaviours. Birds that are not adapting well display a range of abnormal behaviours, including:

Figure 2.7 **Birds in an enriched environment displaying normal plumage.**

Figure 2.8 **A galah housed without any enrichment.**

- Feather-damaging behaviour
- Stereotypic behaviour (repetitive behaviour that serves no purpose)
- Aggression towards other birds and people
- Constant screaming
- Poor socialisation skills, e.g., chewing a cage mate's feathers
- Excessive sleeping

A key concept in environmental enrichment requires that owners and veterinarians know 'species-typical' information. It is then important to select enriching strategies that are behaviourally relevant and physically feasible for the bird. Designing an enrichment plan that is incompatible with the animals' normal behaviour, physical attributes or its existing environment may not only be fruitless but may also cause frustration and potential harm.

There is a tendency to think that environmental enrichment means providing some toys or maybe some branches to chew on. This is only a small part of the process. An enrichment plan can include some or all of the following:

- Foraging enrichment
- Physical enrichment

- Sensory enrichment
- Social enrichment
- Occupational enrichment

Foraging enrichment
Foraging is the act of searching for and finding food. As mentioned earlier, wild birds can spend up to 80% of their day foraging and feeding, most actively in the morning and the evening. Foraging therefore has great social and behavioural importance, but is one of the most severely constrained classes of behaviour in captive parrots; obviously there are potential implications of this common behavioural deficit for bird welfare. Given that birds engage in four basic behaviours on a daily basis – foraging, socialising, grooming and resting – removing the ability to forage (by feeding the same food in the same bowl at the same time) leaves a gaping hole that has to be filled with the other behaviours. A bird that lives alone without other birds to socialise with may therefore start to over-groom (feather damaging behaviour) or sleep excessively. Worse it may develop stereotypic behaviours such as screaming, pacing, and biting the wire of its enclosure.

Foraging enrichment therefore seeks to prevent or treat these problems. It requires the bird(s) to chew and sort through, manipulate, and/or open objects to get to food. It should reflect the bird's natural foraging behaviour and can be increased in complexity as the bird's skill levels increase. Examples of foraging behaviour can include:

- Scattering the food over the floor of the cage or aviary
- Scattering the food in a sand pit or kitty litter tray filled with wooden blocks or recycled paper kitty litter
- Placing the food in small cardboard boxes or paper parcels that have to be chewed open to access the food
- 'Baffle cages' – wire-framed boxes that require beak dexterity to access
- Multiple food dishes around the cage or aviary, some with food, some without
- Covering the food dishes with paper or cardboard that the bird has to chew through to access the food

The bird will have to be taught how to use some of these foraging tools, e.g., leaving a cardboard box open till the bird learns there is food in it, and then gradually closing it.

Figure 2.9 **Physical enrichment.**

Physical enrichment
Physical enrichment ranges from objects placed in the bird's environment (such as toys, swings, ladders, mirrors, etc.) to the environment as a whole (e.g., the space available for the bird to engage in locomotory behaviours such as flying, running, or swimming) (**Fig. 2.9**). There are some general guidelines to follow when considering safe enrichment items for birds:

- If the item is constructed of synthetic components, use sturdy and large enough materials to prevent ingestion. Avoid cotton or natural fibres, as these are frequently eaten and can lead to gut obstructions.
- For multiple birds in a cage, provide multiple enrichment devices to reduce item guarding and aggression.
- For birds fearful of new items, slowly introduce the enrichment to the bird's cage (or the bird to the enrichment area).

Figure 2.10 **Sensory enrichment.**

Sensory enrichment

Sensory enrichment utilises the bird's senses such as sight, hearing, smell, and touch. Providing 'a room with a view' (**Fig. 2.10**), background noises, videos, toys, etc. can improve a bird's welfare. However, it must be used with caution, as some birds may become visibly stressed with loud sounds, certain images, or a lack of security/privacy. Placing a bird's enclosure in the middle of the family room can certainly provide sensory enrichment, but the bird must have the ability to 'get away from it all' when the family activities become too much.

Social enrichment

Social enrichment is the social interactions between birds, and between birds and people (**Fig. 2.11**). It can be indirect, where the bird can see or hear other animals, such as when a bird is in an outdoor cage or aviary. Direct social enrichment includes cage mate pairing, social rooms or flights with numerous birds interacting, and allowing contact between enclosures. Done well, it allows the birds to perform 'species-specific' behaviours (which is likely high when compatible birds are housed together), but it has the potential to have drawbacks and unwanted behaviours such as cage mate aggression. Social enrichment may need to be carefully chaperoned until it is clear the birds can safely intermingle and there is an escape mechanism in the event of aggression.

Occupational enrichment

Occupational enrichment includes items that elicit activities including problem-solving, learning, and choosing and controlling some feature in the bird's environment. (This is different from the use of puzzle foraging toys that do require problem-solving.) Rather these may include items that give birds choices about how they spend their time. For example, although it comes with certain risks, free flight in the house (or even outside) can provide occupational enrichment.

Aviary birds

Aviculture is the practice of keeping birds in captivity, normally within the confines of an aviary, for hobby, display, research, conservation, or income purposes. It is a popular hobby around the world, with operations ranging from a small backyard aviary with a handful of budgerigars to large commercial operations producing birds for the pet market or other hobbyists. Veterinary involvement in aviculture is, unfortunately, predominantly limited to 'fire brigade' medicine – diagnosing and treating disease in individuals and flocks. Aviculturists, perhaps unaware of what veterinarians have to offer, rarely seek advice from them or engage in preventative management and medicine. Aviculturists are not solely to blame for this state of affairs – veterinarians must accept responsibility for not educating clients about preventative medicine and not promoting it to the hobby. Both parties must move towards each other to promote the better management of

Figure 2.11 **Social enrichment.**

avicultural collections and the health of the birds within those collections.

Location and size of the aviary

Local council regulations often stipulate where a backyard structure such as an aviary can be built and how big it can be. Although not an obvious veterinary concern, it can impact on suggestions for alterations and extensions. Similarly, another consideration is the amount of land that is available. Ideally the open side of aviaries, pens, and lofts should face the equator, i.e., north in the Southern Hemisphere and south in the Northern Hemisphere. This maximises exposure to direct sunlight throughout the day. This helps to keep enclosures dry and allows the birds to engage in normal daytime activities. The closed side of the aviary, on the other hand, should face towards the direction of prevailing weather, offering some protection against rain and wind.

Enclosure design and construction

Basic considerations for aviary design include:

- The individual requirements of the species held need to be met by the size of the aviary and by providing shelter sites, perches, feeding areas, and access to water within the cage.
- Common to all good cage design is the recognition that access to all parts of the cage is necessary for cleaning.
- Good visibility also contributes to good husbandry as it allows the health of occupants to be regularly assessed and any problems recognised early.

The aviary frame can be of wood, metal, or polyvinyl chloride (PVC) construction. While wood is the cheapest and easiest to work with, it has several disadvantages: the birds can chew it, it is often treated with preservatives that can be toxic, it cannot be disinfected, and it can harbour external parasites such as red mite (*Dermanyssus* spp.). Metal frames require specialised construction techniques, while PVC frames can be lightweight and prone to unwanted movement.

There are several requirements for aviary wire (Table 2.1): it must be strong enough to withstand chewing by the species in the enclosure, it must be small enough to prevent wild birds entering the enclosure, it must be rust proof but not so heavily or poorly galvanised that the zinc poses a toxic threat to the birds, and it should be 'see through' (this can be achieved by painting it black – the silver colour of wire reflects light back into the eyes of the viewer).

Aviary roofs can be either complete or partial, i.e., the roof is either solid across the entire length or it is partially solid with the rest being wire. The latter design allows sunshine and rainfall to reach the birds, allowing a more natural effect, but the disadvantage is cats, wild birds, possums, and rats running along the top. Some aviculturists reach a compromise by having a small, wire-roofed 'porch' or landing where the birds can get into the rain, and then withdraw into cover if a threat presents.

Aviaries can be either full flight or suspended, depending on the floor. Full flight aviaries have solid floors, whereas suspended aviaries have wire floors raised above the ground (**Figs. 2.12, 2.13**). Dirt floors are cheap, but they are impossible to clean and allow access to the aviary by mice, rats, or snakes. Concrete is expensive and permanent (the aviary can't be moved), but is easy to clean and maintain. Sand and gravel are used by some as a compromise between concrete and dirt. This has to be raked regularly to turn it over and replaced every few months. Weighing up against hygiene is environmental enrichment – some birds, especially Australian species, like to forage for food on the ground, and this is of great psychological benefit to them. Suspended aviaries, while easy to clean, are becoming less

Table 2.1 **Suggested aviary wire sizes for different sized birds**

BIRD SIZE	MESH APERTURE (MM)	WIRE THICKNESS (MM)
Small (20–80 g)	12 × 12	0.7–0.8
Medium (80–200 g)	12–25 × 12	1.24
Medium–large (200–500 g)	12–25 × 12	1.6–1.8
Large (>500 g)	25–75 × 25–50	1.24–2.00

Figure 2.12 **An outdoor aviary which allows birds full flight.**

Figure 2.13 **Example of an outdoor suspended aviary.**

Figure 2.14 **An outdoor aviary with safety porch used to prevent bird escape.**

popular for some larger species as their impact on the birds' psychological wellbeing has become clearer.

Regardless of the aviary design, at some stage someone will have to enter it – to clean it, to recover a bird, to check nest boxes, and so on. This means opening a door, thus providing an escape route for a bird. It is never a good idea to allow a captive bird to go free; not only does it lack the skills needed to survive, but it may also be harbouring disease that could be devastating to native birds. It can also be financially disastrous – some birds are worth many thousands of dollars. For these reasons, responsible aviculturists often have a safety porch or walkway so that they can always maintain a closed door between the bird and the outside world (**Fig. 2.14**).

Aviary furniture

It is important to encourage aviary birds to exercise. To this end, perches should not be so numerous that the bird does not see the need to fly. This can be done by placing only one perch at each end of the aviary or by running a perch along one wall. As with pet bird cages, they should be positioned so that they do not sit above food and water dishes and should be appropriately sized for the birds been kept. Natural non-toxic branches are ideal.

The design of nesting sites varies between species. Canaries, many finches, and pigeons prefer an open nest, whereas parrots prefer a deep hollow.

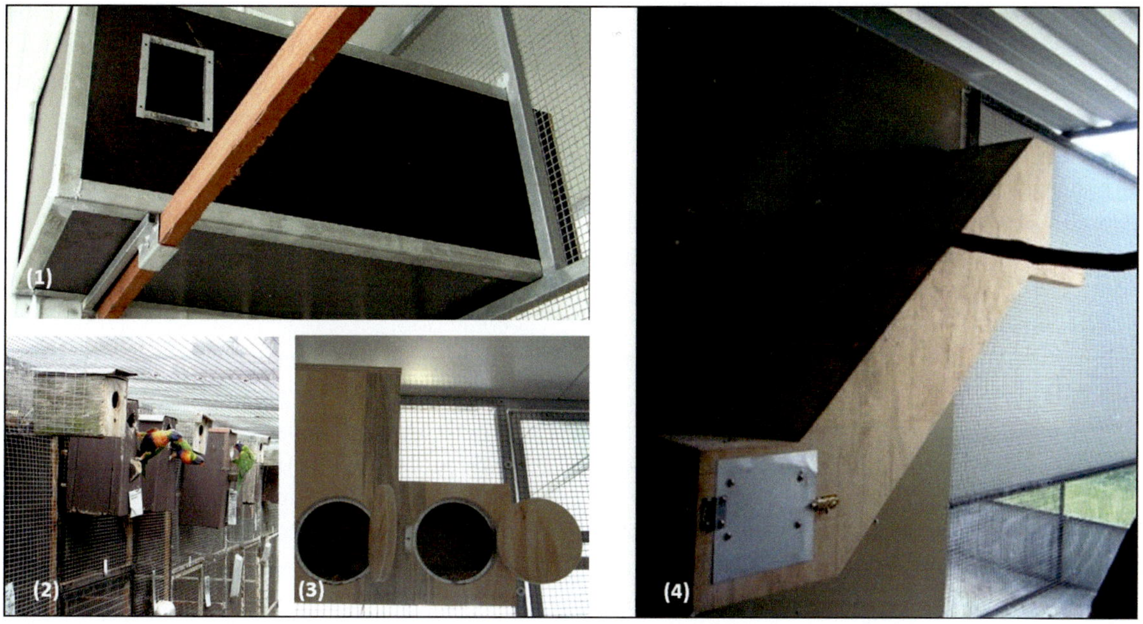

Figure 2.15 **A variety of nest boxes: (1) a horizontal box, (2) A-frame boxes, (3) L-shape box, (4) Z-shape box.**

Aviculturists have come up with numerous designs for nest boxes: A-frames, L-frames, Z-frames, vertical boxes, and horizontal boxes (**Fig. 2.15**). The preferred design varies between species, and often between individuals, and so it is a sound practice to offer birds a choice of nest box designs and locations, allowing them to select their preferred box (the other boxes can be removed). Nest boxes should be positioned so that the bird feels safe entering and leaving it. In particular, sun shining directly into the entrance hole will frequently discourage birds from using them. A privacy shelter frequently overcomes this issue. This can be as simple as a partition in front of the nest box, screening it from view.

Nest boxes should be removed, cleaned thoroughly, and left open to ventilate during the winter months. This prevents/minimises fungal contamination, especially with *Aspergillus* spores, which can cause pneumonia in both adult birds and chicks.

Pest control

Bird enclosures are difficult to keep scrupulously clean; food and faeces attract insects and mice, which in turn attract rats and snakes. Small birds attract larger birds – especially raptors, owls, and corvids. Design and management features that can minimise these problems include:

- Fully roofing aviaries and ensuring there are no overhanging branches.
- Suspended aviaries minimise pest problems, but this may be compromising the birds' psychological welfare.
- Concrete floors are easily cleaned and maintained, but can be expensive to build.
- Fine mesh wire, sometimes doubly wired, can help to restrict access by rats and snakes. Mice, however, can get through almost anything.
- Rat walls – concrete barriers below ground level to a depth of 30 cm – can stop rats and mice burrowing into an aviary.
- Regular cleaning of the aviary floor and avoiding the use of feed hoppers – large dishes of feed that are topped up when empty – minimise the amount of waste that attracts pests.
- Careful use of mice/rat traps and baits can assist, but care must be taken to ensure that the birds do not become inadvertent victims.

Management

Many people are aware of the Five Freedoms used to guide and assess animal welfare for over 60 years. However, these 'freedoms' do not always result in good outcomes. For example, if a bird is free from hunger, what would motivate it to forage for food? This recognition, that emotional needs are equally important as physical needs for animals, has given rise to the concept of the Five Domains.

The Five Domains cover nutrition, environment, health, behaviour, and mental state. This newer approach allows a distinction to be made between the physical and functional factors that affect an animal's welfare and the overall mental state of the animal arising from these factors.

- *Nutrition*: Birds have access to clean water and enough food in a balanced diet. Available food is nutritious, appropriate for the species and individual bird, and varied.
- *Environment*: Birds are comfortable and safe and provided with opportunities and choices to engage in natural and rewarding behaviour.
- *Health*: Birds are fit and healthy, and injuries and illnesses are prevented or immediately and appropriately treated.
- *Behaviour*: Birds are able to express natural and rewarding behaviour, and are able to engage in free movement and play, and have positive contact with birds and people to form social bonds.
- *Mental or affective state*: Nutrition, environment, health, and opportunity for expression of natural behaviour all influence an animal's mental state. Each of these can provide positive and pleasurable experiences that enhance a bird's mental state.

The following principles were originally developed to help aviculturists and veterinarians make a clear distinction between avicultural management (flock health) and the management of an individual pet bird. These principles are:

1. *Aviculture is farming*: Farming, by definition, is productivity-oriented animal stewardship. Many correlations can be drawn between the goals of the dairy, pork, poultry, and beef industries and aviculture.
2. *Productivity is the lifeblood of the farm*: The product from the investment in establishing a collection and effort to maintain it is what keeps the farm viable. Without production, numerous difficulties are experienced by the farm.
3. *Time and cost must always be justified*: Justification for time and cost in each avicultural effort varies immensely between individual owners. Regardless of the individual farm's manner of justification, there must always be thought applied to this principle. Physical time and expense invested in new management efforts must not compromise current maintenance efforts. Otherwise, ramifications from those distractions can pose a significant threat to the overall effectiveness of the farm.
4. *The flock always takes precedence over the individual*: Flock health is oriented to the group rather than the individual. Individual diseased birds are still diagnosed and treated, but the emphasis is always carried towards prevention of the occurrence of the problem(s) in the future. No single bird can be prioritised ahead of the entire collection.
5. *Culling and replacement are fundamental for improvement*: The concept of culling and stock rotation is a fundamental part of any progressive animal production-oriented operation. Based on the aviculturist's goals and objectives, those individual birds not meeting desired goals should be culled by sale or physical removal. When working with particularly rare or valuable species, culling should be a final management step when efforts to identify limiting factors to production have failed. (Note that culling does not include relocation within the aviary or further evaluation to determine the cause of poor production.)

Although these principles remain relevant in some circumstances, the level of veterinary medicine and standard of care for avian patients has increased over the years, such that ante-mortem diagnoses and individualised medication/treatment regimens allow for earlier diagnoses of disease in a flock, as well as better treatment outcomes. The modern avian veterinarian may be called upon to manage avian flock health where some of these principles may not be applicable, because the focus of the client may not be on breeding from their avian collection. Therefore, with the popularity of large backyard pet bird aviaries, it may be more useful for veterinarians involved with the healthcare of these collections to draw experience from the management of flock health in zoological exhibits. In these collections, further considerations to the corresponding principles may include:

1. Pet bird flock owners may place more emphasis on the overall presentation of the institution, including the numbers and species of birds in the exhibit, use of natural substrate and perches, and aesthetically pleasing food items.
2. Productivity is less important to the pet bird flock owner. Each owner may have different

needs and goals, e.g., preferences of certain favourite species within the aviary. A disproportionate amount of resources may be dedicated to such species, simply based on personal preference.

3. The time–cost justification is less applicable for pet bird aviaries. The compliance to veterinary recommendations is usually more associated with the capabilities and willingness of each pet bird aviary owner to follow them.
4. The health of some individual birds may take precedence over the health of the flock. Owners may have preferences for certain individuals and/or species of birds and may be more willing to go the extra mile for them over other birds in the flock.
5. As productivity is not an important factor for pet bird aviaries, euthanasia for improvement of flock productivity may not be relevant. However, euthanasia of birds due to poor quality of life may still be applicable when some birds in the flock become geriatric or incompatible with others.

The Closed Aviary Concept

The fundamental concept behind maintaining a healthy avicultural collection is the Closed Aviary Concept. It is a system of management principles by which aviculturists should be able to maintain their aviaries in a disease-free state. The basis of these principles is to firstly establish a disease-free flock, and then to maintain it that way by preventing the introduction and spread of disease.

The first step is to establish that the flock is healthy. This is usually accomplished through an inspection of the aviary by an avian veterinarian. This visit, discussed later, focuses on two key areas: (1) the aviary infrastructure and management; and (2) the birds – their age, health, and productivity.

Once the aviary has been certified disease-free by the veterinarian, it can be maintained that way using the following steps:

- The aviary should be divided into five geographically separate areas: food storage, breeding birds, nursery and juveniles, isolation, and quarantine (**Fig. 2.16**). The only way a bird enters or re-enters the aviary is through the quarantine area. Quarantine is not an isolation area (which is used for sick birds in the collection). Quarantine is for recently purchased birds and those returning from shows, breeding loans, and other trips to the outside world. During the 30- to 45-day period in quarantine, all birds must pass diagnostic tests to be accepted into the collection. Depending on the aviculturist's resources and previous disease problems, these tests may be as simple as casual observation or

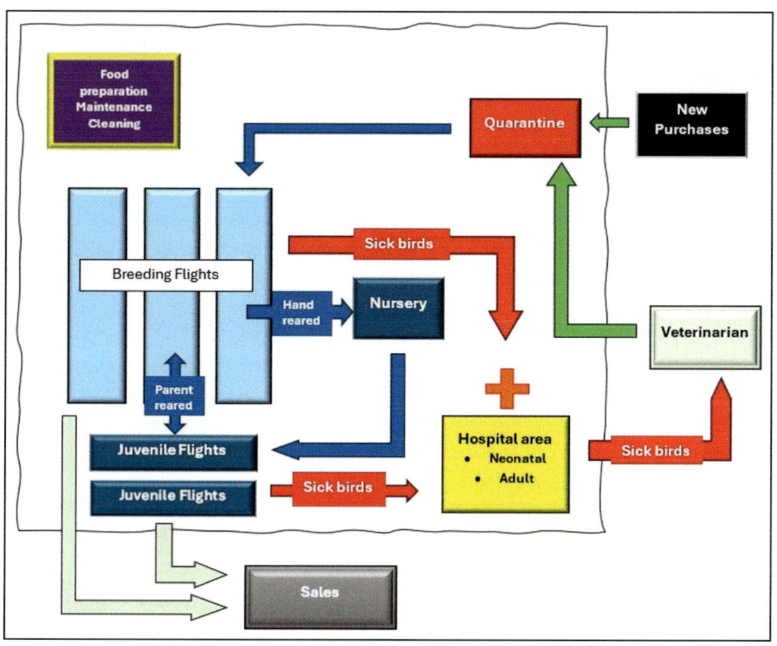

Figure 2.16 **Diagram illustrating the closed aviary concept showing appropriate bird flow between five geographically separated areas.**

as complex as polymerase chain reaction (PCR) tests, cultures, and blood tests.

After passing quarantine, the bird may enter the breeding flock with the ultimate goal of producing offspring. The offspring will be housed in the nursery and juvenile bird area, and eventually be sold or retained as breeding stock.

Occasionally disease may strike even the closed aviary. In these cases, affected birds are placed in isolation until they are recovered and have passed the same diagnostic tests used in quarantine. Isolated birds should not be housed with quarantined birds as they are still members of the closed aviary.

- All supplies (food and water), people, and equipment entering the aviary should be clean and free of disease. The person feeding should wear protective clothing that is not worn outside the aviary. Food should be stored in a clean, separate area that is hygienic and rodent-proof.
- Human traffic flow should move in a one-way direction from the outside to the feed storage, then to the breeding birds, the nursery and juveniles, then to the isolation area, and finally to the quarantine area. This pathway follows from the 'cleanest' birds to the 'dirtiest' birds and prevents unnecessary cross-contamination.

While most of the principles of a closed aviary system remain applicable, the emergence of infectious diseases that can remain dormant in asymptomatic carriers, e.g., avian ganglioneuritis, meaning that even extensive testing of new birds entering the collection and a standard 30-day quarantine period may not be sufficient. Instead, there is emerging evidence that a risk-based quarantine process may be as good as standard quarantine practices. However, in the context of private pet aviary owners, this requires an evaluation of the importing facilities' practices and collection health status. This may be an area in which avian veterinarians need to be more involved. For example, if a veterinarian consults for several pet bird aviary owners, the veterinarian may be best placed to ascertain if the transfer of birds carry certain disease risks. However, such practices require careful consideration of the veterinarian–client–patient confidentiality issues, but through appropriate and careful communication among all stakeholders can still be possible.

The aviary visit

The aviary visit is a key element of flock health. It serves two purposes: to establish a baseline for the flock's health, and then to monitor the flock's status and improvements. It can be divided into three components: the initial assessment of the client and the collection; the physical visit; and the preparation of a report for the client.

The initial client assessment

It is not an uncommon experience for a veterinarian, after been asked to give advice, to become frustrated at the client's perceived unwillingness to make fundamental management and structural changes to their collection in order to follow that advice. At the same time the client becomes frustrated with the veterinarian's perceived lack of understanding of the realities of their circumstances. This shared frustration usually arises because, in the early stages of the collection assessment, the veterinarian failed to carry out an appreciation of the aviculturist's desires, plans, abilities, and limitations. This must be addressed before any further assessment or planning is performed.

This initial assessment can be done either face to face on a visit or (ideally) through a questionnaire completed prior to this visit. Areas that should be explored include:

- What are the client's goals or objectives?
- What does the client hope to achieve through this assessment process?
    - o Identify major problems from the client's perception.
    - o What does the client expect a veterinary inspection and advice to achieve?
- What is the client's level of experience and expertise in bird keeping?
- A description of the collection:
    - o Type of birds
    - o Numbers held
    - o Age and sex of each individual bird
    - o Grouping of the birds (pairs, colonies)
- A description of the collection infrastructure:
    - o Aviary layout (a map is very useful)
    - o Cage construction
    - o Food storage
    - o Quarantine and isolation facilities
- Management:
    - o Diet
    - o Prophylactic treatments

- Daily management, including traffic flow around the collection
- Purchasing policy and quarantine process
- What records are available?
- Collection performance over the last 3–5 years:
  - Breeding successes or failures
  - Losses and medical problems
  - Other issues

After collating this information, the veterinarian should be able to analyse the data so that a clear picture of the client's abilities, expectations, and limitations emerges, as does an idea of the current situation within the flock. With this information in hand, the next step is the aviary visit.

The visit

The visit is for the collection of information and data, not for decision-making and planning. It is important that both the veterinarian and the client understands that it is only part of the collection assessment. Before physically walking through the collection the veterinarian and the client should sit down and discuss the answers to the questionnaire and the veterinarian's assessment of that data. It is important that both parties agree on the information and interpretation, and are clear on each other's intentions, abilities, and limitations.

While inspecting the collection, the veterinarian needs to have three things in mind:

- The information already gathered in the initial assessment
- The principles behind avicultural management
- The Closed Aviary Concept

The inspection can also be broken down into three components: the initial walkthrough, biosecurity assessment and monitoring, and an inspection of the records.

The initial 'walkthrough' inspection focuses on:

- Layout of the complex
- Individual cage design and construction

1. Flooring
2. Roofing
3. Frame
4. Wire
5. Furniture

- Water sources
- Food sources and storage
- Rodent and pest proofing and control
- The nursery, including the incubation, hand-rearing and juvenile holding facilities
- Biosecurity measures, such as cleaning, disinfection, visitor limitations, boots and sometimes protective garments, and equipment usage

Biosecurity assessment and monitoring utilises tests to evaluate the flock for the presence of disease, and may include:

- Faecal examinations
- Cultures
- DNA tests for diseases such as psittacine beak and feather disease (PBFD), polyomavirus, and chlamydia
- Blood tests (haematology, biochemistry)
- Necropsies

The collection's records are also examined. These records should include:

- Breeding records

1. Identity and pedigree
2. Productivity
3. Incubation
4. Paediatrics
5. Overall productivity analysis

- Medical records
- Sample contractual agreements between the aviculturist and purchasers/sellers of birds

The final step – the report

To complete the aviary visit, a written report should be prepared for the aviculturist. This is the most important part of the process and, while it needs to be prepared promptly, it should be prepared thoughtfully and thoroughly. It should include the following:

- Date of the visit and qualifications of the veterinarian
- Background and objectives of the collection
- Description of the collection, including the species and number of birds held
- Aviary design and layout, including a map and aviary construction
- Biosecurity measures

- Nursery management
- Diet, including food storage and preparation
- Individual bird health and results of diagnostic testing
- State of record keeping and an analysis of the productivity of the collection
- Summary of the findings
- Recommendations
- Follow-up

Follow-up
It is important to realise that the visit is not the final step in the process of maintaining a healthy collection. It is suggested that an annual visit be conducted, with subsequent reports, using the original report as a baseline for change and improvement.

NUTRITION

Poor nutrition is the underlying cause of many of the health problems seen in both pet and aviary birds; in fact, most avian veterinarians consider malnutrition to be the major problem affecting the health of birds today. For example, genetic selection of the budgerigar has produced a bird that is phenotypically distinct from its ancestors, e.g., the 'wild' budgerigar usually weighs 25–30 g, whereas 'show' budgerigars can weigh as much as 60 g. Despite these induced genotypic changes, it is still believed by many that the diet of the captive budgerigar should replicate the original native Australian diet. This has led to the commonly-held belief that the budgerigar is a 'seed-eater' and that an all-seed diet is appropriate.

Much of the seed in commercial mixes is in its immature state of growth, and even when matured, it does not provide the same nutrient profiles as wild seeds. Many seeds contain appropriate total protein but lack essential amino acids like lysine and methionine (see Table 2.2). Even for granivorous species, such as the budgerigar, a standard commercial mix is an imbalanced diet because seeds from domestic plants are more concentrated in energy and lower in protein and many other nutrients than seeds available in the wild.

Common problems associated with seed diets include:

- Weight gain/obesity, with associated diseases such as liver disease, cardiovascular disease, and diabetes
- Lipoma formation on the body of budgerigars
- Vitamin A deficiency resulting in skin and feather changes, respiratory problems and renal disease
- Vitamin D deficiency resulting in skeletal abnormalities
- Calcium deficiency, also resulting in skeletal abnormalities as well as muscle weakness

For this reason, it is important that veterinarians treating birds have a sound and practical understanding of the nutrition of pet and aviary birds.

Nutrients

Nutrients are the dietary constituents that provide energy to fuel metabolism and provide precursors for macronutrients (energy, protein, lipids, carbohydrates, and water) and micronutrients (vitamins and minerals), and structural and functional macromolecules. Bird owners frequently feed seed, vegetables, fruit, supplements, and treats ('human food'). Some feed a mix of these foodstuffs, hoping to achieve a balance of nutrients. Others, however, either lacking knowledge or simply frustrated by the selectivity of their bird, feed only one or two food items. The popularity of commercially manufactured formulated diets (which removes the bird's ability to selectively eat an unhealthy diet) is rising, but by no means is it universally accepted.

Energy

Energy, while not a nutrient itself, is a function of other macronutrients and is provided by the metabolism of carbohydrates, fat, and protein. The energy requirement for a bird is more variable than any other nutrient (it varies with body size, activity patterns, environmental temperature, plumage cover, age, and physiologic state) and generally a bird eats enough to satisfy that requirement. Energy requirements also depend on the bird's activity levels and physiological status: in cold weather it is increased threefold and for flight by 11–20 times the basal metabolic rate (BMR).

Food intake fluctuates with environmental temperature, the bird's activity levels, and the energy concentration of the diet.

The macronutrient profile of a diet (protein, fat, and carbohydrate) determines its energy density, with each providing a potential energy source. Because each of these nutrients has a different energy

Table 2.2 Essential amino acids that are dietary requirements for birds

ESSENTIAL AMINO ACIDS
Arginine
Isoleucine
Leucine
Phenylalanine
Methionine
Threonine
Tryptophan
Lysine
Valine

potential, the gross energy content of a diet depends on the relative proportions of these components.

Not all of the energy contained in a food is digestible; some is lost in faeces. The amount of energy that is digested and absorbed is known as the digestible energy. Some of this is then lost in the urine and intestinal gas; the remainder – the metaboliasable energy – is available for the bird's use. Growth and reproduction also increase energy requirements.

The bird's food intake will affect the intake of other nutrients. The requirement for these nutrients remains constant, so:

a. If there is an increased energy requirement, leading to an increased food intake, the required levels of other nutrients in the diet will decrease.
b. If there is a decreased energy requirement, leading to a decreased food intake, the required levels for other nutrients in the diet will increase.

So, as the energy concentration of a diet changes, the concentration of other nutrients changes proportionally. In evaluating a diet, the energy requirements of the bird and the energy concentration of the food must be specified.

Protein

Dietary protein is utilised as a source of amino acids and is required for anabolic processes such as growth, reproduction, and moulting, and as a source of energy. Any excess is excreted as uric acid.

Amino acids are used predominantly by birds to synthesise protein for the bird's use. The amount of these proteins required are:

- High during hatching and growth
- Decline gradually as adult size and weight are reached
- Increase during reproduction
- Decline after the breeding season
- Increase during moulting

These synthesised proteins contain 20 amino acids; however, birds are unable to synthesise nine of them. These amino acids (known as essential amino acids) have to be provided through the diet.

In addition to these essential amino acids, histidine, glycine, and proline can be synthesised but not in sufficient quantities for a growing bird, and so are considered essential amino acids for young birds. The remaining amino acids, known as non-essential amino acids, are synthesised by the bird if there is sufficient dietary nitrogen.

The requirement for some amino acids can be met by either that amino acid itself or the precursor of another essential amino acid. For example, tyrosine can be synthesised but only if sufficient phenylalanine is provided, cysteine requirements can be met by methionine, and glycine requires serine. These synthesised amino acids are known as conditionally essential amino acids.

Therefore, the diets must be high enough in protein to:

- Meet essential amino acid requirements
- Supply the nitrogen needed to synthesise the non-essential and conditionally essential amino acids
- Through metabolic processes give rise to glucose, fat, energy, and several vitamins

The 'ideal' protein is one which contains the exact requirements for a balance of essential amino acids and sufficient nitrogen to allow the synthesis of non-essential amino acids. The greater the deviance in the amino acid balance from that of an ideal protein, the higher the level of protein that will be required to meet the bird's amino acid requirements.

Protein quality is a function of its quantity, its digestibility, and the balance of amino acids. The amount and balance needed to meet a bird's requirements is rarely met by one food. The amino acid that is deficient in a food is known as the limiting amino acid; excess amounts of other amino acids do not compensate for the deficiency of the limiting amino acid. Protein complementation occurs when two or more foods correct their respective deficiencies.

Once consumed, amino acids are either used for anabolic processes or quickly catabolised for energy. Therefore, the correct amount and ratio of amino acids must be consumed within a relatively short time frame. Food containing complementary amino acids must be consumed during the same meal.

The requirements for protein will be determined by the needs for maintenance, growth, reproduction, and moulting:

- *Maintenance requirement* is the level required to meet the requirements of a bird that is not growing, moulting, or reproducing. Levels below this will sustain life, but muscle mass and function will be impaired. The more active a bird is, the lower the percentage of protein in the diet becomes – this is because the bird's energy demands increase and more food (and therefore protein) is consumed. The actual protein requirement (in mg/kg BW) does not vary.
- *Growth requirements* are much higher than maintenance by up to 15 times. Energy requirements, however, are similar to maintenance, therefore the diet must be higher in protein as a percentage of the diet.
- *Reproduction requirements* for egg-laying birds must meet normal maintenance, growth of the oviduct, and egg proteins. Birds that lay large numbers of eggs in a short period (e.g., daily) have much higher requirements than those that lay only occasionally. Protein requirements for oviduct growth will increase prior to egg laying, and the requirements for egg protein will increase the day before laying.
- *Moulting requirements* reflect an increased need for protein to synthesise new feathers, feather sheaths, and epidermal blood vessels. Feathers and feather sheaths are 90% protein. In particular, feathers require cysteine, valine, and leucine. Because the feathers are growing continuously but the bird eats only during the day, the necessary amino acids are mobilised from tissue proteins rather than the diet. Moult is energetically expensive due to loss of feather insulation, the cost of feather synthesis, and increased body protein synthesis. This increase in energy requirements means more food taken in and therefore more amino acid, i.e., there is no need to increase the percentage of protein in the diet.

Lipids

Dietary lipids supply energy, essential fatty acids (EFAs), and pigments as well as carrying fat-soluble vitamins.

- *Energy*: Lipids are digested, absorbed, and deposited in tissues with little or no modification, and serve as an energy source between meals and during embryonic development.
- *Fatty acids* are synthesised in the body into polyunsaturated fatty acids, needed for cell membranes and yolk lipids, and as precursors for some hormones, e.g., prostaglandins. They also act as precursors for eicosanoids, which regulate nearly every physiological mechanism, including egg laying, embryonic development, immunity, and bone growth. Deficiencies can be identified as slow growth, fatty liver, impaired spermatogenesis, and fewer and smaller eggs. Embryonic mortality may also rise. Linoleic and α-linolenic acids cannot be synthesised in the body, making them essential fatty acids. They are available in plant material, seeds, some fruits, and animal prey. Vegetable oil can be used as a supplement if a deficiency is suspected.
- *Carotenoid pigments*: Much of the colour in the plumage of birds is due to dietary pigments known as carotenoids.

There is no dietary requirement for triglycerides, but they are a valuable source of energy when food is limited, and they can improve the palatability of the diet. They also facilitate the absorption of fat-soluble vitamins from the diet. Triglycerides and other lipids can also be synthesised in the body from carbohydrates and amino acids. They are stored in adipose tissue and utilised when energy intake is limited.

The amount of body fat is affected by:

- The level of metabolisable energy intake relative to energy expenditure
- The ratio of dietary protein to fat – high protein usually results in a muscle anabolic effect with little fat deposition; as the level of protein decreases, the amount of fat deposition increases
- The balance of amino acids in dietary protein – poorly balanced proteins often increase fat deposition

Dietary fat has little effect on fat deposition, so long as the ratio of protein to metabolisable energy in unchanged.

Carbohydrates

Carbohydrates are either digestible (starch and sugars) or indigestible (fibre). Digestion of starch and sugars within the gastrointestinal tract results in the production of glucose, the primary carbohydrate absorbed across the intestinal wall. In the absence of gluconeogenic precursors such as fatty acids and amino acids, this glucose is used within the body as an energy source and a precursor for non-essential amino acids, vitamin C, and other metabolites. For budgerigars, carbohydrates are considered an essential nutrient.

When excessive amounts of carbohydrates are fed, a small amount of glucose is stored in the liver and muscles as glycogen; once these pools are replete, the remaining excess is converted to triglycerides and stored as fat. Glycogen can be quickly mobilised between meals to maintain a constant level of glucose in the blood. This reserve is depleted within a few hours, with the bird then reliant on gluconeogenesis (the metabolic synthesis of glucose) via lipolysis (mobilisation of fatty acids) and protein degradation (catabolism) to maintain a resting (but high) blood glucose. Wild parrots feed mainly in the early morning and late afternoon, roosting in between meals. This feeding schedule allows body reserves to be utilised as an energy source between meals. Captive birds, often fed *ad lib*, have a more constant intake of carbohydrates and therefore do not resort as much to gluconeogenesis – leading to increased body fat and obesity.

Indigestible carbohydrates (fibre) are composed of non-starch polysaccharides and lignin, found in the plant cell wall. Parrots, lacking a caecum to ferment this fibre, do not utilise it as an energy source.

Vitamins

Vitamins are unrelated organic molecules required in small but finite amounts in the diet to facilitate many of the chemical reactions involved in bodily function. Most are essential nutrients, i.e., the bird is unable to synthesise them in the body and therefore they must be provided in the diet, often as precursors or coenzyme complexes bound to protein, requiring digestion and metabolic transformation to become available to the bird. There are a number of complex nutrient interactions that may ultimately lead to imbalances in vitamin metabolism. It is therefore inappropriate to assess nutrients in isolation; they must be put into context with the entire composition of the bird's diet.

Vitamins are broadly classified into fat-soluble and water-soluble vitamins. The fat-soluble vitamins (A, D, E, and K) are usually found in the diet associated with lipids and are absorbed along the same pathways. They can be stored for some time in the body, so a constant supply is not required. Water-soluble vitamins are absorbed along different pathways as either free forms or bound to protein. They are generally not stored and so a constant dietary supply is required.

Vitamins are generally the least stable of the nutrients, diminishing in food over time, when the food contains minerals or antioxidants, or when the food is subjected to high temperatures, ultraviolet radiation, or high humidity. Water-soluble vitamins (except vitamin C) are more stable than fat-soluble vitamins.

Vitamin A: Vitamin A is a fat-soluble vitamin which can be formed from the vegetable pigment beta-carotene. Seeds contain no vitamin A and are generally poor sources of beta-carotene. It is therefore not uncommon to observe hypovitaminosis A in psittacine birds fed restricted diets of single, unsupplemented seeds. However, it must be remembered that an excess of vitamin A added as a supplement can also be harmful. In the budgerigar, for example, only 12 µg/bird would be needed daily with a maximum tolerance limit of 750 µg/bird. It is therefore important that vitamin supplements (especially those containing fat-soluble vitamins) are not used indiscriminately for addition to a parrot's diet.

Vitamin D: Vitamin D is essential for regulation of calcium absorption from the digestive tract and deposition/absorption of calcium from bone. Birds exposed to sunlight will synthesise adequate amounts of vitamin D from sterols present in their skin. Alternatively, the diet must be supplemented.

Vitamin E (tocopherol): Vitamin E is essential for cell membrane stability, giving protection against oxidative damage, and it also influences a number of physiological activities through its effects on EFA eicosanoids in the inflammatory pathway. Vitamin E is found in many plants, especially leafy vegetables. Although seed is high in γ-tocopherol, this form is poorly absorbed compared to the α-tocopherol found in leafy plants.

Vitamin K: Vitamin K is found in leafy plants but is deficient in seed and fruit. It forms part of the complex involved in the coagulation of blood. Low dietary levels are required, partly because it is found in the faeces of the bird and intake is supplemented by coprophagy.

Thiamine (vitamin B1): Although thiamine is found in seed, the highest concentration is in the husk, which is not eaten by the bird. Thiamine content also decreases with time if the seed is stored too long. Tannic acid, found in many seeds, fruits, and nuts, binds to thiamine, making it unavailable. Plants high in protein usually have high levels of thiamine.

Riboflavin (vitamin B2): This vitamin is found in leafy plants and fruits. Like thiamine, riboflavin is found in the seed husk and is not consumed by budgerigars. Soaking the seed may remove the riboflavin from the seed due to its extreme water solubility.

Niacin: Although found in seed, niacin in this form has a poor bioavailability except in green, milky grain. Food from animal sources is usually rich in niacin, otherwise supplementation is required.

Vitamin B6: This is another water-soluble vitamin found mainly in the husk of the seed. Supplementation is usually required, although nuts are a good source.

Pantothenic acid: This vitamin is similar in its sources and bioavailability to vitamin B6.

Biotin: Fruit, legumes, and leafy plants are good sources of biotin, but it is unavailable from seed. Coprophagy may provide a dietary source.

Folates: This is readily found in leafy plants, legumes, and citrus fruit, but may be unavailable from seed.

Vitamin B12: Vitamin B12 is not found in seed, although coprophagy may provide an adequate source.

Choline: This vitamin is found in seed and remains stable under most storage conditions. It may also be synthesised by the bird.

Vitamin C: Fruits, vegetables, and many herbs are a rich source of vitamin C, but it is not found in seed. However, it is synthesised from glucose and is not an essential vitamin.

Minerals

Minerals are utilised in the structure of the skeleton, maintenance of body pH, water balance, and are involved in many cellular processes. At least 13 minerals are required for the optimal health of birds and are divided into the macrominerals (calcium, phosphorous, sodium, potassium, chloride, and magnesium) required in large amounts for structural or osmotic functions, and the trace minerals (copper, iodine, iron, manganese, selenium, and zinc) required for other purposes and at much lower levels.

The level of minerals in a food item may vary depending on the soil the plant was grown in, and some minerals may not be as digestible as others. For these, and other reasons, the value of a food as a mineral source will be determined by the concentration and chemical form of the mineral, as well as a variety of food-specific factors (such as fibre content, digestibility, presence of chelators, etc).

Interactions may occur between various minerals at one of three points:

- In the intestine, impacting absorption
- In the body through competition to binding sites, impacting metabolism
- In the liver and kidney, impacting excretion

Therefore, excessive concentrations of one mineral in the food may lead to dietary imbalance of other minerals.

Requirements for many minerals will be determined by the bird's physiologic status, e.g., the requirements for calcium are greater in growing chicks and breeding hens. Macrominerals are often present in food items at levels close to known requirements and are absorbed at a level proportional to need. Trace minerals, on the other hand, are often in excess in the diet but digestion and absorption are actively minimised to prevent toxicosis.

Mineral	Dietary source	Absorption	Interactions	Metabolism
Calcium	Shell grit Cuttlebone Seed	Regulated by parathyroid hormone and vitamin D3	Often bound to phytic acid in seed and oxalates in plants Interacts with phosphorous and free fatty acids	Utilised for skeletal structural strength, muscle metabolism, and eggshell formation
Phosphorous	Seed Inorganic forms often unavailable	Regulated by parathyroid hormone and vitamin D3	Often bound to phytic acid in seed Interacts with most other minerals	Utilised for skeletal structural strength
Electrolytes	Charcoal Plants Salt blocks	Regulated by osmotic conditions in the bird	Sodium is often low in seed, but leafy plants are usually adequate	Utilised in pH balance of the body
Magnesium	Seeds and leafy plants	Actively absorbed across the gut wall	Often bound to phytic acid and fat in seed Interacts with phosphate	Skeletal structural strength
Copper	Seeds (marginal)	Absorbed across GI wall	Often bound to phytic acid, fat, and tannins in seed and diet Interacts with phosphorous and calcium	Enzyme component
Iron	High in many foods except fruit	Absorbed across GI wall	Often bound to phytic acid, fat, and tannins in seed and diet Interacts with phosphorous, calcium, zinc, and copper	Haemoglobin
Zinc	High in plants but low in seed	Absorbed across GI wall	Interacts with phosphorous, calcium, iron, and copper	Required for all physiological processes
Manganese	Leafy plants	Absorbed across GI wall	Often bound to phytic acid, fat, and tannins in seed and diet Interacts with phosphorous, calcium, zinc, and copper	Bone structure
Iodine	Variable in plants, depends on iodine content in soil	Both organic and inorganic forms are readily absorbed	None	Thyroid function
Selenium	Seed	Readily absorbed	High levels of vitamin E reduce the requirement for selenium	Essential component of at least enzymes

Seed

Captive parrots are commonly fed a predominantly or all-seed diet, believed by many to be a 'natural diet' – parrots are, after all, 'seed-eaters' (**Fig. 2.17**). This belief fails to take into consideration that there are some 365 species of parrots, found naturally on four continents and numerous islands. They have evolved in a myriad of geological and climatic conditions, with a wide range of different foodstuffs available to them. The natural diet of wild parrots includes many different plants and plant parts, as well as invertebrates and sometimes even meat from animal carcasses. Their diet changes based on seasonal availability, and their requirements change according to their life stage and physiological status.

It is believed that parrots evolved in Gondwanaland some 50 million years ago. However, agricultural crops were not planted in Australia until some 200 years ago, meaning the grains found in commercial seed mixes are not indigenous to the natural habitat of Australian parrots (**Fig. 2.18**). In fact, many of these grains originate from the American continent and thus are not natural food items for African or Asiatic species either. Although many wild parrots have adapted to, and taken advantage of, the introduction of agricultural crops into their habitat, they

Figure 2.17 **Example of a commercially available all-seed, high-fat bird food mix.**

do not feed exclusively on these crops, browsing on a variety of food items and, perhaps more importantly, foraging over large areas of land to find enough food. Field studies have indicated that most wild parrots spend more than 50% of their day foraging for food, which burns up much of the excess energy contained in agricultural grain. It has been estimated that the energy requirement of a wild bird is more than 1.5 times that of a bird in captivity.

Not only are seed diets unnatural, but they are also nutritionally unbalanced. The fat content varies between 20% and 50%, with some nuts even higher. Although ample carbohydrates are present, they are often surplus to the bird's energy requirements. Total protein is often suitable for adult birds (although not for growing chicks), but the amino acid composition is not suitable for maintaining long-term health. Seed is deficient in vitamin A and beta-carotene (its precursor), vitamin D, and vitamin K. It usually contains adequate amounts of vitamin E (if fresh), and although certain seed types may be marginal in individual B vitamins, if a mix of seeds is offered, requirements for these vitamins may be met. Most seeds have an excess of phosphorous but are deficient in other minerals, including calcium, manganese, iron, and zinc. Iodine and selenium may also be deficient, depending upon the geographical region of production. The high-fat content can amplify the effects of low calcium through saponification in the intestinal tract, making the calcium unavailable for absorption.

In summary, dry seed diets have excessive fat, carbohydrates, and phosphorous; marginal protein; adequate vitamin E; and are deficient in in amino acids, calcium, available phosphorous, sodium, manganese, zinc, iron, riboflavin, pantothenic acid, choline, niacin, and vitamins A, D, K, and B12. While it can sustain life, dry seed is not suitable for long-term health, growth, or reproduction.

In an attempt to improve the nutritional value of seed, some bird owners soak and/or sprout the seeds. This process increases the water content (diluting the energy), lowers the fat content, and, if sprouted, increases the protein and vitamin content. These advantages must be weighed against the risks of bacterial or fungal overgrowth during the process, resulting in fermentation of the seed. Some aviculturists minimise this risk by adding chlorhexidine (or even household bleach) to the mix. Thorough rinsing is necessary before feeding.

Vegetables

In an attempt to balance a seed diet and simulate the diet of a wild parrot, many owners add vegetables to their birds' diets. However, many of these domestic vegetables are deficient in energy, calcium, and other nutrients, and are often higher in water than their wild counterparts. While people can eat a balanced vegetarian diet, it must be remembered that they grow very slowly and have a low reproductive output compared to birds. Birds fed an exclusively vegetable diet will display signs such as poor growth, loss of condition, and poor reproductive success.

On the plus side, vegetables fed as a supplement to a seed diet can dilute the energy content while providing some nutrients not found in seeds (**Fig. 2.19**). It also provides a form of enrichment when fed in large pieces. This does, however, require that the bird eat the vegetables – when given free choice, many birds will eat seed in preference to vegetables, often to the complete exclusion of the vegetables. Feeding strategies, such as limiting access to seed to a short period morning and night, may be needed to overcome this dietary preference.

Fruit

Many of the preceding comments regarding the feeding of vegetables apply to fruit as well, although fruits tend to be higher in energy and water than vegetables. Many arid zone Australian parrots, e.g., some cockatoos, cockatiels, and budgerigars, have not evolved with fruit as part of their diet and show little interest in it. Other, more tropical, parrots such as eclectus, lorikeets, and Asiatic, South American,

Figure 2.18 **Australian native foods available for parrots, including (A) termites and other small insects, (B) green seeding grasses, (C) seed and grit, (D) casuarina cones, and (E) *Grevillea* flowers.**

and African parrots relish fruit in their diet. Some fruits, such as avocado, may be poisonous to birds and should be avoided. Citrus fruits can promote iron absorption, leading to iron storage disease in susceptible species (mynahs, toucans, and lorikeets); they should be fed in only minimal amounts, if at all. In general, fruit should be restricted to a small component of the diet (<10%) or even regarded as a treat rather than a regular food item.

Supplements

The diets consumed by free-living birds can rarely be duplicated in captivity because the vast range of seeds and other food items are not usually available in sufficient quantities. Even if these food items, or very similar ones, could be obtained, they may still not be nutritionally adequate. This is because wild birds eat large amounts of food to meet their energy requirements, but their intake of amino acids,

Figure 2.19 **A cockatiel eating a mix of vegetables, pellets, and seed.**

vitamins, and minerals remains constant. Captive birds eat relatively less food, and if eating a 'wild' diet, they will become deficient in these nutrients. For this reason, supplementation is necessary for birds fed a diet based on seed and vegetables.

As previously mentioned, the level of calcium in seed diets is low, and even in vegetables with reasonable quantities, the availability can be extremely variable. Calcium found in plants is often less digestible than other sources because of the formation of poorly digestible complexes with phytic and oxalic acid, or saponification when fed with high fat diets. In the wild, many birds supplement their diet with mollusc shells, eggshells, and calciferous grit. In captivity, a calcium supplement is required when seed and vegetables are fed as the predominant diet. This supplementation is available in several forms: cuttlebone, shell grit, calcium powder, and calcium syrup, each with its own advantages and disadvantages. Cuttlebone is relished by many smaller parrots but destroyed by larger parrots with little ingestion. Shell grit is readily consumed by all parrots but, if overeaten, may lead to a grit impaction of the proventriculus and ventriculus. (This can be prevented by adding only a small amount to the food every few days, rather than providing a separate bowl of grit.) Calcium powder is often unpalatable and has to be mixed in well with the food. Calcium syrup often has sugars added to improve its palatability, but this can lead to bacterial overgrowth if food and water dishes are not washed daily. Care must be taken not to oversupplement calcium; in some species such as budgerigars, excessive levels can be toxic.

Similarly, vitamin and other mineral supplementation is also required for birds on seed and/or vegetable-based diet. Many birds appear to be able to balance energy, amino acid, and calcium levels in their diets by selecting among dietary items, but there is little evidence they can select for adequate intake of many other nutrients. Vitamin and mineral supplements are frequently offered in the form of powder and liquids to be added to the food or water. The effect is often lost when added to a dry seed mix, as the vitamin coatings on the husk are lost when the bird removes and eats the kernel. Vitamin–mineral pellets added to a seed mix are often of little value, as birds often ignore them. Vitamins and minerals added to the water may not achieve its intent, as these solutions are often very unstable in water and rates of water consumption are extremely variable. Some bird owners will attempt to compensate for this variable intake by oversupplementing the diet with vitamins and minerals. Unfortunately, this practice may result in excessive intake and subsequent toxicosis. Vitamin A and D oversupplementation, in particular, have been reported in parrots with sometimes fatal effects.

Formulated diets

In an attempt to overcome the inadequacies and deficiencies posed by feeding seed and vegetables, formulated diets have been developed for birds as they have for other species (**Fig. 2.20**). The processing binds the dietary ingredients together so that the bird cannot select individual components. These ingredients are typically ground grains (such as corn) and ground legumes (such as soybean meal or peanut meal), with vitamins, minerals, vegetable oil,

Figure 2.20 **Formulated diets, supplemented with vegetables and fruit, provide a more balanced diet for most parrots.**

and purified amino acids added to meet known (or estimated) requirements. These ingredients are then either pelletised or extruded. Pellets are made by heating the mix to 70–80°C and passing it through holes using a roller. When the mix emerges as cylindrical particles of a constant diameter, a turning knife cuts the pellets at pre-set lengths. Extruded diets, on the other hand, are made by forcing the mix, under pressure and at high temperatures (between 90°C and 180°C), through an extruder. Water is added in the extruder as steam for a 'dwell' time of anywhere from 30 seconds to several minutes. The food will take on the shape of the holes in the extruder plates. The water in the mix instantaneously evaporates when it leaves the extruder, along with the intracellular water in the plant cells and bacteria, rupturing the plant cells and killing infectious agents if present. The resultant extruded diet is hydrolysed and sterile but may have lower nutritional value. This last problem is overcome by using more nutrients in the pre-mix before the extrusion to compensate for expected losses. Extruded diets are usually more palatable than pelleted diets.

Many aviculturists and pet bird owners resist the concept of formulated diets, arguing that these diets are unnatural, too expensive, boring for the bird, or that their birds just won't eat them. The first two arguments can be dealt with through client education; the third can be overcome by adding vegetables and fruit to the diet and the use of foraging activities to make the bird 'work' for its food.

The last argument, that the birds won't eat the new diet, is more problematic. Many birds are neophobic (scared of new things); others simply don't recognise the new diet as food. Forcing the birds to convert to a new diet by withholding other foods does not work in many cases; small parrots, in particular, may starve to death if this approach is attempted. There are a few simple guidelines to think about when planning to convert a bird to a healthier diet:

- If there is more seed in the cage than the bird will eat in a day, it will not touch new foods.
- Foods offered in a deep dish are unattractive to most birds. Conversely, foods offered in a flat dish or on a flat surface such as a tabletop will often be explored, picked up, and eaten.
- Birds learn by mimicry; if they see other birds (or their owner) eating the new foods, they are more likely to try it themselves. If they are closely bonded to the owner, they may accept new foods when offered to them by hand.
- Conversion may take a few minutes or a few months.

Strategies utilising these guidelines include: hand feeding the bird with the new foods, giving plenty of praise and encouragement while doing so; if the bird comes out of its cage, placing the new foods on a tabletop and encouraging the bird to pick them up and try them; placing the food on a flat dish at the bottom the cage and removing all seed for 6–8 hours; and putting the bird beside another bird that already eats the formulated diet.

If these strategies do not work, a more calculated approach may be required:

1. Determine how much seed the bird eats in a day. This is done by measuring how much seed is placed in the cage in morning, and then how much is left the next day. The difference between the two amounts is the amount eaten. Repeat this exercise for 2–3 days to get an average daily seed consumption.
2. Once it is known how much seed the bird eats each day, provide only that amount of seed, mixed with an equal amount of pellets plus vegetables made into a mash.
3. Feed this mash fresh each day, but each day remove a small amount of seed. This makes the bird slightly hungrier each day and, hopefully, as it searches through the mash looking for the seed, it will start to eat the pellets and vegetables. Over a period of several weeks, seed can be eliminated from the diet.

During the conversion process the owner should weigh the bird each day. If there is more than a 10% weight loss, or if the bird starts to look unwell, it should be offered seed immediately. If the bird returns to normal weight and behaviour, it should be given a few days to recuperate before starting the process again.

Lorikeets

Lorikeets and lories are brightly coloured, active nectarivorous parrots native to Australia and Oceania. Long popular as aviary birds, they are becoming more popular as pets but have unique dietary requirements. In the wild these birds are high-energy users, flying swiftly and powerfully through their habitat and constantly moving from flower to flower while feeding. Their natural diet (nectar, pollen, manna, honeydew, lerp, insects, and fruit) reflects this high energy usage, and their anatomy reflects this high energy diet (brush tongue; small, relatively weak ventriculus; and a rapid gastrointestinal transit time). Nectar is a sugar-rich, liquid food source, but is low in amino acids, vitamins, and trace minerals. Pollen protoplasm is composed of highly digestible protein and contains a diverse amino acid profile, and is therefore a good food source for lorikeets. Manna is a sugary exudate from damaged eucalypt leaves or woods; honeydew is the sugary excretion of nymphal stages of aphids, coccids, and insects belonging to the family Psyllidae (plant lice); and lerp is a waxy material secreted as a protective scale by psyllids. However, given the high carbohydrate and low protein content of these latter foods, it is likely that they are used only as alternatives to nectar. Nectarivorous birds almost certainly obtain a good supply of protein from insects.

Attempts to replicate this diet in captivity have been met with varying degrees of success. Many of these artificial diets are high in sugar and do not account for the decreased energy output associated with captivity. As well, many do not match the amino acid composition of pollen and insects; the result is often obesity, poor growth, and decreased reproductive performance. Several attempts have been made to produce lower energy diets for lorikeets, but palatability remains an issue for many birds (**Fig. 2.21**).

Generally captive lorikeets are fed a small amount of liquid formula once or twice daily and *ad lib* dry food, supplemented with fruit, vegetables, and sometimes sprouts. Some larger lorikeets, e.g., the rainbow lorikeet (*Trichoglossus haematodus*), will consume seed, but this should not be considered a suitable diet for these birds. More research needs to be done to determine the ideal diet of captive lorikeets and lories; veterinarians should not assume that commercially manufactured diets are nutritionally adequate.

Figure 2.21 **A dry nectar mix prepared for lorikeets.**

FURTHER READING

Gartrell BD. The nutritional, morphologic, and physiologic basis of nectarivory in Australian birds. *Journal of Avian Medicine and Surgery*, 2000; 14(2): 85–94.

Harper EJ, Skinner ND. Clinical nutrition of small Psittacines and Passerines. *Seminars in Avian and Exotic Pet Medicine*, 1998; 7(3): 116–127.

Kalma ID, Janssen GPJ, Moons CPH. Guidelines and ethical considerations for housing and management of psittacine birds used in research. *ILAR*, 2010; 51(4): 409–423.

Koutsos EA, Matson KD, Klasing KC. Nutrition of birds in the order Psittaciformes: A review. *Journal of Avian Medicine and Surgery*, 2001; 15(4): 257–275.

McLean KM, Schook MW, Pye GW. Comparison between standard zoo quarantine practices and risk-based management of animal transfers: A retrospective analysis of avian acquisition morbidity and mortality (2013–2018). *Journal of Zoo and Wildlife Medicine*, 2021; 51(4): 1017–1020.

Speer BL. Avicultural medical management. An introduction to basic principles of flock medicine and the closed aviary concept. *Veterinary Clinics of North America: Small Animal Practice*, 1991; 21: 1393–1404.

Stahl S, Kronfeld D. Veterinary nutrition of large psittacines. *Seminars in Avian and Exotic Pet Medicine*, 1998; 7(3): 128–134.

CHAPTER 3

HANDLING AND GROOMING BIRDS

HANDLING

General

At some time during every examination, the veterinarian must handle, and sometimes restrain, the patient. Traditionally, it was considered acceptable to put on a heavy pair of gloves, pin the bird to the tabletop or the side of the cage, and then drag it, struggling and screaming, into a position where it could be better examined. Undoubtedly, this stress, imposed on an already compromised patient, contributed to the myth that birds were 'soft' animals, prone to dying while just being examined!

Just as it is inappropriate to muzzle and grapple with every dog and cat that comes into a veterinary surgery, it is inappropriate to use heavy-handed restraint on a companion bird. Many of these birds have learnt to trust humans and regard them affectionately. Destroying this trust through aggressive catching and handling techniques can adversely affect the bond between owner and bird. This relationship must be preserved, and handling techniques for closely bonded birds should emphasise minimal stress and fear.

Restraint can be achieved by using one of the following methods, or a combination of two or more:

a. Physical restraint – a combination of physical forces and tools to limit the movement of a bird
b. Chemical restraint – the use of tranquilisers, sedatives, and anaesthesia
c. Psychological restraint – behavioural conditioning using positive reinforcement

No form of restraint should ever be taken lightly, as each restraint has some effect on the behaviour, life, or other activities of the bird. Success depends on the handler's knowledge of the bird, the handler's experience, the application of the correct technique, and the availability and application of the appropriate tools/chemicals.

Basic factors to be considered before attempting to restrain or handle a bird are:

a. Is it safe for the person doing the handling?
b. Is it safe for the bird?
c. Is it possible to accomplish the intended procedure by utilising the suggested restraint method?
d. Can constant observation and attention be given to the bird following the restraint until it is fully recovered from the physical effects of the restraint?

Safety

Nearly all birds will have some means of resisting restraint, and these may inflict physical pain on the person handling them. Parrots can inflict painful bites, and the grasp of some (such as lorikeets) will be both painful and difficult to disengage. Poultry and waterfowl will flap their wings wildly in the handler's face. While raptors have a menacing beak, their talons are much more dangerous. The handler should identify how the bird is likely to defend itself and take appropriate steps to avoid injury to both themselves and the bird.

Preparation

Preparation will minimise both the time necessary for any procedure and the stress the bird will endure. A plan of action should be known and communicated to the team. Equipment needed for the planned procedure should be ready and available. Staff should be free of distractions such as ringing phones or other interruptions during the restraint. Housing and transport containers should be ready to receive the animal either during handling or immediately after.

Wherever possible, take the time to remove perches, dishes, and toys from the cage. Avoiding

DOI: 10.1201/9781003272519-3

these obstructions while trying to catch a small bird will decrease restraint time and stress. Anticipate the bird escaping from either your grip or the cage – ensure windows and doors are closed, ceiling fans are off, and staff outside the room know what you are doing and don't walk in unexpectedly.

Removing the bird from a cage

If the bird is presented in a cage or carrier, it is often more appropriate to ask the owner to remove the bird from its cage. Doing so will often relax the bird in a new environment. If the owner is unwilling (or unable) to do so, the clinician should first offer the bird the opportunity to step out of the cage by itself. If it is unwilling to do so, an attempt should be made to bring the bird out on the clinician's hand (if the cage door is large enough) or on a stick (if the bird accepts this readily). A hand is slowly and gently introduced into the cage, with the back of the hand to the bird. If there is no aggression, the forefinger or hand is extended and placed under the bird's chest. A tame bird will usually step onto the finger or hand. If necessary, one toe is restrained by gently pressing on it with a thumb against the finger, keeping the bird steady, and it is gently brought out of the cage. During this procedure, the clinician should keep talking to the bird, praising it and maintaining eye contact. When the bird is out of the cage, one should continue to talk to it and praise it, scratch its ear, and generally make a fuss of it (**Fig. 3.1**).

If the bird will not step onto a hand, it is best to use a small hand (or paper) towel to gently envelop and then restrain the bird. The bird is shown the towel and allowed to get used to it. If possible, the clinician should slowly envelop the bird in the towel from below; an approach from above is potentially a very intimidating experience for a pet bird. The clinician should keep talking in a friendly voice and maintain eye contact.

Figure 3.1 **Gentle handling of a pet juvenile blue and gold macaw (lutino mutation).**

Birds that are not tame can be caught using a towel as described earlier. These birds will rarely stay still during capture, so a quick capture is the best approach. A few seconds spent removing obstacles in the cage (food and water containers, perches, toys) is a worthwhile endeavour, as it will reduce the bird's opportunities to avoid capture.

Further handling

Once the bird has been removed from the cage, the next step will be determined by how tame it is. Very tame birds can be placed on a T-perch for further examination and weighing; less tame birds may need to be examined while restrained in the towel (**Fig. 3.2**).

As the oils on human skin can be detrimental to the feathers of many species; a light dusting of talcum powder on the clinician's hands is appropriate before beginning an examination. This is particularly necessary for those birds that do not produce powder down (e.g., Amazons, eclectus parrots, and lorikeets). Handled frequently without talcum powder, the green feathers on these birds begin to acquire a black discolouration that can become quite unsightly (**Fig. 3.3**).

For small parrots, the entire body can be supported in the palm of the hand while the head is extended to allow adequate restraint. This can be done by holding the head between the index and middle fingers and extending these fingers (and the bird's head and neck). Alternatively, the bird's head can be restrained by holding the bird's mandible with the thumb and middle finger on either side and placing the index finger on the top of the bird's head (three-fingered grip). The bird's body should always be supported, and the wings and feet controlled with the other fingers or hand. The keel should be allowed to move freely for respiration, and no pressure should be placed on the crop (**Fig. 3.4**).

Medium-sized birds can be restrained using a three-fingered grip, as described earlier, while the bird is wrapped in a towel or held close to the handler's body to help prevent the wings from flapping.

Large psittacine birds can be restrained by holding the head just underneath the beak. Take care not

Figure 3.2 **Using a towel to restrain a green-winged macaw.**

to apply pressure to the bare facial skin of macaws, as it bruises easily (**Fig. 3.5**). Again, the body can be wrapped in a towel, if needed, to control the wings.

Sedation with midazolam and butorphanol, which may be given intranasally, has become popular for use with companion parrots. Based on different species' tolerance to these drugs, beginning with a low dose and having reversal drugs available are advisable.

At all times, the clinician must be aware of the bird and how it is handling the stress of restraint and examination. As many birds are presented for evaluation of an illness, and they have been ill for some time, even being restrained can add extra burden to an already overburdened body. Collapse and death are, unfortunately, not uncommon with critically ill birds. (If the bird appears critically ill when first presented, it is often worthwhile to have oxygen and mask on hand to assist with resuscitation if the bird collapses.) If there is any doubt as to the bird's ability to cope with the stress, it should be immediately returned to a perch or the cage and be allowed to regain its composure before proceeding.

Figure 3.3 **A pet Malabar parrot (blue mutation) with hand oil discolouration across its wings and back. (The bird has had its right leg amputated.)**

Once the procedure is complete, the bird can be gently replaced in its cage the same way it was removed. The whole procedure should be conducted with a minimum of stress, noise, and excitement.

GROOMING

Clients will frequently present their birds for grooming, namely, wing trimming, nail trimming, and beak trimming.

Wing trimming

Wing trimming is a grooming technique designed to keep birds safe while indoors. It is not a means of preventing flight; rather, it lessens the bird's ability to gain height and speed while in flight. It does not mean that the owner can take the bird outside without restraint (also see **Table 3.1**).

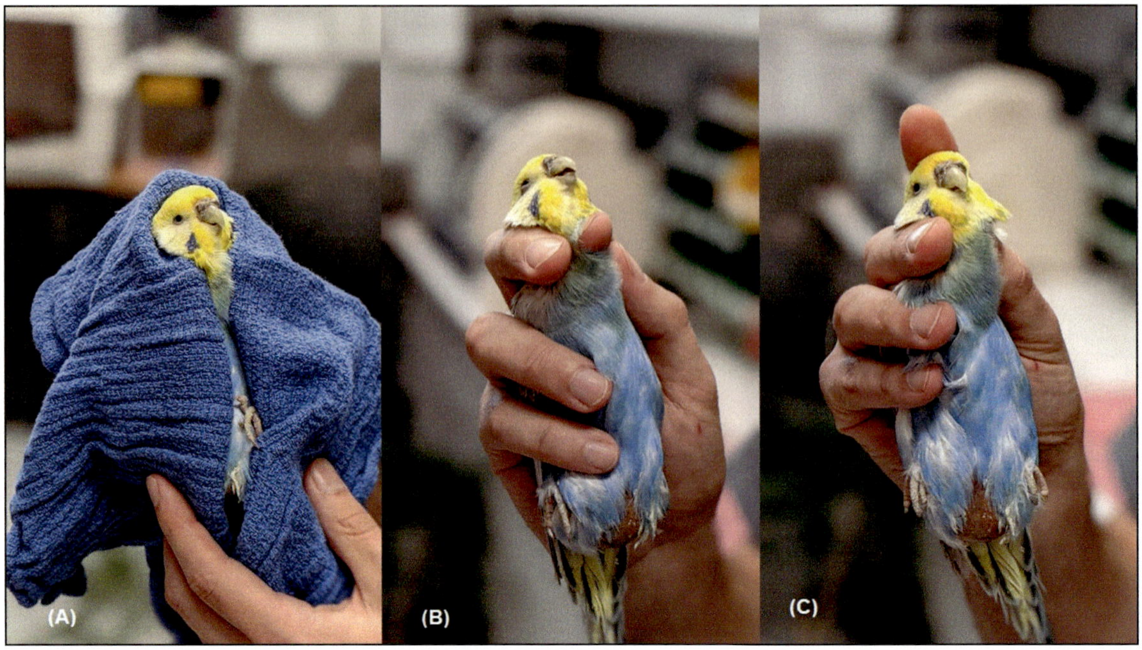

Figure 3.4 **Restraining a budgerigar: (A) towel restraint, (B) two-fingered grip, and (C) three-fingered grip.**

Figure 3.5 **Facial bruising in a blue and gold macaw after heavy-handed restraint.**

Some simple rules for wing trimming:

a. Do not trim a bird's wings until it has learnt to fly. It is only by flying that the bird can learn how to land – the most difficult part of flight. More birds are injured by clumsy landings than anything else associated with flight.
b. Trim only the primary feathers – the long feathers coming off the metacarpus from the carpus to the wing tip. The bird needs its secondary and tertiary feathers to slow down and land safely (**Fig. 3.6**).
c. Trim the same number of feathers on both wings – one-winged clips unbalance the bird and make a bad landing more likely.

Table 3.1 **The impact of wing trimming**

Negative	Positive
May lead to physical trauma and psychological impact (broken blood feathers/beak tips, chewed feathers, fractures (keel/legs/toes/beak), loss of confidence	Cannot escape through an open door or window
Leads to learned helplessness, obesity, cardiovascular disease, mutilation, pododermatitis (bumblefoot), obsessive compulsive disorder	Avoids impacts with ceiling fans, sinks of hot water, cooking surfaces, etc.
More susceptible to predation	Can free range around the house, a form of enrichment
Cannot respond physically to a stressor in a positive way	Owner has more control as the bird cannot 'escape' training sessions
Limits learning and exploration, and prevents normal behaviour	Avoids high velocity impacts with walls, windows, or mirrors

Figure 3.6 **Correct trimming of the primary feathers in a cockatiel.**

d. Never guarantee the effect of a wing trim – the trimmed feathers will moult out and be replaced, and when this will happen cannot be predicted.
e. Trim 5 mm below the edge of the coverts – this avoids cutting through a 'blood quill', a newly erupted feather with blood vessels and nerves in the shaft.

Common injuries associated with poorly performed wing trims include sternal trauma, tail split injuries, and wingtip trauma.

Imping

Imping is the process of repairing a broken feather by grafting or splicing part of another feather to the shaft of the broken feather, to allow a rapid return to full flight ability without having to wait for the natural moult of the broken feather. This is indicated for:

a. Pet birds with physical injuries (tail split injuries, sternal trauma, or wing tip damage) resulting from falling heavily to the ground due to a severe wing trim
b. Behavioural problems possibly associated with a lack of confidence because of an inability to fly
c. Wild birds with damaged feathers preventing an early release

In many cases, it is preferable to plucking the broken feather, as this may result in damage to the feather follicle and a subsequently dystrophic feather. It may take up to 6 weeks or more until the new feather grows completely.

The equipment required includes:

a. Donor feathers – ideally the same feather from a bird of the same species, colour, and size.
b. A sharp cutting instrument that will not splinter feather shafts when cutting, e.g., a box cutter knife.
c. Lightweight but strong small implants ('imping needles') to put inside the shafts of the two pieces of feather to join. Examples include bamboo skewers, the wooden handle from a Q-Tip, aluminium knitting needles, or fine metal needles for very small-diameter feather shafts.
d. Suitable fast-setting glue (5-minute setting time, rather than instant setting). Araldite is good, not instant setting (not superglue). A few minutes setting time allows final adjustment of the orientation of the feather pieces before the glue sets firmly.
e. Small strips of paper to put under the joint when gluing, to stop the glue from getting onto surrounding feathers.
f. A nail file and buffer.

The procedure is usually done under general anaesthesia because of the time required, especially if multiple feathers require imping. Cut the damaged feather shaft with a clean transverse cut, to leave at least 1 cm protruding from the skin (ideally at the base of the feather vane) to obtain the maximum length of exposed shaft. Place a strip of paper under the cut end to protect the surrounding feathers from glue spills later.

Select the appropriate donor feather and bamboo or imping needle. Measure the donor feather against the adjacent feathers (or the equivalent feather on the other wing) to determine the length required for the donor portion. Cut the donor feather shaft transversely, but not too short, because trimming may be required once fitted to the bird's feather stump.

Finally, measure and trim the donor feather against the feather stump. Adjust/trim so the diameter of the cut end of the donor shaft will match that of the feather stump, and when attached to the feather stump it will give the correct feather length. Clean out the inside of the donor shaft and feather stump with a reaming tool, piece of wire/drill bit, to remove loose keratin.

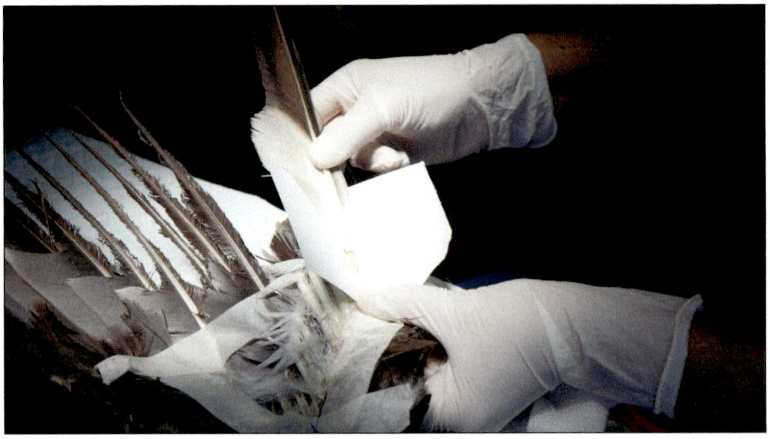

Figure 3.7 **The use of appropriately sized dowel to fit both shafts tightly, then affixed with an epoxy resin.**

Insert one end of the imping needle into the donor shaft to check the fit. Trim the ends of the imping needle so it will fit snugly into the shafts of both pieces of feather. Mix the glue. Glue half the imping needle into the donor feather shaft first, then insert and glue the other part of the imping needle and donor feather onto the feather stump (**Fig. 3.7**).

Before the glue sets, rotate the feather shaft to align the feather properly with the plane of the other feather vanes. When all feathers are glued satisfactorily and the glue has dried, use the nail file and buffer to remove any excess glue from the shaft of the feather. Ensure the glue has fully dried before the bird is recovered from the anaesthetic (**Fig. 3.8**).

Nail trimming

The nail consists of a hard keratin dorsal plate and a softer keratin ventral plate encasing the last phalange and the growth area of the nail (the nail bed). The dorsal plate grows faster than the ventral plate, giving the nail its characteristic curved shape. The nails are normally trimmed naturally by perching on branches of varying diameters, shapes, and textures.

Lack of opportunity to perch naturally can result in overgrown, sharp nails. Trimming a bird's nails is usually performed to avoid scratches and lacerations to the owner's hands and arms while handling it. Occasionally, birds with malnutrition and/or liver disease will develop overgrown nails, requiring trimming. In these cases, the nails are often deformed and soft or flaky. Birds with nails like this should be investigated for underlying problems.

The nails can be trimmed with either small 'cat' nail trimmers (**Fig. 3.9**) or, in larger birds, a hobby grinder (**Fig. 3.10**). The nail is trimmed at a level a few millimetres below the ventral plate. If bleeding occurs, it can be stopped by applying digital pressure for a few minutes or by pressing the nail into dry potassium permanganate powder (Condy's crystals) or a wet bar of soap.

Figure 3.8 **A completed set of imped primary feathers.**

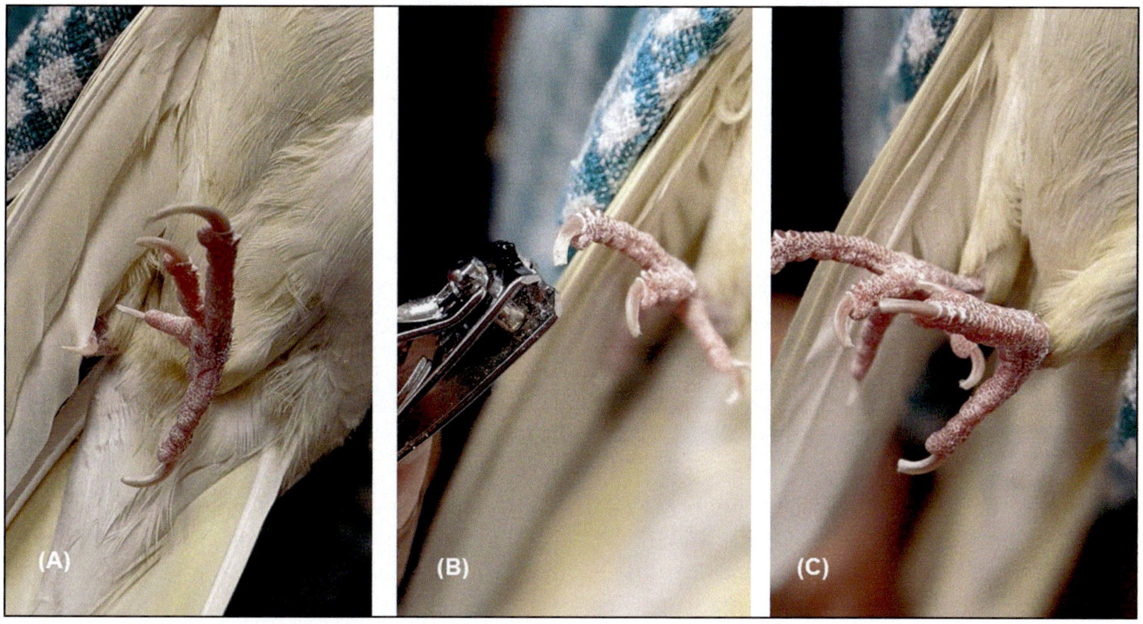

Figure 3.9 **(A-C) Trimming the nails on a cockatiel using nail clippers.**

Figure 3.10 **Trimming the nail on a macaw with a hobby grinder.**

Beak trimming

Although many clients will present their bird for a 'beak trim', it is important to emphasise that this is not a routine grooming procedure. The normal beak grows and wears down at the same rate. Birds will normally keep their beak in a normal shape and length by chewing on branches and other hard wooden structures. Most normal, healthy birds do not need a beak trim. Overgrown beaks can result not only from a lack of such chewing opportunities but also from abnormal keratin growth due to anatomical abnormalities (malocclusion) and disease conditions, such as *Cnemidocoptes* infection, chronic liver disease, malnutrition, and infectious disease conditions including psittacine beak and feather disease (PBFD), neoplasia, and bacterial or fungal infections. Birds presented for the first time for a beak trim should be thoroughly evaluated for such problems, using radiography of the beak and plasma biochemistry analysis and specific disease testing (PCR, biopsy, culture).

It must also be remembered that some birds, e.g., the long-billed corella, have a naturally long beak. It is not inappropriate to check what length a normal beak in an unfamiliar species should be.

It is also important to note the gross and microanatomy of the beak before beginning to trim the beak (see Chapter 12). Nerve-rich papillae extend through bone into the rhamphotheca, providing sensory input from the beak tip. When trimming a beak, the clinician must give careful consideration to the extent of these dermal papillae, as damage to them (e.g., from using clippers or a rotary grinding tool) can result in pain, increased sensitivity, or loss of object and food manipulation abilities, feeding, preening, and social behaviours.

Low-stress handling and restraint are essential components of these procedures. The potential for stress, causing escalating escape and avoidance behaviours, is very real. Sedation should be considered for some birds and, in severely overgrown beaks, especially those with malocclusion, general anaesthesia may be required to achieve a good result without unduly stressing the bird.

Although small birds, e.g., cockatiels and budgerigars, can be trimmed with small nail clippers, larger

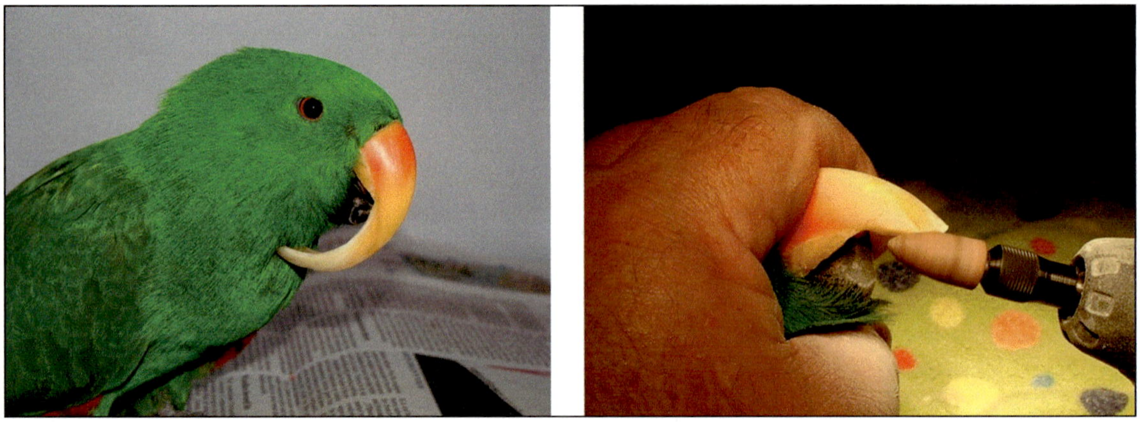

Figure 3.11 **Trimming an overgrown beak in an eclectus parrot.**

birds usually require trimming with a hobby grinder to both shape and shorten the beak (**Fig. 3.11**). Care must be taken with soft or flaky beaks not to inadvertently damage the beak through overheating with a hobby grinder or trimming down to the dermis.

Trimming a beak with a hobby grinder produces a lot of fine dust; personal protective equipment such as safety glasses and a face mask should be worn.

Keratin structures that should be evaluated and may require attention include the rostral and lateral tomia of the rhinotheca, the rostral and lateral tomia of the gnathotheca, the culmen, and the inner occlusal ledge of the rhinotheca (see Chapter 12). Excessive keratin in these areas can be removed, shortened, or reshaped through careful use of a hobby grinder or other tools. The optimal use of this tool requires the selection of the proper size and shape of bits, manual skills in handling and using the tool, and appreciation of the potential for heat that can be generated with excessive pressure or contact time. Short bursts of bit-to-keratin contact will help minimise the heat generated, as opposed to continual grinding pressure throughout the procedure.

Following completion of the procedure, a targeted enrichment strategy should be employed to encourage the bird to use its beak in a normal manner. The provision of fresh browse or other wooden objects to chew is an example of this strategy, and it should be incorporated as a part of the procedure and its follow-up.

Leg band removal

Leg bands are used by breeders to identify individual birds, a technique that predates microchipping and is still in common use. These rings can be made of plastic, aluminium, or stainless steel, and are either closed (i.e., no join) or split/open (a join is present). Split rings can be placed on any bird at any time, but closed rings have to be applied when the bird's foot and leg are small enough for the ring to fit over it. Positioned correctly, the band sits over the podotheca (scaled, unfeathered part of the leg).

Split rings are often malleable to allow them to be spread open and then closed when positioned. Unfortunately, this malleability allows a parrot to crush the band with its beak, either flattening it or allowing the two ends to override. This creates a tourniquet effect, cutting off the blood supply to the lower limb.

All leg bands can become trapped on cage wire or toys, causing the bird to struggle and often resulting in fractures of the limb. Leg bands can also cause constriction of the lower extremity if swelling of the limb occurs for any reason (e.g., trauma or scaly leg mites).

For these reasons, most avian veterinarians recommend that leg bands be removed and replaced with a microchip at an early age. The principle of removal is to cut the ring into halves, allowing the ring to simply fall off. Aluminium and plastic rings can often be cut with small scissors or specialty ring cutters (sometimes available from jewellers). Stainless steel bands often have to be cut with a rotary tool equipped with a cutting wheel (e.g., a Dremel®) (**Fig. 3.12**). Care must be taken when removing the ring to avoid placing too much torque on the rings, as a slip or sudden success may result in the leg snapping due to the force. Care also needs to be taken if there is excessive swelling around the ring, as this may indicate circulatory compromise of the underlying bone. In these cases, it is not unusual for the lower extremity to come off when the band is removed.

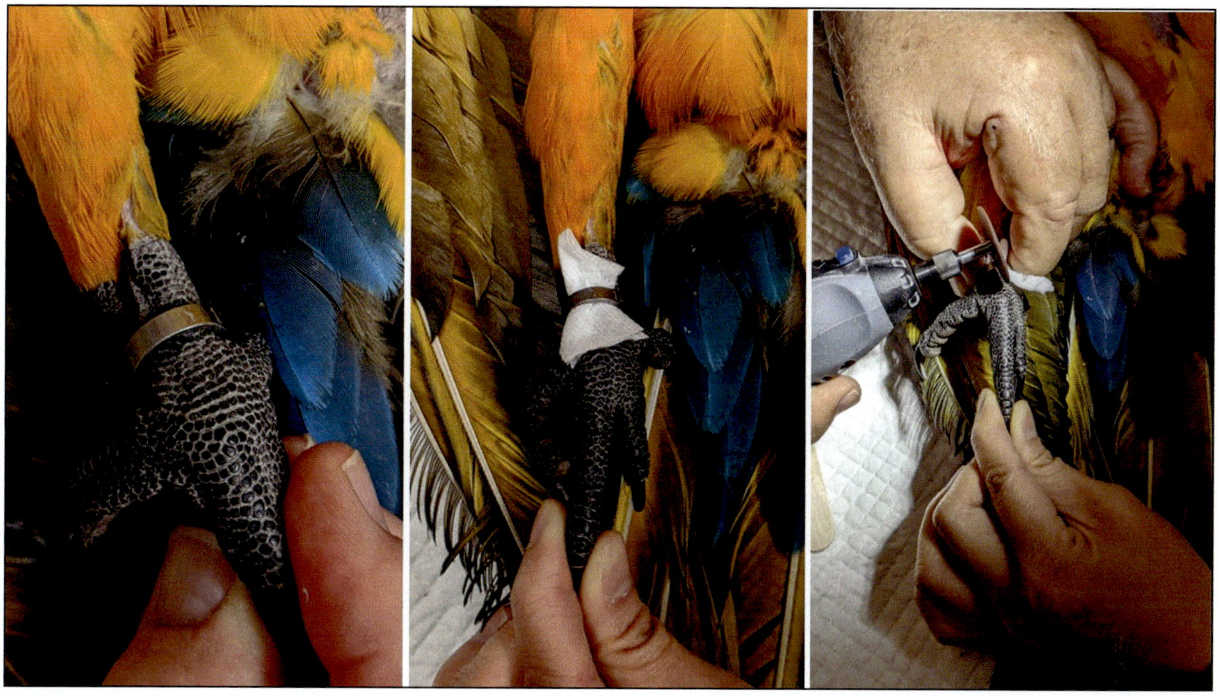

Figure 3.12 **Removal of a stainless-steel ring using a cutting wheel on a hobby grinder. The strip of gauze is soaked in water to protect the skin from heat and lacerations.**

Microchipping

In the last 15 years, microchip implantation has become a popular method of permanent identification in dogs, cats, and laboratory animals. A battery-free transponder or microchip is inserted into the subcutaneous space or muscle tissue. When read by the appropriate scanner, the microchip provides an identification number that can be associated through a database with information about the animal and the owner.

Due to the size of the needle and microchip, it is often safer and less stressful to anaesthetise the bird for implantation. In large birds (>1 kg), the topical application of a local anaesthetic cream (e.g., Emla®) and sedation (if needed) is recommended. Prior to implantation with a microchip, the bird to be implanted should be thoroughly scanned to ensure a microchip is not already in place. The microchip should also be scanned to ensure it is functioning and that the number corresponds to its accompanying documentation. (A second scan should follow-up the procedure to ensure successful implantation.)

By convention, microchips are usually inserted into the left pectoral mass, although the interscapular space is often used in birds larger than 1 kg or very small birds (<50 g). Small birds (<1 kg) are usually anaesthetised for the procedure, while larger birds can be implanted under local anaesthesia. Ostriches and other ratites should be implanted in the pipping muscle at the back of the head shortly after hatching, or subcutaneously in the left thigh at the feather line if they are older than 2–3 days.

Implanting should be performed in such a manner as to minimise the risk of the transponder inadvertently falling out. After prepping the skin with an alcohol wipe, the needle is inserted in a craniocaudal direction (thus placing the microchip at the bottom of the pocket created by the needle) at a shallow angle. The insertion wound can be closed with a drop of tissue glue, and the bird scanned to confirm placement.

It is vital that the details of the bird and microchip are recorded on an appropriate database as soon as possible. Some clients are of the view that microchips can be used to track an escaped bird – they must be left with no doubt that the microchip is only used for identification once the bird has been recovered.

CHAPTER 4

THE PHYSICAL EXAMINATION

UNDERSTANDING THE MASKING PHENOMENON

A common misconception held by many bird owners (and many veterinarians) is that birds are not very resistant to illness. To the novice, it often appears that birds show signs of illness one day, are at the bottom of their cage the next, and dead the day after. This misconception has stemmed from two sources.

Firstly, many of the birds seen in practice are only a few generations descended from wild birds. As such, they still retain many of the protective instincts inherited from their forebears. As many of the species kept as companions are relatively low on the food chain, quite a few of these instincts have been developed to avoid drawing the attention of predators. One such instinct is often known as the masking phenomenon. Predators are naturally drawn to prey that looks or behaves differently from others. Unusual colouring, weakness, and lameness can single out a bird and make it a draw card for a predator. A natural instinct is therefore to avoid looking 'different': a sick bird will make a determined effort to look healthy, even when there are no predators around. The classical 'sick bird look' signs we usually associate with illness (fluffed up, eyes closed, lethargic) only develop when the bird is incapable of masking these signs (**Fig. 4.1**). Many of the patients presented to veterinarians are well past the initial stages of their illness and are now decompensating rapidly.

There are many subtle changes in a bird's behaviour or appearance, however, which are indications of a health problem. While these signs are often discernible by experienced owners and veterinarians, they are just as often overlooked by those with less experience. Missing these early signs, when combined with the bird's efforts to mask obvious clinical signs, invariably leads to the late detection of illness and the presentation of the bird *in extremis*. It is important that veterinarians learn to recognise these early signs of illness, and then educate their clients so that illnesses can be detected before becoming too advanced.

The masking phenomenon and the ease with which early clinical signs can be overlooked highlight the importance of regular health examinations for the companion bird. Long-standing conditions such as malnutrition can be detected and corrected before the bird begins to decompensate and show signs of overt illness.

EXAMINATION ROOM EQUIPMENT

Appropriate equipment for use in the examination room includes:

- A supply of freshly laundered towels in different sizes (or paper towels) for restraining birds.
- Scales capable of weighing in grams, preferably with a detachable T-perch (to allow birds to perch on while being weighed), and a container in which to weigh smaller birds.
- Mouth gag of a suitable size and material for the species being examined.
- A training perch for the bird to perch on while being examined.
- Clinical equipment includes items such as a stethoscope, a focal light source, magnifying loupes, needles and syringes, blood collection bottles, and culture swabs.
- Alcohol for wetting feathers down for a closer examination of the skin and underlying fat and muscle.
- Treats to reward pet birds and make the experience more enjoyable (or at least, less stressful).

The use of heavy gloves to catch and restrain birds should be discouraged. With these gloves on, the

DOI: 10.1201/9781003272519-4

Figure 4.1 **The masking phenomenon. (A) A sick cockatiel taken from its carrier. (B) The same bird a few minutes later, unable to hide signs of illness any longer.**

clinician cannot be sensitive to the small movements of the bird and can easily hurt or even kill the patient.

Ensure the room is escape-proof and that clinic staff do not enter the room unexpectedly. Avoid stressful sights and sounds such as dogs, cats, and other potential predators.

HISTORY-TAKING

The collection of an accurate and thorough history of a patient is as crucial to making a diagnosis as the physical examination and appropriate diagnostic testing. A good history can alert the clinician to likely problems and allow him/her to focus on likely possibilities and refine the rest of the diagnostic approach. History-taking can be divided into:

- Background history:
  - Signalment
  - Origin of the bird
  - Husbandry
  - Nutrition
  - Behaviour
  - Reproductive history
  - Previous medical history
- Presenting problem

Signalment

The first step in obtaining a history is to gain as much information about the bird itself as possible. A good receptionist or technician can often obtain such information, but the clinician needs to be familiar with birds in general, as often clients are not aware of some basic facts about their bird (such as species, sex, and age). Things that need to be ascertained include:

- The species of bird. The clinician needs to be able to recognise common species and have access to literature that will enable him/her to identify other species. Be aware of local names that may differ from those in the literature. Knowing the species, its behaviour, and its dietary requirements can offer the clinician vital clues to likely problems.
- The age of the bird can offer other clues. As a general rule, juveniles are more likely to suffer from infectious diseases and nutritional deficiencies, while adult birds are more likely to suffer from neoplasia, chronic malnutrition, and degenerative conditions.
- The sex of the bird. Many parrots are sexually monomorphic and require DNA or surgical sexing. Do not accept the owner's assertion of their bird's sex unless they have proof of sex identification (e.g., a history of egg laying or a certificate of sex identification that can be correlated with this bird). Knowing the sex of a bird can be vital, as many conditions, including behavioural problems, can be linked to the bird's sex. For example, coelomic hernias are almost non-existent in male birds, while being relatively common in females; yolk-related peritonitis is obviously only seen in hens; cocks do not become egg-bound.

Origin of the bird

- How long have the clients owned this bird? Birds that have been in the owner's possession for many years, with no recent exposure to other birds, are less likely to have infectious diseases. Recently obtained birds are more likely to have been in close contact with other birds and, as such, have possibly been exposed to infectious diseases.
- Where did the owner obtain the bird? With experience, the clinician will be able to identify 'problem sources' of birds in the local area (e.g., a certain breeder or pet shop). Developing a working knowledge of the quality of the sources of pet birds in your area can be a key element of patient evaluation in your practice.
- Is the bird aviary-bred or wild-caught? Wild-caught parrots, although less common in recent times, still appear occasionally in the pet market. The behaviour and diseases of such a bird may well be linked to its source.
- Is the bird hand-reared or parent-reared? The health of juvenile birds is dependent to a large extent on the health and nutrition of the parents. Hand-reared birds have two additional factors to be aware of, namely, the quality of the hand-rearing formula being fed and the skill of the person doing the rearing. Parent-reared birds tend to be more difficult to tame unless they are handled on a regular basis before fledging.
- Hand-reared birds, on the other hand, while usually more closely bonded to people, often have behavioural disorders associated with poor socialisation skills, especially if reared in isolation. If a large number of chicks from different sources are reared in one facility, there is a higher probability of the spread of diseases such as polyomavirus and chlamydia. All of these factors need to be assessed during the history collection, especially when examining juvenile birds.

Husbandry

- How has the bird been managed? Is it an aviary bird, a companion bird, or a zoological specimen? An assessment of a companion bird's husbandry must include the cage, the environment around the cage, and the bird's interaction with its environment.
- Ascertain whether the cage in the examination room is the bird's permanent cage or simply a transport cage. If the latter, ask the owner to describe the permanent cage or to bring photographs or video images.
- Where is the cage located in the house? Is it exposed to toxins such as burning Teflon®, cigarette smoke, or household plants? Does the bird get any privacy? Is it kept isolated, away from family activities? Does it get a good night's sleep, or is it forced to stay up with its owner watching television all night? Does the bird get access to direct, unfiltered sunlight on a daily basis? If the bird is kept outside, is it safe from predators (including wild birds) and from vectors of diseases such as *Sarcocystis*?
- Does the bird come out of its cage? How long does it stay outside the cage each day? Are its wings clipped? Is it supervised when out of the cage? Does it interact with other animals and birds?
- Does the client have any other birds? Any other pets?

See Chapter 2 for a discussion on the husbandry of both companion and aviary birds.

Nutrition

Underlying many health problems in pet birds is the common thread of malnutrition. For many generations, bird owners have accepted as fact that birds eat seed and that is all they need. Birds, like many other animals, have a preference for high-fat diets. Given a choice between seed, vegetables, formulated diets, and fruits, nearly all birds will consume the seed first. Given this preference, it is not surprising to hear many bird owners state, "All he will eat is the seed, so that is all that I give him". See Chapter 2 for further discussion about nutrition.

It is therefore important to ascertain:

- What does the owner feed the bird and what does the bird actually eat? The clinician needs to be aware that there are major species differences in dietary requirements, with, for example, some species having a higher requirement for fat than others. There is no one diet to suit all avian species, any more than there is one diet to suit all mammalian species.

- How much food is being consumed? (Some birds on an excellent diet will still eat too much and gain excessive weight.)
- Is the food prepared fresh daily?
- Are dishes cleaned every day?
- Does the bird dunk food into its water dish, creating a nutrient-rich broth ideal for bacterial contamination?
- Does the bird get any treats such as food off its owner's plate?
- Are vitamin and mineral supplements being offered?

Paediatric cases open another spectrum of questions. What hand-rearing formula is being used? How is it prepared? Are the manufacturer's recommendations being followed? Has anything extra been added to an already balanced diet? At what temperature has it been fed? How much has been fed, and how often? How has the chick been fed: syringe, crop tube, or spoon? How are these utensils cleaned and disinfected?

Behaviour

A behavioural history is becoming increasingly more important as pet birds move out of their cages and more into their owners' lives. Just as countless dogs and cats are euthanised every year because of behavioural problems, many birds suffer the same fate or are transferred from household to household for the same reasons.

As psittacine behaviour is determined to a large extent by the interaction between the bird, its owners, and its environment, questioning must focus on these areas:

- How many hours per day does the bird spend alone?
- What does the bird see and hear when it is alone?
- Does the bird spend time with other birds or other pets?
- Are toys provided for the bird? Does it utilise those toys?

The bird's interaction with its owners (human flock) will need to be clarified:

- Who is the primary caretaker?
- Who does the bird seem to prefer?
- Does the bird dislike anyone?
- How tame is the bird?
- Does it readily step up onto a proffered hand?
- Does it always try to move up onto a person's shoulder? Is this allowed or encouraged?
- Does the bird talk?
- Does it like to be petted? If so, where?
- How does it react to different family members?
- How does it react to strangers?

See Chapter 30 for more discussion on avian behaviour.

Sex of the bird and its reproductive history

Does the owner know the sex of the bird? If not a dimorphic species, how was the sex determined? (See below.) It is important to ascertain if the bird has displayed any reproductive activity. If it is a hen, has it engaged in nesting or territorial behaviour, or has it laid eggs? If a cock bird, has it displayed courtship behaviour towards another bird or even the owner?

Previous medical history

The patient's medical history should be examined. Has the bird been ill before? Who saw it, what did they diagnose, and how was it treated? Has it had remedies supplied by pet shops or breeders? Is the bird being presented for a second opinion or a specialist referral? If appropriate, permission to obtain copies of medical records from the previous (or referring) veterinarian should be requested.

Presenting problem

- What exactly is the problem? This needs to be clarified with the client, as confusion often exists between owners and veterinarians over the precise description of clinical signs.
- When did it start?
- What effect is the problem having on the bird's appetite, thirst, urination, defecation, and behaviour?
- Is this the first time it has happened?
- Are other birds affected?
- Have other treatments been tried? Who prescribed these treatments? Did they work?
- Is this condition progressing, static, or perhaps even improving?

The history-taking described here is not a comprehensive review of every possible question that can be asked of the owner. As the process continues, areas of interest will become apparent, and more questions may be needed to clarify these areas. The clinician must take care to ensure that he/she does not dominate the conversation; rather, they should ask short questions and listen carefully to the client's reply. However, the clinician must be prepared to guide the discussion, as otherwise, some clients may become distracted and lose track of the original question!

Figure 4.2 **A distant exam of these wild cockatoos is sufficient to arouse suspicions they have PBFD.**

THE DISTANT EXAMINATION

Although the history-taking should be done before the bird is handled, the clinician should use this time to gain some insights into the bird and its health by careful observation of the bird, its cage, and its droppings.

The bird

- Most birds, no matter how ill they are, will try to mask their clinical signs when first brought into the examination room. This effort will rarely last long, usually only a minute or two. Avoid disturbing the bird until it has settled; otherwise, valuable clinical signs may be overlooked.
- Watch the bird's breathing. Once the bird has settled in its cage in the examination room, there should be no open-mouth breathing, marked tail bobbing, or marked respiratory effort. There should be no audible respiratory noise. The presence of these signs should alert the clinician to the likelihood of respiratory compromise, and great care must obviously be taken with handling such a patient.
- Look at the bird's posture. Many sick birds are hypothermic and attempt to conserve body heat and energy by fluffing their feathers, sitting still, and sleeping more. These signs are often referred to as the 'sick bird look', but not all sick birds will display these signs. Evidence of a wing droop, lameness, or reluctance to bear weight on one leg indicates a musculoskeletal problem. Spinal deformities can often be detected by an abnormal positioning of the tail. An upright position with a wide-legged stance may indicate egg-binding or a similar space-occupying lesion in the coelom. Birds holding both wings away from their body and panting are usually heat-stressed.
- Examine the plumage. Normally, it should be sleek, well-groomed, and clean. Untidy or dirty plumage can indicate either that the bird is not grooming itself for some reason or that there is a feather dystrophy of some description. Discoloured feathering can reflect a variety of problems, such as psittacine beak and feather disease (PBFD) (**Fig. 4.2**), chronic liver disease, excessive handling with oily hands, or malnutrition.
- Examine the beak and toenails. Overgrown or flaky beaks, or overgrown and twisted nails can be associated with PBFD (in cockatoos), poor husbandry, chronic liver disease, or malnutrition.
- Watch the bird defecate. Look for signs of straining or discomfort, and listen for any accompanying flatulence.
- Observe the bird's behaviour, assessing how tame the bird is and whether it is showing any overt sexual display towards the owner or other people present in the room. Look at how the owner interacts with the bird, as this can give valuable clues to relationships at home.

The cage

(Also see Chapter 2.)

- Is the cage large enough (**Fig. 4.3**)?
- Is it constructed from safe materials?
- Is the floor of the cage solid or wire? Is it covered in grit, sand, or paper?

Figure 4.3 **A cockatoo with badly damaged primary feathers, a consequence of this bird being housed in an inappropriately small cage, leading to its wings striking the bars.**

- What sort of perches and dishes are in the cage, and how are they positioned?
- Are any toys appropriate for the bird and not overcrowding the cage?
- Is the bird offered the opportunity to bathe, or at least to be misted, on a daily basis?

Faecal examination

Birds' droppings are made up of three components: faeces, urates, and urine (**Fig. 4.4**). In a healthy bird, the faecal portion should be formed and homogeneous, with little odour (except for poultry, waterfowl, and carnivorous birds). The colour should range from brown to green. The urates should be a crisp white and slightly moist. Old droppings will have greenish urates as biliverdin leaches out of the faeces. The urine should only extend a couple of millimetres past the dropping. True polyuria should not be confused with so-called excitement polyuria, the excess urine produced by an excited or nervous bird. Lorikeets, due to their liquid diet, will produce large amounts of urine, which should not be mistaken for pathological polyuria. A close examination of the droppings is a valuable starting point for a clinical examination.

Some abnormalities commonly encountered include:

- Diarrhoea (unformed faecal portion).
- Undigested food in faeces.
- Very bulky droppings indicate maldigestion, malabsorption, reproductively active hens, coelomic growth, or pelleted diets.
- Melena or very dark droppings can indicate anorexia or intestinal haemorrhage.
- Malodorous droppings are often associated with bacterial/fungal overgrowth.
- Green urates (biliverdinuria) are often indicative of liver disease.
- Pink/red urates (haematuria/haemoglobinuria) are seen in cases of renal disease, often associated with lead poisoning, especially in Amazon parrots and galahs.
- Yellow urates can be associated with anorexia.
- Orange urates indicate that a vitamin B injection may have been given in the last few hours.
- Thick, pasty urates are seen in dehydrated birds.
- Polyuria (excessive urine – be aware that this is normal in lorikeets and birds eating a lot of fruit).
- Anuria.
- Discoloured urine: similar causes to discoloured urates.
- Fresh blood indicates either a cloacal problem or an oviductal disease.

THE PHYSICAL EXAMINATION

A thorough, systematic physical evaluation of the patient is essential to obtaining clues about the bird's problem and diagnosis. Clinicians should develop a thorough examination protocol that they are comfortable with and use it for every patient regardless of the reason for presentation.

Identification of the sex of birds

To the human eye, many birds are sexually monomorphic (e.g., many South American and African species). Some are initially monomorphic, later

Figure 4.4 **Various forms of faeces from healthy and sick birds. (A) Normal droppings from a healthy bird, with normal urine, urates, and faeces. (B) Haematuria. (C) 'Popcorn' droppings, associated with exocrine pancreatic insufficiency. (D) Green urates, suggestive of liver disease. (E) Whole seed in the droppings, suggestive of upper gastrointestinal problems. (F) Yellow urates and scant faecal production in an inappetant bird. (G) Large, unformed droppings from an egg-laying hen. (H) Malena from a bird with ventricular ulceration. (I) Biliverdinuria from a bird with liver disease.**

becoming dimorphic (e.g., many Asiatic parrot cocks develop a black neck ring at 2–3 years of age). Other species are obviously dimorphic from the time of fledging (e.g., the eclectus parrot). To birds, however, their ability to see in the ultraviolet spectrum of light is believed to make sex identification feasible for them.

Techniques employed by veterinarians and/or aviculturists to identify the sex of their birds include:

- DNA sexing
- Surgical sexing – endoscopic inspection of the gonads (see Chapter 9)
- 'Wait and see' for those species developing dimorphism later in life (**Fig. 4.5**)
- Vent sexing in ratites and poultry involves examining for the presence of a phallus on the ventral floor of the proctodeum.

Weight recording

A vital aspect of avian medicine is the accurate recording of the patient's weight. All birds should be weighed at each visit to the veterinarian and each day while hospitalised. An accurate weight record allows the clinician to quantify the bird's body condition (if normal weights for that species are known), accurately calculate drug dosage rates, and monitor the patient's response to therapy (if it is weighed daily). Over a period of time, the clinician will also develop a working knowledge of expected body weights for different species.

Weights should be recorded to the nearest gram. As such, a veterinary practice dealing with birds will need to invest in a good-quality digital scale, readily available in many electronics or kitchenware stores (**Fig. 4.6**).

Once practised, the weighing of an avian patient is relatively easy to perform. Larger birds can be

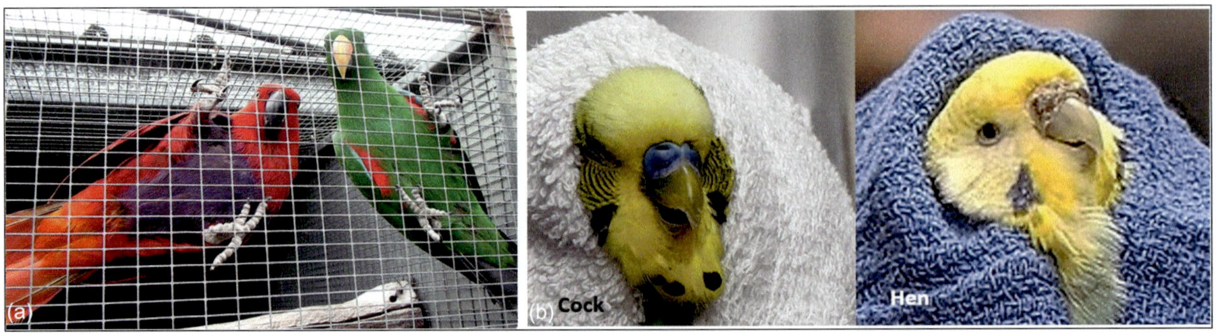

Figure 4.5 **(a) Sexual dimorphism in a pair of eclectus parrots. (b) Sexual dimorphism in budgerigars.**

Figure 4.6 **Using a T-perch on a set of scales allows for easy weighing of most pet birds.**

weighed by having them stand on a T-perch mounted on the digital scale. Smaller birds, such as budgerigars, canaries, and finches may be placed inside a metal or plastic container similarly mounted on the scale.

The patient's weight must be recorded on its medical records to demonstrate to clients their pet's weight gain or loss and to provide part of that patient's minimum database.

Auscultation

When to auscultate is the prerogative of the clinician, but it may be better performed before the bird is handled for too long. The heart rate is usually rapid, although that of some larger pet birds (e.g., cockatoos) can be surprisingly slow compared with stressed or smaller birds. Murmurs, arrhythmias, and bradycardia are occasionally detectable.

Movement of air through the lungs and air sacs can be auscultated over the bird's back, and occasionally friction rubs associated with air saculitis can be detected.

Body condition

Traditionally, a bird's body condition was determined by palpation of the pectoral muscles (**Fig. 4.7**) and allocating a body score based on the muscle and fat coverage of the sternum. This technique fails to take into account that most birds do not store fat in their pectoral region, and they can be carrying significant fat deposits while still having an apparently good body score. Wetting the feathers over the ventral coelom and flanks with alcohol allows visualisation of subcutaneous fat deposits, seen as yellow fat under the skin rather than pinkish-red muscle. Coelomic fat deposits can be evaluated by palpation.

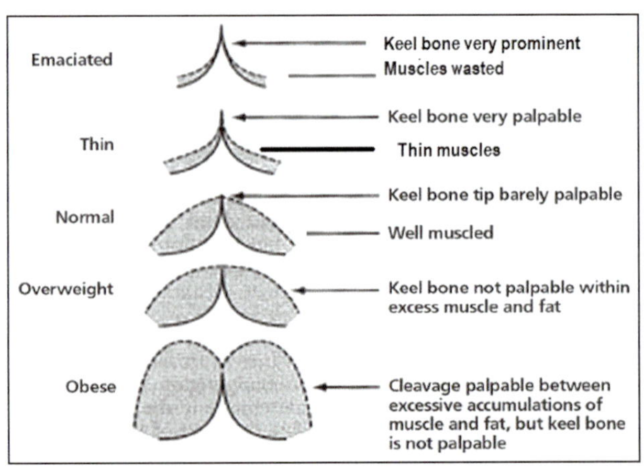

Figure 4.7 **Assessing body condition by pectoral muscle palpation.**

The combination of body weight recording, pectoral muscle and coelomic palpation, and examination of subcutaneous fat allows for a more accurate assessment of body condition than just palpation of the pectoral muscles. Many avian veterinarians prefer not to use a body condition scoring scheme, as there is no international consensus or consistency.

Skin and plumage

The bird's skin and feathering should be examined in detail. Attention should be paid to the following areas:

- Colour of the feathers. Normal colouration of feathers is the result of pigment deposited in the feather keratin and the reflection of light from the feathers. Abnormal colouration of feathers results from anything that affects the pigment deposition or the ultrastructure of the barbs and barbules. PBFD can cause green feathers to turn yellow and blue feathers to turn white. It will also lead to a generalised dirtiness of the feathers, especially in cockatoos. Chronic liver disease and/or malnutrition can cause darkening of feathers (**Fig. 4.8**). As mentioned earlier, frequent handling of green birds by the owner can cause a black discolouration on their feathers. This is not seen in birds with powder down, presumably because the powder keeps the feathers clean.
- Tidiness of the plumage. Birds generally keep their plumage well-groomed and tidy. If the plumage is untidy, with no immediately obvious cause (e.g., recent handling), the clinician should suspect that either the bird is unable to groom itself properly or a generalised feather dystrophy (e.g., PBFD) is present.
- Evidence of feather damage. Chewed and/or broken feathers should lead the clinician to suspect overgrooming, self-mutilation, or malnutrition. Saw-toothed edges can indicate a failure to moult normally, hence old, worn feathers are being retained. It should be noted in cases of feather picking whether the feathers have been bitten off level with the skin or plucked out, or if the shaft has been chewed.
- Evidence of feather dystrophies, retained feather sheaths, retained pulp, haemorrhage in the shaft of feathers, strictures of the calamus, and twisted feathers are indicative of feather dystrophies, often of viral origin (e.g., polyomavirus, beak and feather disease virus).
- Wing clipping (if present). The wings should be examined to determine if the bird's wings have been clipped and, if so, if that clip is appropriate for the species of bird. The quality of the clip should be examined to determine if the cut ends of feathers could be bothering the bird (**Fig. 4.9**).
- Absence or presence of powder down. Powder down is produced by the powder down feathers on the thighs of many species of birds, particularly cockatoos and African grey parrots. It is easily recognised by the presence of a fine white powder on the clinician's hands and clothing after handling the bird. A lack of powder down leads to staining of the feathers and a shiny appearance of the beak and feet. The most

Figure 4.8 **Wing tip trauma in a princess parrot associated with a poorly performed wing trim.**

Figure 4.9 **Darkening of the feather colour in this cockatiel suggests either chronic hepatopathy or hypothyroidism.**

common cause of a loss of this powder is PBFD, but it can be seen in obese birds, elderly birds, and birds that have chewed at their powder down feathers.

- Moulting patterns. Most birds will typically moult heavily twice yearly, in spring and autumn, the 'prenuptial' and 'postnuptial' moults. Outside of these annual moults, there is a steady and progressive turnover of old feathers. Continual heavy moults or the sudden loss of many feathers is abnormal, as is the failure to moult (seen as the retention of worn and broken feathers).
- The presence of stress lines. Stress or disease at the time a feather is growing will lead to a transverse 'break' in the vane of the feather. The presence of many feathers with such stress lines is indicative of a problem in the bird's recent past. The black marks left on green-coloured birds by human skin oils (see earlier) should not be confused with stress lines.
- Birds that are unable to groom properly (e.g., due to spinal disease or wearing an Elizabethan collar) may develop overgrown powder down feathers, retained in their shafts over the thighs. This should not be confused with a feather dystrophy such as that seen with diseases like PBFD.
- The condition of the skin. The presence of erythema, excessive scale, or areas of skin trauma should be noted. This can be done by parting the feathers with a cotton bud or gently blowing on the feathers.
- Areas of trauma. The skin should be thoroughly examined for areas of trauma, especially on the wing tips, sternum, and axillae.
- Flexibility of the feather. The feather of a healthy bird on a good diet should flex, rather than bend, when the tip is drawn down towards the base, and then spring back to a normal position when released.
- Parasites. The presence of parasites on the feathers should also be noted. Microscopic examination of a freshly plucked or molted feather may be needed.

It is important that changes in feathers and skin be recorded in a detailed manner. Veterinarians should familiarise themselves with the descriptive terminology used for the external anatomy of a bird and use that terminology when describing lesions. In addition, it is important that lesions be described accurately (e.g., whether feathers are being stripped, chewed, plucked, or bitten off). Such precise terminology is essential when describing a case to other veterinarians and can assist in developing a clearer picture for all concerned.

The uropygial (or preen) gland is located on the dorsal base of the tail (see Chapter 11). It should be assessed for evidence of impaction, enlargement, inflammation, or trauma.

Head

- Look for asymmetry arising from sinus swellings, exophthalmos, enophthalmos, and trauma.
- Loss of feathers on the head can be due to a variety of conditions. Some cockatiel mutations, especially lutinos, have a bald spot behind the crest. Feather loss in other species can be associated with fungal dermatitis, *Cnemidocoptes* infection, PBFD, or excessive grooming by a cage mate. Feather loss around the eyes can indicate facial rubbing associated with conjunctivitis or sinusitis.
- Matting of the feathers over the crown and nape can indicate that the bird has been vomiting (**Fig. 4.10**).
- The conformation of the beak should be assessed for the presence of congenital or acquired abnormalities such as scissor (wry) beak, prognathism, and bragnathism. Trauma to the beak or localised sinus infections can result in localised anatomical abnormalities (e.g., longitudinal grooves in the keratin). Excessive keratin flaking of the beak can reflect poor nutrition or simply a lack of opportunity to rub the beak on a suitably abrasive surface (such as a cement perch). Overgrowth of the beak can occur with PBFD, *Cnemidocoptes* infection, misalignment of the upper and lower beaks, chronic liver disease, or malnutrition. It is rarely the result of a lack of objects to chew on. (It is important to note that some species, such as the long-billed corella naturally have elongated beaks. This should not be mistaken for an overgrown beak.) Underrunning of the ventral surface of the rhinotheca is commonly seen in cockatoos with PBFD but can also be seen in birds with bacterial or fungal infections in the keratin.
- The cere, the fleshy skin at the top of the beak, is not present in all species. In the budgerigar, cere colour can be used to sex the bird, with

Figure 4.10 **Matting of the feathers associated with vomiting in a budgerigar.**

cocks having a blue cere and hens a brown cere. However, this will vary with the age of the bird, the colour mutation, and the degree of health. Cere hypertrophy, a thickening of the brown cere in the budgerigar hen, may reflect a hyperoestrogenic state.

- The nares, the openings into the sinuses located at the top of the beak, should be symmetrical and dry. They should be open, and the presence of keratin plugs (rhinoliths) is abnormal. Blockage of the nares may result in a subtle movement of the skin over the infraorbital sinuses. This may be the only clue that a blockage exists. Discharge from the nares, often seen as staining and matting of the feathers dorsal to the nares, may be an indication of upper respiratory disease. Previous or chronic episodes of sinusitis may be seen as asymmetrical enlargement of the nares, sometimes associated with a longitudinal groove in the keratin of the beak.
- The eyes should be bright and clean. Ocular discharge and loss or matting of the feathers around the eyes indicate either conjunctivitis or sinusitis. Conjunctival hypertrophy is common in chronic conjunctivitis, especially in cockatiels with *Chlamydia* infection. Focal light, magnification, and fluorescein dye are needed for a detailed ocular examination. See Chapter 13 for a detailed discussion of the ocular examination.
- Examination of the oropharynx can be accomplished by using a gauze bandage or metal gags to open the mouth. The choana (the slit on the roof of the oropharynx) should be free of excessive mucus or discharge and fringed with well-defined papillae. There should be no abscesses or diphtheritic membranes present.
- The ears can be examined by parting the ear coverts with the handle of a cotton bud or similar appliance. The ears should be open and free of discharge or erythema. Visualisation of the tympanic membrane is difficult in most species without the use of an endoscope.

Crop

The crop can be palpated in most birds at the base of the neck, just cranial to the thoracic inlet. It should be carefully and gently palpated to assess if:

- Food is present (i.e., is the bird eating?).
- It feels doughy or fluid-filled, indicating that crop stasis may be present.
- Ingluvoliths or other foreign objects are present.
- The crop mucosa feels thickened.

Care must be taken in critically ill birds not to force fluid or ingesta back up the neck into the oropharynx, as aspiration and death can result.

Body

Palpation of the skin over the trunk occasionally reveals the crackling of subcutaneous emphysema. While this is normal in species such as pelicans, in most species, it is often the result of trauma or infection rupturing air sacs and allowing the escape of air under the skin.

The coelom in the normal bird is concave between the end of the sternum and the pubic bones. The clinician needs to distinguish between internal and external distension of the coelom. Internal distension of the coelom can be due to fat, organ enlargement, ascites, neoplasia, or the presence of an egg. External distension can be due to subcutaneous fat, neoplasia (especially lipomas), or hernias. Radiography and ultrasonography may be required to distinguish between internal and external coelomic distension, and between different aetiologies of both. Coelomic pain or discomfort can occasionally be elicited by careful palpation.

The back should be carefully palpated for evidence of scoliosis, lordosis, or kyphosis. As the notarial vertebrae are predominantly fused, flexibility of the spine cannot be assessed in the same way as it is in dogs and cats.

The carina of the sternum should be palpated for evidence of distortion, trauma, or congenital defects such as splitting. Distortion of the carina, often indicating a history of rickets or other metabolic bone diseases, should lead the clinician to recommend radiographic evaluation of the rest of the patient's skeletal system.

The cloaca can be assessed externally for enlargement and dilation (often indicative of reproductive behaviour in a hen), prolapse, ulceration or inflammation around the mucocutaneous junction, and the presence or loss of sphincter tone. Gently everting the cloaca can give a cursory examination of the mucosa, possibly revealing papillomas in susceptible species. A more thorough evaluation of the cloaca requires endoscopy.

The area between the cloaca and the tail should be assessed for splitting of the skin (**Fig. 4.11**). This is commonly seen in pet parrots (especially cockatiels) and is associated with a poor diet and a poor wing clip. The wing clip causes the bird to land awkwardly, pushing its tail up as it does so. Malnutrition causes the skin to lose its elasticity. The result is that the skin in this area splits, leading to bleeding and feather picking in this area. It is also a common area for hernia in hens, often containing the oviduct or even cystic ovaries, and for lipomas in obese birds.

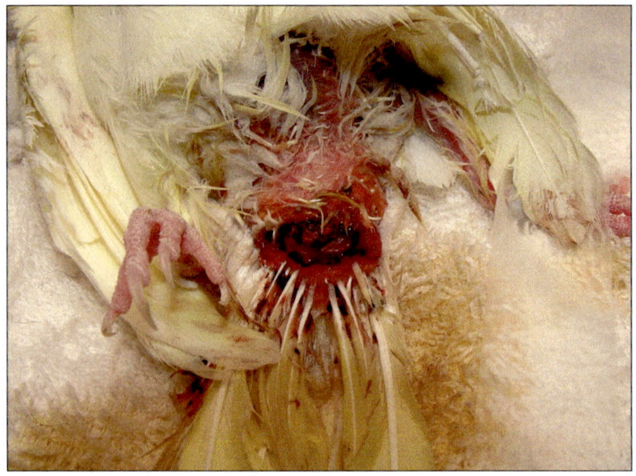

Figure 4.11 **Tail split injury in a cockatiel, associated with a poor wing trim and a heavy fall to the ground.**

Wings

Each wing should be carefully extended and flexed to assess mobility and compared with the other wing. The bones and joints should be palpated for swelling or crepitus. Recent trauma may be evident as greenish discolouration of the soft tissue. This is bruising and should not be mistaken for infection or tissue death. If the cause of a wing droop is not detectable after careful palpation, radiography is required to assess the pectoral girdle. The bones of this girdle are covered in strong muscles, and fractures are often not detectable by palpation alone. The patagium should be evaluated for loss of elasticity, trauma, or scarring and for the presence of a tattoo (indicating the bird has been surgically sexed).

Legs

Each leg should be carefully palpated to detect abnormalities such as fractures, swelling, bruising, healed calluses, and angular deformities of the long bones. Each joint should be extended and flexed to assess mobility and range of movement. Joints should also be examined for swelling and deposition of chalky white uric acid crystals (i.e., articular gout). Each leg should be compared with the other in all aspects (e.g., symmetry, length, swelling).

The toes should be examined for abnormalities, including:

- Missing digits or nails
- Annular constrictions
- Swelling of interphalangeal joints, occasionally with the deposition of uric acid crystals
- Avascular necrosis
- Excessive thinness, especially in neonates
- Abnormal position and conformation of the toes
- Excessively long or twisted nails

The plantar surface of each foot should be examined, and the condition of the metatarsal pad and digital pads noted. Abnormalities seen here include loss of definition of the epidermis (seen as a shiny, reddened surface), swelling, erosions, ulcers, and scabs. Pododermatitis (bumblefoot) is common in raptors but can be seen in any bird where unilateral lameness causes more weight-bearing on the unaffected leg. This in turn can lead to pressure necrosis, infection, and subsequent pododermatitis. Consequently, in cases of unilateral lameness, the opposite leg should always be closely examined.

In some birds, especially overweight specimens, calluses and dermatitis can also be seen on the caudal aspect of the tarsometatarsus. This may be the result of improper perches and a bird that cannot perch normally because of its weight.

Occasionally, due to the nature of the injury or the disposition of the bird, a full examination may require general anaesthesia. If this is the case, radiographs can be taken at the same time, minimising handling of the conscious (and therefore stressed) patient.

Neurological assessment

Birds presented for any of the following problems require a thorough neurological assessment:

- Abnormal posture or behaviour
- Abnormal mentation
- Paresis or paralysis of any or all limbs
- Fractures of limb bones
- Weakness or inability to grip with one or both feet
- Other neurological abnormalities

The assessment should be performed methodically and logically. The order in which the assessment is performed depends on the clinical condition and cooperation of the patient. The degree of cooperation should be considered when interpreting responses to neurological tests.

The first part of the neurological assessment is to step back and observe the bird's posture, mentation, flight, gait, and behaviour. (This is discussed in more detail in Chapter 26.) The bird is then caught and examined physically. In particular, the tone of the wing and leg musculature is evaluated. Muscle atrophy is suggestive of reduced innervation. The skeletal system is palpated for evidence of crepitus or other abnormalities.

A systematic evaluation of the nervous system is then conducted.

Cranial nerve assessment

- The menace response, evaluating cranial nerves II and VII, can be provoked by bringing the hand towards each eye. Normal responses include eye-blink, pulling away of the head, or aggressive action of the beak.
- The pupillary light response evaluates cranial nerves II and III. It is often considered unreliable in birds because, although there is complete decussation of the optic nerves at the chiasma and therefore no consensual pupillary light response, the thin bones of the avian skull can allow light shone in one eye to stimulate a response in the other eye.
- Pupil size. The pupils should be symmetrical. If not, consider intraocular inflammation, ocular structural lesions, or sympathetic neuropathy affecting cranial nerve III.
- A symmetrical eye blink normally occurs when the cornea is gently touched with a moist cotton swab. If not, consider an abnormality affecting cranial nerves V and VII.
- A symmetrical eye-blink should be elicited by touching each side of the face or the lateral canthus. This evaluates cranial nerves V and VII.
- The tongue should be positioned and move normally. Abnormal tongue movement or a deviated tongue indicates damage to cranial nerves IX to XII.
- Reduced beak strength when biting or eating indicates possible damage to cranial nerves V, IX, X, XI, and XII.
- Nystagmus, the periodic, rhythmic, and involuntary movement of both eyeballs in unison in either a vertical, horizontal, or rotary direction, indicates damage to cranial nerve VIII, a cerebellar lesion, or increased intracranial pressure.
- Strabismus, an involuntary deviation of one eye, can occur with damage to cranial nerves III, IV, or VI and vestibular lesions.
- Horner's syndrome (enophthalmos, upper eyelid ptosis, slight elevation of the lower lid, piloerection of feathers on the affected side of the head, pupil constriction, and narrowing of the palpebral fissure) has been reported with intracranial

lesions or lesions affecting the cervical sympathetic tract or brachial plexus. Note that miosis and third eyelid protrusion are not as obvious in birds as in mammals because of the presence of striated muscle.
- Torticollis, a cervical muscle contraction producing neck torsion, has been seen with lesions involving cranial nerve XI.

Peripheral nerve assessment

Spinal reflexes, although difficult to assess objectively in birds, can help determine if a lesion is centrally or peripherally located. In most situations, determining if a reflex is present or absent is sufficient, and the symmetry of response should receive particular attention. Spinal reflexes require only the reflex arcs to be intact; no other parts of the central nervous system (CNS) are involved. Spinal reflex tests include:

- Body balancing. With the wings held close to the body, suspend the bird vertically and head down. Then quickly rotate it up to a horizontal position and observe the fanning of the tail feathers. Dip the bird forward again, back to the vertical position, and observe the tail flick up. This reflex functions to maintain balance.
- Wing withdrawal. Lightly touch a wing and observe it being pulled away. This is a segmental reflex which, if present, indicates that the reflex arc and associated spinal cord segment in the cervico-notarial cord are intact. Damage may still exist higher in the CNS.
- Leg withdrawal. Lightly touch a leg and observe it being pulled away. This is again a segmental reflex; if present, it indicates that the reflex arc and associated segment in the notarial spinal cord are intact.
- Vent reflex. Touch the vent mucosa with a fine object and observe it closing tightly. This is also a segmental reflex that indicates that the reflex arc and associated synsacral spinal cord segment are intact.

Reactions require reflex pathways (ascending and descending fibre tracts in the spinal cord and higher centres). It is important to look closely for subtle asymmetrical deficits.

Comparison of spinal reflexes and reactions helps to localise lesions. If a reflex exists, but its corresponding reaction does not, a lesion exists within the CNS rostral to the segment involved in the reflex arc. Reactions include:

- Proprioception (wings). Note the resting wing carriage. Pull one wing out of its resting position and note the time taken for its return. Only normally innervated wings will correct displacement.
- Proprioception (legs). Observe the bird at rest. Knuckling of the foot is quickly corrected in normal birds.
- Unilateral and bilateral wing fanning should be assessed while the bird is moved up, down, and from side to side, while the feet and lower back are restrained. Forceful, rhythmic, fanning out of the wings should be evident and, in the bilateral test, wing movements should be simultaneous. Unilateral wing fanning will occur only in the normally innervated wing, while in the bilateral test, a lesion causing loss of afferent stimulus on one side will be compensated for by afferent stimulus from the other side, and normal wing movements may occur.
- Placing. With wings held close to the body and the legs free, approach a horizontal surface (such as a desktop). As soon as any part of the foot touches the surface, both feet should swiftly position themselves accurately on the surface in order to support the bird's weight.
- The differentiation between pain perception and withdrawal reflex is critical. Movement of a pinched limb does not indicate that the patient is able to feel the stimulus. Some type of conscious recognition of the stimulus is required (e.g., vocalisation, attempts to escape or bite). This part of the examination is best kept until last, so as not to influence the patient's behaviour during other segments of the neurological examination. Loss of pain perception is a poor prognostic sign. Deficits in a particular wing may only be obviously detected if its counterpart is restrained.

FURTHER READING

Clippinger TL, Bennett RA, Platt SR (1996) The avian neurologic examination and ancillary neurodiagnostic techniques. *Journal of Avian Medicine and Surgery* 10(4): 221–247.

Doneley B, Harrison GJ, Lightfoot TL (2006) Maximising information from the physical examination. In: *Clinical Avian Medicine, Vol 1.* GJ Harrison, TL Lightfoot (eds). Spix Publishing, Palm Beach, pp. 153–212.

Greenacre CB, Lusby AL (2004) Physiological responses of Amazon parrots (*Amazona* species) to manual restraint. *Journal of Avian Medicine and Surgery* 18(1): 12–18.

Platt SR (2006) Evaluating and treating the nervous system. In: *Clinical Avian Medicine, Vol 1.* GJ Harrison, TL Lightfoot (eds). Spix Publishing, Palm Beach, pp. 493–518.

CHAPTER 5

SUPPORTIVE THERAPY

Because so many of the patients presented to veterinarians are in critical condition, treatment (and diagnostic testing) has to be done carefully, in limited stages, to avoid 'overwhelming' the patient and preventing the bird from decompensating, collapsing, and dying.

Supportive care, sometimes overlooked, is a critical part of the treatment for a patient that is often more ill than many other species presented to veterinarians. Although obtaining an accurate diagnosis and then applying a specific treatment are essential components of avian medicine, the clinician must not overlook the importance of supportive care. It is usually far better to have a tentative diagnosis in a live patient than a confirmed diagnosis in a dead one! When the masking phenomenon (hiding signs of illness until the bird is decompensating) is combined with an owner's approach of 'waiting a few days to see if he gets better', the result is often a patient that is presented badly dehydrated, in hypothermic shock, and in a catabolic state due to anorexia. Occasionally, some birds will present with severe respiratory compromise, either acute or chronic in onset. In some cases, especially trauma, the patient may be in pain or have experienced significant blood loss.

Obviously, sending such a patient home with some antibiotic powder to place in its drinking water is unlikely to help much. Following a quick assessment to identify immediate life-threatening problems (e.g., uncontrolled haemorrhage or severe dyspnoea), it is often appropriate to commence supportive care before attempting diagnostic testing and/or treatment.

Patients should be assessed on presentation, and a treatment plan should be developed based on addressing identified problems with:

1. Hypothermia
2. Hypovolaemia (dehydration and/or blood loss)
3. Hypoglycaemia due to inappetence
4. Hypoxia
5. Pain

The clinician must recognise the clinical signs of these conditions and deal with them aggressively, often prior to making any diagnostic attempts (**Fig. 5.1**).

HYPOTHERMIA

Birds are particularly sensitive to hypothermia because of their small size and higher surface

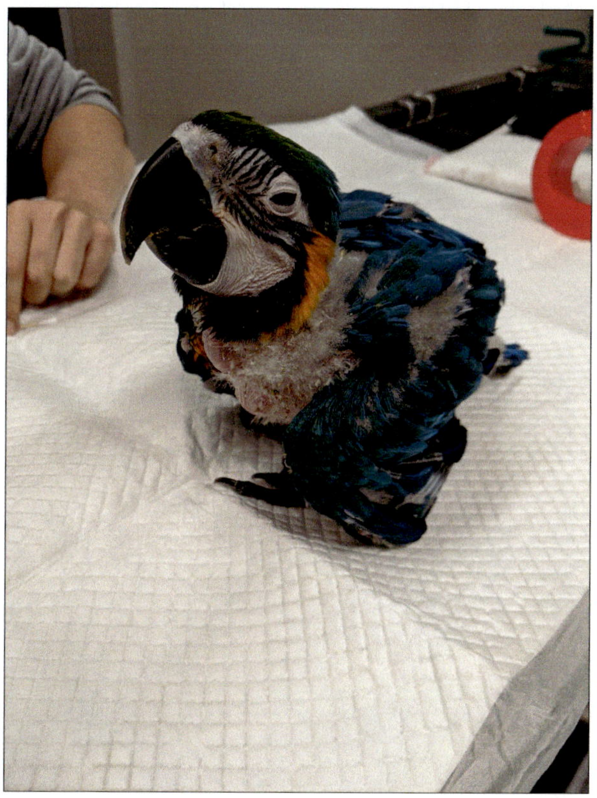

Figure 5.1 **The 'sick bird' look. This blue and gold macaw chick is hypothermic, hypoglycaemic, hypovolaemic, and probably hypoxic. These issues all have to be dealt with to save the patient's life.**

DOI: 10.1201/9781003272519-5

area-to-volume ratio. It is for this reason that sick birds sit still (to conserve energy) and fluff their feathers (to retain body heat). When presented with such a patient, it is prudent to actively warm the bird before attempting any further treatment or diagnosis.

Clinical presentation

The bird's feathers are fluffed to trap body heat, and the bird is lethargic and sleeps a lot to conserve energy. It may be unable to remain on the perch and is often found on the floor of the cage.

Management

Active warming using exogenous heat should be provided by placing the bird in a heated cage, brooder, or incubator, or by placing a heat lamp beside the cage, preferably next to a perch. Ideally, birds should be hospitalised in a heated room. The ambient temperature around the bird should be raised to 30–32°C, with a relative humidity of about 70% (**Fig. 5.2**).

The bird should be monitored for signs of heat stress: panting and wings held away from the body. Care should be taken to ensure that adequate humidity is provided to prevent the bird from dehydrating in a hot, dry environment.

Passive warming, such as covering a cage with a towel or blanket, does not help to keep a bird warm. This technique relies on trapping the bird's body heat and increasing the ambient temperature. Most sick birds cannot generate enough warmth to heat a cage covered with a blanket.

Once the bird has regained its body temperature, as evidenced by increased activity, further treatment and diagnostic efforts can be attempted.

Figure 5.2 **A reptile heating lamp is used to provide a focal heat source for this patient. More complicated ICU cages do a better job but may not be needed in a practice with a small avian caseload.**

DEHYDRATION

Most sick birds can be assumed to be at least 5% dehydrated, if not more.

Clinical presentation

Signs include sunken eyes, strands of mucoid saliva seen in the mouth when the beak is opened, and decreased capillary refill (seen by pressing on the basilic vein on the medial side of the elbow and assessing refill time). There may be wrinkling and/or tenting of the skin on the toes and torso, decreased urinary output, and the urates may appear thick and pasty.

Management

The clinician should aim to correct the calculated fluid deficit over 2–3 days while at the same time providing maintenance fluids (50 ml/kg/day). Fluid volumes to be given can be simply calculated as follows:

- 10% of the bird's body weight in grams = the volume (in ml) to be given daily for 3 days.
- Then reduce to 5–7.5% daily.
- Increase this amount if there are ongoing fluid losses (e.g., diarrhoea, polyuria).
- Divide the total daily requirement into two or three doses.

Fluid therapy may be given via the following routes:

- Orally, if the bird is not moribund or vomiting. In mildly dehydrated patients with a functional gastrointestinal system and the ability to withstand the necessary handling, replacement fluids can be given orally. Tap water, oral rehydration formulae, or lactated Ringer's solution can be delivered by a crop needle or plastic tubing directly into the crop. Hypertonic liquids should be avoided, as these may exacerbate dehydration.
- Subcutaneously, if the bird is not in shock, hypothermic, hypoproteinaemic, or has poor circulation (**Fig. 5.3**).

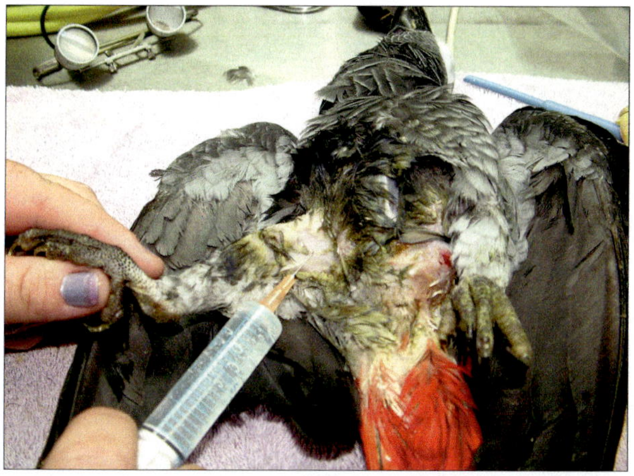

Figure 5.3 **Administration of subcutaneous fluids into the inguinal skin web.**

- Intravenously or intra-osseously. These routes can be difficult to maintain for extended periods in an alert, active patient.

Infusion pumps and syringe drivers are most appropriate for constant rate infusion. While birds may be given up to 100 ml/kg/hour intravenously without adverse effects, rates of between 3 and 10 ml/kg/hour are usually more appropriate. Fluid boluses of 20–30 ml/kg may safely be administered over a 2–3 minute period. If using this technique, the use of a butterfly catheter on the syringe and a Luer injection port in the catheter can minimise handling of the bird.

Guidelines on the selection of the type of fluid to be given are like those for mammalian medicine. Crystalloids, colloids, and whole blood are all used successfully in avian patients (see Table 5.1).

Fluid therapy is an important tool in the resuscitation of a critical patient. A guide to its use is in Table 5.2.

CATABOLISM AND HYPOGLYCAEMIA

Sick birds often do not eat; birds that do not eat, die. It is therefore vital that anorexic or emaciated birds with a functional gastrointestinal tract receive nutritional support.

Clinical presentation

Food may be untouched in dishes and droppings reduced in size and quantity; faeces are often dark (or even black) and urates are small and sometimes yellow. There is wasting of the pectoral muscles and

Table 5.1 **Fluid therapy selection guidelines**

FLUID TYPE	INDICATIONS	EXAMPLES
Replacement crystalloids	Rapid replacement of intravascular volume and electrolytes as seen with shock, haemorrhage, or severe volume depletion (secondary to the losses associated with vomiting, diarrhoea, third body spacing, or excessive diuresis).	• Normal saline (0.9% NaCl) • Lactate Ringer's solution (lactate or acetate) • Normosol R • Plasmalyte A
Maintenance crystalloids	Meet the electrolyte requirements of patients with normal daily electrolyte losses that are unable to maintain adequate fluid or electrolyte intake. They are rarely infused at rates greater than those necessary to meet the patient's maintenance needs.	• 0.45% saline • 0.45% saline + 2.5% dextrose • Lactated Ringer's + dextrose • Plasmalyte 56
Hypertonic crystalloids	Rapidly increase intravascular volume when severe hypovolaemia is present that may lead to impending death. It is contraindicated in severely dehydrated patients or those with uncontrolled haemorrhage.	• 7% saline
Dextrose in water	Used as a vehicle for infusion of other medications and to provide free water in severely hypernatraemic states. Infusion of large volumes of this fluid can lead to dilution of serum electrolytes and/or the development of oedema.	• 5% dextrose
Colloids	Expand intravascular volume by holding and potentially drawing water into the vasculature, therefore indicated for the treatment of hypovolaemia, sepsis, inflammatory conditions such as pancreatitis, and to improve colloid oncotic pressure in patients with hypoproteinaemia secondary to renal or gastro-enteric disease, vasculitis, or burns.	• Hetastarch • Dextrans • Modified gelatin solutions
Whole blood	Severe blood loss or anaemia	• See section titled "Blood loss" on p. 69

Table 5.2 **Fluid therapy in the treatment of a critical patient**

STEP	ACTION
Assess degree of dehydration or hypovolaemia	Physical assessment, blood pressure (if possible), and check PCV (if < 20%, consider a blood transfusion)
Administer crystalloids	1–2 boluses of 10 ml/kg IV/IO
Administer colloids	1–2 slow boluses (over 10 minutes) of 3–5 ml/kg
If patient is not improving	Give third bolus of crystalloids and colloids
If patient still not improving	Give hypertonic saline (7.5%) as a slow bolus (10 minutes) at 5 ml/kg
If no improvement	Maintain IV fluids while waiting for other therapies to take effect

weight loss (all patients should be weighed on every visit so that a record of their normal weight is maintained in their records).

Management

Placing food and water bowls in easily accessible places and offering favourite foods can encourage an ill patient to start eating. Placing food in a dish on the floor of a cage for a patient that is perching (and reluctant to leave the perch) will not encourage eating. The food bowls may need to be placed adjacent to the perch where the bird can easily reach them. Alternatively, spreading the food over the floor of the cage may encourage ground-feeding birds (e.g., many finches and Australian parrots) to browse and pick at the food.

If the patient is not eating but is not vomiting or moribund, crop gavage with an appropriate semi-liquid food can be instituted (**Fig. 5.4**). Foods that will pass easily through a syringe and feeding tube, and are high in energy, easily digestible, and have minimal residue are ideal. For these reasons, parrot hand-rearing formulae are often used for short-term maintenance feeds. If not readily available, pureed vegetable and fruit baby food can be used until a more appropriate replacement is obtained.

The use of appetite stimulants is an emerging area of avian supportive care. Where birds are not eating but are not moribund or vomiting, and crop gavage is considered to be potentially stressful or hazardous, appetite stimulants such as benzodiazepines and capromorelin may be appropriate. Suggested treatment regimens include:

- Capromorelin at 10–12 mg/kg (or lower doses) PO q24h may be effective in increasing food intake with potentially reduced adverse effects.

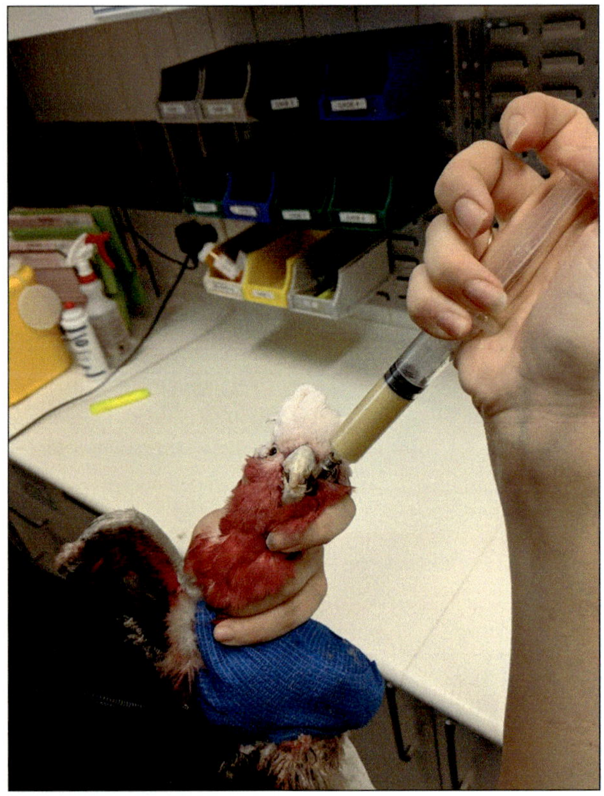

Figure 5.4 **Crop-feeding a galah using a metal feeding tube and parrot hand-rearing formula.**

- Midazolam or diazepam at 1 mg/kg IM will stimulate food intake while causing mild sedation for several hours. Lowering the dose to 0.5 mg/kg will shorten the sedation period but also reduce food intake.

If it is anticipated that the patient will require long-term feeding or if crop gavage cannot be done for any reason, oesophageal feeding tubes can be used to bypass the patient's head and are used in cases such as head trauma where eating or passing a stomach tube is not feasible. Under anaesthesia, curved haemostats are placed in the oropharynx and directed into the oesophagus, so the tip is at a point halfway along the cranial oesophagus on either the right or left side of the neck. The tip is pushed up against the skin (taking care to avoid entrapping the jugular vein) and the jaws are slightly opened. A 3–5 mm scalpel incision is made between the open jaws of the haemostat; they are then closed and pushed through the incision. The tip of the oesophagostomy tube is then grasped with the forceps and drawn rostrally out of the mouth, reversed, and guided into the crop or, if desired, into the distal oesophagus. The exposed end of the tube is then secured to the skin with a Chinese finger trap suture, capped, and the neck lightly bandaged. If the tube becomes blocked with hand-rearing formula, it can often be unblocked by flushing the tube with a pancreatic enzyme extract dissolved in water.

Total parenteral nutrition via an intravenous catheter is not routinely practised in avian medicine currently.

RESPIRATORY COMPROMISE

Respiratory support

Birds, with their high metabolic rates, have a higher oxygen demand than dogs and cats. Conversely, their gas exchange is more efficient; indeed, prolonged oxygen exposure (70–100% for 3–8 days) has been shown to be toxic to birds.

Clinical presentation

Signs include open-mouth breathing, increased respiratory effort (seen as tail bobbing and inspiratory sternal lift), audible respiratory noise, and collapse.

Management

Birds with respiratory distress appear clinically to benefit from a 40–50% oxygen environment. Oxygen may be delivered by mask in the short term or in an oxygen cage.

Oxygen therapy supplied either through a face mask or in an oxygen chamber can help patients with respiratory compromise. Clinicians must be aware that prolonged exposure to 100% oxygen can cause perivascular oedema and increase the degree of respiratory compromise.

In cases of tracheal obstruction (e.g., inhaled foreign body, tracheal stricture, *Aspergillus* granuloma, or diphtheritic plaques), it may be necessary to provide an alternative airway by way of an air sac catheter. This is a 2 mm or larger plastic tube (e.g., a non-cuffed endotracheal tube) placed into the caudal thoracic air sac. With the patient in right lateral recumbency, the left leg is drawn cranially, exposing the left flank. The following landmarks are identified: the caudal rib, the flexor crura medialis muscle, and the pubic bone. A 4–5 mm vertical incision is made in the skin where the flexor crura medialis muscle crosses the caudal rib. Curved haemostat forceps are then used to reflect the flexor muscle dorsally and then dissect bluntly through the coelomic wall. If done correctly, the operator will feel a 'pop' as the haemostats enter the caudal thoracic air sac. The endotracheal tube or other catheter is then advanced through this incision into the caudal thoracic air sac, taking care not to advance it as far as the caudal border of the lung. Patency is determined by observing expired air moving a feather placed over the free end of the catheter. Once satisfied with the placement, the clinician can then use a Chinese finger-trap suture to anchor the catheter to the skin. The free end of the catheter should be long enough so that powder down and feather dust are not freely inhaled; we prefer to leave the catheter as long as the bird's tail, securing it to the tail feathers (**Fig. 5.5**).

BLOOD LOSS

Birds can withstand comparatively greater blood loss than mammals. Clinical signs of anaemia and hypovolaemia do not usually occur until 50–60% of the blood volume is lost. Studies in healthy chickens have shown that after the removal of 30% of the blood volume, the packed cell volume (PCV) returned to normal within 72 hours. The removal of 60% of the blood volume in pigeons did not cause significant clinical signs, and the PCV returned to normal without treatment in 7 days. This response is thought to be the result of:

- An increased capillary surface area within skeletal muscle allows for rapid extravascular fluid resorption to maintain vascular volume
- The ability to mobilise large numbers of immature erythrocytes
- The absence of the autonomic response to haemorrhage contributes to haemorrhagic shock

Birds may be better able to deal with chronic anaemias because the nucleated erythrocyte can

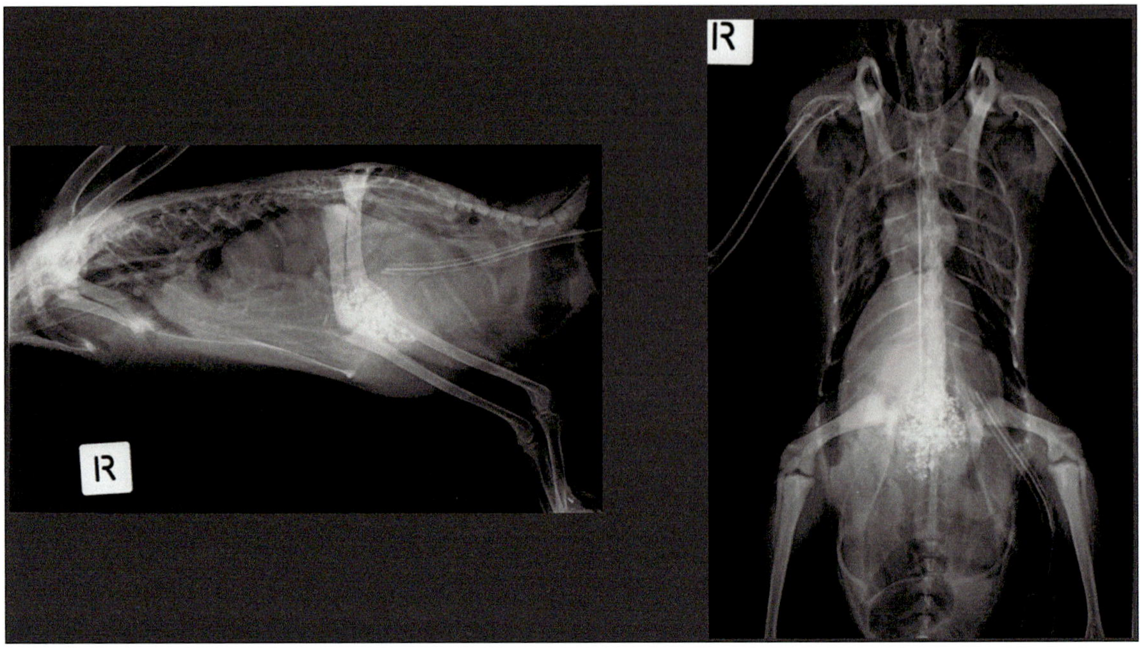

Figure 5.5 **Air sac catheter placed in a duck with a tracheal obstruction.**

produce more haemoglobin in response to anaemia. Anaemias of a chronic nature may, therefore, not require transfusions if the patient is stable and no further blood loss is expected. However, some birds with chronic anaemia may benefit from transfusions with the increased oxygenation of tissues, improved healing, appetite, etc. In addition, a transfusion may further stabilise critically ill birds, while diagnostic testing to determine the cause of the anaemia can be performed.

However, birds that are decompensating may require a blood transfusion to restore circulating blood volume and oxygen-carrying capacity. This includes cases where the PCV acutely falls below 15%, or more than 40% of the blood volume is lost. Transfusion is also indicated when blood loss is associated with collapse, ongoing haemorrhage is present, or there is poor response to conventional shock therapy with crystalloids and colloids (see earlier). The best course of action in acute blood loss in birds may be intravenous administration of balanced crystalloids for mild to moderate haemorrhage (loss of up to 25% of the blood volume) and a transfusion for moderate to severe haemorrhage (greater than 25% loss of blood volume).

Clinical presentation

There may be a history or physical evidence of recent blood loss. Mucous membranes are pale, and the respiratory rate and effort are increased. The bird may be weak and lethargic. The PCV is usually less than 20%.

Management

Many cases of blood loss do not require a transfusion. In mild cases, or when a blood donor is not available, intravenous or intra-osseous colloids or crystalloids, alone or in combination, may be sufficient. When given together, crystalloids are administered at doses of 30–40 ml/kg, while the colloid is administered at 5 ml/kg. With the bird's ability to mobilise immature erythrocytes, it is not uncommon for the PCV to return to normal within 7 days with this therapy alone.

In more severe cases or where significant blood loss is anticipated (e.g., surgery), a blood transfusion may be necessary. Homologous transfusions are ideal, preferably between the same species, but the same genus will give similar results. The half-life of a homologous transfusion is believed to be 6–11 days. Heterologous transfusions, on the other hand, have a half-life of 3 days or less. Transfusion reactions can occur with repeated transfusions, particularly heterologous transfusions.

Blood may be collected into syringes containing an anticoagulant such as sodium citrate (0.1 ml/0.9 ml blood), heparin (0.25 ml/10 ml blood), acid citrate dextrose (0.1 ml/0.9 ml blood), or citrate phosphate

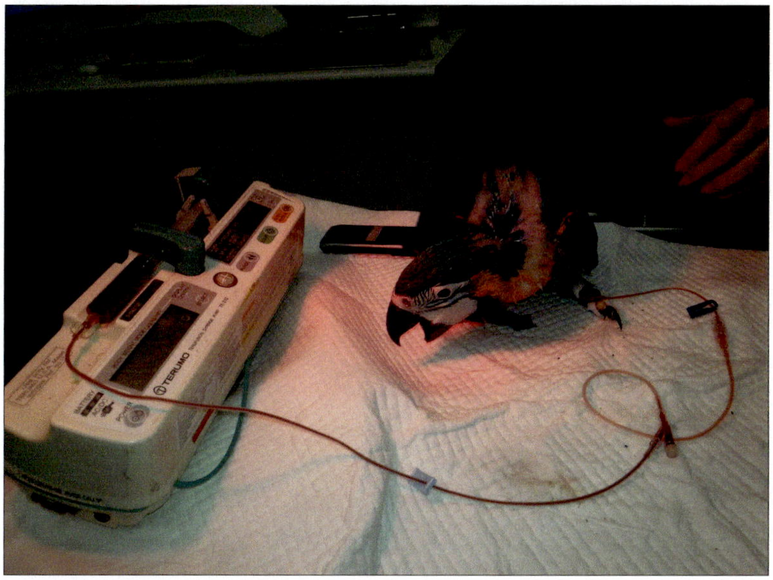

Figure 5.6 **Blood transfusion in a blue and gold macaw chick.**

dextrose (0.1 ml/0.9 ml blood). Up to 10% of the donor's blood volume (1% of its body weight) can be collected. Once collected, the blood is used within 12–24 hours. Nucleated avian erythrocytes are very metabolically active, metabolising fat and protein and consuming 7–10 times more oxygen than mammalian erythrocytes. Consequently, they do not store well.

Transfusions can be given via an indwelling intravenous catheter or intra-osseous catheter into the jugular, basilic, or medial metatarsal veins (**Fig. 5.6**). They can be administered as a constant rate infusion over 1–4 hours or as a slow bolus over 1–5 minutes. (If using a bolus approach, care must be taken to avoid fluid overload.) The amount of blood administered will depend on several factors including the degree of anaemia, size of the donor, and the stability of the patient. As a general guideline, 10–20% of the recipient's blood volume can be administered, and a volume of 1 ml/kg will raise the PCV by 1%. A small blood filter can be used to remove clots or other material from the transfusion.

Transfusion reactions

In mammals, transfusion reactions may result in haemolysis of the donor red cells, fever, urticaria, and anaphylaxis. These signs may be difficult to assess in birds; therefore, death may be the only noted sign. Acute haemolytic reactions are rare and have not been reported in birds. While incompatible crossmatching has been reported in birds in several studies, no visible signs of transfusion reactions were seen in any of the recipient birds, and all birds appeared healthy for 6 months after three separate transfusions.

ANALGESIA

Clinical signs of pain

Birds, unlike domestic mammals, indicate pain in a less obvious manner than many clinicians are accustomed to. They will respond to painful stimuli in one of two ways. Acute, unexpected pain produces a 'fight-or-flight' response, evidenced by excessive vocalisation, wing flapping, and decreased head movement. More chronic pain may result in a 'freeze' response: immobility, closure of eyes, inappetence, and fluffing of feathers. Care must be taken not to misinterpret lack of movement or vocalisation as an indication that the bird is not in pain.

It is wise, therefore, to assume that what would be painful to another species would be painful to humans, and adequate analgesia should thus be provided.

Management

See Chapter 31 for a more detailed discussion of avian analgesia.

Figure 5.7 **A cage cover provides a measure of security to a hospitalised bird.**

HOSPITAL CARE

Hospital care for companion birds requires a veterinary clinic to make accommodations for their differing requirements, often far removed from those needed by dogs and cats.

- *Security*. Most companion birds are 'prey' species, and they do not feel safe or secure when housed near predators (i.e., dogs, cats, birds of prey, and reptiles). Loud noises and constant movement around a cage can be stressful to a sick bird. Ideally, a separate bird ward should be provided, but cage covers can provide some security to a stressed bird (**Fig. 5.7**).
- *Warmth*. As mentioned earlier, hypothermic birds need a focal heat source. Most companion birds, whether sick or healthy, are more comfortable in warmer temperatures. Air conditioning, for example, adds to a bird's energy requirements to maintain its body temperature. If possible, the bird ward should be kept slightly warmer than the rest of the hospital. Care must be taken when heating a room to ensure that humidity levels do not decrease excessively, predisposing a bird to dehydration.
- *Biosecurity*. Many of the diseases for which birds require hospitalisation are infectious. All hospitalised birds are stressed to some degree and therefore likely to be immunocompromised. Attention must therefore be given to prevent aerosol or mechanical transmission of diseases within an avian hospital facility.
- *Feeding*. Most companion birds cannot withstand long periods of food deprivation. Therefore, every effort must be made to ensure that hospitalised birds are eating. Techniques for ensuring patients are receiving nutrition were discussed earlier in this chapter.
- *Psychological care*. Many companion birds are closely bonded to their owners. Hospitalisation of these birds can result in separation anxiety evidenced by anorexia, lethargy, or hyperexcitability. In some cases, feather picking may develop. In these cases, consideration must be given to discharging a patient to home care, on the proviso that the bird is returned to the hospital if it fails to improve when back in familiar surroundings.

FURTHER READING

De Matos R, Morrisey JK (2005) Emergency and critical care of small psittacines and passerines. In: *Seminars in Avian and Exotic Pet Medicine: Emergency and Critical Care*. TN Tully, MA Mitchell, JJ Heatley (eds). 14(2): 90–105.

Graham JE (2004) Approach to the dyspneic avian patient. In: *Seminars in Avian and Exotic Pet Medicine: Emergency Medicine*. AM Fudge, MS Johnston (eds). 13(3): 154–159.

Harrison GJ, Lightfoot TL, Flinchum GB (2006) Emergency and critical care. In: *Clinical Avian Medicine, Vol 1*. GJ Harrison, TL Lightfoot (eds). Spix Publishing Inc, Palm Beach, pp. 213–232.

Jaensch SM, Cullen L, Raidal SR (2001) The pathology of normobaric oxygen toxicity in budgerigars (*Melopsittacus undulatus*). *Avian Pathology* 30(2): 135–142.

Jenkins JR (1997a) Hospital techniques and supportive care. In: *Avian Medicine and Surgery*. RB Altman, et al. (eds). WB Saunders, Philadelphia, pp. 232–252.

Jenkins JR (1997b) Avian critical care and emergency medicine. In: *Avian Medicine and Surgery*. RB Altman, et al. (eds). WB Saunders, Philadelphia, pp. 839–863.

Lichtenberger M (2004) Principles of shock and fluid therapy in special species. *Seminars in Avian Exotic Pet Medicine* 13(3): 142–153.

Lichtenberger M (2006) Emergency case approach to hypotension, hypertension and acute respiratory distress. In: *Proceedings of the Annual Conference of the Association of Avian Veterinarians*, pp. 281–290.

Lichtenberger M, Chavez W, Thamm DH, et al. (2007) Use of hetastarch and crystalloids for resuscitation of acute blood loss shock. In: *Proceedings of the Annual Conference of the Association of Avian Veterinarians*, pp. 103–106.

Paul-Murphy J (2006) Pain management. In: *Clinical Avian Medicine, Vol 1*. GJ Harrison, TL Lightfoot (eds). Spix Publishing Inc, Palm Beach, pp. 233–240.

CHAPTER 6

CLINICAL TECHNIQUES

AFTER the physical examination, a range of diagnostic tests can be employed to move closer to a diagnosis. At the same time, the patient must be treated. Both of these procedures require competent clinical techniques to achieve the best outcome. This chapter describes the clinical techniques used in diagnostic testing and administering treatment. Interpretation of these tests and supportive care are described in later chapters.

DIAGNOSTIC TECHNIQUES

The limited range of physical responses to disease available to birds, and their relatively small size, has increased the requirement for diagnostic testing in order to better diagnose disease and provide appropriate treatment. The full gamut of testing available in small animal medicine is also available for avian practitioners. However, before any diagnostic test can be interpreted, it is vital that the clinician understands how to, and is able to, collect quality samples and position the patient correctly for diagnostic imaging. Diagnostic imaging, endoscopy, and cardiology techniques are discussed in more detail in Chapters 8, 9, and 22, respectively.

Blood collection and handling

There are a number of haematological, biochemical, serological, and polymerase chain reaction (PCR) tests available for avian veterinarians to utilise in the care of their patients. The basic requirement for these tests is to collect a sufficient volume of non-haemolysed blood without endangering the patient. Secondary requirements include storing the collected sample correctly and making fresh blood smears without damaging the relatively fragile nucleated erythrocytes.

Blood can be collected from the right jugular vein (the left jugular vein is accessible but is much smaller) (**Fig. 6.1**), the basilic vein (on the medial aspect of the elbow) (**Fig. 6.2**), or the medial tibiotarsal vein (in large birds) (**Fig. 6.3**). The right jugular vein is usually preferred in companion birds because of the relatively easy access; it lies under an apteryla and, with practice, a sole operator can both restrain the bird and perform the venepuncture. It must be remembered that avian veins have thin walls and tear easily. Coagulation in birds usually relies on extrinsic clotting pathways requiring tissue thromboplastin, rather than the intrinsic clotting pathways utilised by mammals. Care must therefore be taken to prevent accidentally tearing the vein wall, which can lead to a rapidly fatal haemorrhage. It should be noted that occasionally a bird will bleed from the nares after jugular venepuncture. This follows blood from the jugular vein entering the cervicocephalic air sac (an extension of the infraorbital sinus) after inadvertent rupture or tearing during blood collection. This haemorrhage is rarely significant, although anecdotal reports suggest that birds with coagulopathy can suffer a fatal haemorrhage. It is advisable to apply digital pressure to the venepuncture site for at least 30–60 seconds to minimise

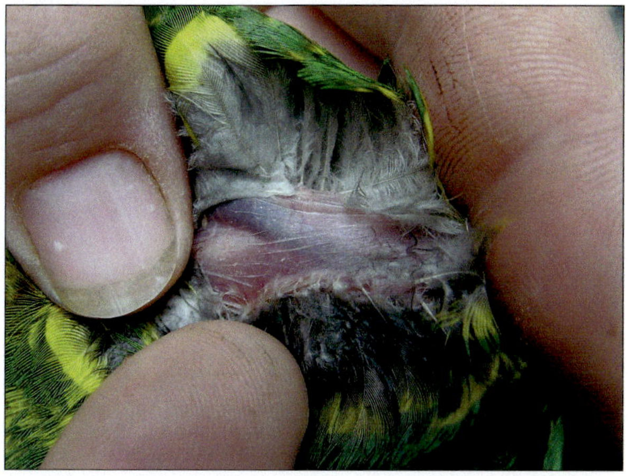

Figure 6.1 **The right jugular vein is usually easily accessed, lying under an apterylae on the right side of the neck, dorsal to the trachea.**

DOI: 10.1201/9781003272519-6

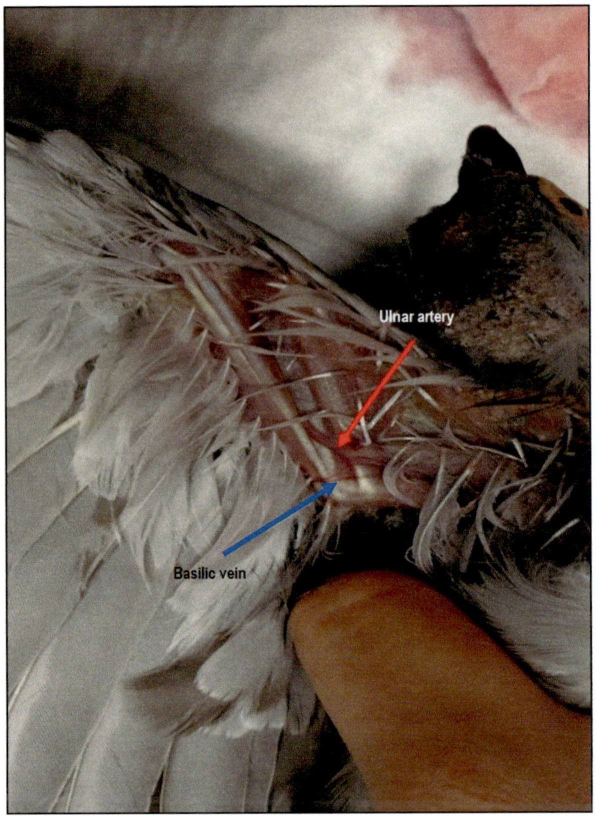

Figure 6.2 **The basilic vein, lying on the medial aspect of the elbow.**

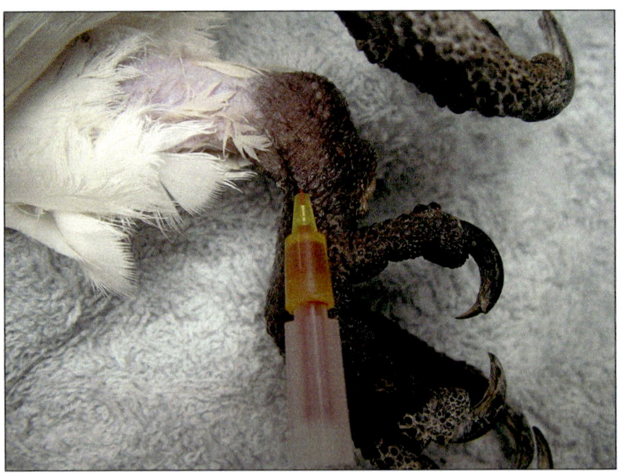

Figure 6.3 **The medial tibiotarsal vein in a cockatoo.**

haematoma formation. Using a 25–29 gauge needle (**Fig. 6.4**) minimises the iatrogenic trauma to the vein, but the smaller the needle bore, the more likely haemolysis is to occur during collection. Once the needle has entered the vein, avoid using excessive pressure to draw back; this will prevent both collapsing the vein and haemolysing the sample. It is sometimes advantageous to use a heparinised syringe for blood collection.

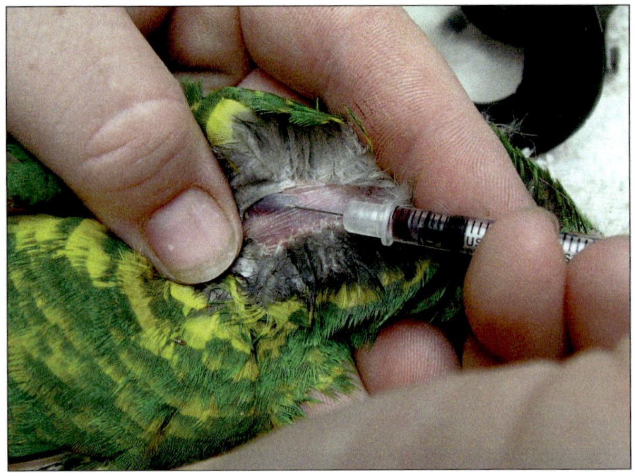

Figure 6.4 **Jugular venepuncture.**

A toenail clip, once advocated as a means of blood collection in birds, is the least desirable method of blood collection because of contamination with debris and uric acid from the bird's droppings, an increased likelihood of the sample clotting, an increased incidence of abnormal cell distributions, and the pain it causes the patient.

Many clients (and veterinarians) are surprised at how much blood can be safely collected from a bird. The blood volume in any given bird is approximately 10% of its body weight. Given that 10% of this blood volume can usually be safely collected for diagnostic testing or other purposes, an amount of blood equal to 1% of a bird's body weight can be safely collected (i.e., 1 ml of blood can be collected for each 100 g of body weight). With the modern laboratory equipment available in both external and in-house laboratories, this is usually a sufficient sample to run both haematological and biochemical assays.

Biochemistry testing is usually best performed on blood that has been collected into lithium heparin. If testing is likely to be delayed for anything more than a few hours, it should be centrifuged, and the plasma decanted and submitted. The use of plain tubes that allow the blood to clot and serum to come off is not advisable. Although serum collected in this manner has no anticoagulant in it, it has been in contact with haemolysing red blood cells for a significant time, which causes more artefactual changes. (Serum is lower in total protein, albumin, and glucose; it is higher in uric acid, potassium, calcium, magnesium, and phosphorous.)

Haematology, on the other hand, is usually best performed on blood that has been stored in sodium EDTA. (Note that some labs prefer lithium heparin for haematology; the clinician should consult with

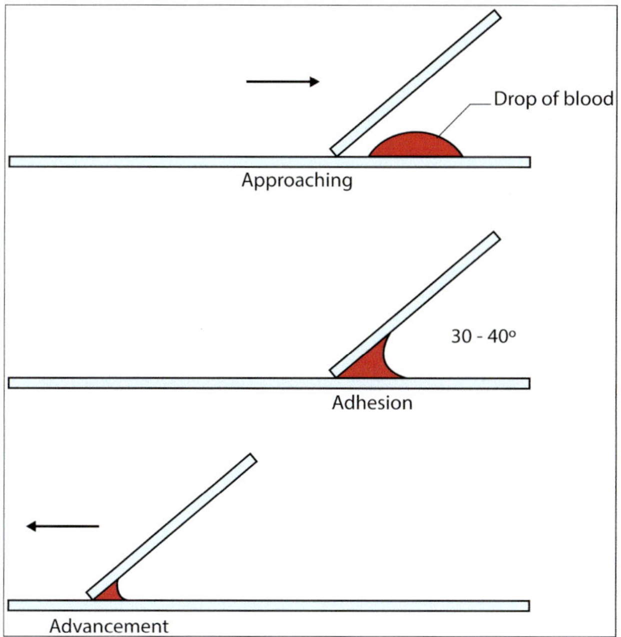

Figure 6.5 **The angled slide 'drag' method: the angled slide is used to drag, rather than push, the blood film along the preparation slide.**

their local laboratory for preferences.) The full range of species-specific effects of each anticoagulant has also not been studied well, so the same anticoagulant should be used when performing serial haematology tests to follow trends in the same patient. Fresh blood smears should always be made before the blood is placed into any anticoagulant. There are several techniques used for preparing blood smears:

- The traditional glass slide angled 'push' method, which often results in significant cell lysis.
- The angled slide 'drag' method: The angled slide is used to drag, rather than push, the blood film along the preparation slide. Cell lysis is much less common than with the previous technique (**Fig. 6.5**).
- The coverslip method: Place a 2 mm diameter drop of blood between two coverslips, allow the drop to spread close to the edges, then slide the coverslips apart horizontally. (If not done gently, significant cell lysis can occur.) The dried coverslips can then be stained and mounted face down on a glass slide using mounting medium, immersion oil, or a drop of adhesive.

Microbiology

Microbiology is an essential tool in avian medicine. It can begin with a Gram stain to determine if bacteria or fungi are present in an area or organ of interest, and in what proportions. The sample is collected, rolled onto a slide, air-dried, and then heat-fixed with a gentle flame. Excessively thick smears should be avoided. The staining procedure for Gram stain is as follows:

- Apply crystal violet solution for 60 seconds, rinse gently with water.
- Apply Gram's iodine solution for 60 seconds, rinse gently with water.
- Decolourise with acetone or ethanol for 2–5 seconds, then rinse gently with water.
- Stain with 0.1% basic fuchsin solution for 60 seconds, rinse gently, and dry.

Gram-positive organisms and yeast stain deep violet, and gram-negative organisms stain red. Gram-negative bacteria are predominantly rod- or coccobacilli-shaped and are usually much smaller than gram-positive rods in birds. Spirochaetes are gram-negative spiral bacteria (**Fig. 6.6**).

Samples collected for culture and sensitivity are collected with sterile swabs, placed in a transport medium, and sent to the laboratory. Common sites cultured in live birds include the cloaca, choana, crop, eye, ear, and skin. Less commonly sampled sites include the sinuses, trachea, joints, air sacs, and the oviduct.

Swabs used to culture the choana, crop, or cloaca should be pre-moistened with either sterile saline or transport medium to minimise iatrogenic damage to mucosal surfaces (**Fig. 6.7**). Cloacal swabs are preferred to faecal swabs, as there is less chance of environmental contamination. The owners should be advised that after the swab has been collected from this site, the next few droppings may have blood on them. This is common and rarely of concern.

Choanal cultures are similarly collected with a pre-moistened swab. Some clinicians prefer to flush the nares with 1–2 ml of sterile saline immediately prior to swabbing the choana; this reduces oral contamination and gives a more representative sampling of the nasal cavity. A mouth speculum or gauze strips are used to hold the mouth open. The swab is gently inserted into the rostral choana.

It must be noted that the sinus and nasal cavity anatomy is such that a choanal culture may not represent the pathology occurring in the sinus. In cases of sinusitis, it is therefore usually preferable to culture a sample from a sinus aspirate. This is done

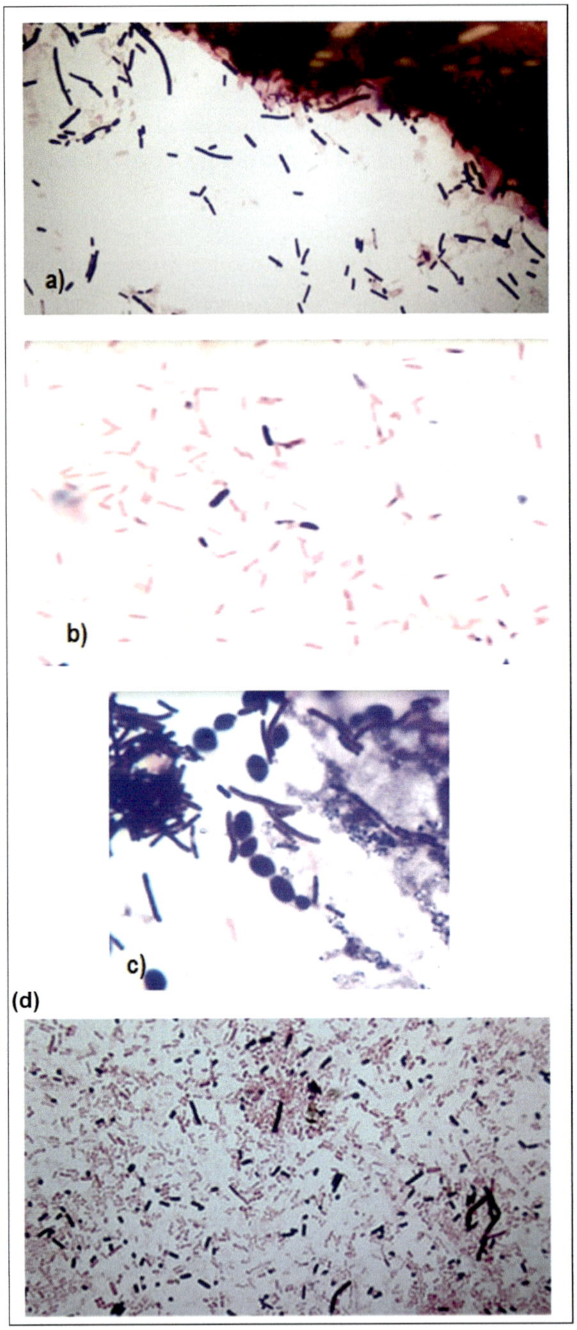

Figure 6.6 **(a) Normal Gram stain in a parrot, with predominantly Gram-positive flora. (b) An overgrowth of Gram-negative bacteria. (c) Gram-positive budding yeast, consistent with Candida spp. (d) A mixed growth of Gram positive and Gram negative rods.**

by inserting a 22–27 gauge needle, attached to a 3–5 ml syringe, midway between the commissure of the beak and the medial canthus of the eye, under the jugal bone. The sinus is more easily entered if the mouth is held open with a speculum during this procedure. Care must be taken to avoid iatrogenic trauma to the eye.

Culture of joints is indicated when bacterial or fungal infection is suspected based on cytology.

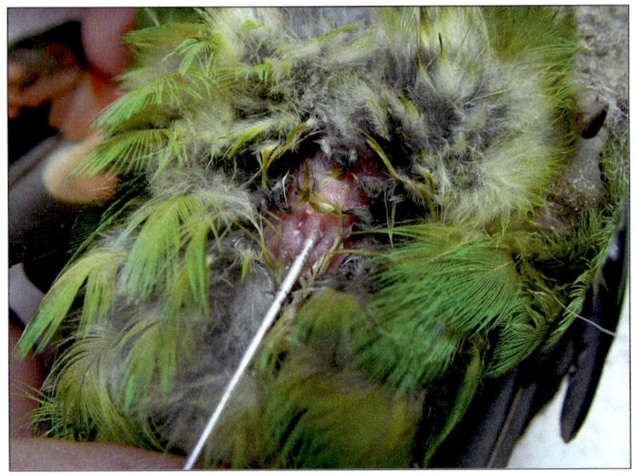

Figure 6.7 **Swabbing the cloaca should be performed using a moistened swab.**

The bird should be anaesthetised unless it is severely depressed. The skin over the joint is prepared aseptically, and joint fluid is aspirated using a small needle. The sample is expressed onto a swab, placed in transport media, and sent to the laboratory.

Tracheal washes are useful for evaluating birds with suspected lower respiratory tract disease. The procedure can be performed on a conscious patient, but care must be taken with birds in respiratory distress. In many cases, however, anaesthesia reduces the chances of undue stress in an already compromised patient, iatrogenic trauma, and contamination of the sample. A sterile catheter, long enough to reach the tracheal bifurcation, is introduced through the open glottis, taking care to avoid contamination through indirect contact with the mouth. Once the catheter is in position, sterile 0.9% saline (0.5–1.0 ml/kg) is quickly infused and immediately reaspirated. (The lower dose rate should be used in larger birds and those with severe respiratory compromise.) Approximately 25–50% of the volume infused should be recovered and can be processed for both culture and cytology.

Cytology

Cytology is a useful tool in avian medicine. Antemortem samples can be collected from the choana, pharynx, sinuses, joints, skin, feather follicles, trachea, lungs, crop, the cloaca, coelom, and the bone marrow. Collection techniques, for the most part, are similar to those described earlier.

Coeliocentesis is indicated in the presence of ascites or other abnormal coelomic fluid. (Often these birds present with dyspnoea, so care should be taken when handling.) The ventral midline is aseptically prepared, and a 21–25 gauge catheter or needle (attached to a syringe) is inserted on the midline immediately caudal to the sternum, directed to the right side of the coelom to avoid the ventriculus. If using a catheter, the stylet is removed immediately when the catheter enters the coelom, and a syringe is attached. Gentle negative pressure is applied until fluid enters the syringe.

Bone marrow cytology is useful for the evaluation of haematopoiesis. Under anaesthesia, the proximal cranial tibiotarsus is aseptically prepared. A small incision is made over the cnemial crest; a spinal or bone marrow needle is inserted into the shaft of the tibiotarsus with gentle, twisting pressure, the stylet is removed, and the sample is aspirated with a syringe and expelled onto a slide. A smear is then made in a manner similar to making a blood smear. For the best evaluation of the bone marrow, a peripheral blood sample collected the same day should also be submitted to the laboratory.

Fluid samples collected for cytology are often dilute, with low cellularity; they must therefore be concentrated before being examined. It is preferable to use a gravity-based cell concentration method that avoids cell distortion. Centrifugation can be used but often causes variable degrees of cellular distortion.

Examination of the feathers and skin

Skin scraping

Skin scrapings are rarely indicated in birds; in fact, the thin skin makes iatrogenic damage (including lacerations) almost probable. The occasional exception is for the diagnosis of *Cnemidocoptes* mites. The characteristic 'honeycomb' appearance of *Cnemidocoptes* mite infection is unmistakable, but a gentle scraping of the lesion can be used to confirm the diagnosis.

Skin cytology

Moist skin lesions can be swabbed, and the exudate rolled onto a microscope slide and stained for examination, e.g., Gram stain or a modified Wright's stain (Diff Quik®). Alternatively, a microscope slide can be pressed directly onto the lesion.

Acetate tape impressions can be used to sample dry or thickened skin. For a tape preparation, firmly and repeatedly press a piece of clear (not frosted) acetate tape to the lesion. The tape is then placed on a microscope slide over a few drops of blue Diff-Quik stain. Examine the sample under 40× to 100× magnification, using the tape as a slide cover.

Care must be taken when interpreting skin samples, as feather dust, melanin granules, and other debris are easily mistaken for yeast or bacteria.

Skin biopsy

The thinness of avian skin makes it difficult to prevent the rolling and contraction of skin biopsy specimens during collection and fixation. This in turn leads to difficulty orienting samples during processing and ultimately interferes with the establishment of a histopathological diagnosis.

To overcome this, non-translucent self-adhesive tape (e.g., Scotch® tape) is attached to biopsy sites before using a standard skin biopsy punch to obtain the biopsy. The specimens are fixed in 10% formalin without removing the tape; remnants of the tape are microscopically evident above the stratum corneum, assuring that none of the stratum corneum is lost during processing. There is no curling or rolling of the specimen, and they can be easily orientated for correct trimming and subsequent histopathological evaluation. The tape technique does not produce any appreciable artefacts.

Feather examination

A dry feather can be placed on a microscope slide and scanned under low power. Examination of the feather shaft under magnification may be needed to diagnose quill mites (*Syringophilus* spp. and others), while the barbs and barbules of the feather vane may reveal lice eggs (nits) and the lice themselves.

Feather pulp cytology

Using a scalpel blade, feather pulp is expressed from the calamus of a feather shaft onto microscope slides. Modified Wright stain (Diff Quik) and Gram staining are then used to examine the pulp for evidence of inflammation or infection. Typically, the cytology reveals a pink-staining proteinaceous background material, purple-staining amorphous globular material, basiloid epithelial cells, and occasional erythrocytes and heterophils. Again, great care must be taken in the interpretation of the samples.

Faecal analysis

Internal parasitism can be assessed by faecal examination.

- Motile protozoa such as *Trichomonas* spp., *Spironucleus* spp., *Cochlosoma* spp., and *Giardia* spp. are best detected by immediate examination of a faecal material suspended in a 0.9% saline solution.
- Protozoan oocysts and nematode eggs are sometimes detected by this method but may require concentration methods such as faecal flotation using hypertonic solutions (zinc sulphate, sodium nitrate, or sugar solutions) or faecal centrifugation. It has been suggested by some researchers that faecal centrifugation consistently detects more parasitic ova than standard faecal flotation techniques. The technique can be performed in most veterinary clinics:
  - Mix 2–5 g of faeces with 10 ml of faecal flotation solution and shake well until dispersed.
  - Strain this fluid into a 15 ml centrifuge tube until a slight positive meniscus forms at the top.
  - Place a coverslip on the tube and centrifuge it in a swinging-head centrifuge at no less than 1200 rpm for 5 minutes.
  - Let it stand for 10 minutes before examining it under the microscope.
- Cestode infections can be difficult to detect by any of the aforementioned means, as proglottid sections are often too heavy to float and may be missed on a fresh faecal smear.
- The detection of *Macrorhabdus ornithogaster* in the faeces is best achieved by concentrating *M. ornithogaster* and separating it from other solid matter in the faeces using the following technique:
  - Homogenise a droplet with approximately 20 times its volume of physiologic saline in a small tube.
  - Let it stand for 10 seconds before examining a small drop of the suspension collected from the meniscus under the microscope.

TREATMENT TECHNIQUES

Once a patient has been stabilised and diagnostic samples have been collected, medication can then be administered based on either a confirmed or suspected diagnosis. The route of administration is determined by a number of factors: the ease with which the patient can be handled, the patient's clinical condition, the pharmacodynamics and pharmacokinetics of the drug itself, and the clinician's experience and preferences. Commonly used administration routes include injection (intramuscular, intravenous, or subcutaneous), oral dosing, topical application, and nebulisation.

Oral medication

In mildly dehydrated patients with a functional gastrointestinal system and the ability to withstand the necessary handling, medications and fluids can be given orally. Oral medications are often surprisingly easy to administer. In large birds (e.g., poultry, waterfowl), tablets can be placed directly into the oropharynx; for smaller birds, tablets can be crushed and suspended in either water or methylcellulose for administration with a syringe (**Fig. 6.8**). When using a syringe, it should be placed over the tongue and the medication delivered in small quantities into the oropharynx. Care must be taken with screaming birds to ensure that the medication is not aspirated. In these birds, and with larger volumes of medication, crop gavage may be more appropriate.

A crop needle (a stainless-steel needle with a stainless-steel ball on the end) or plastic tubing is passed over the tongue and gently guided down the cervical oesophagus and into the crop. Care must be taken not to inadvertently place the needle or tube into the trachea or to force it through the oesophageal/crop wall. It is important that the crop needle be placed into the crop rather than just at the back of the pharynx. Careful palpation of the crop can confirm the placement of the implement before giving the fluids. Failure to do so may result in the fluids flooding the pharynx and oral cavity, possibly leading to aspiration. Care must be taken not to press on the crop

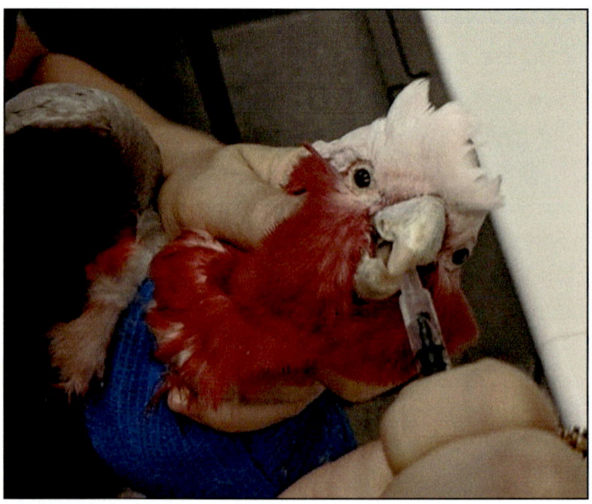

Figure 6.8 **Administering oral medications to a galah.**

while restraining the bird, to overfill the crop, or to fill it too quickly. If fluid appears in the pharynx or comes from the mouth while gavaging, the tube or needle should be immediately withdrawn, and the bird placed back in its cage. Continued handling of the bird is more likely to result in it panicking and aspirating the fluid. As a general rule, the amount given by crop gavage should start initially at 5% of the bird's body weight. It can then be built up to 8–10%. It will typically take about 2–4 hours for the crop to empty completely (thicker mixes, such as hand rearing formula, will take longer) (**Fig. 6.9**).

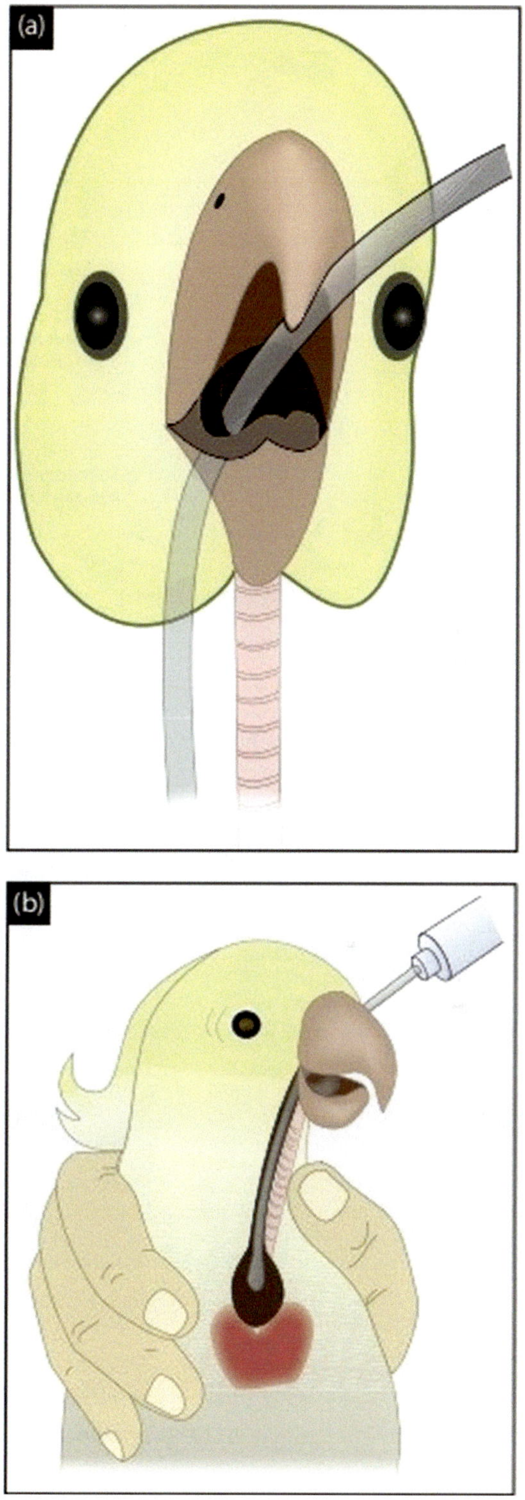

Figure 6.9 **Procedure for the insertion and correct placement of a crop tube for fluid therapy. With the patient facing you and its head restrained, the crop needle is introduced into the bird's mouth on the left side, passed over the top of the tongue and directed to the right side of the oropharynx (a). The tube is gently passed down the oesophagus, into the crop (b).**

Subcutaneous injections

Subcutaneous fluids are not recommended for hypothermic patients or those in shock due to concerns over absorption (peripheral vasoconstriction may lead to poor absorption). They are, however, very useful for most ill patients. They may be given in the inguinal region or the interscapular region; the inguinal region allows for greater volumes to be given with less patient discomfort (**Fig. 6.10**). Volumes up to 10 ml/kg can be given two to three times daily in each site. If oedema is noted prior to subsequent administration, the fluid is not being absorbed (due to poor circulation, excessive volume administered, or hypoproteinaemia) and additional fluid should not be administered by this route.

Intramuscular injections

The pectoral muscles are most commonly used for intramuscular injections because of their large size (**Fig. 6.11**). Other muscles such as the quadriceps, caudal thigh muscles, and gluteal muscles can be used in larger birds. Clinicians need to be aware that many drugs are, at best, tissue irritants when given intramuscularly; caution needs to be taken to avoid giving repeated injections into one site. Care should also be taken with the volume to be injected: a 0.1 ml injection for a 100 g bird is equivalent to a 10 ml injection for a 10 kg dog. If necessary, divide the sternum into quadrants and divide the dose between the quadrants.

Intravenous catheterisation

Fluids may be administered by constant infusion or slow bolus through an aseptically placed intravenous or intra-osseous catheter. The jugular, basilic, and medial metatarsal veins are suitable for intravenous fluid administration. Depending on the size of the

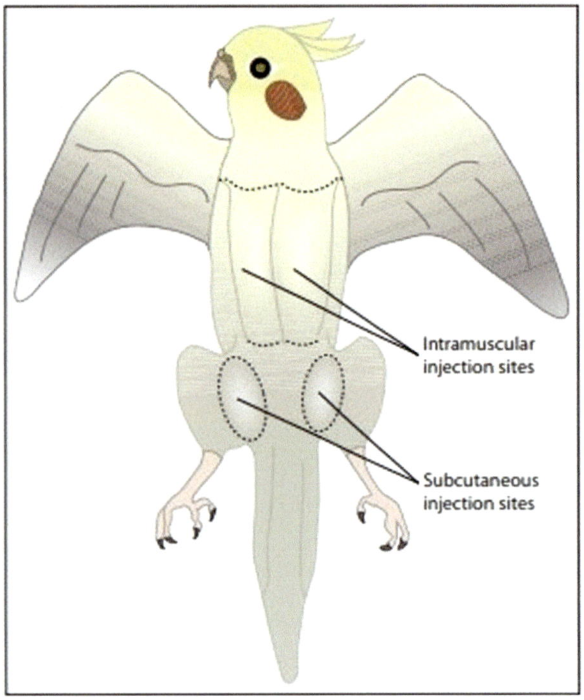

Figure 6.10 **Diagram illustrating suitable sites for the administration of intramuscular and subcutaneous injections.**

Figure 6.11 **Administering an IM injection.**

patient, 22–26 g intravenous catheters can be placed with little difficulty in birds weighing more than 200 g. Secured in place with adhesive tape, tissue glue, or suturing, they can be maintained for several days.

Regional limb perfusion

Distal pelvic limb infections, in particular pododermatitis (bumblefoot) and septic arthritis, are challenging conditions to treat, especially in larger birds, due to repeated mechanical trauma, biofilm formation, difficulty in keeping the affected lesions clean, granuloma formation, and extensive scar tissue. Conventional therapy, such as topical and systemic antibiotic therapy, often fails to reach therapeutic concentrations at the site, and so treatment failures and disease recurrence are common.

Regional limb perfusion (RLP) is commonly used with amikacin and is considered an effective antibiotic delivery technique in horses and other ungulates to treat distal limb infections. Regional limb perfusion of intravenously administered antibiotics in distal limbs can result in high drug tissue concentrations in the target site while concurrent low systemic concentrations remain low. This increases local antimicrobial drug efficacy while reducing systemic adverse effects.

A tourniquet is placed just above the tibiotarsal–tarsometatarsal joint. The medial aspect of the tarsometatarsus is aseptically prepared, the medial metatarsal vein is visualised, and a 25 gauge IV catheter is placed in the vein and secured in place. The drug being used (usually an antibiotic) is infused into the vein; the catheter is removed, and the injection site is covered for 15 minutes. The tourniquet is released after 15 minutes.

Using this technique in well-hydrated patients, it may be feasible to use potentially nephrotoxic drugs (e.g., aminoglycosides) without incurring the renal damage often associated with the effects of the renal portal vascular system.

Intra-osseous catheterisation

For smaller patients, intra-osseous fluids are considered equivalent to intravenous administration, with 50% of the fluid given by this route appearing in the circulation within 30 seconds. Intra-osseous catheters are most commonly placed in the proximal tibiotarsus and the distal ulna. A spinal needle is used to prevent the lumen from plugging with bone and tissue, rendering the catheter useless. Care must be taken to insert the catheter correctly the first time, as subsequent attempts usually result in either a large hole or multiple holes in the bone at the entry site. Leakage will occur through each hole, resulting in discomfort from the deposition of the fluid or drugs into the soft tissues around the bone.

The needle is placed into the tibiotarsus in a manner similar to the normograde insertion of an intramedullary pin. The tibiotarsus is grasped, a small stab incision is made in the skin over the proximal cranial tibiotarsus, lateral or medial to the cnemial crest, and the needle is inserted into the marrow cavity with gentle pressure and a twisting motion of the needle. Following the placement of the catheter, the stylet is removed, and the catheter's position is evaluated to ensure it is within the medullary canal. In larger birds, aspiration will result in blood or bone marrow flashback. (If bone marrow is needed, it can be collected in this manner.) Alternatively, a small amount of saline can be injected and then reaspirated. If the needle is in the medullary canal, this will usually yield blood-tinged fluid. A larger volume of fluid can then be injected; if the needle is outside the bone, the soft tissues will swell. Once placement is confirmed, the catheter should be secured.

For placement in the ulna, the dorsal crest of the distal ulna is palpated with the carpus flexed. The needle is inserted between the distal ulna and the ulnar carpal bone and advanced into the medullary canal. The ulna is held with one hand during insertion to ensure that the needle is inserted at the proper angle. Again, aspiration, flushing, and reaspiration can be used to confirm correct placement. As the ulna in some birds is pneumatic, if air is aspirated, the catheter should be removed and placed in the tibiotarsus. Infusion of a larger volume of fluids into a correctly placed ulnar catheter should result in visible 'blanching' of the basilic vein as the fluid enters the circulation. Once correct positioning is assured, the catheter should be secured with a bandage (**Fig. 6.12**).

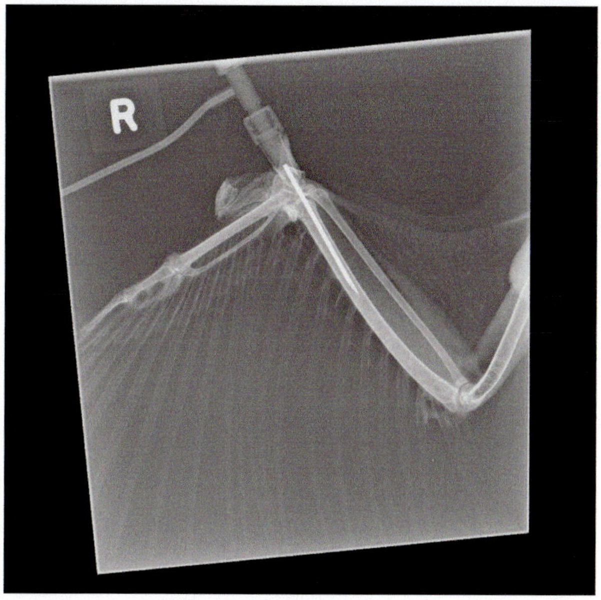

Figure 6.12 **Correct ulnar placement of an intraosseous catheter.**

As the bone is not distensible and the rate at which the fluid enters general circulation is a limiting factor, care must be taken to give intra-osseous injections slowly in order to allow the fluid to enter the circulation without being forced. If the injection is forced, it will exit the insertion hole in the bone, causing discomfort.

Topical medications

While water-based or other drugs can sometimes be used topically in birds, topical application of oil-based medications is contraindicated. Their grooming behaviour usually causes the medication dispersed through the plumage, destroying the insulation and waterproofing capacity of the feathers. The thinness of the avian skin makes transdermal absorption of medication very easy. This can result in systemic absorption of toxic levels of a drug (e.g., the use of topical corticosteroids invariably results in steroidal hepatopathy and other adverse effects).

Nebulisation

Nebulisation is often used in the treatment of respiratory disease, both to deliver medications directly to the site where it is needed and to humidify the respiratory tract. The unique anatomy of the avian respiratory tract means that the nebuliser used should create droplets between 1 and 3 μm in size. Hypertonic saline or diluted acetylcysteine (2% solution in saline) can be used to reduce the viscosity of airway secretions. Antibiotic and antifungal medications can be added to the nebulising solution to deliver the drug directly into the respiratory tract.

FURTHER READING

Campbell TW, Ellis CK (2007) *Avian and Exotic Animal Hematology and Cytology*, 3rd edn. Blackwell Publishing, Ames.

Guzman DSM, Mitchell MA, Gaunt SD, Beaufrère H, Tully TN (2008) Comparison of hematologic values in blood samples with lithium heparin or dipotassium ethylenediaminetetraacetic acid anticoagulants in Hispaniolan Amazon parrots (*Amazona ventralis*). *Journal of Avian Medicine and Surgery*, 22(2), 108–113.

Jaensch SM, Cullen L, Raidal SR (2001) The pathology of normobaric oxygen toxicity in budgerigars (*Melopsittacus undulatus*). *Avian Pathology*, 30(2), 135–142.

Nett CS, et al. (2003) A modified biopsy technique to improve histopathological evaluation of avian skin. *Veterinary Dermatology*, 14(3), 147–151. https://doi.org/10.1046/j.1365-3164.2003.00333.x.

Rosenthal KL, Morris DO, Mauldin EA, Ivey ES, Peikes H (2004) Cytologic, Histologic, and microbiologic characterization of the feather pulp and follicles of feather-picking psittacine birds: A preliminary study. *Journal of Avian Medicine and Surgery*, 18(3), 137–143.

CHAPTER 7

INTERPRETING DIAGNOSTIC TESTS

CHOOSING WHICH TESTS TO USE

Veterinarians treating birds are faced with challenges often not encountered by their counterparts treating dogs and cats. Many birds are presented to veterinarians only when near terminally ill, and a quick answer is often the difference between life and death. Birds are often limited in their range of expression of clinical signs, and many clinicians, through no fault of their own, lack the experience to conduct a thorough physical examination. The combination of these factors has led to an increasing tendency in avian medicine to conduct exhaustive diagnostic tests on patients, often with scant attention being paid to a complete history and a careful physical examination, and with little attempt to refine or focus the diagnostic efforts. The selection of diagnostic tests should be based on a solid understanding of the species in question, the results of a thorough history-taking and physical examination, and a shortened list of probable differential diagnoses based on this workup.

Before proceeding with diagnostic tests, the clinician should first ask:

- Are the test(s) appropriate to the patient (e.g., species, age, sex) and its clinical signs?
- Has the test been validated to ensure that the result obtained is likely to be both accurate and meaningful?
- Are the physical risks to the patient and the cost of the test(s) to the client justified by the likely clinical value of the results?

If the answers to these questions are yes, then diagnostic testing should proceed.

Available diagnostic tests include:

- Haematology
- Blood coagulation
- Clinical biochemistries
- Protein electrophoresis
- Serology
- PCR (polymerase chain reaction)
- Microbiology
- Cytology

HAEMATOLOGY

Haematology, often referred to as the complete blood count (CBC), is an important test in determining many disease states in birds. A CBC involves assessing:

- The erythrocytes, through the determination and assessment of:
  - The haematocrit or packed cell volume (PCV) (**Figs. 7.1, 7.2**)
  - Total erythrocyte count
  - Erythrocyte morphology
  - Reticulocytes
- The leucocytes, through the determination and assessment of:
  - Total white cell count
  - Leucocyte differential count
  - Cellular morphology
- Thrombocyte numbers

Erythrocytes

Elliptical in shape with a centrally located, oval nucleus, avian erythrocytes are usually larger than mammalian erythrocytes with a relatively short half-life (28–45 days) when compared with those of mammals (e.g., 120 days in man), which results in a greater number of immature erythrocytes in the peripheral circulation (**Fig. 7.3**).

DOI: 10.1201/9781003272519-7

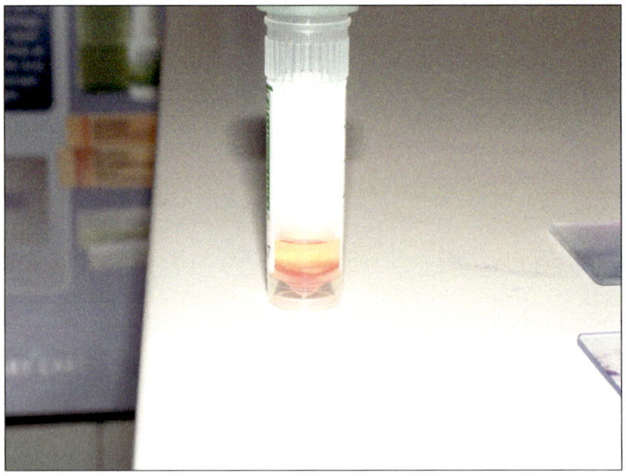

Figure 7.1 **Spun down blood collection showing severe anaemia.**

Cellular responses

Erythropoiesis occurs in the bone marrow (and, in the embryonic chick, the yolk), although ectopic erythropoiesis can occasionally be found on the spleen and liver. It is controlled by several factors including hormonal stimulation (androgens and adrenocorticoids) or oxygen concentration in tissues. Hypoxia stimulates the production and release of erythropoietin from the kidneys, which then stimulates erythropoiesis.

Birds with high metabolic rates have a higher oxygen requirement and therefore require a greater oxygen carrying capacity – and vice versa. The PCV of most parrots lies between 0.4 l/l and 0.55 l/l. A low PCV (<0.35 l/l) can indicate blood loss, anaemia, shock, or haemodilution following fluid therapy. A high PCV (>0.55 l/l) indicates dehydration or polycythaemia (primary or secondary). Primary polycythaemia may be seen in birds with bone marrow dysplasias. Birds kept at high altitudes, or those with chronic respiratory or cardiac disease, may be polycythaemic secondary to low oxygenation of the blood.

Total red blood cell (RBC) numbers in birds are affected by age, sex, environment, hormonal influences, and hypoxia. Because the erythrocytes are larger, red cell numbers are lower in birds than in mammals. RBC numbers tend to be lower in young birds and females. The normal RBC count in birds is 1.5–4.5 × 10^{12} cells/l.

Morphological abnormalities seen in avian erythrocytes include:

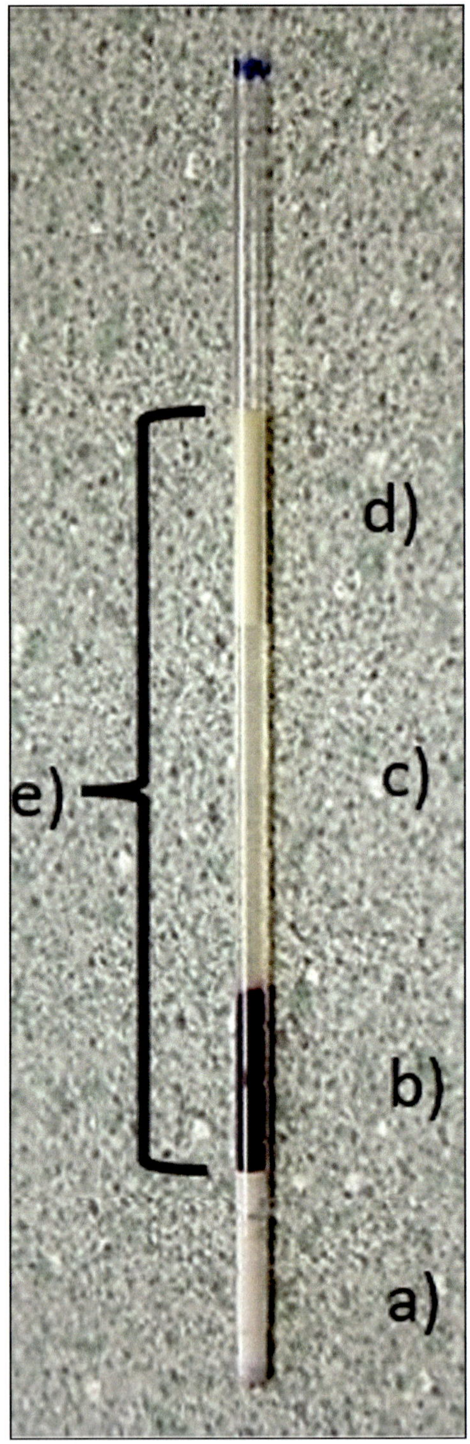

Figure 7.2 **Haematocrit tube after centrifugation: (a) plasticine plug, (b) erythrocytes, (c) plasma, (d) lipaemia, (e) total column of blood.**

- Excessive polychromasia. Polychromasia is an indicator of the patient's erythrocyte regenerative abilities. Although some polychromasia

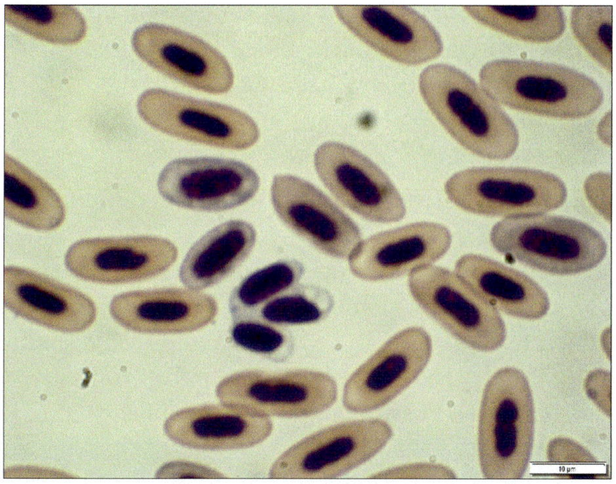

Figure 7.3 **Avian erythrocytes: Elliptical in shape with a centrally located, oval nucleus. They are usually larger than mammalian erythrocytes with a relatively short half-life (28–45 days) when compared with those of mammals (e.g., 120 days in man), which results in a greater number of immature erythrocytes in the peripheral circulation. The cytoplasm of these immature erythrocytes stains more basophilic than that of mature erythrocytes (polychromasia), they are non-uniform in size (anisocytosis), and the nuclear chromatin is less condensed than mature erythrocytes. In most healthy parrots they represent approximately 1–5% of the total erythrocyte count.**

(1–5%) is normal, excessive polychromasia indicates a regenerative response to blood loss or anaemia.

- Reticulocytosis, especially when combined with increased polychromasia, is seen as a regenerative response to blood loss or anaemia.
- Anisocytosis. Variation in the size of avian erythrocytes is occasionally seen in peripheral blood smears as a normal finding. However, the number increases in response to anaemia.
- Poikilocytosis, or variable cell shapes, may represent artefactual error, but is also seen when severe systemic infections affect the bone marrow. Erythrocytes may appear round, elongated, or irregular. The nucleus may vary in appearance, location, and number. Erythrocytes that appear round with oval nuclei are indicative of accelerated erythropoiesis. Binucleated erythrocytes may also indicate abnormal erythropoiesis in association with severe, chronic inflammatory processes and neoplasia. Poikilocytes are susceptible to damage, and therefore have a shorter life.
- Erythrocytic ballooning has been reported to be commonly associated with, although not pathognomonic for, lead toxicosis. It is also seen with 'conure bleeding syndrome'. There are bulges in the normal ellipsoid shape of the erythrocyte, often accompanied by areas of hypochromasia.
- Haemoparasites are occasionally seen in the erythrocyte cytoplasm of wild-caught birds or those exposed to biting insects.

Understanding anaemia

Anaemia, an absolute decrease in erythrocyte numbers, haemoglobin concentration, and PCV, is usually the result of increased loss, increased destruction, and/or decreased production of erythrocytes. It is classified as regenerative (haemolytic, or haemorrhagic) or non-regenerative.

Regenerative anaemia

Regenerative anaemia is characterised by the presence of polychromasia, reticulocytosis, macrocytosis, and anisocytosis **(Figs. 7.4, 7.5)**. It is an

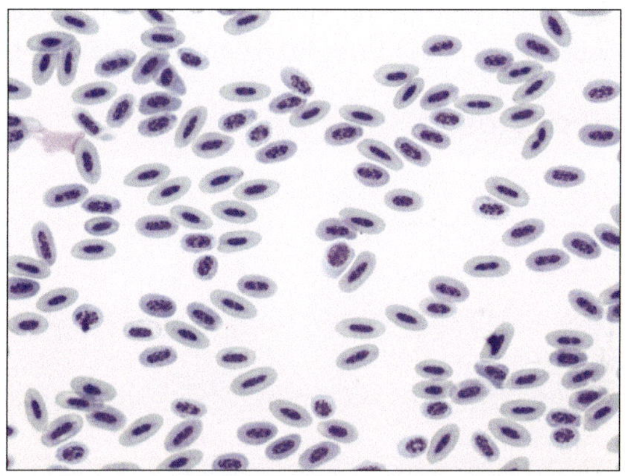

Figure 7.4 **Blood cell sample with polychromasia (variability in erythrocyte cytoplasmic colour) and anisocytosis (variability in erythrocyte size) indicating mild regenerative anaemia.**

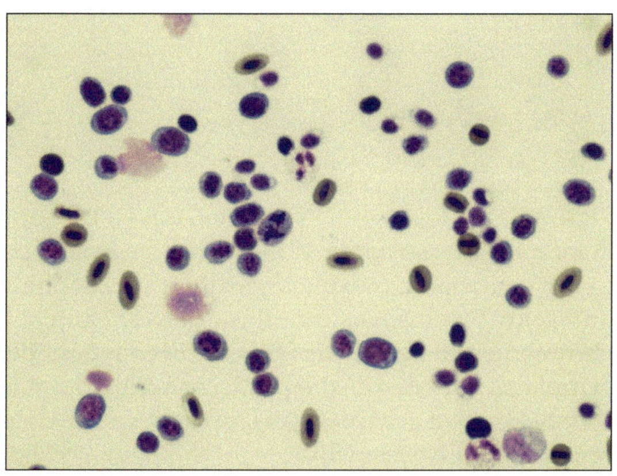

Figure 7.5 **Blood cell sample with mitotic figures, poikilocytosis (variability in erythrocyte shape), and binucleate erythrocytes indicating pronounced regenerative anaemia.**

indication of an appropriate bone marrow response (increased erythrocyte production and release of reticulocytes) to decreased red cell mass through blood loss or destruction.

- *Haemolytic anaemia* is due to increased destruction of circulating erythrocytes. Possible aetiologies include:
  - Haemoparasites (*Plasmodium* spp. and *Aegyptianella* spp.)
  - Bacterial septicaemia
  - Acute toxicosis (oil ingestion, lead, aflatoxicosis)
  - Immune-mediated conditions
- *Haemorrhagic anaemia* results from blood loss due to trauma, gastrointestinal parasitism, coagulation disorders (as in conure bleeding syndrome), organ rupture, or ulceration, aneurysms, and some viral diseases. Following acute haemorrhage, there is often a rapid response during which erythropoiesis increases dramatically. Healthy birds are remarkably adept at dealing with acute blood loss and haemorrhagic shock rarely occurs. Studies have shown that after the loss of even 50% of the blood volume, the PCV returns to normal within 3–7 days.

Non-regenerative anaemia

Non-regenerative anaemia is the most common type of anaemia described in birds. It arises due to a lack of appropriate bone marrow response combined with the relatively short life of avian erythrocytes. It is characterised by a lack of a regenerative response (i.e., little or no reticulocytosis, polychromasia, or anisocytosis). Possible aetiologies include:

- Acute, overwhelming infections: bacterial, viral (e.g., PBFD) or fungal
- Chronic disease including chronic inflammatory conditions, chronic infectious diseases (tuberculosis, colibacillosis, salmonellosis, aspergillosis, chlamydiosis), some viral diseases, hypothyroidism, and neoplasia
- Toxicosis
- Nutritional deficiencies

Leucocytes

TYPE	ROLE
Heterophils (**Fig. 7.6**) 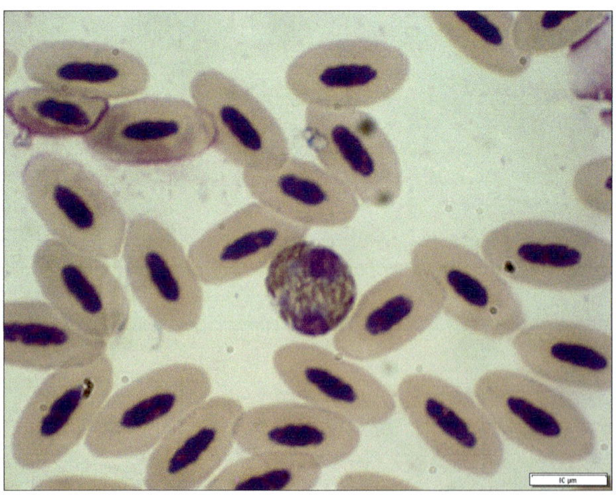 Figure 7.6 **Heterophil: The avian equivalent of the mammalian neutrophil. While it has a similar function to the neutrophil, morphologically it appears quite different. The nucleus contains coarsely clumped chromatin and usually has two to three lobes. The cytoplasm contains eosinophilic, spherical, oval, or spindle-shaped granules. It is, in most species, the predominant white cell.**	Equivalent to the mammalian neutrophil, heterophils are usually the most common leukocyte seen in peripheral blood. They lack lysozymes, which is why birds form caseated, rather than liquid, pus.

Lymphocytes (**Fig. 7.7**)

Lymphocytes are second only to heterophils in frequency in most species, except Amazon parrots and some passerines (canaries) in which the lymphocytes appear to be the predominant leucocyte.

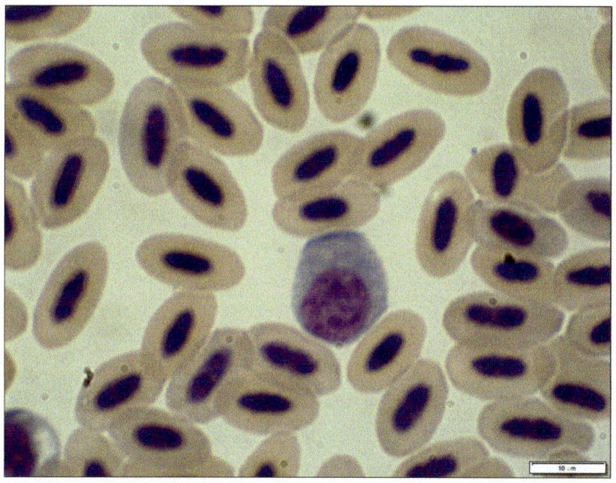

Figure 7.7 **Lymphocyte: Size and shape vary, with small, medium, and large cells that may be round or moulded around neighbouring cells being seen in the same smear. They have a large, round, centrally located nucleus with densely clumped or reticulate nuclear chromatin. The cytoplasm is clear or slightly basophilic. Reactive lymphocytes (i.e., those responding to antigenic stimulation) usually have a more deeply basophilic cytoplasm, vacuoles in the cytoplasm, a clear perinuclear ring, and scalloping of the cell edges.**

Eosinophils (**Fig. 7.8**)

The function of eosinophils is still largely unknown; eosinophilia in pet birds is rare, sometimes associated with marked tissue damage or, occasionally, parasitic infections.

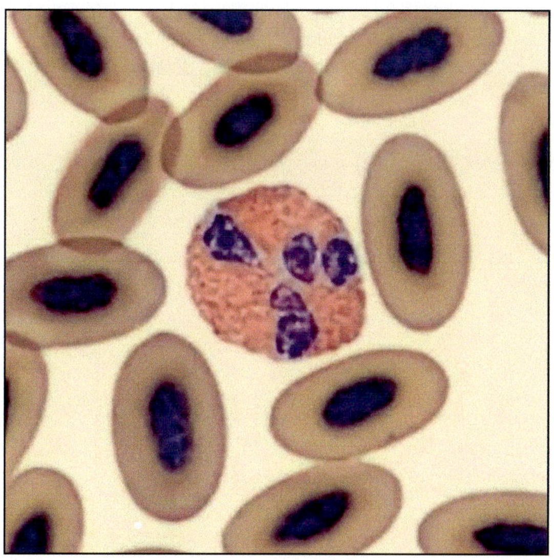

Figure 7.8 **Eosinophil: A round cell with a slightly basophilic cytoplasm (in contrast to the colourless cytoplasm of the heterophil). The granules are usually rounded, although shape and colour may vary greatly between species. The granules are distinctly eosinophilic and brighter in colour compared with the heterophil granules.**

Monocytes
(**Fig. 7.9**)

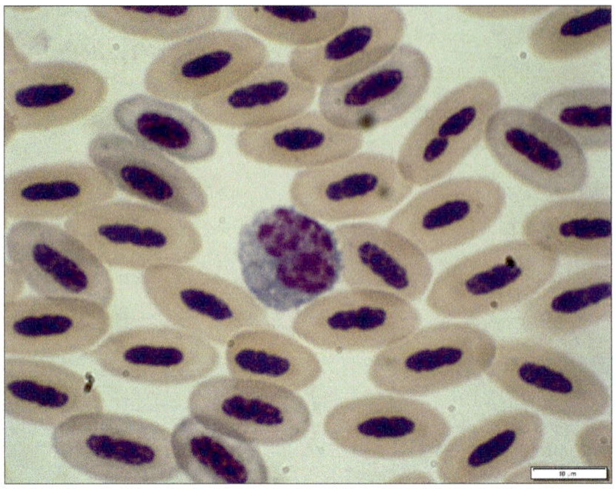

Figure 7.9 **Monocyte: The largest of the mononuclear leucocytes but rarely seen in peripheral blood smears. They spend only a short time in circulation before passing into tissues and becoming macrophages. The eccentric nuclei are either round, elongated, or indented, and the cytoplasm typically stains a blue-grey colour with a reticular or finely granular appearance, with occasional vacuoles. Care must be taken not to confuse them with large lymphocytes.**

Monocytes respond to diseases such as granulomatous diseases, including chlamydiosis, fungal and/or bacterial granulomas, and massive tissue necrosis. The peripheral monocytes are taken out of circulation to become tissue macrophages.

Basophils
(**Fig. 7.10**)

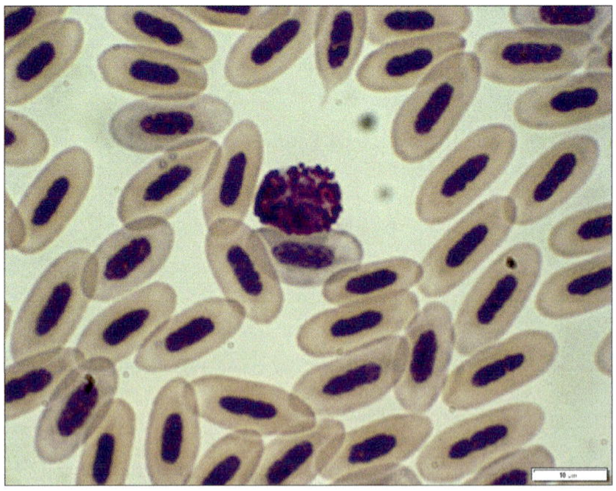

Figure 7.10 **Basophil: Uncommon in peripheral blood smears of birds. They appear as small cells with clear cytoplasm and spherical basophilic granules. The nucleus stains a light blue colour. Care must be taken not to confuse them with immature heterophils.**

Basophilia has been reported in respiratory diseases (e.g., air sac mites in canaries), chlamydiosis, and tissue trauma more than 48 hours old. In birds, basophils appear to play an important role in early inflammatory and immediate hypersensitivity reactions but differ from those in mammals by not contributing to delayed hypersensitivity.

The white blood cell (WBC) count and differential are important tools in assessing a patient's response to disease or injury. The WBC count in birds can be determined using three testing methods:

- An automated count, which has recently become available in some laboratories.
- An estimated WBC is determined from a blood smear by counting all leucocytes in ten high-power (40×) microscopic fields, dividing by the number of fields, and then multiplying this average by 2,000, giving a total WBC/µl.
- The total leukocyte count is best made using a method that involves the staining of avian

heterophils and eosinophils using phloxine B as the diluent. Commercial kits for this purpose are available (Avian Leukopet, Vetlab Supply, USA). A measured amount of blood (typically 25 μL) is diluted with the phloxine B test kit. A Neubauer-ruled hemacytometer is then charged with the blood–phloxine mixture and allowed to stand for at least 5 minutes. Then, the stained heterophils and eosinophils are counted on one side of the chamber (nine large squares).

The total leukocyte count is then obtained using the following calculation:

$$\text{TWBC}/\mu\text{L} = \frac{(\text{eosin-stained cells [counted in 9 squares]} \times 2 \times 1.1 \times 16)}{(\%\ \text{of heterophils} + \text{eosinophils})}$$

Changes in total leucocyte count

Leucocytosis can be normal in young birds, but it can also be due to:

- Stress
- Inflammation, often associated with bacterial and fungal infections
- Trauma
- Leukaemia
- Bleeding into an internal body cavity or neoplasm

Leucopenia can be due to:

- Chronic inflammation or disease, often with an acute decompensatory episode at the time of presentation
- Overwhelming bacterial and viral infections (e.g., PBFD in juvenile parrots, especially young grey parrots)
- Artefacts resulting from poor sample handling (blood clotting) or technique

White cell differential counts are best obtained from fresh blood smears, as cellular morphology can be affected by anticoagulants in blood collection tubes. A differential count is typically obtained by examination of stained smears under high magnification. Both the type and morphology of the white cells seen are recorded. In many cases, the differential count and cellular morphology give more indication of a bird's health status than the total white cell count.

Changes in heterophils

Abnormal changes seen with heterophils include:

- Heterophilia. Seen with inflammatory disease, trauma, toxicoses, and stress. It is considered normal in many young birds.
- Heteropaenia. Seen with overwhelming septicaemia and certain viral diseases (e.g., PBFD in young African grey parrots). Poor smear preparation technique can 'smudge' many heterophils, causing an artefactual heteropaenia.
- Toxic changes (increased cytoplasmic basophilia, vacuolisation, nuclear degeneration, degranulation, or abnormal granules) are seen with severe diseases that affect production and release from the bone marrow. As the severity of the disease increases, the toxicity of the heterophils may also increase as well.
- Immature (band) heterophils have a less distinctively lobulated nucleus and fewer granules, which are round and deeply basophilic. (Care must be taken not to confuse basophils with immature heterophils.) They are produced in response to an overwhelming demand for heterophils and a depletion of the mature storage pool in the haematopoietic tissues. Most commonly associated with overwhelming infections and accompanied by a leucopaenia, a degenerative left shift often carries a poor prognosis.

Changes in lymphocytes

Occasional reactive lymphocytes are a normal finding in avian blood, but large numbers indicate marked antigenic stimulation as seen in severe infections (e.g., severe viral infections, chlamydiosis, aspergillosis, salmonellosis, and tuberculosis).

Lymphocytosis is seen in:

- Chronic antigenic stimulation resulting in a greatly expanded circulating lymphocyte pool. It is seen mainly in patients with chronic inflammation (infectious or non-infectious origin). The most common causes of chronic inflammation are infectious agents (viruses, fungi, parasites, bacteria).
- Physiologic leucocytosis is accompanied by heterophilia. It is often seen in young animals and those not accustomed to handling. Physiologic leucocytosis is mediated by epinephrine release or sudden physical exertion.
- Lymphoproliferative disease (leukaemia, lymphosarcoma) may be accompanied by lymphocytosis that varies from mild to severe (>100,000 cells/μl).

- It can be a normal finding in Amazons and canaries.

Lymphopaenia is seen in:

- Viral infections and diseases that cause bursal damage or bone marrow suppression.
- Some drugs and toxins.
- Corticosteroid release (stress) or administration (iatrogenic).
- Relative to a marked increase in heterophils.

Changes in monocytes

Monocytosis is commonly associated with chronic granulomatous infections such as chlamydiosis, tuberculosis, mycoses, and bacterial granulomatous diseases (e.g., mycobacterial infections).

Because circulating numbers are usually so low, monocytopaenia is not reported.

Changes in basophils

Basophils appear to participate in the initial phase of acute inflammation or hypersensitivity reactions. Like monocytes, their circulating numbers are usually so low that an absolute decrease in circulating numbers is not reported.

Thrombocytes

The function of thrombocytes is unclear; they contain little thromboplastin, so it is unlikely that they initiate clotting. With bacterial infections they tend to increase in numbers and become activated (pseudopodial formation and vacuolation) and tend to aggregate in clumps. They appear to have some phagocytic activity (**Fig. 7.11**).

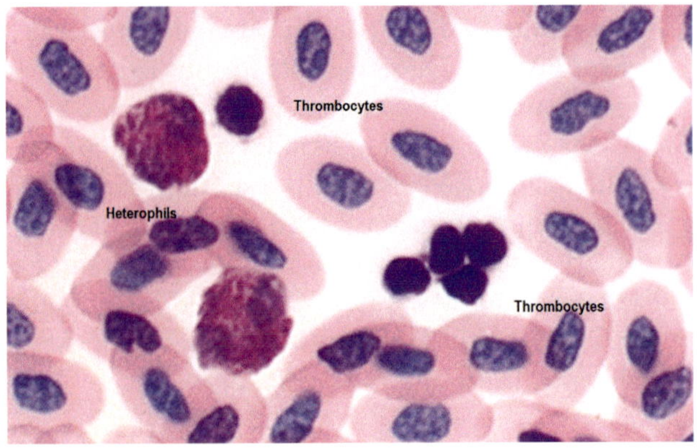

Figure 7.11 **Thrombocyte: Small, oval, nucleated cells that can be differentiated from erythrocytes by their size (they are smaller than erythrocytes) and their nucleus, which is larger, more rounded, and darkly basophilic staining. The cytoplasm is colourless or a faint blue colour with one to two small basophilic inclusions at the poles. Total counts are difficult and not routinely performed as the thrombocytes tend to clump. However, there are typically one to two cells seen per high-power field.**

Thrombocytosis is rarely reported and may arise as in response to thrombocytopenia. Thrombocytopenia may occur due to bone marrow suppression or disease processes causing an excessive demand (e.g., viral diseases such as circovirus, reovirus, or polyomavirus).

ASSESSING BLOOD COAGULATION

Vascular injury results in three events:

- The exposure of blood to tissue thromboplastin (factor III), which activates the extrinsic clotting pathway.
- The exposure of blood to subendothelium, which both activates the intrinsic pathway (which is weak or absent) and stimulates the adhesion and aggregation of thrombocytes to form a plug in the injury.
- Vasoconstriction, through various humoral and local mechanisms, which slows blood flow, reduces blood loss and allows time for thrombocytes to adhere to the injury.

Therefore, the extrinsic pathway is the major means of blood clotting in birds. Following vasoconstriction and thrombocyte adherence, the extrinsic pathway forms a fibrin clot over the thrombocytes and allows the vascular endothelium to begin repairs. This entire process can occur in as little as 10 seconds.

Unfortunately, there are no commercially available tests to evaluate the avian extrinsic pathway. Clinical tests for the assessment of coagulation in birds include:

- Thrombocyte evaluation. There should be one to two thrombocytes per oil immersion field. An estimated thrombocyte count/µl can be performed by counting the average number of thrombocytes in five high-power fields and then using the formula: estimated thrombocyte count per µl = average number of thrombocytes/1,000 × 3,500,000. The normal count is 20,000–30,000 per µl (20–30 × 10^9/l).

- Bleeding time has been used in poultry by pricking the comb with a needle. The normal bleeding time averages 8 minutes.
- Whole blood clotting time, where whole blood is placed in a plain tube, incubated and then inverted until clot forms. This assesses the intrinsic and common pathway and blood will normally clot in 2–10 minutes. This time will be accelerated by contamination with tissue thromboplastin. An alternative technique is to place blood in capillary tubes and break the tubes until a clot forms, normally in less than 5 minutes.
- Prothrombin time (PT), which evaluates the extrinsic and common pathway, can be used but requires avian thromboplastin. Mammalian thromboplastin significantly prolongs PT. While avian PT has been measured in multiple species and demonstrated to be prolonged in birds exposed to anticoagulant rodenticides, evaluation is complicated by the lack of an available source of standardised avian thromboplastin. The use of mammalian thromboplastin results in much longer normal PT times, and analysers utilising mechanical clot detection methods often time out before returning a result.
- Recent evaluations of viscoelastic coagulation indicate it may be useful. It evaluates the formation and lysis of a clot over time, providing information to better assess a patient's true clotting potential. Viscoelastic testing allows the measurement and visual representation of the changing viscoelastic properties of whole blood as coagulation occurs. It provides a rapid assessment of the initiation process, clot strength and stability, along with clot retraction and dissolution.

Figure 7.12 In-house laboratory equipment such as this bench top diagnostic blood analyser enables the rapid diagnosis of critically ill patients. The downside is that there is often a reduced number of analytes reported.

CLINICAL BIOCHEMISTRY

Clinical biochemistry involves the measurement of specific groups of chemicals within the body and the interpretation of the results obtained (**Fig. 7.12**). These chemicals include metabolites, tissue enzymes, electrolytes, minerals, and bile acids.

The levels of these chemicals in the blood can be influenced by either physiological or pathological processes. Physiological variations can be due to age, sex, body fat to muscle ratio, nutritional status, reproductive status, and species. However, pathological processes, including cellular damage or abnormal function of an organ system (or systems), often produce significant changes in blood levels. There are three major causes of abnormal clinical biochemistries:

1. Normal variation between species and individuals
2. Artefacts
3. Pathology

Normal variation between species and individuals

There are over 9,000 species of birds, with major differences in anatomy, physiology, form, and function. Some are carnivorous, some are nectivorous, some are granivorous, and some are omnivorous. It is unrealistic to expect that they would all conform to a relatively narrow range of biochemical values.

Other variations arise between individuals of the same species. These variations occur because of differences in age, sex, diet, husbandry, etc. For this reason, some clinicians recommend establishing a set of normal values for individual birds during annual health examinations, and then using these values as a comparison should the bird become ill.

Artefacts

When interpreting a biochemistry analysis, care must be taken to distinguish between abnormal results due to disease and abnormal results due to other factors. These other factors, referred to as artefacts, can occur for a variety of reasons, including:

- Physiological changes
- Previous therapy
- The clinical condition of the patient
- The collection method
- Storage and transport of the sample

Physiological changes

Stress due to transport and handling of the patient can lead to a release of endogenous corticosteroids, resulting in changes in the haemogram and in blood glucose.

Lipaemia, while occasionally seen in diseases of the liver and reproductive system, can also occur naturally in the reproductively active female (**Fig. 7.13**). Regardless of the cause, lipaemia can cause false elevations in bile acids, protein, calcium, phosphorous, and uric acid. It may also falsely decrease amylase. Postprandial lipaemia is uncommon in pet birds, so fasting will not help; the clinician needs to check with the laboratory if the sample submitted was lipaemic before interpreting these biochemistries.

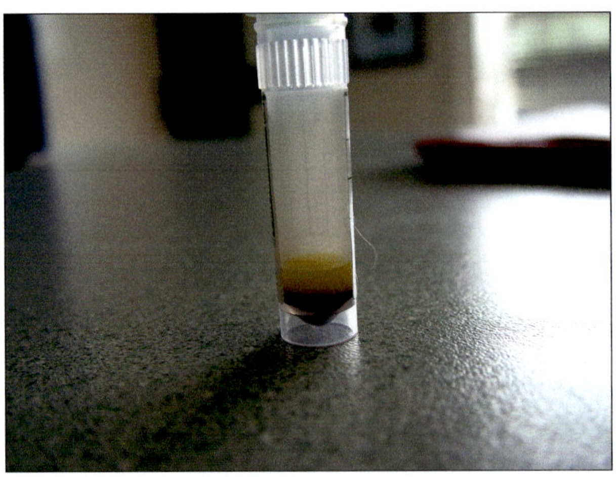

Figure 7.13 **Centrifuged blood sample demonstrating lipaemia. This condition in birds is usually associated with high-fat diets, liver disease, diabetes mellitus, and reproductive activity.**

Previous therapy

Before interpreting biochemistries, the clinician should consider if any treatment given prior to the sample collection could have had an effect on the results. Therapy given by another veterinarian or in an attempt to stabilise a crashing patient can have marked effects. Parenteral fluids can dilute biochemistries; exogenous corticosteroids can markedly elevate aspartate aminotransferase (AST), creatine kinase (CK), and lactate dehydrogenase (LDH); intramuscular injections, particularly of irritant drugs, can do the same.

The clinical condition of the patient

Trauma, starvation, and dehydration can all have marked effects on biochemistries, and need to be considered when interpreting results. Trauma can cause elevations in AST and CK and possibly glucose; starvation can lower glucose and also elevate AST and CK if protein catabolism has begun; dehydration can elevate uric acid.

The collection method

Ideally, sample collection should be performed in such a manner that it has minimal impact on the patient while providing an artefact-free sample suitable for analysis. This usually requires venepuncture to be performed on a minimally stressed patient. Inexperienced clinicians may need to consider gaseous anaesthesia in order to collect a good sample without the bird struggling.

Venepuncture should be performed on a large vein (e.g., the jugular) using a needle that is large enough to minimise haemolysis while being small enough to minimise trauma to the blood vessel wall. Haemolysis can cause elevations in bile acids, LDH, CK, alkaline phosphatase (ALP), potassium, and phosphorous. Glucose and albumin may be decreased. Calcium may be elevated or decreased, according to the methodology used.

Toenail clipping should be discouraged. Not only is it unduly painful, but crush artefacts, and contamination with uric acid and bacteria from droppings on the perch can cause elevations in uric acid and decreased glucose if testing is delayed long enough for bacterial growth to occur in the sample.

Storage and transport of the sample

Blood collected for biochemistry analysis should be placed immediately into a lithium heparin tube. Ideally, miniature tubes as used in medical paediatrics should be used. The sample should be gently rolled or rocked; clotting must be avoided, but haemolysis must be as well. If the analysis is to be performed in-house, it should be processed immediately. If a delay is likely, or if the sample is to be shipped to an outside laboratory, the sample should be centrifuged, and the plasma harvested. Sending whole blood to an outside laboratory can result in decreased glucose (as cell metabolism continues) and haemolysis.

EDTA tubes are unsuitable for biochemistry analysis, but can be used for haematology, lead analysis, and fibrinogen determination.

Biochemistry analysis by organ system

Liver

The detection of liver disease through biochemistry is complicated by the fact that there are no specific 'liver enzymes' that can be evaluated conclusively in each and every case. Liver disease can be broadly classified into three conditions: hepatocellular rupture, decreased hepatic function, and cholestasis. These conditions can occur either separately or concurrently.

Hepatocellular rupture

Hepatocellular rupture releases intracellular enzymes, which then reach elevated levels in the blood. These so-called 'leakage enzymes' include:

- Aspartate aminotransferase. This cytosolic enzyme is found in many tissues in the body, but the highest concentrations are found in skeletal muscle and liver. Significant elevations usually represent either muscular or hepatocellular damage. AST, therefore, must be interpreted alongside CK (released from damaged muscle) to distinguish between the two. In general, an elevated AST with a normal CK indicates hepatocellular rupture. However, CK has a much shorter half-life than AST; a single-point muscle injury (e.g., an injection) 4–7 hours before sample collection could duplicate this biochemistry pattern. Although AST is considered to be the most useful liver enzyme, it cannot be considered in isolation as an indicator of liver disease.
- Glutamate dehydrogenase (GLDH), a mitochondrial enzyme, is the most specific enzyme for the detection of liver disease, but its sensitivity is low. Because it is bound to mitochondria, extensive and severe liver damage is required before elevations are detectable.
- Lactate dehydrogenase is not specific to any tissue; its main advantage lies with a half-life shorter than CK. Persistent elevation in the presence of normal CK is strongly suggestive of liver disease.
- Alanine aminotransferase (ALT) and alkaline phosphatase (ALP) are not considered useful in detecting liver disease in birds. ALT in birds is very nonspecific for the liver, and normal levels have been shown in cases with severe liver damage. ALP elevations are more commonly associated with bone disease in birds.

Decreased liver function

Decreased liver function can occur with any number of liver diseases, not all of which involve hepatocellular rupture. Chronic cirrhosis, amyloidosis, and hepatic lipidosis can all have an adverse effect on liver function without causing any cellular damage. In these cases, a 'liver function test' is necessary to detect the problem. Bile acids serve this purpose well. Produced in the liver, they are excreted in bile into the small intestine where they act to emulsify fat. Most of the bile acids are then resorbed in the small intestine, enter the portal system and are taken up by the liver to be recycled. Elevated levels occur when there is impairment of the liver's ability to remove bile acids from the portal circulation. A two- to fourfold increase in bile acids indicates a significant decrease in liver function. It needs to be noted though that a severely dysfunctional liver (e.g., end-stage cirrhosis) may not be able to produce normal levels of bile acids, leading to low to normal results. Total protein, especially albumin, may also be decreased with decreased liver function.

Cholestasis

Cholestasis occurs when the biliary system is partially or totally obstructed. This can be seen with biliary neoplasia, pancreatic disease, or diffuse swelling of the entire liver. Gamma glutamyl transferase (GGT) is an enzyme found in the cell membranes of the bile ducts. Elevations can be seen in cholestatic disease (e.g., bile duct carcinoma), but it is considered to be a relatively insensitive test for liver disease in parrots. Bilirubin is not produced in birds; they utilise biliverdin instead. There are no commercial assays for biliverdin.

Kidney

The end-product of protein metabolism in birds is uric acid. It is produced in the liver, enters the circulation and is then secreted by renal tubules (>90%) or filtered in the glomerulus (<10%). Significant loss of renal tubules will therefore result in elevations of uric acid. Dehydration is less likely to cause hyperuricaemia because glomerular filtration is relatively unimportant.

At first glance it would appear that uric acid offers a sensitive and specific test for renal disease. There are, however, several confounding factors. Firstly, species differences: carnivorous birds have higher normal uric acid levels than granivorous birds. Secondly, age: juvenile birds may have lower levels than adults. Thirdly, although significant elevations usually indicate renal disease, normal levels do not mean the kidneys are normal: mild increases could indicate early renal disease or dehydration (or both). There must be severe renal damage before uric acid levels begin to rise.

Because of this relative insensitivity of uric acid in detecting renal disease, levels are best interpreted alongside a determination of the bird's water intake and loss and a physical examination. To distinguish renal disease from dehydration, the patient's haematocrit, total protein, and blood urea nitrogen (BUN) should be evaluated concurrently. Dehydration can lead to decreased glomerular filtration rates (GFRs), in turn leading to elevated levels of BUN; this same decrease in GFR can lead to elevations of uric acid without primary renal disease being present. It is therefore prudent, in cases of an elevated uric acid level, to rehydrate the patient over 2–3 days before definitively diagnosing renal disease. Persistent hyperuricaemia after fluid therapy, and with haematocrit, total protein, and BUN returning to normal, confirms a diagnosis of renal disease.

Creatinine is generally accepted as being of little or no value in evaluating renal function in birds. Phosphorous elevations are usually not seen in birds with renal disease.

Reproductive system

Clinical biochemistries can tell the clinician little about the male reproductive tract; they can, however, reveal something about the activity of the female reproductive tract. Oestrogen, produced by developing follicles, induces the production of calcium-binding protein and vitellogenesis in the liver. The net result of this activity is an increase in circulating total protein, calcium, triglycerides, and cholesterol. The serum may appear lipaemic.

Radiographic evidence of hepatomegaly and increased long bone density can confirm reproductive activity. It should be noted, though, that normal calcium and protein do not reflect a lack of reproductive activity. Elevated triglyceride levels can be due to dietary factors, liver disease, ovarian activity, pancreatic disease, or causes not yet understood in parrots (see "Lipids" section).

Gastrointestinal system

Gastrointestinal disease typically only gives nonspecific results with clinical biochemistry. Elevations of CK, AST, and LDH are not uncommon, and are not specific to the intestinal tract. Electrolytes may give more information and should be evaluated with an understanding of the patient's appetite and thirst, hydration status, previous or current therapy, and pathologic processes (i.e., gastrointestinal or renal disease) which may alter electrolyte concentrations.

- Sodium must be interpreted with the knowledge of the patient's hydration status. It may be elevated with decreased water intake or dehydration through renal disease, vomiting or diarrhoea. Sodium may also be lost through the gastrointestinal tract or the kidneys. Other causes of hyponatraemia include overhydration, end-stage liver disease, and congestive heart failure.
- Chloride is interpreted alongside sodium. It may be elevated with vomiting or regurgitation, although this is uncommon; low levels are usually associated with regurgitation or vomiting, renal disease, congestive heart failure, and other conditions which cause water retention.
- Potassium may be decreased with vomiting/diarrhoea and elevated with dehydration, haemolysis, tissue damage, or poor sample handling.

There are many other possible causes of electrolyte disturbance, and our understanding of avian electrolyte balance is still in the very early stages.

Amylase and lipase have been proposed as useful parameters in the detection of pancreatic disease. There is still considerable discussion of the incidence of pancreatic disease and the specificity of these enzymes. Significant elevations of these enzymes, when accompanied by clinical signs of gastrointestinal dysfunction (vomiting, ileus, diarrhoea, coelomic pain) should lead the clinician to consider pancreatic disease as a differential diagnosis. However, normal levels do not preclude a diagnosis of pancreatic disease, nor do abnormal levels confirm such a diagnosis.

Blood glucose

Glucose is an essential energy source for nearly every cell in the body. Blood levels are governed by its intake, absorption, the interactions of hormones controlling carbohydrate metabolism (insulin, glucagon and somatostatin), the body's metabolism, its ability to store glucose, and its excretion. As disorders of glucose metabolism involve so many organ systems, it is treated here as a separate entity.

Hyperglycaemia may be a normal physiological process, (e.g., in juvenile birds). However, elevated levels are usually related to increased production or release (e.g., stress) or failure of tissues to take it up out of the blood (diabetes mellitus). Iatrogenic hyperglycaemia occurs when corticosteroids are administered or intravenous dextrose is given. Female reproductive disease may also elevate blood glucose, but this may be an indirect result due to inflammation affecting the endocrine pancreas.

Hypoglycaemia may result from poor handling of blood samples (i.e., artefactual rather than factual), or with decreased food intake (starvation, anorexia), increased glucose usage (septicaemias, neoplasia, and multi-organ failure), or decreased production (liver disease).

Lipids

Lipid-associated disorders in birds include adipocytic tumours (lipomas, haemangiolipomas, myelolipomas, liposarcomas), hepatic lipidosis, xanthomas, and atherosclerosis. They are increasingly recognised as the effects of chronic malnutrition become obvious.

Cholesterol and triglycerides are lipids synthesised in the body and used for a variety of metabolic processes. As they are insoluble, they are transported through the body bound to lipoproteins – high-density lipoprotein (HDL), low-density lipoprotein (LDL), and very low-density lipoprotein (VLDL).

Cholesterol, synthesised in the liver, plays a major role in the production of bile acids, steroid hormones, and vitamin D. It also provides stability and flexibility to cellular plasma membranes. Triglycerides, fatty acids attached to glycerol, are synthesised in the liver and intestinal mucosa and stored in fat. They are used as an energy source for metabolic processes.

VLDLs transport cholesterol, triglycerides, and protein around the body. Once in circulation, triglycerides are released to cells for energy; as the triglycerides are removed, the cholesterol component increases until it is higher than the triglyceride component, forming LDL. This is then taken up by cells where the contents are either stored, used for cell membrane structure, or converted into other products such as steroid hormones or bile acids. Excess cholesterol in the cells and blood vessels is taken up by HDL, transporting it back to the liver for subsequent conversion to bile acids.

So, while cholesterol and triglycerides are necessary for normal body functions, high levels (associated with high fat diets, lack of exercise, obesity, and reproductive activity) may predispose birds to disease.

Elevations in cholesterol concentrations have been reported to be associated with a variety of disease conditions in birds, including hepatic lipidosis, atherosclerosis, hypothyroidism, bile duct obstruction, and diabetes mellitus. Decreased plasma cholesterol concentrations have been associated with some cases of hepatic disease (such as aflatoxicosis), decreased dietary fat, *E. coli* endotoxaemia, and spirochaetosis.

Elevated triglycerides are frequently associated with reproductive activity in female birds. Very high levels that are persistently elevated have been associated with ovarian pathology.

Elevated VLDL, LDL, and HDL concentrations, especially when combined with elevated cholesterol and triglycerides, are considered to be likely markers for atherosclerosis and hepatic lipidosis. More research needs to be done in this field. Most birds transport cholesterol mainly in the form of HDL (unlike humans, where LDL is predominant). Therefore, higher HDL values may not be associated with decreased risks of developing atherosclerosis as it is in humans – in fact, the converse may be true.

PROTEIN ELECTROPHORESIS

Protein electrophoresis (EPH) – the measurement of different fractions of plasma protein – can take the clinician beyond the white cell count in differentiating a bird's response to disease and stress. Used in conjunction with haematology, it is a valuable tool in the assessment of a bird's health status.

1. Plasma protein is composed of two major fractions: albumin and globulins. Albumin is a reserve protein and carrier of other molecules, while globulins include the inflammatory proteins, clotting proteins, and immunoglobulins. A range of physiological and disease factors affect the levels of protein in the blood, including the bird's age, sex, diet, hydration status, haemorrhage, inflammation, and tissue repair. Albumin and globulin are further broken down into smaller fractions: *pre-albumin* is often included in measurements of albumin. Its

occurrence is variable, with different researchers reporting its absence or presence in birds of different ages and sex. It has a role in thyroxine binding and transport.

2. *Albumin* is produced in the liver and serves as the major storage reservoir of proteins. It contributes 75% of osmotic activity of plasma. Its role is that of a carrier of minerals, vitamins, fatty acids, and thyroid hormones.
3. *Globulins* are composed of five fractions.
   a. *Alpha 1 and 2 globulins* are produced in the liver. They include:
      - Macroglobulins – inhibit proteolytic enzymes
      - Lipoproteins – lipid transport
      - Haptoglobin – haemoglobin transport
      - Ceruloplasmin – copper transport
      - Amyloid A – acute inflammatory protein
   b. *Beta 1 and 2 globulins* are also produced in the liver. They include:
      - Complement – enzymatic proteins involved in antigen–antibody interactions
      - Haemopexin – haem-binding protein
      - Ferritin – iron storage protein
      - Fibrinogen – clotting factor I; nonspecific indication of inflammation
      - C-reactive protein – produced in response to tissue injury, inflammation, or infection; immunosuppressive; promotes phagocytosis, inhibits platelets, activates complement
      - Lipoproteins
      - Some immunoglobulins (IgM, IgA)
      - Transferrin – a gamma globulin in chickens (beta globulin in mammals)
   c. *Gamma globulins* are the immunoglobulins, produced by B-lymphocytes. They include:
      - IgA – found in the intestinal tract
      - IgM – first immunoglobulin produced
      - IgG – supersedes IgM after 10–14 days; also known as IgY
      - IgN – found in some waterfowl

EPH uses electrophoresis, the movement of charged particles suspended in a fluid on various media (e.g., paper, gel), under the influence of an applied electric field. It is widely used in the separation of plasma protein fractions and is therefore valuable in the study of diseases in which the serum and plasma proteins are altered.

Submission of samples for protein electrophoresis

Plasma must be submitted for EPH, as serum is produced by the effect of clotting activation, which reduces fibrinogen and other globulin levels in the sample. Collect the blood into lithium heparin, centrifuge, remove the plasma, and submit.

Interpretation of results

As with all diagnostic tests, EPH must be interpreted in conjunction with a good history, clinical examination, and other diagnostic tests. It is not an indicator of specific disease, as earlier work had suggested. Rather, it should be used as an indicator of debility and inflammation. Not all diseases will result in changes in an EPH, e.g., neoplasia rarely causes any changes. Note that reference intervals vary, often significantly, between laboratories.

<table>
<tr><th rowspan="2">FRACTION</th><th rowspan="2">GENERIC REFERENCE INTERVALS (G/DL)</th><th colspan="2">POSSIBLE CAUSES OF CHANGES</th></tr>
<tr><th>INCREASE</th><th>DECREASE</th></tr>
<tr><td>Albumin</td><td>• 1.2–3.2</td><td>• Dehydration
• Reproductive activity</td><td>• Decreased synthesis
• Liver disease
• Poor diet
• Increased loss
• Protein-losing enteropathies and nephropathies
• Intestinal parasites
• Increased usage
• Chronic inflammatory disease</td></tr>
<tr><td>Alpha 1 globulins</td><td>0.8</td><td>• Acute inflammation or infection</td><td></td></tr>
<tr><td>Alpha 2 globulins</td><td>0.8</td><td rowspan="2">• Acute infection/inflammation
• Reproductively active female
• Acute infection in a chronically ill bird</td><td rowspan="3">• Chronic-active inflammation</td></tr>
<tr><td>Beta globulins</td><td>0.57</td></tr>
<tr><td>Gamma globulins</td><td>0.57</td><td>• Chronic inflammation or infection
• Acute infection in a chronically ill bird</td></tr>
</table>

Figure 7.14 The ImmunoComb® *Chlamydia psittaci* test is an example of an in-house serological test. It only detects IgG, meaning that it may not detect acute cases where only IgM is produced.

SEROLOGY

Serological tests are designed to detect antibodies (usually IgG or IgM) in serum or plasma (**Fig. 7.14**). The results can be affected by a range of factors:

- Host factors
  - Persistence of antibodies post-infection. Serial tests, weeks apart, are needed to assess what is happening in the patient.
  - Lack of antibody post-infection. If a test only assesses IgM, chronic cases with only IgG will be missed. Similarly, some animals that fail to seroconvert due to immunosuppression will test negative.
  - Changes in antibody titre. A single sample only indicates prior infection. Detection of active infections requires a measurement of rising titres (usually fourfold) over a 4–6-week period.
  - Species-specific reaction to certain testing methods.
- Antigen factors
  - Prepatent period. Usually 1–2 weeks. If testing during the prepatent period, a false negative may be obtained.
  - Antigen-specific reaction to certain testing methods.
- Assay factors
  - Laboratory error.
  - Poor test selection.
  - Lack of reagents.
  - Cross-reaction between antibodies may cause false positives.
  - Sensitivity and specificity of test.

A variety of serological tests is available today. The selection of the best test requires a clinician to have an understanding of the principles of immunology and the interaction between host and pathogen. He/she also needs to have an understanding of the types of tests available and the advantages and disadvantages of each.

Using serology

Serology has several advantages that make it unlikely its use will ever disappear:

- PCR and other antigen-detection tests can sometimes be too accurate, detecting extremely low levels of antigen (or portions of antigen) that may not be significant. At other times, the antigen may be extremely difficult to detect: it may shed only intermittently or at very low concentrations. In both these cases, serological evaluation of the bird (or birds) may better detect the presence of significant levels of antigen. Serological assays of a flock may be more sensitive in detecting a pathogen than more direct antigen-detection tests.
- Its relatively low invasiveness. Direct testing for an antigen that is only intermittently shed (if at all) often requires the collection of a tissue sample, either by autopsy or biopsy. Serology, on the other hand, simply requires a blood sample. This is obviously much less invasive.
- Serological tests, especially when performed in large numbers, can be significantly cheaper than more direct pathogen-detection tests.

Although at first glance these appear to be almost overwhelming advantages, other factors play key roles in the selection of an appropriate serological test. These factors can be categorised into three groups, discussed next.

Characteristics of the test

The selection of the test is of paramount importance. Is this test appropriate for this species and this disease? Antibodies are not necessarily consistent across species; a test that works well in poultry may not work well, or at all, in parrots.

The sensitivity and specificity of a test will often determine its usefulness. Sensitivity is a measure of a test's ability to accurately detect antibodies in an infected bird. A highly sensitive test will hopefully detect most of the antibody-positive birds in a population. As such, it is useful as a screening test during the early days of a disease outbreak, where it is important to identify quickly potentially infected birds. Once these birds have been removed from a population, it becomes more important to be sure that the remaining birds that have tested negative are, in fact, not infected. At this time a highly specific test is desirable. Specificity is the proportion of truly non-infected birds that test negative. A clinician using a serological test should therefore be aware of the published sensitivity and specificity of a test.

Another test characteristic to be considered is its complexity to perform. The more complex a test, the more susceptible it is to operator error leading to erroneous results. Errors can occur in the collection, storing, and transport of the sample to the laboratory, and then again in carrying out the test; equipment can be faulty, the operator may be inexperienced, or reagents may be out of date. There may also be error in the interpretation of the results, or even dissension about the significance of the results.

In some cases, a test requires specific reagents, which may not be readily available (e.g., the haemagglutination inhibition test for PBFD requires specially prepared galah erythrocytes, readily available in Australia, but harder to obtain elsewhere; however, goose erythrocytes can sometimes be used). This limits its availability and usefulness elsewhere in the world.

Characteristics of the antigen and the disease

Some antigens/diseases are not readily detectable by some tests due to the behaviour of the antigen. Some viruses, for example, do not agglutinate erythrocytes, making haemagglutination inhibition tests unsuitable.

Some pathogens, such as viruses like the influenza A virus, undergo antigenic variance wherein the surface proteins of the virus change and prevent immediate detection. Other pathogens, such as *Mycobacterium* spp. and *Aspergillus* spp., seem not to provoke an antibody response in some patients for, as yet, undetermined reasons. In these situations, false-negative test results may occur.

From an epidemiological point of view, there will always be false-positive and false-negative results for any test. The clinician must therefore interpret results with care; if a disease is of a low incidence in a species or in a geographical area, is a positive result truly significant or even accurate? And vice versa. As with any laboratory test, the results must be interpreted as a part of the diagnostic test, not as a sole entity.

Host factors

Some clinicians seem to fall into the trap of forgetting that there is a patient attached to a laboratory test result. How the host is responding to the challenge of a pathogen has a major impact on the interpretation of a serological test (or any other test). The interaction between the host and the pathogen must be understood by the clinician in order to accurately assess the patient's status.

In the initial stages of infection (the prepatent period) there will be a delay while the immune system firstly recognises the presence of an antigen and then mounts a response to it (usually in the form of IgM). In this prepatent period (usually 1–2 weeks), a serological assay that only detects IgG will give a false-negative result. On the other hand, direct antigen-detection tests may have a prepatent period of only 1–2 days, and are therefore more reliable in detecting early infection.

With appropriate treatment or an effective immune defence (or both) the host will inactivate/kill the pathogen and then clear it from the body. During this process there is a rising antibody titre. After the pathogen has been cleared, these titres should start to decline. The rate of this decline is dependent on many factors (e.g., the strength of the immune system, the persistence of the antigen, the degree of response to a specific pathogen). A single-point serological assay is unable to distinguish between rising and declining antibody titres; serial tests, 2–3 weeks apart, are needed to assess what is happening in the patient. For example, polyomavirus infection in a bird may result in detectable antibody titres for the life of the bird, even though the bird cleared the virus from its body after a few weeks or months. So while a single serological test may indirectly detect that a pathogen was/is present in the patient, it cannot (by itself) detect the ongoing presence of disease.

The serological detection of pathogens that are ubiquitous in the environment is fraught with error. Most birds, for example, are exposed to *Aspergillus* spp. at some time; relatively few develop aspergillosis. The detection of antibodies to *Aspergillus* does not, therefore, constitute a diagnosis of aspergillosis; rather, it indicates the patient has been exposed to the pathogen and has mounted an immune response to it.

Despite the presence of a pathogen actively causing disease, some birds fail to produce antibodies. This immunosuppression can be due to a number of factors. Concurrent disease frequently causes immunosuppression: viruses such as PBFD, polyomavirus, pox and herpes, bacterial and parasitic infections, and aflatoxicosis have all been implicated in immunosuppressed patients. Certain drugs such as tetracycline, tylosin, and gentamicin are known to decrease antibody production. Environmental stress (inadequate temperatures, humidity, noise, poor nutrition) is known to suppress both the bursa and the thymus, possibly through corticosteroid production by the adrenal gland. Immunosuppressed birds, for whatever reason, may give a false-negative result.

A final consideration is that antibodies produced in response to a particular pathogen may be difficult to distinguish from antibodies produced against another pathogen. Antibodies produced by vaccination may be indistinguishable from those produced by natural infection. In either case, false-positive results can be produced.

Scenarios involving the use of serology

Disease outbreaks

Serology is often employed in the face of an outbreak of an infectious disease. Serology has the advantages of being relatively simple (most labs can perform it), quick, and cost-effective, particularly when compared with other (more direct) tests such as DNA/PCR and histopathology. Large numbers of birds can be tested quickly, giving an indication of the incidence of the disease in a given population. In the early stages of an outbreak, high sensitivity is needed; all infected (positive result) birds need to be identified as soon as possible. Sensitivity can be increased by using known high-sensitivity tests; increasing the sample size of the population being tested; and parallel testing, using several different tests to increase the likelihood of an infected bird being detected.

Once the incidence of a disease has been determined, steps can be taken to control it. This will usually require the removal of birds that have tested positive and further testing of those that have tested negative. This introduces the problem of false-negative results (seronegative birds that are latent carriers of a disease). At this time highly specific tests are needed, and may require the employment of a different serological test or tests. As the incidence of a disease approaches zero, the detection of these latent carriers becomes even more difficult, and complete eradication may not be possible without complete culling.

Specific pathogen-free flocks

One of the cornerstones of successful farming is maintaining a disease-free flock. This is no different with birds, whether it is an intensive poultry flock, a zoological collection, or an avicultural facility. The development of a specific pathogen-free (SPF) flock requires the testing of all members of that flock, preferably before they join it.

Veterinarians advising clients on a SPF protocol encounter similar problems to those faced when dealing with an outbreak of disease. No test can hope to be 100% sensitive and 100% specific. Therefore, parallel testing (several different tests, not necessarily all serological) and serial testing (repeated testing over a period of time) may help to detect those birds that are latent carriers and test negative the first time. While this can add considerably to the cost of new birds, it must be weighed against the costs of remedying an outbreak of disease if it escapes detection and is introduced.

The individual bird

While serological tests are extremely useful in screening flocks, they often lack the sensitivity and specificity required for screening an individual bird. While relatively inexpensive, they may lack the accuracy of a more direct antigen-detection test. However, these more direct tests only detect the presence of a pathogen, not necessarily the presence of a disease. Combining the two tests can lead to a better screening of an individual. By looking for the pathogen, and the body's immune response to it, a better understanding of the bird's status can be achieved. When combined with ancillary tests such as haematology and biochemistry (which can reflect the effect the pathogen is having on the bird), a more complete picture of the host–pathogen interaction becomes apparent.

PCR TESTING

PCR is the amplification of a specific fragment of nucleic acid (DNA) from minute quantities of source DNA material. It is rapidly becoming one of the most widely used techniques in molecular biology because it is quick, inexpensive, and simple.

The specificity of PCR depends on the primers used. As a general rule, most PCR assays are highly specific, but a quantitative PCR assay may have variable degrees of sensitivity and may fail to detect very low quantities of DNA. In an attempt to increase the sensitivity of PCR, 'real-time PCR' has been developed. This assay allows the laboratory to view the increase in the amount of DNA as it is amplified by characterising (through the use of fluorescent markers) the point in time during cycling when amplification of the target DNA is first detected, rather than the amount of target produced at the end of the assay. The higher the starting copy number of the nucleic acid target, the sooner a significant increase in fluorescence is observed.

While PCR is an accurate test for the detection of target DNA, results must be interpreted with caution. Firstly, the sensitivity and specificity of the test means that false positives can easily occur if the sample is contaminated during collection, processing, or testing. Secondly, a problem sometimes arises with an overdiagnosis of a disease solely through the detection of a pathogen, without necessarily diagnosing the presence of a disease process. Clinicians must therefore exercise caution in the use of PCR, using it as a tool for obtaining a diagnosis, rather than as a diagnosis in its own right.

MICROBIOLOGY

Over the last several decades there has been a quantum shift in thought about the significance of bacterial and fungal cultures taken from birds. The concept of bacterial and fungal infections being primary diseases has largely been replaced by a better understanding of how pathogens can take advantage of an immunosuppressed, malnourished patient. In other words, clinicians, instead of asking 'What infection does this bird have?' are now asking 'Why does this bird have an infection?' This change in thought has led to better diagnoses, better treatments, and better survival rates.

As part of this evolving thought process, clinicians have reviewed many of the tests that were commonly used to diagnose disease, and are questioning the correct use and interpretation of tests such as Gram stains and cultures. It is not appropriate to base a diagnosis of health or disease in a bird solely on a Gram stain and/or culture. Questions that need to be asked when selecting this test include:

- Does the clinical database (history, physical examination, haematology, and biochemistries) collected on this bird suggest a bacterial or fungal infection?
- Is the site the sample has been collected from likely to give significant information about the patient's problem?
- What do the culture or stain results mean?

As veterinary skills and diagnostic equipment have improved, clinicians are now recognising that many diseases in birds are not caused by infectious agents. Nutritional, metabolic, endocrine, and traumatic conditions, amongst others, can all give rise to the classic 'sick bird look'. Neither bacterial cultures nor antibiotics are likely to assist in the diagnosis and treatment of these patients. A sound clinical database provides the clinician with the best information from which to direct further diagnostic and therapeutic steps.

While a Gram stain or culture from the faeces of a bird with gastroenteritis is likely to provide valuable information, it does not constitute a diagnosis in its own right. Nor will it assist in the diagnosis of a bird with a respiratory problem or a traumatic injury, or even an assessment on a bird that is otherwise normal and healthy. But there still persists a school of thought that insists a faecal Gram stain should be performed on all birds, regardless of why they have been presented. Cultures (**Fig. 7.15**) and

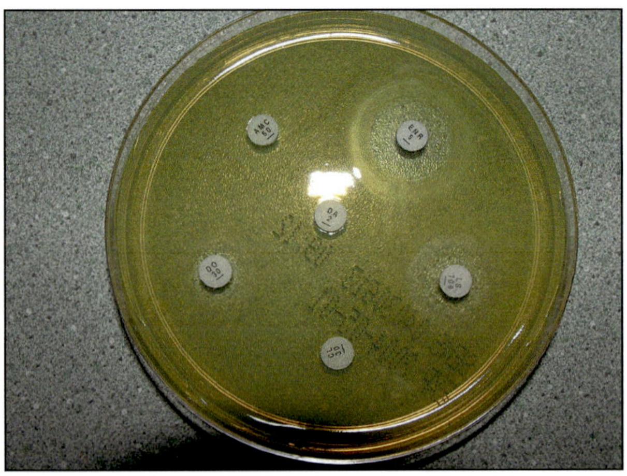

Figure 7.15 **A culture plate used to investigate antibiotic sensitivity. It should be remembered that although this test indicates the most appropriate antibiotic to use, it does not answer the clinician's primary question, 'Why does this bird have an infection?'**

Gram stains are useful diagnostic tools, but the clinician must evaluate if their use is justified for each patient.

Normal faecal bacteria cultured from healthy birds include gram-positive bacilli (*Lactobacillus* spp., *Bacillus* spp., *Corynebacterium* spp., and *Streptomyces*) and gram-positive cocci (*Staphylococcus epidermidis*, *Streptococcus* spp., *Aerococcus* spp., and *Micrococcus* spp.). Older birds tend to have more *Corynebacterium* and less *Lactobacillus* than juvenile birds. *Escherichia coli* is commonly recovered from cockatoos but less commonly in other species. Other gram-negative bacteria occasionally found in clinically normal birds include *Enterobacter*, *Klebsiella*, *Citrobacter*, *Pasteurella*, and *Moraxella* spp. While some *Pseudomonas* spp. have been recovered from healthy birds, *P. aeruginosa* is rarely recovered from healthy birds. Isolation of *Proteus*, *Salmonella*, *Pseudomonas*, *Klebsiella*, *Listeria*, *Erysipelothrix*, and haemolytic *Staphylococcus aureus* is clinically significant in sick birds.

Cultures from other sites (eyes, ears, choana) usually show similar results.

A heavy growth of a pure culture of any bacteria can reflect bacterial overgrowth and warrants further investigation.

Candida albicans is often detected by Gram stain in the droppings of birds eating yeast-containing feedstuffs (e.g., bread, biscuits). Unless the yeast are numerous and budding, or can be cultured in heavy numbers, they have little or no clinical significance. Development of pseudohyphae, seen on Gram staining, is significant and may indicate invasion of the intestinal mucosa by the yeast.

Culture of *Aspergillus* spp. or *Cryptococcus* spp. from a healthy bird is likely to reflect environmental contamination rather than infection.

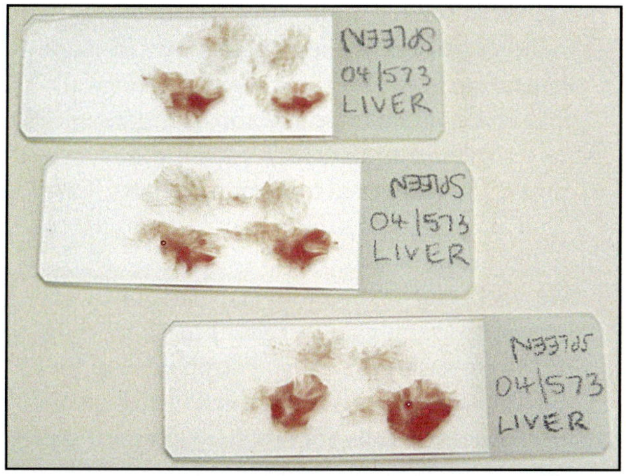

Figure 7.16 **Smears of liver and spleen tissue will be air-dried and then fixed in methanol ahead of staining. Note: It is always advisable to prepare more than one smear so that different stains can be applied. (Photo courtesy S Raidal.)**

Figure 7.17 **Stained liver tissue smear demonstrating how artefacts can be confused with pathology. In this case a talc crystal (identified) is from a disposable glove (Diff Quik, 100×). (Photo courtesy S Echols.)**

CYTOLOGY

Cytological samples (**Figs. 7.16, 7.17**) can be obtained from:

- Fine needle aspirates of masses or organs
- Coeliocentesis
- Sinus aspirates
- Crop washes
- Arthrocentesis
- Tracheal and air sac washes
- Contact smears from skin and internal organs
- Scrapings from the palpebral conjunctiva, cornea, oral cavity or tissues that normally yield poorly cellular samples

Assessment of a cytological sample requires the clinician to firstly identify both the cell types and the cellular response that is occurring in the sample.

Classification of cell types

- Haemic cells are those cells found in the blood and the haematopoietic tissues.

- Epithelial cells typically exfoliate easily and are found in clusters or sheets. They vary in shape depending upon their origin and can be oval, cuboidal, columnar, or polygonal. They typically have an abundant cytoplasm, small round to oval nuclei, and distinct cytoplasmic margins. Cells from secretory epithelium may contain cytoplasmic granules or vacuoles.
- Mesenchymal cells tend to exfoliate poorly and normally occur as single cells. They have indistinct cytoplasmic margins. Fibroblasts are the most frequently encountered cell of this group.
- Nervous tissue cells are rare in cytological specimens. They may be seen as deeply basophilic, stellate cells with cytoplasmic projections.

Classification of cellular responses

The goal of cytology is to classify the cell response into one of the basic cytodiagnostic groups. These groups include inflammation, tissue hyperplasia, benign neoplasia, malignant neoplasia, and normal cellularity.

Inflammation

A diagnosis of inflammation is made when an increased number of inflammatory cells is detected in the sample. The inflammatory cells of birds are heterophils, lymphocytes, plasma cells, and macrophages. Eosinophils may be included in the list of inflammatory cells; however, eosinophilic inflammation is either extremely rare or difficult to detect based on routine cytological methods. Heterophils and eosinophils may be difficult to differentiate based on routine cytological methods. Avian inflammatory responses are classified as heterophilic, mixed cell, or macrophagic:

- Heterophilic inflammation is present when heterophils make up >70% of the inflammatory cells and indicates an acute inflammatory response. Degenerate heterophils indicate a toxic microenvironment, usually caused by microbial toxins. If bacterial phagocytosis can be demonstrated, the cytodiagnosis of septic heterophilic inflammation can be made. If only extracellular bacteria are found, they may represent either normal flora or contaminants of the sample.
- Mixed-cell inflammation is represented by the presence of heterophils and mononuclear leucocytes. Heterophils usually make up at least 50% of the inflammatory cells. Mixed-cell inflammation usually represents an established, active inflammation.
- Macrophagic inflammation is indicated by the predominance of macrophages (>50%) in the inflammatory response. This type of inflammation does not necessarily imply chronicity, but may be suggestive of a number of aetiologies (e.g., intracellular pathogens). Macrophagic inflammation is common to certain avian diseases including tuberculosis, chlamydiosis, foreign body reaction, mycotic infections and cutaneous xanthomatosis. Multinucleate giant cell formation is often associated with macrophagic inflammation. Giant cells can appear within hours of the onset of some inflammatory responses and, unlike in mammals, their presence does not imply chronic inflammation.

Tissue hyperplasia or benign neoplasia

Tissue hyperplasia resulting from cellular injury or chronic stimulation is difficult to differentiate from benign neoplasia on cytology alone. Cells from hyperplastic tissue appear mature and do not exhibit much pleomorphism.

Malignant neoplasia

Malignant neoplastic cells show varying degrees of nuclear pleomorphism. There is an increase in nuclear size, which is reflected by an increased nucleus-to-cytoplasm ratio. Multinucleation may be present. The nuclei often have coarse, hyperchromatic chromatin and large or multiple (more than five) nucleoli. The cytoplasm shows increased basophilia, decreased volume and variability in the staining and it may have abnormal vacuolation or inclusions. There is often variation in cell margins. The cell types seen can be grouped into four basic classifications:

- Carcinomas – malignancies of the epithelial cells. Adenocarcinomas are frequently seen in birds. Cytological evidence of adenocarcinomas includes epithelial cells that tend to form giant cells, have cytoplasmic secretory vacuoles, and tend to occur in aggregates.
- Sarcomas – malignancies of mesenchymal cells. Fibrosarcomas are the most frequently encountered sarcomas of birds. The cells tend

to exfoliate poorly. They are abnormal appearing fibroblasts, which are spindle-shaped cells that typically exfoliate as single cells. Abnormal fibroblasts show increased cellular size and nuclear: cellular ratios and nuclear and cellular pleomorphism. Other mesenchymal cell neoplasms such as chondromas, chondrosarcomas, and osteogenic sarcomas may produce a heavy eosinophilic background material that can be seen on the microscopic sample.

- Discrete or round cell neoplasms. The only common neoplasm of this type is lymphoid neoplasia. Cellular features include a marked increase in the number of lymphoblasts, nuclear and cellular pleomorphism, an increase in cytoplasmic basophilia and mitotic figures, and abnormal or multiple nucleoli.
- Poorly differentiated neoplasms produce cells having features of malignancy, but the cells are difficult to classify as carcinomas or sarcomas.

Cytology of commonly sampled fluids and tissues

Coelomic fluids

Coelomic fluids can be classified as transudates, modified transudates, exudates, haemorrhagic effusions, or malignant effusions.

Transudates are odourless, transparent fluids characterised by total cell count <1 × 10⁹/l (<1,000 cells/mm³), a specific gravity <1.020, and a total protein <30 g/l. Transudates do not clot. The cell types are primarily macrophages and occasional mesothelial cells. Transudates occur as a result of oncotic pressure changes or other circulatory disturbances. The common causes of coelomic transudates in birds include hepatic cirrhosis, cardiac insufficiency, and hypoproteinaemia.

Modified transudates have total cell counts of 1–5 × 10⁹/l (1,000–5,000 cells/mm³). The mononuclear leucocytes predominate, with occasional mesothelial cells and rare heterophils. The mesothelial cells tend to be round or oval with increased cytoplasmic basophilia. Multinucleation, cytoplasmic vacuolation, and mitotic activity are often associated with reactive mesothelial cells. Care should be taken not to mistake these cells for malignant neoplasia. Transudates and modified transudates are commonly found in the coelomic cavity of mynah birds suffering from haemochromatosis.

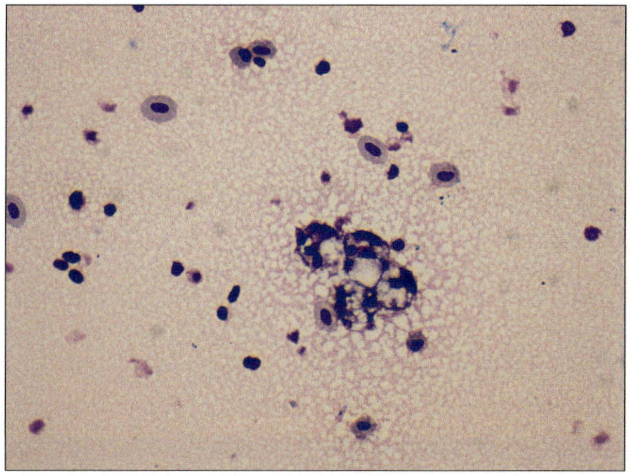

Figure 7.18 **Stained sample of egg yolk from a budgerigar reveals coelomitis (Diff Quik, 40×). (Photo courtesy S Echols.)**

Exudative effusions are characterised by high cell counts >5 × 10⁹/l (>5,000 cells/mm³), a specific gravity >1.020, and protein content >30 g/l. The majority of cells are inflammatory cells. Acute exudative effusions demonstrate primarily a heterophilic inflammatory response; however, macrophages quickly move into the fluid, creating a mixed-cell inflammatory response within a few hours of onset. Lymphocyte and plasma cells are often seen in long-standing exudative effusions. Lesions often associated with exudates include septic peritonitis, egg-related peritonitis, and coelomic malignancies (**Fig. 7.18**).

Haemorrhagic effusions are identified by the presence of erythrocytic phagocytosis. Thrombocytes usually disappear rapidly in haemorrhagic effusions. Iron pigment or haemosiderin crystals found in macrophages are also indicators of haemorrhagic effusions. Iron pigment appears grey to blue-black using Wright's stains. Haemosiderin appears as diamond-shaped, golden crystals within the macrophage cytoplasm.

Malignant effusions have features of either exudative or haemorrhagic effusions, but contain cells compatible with malignant neoplasia. Cystadenocarcinomas of the ovary are a common cause of malignant effusions in older females. These cells often form cellular aggregates of balls or rosettes and have cytoplasmic secretory vacuolation.

Urate peritonitis is a rare effusion that can occur when urinary fluids leak into the coelomic cavity. The acute lesion is poorly cellular, but it contains a marked number of sodium and potassium urate crystals. Urate crystals are spherical and have a spoke-wheel appearance. If the bird survives long enough, inflammatory cells will migrate into the fluid.

Cytology of the alimentary tract

Oral cavity

The oral cavity is easily sampled when lesions are visible. Normal cytology of the oral cavity shows occasional squamous epithelial cells, varying amounts of background debris and extracellular bacteria represented by a variety of morphological types. The differential diagnoses for common oral lesions include septic stomatitis, candidiasis, trichomoniasis, and squamous cell hyperplasia. *Alysiella filiformis*, a gram-negative small paired coccobacillus that forms ribbon-like chains, is normally associated with squamous epithelial cells.

- Smears made from a bacterial abscess reveal either a heterophilic or mixed-cell inflammation with bacterial phagocytosis.
- Candidiasis is evidenced by large numbers of typical organisms. *Candida* can be a normal inhabitant of the upper alimentary tract, so low numbers of organisms do not produce inflammation. An inflammatory response often occurs when the infection has involved the mucosa. The presence of hyphae formation suggests a potential systemic invasion by the yeast.
- Trichomoniasis is best diagnosed by observing the movement of the piriform flagellate protozoa in a wet mount preparation. It is important to recognise these organisms in a stained cytological sample if wet mount preparations are not part of the routine examination or if trichomoniasis is not suspected. Trichomonads appear as basophilic, piriform cells with flagella on Wright's stained smears (**Fig. 7.19**). The cell nucleus usually stains more eosinophilic than most cell nuclei. An eosinophilic axostyle can often be seen as a straight line running from the nucleus to the opposite pole of the cell. An inflammatory response is usually found associated with trichomoniasis.
- Squamous cell hyperplasia and metaplasia lesions may grossly resemble bacterial, yeast, or protozoal infections, but the cytological appearance is very different (**Fig. 7.20**). Normally, squamous epithelial cells exfoliate as single cells or small sheets. With squamous hyperplasia associated with vitamin A deficiency, smears with large numbers of cornified squamous epithelial cells that exfoliate in large sheets or aggregates are common. The cytology resembles that of the vaginal cytology of a dog in oestrus. Inflammatory cells are not seen unless there are secondary infections with bacteria, yeast, or protozoa.

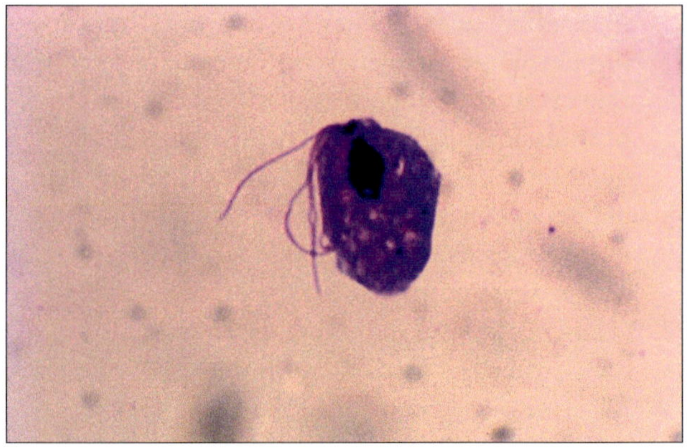

Figure 7.19 **Staining reveals a *Trichomonas* spp. within a tissue smear from a bird with bacterial pharyngitis. (Photo courtesy R Schmidt.)**

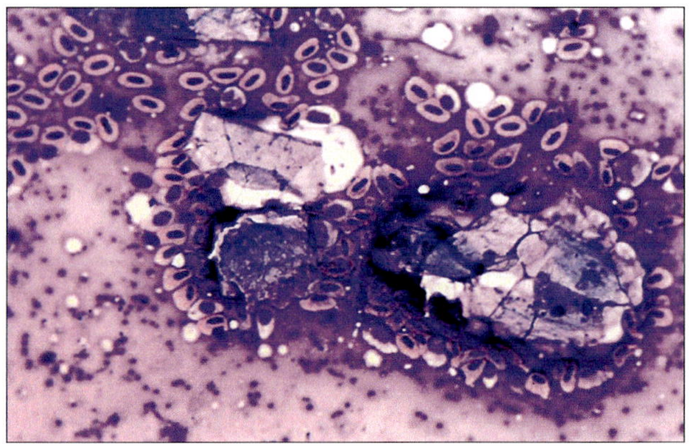

Figure 7.20 **Stained tissue smear from an Amazon parrot reveals squamous metaplasia with secondary bacterial invasion due to vitamin A deficiency. (Photo courtesy R Schmidt.)**

Oesophagus and crop

Cytological evaluation of the oesophagus and crop is indicated in birds with clinical signs of regurgitation, vomiting, delayed crop emptying or other suspected oesophageal and crop disorders. Normal cytology reveals occasional squamous epithelial cells and a variable amount of background debris and extracellular bacteria. An occasional yeast is accepted as normal. Also, some foods contain yeast as a source of supplemental B vitamins, so this source of large numbers of non-budding yeast must be ruled out. The presence of many bacteria represented by one morphological type, even without inflammatory cells, may indicate a problem. This may be a common finding in peracute ingluvitis. A pH >7 is also suggestive of acute or peracute ingluvitis. *Capillaria* ova may be detected in cytological samples from the oesophagus or crop of some birds with capillariasis. These ova are double operculated and may not stain.

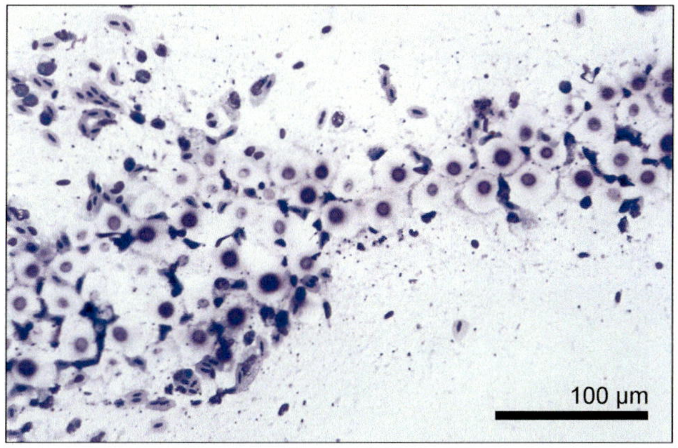

Figure 7.21 **Fungal spores revealed in a stained sample from a bird who has nasal cryptococcosis. (Photo courtesy S Raidal.)**

Cloaca

Cloacal cytology is indicated whenever a disorder of the lower intestinal tract, reproductive tract, urinary tract, or cloaca is suspected. Normal cytology reveals a few non-cornified epithelial cells, extracellular bacteria, background debris, and urate crystals. Abnormal findings would include the presence of inflammatory cells, large numbers of yeast, or a uniform population of bacteria.

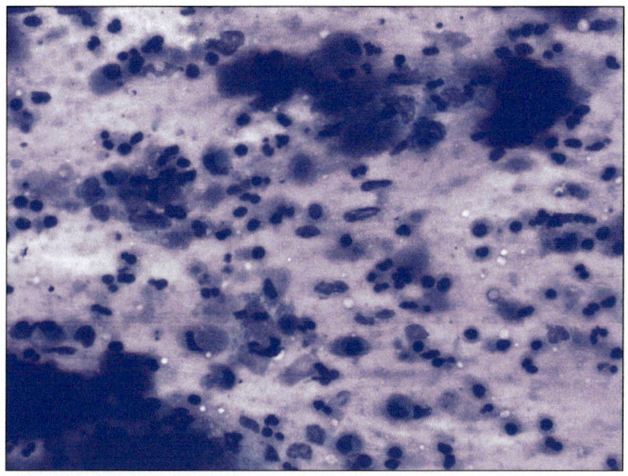

Figure 7.22 **Tissue smear from an umbrella cockatoo once stained reveals evidence of purulent tracheitis (Diff Quik, 40×). (Photo courtesy S Echols.)**

Cytology of the respiratory tract

Nasal and infraorbital sinuses

Normal cytology reveals occasional non-cornified squamous epithelial cells and low numbers of extracellular bacteria with little background debris. Evidence for periorbital sinusitis is provided by the presence of inflammatory cells in the aspirate. Lesions with a bacterial aetiology are indicated by septic, heterophilic, or mixed-cell inflammation. Mycotic lesions often reveal either mixed-cell or macrophagic inflammation, with the presence of fungal elements such as yeast, hyphae, or spores (**Fig. 7.21**). Chlamydial sinusitis often reveals mixed-cell or macrophagic inflammation. Chlamydial inclusions appear as small, blue-to-purple spherules, often in dense clusters, within the cytoplasm of macrophages when stained with Wright's stain.

Trachea

Normal cytology from a tracheal wash consists of a few ciliated respiratory epithelial cells and goblet cells.

Septic tracheobronchitis reveals inflammatory cells showing bacterial phagocytosis (**Fig. 7.22**). Degenerative respiratory epithelial cells show a loss of cilia, cytoplasmic vacuolation, and karyolysis. There is often increased mucin formation, which causes an increased thickness to the non-cellular background. Mycotic lesions involving the trachea, syrinx, and bronchi may reveal fungal elements on the tracheal wash. Aspergillosis is characterised by thick, septate hyphae that branch at 45° angles. Mycotic lesions usually reveal mixed-cell or macrophagic inflammation (**Fig. 7.23**).

Air sacs

Normal air sac samples are poorly cellular with the presence of a few non-cornified epithelial cells. Bacterial infections show the typical septic inflammatory patterns. Chlamydial and mycotic lesions demonstrate mixed-cell or macrophagic inflammation with the presence of chlamydial inclusions or

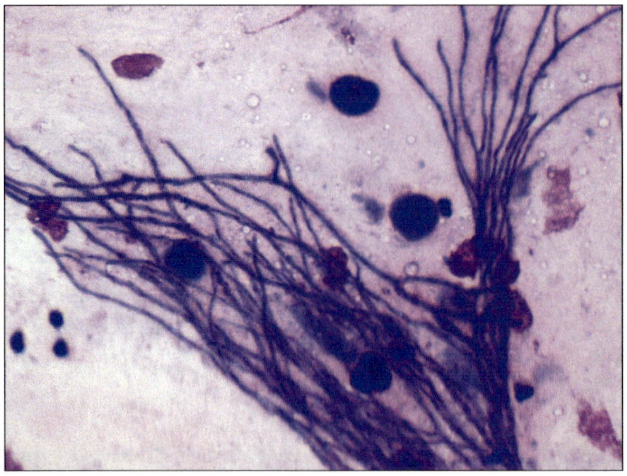

Figure 7.23 **Fungal hyphae clearly visible after staining in a sample from an Amazon parrot with fungal tracheitis (Diff Quik, 40×). (Photo courtesy S Echols.)**

fungal elements, respectively. Neoplastic lesions of the respiratory tract of birds are rare.

Skin

Normal skin samples typically contain squamous epithelial cells, debris, and extracellular bacteria.

- Bacterial infections involving the skin are usually associated with a heterophilic or mixed-cell inflammation (**Fig. 7.24**) (bacterial phagocytosis must be demonstrated to detect a septic inflammatory lesion). Care must be taken not to confuse basophilic-staining powder down for bacteria or yeast.
- Cutaneous xanthomatosis is a unique condition of birds caused by an excessive accumulation of lipids in the skin. It is a macrophagic inflammatory response, with multinucleated giant cells and cholesterol crystals observed on the cytological specimen. (Cholesterol crystals appear as angular, translucent crystals that vary in size and shape.)
- Subcutaneous lipomas produce a cytological specimen that appears 'greasy' on the unstained slide. The cytology reveals numerous lipocytes, which vary in size. Fat droplets usually partially dissolve in the alcohol-based stains, but are easily seen in water-soluble stains such as new methylene blue. Special fat stains such as Sudan IV can be used to demonstrate the fat droplets.
- Feather cyst cytology may reveal RBCs and erythrocytosis in early lesions. More chronic lesions develop a caseous exudate with a mixed-cell inflammation.
- Cutaneous and subcutaneous malignancies are rare in birds. Lymphoid neoplasia produces a highly cellular sample of immature lymphocytes. Cutaneous melanosarcomas have also been found in birds.
- Avian poxvirus lesions reveal clusters of squamous epithelial cells that contain large eosinophilic cytoplasmic vacuoles. The large cytoplasmic vacuoles found in the affected squamous cell push the cell nucleus to the cell margin.

Cornea and conjunctiva

Normal conjunctival scrapings provide poorly cellular samples with little background material. The cells may contain intracytoplasmic pigment granules. Normal cytology of the cornea is also poorly cellular and consists of occasional non-cornified squamous epithelial cells.

Inflammatory lesions involving the cornea and conjunctiva reveal inflammatory cells (**Fig. 7.25**)

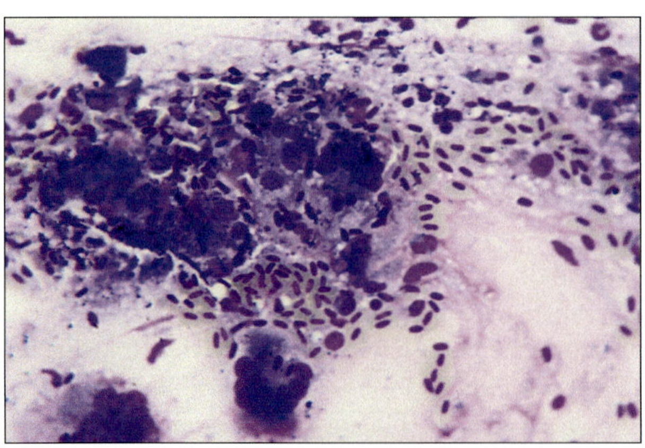

Figure 7.24 **Stained tissue smear shows presence of erythrocytes, leucocytes, and epithelial cells indicative of pododermatitis. (Photo courtesy R Schmidt.)**

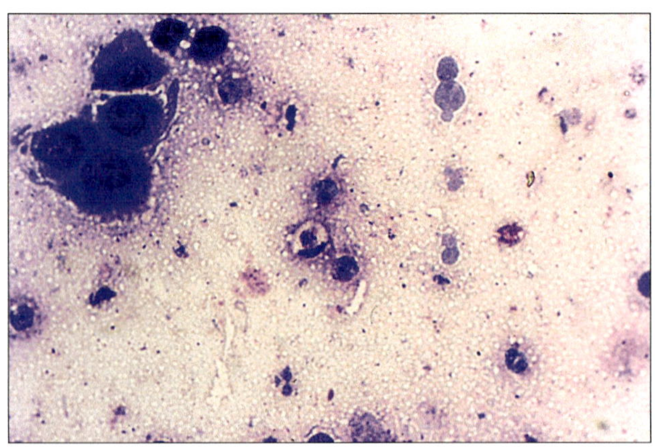

Figure 7.25 **Stained conjunctival smear reveals epithelial cells, leucocytes, and encapsulated yeast cells (*Cryptococcus* spp.) (identified) in the centre of the image. (Photo courtesy R Schmidt.)**

and increased numbers of exfoliated epithelial cells. Chronic lesions may also reveal the presence of cornified squamous epithelial cells that are not normally found in the conjunctiva or cornea.

Synovial fluid

Normal synovial fluid is poorly cellular. The cells are mononuclear cells, representing either synovial lining cells or mononuclear leucocytes. The background of normal synovial fluid consists of a heavy, granular, eosinophilic substance representing the mucin in the fluid.

An increase in the inflammatory cells and change in the colour, clarity, and viscosity of the fluid is indicative of inflammatory joint lesions. There may be a decrease in the granular eosinophilic background material, suggesting a decrease in mucin content. Erosion of the articular cartilage may result in the presence of multinucleated osteoclasts in the synovial fluid. Spindle-shaped fibroblasts suggest erosion into the fibrous layer of the articular capsule. Septic joint lesions may demonstrate bacterial phagocytosis by leucocytes. Traumatic arthritis also results in increased numbers of inflammatory cells.

Articular gout produces a cream yellow-coloured deposit in affected joints. Cytology reveals numerous, needle-shaped crystals. Inflammatory cells are often present and the mucin content is often reduced, as reflected in the reduction in the amount of eosinophilic granular background.

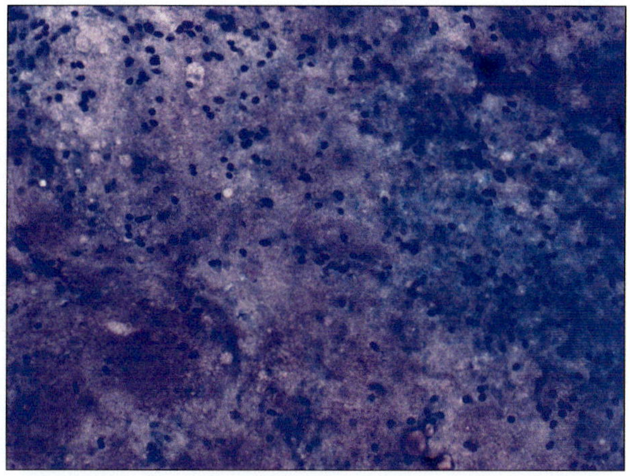

Figure 7.26 **Stained liver tissue smear from an African grey parrot indicates hepatic necrosis (Diff Quik, 20×). (Photo courtesy S Echols.)**

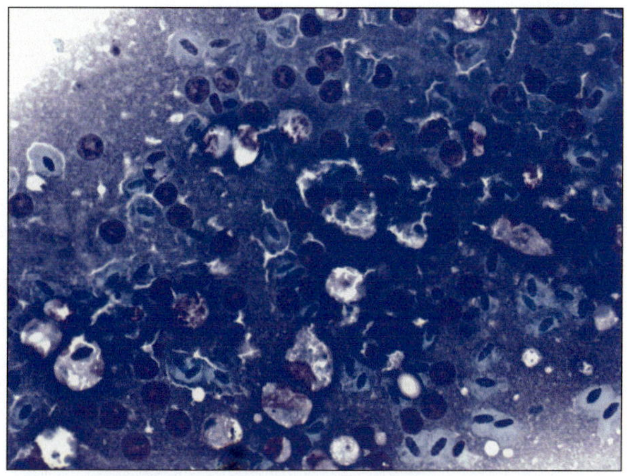

Figure 7.27 **Stained liver tissue smear from an African grey parrot indicates granulomatous hepatitis (Diff Quik, 400×). (Photo courtesy S Echols.)**

Cytology of internal organs

Liver

Cytological samples are usually highly cellular, with a predominance of hepatocytes, erythrocytes, and free nuclei. Hepatocytes are large epithelial cells that occur in sheets or clusters or as single cells. Normal haematopoiesis is occasionally found because the liver is a common location for ectopic haematopoiesis. Macrophages containing haemosiderin are occasionally seen.

Inflammatory lesions of the liver reveal numerous mature heterophils and an increase in the number of macrophages and plasma cells (**Figs. 7.26, 7.27**). It is important not to confuse normal ectopic granulopoiesis with heterophilic inflammation. (If developing stages of the heterophils can be found, the cytology is representative of granulocytopoiesis. If the heterophils are mature cells, the cytology indicates inflammation.)

Avian tuberculosis produces a macrophagic inflammatory response in the liver. The cytology reveals numerous macrophages and multinucleated giant cells (**Fig. 7.28**). The background of the smear and the macrophages may contain numerous bacterial rods that do not stain the waxy cell wall produced by mycobacteria.

Chlamydiosis often results in mixed-cell or macrophagic inflammation in the liver, with a marked increase in the number of plasma cells. Small, blue-purple, intracytoplasmic inclusions suggestive of chlamydial elementary and initial bodies may be seen in macrophages.

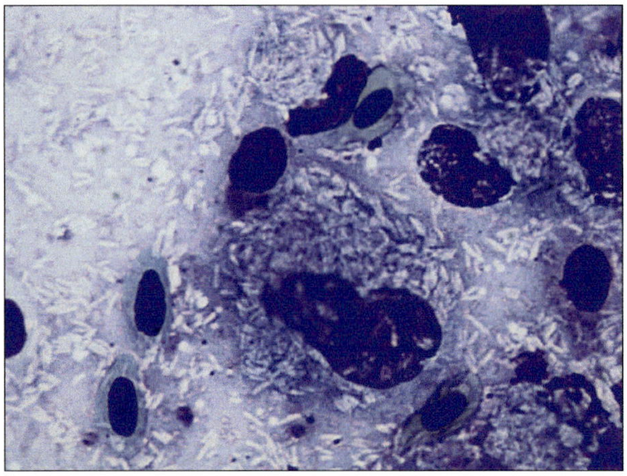

Figure 7.28 **Stained liver tissue smear from a bronze-winged pionus indicates granulomatous hepatitis due to a mycobacterial infection (Diff Quik, 100×). (Photo courtesy S Echols.)**

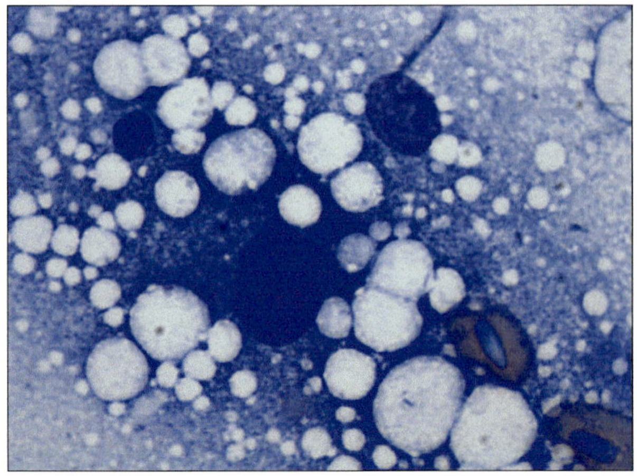

Figure 7.30 **Stained liver tissue smear from an umbrella cockatoo showing hepatic lipidosis (Diff Quik, 100×). (Photo courtesy S Echols.)**

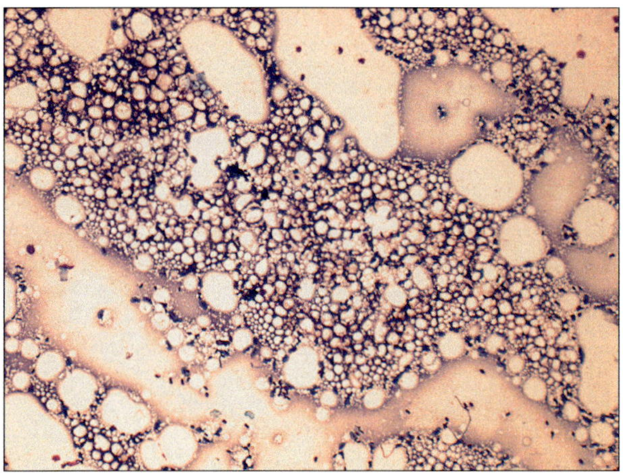

Figure 7.29 **Stained liver tissue smear from a cockatiel indicate hepatic lipidosis. (Diff Quik, 10×). Note the enlarged hepatocytes that contain round, cytoplasmic vacuoles. (Photo courtesy S Echols.)**

Hepatic lipidosis reveals enlarged hepatocytes that contain round, cytoplasmic vacuoles (**Figs. 7.29, 7.30**).

Occasionally, parasites may be found on hepatic imprints. Those commonly seen are schizogony of *Haemoproteus* and *Leukocytozoon* and sporozoites of *Atoxoplasma* and microfilaria.

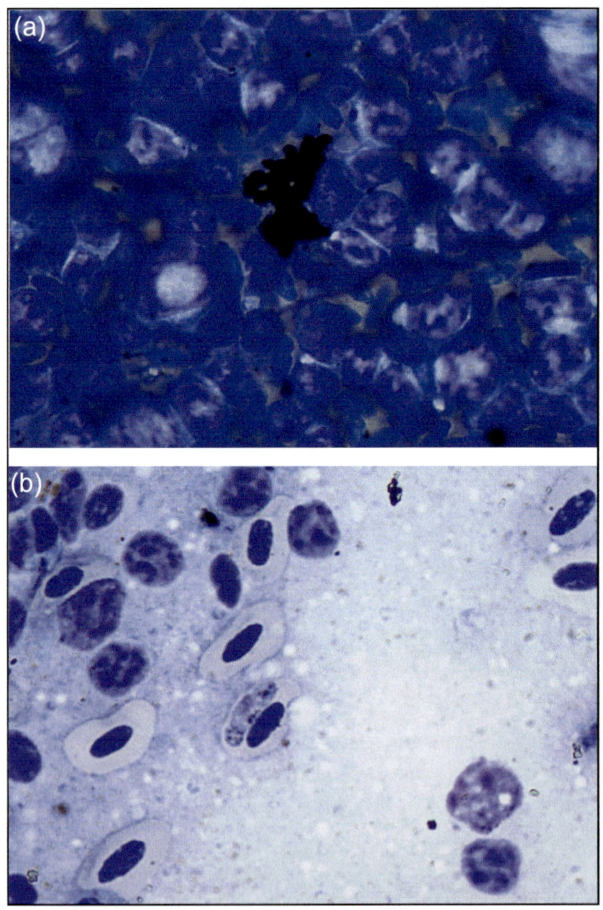

Figure 7.31 **Stained splenic tissue smear from a pigeon with haemoparasitism revealing haemosiderosis (a) and reactive splenitis (b). Note the haemoparasite located within the erythrocyte (b). (Diff Quik, 100×). (Photos courtesy S Echols.)**

Spleen

Normal cytology shows a marked number of erythrocytes and lymphocytes, reflecting the cytology of a lymphoid tissue. Chlamydial infections often cause a marked increase in the number of splenic plasma cells. Macrophages often demonstrate intracytoplasmic chlamydial inclusions. See **Figs. 7.31, 7.32**.

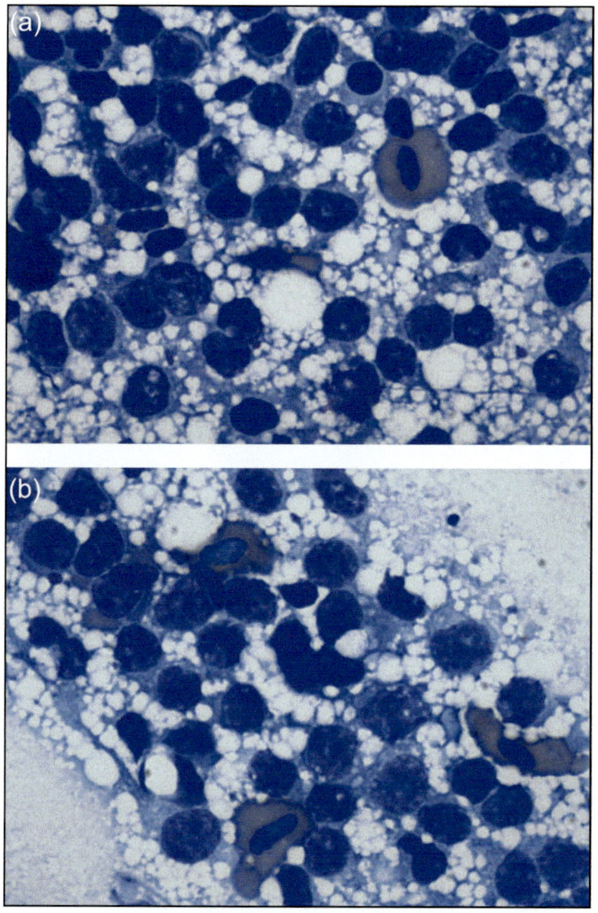

Figure 7.32 **Stained splenic tissue smears (a) and (b) from an umbrella cockatoo indicates hepatic lipidosis (Diff Quik, 100×). (Photos courtesy S Echols.)**

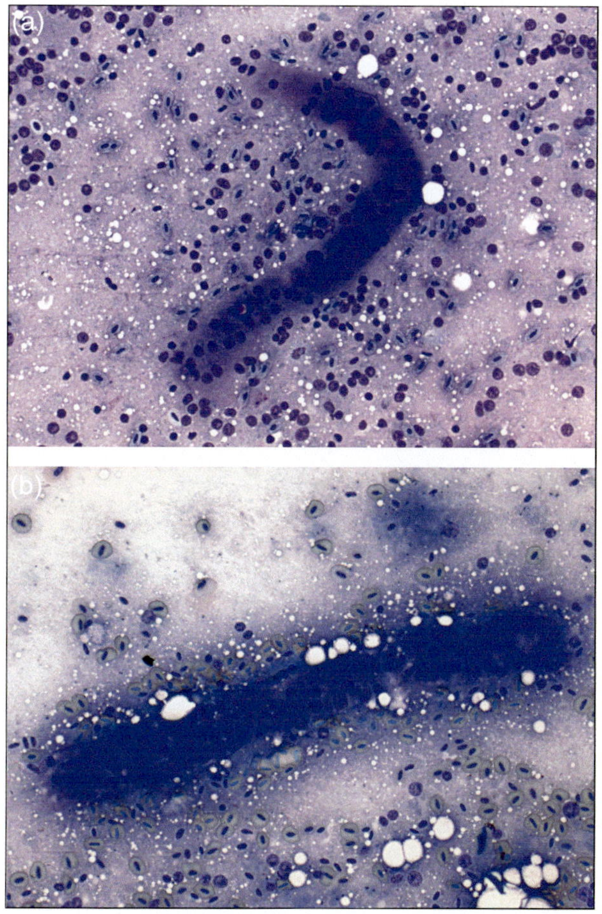

Figure 7.33 **Stained tissue smears from an umbrella cockatoo show a normal renal tubule (a) and an abnormal renal tubule due to hepatic lipidosis (b) (Diff Quik, 20×). (Photos courtesy S Echols.)**

Kidney

Normal kidney produces a highly cellular sample that contains numerous epithelial cells with an abundant, slightly basophilic cytoplasm and slightly eccentric, round-to-oval nuclei (**Fig. 7.33a**). Abnormal cytology (**Fig. 7.33b**) may include inflammatory cells or the presence of neoplastic cells:

- Epithelial cells from renal adenomas show increased cytoplasmic basophilia, slight pleomorphism and occasional mitotic figures.
- Renal adenocarcinomas produce epithelial cells having features of malignant neoplasia.
- Nephroblastomas produce poorly differentiated epithelial and mesenchymal cells.

FURTHER READING

Beaufrere H (2022) Blood lipid diagnostics in psittacine birds. *Veterinary Clinics of North America: Exotic Animal Practice*, 25: 697–712.

Beaufrère H, Cray C, Tully T (2014) Association of plasma lipid levels with atherosclerosis prevalence in Psittaciformes. *Journal of Avian Medicine and Surgery*, 28(3): 225–231.

Campbell TW, Ellis CK (2007) *Avian and Exotic Animal Hematology and Cytology*, 3rd edn. Blackwell Publishing, Ames.

Fudge AM (1997) Avian clinical pathology. In: *Avian Medicine and Surgery*. RB Altman, SL Clubb, GM Dorrestein, K Quesenberry (eds). WB Saunders, Philadelphia, pp. 142–157.

Fudge AM (2000) *Laboratory Medicine: Avian and Exotic Pets.* WB Saunders, Philadelphia.

Gerlach H (1994) Bacteria. In: *Avian Medicine: Principles and Application.* BW Ritchie, GJ Harrison, LR Harrison (eds). Wingers Publishing, Lake Worth, pp. 949–983.

Harr KE (2006) Diagnostic value of biochemistry. In: *Clinical Avian Medicine, Vol 2.* GJ Harrison, TL Lightfoot (eds). Spix Publishing Inc, Palm Beach, pp. 611–630.

Hochleithner M (1994) Biochemistries. In: *Avian Medicine: Principles and Application.* BW Ritchie, GJ Harrison, LR Harrison (eds). Wingers Publishing, Lake Worth, pp. 223–245.

Keller KA, Guzman DSM, Aciern, MJ, et al. (2015) Thromboelastography values in Hispaniolan Amazon parrots (*Amazona ventralis*): A pilot study. *Journal of Avian Medicine and Surgery*, 29(3): 174–180.

Ritchie BW (1995) Diagnosing viral infections. In: *Avian Viruses: Function and Control.* Wingers Publishing, Lake Worth, pp. 83–104.

Samour J (2006) Diagnostic value of hematology. In: *Clinical Avian Medicine, Vol 2.* GJ Harrison, TL Lightfoot (eds). Spix Publishing Inc, Palm Beach, pp. 587–610.

CHAPTER 8

DIAGNOSTIC IMAGING

RADIOLOGY

Radiology has long been a valuable diagnostic tool in avian medicine and is now being supplemented by other diagnostic imaging modalities such as ultrasound, computed tomography (CT), and magnetic resonance imaging (MRI). The increasing availability and decreasing cost of these advanced diagnostic imaging modalities circumvent some of the problems traditionally associated with low clinician 'comfort level' in avian radiology. For example, a lack of training and experience in radiology, as well as the apparent difficulty in obtaining good-quality images, is less of a problem with CT/MRI due to the ability to render 3D images, which increase confidence in visualisation of lesions. The short scan time also decreases the anaesthetic risk for small patients, although there is still an existing risk associated with intravenous contrast administration. However, digital radiography is still expected to remain the first-line diagnostic imaging modality most available to clinicians in the near future, so the basic principle – do it right the first time – is as true in birds as it is in other species.

OBTAINING HIGH-QUALITY IMAGES WITH RADIOLOGY, FLUOROSCOPY, AND CT

The goal in diagnostic imaging is to obtain high-quality images which are suitable for clinical purpose, using the minimum radiation dose to the patient and veterinary staff. Image quality is difficult to define because it is subjective in its nature, but a high-quality image enables the observer to extract information from the image that assists in making an accurate diagnosis.

Obtaining a high-quality image requires the following:

1. Equipment and techniques are used to provide:
   - Good contrast, i.e., the difference in density between two adjacent structures. High radiographic contrast is obtained where density differences are distinct (black to white). Low radiographic contrast occurs where adjacent regions have a low-density difference (dark grey to light grey). The greater the contrast, the more visible features become.
   - Appropriate exposure, which is determined by the amount of radiation energy used relative to the patient's body size. When insufficient energy is used, most of the radiation is absorbed by the patient and does not reach the receiver, producing an underexposed (white) image. Conversely, excessively high energy radiation produces an overexposed image, i.e., too dark to provide detail.
2. The patient is positioned so that the relevant part of the body is the centre of the image focus.
3. The patient must not move during the imaging. Movement – even rapid breathing – blurs the imaging and reduces the sharpness of the image. Scatter of the radiation beam at the periphery of the film will also reduce image sharpness if the beam is not collimated.

Equipment

Good avian radiography requires short exposure times (to minimise the effect of 'blurring' due to the patient's relatively rapid respiratory rate) and a combination of high-detail screens and films. These requirements mean that a relatively powerful X-ray machine is needed, with the ability to produce a high mA with a short exposure time. Most X-ray machines used in veterinary practices will fulfil these requirements.

DOI: 10.1201/9781003272519-8

Digital radiography (DR) is now available in many private practices. With the ability to obtain and manipulate high-definition images as well as storing them electronically, high-definition digital radiography offers major advancements in avian radiology. The authors finds that, for small patients (<50 g), a combination of digital plates and a dental X-ray machine achieves excellent images that were previously difficult to achieve using film, or even using DR or computed radiography (CR) equipment with conventional X-ray tube or machines. For birds over 100 g, DR or CR equipment with conventional X-ray tube or machines produce high-quality images. Having a Picture Archiving and Communication System (PACS) that allows the clinician to more closely examine images, measure parts of them, annotate them, and then communicate these images with colleagues and clients via email or CD makes for a huge leap in the quality of avian diagnostic imaging. PACS is also essential for storage and sharing of CT/MRI images.

If using film, non-screen film gives good detail but requires a higher exposure time (and therefore movement artefacts are likely). Mammography and extremity film (single emulsion film), using high-detail, rare earth cassettes will provide excellent results but require higher peak kilovoltage (kVp) and milliamperes (mA).

Restraint and positioning

Whether using conventional or digital radiography, positioning of the patient is absolutely critical for the accurate interpretation of an exposed radiograph. Correct positioning is best obtained under anaesthesia or sedation; only very tame (or very sick) patients may be positioned conscious in a plexiglass restraint device or similar restraints. The time of anaesthesia is generally brief (less than 10 minutes). If the patient's clinical condition precludes anaesthesia, positioning may be achievable using sedation with a benzodiazepine and/or opioid. This can then be reversed if needed.

Manual restraint is often used by inexperienced clinicians in the mistaken belief that anaesthesia in a sick bird is too dangerous. Not only is anaesthesia less stressful, and therefore safer, for the bird, manual restraint places the operator at an unacceptable risk of exposure to radiation.

The patient should be positioned in a symmetrical, straight fashion. If the bird is rotated, the radiograph will be difficult to interpret and may not be diagnostic. It is important to take the time to get the positioning correct so that the procedure does not have to be unnecessarily repeated.

Figure 8.1 **Lateral positioning.**

Good radiographic technique requires that a minimum of two orthogonal views (at 90° from each other) be taken for every study. By capturing the same region from different angles, orthogonal views help minimise the effect of tissue superimposition and allow for better visualisation of anatomical details and potential pathology. Several views are routinely used in avian medicine:

a. The lateral view is taken with the patient in lateral recumbency, with the wings extended dorsally and the legs pulled caudally. The limbs can be held in position with adhesive tape (**Fig. 8.1**). When examining the resultant image, the clinician should first assess that the acetabulae are superimposed, as should the coracoids (**Fig. 8.2**). In some cases, left- and right-sided views may be needed.
b. The ventrodorsal view is taken with the patient in dorsal recumbency with the wings extended laterally and the legs caudally alongside the tail, with the femurs ideally parallel to the spine and each other (**Fig. 8.3**). The keel should be superimposed over the vertebrae – a thin air space in the carina (seen as a thin black line) is used to check this (**Fig. 8.4**).
c. To evaluate the pectoral girdle, a caudoventral-craniodorsal oblique radiographic view made

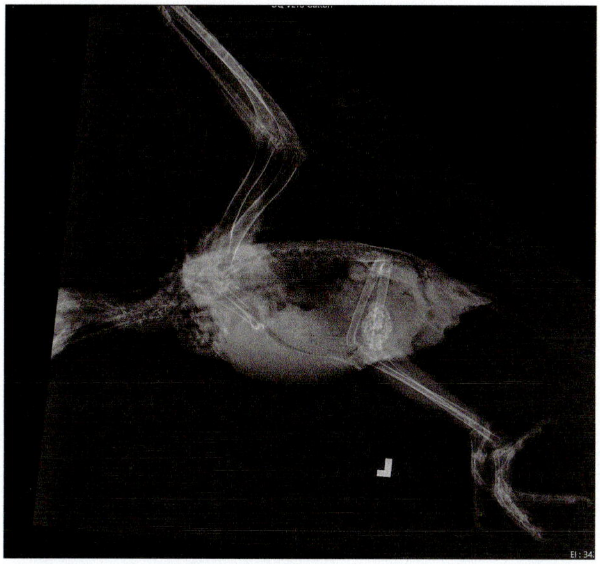

Figure 8.2 **Lateral radiograph.**

Figure 8.3 **Ventrodorsal positioning.**

at 45° to the frontal plane (the 'H-view') may be indicated. It is used to examine the coracoid, clavicle, and scapula without these bones being superimposed upon one another. This image is acquired by placing the patient in the same position that would be used for a standard whole-body ventrodorsal view. It is important that the patient's head and feet are aligned with the length of the table. This technique requires a machine that will allow for the head of the X-ray machine to rotate to 30°–45°. The exact angle depends on the species of bird being examined. Rotate the head of the machine to the desired angle and then move the machine caudally in relation to the patient (**Fig. 8.5a**). The beam should be aimed at the shoulders of

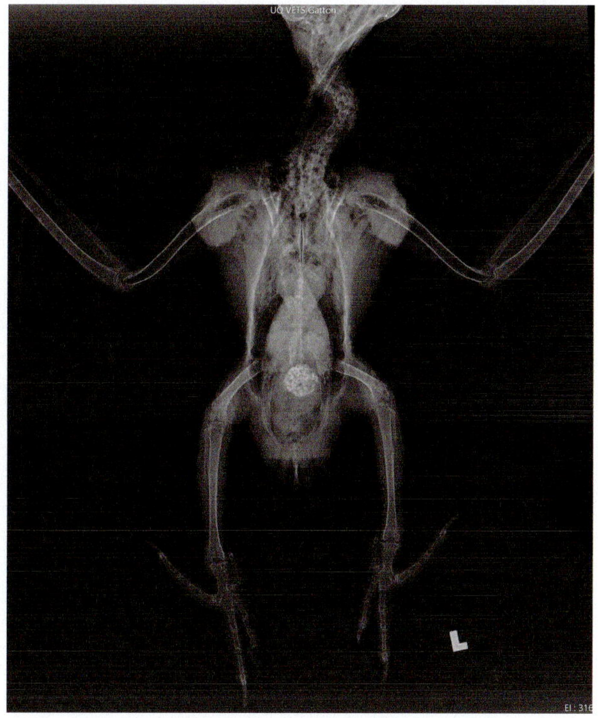

Figure 8.4 **Ventrodorsal radiograph.**

the patient passing through it in a caudal–ventral/cranial–dorsal angle (**Fig. 8.5b**). If well-positioned, all of the pectoral girdle is visible without superimposition (**Fig. 8.6**).

d. Lateral and anterior–posterior (cranial–caudal) leg (**Fig. 8.7**) or posterior–anterior (caudal–cranial) wing radiographs should be taken. This may require the use of a horizontal beam directed at a vertically-positioned plate (**Fig. 8.8**).
e. A quick 'bird in a box' technique can be used for birds too sick to handle the stress of being restrained or anaesthetised. The bird is placed in a cardboard box or paper bag in the standing position, and dorsoventral and horizontal lateral images are taken (**Fig. 8.9**). This technique is only helpful for checking the patient for an egg, large coelomic mass/fluid, or metal foreign body – it is of no value in examining organ size or location.

Contrast

Contrast studies can add greatly to the diagnostic value of radiology. It is often necessary to contrast one tissue or structure from its surroundings or to provide greater detail of that tissue/structure.

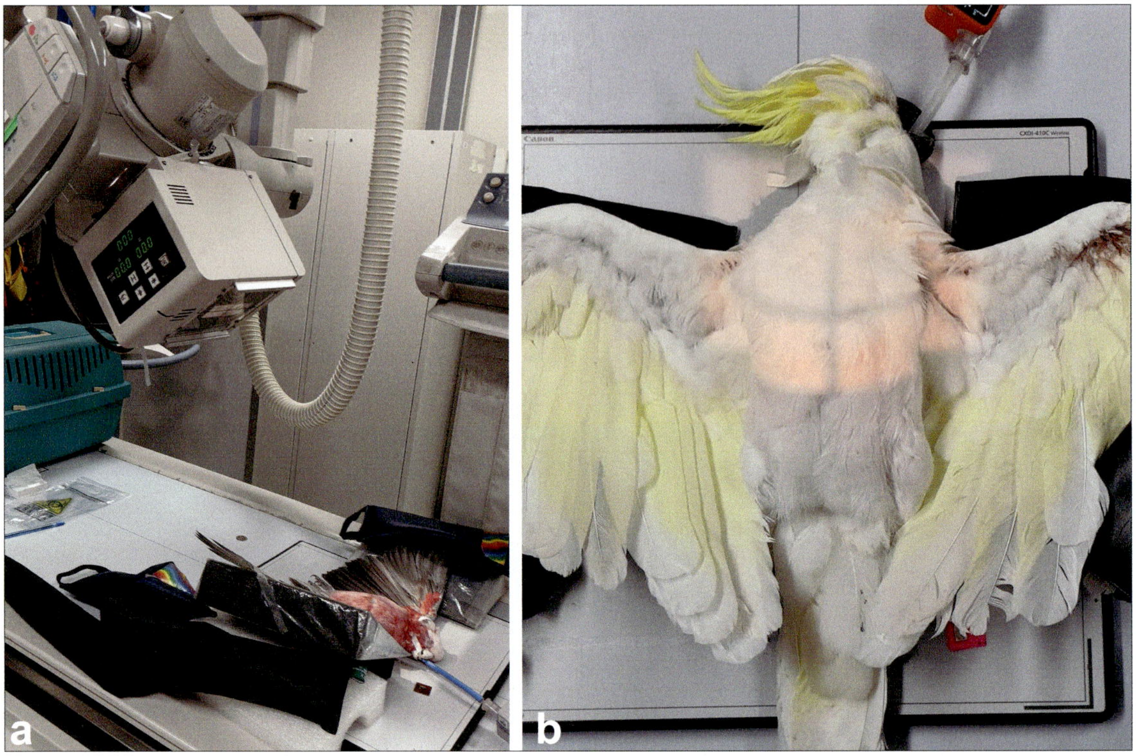

Figure 8.5 **(a and b) H-view positioning.**

Contrast media such as barium, iohexol, and diatrizoate have all been used in avian radiology.

Intravenous contrast

Intravenous contrast media can be used to help distinguish between soft tissue structures in the coelom, provide more information about lesions and abnormal vascularisation, and to highlight areas of inflammation. Most commonly used in CT, it can be used in radiography as well. Because of the short duration of effect, especially when combined with the patient's rapid heart rates and high cardiac output, it is generally recommended to administer 1–2 ml/kg of iohexol, an IV non-ionic iodinated contrast medium (Omnipaque 240, GE Healthcare Canada Inc., Mississauga, Ontario), at a rapid rate (over a few seconds) followed by rapid scanning techniques to enhance image contrast. (This requirement can be hampered by the viscosity of the media. While many have a low viscosity that can easily pass through the small gauge IV catheters, some may require dilution with sterile normal saline.) The same dose can be administered intraosseously with the same effect as intravenous administration.

While generally considered safe in mammals, deaths in small birds (<150 g) immediately after contrast medium administration justifies caution in its use in these birds. Potential causes of death following contrast medium administration may be the osmolality of the contrast medium and rapid shifts in the plasma osmolality, or the rapid expansion of the plasma volume may lead to baroreceptor activation and bradycardia.

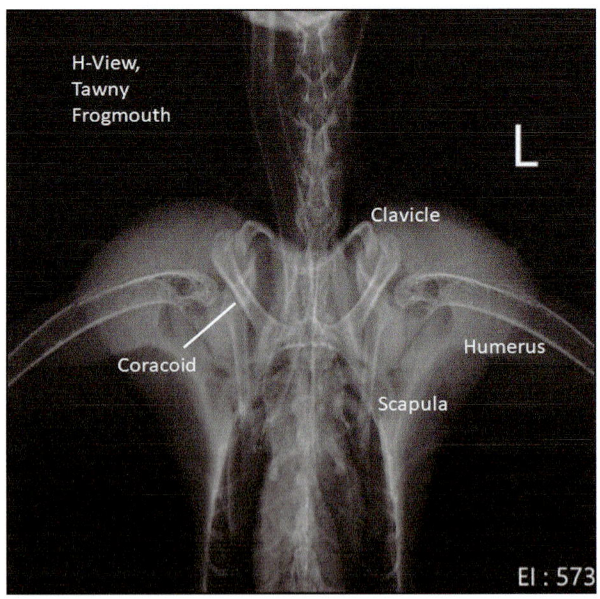

Figure 8.6 **H-view radiograph.**

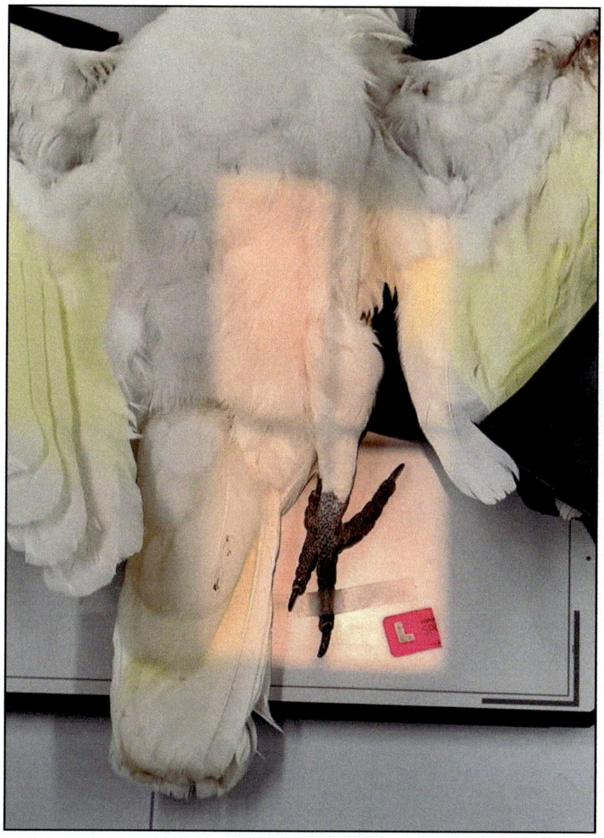

Figure 8.7 **Focused view positioning, pelvic limb.**

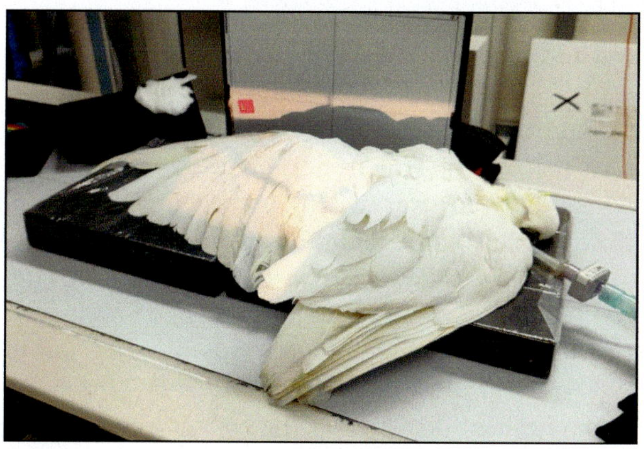

Figure 8.8 **Focused view positioning, caudocranial wing.**

Gastrointestinal studies

Gastrointestinal contrast studies (**Fig. 8.10**) utilising iohexol or diatrizoate are often comparable to those using the more routinely recommended barium sulphate. Barium and iohexol can be used undiluted or diluted with 1:1 with water. (Higher dilutions may result in poor contrast.) Diatrizoate has a high osmolality; this hypertonicity may cause dehydration in small birds. Care should therefore be taken when using diatrizoate, but it may be safer than barium if a gastrointestinal perforation is suspected.

The three most common mistakes made when using contrast media are:

1. Failing to take plain survey images immediately before contrast administration. This can rule out other disease processes, to adjust the unit's settings and the patient's positioning, to confirm adequate preparation of the abdomen (colon and bladder), and to use as a pre-contrast comparison.
2. Using an insufficient volume of contrast medium. Too small a volume does not distend the lumen sufficiently to appreciate detail.
3. Taking too few images after oral contrast medium is given.

The dose is based on the estimated size of crop volume (approximately 25–30 ml/kg for parrots). In addition to time-zero survey radiographs, additional orthogonal images are typically taken at 5, 15, and 30 minutes, then hourly afterwards till the study is complete. It is recommended to fast a parrot for 3–4 hours before a gastrointestinal contrast study (if fasting fails to empty the crop, the contents should be removed with a large bore feeding tube before administering the contrast media), but transit time does not appear to be affected by anaesthesia. Transit time for barium (crop to cloaca) in a healthy bird is approximately 3 hours; using iohexol it may be as short as 1 hour.

A double contrast study gives superior mucosal detail of the upper gastrointestinal tract. Using 100% barium suspension or concentrated non-ionic water-soluble contrast medium, 25–50% of the crop volume is infused as a positive contrast medium, followed by inflation of the crop with air (100–200% of estimated crop volume) to achieve moderate distention. The feeding tube is removed, and the crop massaged distally while digital pressure is maintained on the cervical oesophagus. Radiographs are taken immediately after massaging the crop and then every 15–30 minutes to document the passage of contrast medium throughout the gastrointestinal tract. The study is terminated when contrast medium reaches the cloaca.

Cloacal contrast

Radiographic contrast medium studies of the cloaca can be challenging. A large bowel follow-through on

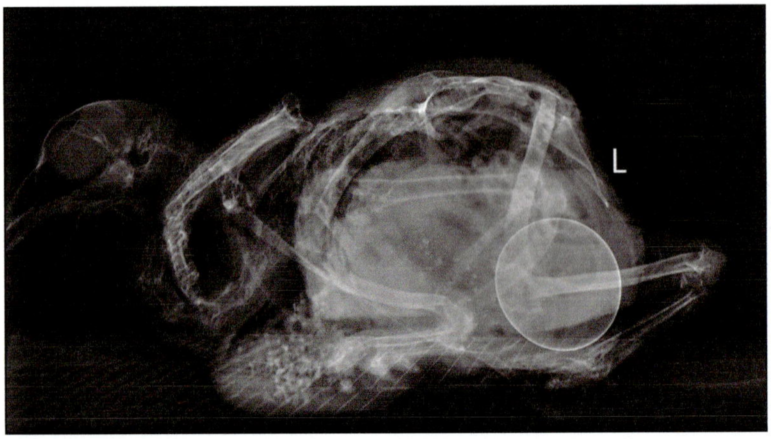

Figure 8.9 **'Bird in a box' radiograph.**

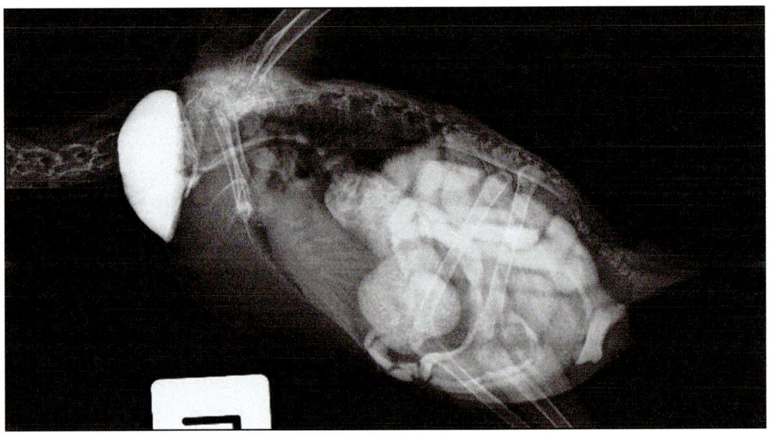

Figure 8.10 **Gastrointestinal contrast study showing dilated intestinal loops in an Alexandrine parrot.**

a gastrointestinal study may not produce sufficient cloacal distention, but it may help to assess for large bowel disease. A cloacogram under anaesthesia may be more definitive. A temporary ventoplasty, to assist in contrast medium distention, may be required. After taking survey radiographs, the cloaca is gently irrigated with sterile saline to remove faecal material and other debris. The ventoplasty sutures are then loosely tightened, and an iodinated contrast medium is then injected through the vent into the cloaca, using a soft, primed catheter (to avoid air bubbles). When the contrast medium begins to reflux, tighten the sutures further and take the first image. Inject a small volume of contrast medium just prior to each radiographic exposure to allow for leakage of the contrast agent. Ensure there is adequate absorbent material around the vent to minimise the effects of contrast medium leakage or wet feathers.

Urography

The patient should be fasted for about 2 hours before the contrast medium is given. It may be necessary to aid emptying the gut by introducing liquid paraffin into the cloaca. For urography, the organic iodine compounds are warmed to body temperature and slowly injected intravenously under anaesthesia, usually into the basilic vein. The dose is 2 ml per kg body weight, as a 70–80% solution of organic iodine compound or containing 300–400 mg of iodine per ml (800 mg/kg) intravenously. The quality of the contrast obtained depends on the concentrating capacity of the kidneys, on the preparation used, and on the iodine concentration of the medium. The aorta, heart, and pulmonary arteries are demonstrated 10 seconds after the injection, kidneys and ureters are shown 30–60 seconds after injections, and the cloaca and terminal gut are visualised 2–5 minutes after the injection (**Fig. 8.11**). Urography must not be performed in dehydrated or severely renal compromised birds.

Sinography

To visualise the infraorbital sinuses and nasal cavity, iodinated compounds are instilled directly into the

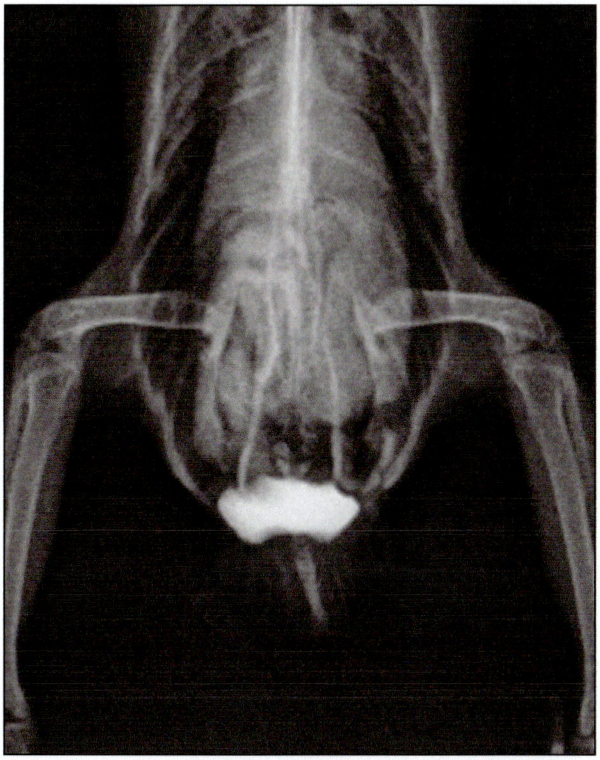

Figure 8.11 **Contrast outlining ureters and cloaca.**

sinuses to demonstrate the flow of contrast medium into the nasal cavity. From 0.1 to 1.0 ml of 15–20% iodine agent is installed directly in the nares or in the sinus. After taking the radiographs, the contrast medium should be flushed out with sterile saline to minimise local reaction after radiography has been performed. (Side effects have been described in some birds, including oedema and periorbital swelling.)

Normal anatomy

As with any species, it is important to become familiar with the normal gross and radiographic anatomy of birds (**Figs. 8.12, 8.13**). Reviewing all systems with a systematic approach in all studies will increase the diagnostic yield of each study, while improving the clinician's comfort and familiarity in interpreting avian radiographic images.

Musculoskeletal system

Reviewing the radiograph should include the following: the conformation of the skeleton; the density and thickness of the bone cortices; the radio-opacity of the medullary cavities, including that of the pneumatic bones (humerus and femur); and the presence of internal struts in the humerus and other long bones (**Fig. 8.14**).

Cardiovascular system

In general, a bird's heart is located in the cranioventral portion of the coelomic cavity, with its long axis slightly to the right and dorsal to the sternum, extending from the second rib to the fifth and sixth ribs. The apex is superimposed by liver lobes, making the heart–liver interface indistinct. In the

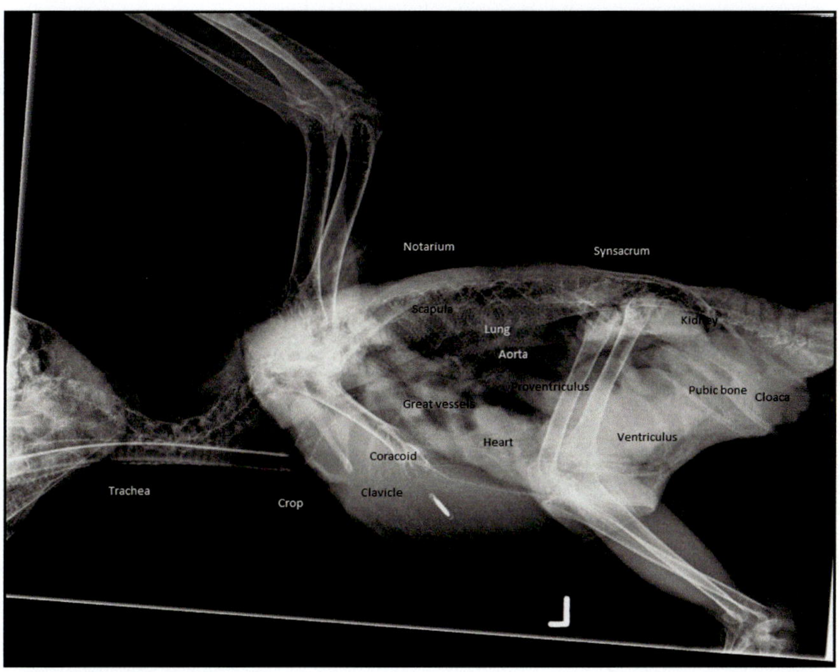

Figure 8.12 **Normal anatomy, lateral view.**

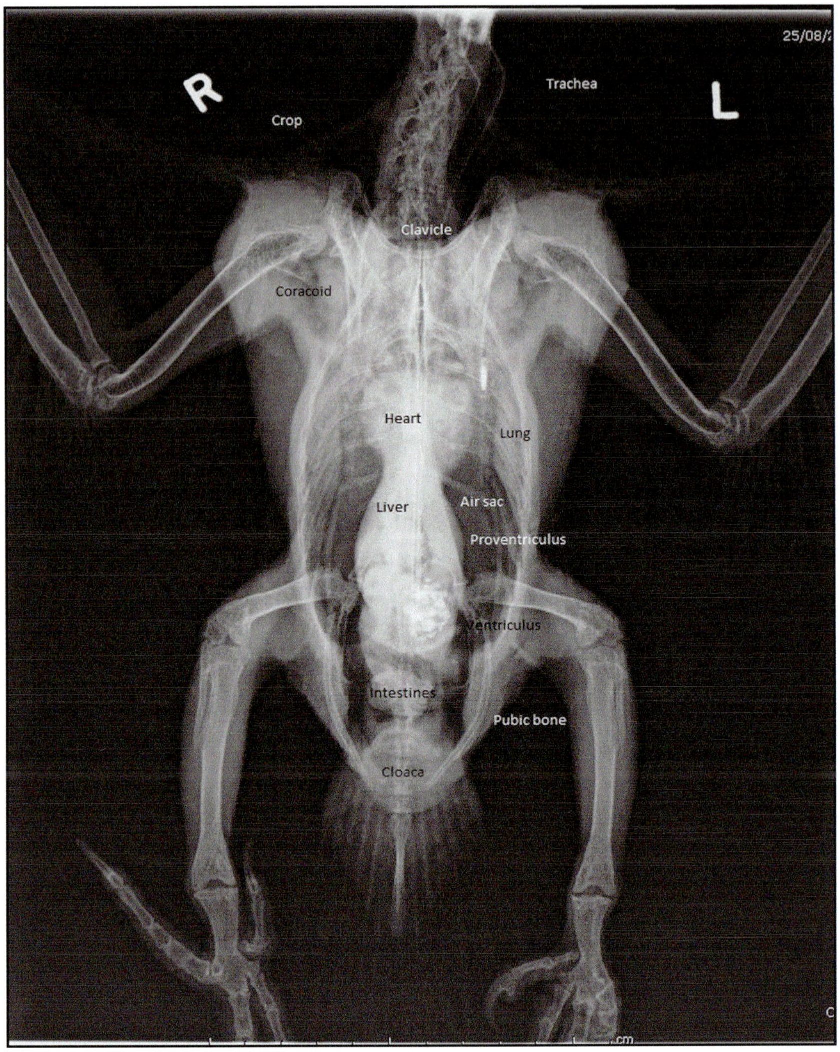

Figure 8.13 **Normal anatomy, ventrodorsal view.**

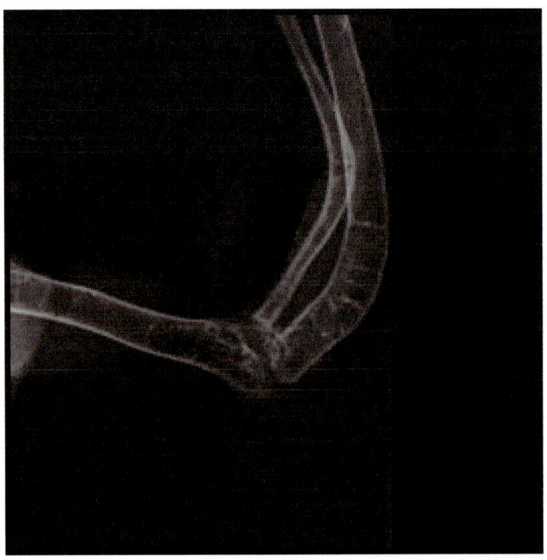

Figure 8.14 **Increased number and size of internal bone struts in a bird with metabolic bone disease.**

ventrodorsal view, this interface (often referred to as the cardiohepatic shadow) forms an 'hourglass' silhouette, with the narrower waist seen at the junction between the hepatic and cardiac silhouettes. This superimposition makes the measurement of cardiac silhouette length, from the lateral view, impractical. The shape of the heart on the ventrodorsal (VD) view should be symmetrical, broader at the base than the apex, and, as previously mentioned, there should be no clear distinction between the heart and the liver (the cardio-hepatic shadow). On a lateral view, the aorta can usually be visualised leaving the base of the heart and coursing caudally, below and approximately parallel to the spine.

The size and shape of the heart is therefore best assessed in a VD view. The maximum widths of the cardiac silhouette and the ribcage at the same point

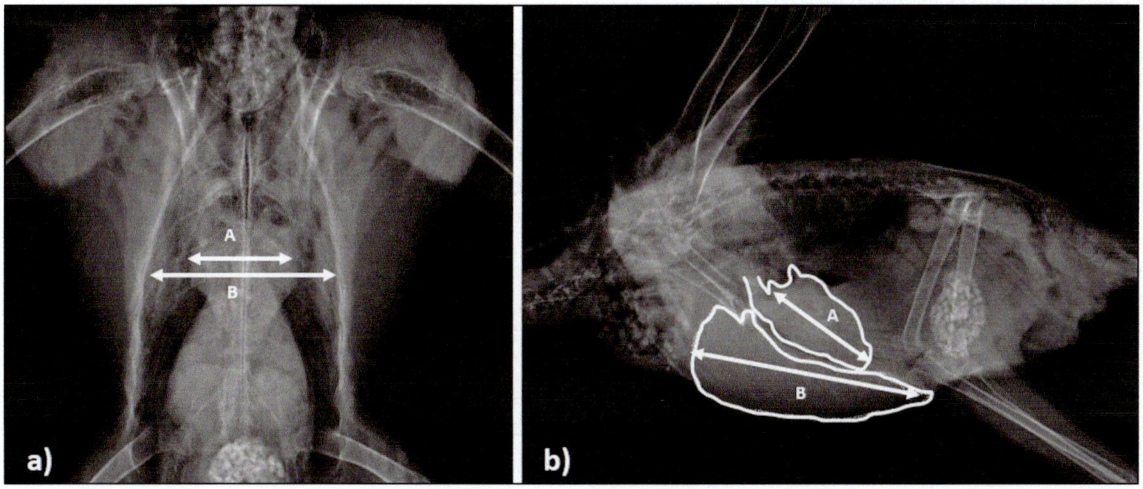

Figure 8.15 **Heart size measurement. (a and b) See text.**

are measured. In many small to medium-sized parrots, the maximum width of the cardiac silhouette should be equal to 51–61% of the maximal width of the rib cage. There is, however, some variation – one study using blue-fronted Amazon parrots (*Amazona aestiva*) found a normal ratio of 39–45%. The relationship between heart size and body size has to be therefore individually established for each species being evaluated (**Fig. 8.15a**).

On the lateral view the length of the heart (apex to base) should be 35–40% the length of the sternum (cranial carina to caudal sternum) (**Fig. 8.15b**).

Respiratory system

The proximal trachea can be readily visualised. The distal trachea, within the rib cage, and the syrinx can be difficult to evaluate due to the overlying ribs, soft tissues and great vessels. Thorough interpretation of the lungs requires orthogonal views because a significant proportion of the lungs are covered by the cardiac silhouette on the VD view. On the lateral view, the right and left lung fields are superimposed rendering the localisation of unilateral lesions impossible. Careful consideration of both views may allow for the detection and localisation of pulmonary lesions. The normal radiographic appearance of the avian lung is somewhat honeycomb-like (due to the appearance of end-on parabronchi). The air sacs can be visualised around the pectoral joints (clavicular) and caudal to lungs, lateral to the liver (thoracic and abdominal). Healthy air sacs have little or no radio-opacity, with no lines visible between adjacent air sacs.

Digestive tract

The crop is usually empty, although small amounts of ingesta are sometimes present. Gas or fluid should not be present. The oesophagus passes over the dorsal aspect of the heart and proximal liver and becomes the proventriculus about one-quarter of the way along the dorsal hepatic surface. In health, a narrow air sac space can often be identified between the proximal portion of the dorsal hepatic silhouette and the oesophagus/proventriculus. The oesophagus and proventriculus typically course on an approximately 40–45° angle from the notarial vertebrae before entering the ventriculus (or gizzard). The ventriculus is seen as a spherical organ, often filled with slightly radio-opaque grit. On a VD view it lies on the left side of the spine, between the coxofemoral joints. On a lateral view it lies on the ventral coelomic floor between the stifle joints of the extended legs. On the dorsal side of the proventricular–ventricular junction lies the spleen. The intestinal tract lies caudal to the ventriculus, with individual loops usually indistinguishable. It terminates in the cloaca, usually visible as a small dome-shaped soft tissue density at the base of the tail.

Liver

The heart and liver are contrasted against the thoracic and abdominal air sacs. In general, the silhouette is of an hourglass shape on the VD view with the hepatic waist being marginally wider than that of the heart. The lateral margins of the liver should not extend beyond a line joining the pectoral and coxofemoral joints on each side (**Fig. 8.16**). The

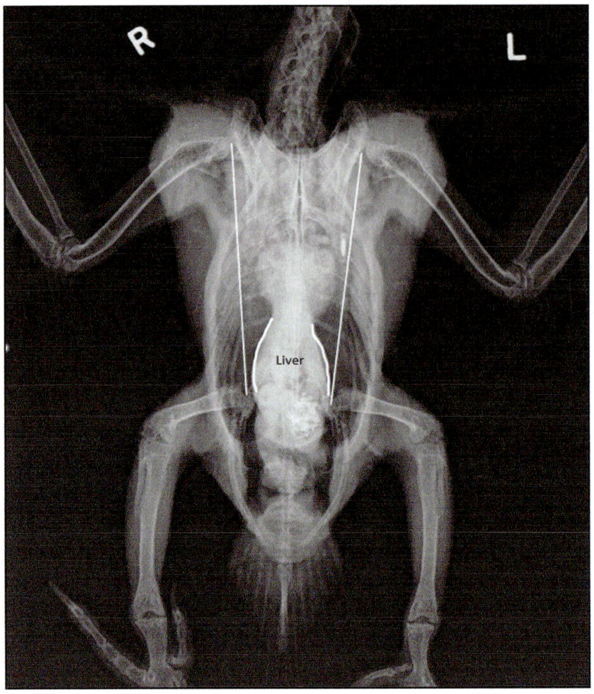

Figure 8.16 **Measuring liver size.**

proventriculus lies on the left side and sometimes lateral to the liver; contrast media may be required to differentiate it from the liver. On the lateral view, the liver is caudal to the heart and fills approximately one-third of the ventral coelom, tapering caudally and ventrally at about 15–20° from the heart base. There is often a small space between the dorsal liver and the ventral proventriculus.

Urogenital system

The kidneys are best assessed on the lateral view, despite superimposition. They are difficult to assess on the VD view because of overlying bone and soft tissues (although the cranial margin of the cranial division of the kidneys can occasionally be visualised). The kidneys are surrounded by the abdominal air sac, including dorsal diverticula of the abdominal airs sacs which lie between the kidneys and the synsacrum. On the lateral view, with both acetabulae superimposed, the kidneys should extend below a line drawn through the ventral acetabular rim, parallel to the spine (**Fig. 8.17**). The ovary or testes may be visible immediately cranial and slight ventral to the cranial division of the kidney. The oviduct is generally not distinguishable from the intestinal tract, unless it contains an egg.

Radiographic abnormalities

Musculoskeletal system

Abnormalities seen include:

- Congenital abnormalities such as bifid sternum are occasionally seen. This develops as a failure of the two halves of the sternum to close during embryonic development.
- Developmental abnormalities such as kyphosis, scoliosis, bowing of the long bones and tibiotarsal rotation are often seen in juvenile birds on a poor diet, or whose parents were on a poor

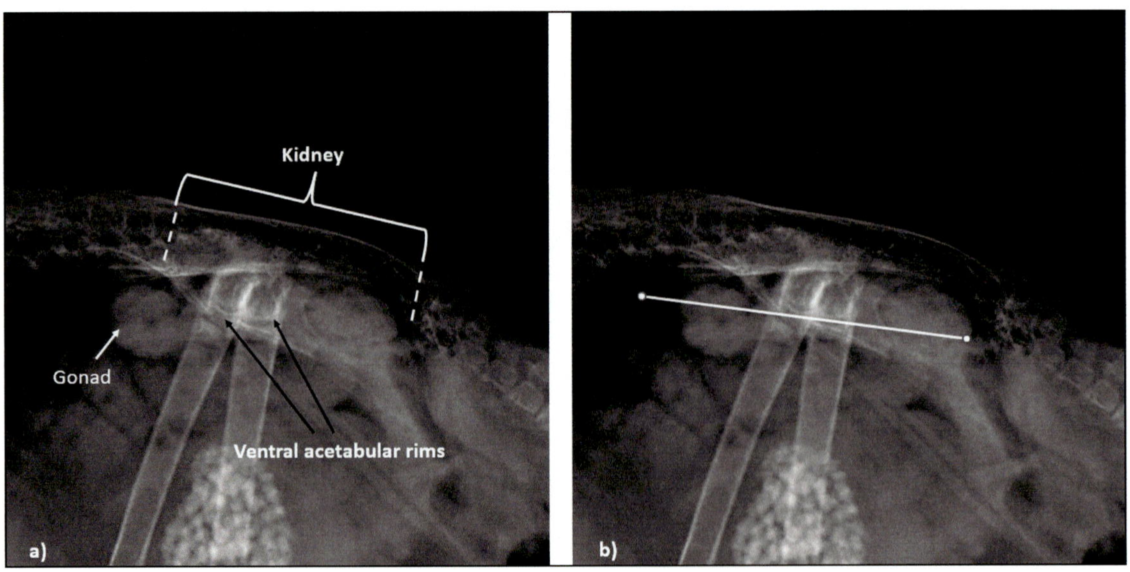

Figure 8.17 **Renal margins: (a) regional anatomy, (b) line through ventral acetabular rims.**

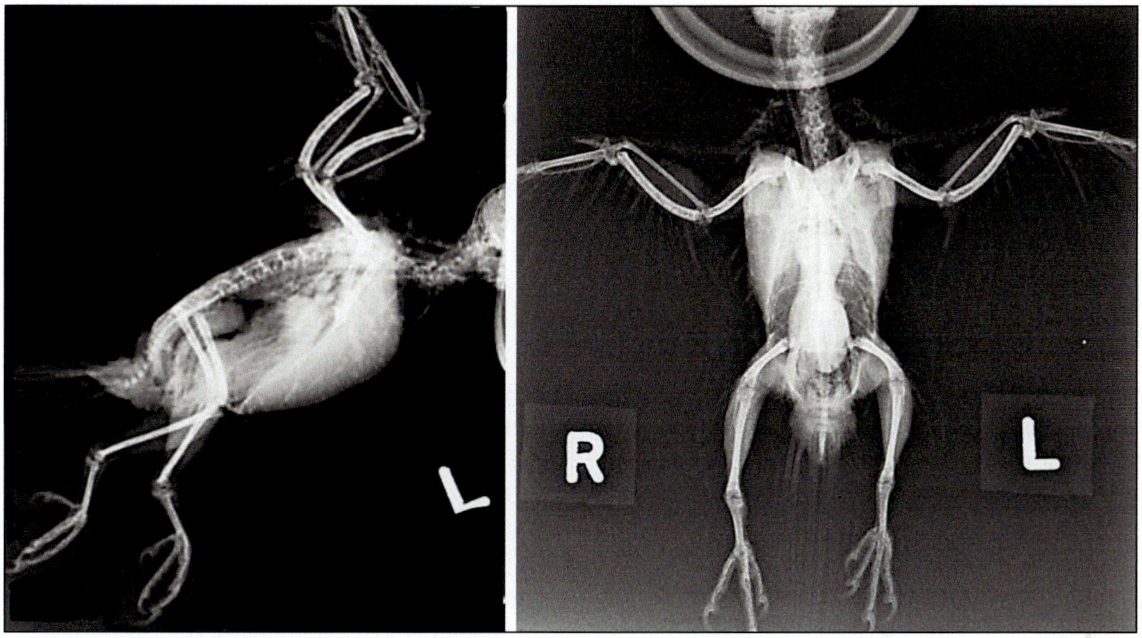

Figure 8.18 **Polyostotic hyperostosis in a male budgerigar with a Sertoli cell tumour.**

diet. Metabolic bone disease (usually nutritional secondary hyperparathyroidism) is characterised by these changes, as well as decreased cortical thickness and the presence of large numbers of internal struts in the metaphyseal regions of the long bones.

- Fractures and dislocations. These are commonly seen in older birds, malnourished birds, or those that have been subject to trauma. Care must be taken to ensure that subtle changes are not overlooked when obvious fractures or dislocations are seen. Coracoid fractures in particular, can be easily overlooked.
- Arthritis is common in older birds although septic arthritis can occur at any age. The amount of bony proliferation with degenerative joint disease in birds is less obvious radiographically than in mammals. Septic arthritis in birds often presents radiographically as an enlarged joint space and requires arthrocentesis to confirm.
- Osteomyelitis typically presents as osteolysis and periosteal reaction. It is distinguished from osteosarcoma by its diffuse nature, often affecting several adjacent bones.
- Neoplastic changes in the bones usually appear lytic, although periosteal reactions and remodelling of the bone can be frequently seen. Osteosarcomas in bird do not appear to metastasise as quickly as they do in mammals.
- Polyostotic hyperostosis. Under the influence of oestrogen, calcium is laid down in the medulla of the long bones as a reservoir for egg formation. This gives the bones a patchy or solid radio-opaque appearance. This condition is also seen in oestrogen-secreting Sertoli cell tumours (**Fig. 8.18**).
- Osteopaenia is seen as a reduced radio-opacity of the bones, often with a patchy appearance in the medullary cavity.

Cardiovascular system

Abnormalities seen include:

- Cardiomegaly is occasionally seen, but it requires determination of the heart width:rib cage width ratio (see earlier) to distinguish absolute cardiomegaly from a relative cardiomegaly seen with microhepatica (see later).
- Pericardial effusions can be seen as a globoid cardiac silhouette, but ultrasound is needed to definitively visualise fluid in the pericardial sac.
- Atherosclerosis can be seen as increased radio-opacity of the great vessels, especially the aorta. In more advanced cases these vessels can demonstrate a patchy mineralised appearance.

Respiratory system

Abnormalities seen include:

- Pneumonia and pulmonary contusions can result in a generalised decrease in the prominence of the parabronchi.

- Air saculitis has a spectrum of severity and can present with variably radio-opaque lines highlighting thickened air sacs, through to near complete loss of coelomic detail.
- Tracheal foreign bodies can occasionally be visualised, although care must be taken not to confuse ingesta in the crop with a tracheal foreign body.
- Hyperinflation of the air sacs occurs with chronic respiratory disease and is seen as relatively overinflated air sacs, extending down around the cloaca and caudal intestinal tract.
- Subcutaneous emphysema occurs with air sac rupture associated with either trauma or chronic infection. Air can be seen under the skin anywhere on the body.

Digestive system

Abnormalities seen include:

- Crop stasis. Although it is common to have a small amount of ingesta in the crop, large quantities of fluid and ingesta distending the crop are considered abnormal.
- Proventricular dilation. The proventriculus is usually empty and compact when examined radiographically, with a clearly defined isthmus where it enters the ventriculus. The presence of gas, ingesta, or grit in the proventriculus is abnormal, especially if the proventriculus is greatly distended.
- Ventricular impaction. While it is common to see a small amount of radio-opaque material in the ventriculus, it should not be tightly compacted or extending cranially into the proventriculus or caudally into the duodenum.
- Intestinal ileus. Large, gas-filled loops of intestine are suggestive of intestinal ileus.
- Cloacal impaction. The cloaca should be visible as a soft tissue density dome-shaped termination of the rectum. Distension or the presence of radio-opaque material is abnormal.
- Coelomic effusion. Loss of detail in the coelom is suggestive of fluid. On a lateral view the ventral coelomic wall between the end of the sternum and the pubic bones should be slightly concave; a convex ventral wall is suggestive of coelomic distension due to effusion, organomegaly, or excessive coelomic fat.

Liver

Abnormalities seen include:

- Hepatomegaly. An enlarged liver extends laterally past the lines drawn between the shoulder and acetabulum (**Fig. 8.16**). On a lateral view the space between the dorsal liver and the ventral proventriculus disappears, and the proventriculus may be pushed dorsally and the ventriculus caudally.
- Microhepatica. It is normal in some species, e.g., macaws, to see a liver that is much smaller in appearance than expected. This is usually not a clinical condition.
- Traumatic injuries. Blunt force trauma, e.g., collisions with windows and cars, can lead to rupture of the air sacs and air leaking out into the coelom. This is seen as a separation of the liver ventrally from the sternum, cranially from the liver, and caudally from the ventriculus.

Spleen

Splenomegaly is readily apparent on the lateral view and occasionally on the VD view. While it usually suggests an infectious process, splenomegaly can also be seen with leukaemic diseases such as lymphoma or lymphosarcoma.

Urogenital system

Abnormalities seen include:

- Nephromegaly. The kidneys are normally surrounded by air and lie above a line through the ventral acetabular rim and parallel to the spine. Extension ventral to this line and loss of the surrounding air space is suggestive of nephromegaly.
- Renal mineralisation, seen as increased radio-opacity of the kidneys, is uncommon.
- Urolithiasis is rarely seen. Radio-opaque densities lying between the caudal division of the kidney and the cloaca are very suggestive of uroliths. Other findings include an atrophic ipsilateral kidney with a normal to hypertrophic contralateral kidney. In some cases, a dilated ureter obstructed with one or more urate stones may be visible radiographically.
- Testicular neoplasia may present as increased mass effect in the area of the cranial division of the kidney. It must be distinguished from enlargement associated with normal seasonal hormonal activity. The degree of this

enlargement is somewhat species-specific, e.g., in ducks there is a marked enlargement. If a Sertoli cell tumour is present, the resultant hyperoestrogenism may produce polyostotic hyperostosis.

- Ovarian enlargement will also produce an increased mass effect in the area of the cranial division of the kidney. Large cysts may cause a loss of coelomic detail and must be distinguished from a coelomic effusion by ultrasound.
- Egg binding is usually readily apparent as an egg in the caudal coelom. If it has been present for more than a few days, the shell may become thickened and roughened in appearance. Soft-shelled eggs may be more difficult to visualise, requiring ultrasound to detect.
- Retained eggs present radiographically as either very thick-shelled eggs or collapsed radio-opaque shapes that are not readily recognisable as an egg.

ULTRASOUND

Ultrasonography remains less commonly used in avian medicine. The best images it obtains are those of organs that are largely composed of fluid and soft tissue. The sonogram produced by the reflection of high frequency sound waves generated by a transducer is based on the unique echoes produced by different tissues. Bone completely absorbs the sound waves, while air reflects them (preventing passage through to deeper structures). The presence of air sacs therefore limits the usefulness of ultrasound in birds. However, it can be used to give some definition of the internal structure of tissues and organs, including examination of small organs such as the gall bladder, spleen, and pancreas; examination of fluid filled lesions; and to measure cardiac function. It is a valuable tool in assessing coelomic distension in avian patients (fluid vs. organomegaly), although often limited by the size of the patient in comparison with the footprint of the ultrasound probe, as well as the presence of air sacs and the large ventrally located keel bone. Together, these factors may limit available acoustic windows and limit the visualisation of internal organs.

The focal range (depth) of the transducer is inversely related to its frequency as the result of attenuation (the loss of acoustic energy as the wave travels). High frequency transducers provide a more detailed picture, but the focal range is less.

As a rule, the smaller the patient, the higher the frequency that will be required. For very small patients (<50 g), a frequency of between 7 MHz and 12 MHz is often suitable, while larger patients may require a transducer with a frequency of 3.5–7 MHz. Lower frequencies can be used to localise an area of interest, and then a higher frequency transducer can be used to examine it in detail.

Three types of transducers are commonly used in practice:

- Linear – higher frequency transducers with an often-large rectangular footprint (surface that touches the patient's skin), which may be impractical on very small patients.
- Curvilinear – lower frequency transducers with a convex footprint.
- Phased-array (also known as sector) – low frequency transducers that generate an image from a point and are good for getting between ribs.

Because of their smaller footprint, available frequencies, and the sector image, curvilinear probes are often preferred.

Ultrasonic colour Doppler combines anatomical information derived using ultrasound, with velocity information obtained from doppler technology. It generates a colour map depicting movement of blood (Doppler mode) superimposed on greyscale images of tissue anatomy (B mode). The most common use of the technique is to image the movement of blood through the organs, blood vessels, and the heart. Blood flow towards the probe is coloured red, while flow away from the probe is coloured blue. Colour Doppler can be used to identify abnormal vascularity, such as hyperaemia or neovascularisation, or to determine if an abdominal mass is avascular (e.g., an abscess) or vascular (e.g., neoplasia).

Scanning

The patient should be fasted, if possible or safe, prior to the procedure.

The most common acoustic windows used are the ventromedial coelom (caudal to the sternum) and right lateral coelom, behind the last rib (**Fig. 8.19**). (The right side is used to avoid interference from grit in the ventriculus.) Any air between the patient's skin and the transducer will produce artefacts that make interpretation difficult, if not impossible. It is therefore vital to prepare the skin appropriately.

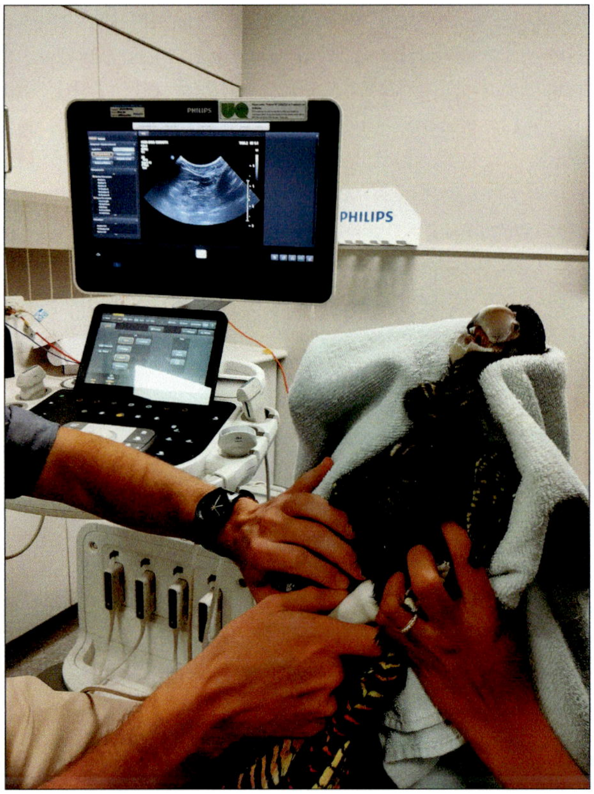

Figure 8.19 **Restraining a bird for ultrasonography.**

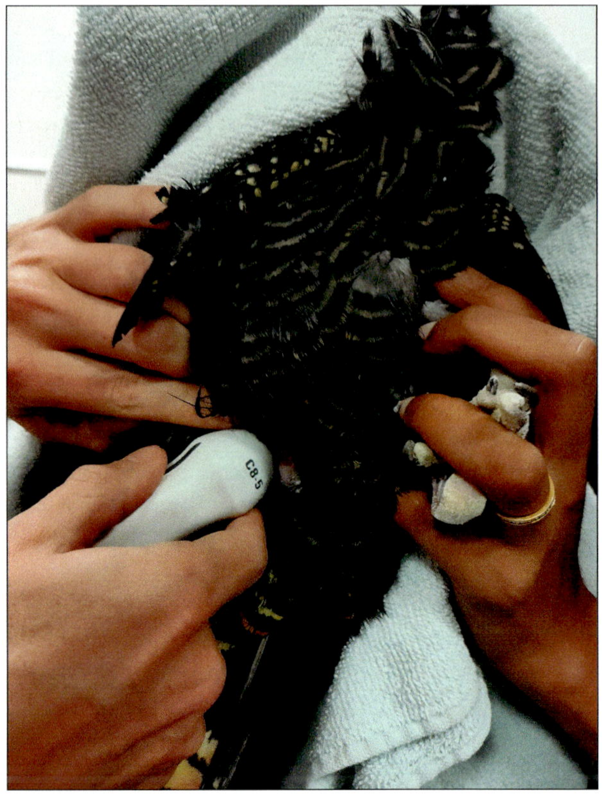

Figure 8.20 **Applying the transducer to the skin.**

With the patient restrained or anaesthetised in an upright, dorsal, or left lateral position, the feathers are parted (to reveal apterylae) or plucked and ultrasound gel applied to the skin (**Fig. 8.20**). (Be aware that ultrasound gel can affect radiographic images. If both ultrasound and radiology are planned, radiology should be performed first.)

In some cases, the use of stand-offs should be considered. These are gel- or water-filled pads, e.g., the fingers of a disposable glove. They act as an acoustic interface, allowing better sound wave penetration and enhancing contact with the skin. After placing coupling gel on the skin, a stand-off is placed over the area of interest and the operator scans through the stand-off.

A systematic approach to imaging be used to ensure that the location, echotexture, measurements, outline, number, and size (LEMONS) of any findings are considered. Whenever possible each organ should be scanned in two planes: transverse and sagittal/longitudinal.

Using the ventromedial window, the liver and heart are first examined in two planes (vertical and horizontal). The transducer is then directed to the left of the bird's body to examine the gastrointestinal tract from the proventriculus to the cloaca. Using a high frequency transducer, the wall of the tract and peristaltic movement can be assessed. The pancreas, lying against the ventral coelomic wall, is often difficult to see even using a spacer. The kidneys and gonads, surrounded by air sacs, are difficult to visualise unless enlarged. Mature active gonads, especially ovarian follicles, may be seen; immature gonads are difficult to identify.

Interpreting the sonogram

The sonogram is based on the unique echoes produced by different tissues and their compressibility. The appearance of tissue on the sonogram represents its echogenicity compared with surrounding structures (**Fig. 8.21**).

- Structures that return minimal echo, such as serous fluids, blood, and urine, will appear black on the sonogram and are anechoic.
- Tissue that reflects more waves than neighbouring tissues (e.g., air, bone) will appear whiter, or hyperechoic, on the sonogram.
- Tissue that reflects ultrasound beams less than the adjacent tissues (e.g., muscle vs. bone) appear darker, or hypoechoic, on the screen.
- Tissues of the same density are isoechoic, i.e., difficult to distinguish between.

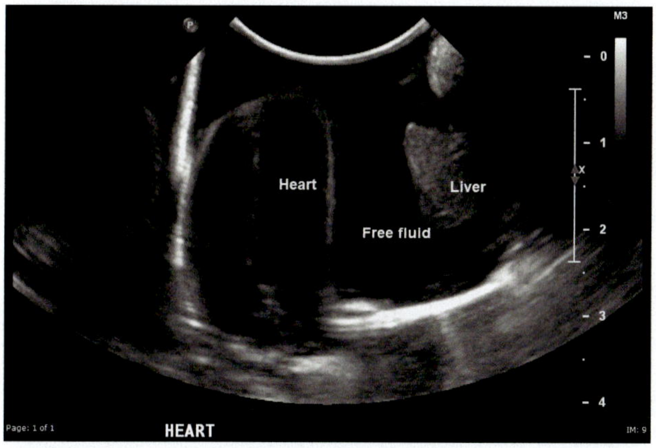

Figure 8.21 **Ultrasound of dilated cardiomyopathy and heart failure in an Alexandrine parrot.**

Ultrasonography therefore cannot be used to scan gas-filled structures (e.g., avian air sacs, mammalian lungs) or bony tissues, as the sound beam is totally reflected at soft tissue–gas interfaces and absorbed at soft tissue–bone interfaces. Intestinal gas, air sacs, and bone also 'shadow' any other organs beyond them.

The clinician must appreciate the wide variation in anatomy between the various species that may be encountered. Care must be taken not to overinterpret normal structures in an unfamiliar species, and whenever possible, known standards should be consulted. Sonographic lesions are sometimes quite characteristic of a given disease process, but more often the changes are nonspecific. Although ultrasonography can be quite sensitive to detect disease, the changes are not specific for a given disease in most cases unless a characteristic change in anatomic presentation and echogenicity are detected.

Organ	Normal	Pathology
Liver	The parenchyma has a homogeneous, finely granular appearance with sharp edges. The gall bladder, if present, lies to the right of midline.	Blood vessels appear enlarged, with increased wall density, with severe hepatomegaly. Hepatic lipidosis produces hepatomegaly with a diffuse increase in echogenicity. Neoplasia causes focal or diffuse changes in echogenicity. Abscesses, granulomas, or necrosis also produce focal or diffuse changes in echogenicity. Haematomas produce focal, hypoechoic lesions.
Spleen	Can be seen on the lateral approach. Its size and shape vary between species. It has a slightly more granular and echodense appearance than the liver.	
Kidneys	Normal kidneys are difficult to visualise.	Enlargement is readily seen in some patients. Neoplasia causes a non-homogeneous parenchyma Renal cyst are clearly defined anechoic areas.
Reproductive tract	Both approaches can be used. Inactive gonads are difficult to detect, while active follicles appear as anechoic areas. The oviduct can be difficult to distinguish from the intestines unless an egg is present. It lacks peristalsis, a feature which may assist in differentiating the two tracts.	Eggs are easy to detect if shelled. Prior to this, they can be seen as the hyperechoic yolk surrounded by the hypoechoic albumen. Ovarian cysts and gonadal neoplasia are also detectable. Metritis and pyometra can be detected readily. Free fluid in the coelomic cavity associated with yolk peritonitis is usually obvious.
Gastrointestinal tract	Structure and motility can be visualised. The proventriculus is often only visible if enlarged. The ventriculus is identifiable by the presence of shadowing arising from insoluble grit in the lumen. Layers within the intestinal walls can be recognised. The duodenal loop is easy to identify, as it lies close to the body wall. A spacer may be required. The cloaca has a narrow, hypoechoic oval form with a regular edge. The presence of faecal material can make visualisation difficult, but retrograde flushing from the vent can help to remove this material.	Proventricular dilation can be seen, e.g., PDD and lead toxicosis. Thickening of wall (>2 mm) can be seen with enteritis. Cloacal concretions, papillomas, and dilation can be seen.
Pancreas	The pancreas is observed as a hyperechoic structure lying between the loops of the duodenum, with a blood vessel through the centre of its structure.	
Heart	Only B mode and Doppler flow have been used successfully in birds.	Cardiomegaly and pericardial effusion are readily differentiated. Mitral valvular disease has been reported.

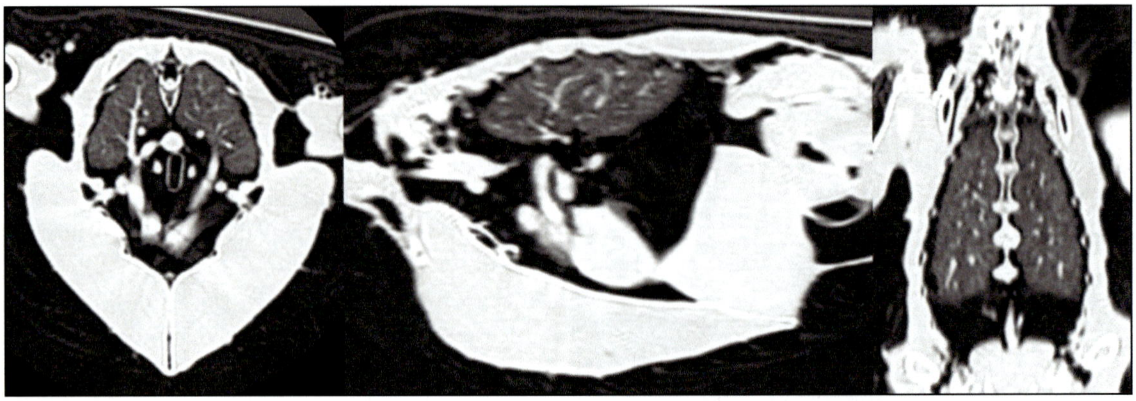

Figure 8.22 **Blue and gold macaw. CT scan images of the lungs and air sacs in lung windows.**

COMPUTERISED TOMOGRAPHY (CT)

Avian anatomy, with an often-complex overlay of bone and soft tissue structures, can make radiographic assessment sometimes quite difficult. The use of new technologies such as CT, fluoroscopy, and magnetic resonance imaging (MRI) can assist with the non-invasive assessment of internal organ structure and function.

CT makes use of computer-processed X-rays, produced by an X-ray tube rotating around the patient and opposite rows of detectors, to produce tomographic images ('slices') of the patient. Increasing the number of sensors increases the resolution, e.g., a 16-slice scanner has 16 sensors. Current technology can now take 16, 32, 40, 64, or 128 slices. The data is then processed by a computer to generate images in several planes (i.e., transverse, sagittal, and coronal) and even three-dimensional reconstructions. Its advantages over conventional radiography include the elimination of superimposition, improved detail, and the ability to create a 3-D reconstruction for planning complex surgeries.

New developments in the field of CT include:

- The use of micro-CT to produce even higher resolution images and 3D reconstructions of small patients, such as birds. Such technology is becoming more available to most clinicians and can be used to gain a more detailed understanding of avian anatomy and disease processes.
- Cone-beam computed tomography (CBCT), widely used in human medicine for head and neck imaging (such as dentistry), has increasingly been used in veterinary medicine. It involves the use of a C-arm rotating around a patient placed on a small imaging platform (40 cm), capturing slices using a cone-shaped X-ray beam.

  However, this technology has two major drawbacks: first, it is very slow when compared to conventional CT (one rotation around the patient can take 15–30 seconds), making movement artefacts highly likely; second, while it has high spatial resolution, it also has poor contrast resolution. This makes CBCT ideal for hard tissue imaging, e.g., bones and teeth, but less suitable for abdominal imaging.

Examples of organs that are more easily visualised on CT images compared to radiographic images, especially with intravenous contrast enhancement, include the lungs (**Fig. 8.22**), spleen, proventriculus, ventriculus, and kidneys. Lesions of the spinal column that may not be obvious on radiographs can also be detected on CT images.

Exposure to radiation must be minimised by limiting the area to be examined and the number of studies performed in a short period of time. Although only short scanning times are required with modern equipment, it is vital the patient does not move during the procedure. Deep sedation or general anaesthesia is therefore usually required. Although scanning times are short, it does take some time to format the images and more time to interpret them.

Intravenous contrast media can be used to help distinguish between soft tissue structures in the coelom, provide more information about lesions and abnormal vascularisation, and to highlight areas of inflammation (**Fig. 8.23**).

FLUOROSCOPY

Fluoroscopy uses an image intensifier that amplifies low-level ionising radiation to allow the capture

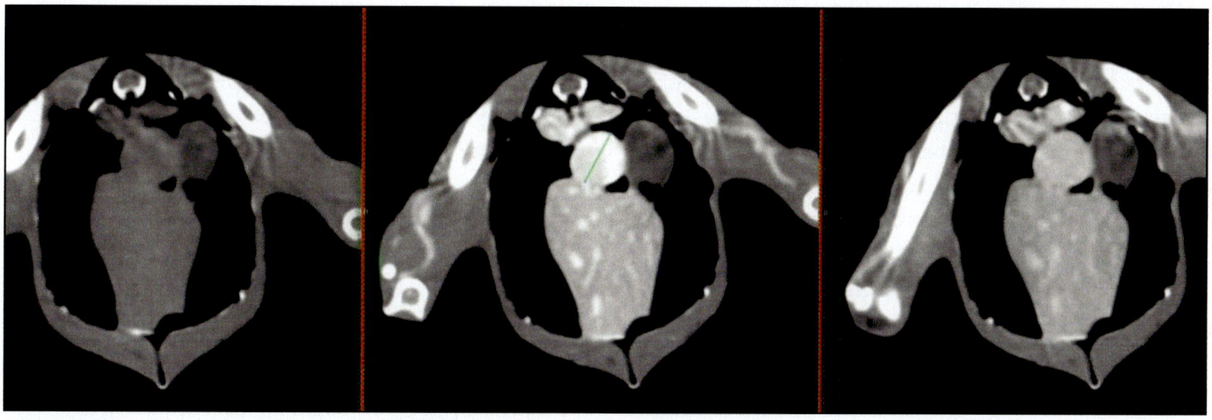

Figure 8.23 **Blue and gold macaw. CT scan images of the spleen in soft windows pre-contrast, post-contrast arterial phase, and post-contrast venous phase (the left side of the coelomic cavity is on the right side of the images).**

of real-time motion using a closed charge-coupling video imaging device. It uses the continuous emission of X-rays, captured on a screen, to produce a dynamic image that provides real-time visualisation of body structures and function. In contrast to radiographic gastrointestinal contrast studies (which require multiple episodes), anaesthesia is often not required; the patient can often be restrained by placing it in a closed cardboard box fitted with a wooden perch (**Fig. 8.24**).

Fluoroscopy lacks the sharpness and contrast of digital radiographs but can be paired with digital imaging (e.g., via a horizontal beam through the restraint box) if more detail is required.

This technique is particularly useful for studying gastrointestinal motility, especially in diseases such as proventricular dilatation disease. By gavaging the patient with contrast media such as barium and then placing it in a cardboard box, the passage of the contrast media through the gastrointestinal tract can be watched in 'real time' (**Fig. 8.25**). Care must be taken with the length of time utilised for a study, as radiation burns can occur with prolonged exposure.

Figure 8.24 **Fluoroscopy machine in place. The patient is in the cardboard box; the towel is to prevent escape.**

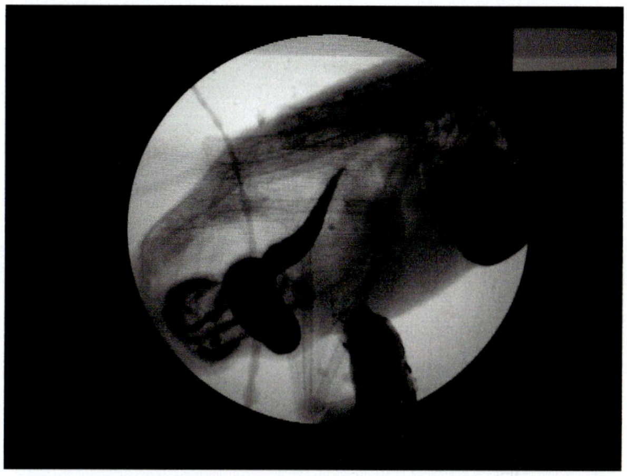

Figure 8.25 **The resultant image showing contrast moving through the gastrointestinal tract.**

MAGNETIC RESONANCE IMAGING (MRI)

Like CT, MRI displays objects in slices. Unlike CT, MRI does not utilise radiation but rather uses a powerful magnetic field in combination with high-frequency radio waves to align and spin protons (nuclei of hydrogen atoms) in the body, and then obtain signals resulting from these protons. Because most hydrogen atoms in the body are trapped in water molecules, the MR image produced represents the different types, distributions, and volumes of water in various tissues, allowing differentiation of soft tissue types. Images are composed of a series of 2D images or 'slices' that may be viewed alone or in 3D reconstructions. The use of intravenous contrast agents such gadolinium is considered a routine procedure for MRI scans to enhance contrast and compare pre- and post-contrast images.

This ability to differentiate soft tissues and the wide variety of ways in which image contrast (weighting) can by generated for visualising the body's anatomy and physiology makes MRI a superior imaging tool for soft tissues. However, CT is more useful for imaging bony structures, has better resolution, a much shorter investigation time, and is currently less expensive.

The strength of the magnetic field used in MRI, the magnetic flux density, is expressed in teslas (T). In clinical veterinary medicine, MRI units typically range from 0.2 to 3 T.

By changing the settings of the MRI scanner (e.g., pulse repetition time and echo time) different tissue weightings are obtained and subsequently used to analyse the tissue composition. The three types of weightings commonly used are as follows:

- T1-weighted images (longitudinal relaxation or simply T1) relate to the rate of recovery of the proton spins after each perturbation (called excitation) required to elicit an observable signal. Therefore, tissues with long T1 (e.g., water) appear hypo-intense.
- T2-weighted images (transverse relaxation or T2), relate to the rate of decay of the observable signal. Therefore, in contrast to T1, tissues with long T2 appear hyper-intense.
- Proton-density-weighted (PDW; spin-density-weighted) images. Contrast between tissues is primarily dependent on the number of available spins (protons) (i.e., brightness is highest in tissues with a high-protein density). These images minimise the T1 and T2 effects, therefore resulting in less contrast than T1 or T2 images. Although these images may be useful for depicting normal anatomy, they are used less often.

Because each type of tissue has its own characteristics for T1, T2, and PDW, the images obtained from these three weighting types can be combined to identify the origin of the examined tissue.

Limiting factors in the use of MRI use are:

- A full investigation may take at least 45 minutes, requiring extended anaesthesia times with limited access to the patient during the procedure.
- The small size of most exotic patients in combination with the low spatial resolution of most systems (less than 0.5 T) limits the diagnostic value of MRI in small patients, especially since the minimal slice thickness needs to be 1–3 mm to achieve a usable signal-to-noise ratio with these machines.
- Imaging may be hindered by the high respiratory and heart rates or by the presence of artefacts caused by the presence of air sacs and/or a transponder.

FURTHER READING

Echols S. The grey parrot anatomy project. http://www.avianstudios.com/the-grey-parrot-anatomy-project, accessed December 2014.

Helmer P. (2006) Advances in diagnostic imaging. In: Harrison GJ, Lightfoot TL (eds). *Clinical Avian Medicine, Vol 2*. Spix Publishing; 653–659.

McCready JE., Gardhouse SM, Appleby R, Beaufrère H. (2021) Mortality rate of birds following intravenous administration of iodinated contrast medium for computed tomography. *Journal of the American Veterinary Medical Association*, 259(1): 77–83.

Silverman S, Tell LA. *Radiology of Birds: An Atlas of Normal Anatomy and Positioning*. Elsevier Health Sciences, 2010.

Veladiano IA, Banzato T, Bellini L, et al. (2016) Normal computed tomographic features and reference values for the coelomic cavity in pet parrots. *BMC Veterinary Research*, 12: 182. https://doi.org/10.1186/s12917-016-0821-6.

CHAPTER 9

ENDOSCOPY

Endoscopy – the examination of the interior of body organs, joints, or cavities through an endoscope – had its beginnings in the early 1900s when lighted telescopes were first used to examine inside a human patient. In avian medicine, rigid endoscopes were first used in the 1970s and 1980s to determine the sex of parrots, many of which are sexually monomorphic (i.e., there are no distinguishing physical differences between the sexes). Examination of the gonads, located between the lung and kidney, allowed a rapid and accurate identification of sex and revolutionised the breeding of many species. It was quickly realised that the unique avian anatomy, with its air-filled body cavities, made birds an ideal candidate for endoscopy. Insufflation is not required, reducing the costs and technical difficulty involved. Techniques and instruments have been developed to take advantage of this, and endoscopy is now a widely utilised diagnostic tool in avian medicine around the world.

EQUIPMENT

Modern rigid endoscopes utilise a series of glass rods separated by small air spaces contained within a rigid telescope. This rod lens system, often referred to as the Hopkins system (**Fig. 9.1**) after its designer, is considered to give better light transmission, image resolution, magnification, and a wider field of view than the older convex glass lens systems. Most practitioners use a 2.7 mm rigid endoscope with a 30° oblique view (**Fig. 9.2**). This allows a much wider field of view than the more common 0° scopes (**Fig. 9.3**). Smaller endoscopes (1.9 mm) are available for smaller patients (<100 g), but the loss of optical detail limits their use.

Despite its rigidity, the endoscope is susceptible to damage if excessively flexed or torqued. For this reason, it is recommended to use either an examination and protective sheath or an operating sheath.

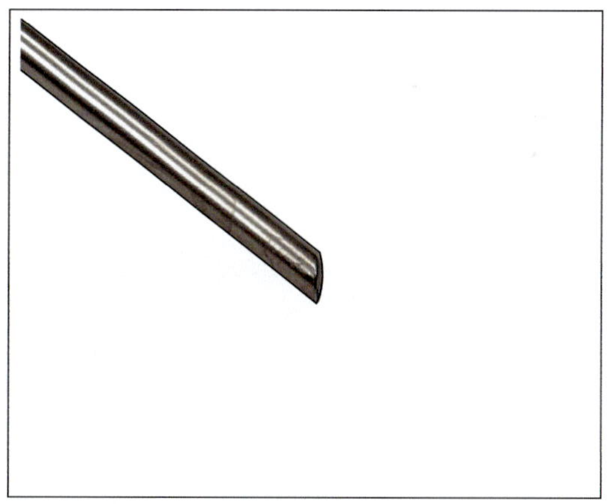

Figure 9.2 **The 30° oblique tip.**

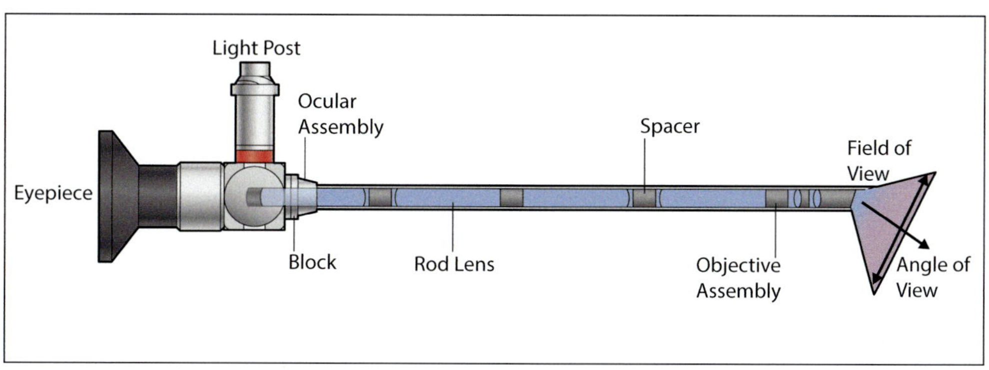

Figure 9.1 **The 2.7 mm rigid endoscope using the Hopkins rod lens system.**

DOI: 10.1201/9781003272519-9

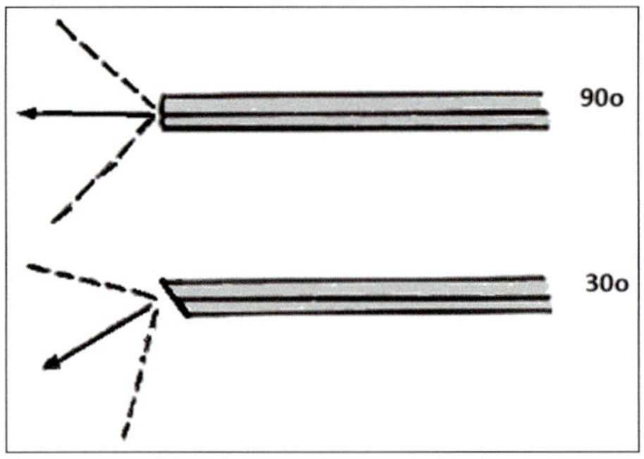

Figure 9.3 **The viewing angles 30° compared to 90°.**

The sheath slides over the telescope and locks into place. The operating sheath provides two stopcocks (to allow irrigation or insufflation) and an operating channel for the use of endoscopic instruments (**Figs. 9.4, 9.5**).

Connecting the telescope to the light source is a light cable (also known as a fibre optic light guide). This is composed of bundled fibre optic cables that transmit light from the light source to the telescope. The fibre optic cables are flexible but do not readily withstand twisting forces.

The use of cold light sources (i.e., light generated without the use of incandescence) minimises the generation of heat at the tip of the telescope. The most commonly used light sources are either tungsten-halogen or xenon. Xenon light sources provide a more intense, whiter, and cooler light than tungsten-halogen. Not only does this give better illumination while viewing inside the body, but it is essential if cameras and video documentation are to be used. Smaller LED light sources are now also available but do not provide the degree of illumination provided by the xenon light source.

Figure 9.5 **The instrument channel for the sheath can be seen above the lens.**

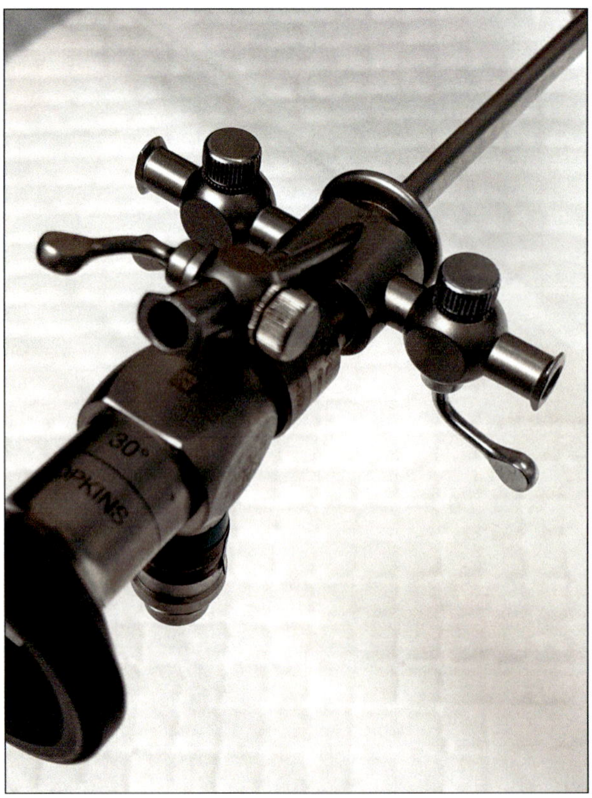

Figure 9.4 **Operating sheath for the 2.7 mm endoscope.**

Although the endoscope can be used with the operator viewing the image through the telescope's eyepiece, this can lead to both eye and back strain. The use of an endoscopic camera and monitor provides better magnification of the operating field while being more ergonomic for the operator. A three-chip digital camera gives a high-quality picture while being light and easily used while attached to the endoscope. Viewing the image on a high-quality monitor improves both the image seen and the operator's ergonomics. Video documentation is desirable as it allows better documentation of procedures and is a valuable tool for educating both veterinarians and clients.

A variety of endoscopic instruments are available for use, including biopsy forceps, grasping forceps, scissors, aspiration/injection needles, and radiosurgical probes. Two sizes are generally used: 5 Fr (for use with the 2.7 mm endoscope) and 3 Fr (for use with the 1.9 mm endoscope). The biopsy forceps are used for collecting tissue biopsies. The cutting edge is sharp, and excessive force to close the forceps is usually not warranted and may result in crushing artefacts. The grasping forceps are used to manipulate tissue and remove small foreign bodies. The scissors are most frequently used for incising fibrous capsules (e.g., over the liver) or air sacs. The fine aspiration/injection needle is used to inject drugs into lesions, irrigate a small area, or aspirate material. The use of radiosurgery probes allows the use of radiosurgery during endosurgical procedures.

Caring for the equipment

Handling

Rigid endoscopes are fragile and must be handled with care during transport, use, and cleaning. Each telescope contains glass fibre optic bundles and a series of rod lenses; excessive forces applied to these fragile components can damage the telescope to the point of being unusable. When not in use, the telescope should be stored in the padded telescope case provided by the manufacturer. When out of the case, the plastic telescope guard should be placed over the telescope until ready to use and set down on a flat surface to avoid accidental flexion and possible fracture of the optics.

Rigid endoscopes should always be picked up by the eyepiece, never by the shaft. Be aware of the torsional forces applied to the shaft of the telescope when in use. Avoid overflexion by partially supporting the long axis of the telescope with one hand while in use. The appearance of a crescent on part of the image indicates that the shaft is in imminent danger of being overflexed.

The light cable consists of fibre optic bundles; although flexible, these bundles are susceptible to damage if twisted or kinked. Careful handling and coiling of the cable when not in use are essential to prevent broken fibre optics. Broken fibre optic cables will appear as black dots in the viewed image.

Endoscopic instruments must be carefully cleaned after use and tested to ensure they are opening and closing smoothly.

Cleaning

Disconnect the light cable from the telescope and remove any light cable adaptors. Remove any residual blood, protein material, and contaminants with a sponge, soft cloth, or a cotton cloth applicator using a neutral pH enzymatic cleaning solution, and then place the telescope in a flat plastic container and soak it with the same cleaning solution for no more than 45 minutes. Never use an ultrasonic cleaner. Rinse thoroughly in distilled water to remove any residual cleaning solution.

Clean the lenses (tip and eyepiece) and the fibre optic inlet post with alcohol wipes or cotton tip applicators soaked in 70% alcohol to remove any residue or film. When satisfied they are clean, wipe these optical elements with a lint-free soft cloth and then dry the entire telescope with a lint-free soft cloth or filtered, compressed air.

Sterilisation/disinfection

Rigid telescopes may be sterilised or disinfected following the manufacturer's recommendation. If cold sterilising, do not soak for more than 45 minutes. Some telescopes can be autoclaved, although this may affect their longevity. Only use an autoclave after confirming with the manufacturer that the telescope is suitable for this process. Ensure that the telescope is placed in a suitable container and is not in contact with other metal items (including the container). Do not wrap the telescope in a towel, as this may leave residue and stains. After autoclaving, allow the telescope to slowly cool to room temperature – abrupt temperature changes may cause the glass rods to fracture.

Operating the endoscope

Endoscopy equipment should be stored in a mobile vertical tower with the monitor on top at the operator's eye level. This allows the equipment to be positioned so that the operator, sitting or standing facing the patient, is also squarely facing the monitor. Such a position reduces operator fatigue and avoids twisting the operator's back or neck during lengthy procedures.

Wherever possible, the endoscope should not be used without the protective or operating sheath over the shaft. After making a suitable entry incision, the shaft is held and supported by the inferior hand, while the dominant hand holds and controls the body of the telescope and camera in a pistol grip

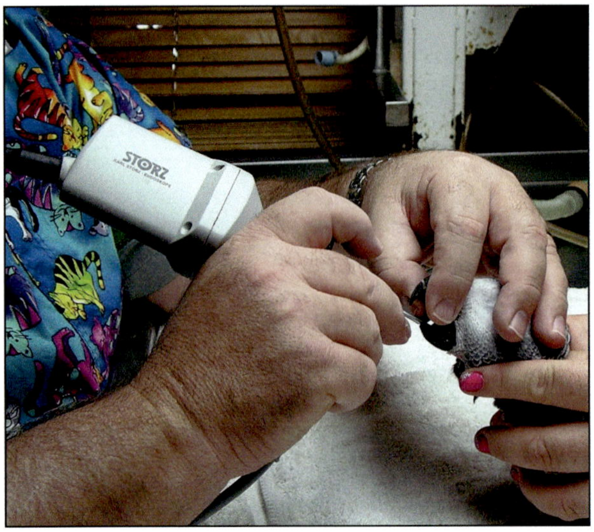

Figure 9.6 **The pistol grip used to operate the endoscope.**

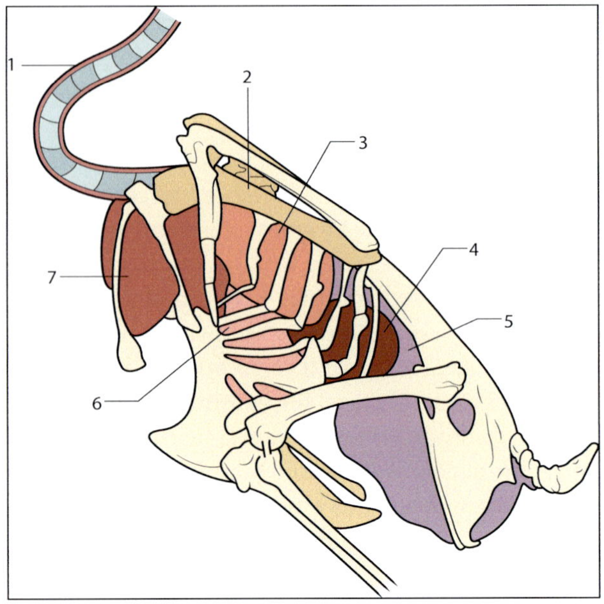

Figure 9.7 **The anatomy of the air sacs facilitates endoscopic examination. 1. Cervical air sac; 2. Extension of the clavicular air sac into the humerus; 3. Lungs; 4. Caudal thoracic air sac; 5. Abdominal air sac; 6. Cranial thoracic air sac; 7. Clavicular air sac.**

(**Fig. 9.6**). This position is maintained as the endoscope is introduced and the body cavity is examined. If it is necessary to use instrumentation, the inferior hand is closed around the shaft, with the thumb running along the shaft towards the operator. This hand is now taking the full weight of the telescope and camera. The dominant hand is free to introduce and use instruments.

ENDOSCOPIC PROCEDURES

Coelioscopy

It is still common that in avian medicine many diagnoses are made only to the organ level (e.g., 'liver disease' or 'kidney disease'), often with a presumed generic cause (e.g., 'fatty liver'). Prior to undertaking any endoscopic procedure, a clinical workup should locate the area/organ of interest, and this information is used to determine which approach is indicated for the patient. For example, serum biochemistry may suggest that liver or kidney disease may be present; radiology may demonstrate a change in internal organ size and location, or a lesion may be visible in one quadrant of the body.

General approach

See later in the chapter for a description of the specific approaches commonly used. The left lateral approach is the most commonly used approach and is used to visualise the liver, lung, kidney, gonad, and spleen. The right lateral approach is used to evaluate the pancreas, and the ventral approach is used if both lobes of the liver need to be examined or sampled, or if the bird has free fluid in the coelom (this approach does not enter through the air sacs) (**Fig. 9.7**).

In order to enter the air sacs, the tip of the endoscope is used to create an opening in the associated air sac membranes using a sharp twisting motion. These membranes are normally thin and transparent, allowing visualisation of viscera on the other side. If air saculitis is present, the wall can be thick, irregular, and opaque. Care should be exercised when penetrating these air sacs, as more force may be needed, coupled with poor visualisation of viscera on the other side. Endoscopic scissors can be used to make a small incision in the air sac if needed.

Biopsy collection

Indications

Endoscopy not only allows the clinician to examine internal organs but also to collect tissue biopsies for histopathology or culture. Of all the antemortem diagnostic tests available, this is the most important since neither biochemistry, haematology, radiography, nor direct or endoscopic visualisation provides reliable predictive information. Endoscopy offers a minimally invasive technique to collect these biopsies. Even if the target organ appears grossly unremarkable, a biopsy is still recommended – preferably earlier in the disease course, before permanent damage develops, and while the patient is more stable, with less systemic risk, to facilitate both a safer anaesthetic and a better recovery.

Technique

Manipulating the endoscope as described earlier, the biopsy forceps are introduced into the operating channel until they are just visible in the operating field. They are then advanced slightly to ensure that the spring-loaded cups can open and close without interference from the operating sheath. The operator should avoid the temptation to advance them any farther; instead, the entire system is advanced as one unit to the organ of interest. A suitable area is selected, and the biopsy cups are opened and then closed firmly on the selected site. If the cups do not cut through the tissue cleanly, a short sharp tug is used to collect the sample, rather than dragging the organ away from its normal position.

With the cups closed, the biopsy forceps are then drawn back into the sheath, and either the forceps or the entire unit is withdrawn from the patient. The tissue biopsy is then removed gently from the cups; the authors prefer to irrigate with a fine stream of sterile saline so that the sample is 'washed' onto a piece of gauze swab in a tissue cassette. The sample can then be placed in formalin or culture medium.

Endosurgery

Avian endosurgery is a minimally invasive surgical technique that may be single-, double-, or triple-entry in nature.

- Single-entry techniques are similar to those employed for diagnostic evaluations and biopsy. They rely on the use of a single instrument inserted through the operating channel of the sheathed telescope. While this system has proven extremely effective for diagnosis, it has obvious drawbacks when attempting to perform endosurgical procedures because surgical instrumentation (e.g., forceps, scissors, retractors, radiosurgery devices, etc.) must be used independently within the endoscopy field.
- Double- and triple-entry techniques utilise two or three separate devices simultaneously. It is necessary to triangulate (coordinate) the surgical instruments with the telescope, but they are no longer restricted to the single plane within the operating sheath.

Procedures such as gonadectomy, intestinal surgery, mass resection/removal, etc., can be performed endosurgically but require training and experience.

ENDOSCOPIC APPROACHES

Several approaches are used in birds; the more commonly used ones are described next.

Tracheal approach

The tracheal approach is used for the examination of the trachea and syrinx for foreign bodies, diphtheritic plaques, and tracheal strictures. It is important to note that for anything other than a quick examination of the trachea and syrinx, the patient should have an air sac catheter placed to provide anaesthesia and respiratory support during the procedure. Under anaesthesia, the patient is placed in either dorsal or ventral recumbency, and the head and neck are extended. The endoscope is introduced over the tongue and through the glottis into the trachea (**Figs. 9.8, 9.9**). It is important that the patient is in a surgical plane of anaesthesia and that the trachea is kept extended and straight to avoid iatrogenic trauma. It is also important to keep in mind that the tracheal cartilage in birds is a complete cartilage ring – and therefore the trachea cannot be stretched from the inside – and that the lumen of the trachea narrows as it passes caudally to the syrinx. If the syrinx cannot be visualised, it is sometimes helpful to withdraw the endoscope, reverse the patient's positioning (i.e., dorsal to ventral and vice versa), and then reintroduce the endoscope. It is not uncommon

Figure 9.8 **The tracheal approach.**

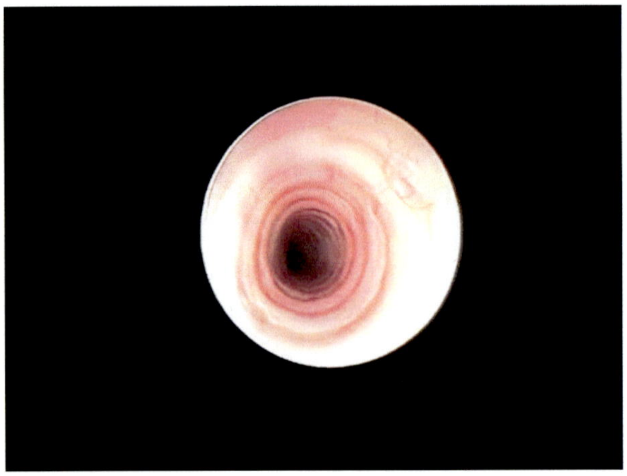

Figure 9.9 **The trachea viewed endoscopically.**

to see some iatrogenic trauma and bleeding during tracheoscopy, but this rarely causes significant complications.

Left coelomic approach

The standard positioning is to place the patient in right lateral recumbency with the wings extended and restrained. The left foot is pulled forward and secured to the neck (**Fig. 9.10**). The left flank behind the leg is then aseptically prepared – it is rare to have to remove more than a few feathers (**Fig. 9.11**). The following landmarks are identified: the caudal rib, the *flexor crura medialis* muscle, and the pubic bone (**Fig. 9.12**). A 4 mm vertical incision is made in the skin where the *flexor crura medialis* muscle crosses the caudal rib (**Fig. 9.13**). Curved haemostat forceps are then used to reflect the flexor muscle dorsally and then blunt dissect through the coelomic wall (**Fig. 9.14**). If done correctly, the operator will feel a 'pop' as the haemostats enter the caudal thoracic air sac.

Figure 9.10 **A galah is positioned for a left flank approach. The wings are extended and secured, and the left leg is brought forward and secured to the neck.**

Figure 9.11 **The left flank prepped for endoscopy. Minimal feather removal is required.**

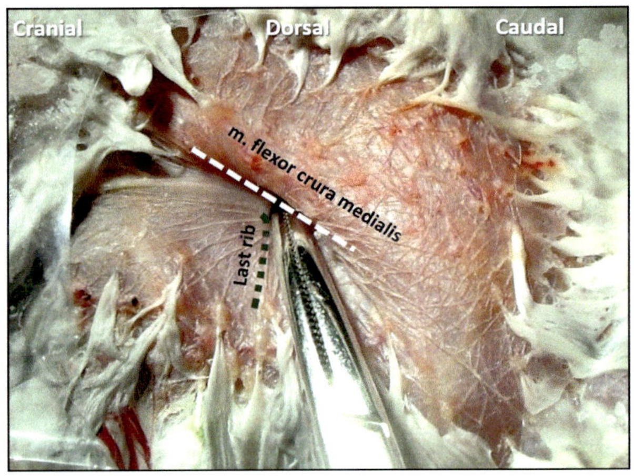

Figure 9.12 **Landmarks for a left flank approach.**

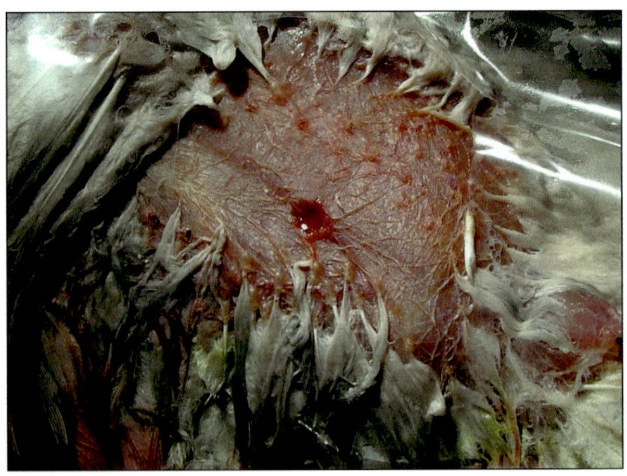

Figure 9.13 **A skin incision is made where the last rib crosses under the *flexor crura medialis*.**

Once the coelom has been entered, the haemostats can be withdrawn, and the endoscope introduced (**Fig. 9.15**). If entry has been performed correctly,

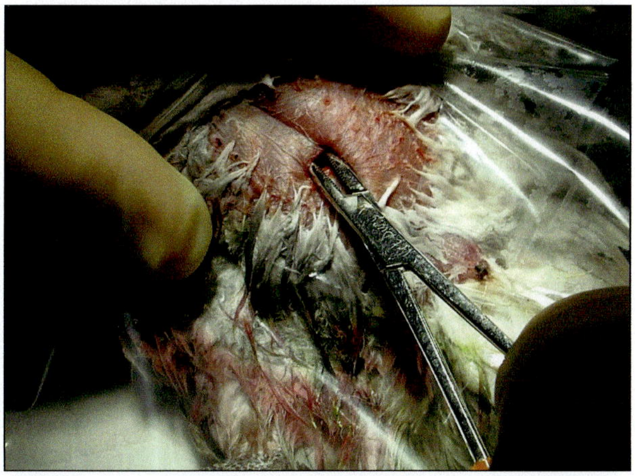

Figure 9.14 **Blunt dissection through the body wall.**

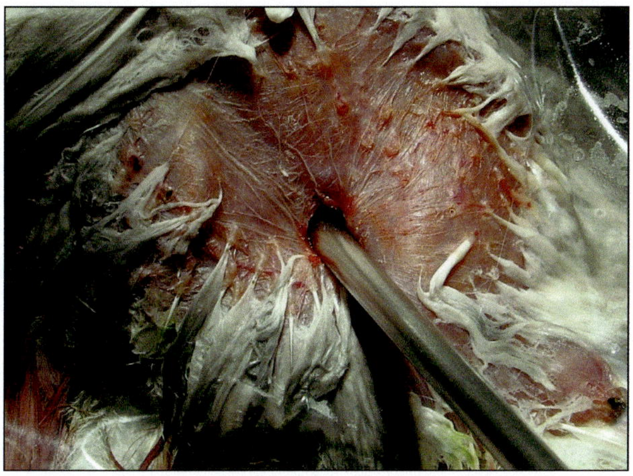

Figure 9.15 **The rigid endoscope is introduced through the incision into the caudal thoracic air sac.**

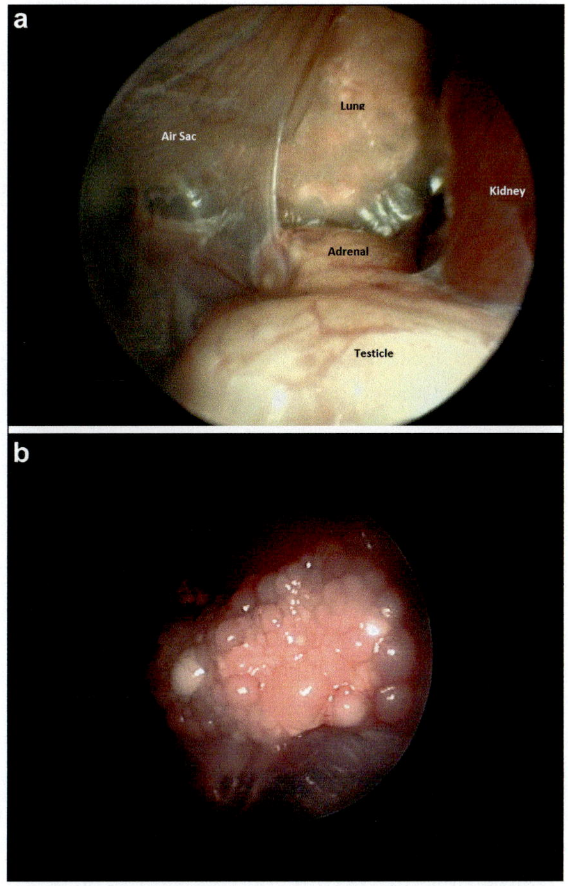

Figure 9.16 **(a) The view seen in the abdominal air sac, looking cranially. (b) The ovary.**

the endoscope will now be in the caudal thoracic air sac – this can be confirmed by a clear view of the lung and ostium, with air sac membranes visible on the left and right of the view (**Fig. 9.16a**). The endoscope can be advanced to the lung and, in some patients, can enter the lung through the ostium. The air sac membranes on either side of the caudal thoracic air sac can be penetrated either with endoscopic scissors or by placing the tip of the endoscope against the membrane and twisting sharply. Going to the left allows entry into the cranial thoracic air sac; to the right lies the abdominal air sac. The kidney, adrenal gland, gonad (**Fig. 9.16b**), and oviduct/ductus deferens are visible within the abdominal air sac, against the dorsal wall; the spleen and proventriculus are located ventrally, below the gonad–lung–adrenal triad. Within the cranial thoracic air sac lie the liver, proventriculus, and, farther cranially, the heart.

Once the procedure is completed, the endoscope is withdrawn. Tears in the air sac membranes heal well without any requirement for surgical closure. The muscle and skin incisions can be closed with either sutures or tissue glue if the operator feels it is advisable, but are often left to heal by secondary intention (which occurs within a few days).

If necessary, the same procedure can be performed on the right flank of the bird.

Ventral coelomic approach

The ventral coelomic approach is used to access the liver for examination and biopsy. With the bird in dorsal recumbency and elevated cranially, the area over the caudal sternum and ventral coelomic wall is plucked and prepared aseptically. A small midline skin incision is made 2–3 mm caudal to the sternum; the linea alba is identified, lifted, and carefully incised. The view may be obstructed by coelomic fat. The endoscope is introduced and directed cranially to the left and right of the midline. The left and right ventral hepatic peritoneal cavities, formed by peritoneal partitions, are identified and entered after either sharp or blunt dissection. The right liver lobe is larger than the left and is often the easiest to visualise. Once the procedure is completed, the linea alba and skin are closed in separate layers.

Choanal approach

Introducing the endoscope through the choanal slit allows visualisation of the left and right internal nares, the nasal cavities, and nasal turbinates. Saline can be flushed through the external nares to provide a better examination and to flush any foreign bodies out – or at least into the field of view. The patient should be intubated before this is performed.

Cloacal approach

Introducing the endoscope into the cloaca while insufflating with saline allows excellent visualisation of the three chambers of the cloaca, the ureteral openings, the opening into the oviduct or the ductus deferens, and – in young birds – the cloacal bursa. After the endoscope is inserted into the cloaca, the vent is occluded while warmed saline is flushed into the cloaca, 'ballooning' the chambers for examination. Care should be taken to release this saline periodically to prevent retropulsion of excessive amounts of saline into the rectum and the rest of the gastrointestinal tract.

Approach to the upper gastrointestinal tract

The oesophagus can be examined by passing the endoscope past the glottis in an anaesthetised patient. The crop can be examined in birds with crops (parrots, pigeons, doves, gallinaceous birds, and some passerine birds) by passing the tip of the endoscope through the cervical portion of the oesophagus. The bird should be fasted for several hours before the procedure to increase visualisation when possible. Insufflation of the crop with air or water will increase visualisation. An ingluviotomy is usually needed to examine the proventriculus and ventriculus. The bird is anaesthetised, intubated, and placed in dorsal recumbency. The cervical oesophagus or pharynx should be packed off with gauze swabs if possible. A small incision is made into the crop, and the endoscope is passed through the distal oesophagus into the proventriculus. The ventriculus may be entered by passing the endoscope through the proventricular–ventricular junction. Insufflation with saline will be necessary; again, care should be taken to remove some of this saline to minimise the amount passing down into the duodenum. When the procedure is completed, the crop is closed with a two-layer closure.

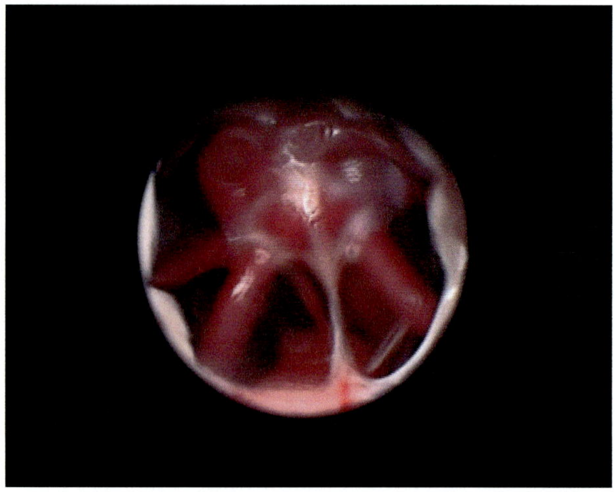

Figure 9.17 **The view in the clavicular air sac, including the base of the heart and the great vessels.**

Clavicular approach

The clavicular approach is used for visualisation of the clavicular air sac, syrinx, heart, brachiocephalic trunks, internal carotid, thyroid, and parathyroid (**Fig. 9.17**). The bird is positioned in dorsal recumbency. The point of entry is in the thoracic inlet, ventral to the trachea and oesophagus, after reflecting the crop. The skin is incised and dissected between the crop and clavicle to access the clavicular air sac. The thyroid gland is located ventral to the carotid artery and has a small, red-brown, ellipsoidal appearance. It can be biopsied from this approach. The clavicular air sac is evaluated for the presence of cranial coelomic masses.

FURTHER READING

Divers SJ. (2005) Minimally invasive endoscopic surgery of birds. *Journal of Avian Medicine and Surgery*, 19(2): 107–120. https://doi.org/10.1647/2004-002.

Divers SJ. (2010) Endoscopy equipment and instrumentation for use in exotic animal medicine. *Veterinary Clinics of North America: Exotic Animal Practice*, 13(2):171–185.

Divers SJ. (2010) Avian diagnostic endoscopy. *Veterinary Clinics of North America: Exotic Animal Practice*, 13(2): 187–202.

Hung CS-Y, Sladakovic I, Divers SJ. (2020) Diagnostic value of plasma biochemistry, haematology, radiography and endoscopic visualisation for hepatic disease in psittacine birds. *Veterinary Record*, 187(3): 93–100.

CHAPTER 10

DIFFERENTIAL DIAGNOSES

Listed by clinical signs are the differential diagnoses that the clinician should consider when examining a patient. Although every effort has been made to make this list as exhaustive as possible, it would be difficult, if not impossible, to guarantee that every possible differential diagnosis is included.

In many cases, the list goes only as far as describing the organs likely to be involved. Clinicians are referred to the chapters dealing with those organs for a more comprehensive discussion of diseases that could affect them.

CHANGE IN DROPPINGS

Diarrhoea

- Enteritis – bacterial, fungal, viral, or toxic (Chapter 19, Disorders of the gastrointestinal tract)
- Intestinal parasites (helminths, protozoa) (Chapter 19, Disorders of the gastrointestinal tract)
- Avian ganglioneuritis (Chapter 19, Disorders of the gastrointestinal tract)
- Chlamydiosis (Chapter 20, Disorders of the liver)
- *Macrorhabdus ornithogaster* (previously known as megabacteria or avian gastric yeast) (Chapter 19, Disorders of the gastrointestinal tract)
- Liver disease (Chapter 20, Disorders of the liver)
- Zinc and lead poisoning zinc (Chapter 20, Disorders of the liver)
- Intestinal foreign body (Chapter 19, Disorders of the gastrointestinal tract)
- Change in diet (Chapter 19, Disorders of the gastrointestinal tract)
- Normal in lorikeets (Chapter 4, The physical examination)

Change in colour of the faecal portion

- Black: anorexia, melena, foreign body (Chapter 4, The physical examination)
- Pale: exocrine pancreatic insufficiency, maldigestion due to *Macrorhabdus* or other gastrointestinal diseases (Chapter 21 Disorders of the pancreas; Chapter 19, Disorders of the gastrointestinal tract)
- Brown: pelleted diet (Chapter 4, The physical examination)
- Dietary change (e.g., purple with berries) (Chapter 4, The physical examination)

Enlarged faecal portion

- Egg laying, broody (Chapter 4, The physical examination)
- Space-occupying lesion in coelom (Chapter 4, The physical examination)
- Exocrine pancreatic insufficiency (Chapter 21, Disorders of the pancreas)
- Formulated diets (Chapter 4, The physical examination)
- Malabsorptive diseases (Chapter 4, The physical examination)
- Hernia (Chapter 4, The physical examination)

Frank blood in the droppings

- Oviductal disease (Chapter 27, Disorders of the reproductive tract)
  - Egg binding
  - Salpingitis
  - Metritis
  - Neoplasia
- Cloacal disease (Chapter 19, Disorders of the gastrointestinal tract)
  - Internal papilloma disease
  - Cloacoliths
  - Cloacitis

DOI: 10.1201/9781003272519-10

  - Cloacal neoplasia
  - Cloacal prolapse
- Warfarin toxicosis
- Severe enteritis (Chapter 19, Disorders of the gastrointestinal tract)

Whole seed in the droppings

- Increased intestinal motility (Chapter 19, Disorders of the gastrointestinal tract)
- Any condition affecting proventricular and ventricular function (Chapter 19, Disorders of the gastrointestinal tract)
  - Avian ganglioneuritis
  - Candidiasis
  - *Acuaria* (gizzard worm) infection
  - *Macrorhabdus*

Change in the colour of the urates

- Green: hepatic disease, post-trauma (metabolism of haemoglobin) (Chapter 20, Disorders of the liver)
- Pink/crimson: renal disease, lead poisoning (Chapter 4, The physical examination)
- Yellow: anorexia, early or resolving hepatic disease (Chapter 5, Supportive therapy)
- Orange: vitamin B injection, doxycycline injection (Chapter 4, The physical examination)

Malodorous droppings

- Clostridial overgrowth (Chapter 19, Disorders of the gastrointestinal tract)
- Cloacoliths (Chapter 19, Disorders of the gastrointestinal tract)
- Coccidiosis (Chapter 19, Disorders of the gastrointestinal tract)
- Yeast enteritis (Chapter 19, Disorders of the gastrointestinal tract)
- Faecal retention, e.g., hernia, broody behaviour (Chapter 4, The physical examination)

Polyuria

- Stress or fear, especially on initial presentation ('stress polyuria') (Chapter 4, The physical examination; Chapter 24, Disorders of the urinary system)
- Renal disease (Chapter 24, Disorders of the urinary system)
- Diabetes mellitus (Chapter 21, Disorders of the pancreas)
- Diabetes insipidus (Chapter 24, Disorders of the urinary system)
- Hepatic disease (Chapter 20, Disorders of the liver)
- Pancreatic disease (Chapter 21, Disorders of the pancreas)
- Hyperadrenocorticism (Chapter 24, Disorders of the urinary system)
- Heavy metal toxicosis (Chapter 24, Disorders of the urinary system; Chapter 26, Disorders of the nervous system)
- 'Phosphate flush' due to an all-seed diet. (Chapter 24, Disorders of the urinary system)
- Psychogenic polydypsia, usually seen in juvenile birds (Chapter 24, Disorders of the urinary system)
- Pituitary adenoma in budgerigars (Chapter 26, Disorders of the nervous system)
- Normal in lories and lorikeets (Chapter 4, The physical examination)
- Normal in birds being hand-fed or crop-fed (Chapter 29, Paediatrics)
- Hens during broody behaviour (Chapter 4, The physical examination)

APPETITE AND THIRST

Increased appetite

- Diabetes mellitus (Chapter 21, Disorders of the pancreas)
- Disorders of digestion, including *Macrorhabdus* and endoparasitism (Chapter 19, Disorders of the gastrointestinal tract)
- Jaw injuries (the bird is actually not eating, but rather is attempting to do so) (Chapter 12, Disorders of the beak and cere)
- Exocrine pancreatic insufficiency (Chapter 21, Disorders of the pancreas)
- Starvation (check if there is food, and not just husks, in the food dish)

Decreased appetite

- Nonspecific sign of illness
- Mouth or jaw injuries (Chapter 12, Disorders of the beak and cere)

Polydypsia

- Usually secondary to polyuria (see earlier) (Chapter 24, Disorders of the urinary system)
- Formulated diets may cause a moderate increase in thirst in some birds
- Psychogenic polydypsia, occasionally seen in weaning cockatoo chicks (Chapter 24, Disorders of the urinary system)

Decreased thirst

- Nonspecific sign of illness
- Sufficient water in green foods and vegetables

VOMITING

- Ingluvitis, e.g., trichomoniasis, bacterial and yeast infections (Chapter 19, Disorders of the gastrointestinal tract)
- Generalized illness, e.g., renal disease (Chapter 24, Disorders of the urinary system)
- Ileus, associated with heavy metal toxicosis or other gastrointestinal disease (Chapter 19, Disorders of the gastrointestinal tract; Chapter 24, Disorders of the urinary system)
- Foreign body obstruction (e.g., from chewing on substrate or cotton fibre toys) (Chapter 19, Disorders of the gastrointestinal tract)
- *Macrorhabus* (Chapter 19, Disorders of the gastrointestinal tract)
- Motion sickness
- Behavioural regurgitation, e.g., after crop feeding or handling (Chapter 29, Paediatrics)
- Extramural obstruction, e.g., neoplasia pressing on the distal oesophagus (Chapter 19, Disorders of the gastrointestinal tract)

WEIGHT LOSS

- Nonspecific sign of illness
- Response to controlled diet (Chapter 2, Husbandry and nutrition)
- Increasing fitness due to increasing exercise
- Normal weaning response (if less than 10% weight loss) (Chapter 29, Paediatrics)
- Malabsorption/maldigestion (Chapter 19, Disorders of the gastrointestinal tract; Chapter 21, Disorders of the pancreas)

POSTURE

Fluffed, immobile, eyes closed, both legs on perch

- Nonspecific sign of illness (Note: Some sick birds will make an effort to appear normal when being examined but can rarely maintain this for more than a minute or two before reverting to the 'sick bird look'.) (Chapter 4, The physical examination)

Fluffed, immobile, eyes closed, head tucked under one wing, only one leg on perch

- Sleeping (Chapter 4, The physical examination)

Tail pointing down, perpendicular to cage floor, while wings remain at a normal angle

- May be increased respiratory effort as evidenced by tail bobbing (Chapter 23, Disorders of the respiratory tract)
- Spinal kyphosis (Chapter 17, Disorders of the spine)
- Caudal coelomic mass, e.g., neoplasia, pyometra, egg binding (Chapter 27, Disorders of the reproductive tract)

Tail bobbing up and down in an exaggerated movement

- Respiratory disease: tracheal obstruction, pulmonary disease, severe air sac disease (Chapter 23, Disorders of the respiratory tract)
- External pressure on air sacs: internal organ enlargement, egg binding, ascites, obesity, kyphosis compressing coelomic space (Chapter 17, Disorders of the spine; Chapter 27, Disorders of the reproductive tract)

Head held down but bird is looking up, wings spread out, tail spread

- Reproductively active hen: courting posture
- Heat stress (Chapter 5, Supportive therapy)
- Egg binding (Chapter 27, Disorders of the reproductive tract)

Wing droop

- Bilateral: generalized weakness, heat stress, bilateral wing injuries (Chapter 4, The physical examination)
- Unilateral: wing injury; as a rule, the lower the droop, the more distal the injury (Chapter 16, Disorders of the pectoral limb)

Sitting on the floor of the cage, body upright, perhaps panting

- Leg injuries (Chapter 15, Disorders of the pelvic limb)
- Egg binding (Chapter 27, Disorders of the reproductive tract)
- Renal disease causing paresis (Chapter 26, Disorders of the nervous system)
- Spinal disease causing paresis (Chapter 26, Disorders of the nervous system)
- Heavy metal intoxication (Chapter 26, Disorders of the nervous system)
- Clenched leg paralysis (lorikeet paralysis syndrome, Barraband paralysis syndrome) (Chapter 26, Disorders of the nervous system)

Neurological signs (ataxia, fitting, paralysis/paresis, tremor)

- Weakness due to illness
- Agonal signs
- Lead poisoning (Chapter 26, Disorders of the nervous system)
- Head trauma (Chapter 26, Disorders of the nervous system)
- Hypocalcaemia (Chapter 26, Disorders of the nervous system)
- Hypoglycaemia (Chapter 29, Paediatrics)
- CNS lesions (e.g., congenital disease, proventricular dilatation disease, encephalitis, neoplasia) (Chapter 26, Disorders of the nervous system)
- Cardiovascular disease (Chapter 22, Disorders of the cardiovascular system)

Bilateral leg paresis/paralysis

- Spinal disease: trauma, neoplasia
- 'Obturator' paralysis in hens that are egg-bound or have recently laid eggs
- Hypocalcaemia (Chapter 26, Disorders of the nervous system)
- Barraband paralysis syndrome (Chapter 26, Disorders of the nervous system)
- Lorikeet paralysis syndrome (Chapter 26, Disorders of the nervous system)
- Bilateral leg trauma (Chapter 26, Disorders of the nervous system)
- Renal swelling (Chapter 26, Disorders of the nervous system)
- Heavy metal poisoning (Chapter 26, Disorders of the nervous system)

Unilateral leg paresis/paralysis

- Fractured leg (Chapter 15, Disorders of the pelvic limb)
- Soft tissue trauma (Chapter 26, Disorders of the nervous system)
- Renal enlargement (e.g., neoplasia) pressing on the sciatic nerve, predominantly seen in budgerigars (Chapter 26, Disorders of the nervous system)
- Neoplasia (Chapter 26, Disorders of the nervous system)

One wing held out to the side, resting on the floor or perch

- Check for leg trauma or unilateral leg paralysis/paresis on that side (Chapter 15, Disorders of the pelvic limb)

FEATHERS AND SKIN

Generalized feather loss

- Psittacine beak and feather disease (PBFD) (Chapter 11, Disorders of the skin and feathers)
- Polyomavirus (Chapter 11, Disorders of the skin and feathers)
- Feather-damaging behaviour (Note: If this is self-inflicted, the head will appear normal.) (Chapter 11, Disorders of the skin and feathers)
- Extreme age (Chapter 11, Disorders of the skin and feathers)
- Obesity
- Normal lack of feathers along apterylae (Chapter 11, Disorders of the skin and feathers)
- Excessive mutual grooming by cage mate (Chapter 4, The physical examination)

- Bald patch on crown of head, behind the crest, is normal in lutino cockatiels (Chapter 11, Disorders of the skin and feathers)
    - Obesity
    - Spinal abnormalities
    - Elizabethan collar to prevent feather picking
    - Beak malformations

Feathering gradually darkening or becoming 'greasy-looking'

- Liver disease (Chapter 4, The physical examination)
- Malnutrition (Chapter 4, The physical examination)
- Hypothyroidism (Chapter, 11 Disorders of the skin and feathers)
- Age (Chapter 4, The physical examination)

Broken primary feathers on wings and/or tail

- Heavy falls, often associated with poorly done wing clips (Chapter 4, The physical examination)
- Small caging combined with wing flapping or inability to perch steadily (Chapter 4, The physical examination)
- Cage mate aggression (Chapter 2, Husbandry and nutrition)
- Malnutrition (Chapter 4, The physical examination)
- Feather-damaging behaviour (Chapter 11, Disorders of the skin and feathers)

'Stress lines' (horizontal breaks in the feather vane)

- Physiological stress or illness at the time the feather emerged through the skin; if severe, could indicate generalized disease in the bird (Chapter 4, The physical examination)
- Fenbendazole use if administered while feathers were growing (Chapter 11, Disorders of the skin and feathers)

Long straw-like feathers over the thighs or other feathers still encased in keratin sheaths

- 'Straw feathers' in canaries, a lethal genetic disorder (Chapter 11, Disorders of the skin and feathers)
- Inability to groom properly (Chapter 4, The physical examination)
    - Illness

Abnormally coloured feathers

- Red feathers on African grey parrots (Chapter 11, Disorders of the skin and feathers)
    - PBFD
    - Chronic feather plucking
    - Malnutrition
    - Liver disease
    - Normal genetic variation
- Abnormally coloured eclectus parrots (Chapter 11, Disorders of the skin and feathers)
    - PBFD
    - Polyomavirus
    - Nutritional issues, especially with hand-rearing formulae
    - Liver disease
    - Thyroxine medication
- Green feathers turning black: saprophytic fungal growth resulting in grossly visible opaque black discolouration. This fungal growth is usually due to the oils from human hands left on the feathers after petting or holding the bird. This is not seen in birds with powder down (e.g., cockatoos and African greys) because the powder keeps the feathers clean. (Chapter 3, Handling and grooming birds)
- Galahs (rose-breasted cockatoos) with dark grey and pink feathers (Chapter 11, Disorders of the skin and feathers)
    - Chronic liver disease
    - Hypothyroidism
- Lutino cockatiels becoming a deep 'buttercup' yellow; feathers have a 'greasy' appearance (Chapter 11, Disorders of the skin and feathers)
    - Chronic liver disease
    - Hypothyroidism
- Green feathers turning yellow, blue feathers turning white (Chapter 11, Disorders of the skin and feathers)
    - PBFD
    - Malnutrition
- Young white cockatoos with dirty feathering (Chapter 11, Disorders of the skin and feathers)
    - PBFD
    - Feeding, e.g., charcoal blocks

Ragged-looking plumage without pruritus

- Delayed moult (Chapter 11, Disorders of the skin and feathers)
  - Malnutrition
  - Endocrine disorders
  - Abnormal diurnal rhythm
  - Excessive egg production
- Incompatibility of cage mates (e.g., aggressive plucking is common in zebra finches, red-eared waxbills, golden-breasted waxbills, and orange-cheeked waxbills) (Chapter 11, Disorders of the skin and feathers)
- Improper housing
- Dermatophytosis (Chapter 11, Disorders of the skin and feathers)

Continued growth of flight, tail, and contour feathers in budgerigars unable to fly

- 'Feather duster' syndrome (Chapter 11, Disorders of the skin and feathers)
- Associated with a lethal recessive gene
- Some reports of association with herpesvirus

Pruritis

- Lice and mites (Chapter 11, Disorders of the skin and feathers)
- Malnutrition (Chapter 11, Disorders of the skin and feathers)
- Dermatitis (Chapter 11, Disorders of the skin and feathers)
- Polyfolliculosis (Chapter 11, Disorders of the skin and feathers)
- Feather inclusion cysts (Chapter 11, Disorders of the skin and feathers)
- Allergic dermatitis (Chapter 11, Disorders of the skin and feathers)

Feathers missing on the head

- Cage mate aggression (Chapter 11, Disorders of the skin and feathers)
- Dermatitis (Chapter 11, Disorders of the skin and feathers)
- Folliculitis (Chapter 11, Disorders of the skin and feathers)
- Cutaneous neoplasia (Chapter 11, Disorders of the skin and feathers)
- Dermatophytosis (Chapter 11, Disorders of the skin and feathers)
- Cutaneous candidiasis (Chapter 11, Disorders of the skin and feathers)
- Trauma (Chapter 11, Disorders of the skin and feathers)

Greenish discoloration of skin

- Bruising (Chapter 4, The physical examination)

Flaky, dry skin

- Malnutrition (Chapter 11, Disorders of the skin and feathers)
- Lack of bathing opportunities (Chapter 11, Disorders of the skin and feathers)
- Dermatophytosis (Chapter 11, Disorders of the skin and feathers)
- Cutaneous candidiasis (Chapter 11, Disorders of the skin and feathers)

Feather cysts

- Retained feathers: often pruritic; usually seen on wings or over sternum (Chapter 11, Disorders of the skin and feathers)
  - Traumatic, often associated with incorrect housing
  - Incomplete removal of a broken 'blood feather'
  - Overpreening by bird
  - Secondary to folliculitis
- Benign follicular tumours, commonly seen in canaries, but can be seen in other species; usually not pruritic and can occur anywhere on the body (Chapter 11, Disorders of the skin and feathers)
- Mycobacterial granulomas in the skin can be confused with feather cysts (Chapter 16, Disorders of the pectoral limb)

Intense pruritus and self-mutilation

- Polyfolliculosis (Chapter 11, Disorders of the skin and feathers)
  - Chronic condition most commonly seen in budgerigars and lovebirds
  - Newly emerging feathers have short, stout quills with retained sheaths
  - Usually intensely pruritic

- Unclear whether it is a primary problem or secondary to folliculitis
- Self-mutilation of African lovebirds (*Agapornis* spp.) (Chapter 11, Disorders of the skin and feathers)
  - Seen on the shoulder region and prepatagial membrane or, less commonly, in the inguinal region, chest, back, base of tail, and around the cloaca
  - Can be unilateral or bilateral
  - Intensely pruritic
  - Cause still undetermined
  - Axillary mutilation (Chapter 11, Disorders of the skin and feathers)
- Superficial chronic ulcerative dermatitis (SCUD)
  - Intense pruritus and self-trauma to the axillae and patagial membranes
  - Reported in grey parrots, cockatiels, galahs, and other cockatoo species
  - Cause undetermined

Enlarged uropygial (preen) gland

- Hyperplasia (Chapter 11, Disorders of the skin and feathers)
- Neoplasia (Chapter 11, Disorders of the skin and feathers)
- Blockage with keratin plug (sometimes associated with hypovitaminosis A) (Chapter 11, Disorders of the skin and feathers)
- Infection (Chapter 11, Disorders of the skin and feathers)

Yellow subcutaneous deposits

- Fat (Chapter 11, Disorders of the skin and feathers)
- Lipoma (Chapter 11, Disorders of the skin and feathers)
- Xanthoma (Chapter 11, Disorders of the skin and feathers)

Feather-damaging behaviour

- Physical problems (Chapter 11, Disorders of the skin and feathers)
  - Dermatitis/folliculitis
  - Underlying painful lesions (e.g., arthritis, neoplasia, internal organ disease, or enlargement [if painful or uncomfortable])
  - Malnutrition
  - Nicotine sensitivity
  - Reproductive-associated feather picking
  - Giardiasis in cockatiels
- Psychological causes (Chapter 30, Behavioural problems)
  - Fear
  - Boredom
  - Insecurity
  - Anxiety
  - Attention-seeking behaviour
  - Sexual frustration
- Combination of any of the above

WINGS

Blood on feathers

- Wing tip trauma, often associated with excessively severe wing trimming and improper housing (Chapter 3, Handling and grooming birds)
- Broken 'blood feather' (Chapter 11, Disorders of the skin and feathers)
- Trauma to bone, muscle or skin (Chapter 11, Disorders of the skin and feathers)

Wing drooping

- Broken bones (Chapter 4, The physical exam)
- Muscle damage (Chapter 4, The physical exam)
- Weakness (Chapter 4, The physical exam)
- Respiratory disease (Chapter 31, Analgesia and anaesthesia)

Swellings

- Neoplasia (Chapter 11, Disorders of the skin and feathers)
- Healing/healed broken bone (Chapter 33, Orthopaedic surgery)
- Soft tissue trauma (Chapter 11, Disorders of the skin and feathers)
- Feather cyst (Chapter 11, Disorders of the skin and feathers)
- Granuloma (Chapter 11, Disorders of the skin and feathers)
- Air sac rupture with subcutaneous emphysema (Chapter 8, Diagnostic imaging)

Wings held away from the body

- Heat stress (Chapter 4, The physical examination)
- Behavioural
    - Courtship
    - Fear
    - Aggression

Green discolouration

- Bruising (Chapter 4, The physical examination)

FEET AND LEGS

Limping

- Injury to bone, muscles, or joints (Chapter 15, Disorders of the pelivc limb)
- Pododermatitis (bumblefoot) (Chapter 15, Disorders of the pelvic limb)
- Hypocalcaemia (Chapter 26, Disorders of the nervous system)
- Pre-existing deformity (e.g., developmental varus/valgus deformity) (Chapter 15, Disorders of the pelvic limb)

Swollen joints

- Articular gout (Chapter 15, Disorders of the pelvic limb)
- Arthritis (Chapter 15, Disorders of the pelvic limb)
    - Degenerative
    - Infectious
- Neoplasia (Chapter 15, Disorders of the pelvic limb)
- Trauma (Chapter 15, Disorders of the pelvic limb)

Missing nails or toes

- Aggressive cage mates (Chapter 15, Disorders of the pelvic limb)
- Unsafe caging or cage furniture (Chapter 15, Disorders of the pelvic limb)
- Ergotism (Chapter 15, Disorders of the pelvic limb)
- Toe constriction due to fibrous band formation (constricted toe syndrome) or foreign bodies (e.g., cotton thread) (Chapter 15, Disorders of the pelvic limb)
- Frost damage (Chapter 15, Disorders of the pelvic limb)

Overgrown nails

- Chronic liver disease (Chapter 15, Disorders of the pelvic limb; Chapter 20, Disorders of the liver)
- Inadequate/unsuitable perches (Chapter 3, Handling and grooming of birds; Chapter 15, Disorders of the pelvic limb)
- Malnutrition (Chapter 3, Handling and grooming birds; Chapter 15, Disorders of the pelvic limb)
- PBFD (Chapter 15, Disorders of the pelvic limb)
- Abnormal toe anatomy preventing normal perching, e.g., articular gout, chronic trauma (Chapter 15, Disorders of the pelvic limb)

Abnormal shape or direction of the legs

- Coxofemoral subluxation: splay leg (Chapter 15, Disorders of the pelvic limb)
- Nutritional secondary hyperparathyroidism (Chapter 16, Disorders of the pectoral limb)
- Incorrectly aligned healed fractures (Chapter 33, Orthopaedic surgery)
- Joint luxation (Chapter 15, Disorders of the pelvic limb)
- Slipped tendon (the gastrocnemius tendon 'slips' out of the groove on the back of the hock); associated with trauma, poor diet, poor conformation, and perhaps a genetic influence (Chapter 15, Disorders of the pelvic limb)
- Angular limb deformity (rotated femur and/or tibiotarsus) (Chapter 15, Disorders of the pelvic limb)

Hyperkeratosis of the scaled part of the leg

- *Cnemidocoptes* (Chapter 15, Disorders of the pelvic limb)
- Nutritional deficiency, especially zinc and/or biotin (Chapter 15, Disorders of the pelvic limb)
- Dermatophytosis (Chapter 11, Disorders of the skin and feathers)

Self-mutilation of foot and toes

- Osteomyelitis (Chapter 15, Disorders of the pelvic limb)
- Myositis/tendonitis (Chapter 15, Disorders of the pelvic limb)
- Pododermatitis (Chapter 15, Disorders of the pelvic limb)
- Neuralgia (Chapter 15, Disorders of the pelvic limb)
- Necrosis of the extremities (see earlier section "Missing nails or toes") (Chapter 15, Disorders of the pelvic limb)
- Leg ring constriction (Chapter 15, Disorders of the pelvic limb)

BEAK

Overgrown beak

- Sometimes with bruising present in the keratin of the beak

- Liver disease (Chapter 20, Disorders of the liver)
- *Cnemidocoptes* (Chapter 12, Disorders of the beak and cere)
- Lack of occlusal wear (Chapter 3, Handling and grooming of birds)
- Lack of chewing opportunities (Chapter 3, Handling and grooming of birds)
- Hypothyroidism (Chapter 11, Disorders of the skin and feathers)

Beak twisted to the left or right

- Scissor beak (Chapter 12, Disorders of the beak and cere)
  - Congenital
  - Acquired

Upper beak inside the lower beak

- Prognathism (Chapter 12, Disorders of the beak and cere)

Inability to close beak properly

- Subluxation of the palatine bone due to hyperextension of the maxilla in macaws (Chapter 12, Disorders of the beak and cere)
- Hyperextension of the mandible (Chapter 12, Disorders of the beak and cere)
- Fractured jaw (Chapter 12, Disorders of the beak and cere)

Flakes of keratin on the beak

- Malnutrition
- Lack of an abrasive surface in the cage to groom beak on (Chapter 4, The physical examination)

White, crusty, honeycombed lesions on the beak

- Sometimes also on feet and vent
- Scaly face mite (Chapter 12, Disorders of the beak and cere)

Nares unequal in size

- Chronic respiratory disease (Chapter 12, Disorders of the beak and cere; Chapter 23, Disorders of the respiratory tract)
- Cere abscess (Chapter 12, Disorders of the beak and cere)
- Neoplasia (Chapter 12, Disorders of the beak and cere)
- Rhinolith (Chapter 12, Disorders of the beak and cere)

Thickening and hypertrophy of cere of budgerigar hens

- Cere hypertrophy associated with hyperoestrogenism (Chapter 12, Disorders of the beak and cere)

Nares blocked or staining/matting of feathers above nares

- Chronic respiratory disease (Chapter 12, Disorders of the beak and cere; Chapter 23, Disorders of the respiratory tract)
- Choanal atresia (African grey parrots) (Chapter 23, Disorders of the respiratory tract)

EYES

Feather loss around eyes

- Rubbing of the face against a perch or sides of cage (Chapter 13, Disorders of the eye)
  - Conjunctivitis (Chapter 13, Disorders of the eye)
  - Sinusitis (Chapter 13, Disorders of the eye; Chapter 23, Disorders of the respiratory tract)
  - Ocular and periocular pain due to other causes (e.g., neoplasia, avian poxvirus) (Chapter 11, Disorders of the skin and feathers; Chapter 13, Disorders of the eye)
  - Blocked nasolacrimal duct
  - Choanal atresia (Chapter 23, Disorders of the respiratory tract)
- Overgrooming by companion bird (Chapter 13, Disorders of the eye)

Eyelid abnormalities

- Congenital (Chapter 13, Disorders of the eye)
  - Cryptophthalmos – failure of eyelid formation, resulting in fusion of the eyelid margins; occasionally seen in cockatiels
  - Blepharophimosis – narrowing of the palpebral fissures without fusion of the eyelid margins; occasionally seen in all species.
- Acquired (Chapter 13, Disorders of the eye)
  - Symblepharon – adhesion of the eyelid(s) to the globe; can be seen as a sequela to severe conjunctivitis
  - Acquired blepharophimosis; occasionally seen after conjunctivitis or other ocular inflammatory conditions
  - Scarring and deformity; occasionally seen after avian poxvirus infections

Thickening and hyperaemia of the conjunctiva

- Often associated with epiphora or ocular discharge

- Conjunctivitis (Chapter 13, Disorders of the eye)
  - Chlamydiosis
  - Mycoplasma
  - Mycobacteria
  - Parasites ('eye' worms): *Oxyspirura mansoni*, *Ceratospira* spp., and *Thelazia* spp.
  - Other chronic infections
  - Allergic, especially secondary to nicotine
- Sinusitis (Chapter 13, Disorders of the eye; Chapter 23, Disorders of the respiratory tract)
- Avian poxvirus (Chapter 11, Disorders of the skin and feathers; Chapter 13, Disorders of the eye)
- Neoplasia (Chapter 13, Disorders of the eye)

Exophthalmos

- Sinusitis (Chapter 13, Disorders of the eye; Chapter 23, Disorders of the respiratory tract)
- Associated with stunting or nutritional secondary hyperparathyroidism in chicks
- Retrobulbar neoplasia, especially lymphoma
- Pituitary adenomas in budgerigars

Enophthalmos

- Dehydration (Chapter 29, Paediatrics)
- Sinusitis in macaws (Chapter 23, Disorders of the respiratory tract)
- Microphthalmia (Chapter 13, Disorders of the eye)

Corneal changes

- Usually multifocal, white, glistening, raised, non-inflammatory corneal lesions; lipid or cholesterol deposits associated with high-fat diets and obesity
- Stromal abscess – associated with severe pain; white material within the corneal stroma
- Keratitis or corneal ulceration
  - Primary: bacterial, fungal or viral (avian poxvirus)
  - Secondary: trauma or exposure due to eyelid deformity or malformation
- • Mass on the cornea
  - Dermoid
  - Staphyloma – uveal herniation into a weakened, distorted area of the cornea
  - Descemetocoele

Hyphema

- Trauma

- Warfarin toxicosis
- Neoplasia

Cataracts

Similar aetiology to mammals; can be inherited or acquired.

FACE

Swellings on the face

- Sinusitis (Chapter 13, Disorders of the eye; Chapter 23, Disorders of the respiratory tract)
- Neoplasia (Chapter 11, Disorders of the skin and feathers; Chapter 12, Disorders of the beak and cere)
- Insect bite
- Trauma

Matting/staining of feathers below and caudal to eye

- Otitis externa

Matting of the feathers over face and head

- Vomiting

Small nodules on the facial skin of macaws

- Macaw acne – ingrown feather follicles that produce a reactive nodule

BODY

Prominent keel bone ('going light')

- Chronic illness, resulting in catabolism of the muscle
- Inability to fly causes the muscles to atrophy due to disuse (Note: This degree of muscle wastage is usually not as severe as seen with illness.)
- Poor/inappropriate diet/husbandry
- Endoparasitism

Twisted keel bone

- Nutritional secondary hyperparathyroidism
- Genetic abnormality

Split keel bone (non-traumatic)

- Bifid sternum – genetic abnormality where the two halves of the sternum fail to fuse correctly

Ulcerative lesion on the cranial end of the keel

- Chronic self-mutilation and trauma resulting from falling heavily to the ground; most commonly seen in birds with overly severe wing trims
- Ulcerating lipoma or 'fat abscesses'

Overabundance of pectoral muscle mass (i.e., 'cleavage') along keel

- Obesity

Enlarged coelom

- Increased space between the sternum and the pubic bones

- Solid enlargement
  - Obesity
  - Hepatomegaly (Chapter 20, Disorders of the liver)
  - Internal neoplasia (e.g., renal, gonadal)
  - Egg binding (Chapter 27, Disorders of the reproductive tract)
  - Oviductal enlargement in breeding season (Chapter 27, Disorders of the reproductive tract)
  - Normal in young birds still being fed (Chapter 29, Paediatrics)
  - Intestinal parasitism (severe) (Chapter 19, Disorders of the gastrointestinal tract)
- Fluid enlargement
  - Yolk-related peritonitis (Chapter 27, Disorders of the reproductive tract)
  - Ovarian cyst (Chapter 27, Disorders of the reproductive tract)
  - Neoplastic cyst
  - Ascites associated with heart and/or liver disease (Chapter 20, Disorders of the liver; Chapter 22, Disorders of the cardiovascular system)
  - Neoplastic effusion

- Hernia

Subcutaneous emphysema

- Trauma
- Post-endoscopy air leakage
- Rupture of cervicocephalic air sac due to trauma or air saculitis (Chapter 23, Disorders of the respiratory system)

Split in skin between vent and tail

- Heavy landing associated with overly severe wing trim

Subcutaneous masses

- Lipomas (Chapter 11, Disorders of the skin and feathers)
- Xanthomas (Chapter 11, Disorders of the skin and feathers)
- Fat deposits
- Hernias
- Abscesses
- Feather cyst, (Chapter 11, Disorders of the skin and feathers)

PAEDIATRICS

Crop not emptying or slow to empty

- Generalized illness with ileus (Chapter 19, Disorders of the gastrointestinal tract)
- Bacterial or yeast ingluvitis (Chapter 19, Disorders of the gastrointestinal tract)
- Hand-rearing formula mixed incorrectly; incorrect temperature or consistency (Chapter 29, Paediatrics)
- Foreign body obstruction (Chapter 19, Disorders of the gastrointestinal tract)

Cervical emphysema

- Ruptured cervicocephalic air sac, often due to rough handling while being fed
- Gulping air while feeding (in this case the air is in the crop, not under the skin)

Erythematous skin

- Dehydration
- Heat stress
- Generalized illness

Pallor

- Cold stress
- Illness
- Anaemia

Overly large head

- Stunting – reduced rate of growth due to any cause (Chapter 29, Paediatrics)

Bruising on the skin

- Severe bacterial or viral infection, especially polyomavirus
- Injections (e.g., enrofloxacin)
- Trauma from parents or siblings

Feathers not growing normally

- Stunting – reduced rate of growth due to any cause (Chapter 29, Paediatrics)
- Polyomavirus (Chapter 11, Disorders of the skin and feathers)

Swollen toes

- Constricted toe syndrome
- Bedding or thread wrapped around toe, acting as a tourniquet

Thin toes

- Stunting – reduced rate of growth due to any cause (Chapter 29, Paediatrics)

Vomiting

- Ingluvitis (Chapter 19, Disorders of the gastrointestinal tract)
- Generalized illness, e.g., renal disease
- Ileus, associated with heavy metal toxicosis or other gastrointestinal disease
- Foreign body obstruction (e.g., from chewing on substrate or cotton fibre toys)
- Weaning, particularly South American species

Refusing to eat

- Weaning (Chapter 29, Paediatrics)
- Nonspecific sign of illness
- Inappropriate husbandry

Reddened skin or scab over the crop

- Crop burn (Chapter 19, Disorders of the gastro-intestinal tract)
- Trauma from crop needle

FURTHER READING

Doneley B, Harrison GJ, Lightfoot TL. (2006) Maximizing information from the physical examination. In: *Clinical Avian Medicine, Vol 1.* GJ Harrison, TL Lightfoot (eds). Spix Publishing Inc, Palm Beach, pp. 153–212.

CHAPTER 11

DISORDERS OF THE SKIN AND FEATHERS

ANATOMY AND PHYSIOLOGY

Avian skin is attached to both the underlying muscles and, more tightly, to the skeleton. It consists of the thin epidermis (only 10 cell layers deep) and the thicker underlying dermis (**Fig. 11.1**). Feather follicles originate from the dermis. True glands are absent from much of the skin, although epidermal cells may secrete a lipoid sebaceous material. The external ear has glands that secrete a waxy material, but the only other gland found on the skin is the uropygial gland. This gland, absent in ratites,

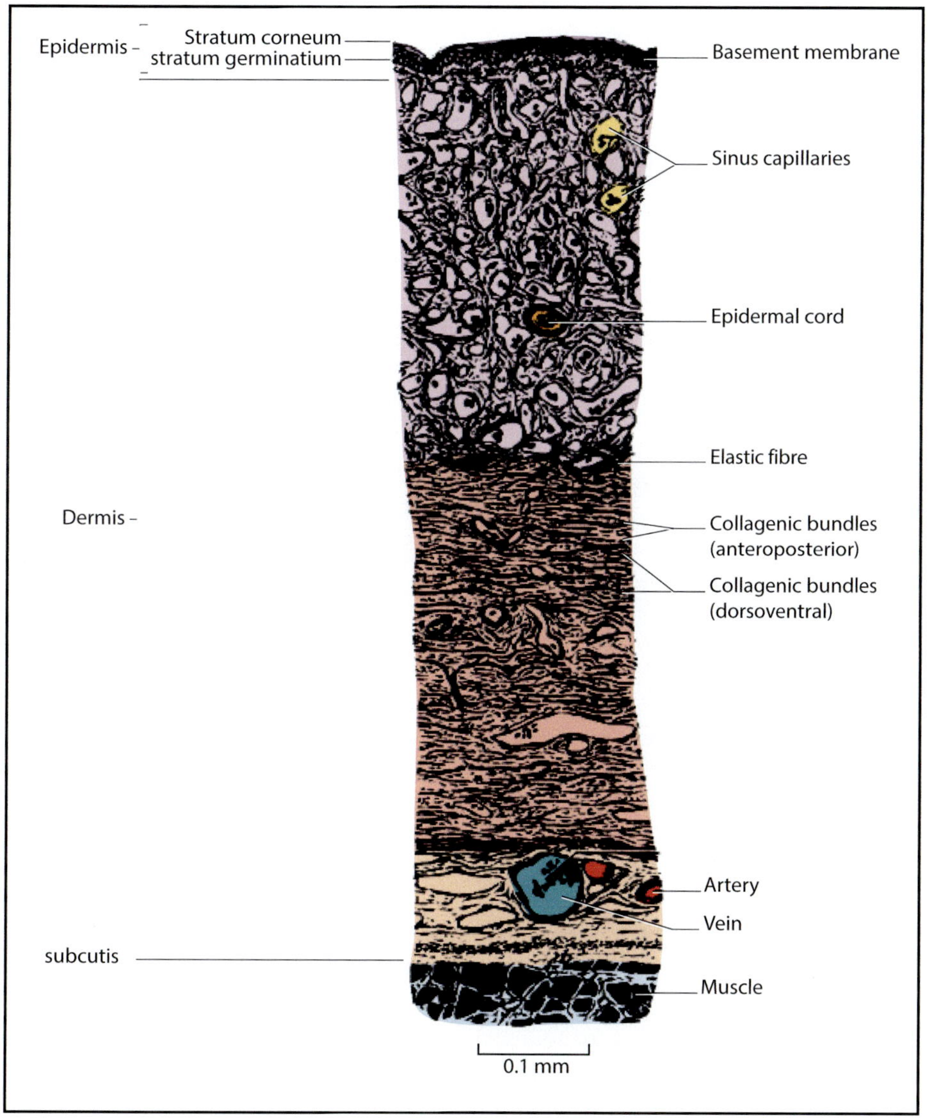

Figure 11.1 **Microscopic structure of avian skin.**

DOI: 10.1201/9781003272519-11

many pigeons, woodpeckers, and some parrots (*Amazona*, *Anodorhynchus*, and *Cyanospitta*), is a bi-lobed compound tubuloalveolar holocrine gland on the dorsum of the tail. It secretes a lipoid sebaceous material containing lipids, fatty acids, waxes, and 7-dehydrocholesterol (a vitamin D3 precursor). It is spread over feathers during grooming, which may assist in waterproofing, cleaning, and reducing skin infections. It is thought it may also have a role in sex identification; uropygial secretions may play a role in ultraviolet reflectivity, which in turn may be used by birds for sex differentiation. It also is likely to play a role in vitamin D3 synthesis.

Scales, raised areas of highly keratinised epidermis separated by folds of less keratinised skin, cover the non-feathered part of the leg, known as the podotheca (**Fig. 11.2**). The claws, which enclose the terminal phalanx of each digit, are made up of two plates: the strongly keratinised dorsal plate enclosing the dorsal and lateral aspects of the phalanx, and the softer ventral plate forming the sole of the claw. The dorsal plate grows faster than the ventral plate; therefore, the nails curve downwards.

The unique structure of avian skin is the feathers. These arise from feather follicles, arranged in tracts around the body known as pterylae. The featherless skin between these tracts is called apterylae. Each follicle is a cylindrical pit in the skin, lined with epidermis and dermis. At the base of the follicle is the dermal papilla, a small mound of dermis that enters the proximal shaft of the feather (the calamus) through a small hole known as the inferior umbilicus. The epidermis covering the dermal papilla is continuous with the calamus and with a thin layer of epidermis lining the follicle (**Fig. 11.3**).

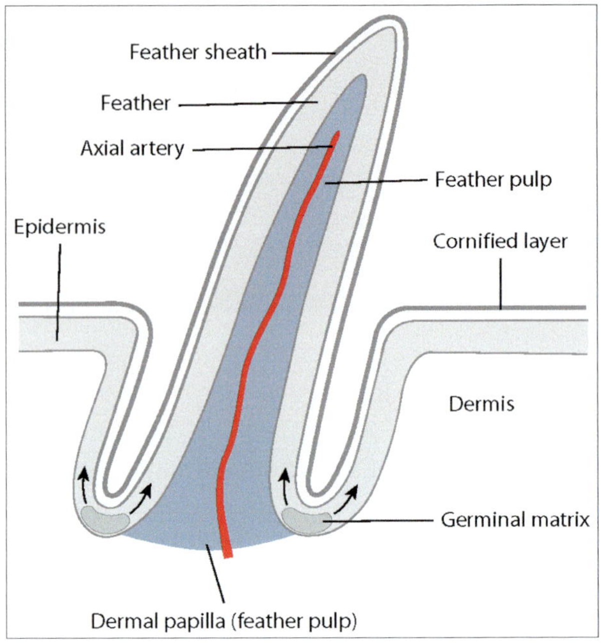

Figure 11.3 **The feather follicle.**

Each feather shaft consists of the calamus, embedded in the follicle, and the rachis, the main shaft beyond the calamus. They are distinguished by the distal (superior) umbilicus, a small opening into the shaft found at the junction of the rachis and calamus. Occasionally, there is an after feather (the hypopenna), a small extra feather on the rim of the distal umbilicus (**Fig. 11.4**).

Coming off the rachis are the barbs, and coming off the barbs are the barbules, filaments that interlock to form the vane (**Fig. 11.5**). This is the pennaceous region. The vanes are asymmetrical, with

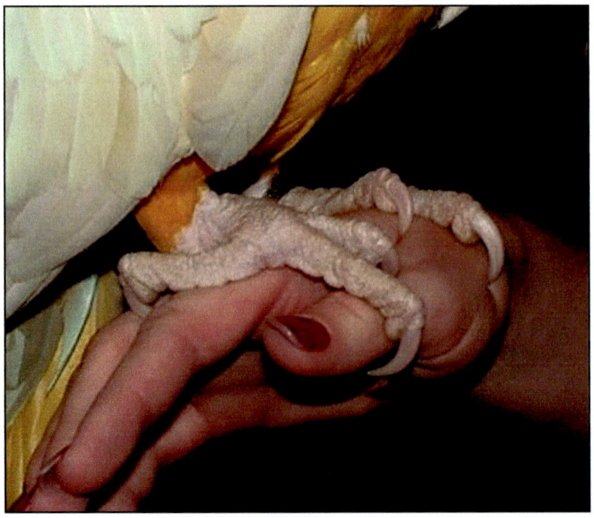

Figure 11.2 **The podotheca and curved nails of a lutino blue and gold macaw.**

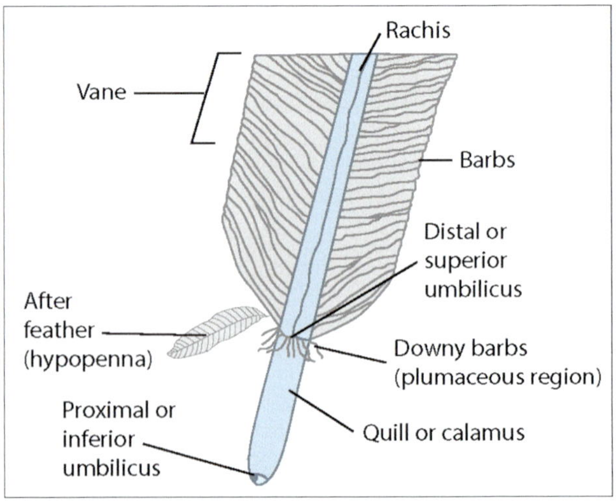

Figure 11.4 **The calamus region of a feather.**

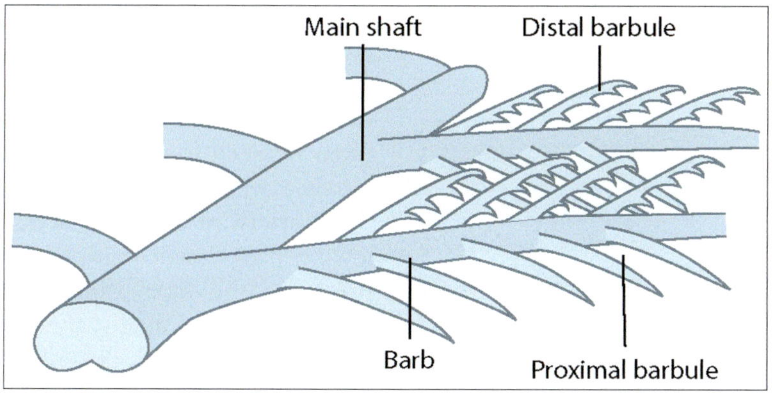

Figure 11.5 **Barbs and barbules.**

the external vane narrower than the internal vane. On the dorsal wing, the external vane of one feather overlaps the internal vane of the next. Just below the vane is the plumulaceous region, where a few downy barbs fail to interlock (**Fig. 11.6**).

Within the calamus of an immature feather is the pulp, a loose reticulum of mesoderm with an axial artery and vein. This pulp retracts as the feather matures, leaving pulp caps (empty chambers within the calamus).

There are seven types of feathers: contour, semiplume, down, powder down, hypopenna, filoplume, and bristle (**Fig. 11.7**).

- Contour feathers include the flight feathers and the body feathers. The flight feathers on the tail are called the retrices. The flight feathers on the wings are known as the remiges: the primaries (9–11) arise from the periosteum of the metacarpus; the secondaries (6–32) arise from the periosteum of the ulna; and the tertiaries arise from the humeral area. Overlying them are the coverts (**Fig. 11.8**).
- Semiplume feathers have a wholly fluffy vane, with the rachis longer than the barbs. They lie along pterylae margins, acting as insulation.
- Down feathers are also wholly fluffy, but the rachis is either absent or shorter than the longest barb. Distribution varies between species.
- Powder down feathers are structured like down feathers, although some are semi-plumes or contour feathers. They shed a fine waxy powder, which is made up of keratin flakes. This powder forms a waterproofing coat over the contour feathers and may play a role in keeping the bird clean. Powder down feathers are usually grouped in patches (e.g., on the thigh), although some species have them widely distributed. They are

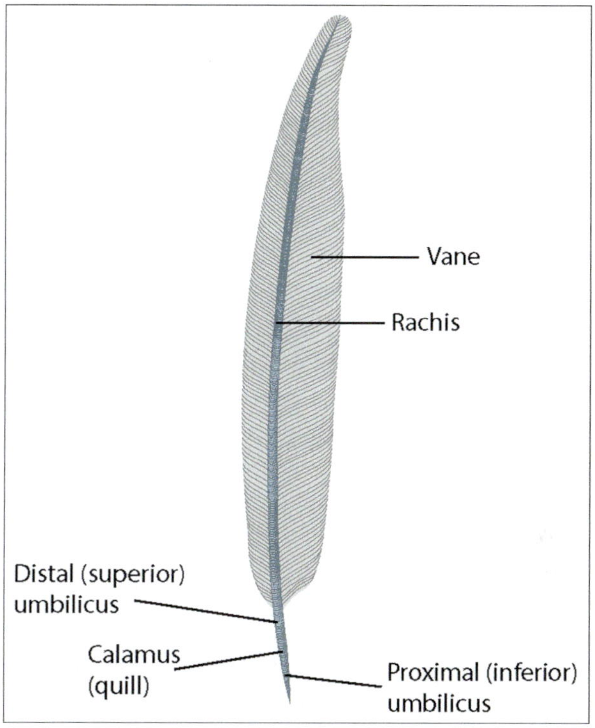

Figure 11.6 **Structure of a feather.**

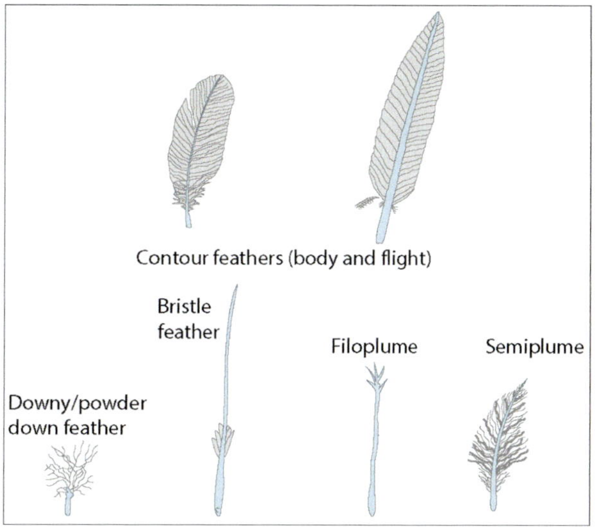

Figure 11.7 **Types of feathers.**

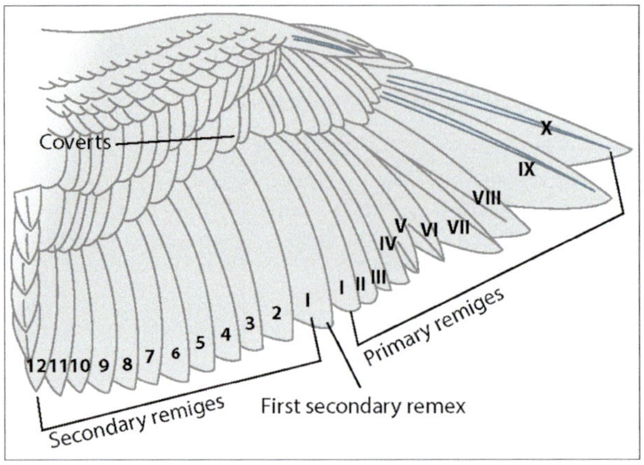

Figure 11.8 **Feathers of the wing.**

found in herons, parrots, toucans, pigeons, and bowerbirds.
- Hypopennae (1–5) are small feathers projecting from the distal umbilicus of pennaceous and plumaceous feathers (afterfeathers). They are usually not associated with retrices or longer remiges.
- Filoplumes have a long, fine shaft with a tuft of short barbs/barbules at the end. They possibly have a sensory/proprioceptive role and are found close to the follicles of contour feathers.
- Bristles have a stiff rachis, with either a few barbs at the proximal end or no barbs at all. They are found around the mouth, nares, and eyes, and possibly have a tactile function.

The colour of feathers is the result of the combination of pigments and feather structure. Carotenoids or psittacins (yellow pigments absorbed from the diet, including reds, oranges, and pinks) create the foreground colour. Melanins are the grey pigments (including black, grey, and brown) that create the background and also the foreground colour. Each feather barb has a cortex (an outer layer) containing either carotenoid pigments (psittacins) or melanin pigments. If melanin is in the cortex, it is known as foreground colour and produces black, greys, dark browns, and chestnut reds. This is the marking seen in many birds. The barb also has a medulla, which only ever contains melanin (background melanin). All of these pigments are distributed in different layers and, when combined with special features of barb structure that affect the passage of light, produce the spectrum of colours seen.

Moulting, the shedding of old, worn feathers and the renewal of plumage, is a regular event. It is controlled by a wide range of factors including thyroid activity, reproductive hormones, photoperiod, body condition, age, and diet. After a series of juvenile moults to attain adult plumage, most birds go on to moult one to two times annually. These moults, often referred to as the prenuptial and postnuptial moults, occur in spring and autumn, respectively. The pattern of moulting is orderly and in the following progression (with some overlap): the inner primaries, the outer primaries, the secondaries and tail feathers, and finally, the body contour feathers. It is usually bilaterally symmetrical and is paced so as to avoid loss of flight capacity at any time.

When it is time to moult an old feather, a proliferation of epidermal cells at the base of the follicle (the epidermal collar) separates the old feather from the dermal papilla and allows it to shed. These epidermal cells then start to group themselves into two series of spiral barb ridges. The tips of these ridges end along a longitudinal line on the ventral aspect of the feather (the seam). On the dorsal side of the feather, the epidermis thickens to form the rachis. Within this structure is the dermal core, consisting of the axial blood vessels, with mesoderm around them. As it grows, the feather emerges from the follicle as a pointed projection with a dermal core and an epidermal cover (sheath). This sheath then progressively ruptures, freeing the barbs that have separated along the seam and allowing the feather to open. Much of the increased grooming activity seen in birds at this time is to remove this sheath.

Because birds lack sweat glands, they rely on evaporative heat loss from the respiratory tract and heat transfer through apterylae (the featherless tracts of skin) to cool their bodies. To do this, many birds hold their feathers close to the body and may extend their wings, exposing the apterylae. Conversely, to retain body heat when ill or cold, they fluff their feathers up to trap body heat against the skin. (Ostriches do the reverse, i.e., they raise their feathers to promote heat loss and hold them close to conserve body heat.)

CONGENITAL DISORDERS

'Feather duster' or 'chrysanthemum' syndrome in budgerigars

Definition/overview

Juvenile budgerigars exhibit continued growth of flight, tail, and contour feathers, with the feathers

Figure 11.9 **A 'feather duster'.**

being soft and curled, giving the bird the appearance of a feather duster (**Fig. 11.9**).

Aetiology

This is believed to be a lethal recessive genetic disorder. There has been discussion about an association with budgerigar herpesvirus, but a 'cause and effect' has not been established.

Clinical presentation

Affected birds have excessively long and curled feathers; they are unable to fly and make barely audible noises. Most die within the first few years of life.

'Straw feather' in canaries

Definition/overview

Feathers fail to emerge from the feather sheath, giving them a straw-like appearance.

Aetiology

This is a lethal disorder believed to be hereditary in canaries.

Differential diagnosis

It must be differentiated from any condition that prevents the bird from grooming properly.

Feather cysts in canaries

Definition/overview

These appear as ingrown feathers forming hard, yellow nodules on the skin of canaries with complex feathering, such as the Norwich, Border, and Gloucester strains.

Aetiology

They may actually be benign neoplasms of the feather follicle. Several dermal papillae form in each follicle, resulting in a tangle of feathers that fail to erupt from the skin. It is believed to be a hereditary condition. Affected canary strains have been genetically selected to produce an extra down-type feather (double buff) that may predispose them to cyst formation. The mode of inheritance is not clear, and other factors (e.g., infections, trauma) may play a role in the development of the condition.

Clinical presentation

The dorsal thoracic area is a frequent site of multiple cysts, but they can occur anywhere on the body and wings. They often occur in several sites along a feather tract. The cysts appear as hard, yellow nodules containing trapped curling feathers. Badly affected birds have irregularly directed feathers all over their bodies. Actively growing feather cysts will have vascular walls and contain blood and gelatinous material. Mature cysts will contain drier keratinous material, and the cyst wall may be more expansive, thickened, and reduced in vascularity.

Management

Treatment of individual cysts can be conservative, simply lancing and expressing the contents of the cyst. Recurrence is common, but surgical excision of the affected follicle is generally curative. When multiple cysts are present, surgical excision of the affected tract may be required. Cysts on the wings may not be amenable to surgery without partial wing amputation. Therefore, conservative therapy may be warranted, as long as the owner is aware of the probability of an ongoing problem.

'Porcupine feathers' in homer and fantail pigeons

A similar condition to 'straw feather' in canaries.

Baldness in lutino cockatiels

A bald patch is present behind the crest feathers on the crown of the head in lutino cockatiels (**Fig. 11.10**). No treatment is required or available.

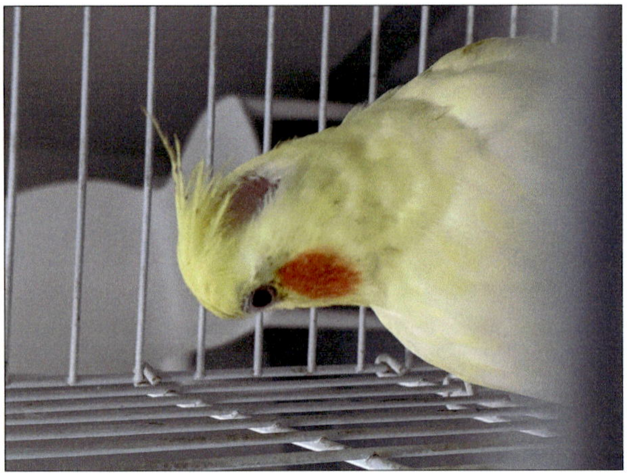

Figure 11.10 **Alopecia in a lutino cockatiel.**

NUTRITIONAL DISORDERS

Malnutrition may affect the skin and feathers in several ways.

Feather quality

- Brittle feathers.
- Abnormal moulting resulting in frayed or damaged feathers.
- Inability to preen, resulting in retained feather sheaths (**Fig. 11.11**).
- Stress bars are breaks in the feather vane due to brief episodes of dysfunction of the epidermal collar associated with the release of corticosteroid hormones during some stress. They can be associated with periods of malnutrition, especially during the weaning process (**Fig. 11.12**).

Feather colour

Colour in birds is the result of a combination of feather structure (affecting the passage or reflection of light), melanin pigments (black, grey, and brown), and carotenoid pigments (yellow and red). Nutrition will affect both the feather structure and the amount and type of carotenoids in the feather. Malnutrition can therefore produce some of the following effects on feather colour:

- Fading and dullness of plumage.
- Red feathering in African grey parrots.
- Abnormal colours in eclectus parrots (**Fig. 11.13**).
- Colour changes (e.g., green feathers turning yellow, blue feathers turning white).

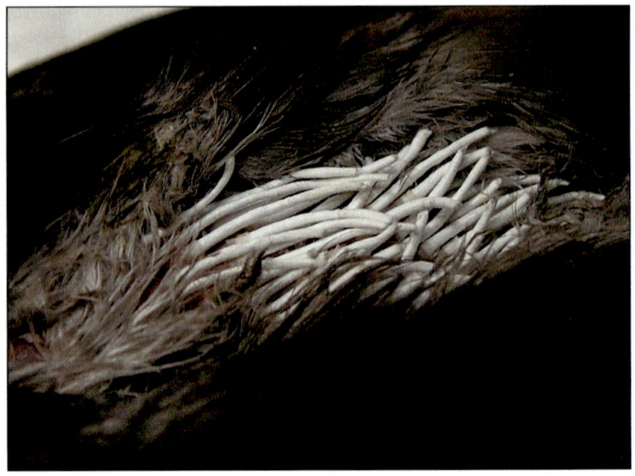

Figure 11.11 **Retained feather sheaths due to an inability to preen.**

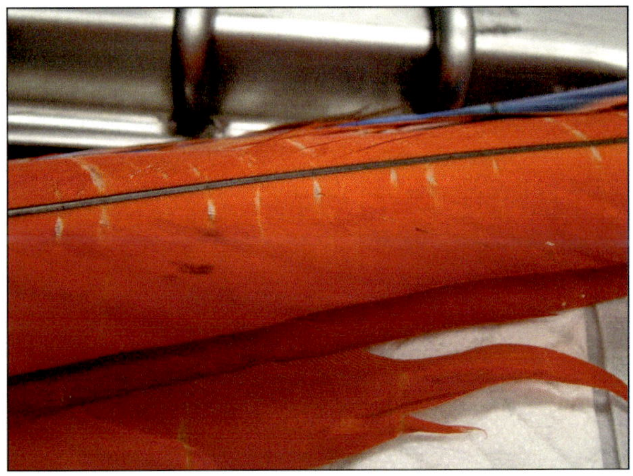

Figure 11.12 **Stress bars.**

Figure 11.13 **Abnormal colours in an eclectus hen.**

It must be remembered that other agents can also affect feather colour, notably genetics and viral diseases such as psittacine beak and feather disease (PBFD).

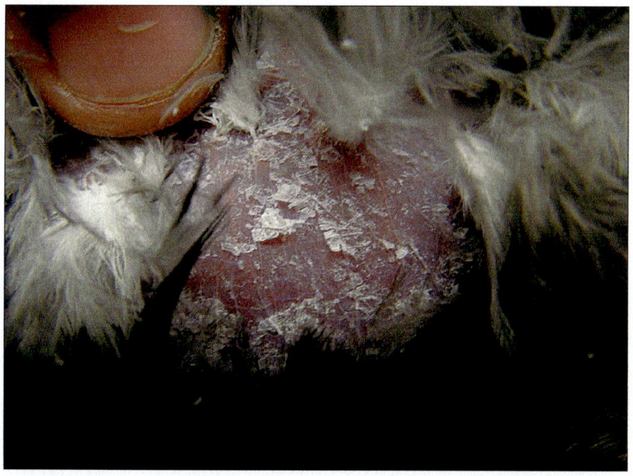

Figure 11.14 **Scaly skin associated with malnutrition.**

Skin changes

- Increased scaliness of the skin (**Fig. 11.14**).
- Subcutaneous fat deposits give the skin a yellowish hue.
- Increased skin fragility causes the skin to tear easily. A common example of this is 'tail split' injuries in cockatiels with poorly done wing clips. The combination of a heavy fall and fragile skin leads to a split in the skin between the vent and the tail.

Management

A dietary assessment to identify and correct nutritional deficiencies is an essential component of the evaluation of any dermatological problem in birds (see Chapter 2).

ENDOCRINE DISORDERS

Hypothyroidism

Clinical presentation

Poultry with thyroiditis demonstrate changes in feather quality and colour. Black, brown, and yellow feathers become red, longer, and more pointed, and have fewer pennaceous barbules than normal. A similar syndrome has been seen in galahs with suspected hypothyroidism. These birds develop long, narrow primary flight feathers and develop a pink-red discolouration of the grey plumage. At the same time, the normally pink feathers deepen in colour intensity (**Fig. 11.15**).

An obese scarlet macaw with confirmed hypothyroidism had non-pruritic feather loss, mild non-regenerative anaemia, mild leucocytosis, heterophilia, hypercholesterolaemia and sparse feathers, and it had not moulted in over a year. It responded well to thyroxine therapy.

Lutino cockatiels with suspected hypothyroidism may show deepening colour intensity and a loss of barbs and barbules in the feathers, giving the bird a 'greasy' deep yellow colour (**Fig. 11.16**). The rhinotheca is often overgrown. Results of liver function tests in these birds are usually normal, although cholesterol levels are often elevated.

Diagnosis

Hypothyroidism is often overdiagnosed. Diagnosis is difficult because normal resting T_4 levels in birds (5–15 ng/ml) are much lower than in mammals and

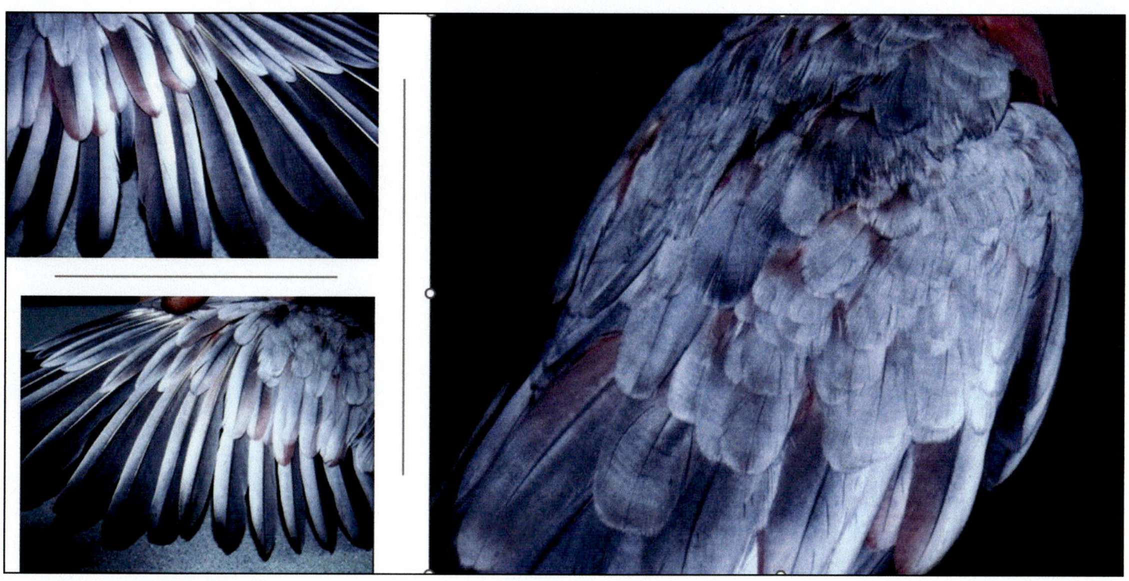

Figure 11.15 **A galah with hypothyroidism.**

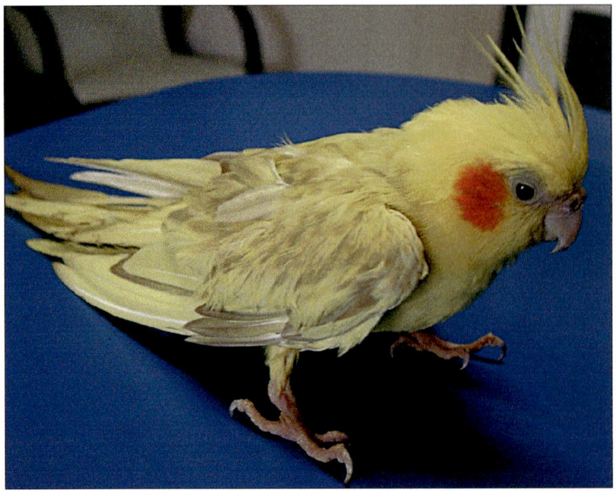

Figure 11.16 **A lutino cockatiel with suspected hypothyroidism.**

are often below the detectable limits of many laboratory techniques and equipment. Additionally, T_3 (0.5–4 ng/ml) may be as important as T_4 in overall thyroid function; therefore, measuring only T_4 may not present an accurate picture of the health of the thyroid gland. The half-life of both hormones is shorter in birds than in mammals, resulting in a significant diurnal variation in a normal bird. There is also a seasonal variation, with higher baseline levels in spring. Diagnosis of hypothyroidism therefore requires demonstration of a failure to respond to thyroid stimulating hormone (TSH); as avian TSH is not commercially available, mammalian TSH has to be used. After obtaining a baseline T_4, the bird is given 1 IU/kg of TSH and T_4 is determined 4–6 hours later. A positive response in a normal bird will see a 2.5-fold or greater increase of the resting T_4 level.

Supporting evidence for a diagnosis of hypothyroidism includes hypercholesterolaemia, elevated triglycerides and liver enzymes, and a mild non-regenerative anaemia. It has been suggested that, in larger birds, an endoscopic biopsy of the thyroid gland may lead to a diagnosis of hypothyroidism.

Management

Despite the lack of laboratory confirmation, some obese birds that demonstrate a lack of weight loss following a rigid diet, accompanied by poor quality feathers and infrequent moults, respond favourably to thyroxine therapy. Thyroid supplementation can be given orally at a starting dose of 0.01–0.02 mg/kg PO q12h. Note that no pharmacologic data exist to confirm the proper administration of thyroid hormone replacement in birds and that response to thyroid supplementation does not confirm the diagnosis.

Delayed moulting

Aetiology

Delayed moult is usually caused by a combination of malnutrition, abnormal diurnal rhythm (especially for companion birds kept indoors), concurrent illness, and endocrine disorders (e.g., hypothyroidism). Excessive egg production may cause abnormal or delayed moulting due to endocrinal effects on feather growth.

Clinical presentation

Birds present with untidy plumage, bald spots, and damaged, brittle feathers. The ends of the feathers often become frayed and lose their colour (**Fig. 11.17**). Other (non-dermatological) signs may include decreased vocalisation in male canaries, excessive or decreased egg production, and general lethargy.

Management

Normal diurnal rhythms should be re-established by ensuring the bird gets at least 12–14 hours of 'sleep time' in a darkened, quiet environment. Simply covering a cage in a busy family area is insufficient to meet the sleep requirements for most birds. Nutritional problems must be corrected.

Figure 11.17 **Worn feather tips in a bird with delayed moult.**

Egg production should be controlled through nutritional, environmental, social, and hormonal manipulation (see Chapter 27).

Trauma

Aetiology

A variety of conditions can result in traumatic skin damage, including self-trauma, sharp objects, and attacks by other animals.

Clinical presentation

Care must be taken to differentiate between a laceration and a normal apteryla. Avian skin sometimes appears transparent, giving the appearance of a wound with underlying structures visible. Most skin wounds will be obvious, with blood matted into the surrounding feathers.

Treatment

Some of these injuries are simple lacerations; others are more complex injuries associated with extensive soft tissue and even bone damage. Considerations when planning repair include:

- Avian skin is closely attached to underlying bony structures but less closely attached to muscle. Loose skin can be found on the neck and the inguinal area, leaving most of the skin on the torso relatively immobile. This makes undermining and mobilising skin to close wounds on the torso and limbs more difficult than similar surgery in many mammalian species.
- Avian skin has a fatty subcutaneous layer, but sutures in this layer often do not hold well.
- Postoperative swelling is not as severe in birds as it is in mammals, and sutures can therefore be tied more tightly and closer together.
- Avian heterophils lack lysozymes, meaning that avian pus is usually caseous and therefore does not drain from a contaminated wound readily. The use of drains in these wounds may not be of significant benefit.
- If primary closure is achieved, sutures can be left in place for 10–14 days. However, the thin epidermis of birds allows dehiscence of primary skin closure if primary healing has not occurred. If a significant scab occurs over the sutures, it is best not to remove them until the scab is ready to slough – this may take several weeks.

Skin laceration repairs in birds, therefore, require that the surgeon:

- Thoroughly debrides and flushes the wound. In most cases, the use of sterile saline is satisfactory as a flushing solution. Chlorhexidine and povidone iodine, even diluted, can be toxic to granulation tissue and should be avoided if possible.
- Mobilises available skin where possible to achieve primary closure without tension on the wound edges.
- Avoids the use of drains unless combined with a flushing or irrigation system.
- Uses minimally reactive absorbable sutures, placed closely together.
- Removes the sutures only when healing is complete.

Where primary closure cannot be achieved, either skin grafts/flaps or healing by second intention is required. Using the same surgical techniques employed in mammals such as dogs and horses, skin grafting and flaps are feasible alternatives in birds. If skin grafting is not employed, good healing can often be achieved by secondary intention. In this situation, avoid the underlying subcutis and muscle drying out by the use of wet or hydrocolloid dressings (e.g., Duoderm®) (**Fig. 11.18**).

Aggressive postoperative antibiosis and analgesia will significantly improve morbidity and mortality rates in these patients. In most cases, if the area is less painful after the surgery, the bird will usually not bother the surgery site, and an Elizabethan collar is unnecessary. These collars are stressful to the bird and reduce its ability to move, eat, and drink normally; they should be used only when necessary.

BACTERIAL INFECTIONS

Definition/overview

Staphylococcus spp. are suspected of being the most common bacterial skin pathogens. They may result in generalised skin infections, which may appear as folliculitis or dermatitis. Localised or multifocal swellings may be abscesses from infected wounds, damaged feather follicles or foreign bodies, or *Mycobacterium* infection.

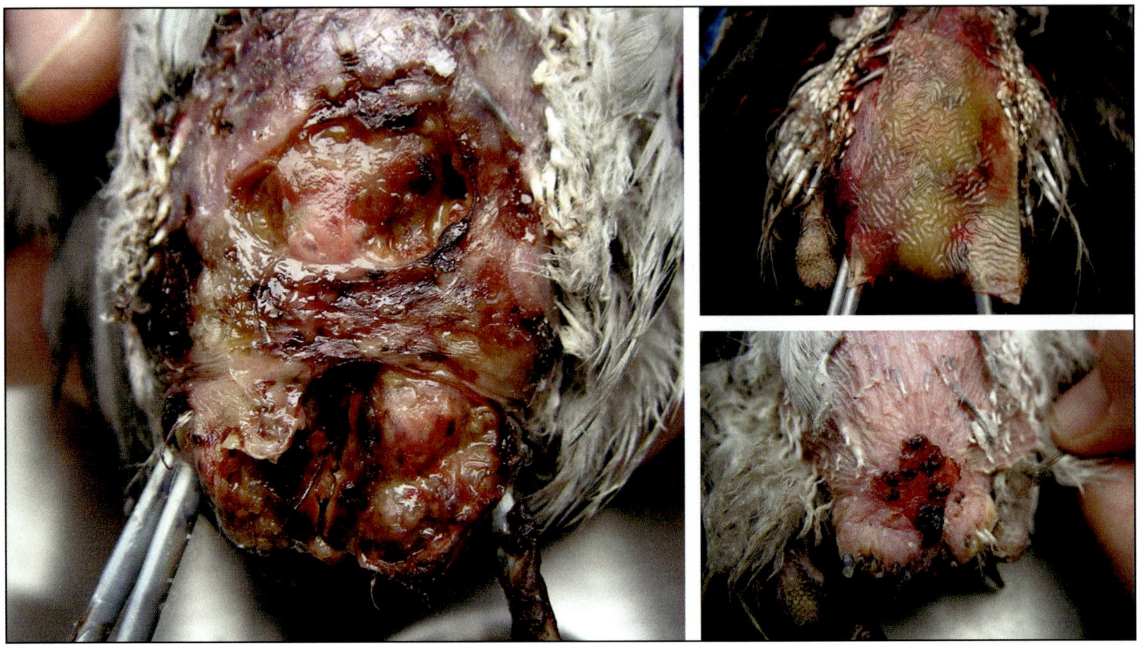

Figure 11.18 **Wound healing by second intention, using a hydrocolloid dressing, showing the improvement over 14 days.**

Clinical presentation

In generalised *Staphylococcus* infections, the skin is usually very pruritic and often erythematous (**Fig. 11.19**). *Mycobacterium* infection causes localised or multifocal lesions that may be wart-like, dry flaky swellings of the skin, granulomatous skin lesions, or raised ulcers.

Diagnosis

Diagnosis requires a skin biopsy (including follicles) and culture. MRSA is becoming a more recognised entity in these cases.

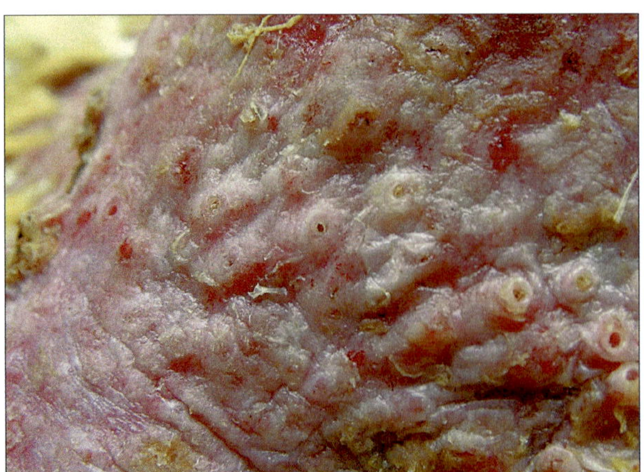

Figure 11.19 **Staphylococcal dermatitis and folliculitis.**

Management

Systemic antibiotic therapy is given as indicated by culture. Washing the bird two to three times weekly with benzyl peroxide or chlorhexidine shampoos can be beneficial by removing a lot of the scales and debris on the skin and reducing the bacterial load. Surgical excision of localised lesions may be curative.

The zoonotic potential of *Mycobacterium* and MRSA must be discussed with the owner before attempting treatment (see Chapter 19).

FUNGAL INFECTIONS

Aetiology

Common fungal pathogens of the avian skin include *Candida albicans*, *Malassezia pachydermatis*, dermatophytes (*Microsporum gallinae*, *M. gypseum*, *Trichophyton verrucosum*, other *Trichophyton* spp.), *Cryptococcus bacillisporus* (formerly *C. neoformans* var. *gattii*) and *C. neoformans* var. *grubii* (formerly *C. neoformans* var. *neoformans* serotype A), and *Aspergillus* spp.

Clinical presentation

Candida causes lesions around the commissures of the mouth and nares and occasionally around feather follicles on the head, back, and ventral abdomen.

Figure 11.20 **Favus in a chicken.**

Malassezia results in generalised pruritus, sometimes with crusting and folliculitis.

Dermatophytes are a slow-spreading infection causing scabs, crusts, and alopecia on the body and thin-skinned areas of the head and upper beak, and a rough, porous appearance of the podotheca. In gallinaceous birds, the characteristic scaly, crusty lesions of the wattle, comb, and legs are known as 'favus' (**Fig. 11.20**).

Cryptococcus has been reported to cause subcutaneous nodules on the head and body and raised nodular lesions on the face and beak.

Cutaneous aspergillosis has been associated with focal ulcerative pruritic lesions.

Diagnosis

Cytology must be used with caution: feather dust stains similarly to yeast organisms and can be nearly impossible to distinguish from yeast bodies. Biopsy and fungal culture may be required to confirm the diagnosis.

Management

Systemic antifungals (itraconazole, fluconazole, ketoconazole, terbinafine) are given orally for 1–3 months. Topical treatment may be carried out using enilconazole wash, clotrimazole wash or cream, or miconazole shampoo.

VIRAL INFECTIONS

Viral infections can cause dystrophic feathers (beak and feather disease virus, polyomavirus, adenovirus, and parvovirus [waterfowl]) or skin lesions (poxvirus, papillomavirus, and herpesvirus).

Psittacine beak and feather disease

Aetiology

Beak and feather disease virus (BFDV), a circovirus, is a non-enveloped, single-stranded DNA virus measuring 14–17 nm.

Pathogenesis

The virus has a minimum incubation period of 21–25 days, but it could be as long as several years. It is shed in faeces, crop secretions, and feather dust (vertical transmission is suspected but not confirmed). Following infection, the virus is disseminated haematogenously, localising in the feather follicle epithelium, the thymus, and the bursa. In cockatoos, it also localises in the growth areas of the beak. The result is feather dystrophy, immunosuppression, and, in cockatoos, overgrowth and underrunning of the rhamphotheca.

Clinical presentation

All parrots are susceptible, but New World parrots and cockatiels appear to be rarely affected. PBFD is more common in juveniles than in adults, but naïve adults are susceptible. There are two forms of PBFD: acute and chronic.

Acute PBFD is seen in juveniles (especially grey parrots) around weaning age. They are lethargic, fluffed, and anorexic. Haematology may show pancytopenia and non-regenerative anaemia. Affected birds often die (within a few days) with severe hepatic necrosis before feather abnormalities develop. Feather lesions have been noted in fledgling birds at 28–32 days old.

Chronic PBFD causes progressive replacement of normal feathers with dystrophic feathers (retained sheaths and blood supply, clubbed appearance, stress lines, constrictions, and abnormal shapes) (**Fig. 11.21**). The degree and location of the feather loss may depend on the state of moult when the bird was initially infected. Typically, lesions develop in the order of powder down, contour, primaries, secondaries, tail, and then the crest. Poicephalus species and lories may only lose tail feathers and primary flight feathers (**Fig. 11.22**). These feathers may then regrow. Neophemas may develop untidy plumage and lose feathers easily when handled. Many parrots

Figure 11.21 **Chronic PBFD, as seen in cockatoos, showing feather loss and dystrophies, as well as beak changes.**

Figure 11.22 **PBFD in a lorikeet, showing only the loss of the remiges and retrices.**

develop feather colour changes: blue feathers become white, and green feathers become yellow (**Fig. 11.23**). Beak lesions, including palatine necrosis, ulceration, elongation, and easily fractured beaks, are seen only in cockatoos. Immunosuppression is common in all species (**Fig. 11.21**).

Diagnosis

Histopathology

Basophilic intracytoplasmic inclusion bodies found in feather follicles and the cloacal bursa are considered diagnostic (**Fig. 11.24**). Care must be taken not to confuse them with herpesvirus or adenovirus. Feathers and skin show multifocal necrosis of epidermal cells, epidermal hyperplasia, and epidermal hyperkeratosis. Diffuse necrosis of epidermal cells is seen throughout the epidermal collar and in the basal and intermediate layers of developing feathers. Beak histopathology shows hyperkeratosis and separation of the cornified outer layer from underlying tissues and bones. Atrophy and focal aggregation of necrotic cells are seen in the thymus and bursa.

Haemagglutination (HA) and hemagglutination inhibition (HI)

Haemagglutination (HA) is the agglutination of washed erythrocytes induced by the presence of a

Figure 11.23 **PBFD in a king parrot (left) and a budgerigar (right).**

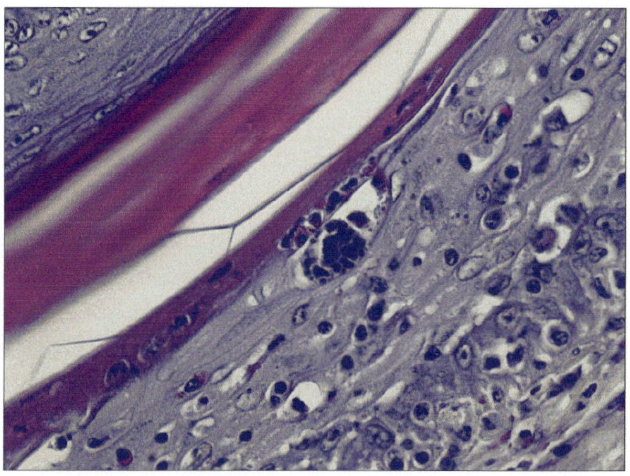

Figure 11.24 **Basophilic intracytoplasmic inclusion bodies found in feather follicles.**

virus in the sample. The result, expressed as a titre, is the highest dilution of the erythrocytes that causes agglutination. The higher the titre, the higher the virus excretion in the patient.

Haemagglutination inhibition (HI) is the inhibition of this agglutination due to the presence of an antibody. The titre is the highest dilution that still inhibits agglutination, i.e., the higher the titre, the higher the level of antibodies. A positive HI titre confirms previous exposure to BFDV infection.

- Birds with chronic PBFD typically have a high HA titre and a low HI titre.
- Birds with high HI titres are generally free of the disease.

PCR

Viral-specific PCR tests are the most sensitive tests for detecting BFDV infection, but they give no indication of whether the bird is latently infected or transiently viraemic, as non-replicating viral DNA may take up to 3 months to clear from the blood. Therefore, PCR should be used with HA/HI, rather than replacing it. PCR can be used alongside viral DNA sequencing and other genotyping methods to determine the source of infection.

If HA/HI is not available, clinically normal birds testing positive on PCR should be retested after 3 months.

Management

At this time, supportive care (e.g., treating secondary infections and providing a good diet) is all that can be done for a bird with PBFD. Although research has been conducted, there is no commercially available vaccine at this time. As this disease is highly contagious, owners should be advised to isolate affected birds from other birds. Companion birds can often live several comfortable years in a single-bird household. In an aviary or multibird household, euthanasia may have to be considered.

Prognosis

Some birds (e.g., lorikeets) appear to mount an effective immune response and apparently recover. However, many of these birds become inapparent carriers. Some birds, despite showing clinical signs, may live for 10–30 years, but most infected birds die within 2 years from secondary diseases related to immunosuppression.

Avian polyomavirus

Aetiology

Polyomavirus is a non-enveloped virus that is relatively environmentally stable. All parrots can be

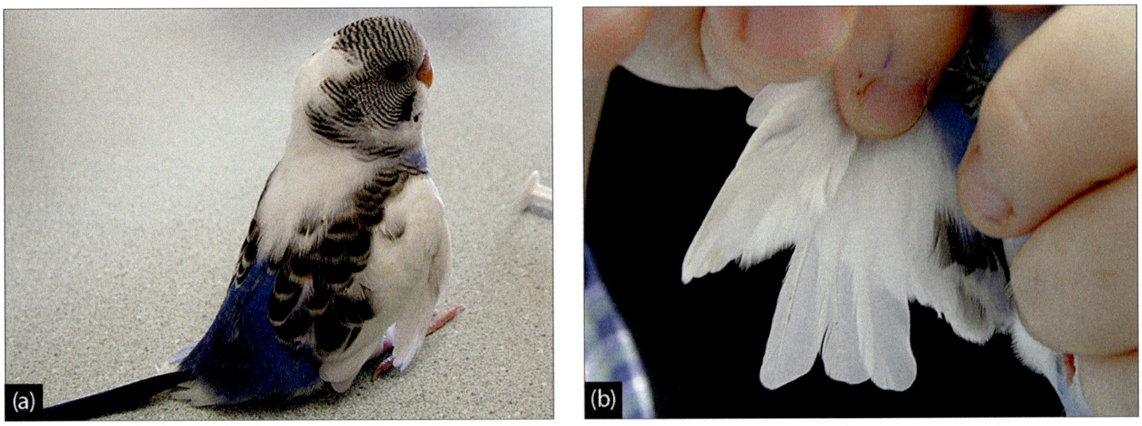

Figure 11.25 **Budgerigar with avian polyomavirus (French moult) (a) showing feather loss on the wings and tail (b).**

affected, but it is most common in macaws, conures, eclectus parrots, and caiques. It is rare in African grey parrots, cockatoos, and cockatiels. Passeriformes, including grass finches, canaries, goldfinches, and greenfinches, can also be infected.

Clinical presentation

In budgerigars (the primary reservoir of polyomavirus), signs include decreased hatchability and embryonic death, abdominal distension, subcutaneous haemorrhages, and feather dysplasia that may resolve after several months (primary and secondary feathers on the wings and tail are lost or fail to erupt – in the budgerigar fancy, these birds are known as runners and the condition is known as French moult) (**Fig. 11.25**). There may be neurological signs if the cerebellum is affected.

In other parrots, the virus typically causes sudden death (especially those <15 days old) 10–14 days after exposure. In larger parrots, most deaths occur between 20 and 140 days old.

In estrilid finches (grass finches), the virus causes acute mortality in 2–3-day-old fledglings, young adults, and mature finches. Survivors may demonstrate poor feather development and misshapen lower beaks.

More details on this virus can be found in Chapter 29.

Avian poxvirus

Aetiology

Avian poxvirus is a large (up to 400 nm), enveloped DNA virus. All avipoxviruses are morphologically similar, but they have differing host specificity.

Pathogenesis

The virus is environmentally stable and can survive for years in dried organic debris. The virus must enter the body through mucous membranes or abraded skin. It is unable to penetrate intact epithelium. It can be transmitted directly through fighting, feather-damaging behaviour, or preening, or it can be indirectly transmitted by blood-sucking insects (e.g., mosquitoes). The virus will either remain at the point of entry, causing localised infection, or spread haematogenously to the liver and bone marrow, producing a systemic infection. The incubation period is 7–9 days in pigeons and 4 days to 3 weeks in canaries.

Clinical presentation

There are four different syndromes:

- Dry pox: discrete scabby lesions on non-feathered parts of the body (e.g., combs, wattles, beaks, eyelids)
- Wet pox: fibronecrotic diphtheritic lesions in the oropharynx
- Septicaemic form in canaries
- Viral-induced neoplastic lesions in the skin and lungs

The presenting syndrome is determined by the strain of the virus, the mode of transmission, and the age, species, and health of the infected bird. Lesions may become infected and painful and may interfere with eating, respiration, and vision.

Diagnosis

Intracytoplasmic inclusion bodies, known as Bollinger bodies, are considered pathognomonic. Virus isolation and PCR can be utilised.

Management
Most localised lesions are self-limiting, healing in 3–4 weeks (although diphtheritic lesions may persist for several months) and producing immunity for 6–12 months or longer. If lesions are infected, antibiotics and gentle cleansing are indicated. Forceful removal of scabs may result in scarring and deformity.

Control revolves around the isolation of affected birds, minimising fighting, and preventing access of biting insects. Sodium hypochlorite is an effective disinfectant. Attenuated live vaccines are available for pigeons and poultry. The fowl pox vaccine is effective in ostriches. There is also a canary pox vaccine available.

Papillomavirus
Aetiology
Papillomavirus is a double-stranded non-enveloped DNA virus.

Clinical presentation
Cutaneous papillomas (small fleshy pedunculated masses originating primarily from featherless areas such as the feet and face) are seen in African grey parrots, canaries, and finches. Their effects are usually simply mechanical, and so the clinical signs are determined by their location.

This is a different disease syndrome from internal papilloma disease seen in many parrots (see Chapter 19).

Diagnosis
Koilocytosis (large dense cytoplasmic perinuclear space with irregular edge) is the hallmark of papillomaviruses. Virus isolation is the gold standard, but papillomavirus viral particles may be rare or absent. Western blot immunoassay and immunohistology are ideal, along with standard histology.

Management
Radiosurgery, cryosurgery, and laser surgery have all been used to remove the papillomas and, possibly, stimulate an immune response to delay or prevent recurrence.

PARASITIC INFECTIONS

Mites
Mites are the most important and commonly encountered external parasite of birds. Adult mites have eight legs; the larval stage has six legs.

Cnemidocoptes pilae
See Chapter 12.

Dermanyssus gallinae (red or roost mites)
This mite lives in the environment, where it can survive long periods without feeding. It takes blood meals at night and leaves the bird during the day. It causes skin irritation, scaliness, and anaemia (especially in juveniles or small birds). Heavy infestations may kill chicks in the nest box.

Ornithonyssus spp. (fowl mite)
These host-dependent, blood-sucking mites remain on the bird, where they congregate at the eyelids and vent. They may cause anaemia, pruritis, and damaged plumage. They may also transmit *Borrelia* and *Lankesterella*.

Epidermoptic mites
This group includes *Myialges* spp., *Epidermotes* spp., and *Microlichus* spp. They may be carried mechanically by the hippoboscid fly. They cause depluming dermatitis with scale formation in passerines.

Quill mites
This group includes *Syringophilus* spp., *Dermoglyphus* spp., *Pterolichus* spp., *Analges* spp., and *Harporhynchus* spp. (**Fig. 11.26**). Their entire life cycle is spent on the host. They can be found by examining the pulp material within a developing or damaged feather. They cause partial or complete loss of the feather.

Figure 11.26 **Harporhynchus mite 'nests' in a wild lorikeet.**

Figure 11.27 **Lice (unknown species) on the primary feathers of a budgerigar.**

Lice

Definition/overview

Lice are six-legged, wingless, crawling external parasites with chewing mouthparts (**Fig. 11.27**). Sucking lice do not infect birds. They are species-specific. They may serve as a vector for some contagious diseases (e.g., eastern equine encephalitis).

Species

There are many hundreds of species of lice found on birds. Just a few examples include:

- Biting lice of parrots include *Neopsittaconirmus*, *Psittaconirmus*, *Eomenopon*, and *Pacifimenopon*.
- Biting lice of pigeons include the Philopteridae and Menoponidae families.
- Biting lice of raptors include *Mallophaga* spp.

Life cycle

Lice usually live on just one host. Eggs are glued to feather shafts and hatch in 4–7 days. Nymphs go through three moults before maturing.

Clinical presentation

Lice feed on skin scales, feather debris, and hair, causing pruritus and poor feather quality.

Fleas

The stickfast flea, *Echidnophaga gallinacea*, is a blood-sucking ectoparasite. It occasionally attaches to the skin of the head of non-poultry birds and causes irritation and blood loss, resulting in depression, anaemia and pruritis on the skin of the head.

Flies

Flies that cause problems with aviary and companion birds include *Pseudolynchia canariensis* (the pigeon fly) and the hippoboscid fly (louse fly, parrot fly). Blood-sucking flies are found on many species of birds. They seldom leave the host but are not host-specific. They lay their eggs off the host, in a secluded spot. They also act as vectors of Haemoproteus.

Ticks

Argas persicus, the fowl tick, is a soft-bodied tick that feeds for 3–7 days on the bird before dropping off. It normally lives in the environment. *Ixodes holocyclus and I. cornuatus*, hard-bodied ticks, cause ascending paralysis.

Management

Sprays and washes can be effective. Natural and synthetic pyrethrins are effective and generally safe. Organophosphate washes are not recommended because of the high incidence of adverse effects. Ivermectin and moxidectin are effective against blood-sucking and keratin-feeding parasites.

NEOPLASTIC AND PSEUDONEOPLASTIC CONDITIONS

Lipomas

Incidence

Lipomas can be seen in any bird but are most common in Amazons, budgerigars, lovebirds, Quaker parrots, and galahs. They usually occur in older birds; there does not appear to be a sex predilection. They are associated with an exclusively, or at least predominantly, all-seed diet, combined with a sedentary lifestyle.

Clinical presentation

Masses are usually located along the main fat deposits on the body (around the crop, the ventral abdomen, thighs, and down the legs). They are occasionally found around the cloaca. They are rarely seen on the back or the wings. These are firm masses, often embedded in subcutaneous fat, with the overlying skin often yellow in colour. The masses may be mobile. Birds are occasionally presented for bleeding due to the lipoma ulcerating.

There is a strong correlation between the diagnosis of a lipoma and the presence of hepatic lipidosis. This correlation needs to be kept in mind when evaluating treatment options for a patient.

Diagnosis

Diagnosis is based on the history and clinical appearance. Lipomas must be distinguished from hernias and xanthomas. Radiography can be useful for differentiating lipomas from hernias.

Management

Lipomas are often an indication of extensive underlying disease processes, including cardiovascular disease, hepatic lipidosis, hypothyroidism, and occasionally, diabetes mellitus. Before starting any treatment, it is important that this information is communicated to the owner and the patient's overall health (as a whole) is determined.

Surgical excision, while curative, is often not the first treatment of choice. The author's preference is to only excise ulcerating lipomas or those that are mechanically interfering with the bird's lifestyle (e.g., interfering with the normal range of motion of the leg). If surgery is decided upon, great care must be taken to monitor for coagulopathies, often not apparent until the postoperative period. In the case of ulcerating lipomas, where surgery is required urgently, fresh whole blood transfusions may be required during and after the procedure (see Chapter 5).

If the lipoma is not ulcerating, the bird should be converted to a formulated diet and its weight monitored. In many cases, the lipoma will shrink to a size where it is either no longer clinically significant (and therefore no further treatment is required) or, if surgery is indicated, it becomes a simpler procedure with fewer postoperative complications.

Some clinicians report that treatment with L-levothyroxine is beneficial in shrinking lipomas. If the decision is made to use this therapy, care must be taken to monitor for adverse side effects such as tachycardia and a hypermetabolic state.

Other neoplasms

Fibromas/fibrosarcomas can occur anywhere on the body. They may be subcutaneous or proliferative with a nodular, red appearance. They tend to be locally invasive and recurrent. Surgical excision has been followed by both radiation and chemotherapy with some success.

Squamous cell carcinomas also may occur anywhere on the body, being most prevalent at mucocutaneous junctions of the head, on the distal wing, and on the phalanges. They also tend to be locally invasive, and radiation therapy has been attempted with some success.

Soft tissue sarcomas, including haemangiomas and haemangiosarcomas, have been reported in several locations in birds.

Lymphosarcoma will occasionally present as hyperkeratotic, alopecic areas of skin, often on the head, or nodular dermal masses. Chemotherapy has been used successfully in some cases.

Xanthomas

Incidence

Xanthomas are locally invasive benign masses consisting of foamy, lipid-laden macrophages, multinucleated giant cells, inflammatory cells, and cholesterol crystals. They can be seen in any bird at any age. They are considered by some authors to be a metabolic problem (hypercholesterolaemia) combined with an inflammatory response following localised trauma and haemorrhage, with resultant deposition of cholesterol triggering an inflammatory response.

Clinical presentation

Xanthomas are generally friable, yellow-coloured, fatty-appearing masses that may be located anywhere on the body, but are often seen on the distal wing, ventral abdomen, and sternum. Rarely, they can be seen in internal organs such as the bones, brain, and liver. Lesions may be pruritic and are often associated with other pathology such as lipomas, hernias, and sites of chronic irritation, e.g., wing tip trauma. Occasionally they will bleed due to repeated self-trauma.

There appears to be a strong correlation between xanthomas and atherosclerosis (a xanthomatous chronic inflammatory vascular disease), a correlation that should be investigated in patients about to undergo treatment.

Diagnosis

Diagnosis is made on the physical appearance of the mass and confirmed by biopsy. Histopathology reveals masses of foamy macrophages and multinucleated giant cells associated with cholesterol clefts. Many birds will have concurrent hypercholesterolaemia, although this is not consistent.

Management

Surgical excision is often difficult or not feasible due to the widespread nature of the lesion, the friability of the tissue and overlying skin, and the vascularity of the lesion, which requires good haemostasis. Severe painful xanthomatosis on the distal wings may require amputation.

Medical therapy involves treating the underlying causes of hypercholesterolaemia (e.g., diet, hepatopathy, hypothyroidism), using NSAIDs (e.g., meloxicam) to control the inflammatory response, and prevention of further trauma to the area. Some may respond to thyroxine supplementation.

TOXIC CONDITIONS

Fenbendazole can damage feathers if given during new feather growth. The feathers that develop during this period appear dysplastic and brittle, breaking easily.

Silver sulphadiazine, used as a topical antibacterial cream, has been reported to cause argyria (e.g., silver poisoning) after excessively prolonged application. It causes a bluish discolouration of the skin, features, and mucous membranes.

CONDITIONS AFFECTING THE UROPYGIAL GLAND

Common conditions affecting the uropygial gland include neoplasia, impaction, and rupture. The gland should always be palpated during a physical examination, noting that it is not always present. In grey parrots (*Psittacus erithacus*) the normal gland is very large – easily mistaken for an abnormality.

Neoplasia of the uropygial gland

Incidence

These tumours are seen in those parrots that possess a uropygial gland (**Fig. 11.28**). They are most often seen in budgerigars, but any species can be affected.

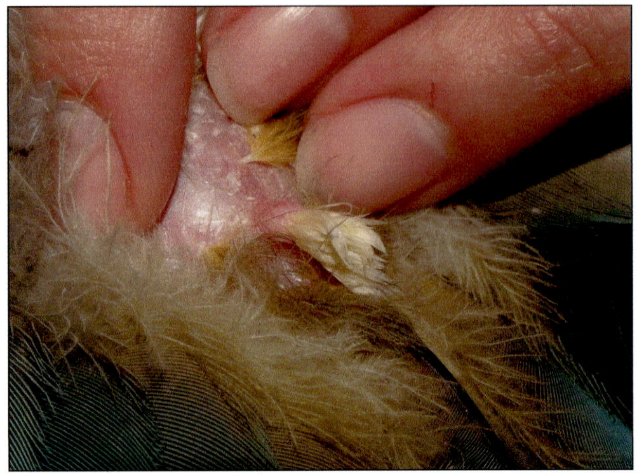

Figure 11.28 **Normal preen gland in a macaw.**

Most tumours are adenomas, carcinomas, and squamous cell carcinomas.

Clinical presentation

There is abnormal swelling of the uropygial gland (**Fig. 11.29**). It will occasionally present as an ulcerated mass with bleeding from the dorsal tail area.

Diagnosis

Tumours must be distinguished from adenitis, hypovitaminosis A with glandular metaplasia and hyperkeratosis, abscessation, and impactions. Biopsy is the best means of achieving a diagnosis.

Management

Management of a neoplastic preen gland can be difficult.

Figure 11.29 **Preen gland carcinoma in a budgerigar.**

- Surgical excision can be achieved in the early stages, but these masses are often very vascular and extensive by the time they are noticed by the owner.
- Tail amputation can be performed but with a guarded prognosis.
- Systemic chemotherapy appears to have little success in treating this condition. Electrochemotherapy (combining electroporation with localised injection of bleomycin, carboplatin, or cisplatin) is a more promising treatment.
- Irradiation with strontium-90 has shown promise as an effective therapy. Surgical debulking may be necessary in some cases. Response time after therapy is directly related to the mitotic activity of the neoplasm (the higher the mitotic rate, the faster the response time).

Impaction

Impaction is associated with hypovitaminosis A, infection, neoplasia, or trauma. The gland, either bilaterally or unilaterally, becomes distended with a caseous-like secretion.

Clinical signs

Many birds show minimal discomfort or pain until the gland becomes quite large. The affected lobes can be very distended or even cystic in appearance.

Treatment

Hot compresses and careful, gentle expression of the gland can temporarily relieve the impaction, but the underlying causes need to be addressed.

In severe cases, or those that cannot be treated, surgical ablation of part (or the whole) of the gland may be required.

Infection and abscesses

Untreated impactions can become infected and go on to form abscesses. These usually are painful and erythematous and may be ruptured on presentation. Surgical ablation of the gland may be required if the abscess fails to respond to debridement, flushing, antibiotics, and NSAIDs.

IATROGENIC TRAUMA

The most common iatrogenic trauma seen in companion birds is that resulting from an excessive wing trim. One-winged trims and/or trimming secondary and tertiary flight feathers reduce the bird's ability to slow down while in flight and control its descent. The result is a heavy fall. Repeated (or even single) heavy falls can result in broken 'blood feathers' (newly erupted feathers with a rich blood supply in the shaft), bruising to the cranial sternum that leads the bird to mutilate the area, often resulting in an ulceration in the skin and muscle, sometimes with an underlying osteomyelitis, and tail splits (splits in the skin between the vent and the tail).

Treatment must be aimed at treating the existing trauma while preventing further trauma. Badly broken feathers may need to be removed. (If more than one feather needs to be removed, general anaesthesia and analgesia are appropriate.) If the feather(s) are not badly broken, it may be possible to leave them *in situ* if the haemorrhage has stopped. Sternal trauma, if mild, may respond to antibiotic and analgesic therapy while preventing the bird from further damaging the area (Elizabethan collar and/or a neck brace). If osteomyelitis is present or the ulceration is deep, surgical debridement may be necessary. Tail splits can be debrided and sutured.

Further trauma can be reduced by restoring flight. This can be achieved by 'imping' new primary feathers onto the stumps of those that have been trimmed and/or removing the stumps of trimmed feathers (under anaesthesia) to encourage the regrowth of new feathers. The bird should be confined in a large cage and not allowed free flight until it is able to fly safely. The perches in the cage should be lowered and the floor padded with soft towels to prevent heavy falls and hard landings within the cage.

MISCELLANEOUS CONDITIONS

Polyfolliculosis/polyfolliculitis

Definition/overview

This is a chronic pruritic condition that can cause multiple short feathers to erupt from one follicle. It is most common in budgerigars and lovebirds (*Agapornis* spp.).

Aetiology

The aetiology remains unclear, but some feel it may be viral or genetic in origin. There is some discussion as to whether this condition is the cause of the bird's intense pruritus or is the result of damage done by the bird in response to the pruritus.

Clinical presentation

Affected birds are presented for pruritus predominantly over the back and neck. Newly emerging feathers have short, stout quills with retained sheaths (**Fig. 11.30**). There appear to be several feathers emerging from a single, enlarged follicle. Histologically, there is the appearance of multiple short feather shafts growing from the same follicle, with a thin layer of epidermis separating the shafts. Chronic inflammation occurs beneath the pulp cap and the feather sheath is thickened.

Management

Surgical excision of affected follicles or feather tracts may be of benefit. NSAIDs such as meloxicam may be of benefit.

Lack of bathing opportunities

In the zebra finch (*Taeniopygia guttata*), restricting bathing opportunities reduced plumage condition through increased lipids and dirt in the short term, and in the long term, bathing deprivation led to increased corticosterone. This increase in corticosterone can subsequently have an effect on moulting and the immune system.

Feather cysts in parrots

Definition/overview

Feather cysts in species other than canaries most often represent an acquired condition, which may be the result of an infection, trauma, or other conditions that interfere with normal feather growth.

Clinical presentation

These cysts usually occur on the wings, tail, or neck (**Fig. 11.31**). Three types of cysts have been described in budgerigars: variably-sized spherical cysts containing caseous material and the remains of

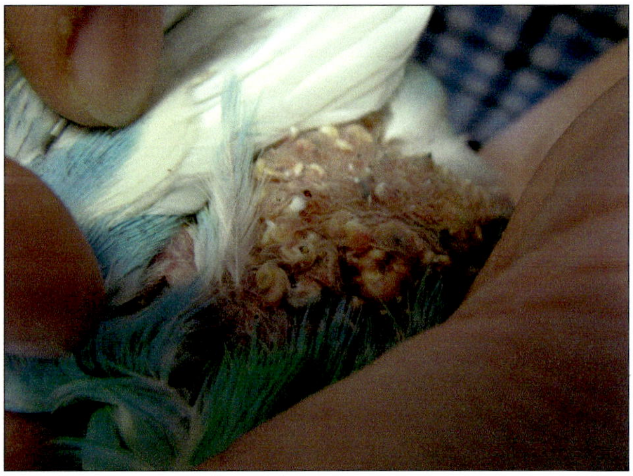

Figure 11.31 **Feather cyst on the wing of a sun conure.**

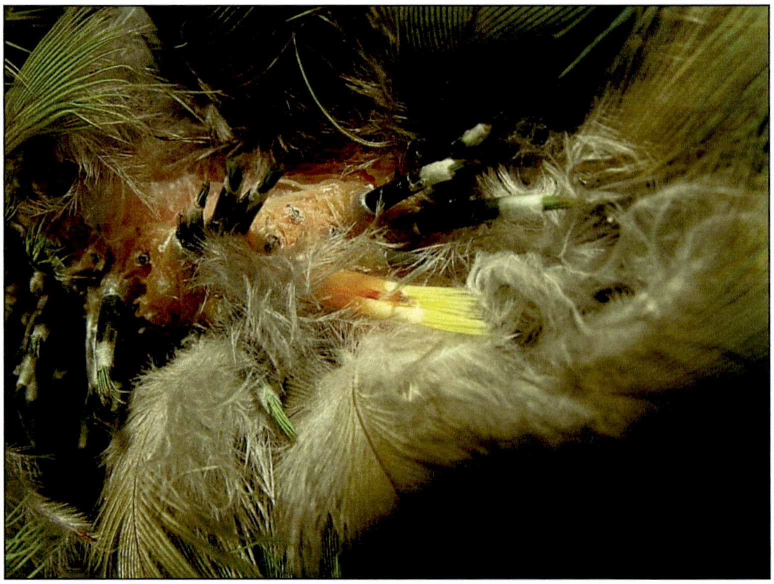

Figure 11.30 **Polyfolliculosis in an *Agapornis* species.**

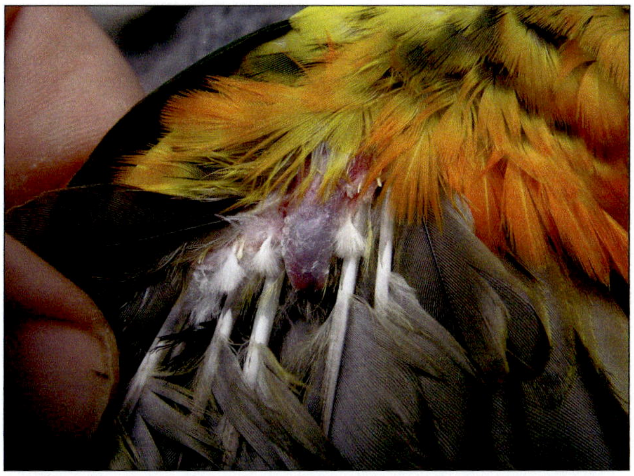

Figure 11.32 **Multiple feather cysts in a budgerigar.**

one or more feathers, encased in a fibrous capsule (**Fig. 11.32**); similar cysts but with the surrounding skin inflamed and thickened; and long, narrow cysts on the wing, often found side by side, with each cyst corresponding to the follicle of one primary feather. They contain caseous material similar to the other types, but they often have a malformed feather protruding from the tip of the cyst.

Management

Surgical excision of affected follicles is usually curative.

FEATHER-DAMAGING BEHAVIOUR AND OTHER SELF-MUTILATING CONDITIONS

Self-mutilation in Agapornis species

Clinical presentation

This is an intensely pruritic common self-mutilation problem affecting the shoulder region and prepatagial membrane. Less commonly, it can involve the inguinal region, chest, back, base of the tail, and around the cloaca. It can be unilateral or bilateral. Polyfolliculosis is found in some cases.

Long-term scarring of the prepatagial membrane often restricts the bird's ability to fly, and the affected area frequently cracks and bleeds when the bird stretches its wings.

Management

Treatment is often unrewarding. Placing an Elizabethan collar until the skin wounds have healed can be of benefit, but recurrence once the collar is removed is common. Antibiotics and anti-inflammatory therapies should be given. (See Chapter 16 for a more comprehensive overview of this problem in other avian species.)

Cockatiel feather mutilation syndrome

Definition/overview

This is associated with intestinal giardiasis and has been reported in cockatiels only. It appears to be less common now than when first reported.

Aetiology

It is postulated that giardiasis causes an intestinal malabsorption syndrome, leading to a vitamin E/selenium deficiency, which in turn leads to dry, flaky skin that can progress to episodes of feather pulling, alopecia, and pruritus.

Clinical presentation

Dry, flaky skin that can progress to episodes of feather pulling, alopecia, and pruritus. Other clinical signs include weight loss, depression, ruffled feathers, chronic diarrhoea, neonatal mortality, cachexia, and weakness.

Diagnosis

Eosinophilia and hypoproteinaemia are commonly reported; the eosinophilia is most likely due to the tissue damage associated with feather-damaging behaviour. Detection of trophozoites (on very fresh faecal smears suspended in normal saline) and cysts (using centrifuged zinc sulphate flotation) is very suggestive that this syndrome is present.

Management

Treatment with nitroimidazoles is usually effective, with a rapid response commonly noted.

Quaker mutilation syndrome

Aetiology

The aetiology is unclear. There is no histopathologic evidence of a pathogen. Insufficient cases have been fully worked up to determine a pattern.

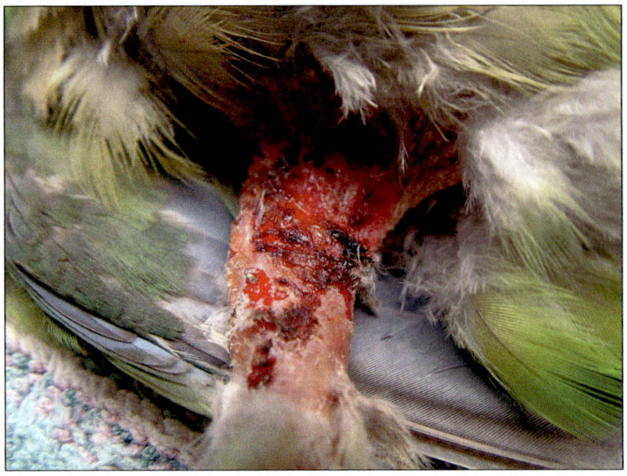

Figure 11.33 **Quaker self-mutilation syndrome.**

Clinical presentation

There is an acute onset of severe self-induced skin trauma, often directed at the neck and sternum and usually unrelated to prior episodes of feather-damaging behaviour (**Fig. 11.33**).

Management

Treatment is directed at preventing further trauma through the use of neck braces and/or Elizabethan collars. Analgesia is provided through the use of NSAIDs, and antibiotic coverage is administered. Euthanasia is often requested because of the severity and recurrence of the trauma.

Feather-damaging behaviour directed towards other birds

Towards nestlings

Some parents will start to pluck feathers from their nestlings while they are still in the nest box. Mild cases involve the removal of feathers, usually from the back, dorsal wings, and head. Severe cases may see the parent(s) remove wing tips or toes, or even kill the nestling.

Two theories exist as to why this behaviour occurs:

- A learnt behaviour – feather pluckers were often feather plucked themselves as chicks.
- A desire to remove the nestlings to prepare the nest box for another clutch of eggs.

Affected chicks are best removed for hand-rearing. They should not be used for breeding if possible.

Towards a bonded mate

Some individuals will pluck feathers from their mate, often at the beginning of the feather season. Both hens and cocks may do the plucking. Usually, it is the feathers on the head and face that are plucked. It is unusual to see this plucking extend to traumatic mutilation. Many of these pairs will breed successfully. They should be separated if mutilation is likely to result in injury.

Towards cage mates

In overcrowded or otherwise stressful living conditions, some birds (particularly finches, budgerigars, *Agapornis* spp., *Neophema* spp., and gallinaceous birds) will pluck the head and back of their cage mates. Occasionally, this behaviour can lead to cannibalism and death. Some birds will also chew and pluck the flight and tail feathers of cage mates, even in good living conditions.

Affected birds and, if possible, the offending bird(s) should be removed. Stocking density and stress factors need to be addressed and corrected. Foraging activities and toys should be provided for birds that appear to be indulging in this behaviour due to a lack of environmental enrichment.

Feather-damaging behaviour directed towards the bird itself

Definition/overview

Feather-damaging behaviour is damage to, or destruction of, a bird's feathering, in this case by the bird itself. It can include a number of behaviours, such as:

- Feather chewing – the bird chews at the ends of its feathers, damaging the vane.
- Feather barbering – the bird chews along the entire length of the feather shaft.
- Feather plucking – the bird pulls a complete feather out.
- Feather picking – the bird bites the feather off at the level of the skin.
- Feather snapping – the bird breaks the feather shaft, leaving the ends of the feathers ending in a V shape.

This is one of the most common and most frustrating conditions that veterinarians can be presented with. The causes of this condition are often multifactorial and cascading in their effects. By the time the bird is presented to the clinician, the original inciting cause may have disappeared or been obscured by other complicating or reinforcing factors.

It must be stressed to the bird's owner that a successful outcome may simply be a reduction, rather than elimination, of the activity. In other words, a feather picker will nearly always remain a feather picker but may pick less.

Aetiology

Feather-damaging behaviour may be due to either physical or behavioural problems. It is simplistic and inaccurate to diagnose 'boredom' or 'fear/anxiety' without a thorough investigation to rule out physical problems first. It is also simplistic to believe that a single aetiology exists for each individual case. In many cases, several factors, both physical and behavioural, have combined to produce the clinical sign of feather-damaging behaviour.

Physical causes of feather-damaging behaviour

Dermatitis: infectious (bacterial, fungal, viral), chemical (e.g., nicotine absorbed from the owner's fingers), or allergic (still unproven, but strong evidence exists for allergic dermatitis to be a factor in some feather-pickers). Allergies are suspected to occur in birds, but a reliable diagnostic test has yet to be developed.

- Folliculitis: bacterial, fungal, viral.
- Malnutrition: an all-seed diet often results in dry, flaky skin that is predisposed to superficial infections, resulting in pruritus.
- Environmental conditions: extremes of humidity, aerosol contamination, and cigarette smoke have been associated with feather-damaging behaviour.
- Heavy metal toxicoses are often implicated as a cause of feather-damaging behaviour. However, all reported cases are anecdotal, and there is no reliable evidence that heavy metal toxicosis causes feather-damaging behaviour.
- Underlying painful lesions: hepatopathy (e.g., chlamydiosis), osteomyelitis, pancreatic disease, renal disease, neoplasia, underlying abscesses.
- Reproductive activity, perhaps through ovarian and oviductal enlargement and liver changes (vitellogenesis), is believed to cause abdominal discomfort that may be seen as feather-damaging behaviour of the thighs and ventral abdomen of some reproductively active hens.
- Parasites – external (mites, lice) and internal (*Giardia*) – are often overdiagnosed as a cause of feather-damaging behaviour. Pet shop employees, in particular, often advise bird owners to treat feather-damaging behaviour in birds for lice. *Giardia* has only been associated with feather-damaging behaviour in cockatiels, yet many case studies report *Giardia* testing in non-cockatiel species.

Psychological causes of feather-damaging behaviour

See Chapter 30 for a list of causes.

Diagnosis

A thorough history of both the bird and the problem is required. In particular, this history needs to focus heavily on the bird's interaction with its environment and the people (and other animals) around it. A detailed physical examination should be performed and a minimum database on the bird's health status gathered (haematology and clinical biochemistries, whole-body radiographs). Based on this physical examination and clinical database, the following additional diagnostic tests may be warranted:

- Skin biopsy
- Pathogen detection through culture or PCR
- Endoscopy

If a physical cause is unable to be determined and the patient's history supports it, a diagnosis of behavioural feather damaging can then be made (**Table 11.1**). At this stage, every effort should be made to incorporate the services of a veterinary behaviourist or a suitably qualified parrot behaviour consultant. See Chapter 30 for a more detailed discussion on causes and management of this behaviour.

Management

Elizabethan collars are rarely indicated unless self-mutilation and/or physical trauma are occurring. In many cases, Elizabethan collars and other restraints may worsen the patient's anxiety state and aggravate the original problem.

Psychotropic drugs may be of benefit as a short-term therapy to break a feather-damaging behaviour 'cycle', but are rarely, if ever, warranted as a long-term solution. They can be used to suppress, modify, or change unwanted behaviour, but should be used in conjunction with behavioural modification. Most exert their effect by enhancing or inhibiting the effects of excitatory or inhibitory neurotransmitters. Examples of drugs that may have a role in short-term therapy can be found in Appendix 1: Formulary.

Table 11.1 Diagnosis of feather damaging behaviour.

TIMING AND NATURE OF FEATHER DIFFERENTIAL DIAGNOSES	DAMAGING BEHAVIOUR	REINFORCING FACTORS
Occurs when owner is not present	Separation anxiety; boredom	Habit; lack of environmental enrichment/foraging activities
Occurs when owner is present but not paying attention to the bird	Attention seeking behaviour	Owner's behaviour (drama, attention); habit
Bird interrupts other behaviour to damage feathers	Obsessive/compulsive disorder; true pruritus	Owner's behaviour (drama, attention); habit
As well as the feather damaging behaviour, the bird exhibits signs of unwarranted fear, anxiety or stress	Generalized anxiety disorders	Owner's behaviour (avoidance); habit; major changes in household
Problem starts at an extremely young age; handfed bird	Improper preening; poor early socialization	Owner's behaviour (drama, attention); habit
Involves primarily remiges and retrices. The feathers are frayed and splintered due to small cage	Anxiety disorders. Iatrogenic: improper wing trim, feather trauma	Owner's behaviour (drama, attention); habit
Overly bonded, sexually mature bird that displays sexual behaviours out of context	Reproductively related	Owner's behaviour (drama, attention); habit

These drugs, if used at all, should be employed as a short-term solution. Behavioural feather damaging is a behavioural problem, and the real solutions lie in behavioural modification. See Chapter 30 for more detailed information.

Prognosis

The goal is to decrease the feather-damaging behaviour; it is unlikely that a permanent cure will be achieved. Regular communication and follow-up evaluations are essential to monitor changes in the bird's behaviour and environment. If the result is a healthy, well-socialised bird that actively engages with its environment and its human companions, the state of its plumage becomes less significant.

FURTHER READING

Briscoe JA, Reisner IR, Rosenthal KL (2004) Incorporating the veterinary behaviourist: A new model for the diagnosis and treatment of a feather damaging pet parrot. In: *Proceedings of the Annual Conference of the Association of Avian Veterinarians*, pp. 293–296.

Chitty J (2005) Feather pulpitis in plucking parrots. In: *Proceedings of the Annual Conference of the Association of Avian Veterinarians*, pp. 259–265.

Clubb SL (2006) Clinical management of feather damaging behaviour associated with inflammatory skin disease in parrots. In: *Proceedings of the Annual Conference of the Association of Avian Veterinarians*, pp. 73–78.

Clubb SL, Buerkle M, Crosta L, Ciembor PG, Latima KS, Garner MM, Ritchie BW (2004) Feather damaging behaviour in Lories and its association with psittacine circovirus. In: *Proceedings of the Annual Conference of the Association of Avian Veterinarians*, pp. 319–320.

Clubb SL, Elmo N, Buerkle M, Crosta L, Enders F (2004) Incidence and characterisation of feather damaging behaviour in a large parrot collection. In: *Proceedings of the Annual Conference of the Association of Avian Veterinarians*, pp. 321–332.

Clubb SL, Garner MM, Cray C, Goodman M (2004) Diagnostic assessment of feather picking behaviour in African Grey Parrots (*Psittacus erithacus*). In: *Proceedings of the Annual Conference of the Association of Avian Veterinarians*, pp. 313–318.

Danylyk I, Oldfield T, Raidal SR (2010) Evaluation and assessment of hypothyroidism in Psittacine birds. In: *Proceedings of the Annual Conference of the Association of Avian Veterinarians and Unusual and Exotic Pet Veterinary Special Interest Group.*

Ferrell ST (2002) Avian integumentary surgery. *Seminars in Avian and Exotic Pet Medicine* 11(3):125–135.

Garner MM (2006) Inflammatory skin disease in feather picking birds: Histopathology and species predispositions. In: *Proceedings of the Annual Conference of the Association of Avian Veterinarians*, pp. 17–20.

Gartrell BD, Rogers L, Alley MR (2005) Eosinophilic dermatitis associated with *Trichophyton asahii* in a cockatiel (*Nymphicus hollandicus*). *Journal of Avian Medicine and Surgery* 19(1):25–29.

Gill JH (2002) Avian skin diseases. *Veterinary Clinics of North America: Exotic Animal Practice* 4(2):463–492.

Greenacre CB (2005) Viral diseases of companion birds. *Veterinary Clinics of North America: Exotic Animal Practice* 8(1):85–106.

Hochleithner M, Hochleithner C (1996) Surgical treatment of ulcerative lesions caused by automutilation of the sternum in psittacine birds. *Journal of Avian Medicine and Surgery* 10(2):84–88.

Johnston MS, Preziosi DE, Morris DO, Rosenthal KL, Rankin S (2004) The role of *Malassezia* in feather destructive behaviour in psittacines. In: *Proceedings of the Annual Conference of the Association of Avian Veterinarians*, pp. 89–90.

Koski MA (2001) Dermatological diseases in psittacine birds: An investigational approach. *Seminars in Avian and Exotic Pet Medicine* 11(3):104–125.

Lightfoot TL, Nacewiz CL (2006) Psittacine behaviour. In: *Exotic Pet Behaviour.* TB Bays, TL Lightfoot, J Mayer (eds). Saunders Elsevier, St Louis, pp. 51–102.

Martin KM (2004) Behavioural approach to psittacine feather-picking. In: *Proceedings of the Annual Conference of the Association of Avian Veterinarians*, pp. 307–312.

Merryman JI, Buckles EL (1998) The avian thyroid gland. Part one: A review of the anatomy and physiology. *Journal of Avian Medicine and Surgery* 12(4):234–237.

Merryman JI, Buckles EL (1998) The avian thyroid gland. Part two: A review of function and pathophysiology. *Journal of Avian Medicine and Surgery* 12(4):238–242.

Nemetz LP (2004) Strontium-90 therapy for uropygial neoplasia. In: *Proceedings of the Annual Conference of the Association of Avian Veterinarians*, pp. 15–20.

Phalen DN (2006) Implications of viruses in clinical disorders. In: *Clinical Avian Medicine, Vol 2.* GH Harrison, TL Lightfoot (eds). Spix Publishing Inc, Palm Beach, pp. 721–746.

Powers LV, Van Sant F (2006) Axillary and patagial dermatitis in African Grey Parrots (*Psittacus erithicus*). In: *Proceedings of the Annual Conference of the Association of Avian Veterinarians*, pp. 101–105.

Raidal SR (1995) Viral skin diseases of birds. *Seminars in Avian and Exotic Pet Medicine* (4):77–82.

Ritchie BW (1995a) Circoviridae. In: *Avian Viruses: Function and Control.* BW Ritchie (ed). Wingers Publishing, Lake Worth, pp. 223–252.

Ritchie BW (1995b) Papoviridae. In: *Avian Viruses: Function and Control.* BW Ritchie (ed). Wingers Publishing, Lake Worth, pp. 127–170.

Ritchie BW (1995c) Poxviridae. In: *Avian Viruses: Function and Control.* BW Ritchie (ed). Wingers Publishing, Lake Worth, pp. 285–312.

Rosenthal KL, Morris DO, Mauldin ES, Ivey ES, Peikes H (2004) Cytologic, histologic and microbiologic characterization of the feather pulp and follicles of feather-picking psittacine birds. *Journal of Avian Medicine and Surgery* 18(3):137–143.

Schmidt RE (2002) Avian thyroid metabolism and diseases. *Seminars in Avian and Exotic Pet Medicine* 11(2):80–83.

Schmidt RE, Lightfoot TL (2006) Integument. In: *Clinical Avian Medicine, Vol 2.* GH Harrison, TL Lightfoot (eds). Spix Publishing Inc, Palm Beach, pp. 395–410.

Schmidt RE, Reavill DR, Phalen DN (2003) Integument. In: *Pathology of Pet and Aviary Birds.* RE Schmidt, DR Reavill, DN Phalen (eds). Iowa State Press, Ames, pp. 177–196.

Seibert LM (2006) Feather-picking disorder in pet birds. In: *Manual of Parrot Behaviour.* AU Luescher (ed). Blackwell Publishing, Oxford, pp. 255–266.

Tully TN, Foil CS, Nett-Mettler C, Columbini-Osborn S, Heatley JJ, Hosgood G (2006) Status of intradermal skin testing in avian species. In: *Proceedings of the Annual Conference of the Association of Avian Veterinarians*, pp. 33–37.

Welle K (2005) Clinical approach to feather picking disorders in pet birds. In: *Proceedings of the Annual Conference of the Association of Avian Veterinarians Australian Committee*, pp. 83–87.

CHAPTER 12

DISORDERS OF THE BEAK AND CERE

ANATOMY AND PHYSIOLOGY

The avian skull (**Fig. 12.1**) consists of the neurocranium (the braincase and the orbit) and the splanchnocranium (the viscerocranium). The splanchnocranium is the skeleton of the face and consists of the mandible, maxilla, jugal bars, quadrate, palate, and hyoid apparatus.

The bones of the splanchnocranium are present at hatching but continue to grow via single ossification centres in the premaxilla and vomer bones and in the paired maxillary, palatine, nasal, jugal, quadratojugal, pterygoid, and quadrate bones. The age at which growth ceases is also undetermined but is likely to occur between fledging and weaning.

The mandible and maxilla are covered in horny keratin (**Fig. 12.2**), called the rhamphotheca; the mandibular rhamphotheca is known as the gnathotheca, and the maxillary rhamphotheca is called the rhinotheca. The dorsal midline of the rhinotheca is the culmen, and the ventral midline of the gnathotheca is the gonys. The cutting edge of both the upper and lower beak is the tomia, while the soft tissue between the mandibular rami is the interramular region. Histologically, the rhamphotheca resembles skin, with the dermis attached to the periosteum of the underlying bone. The epidermis is modified in that the stratum corneum is thickened and hardened, as the cells contain free calcium phosphate and crystals of hydroxyapatite.

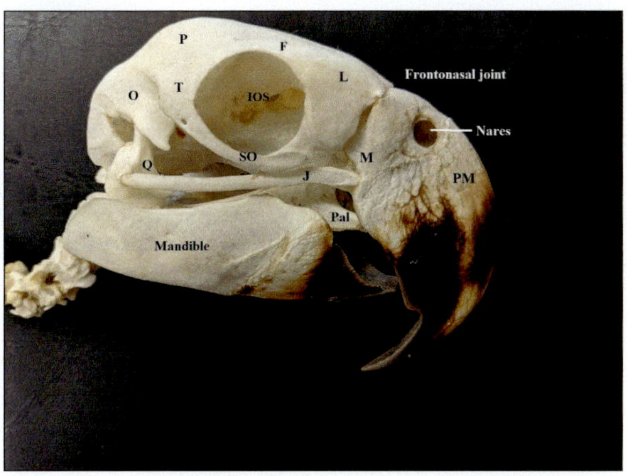

Figure 12.1 **Macaw skull. The bones of the skull are grouped into the neurocranium (braincase orbital bones) and the splanchnocranium (bones of the face). The neurocranium consists of the frontal bone (F), lacrimal bone (L), the suborbital arch (S), the occipital bone (O), temporal bone (T), and parietal bone (P). Between the two orbits is the thin interorbital septum (IOS). The splanchnocranium consists of the mandible, premaxilla (PM), maxilla (M), jugal bars (J), quadrate (Q), palatine (PAL), and hyoid apparatus.**

The neurological innervation of the upper beak is provided by the ophthalmic and maxillary divisions of the trigeminal nerve, while the mandibular division of the trigeminal nerve innervates the lower beak. Nerve-rich papillae extend through the bone into the rhamphotheca, providing sensory input from the beak tip. In the treatment of beak abnormalities (malocclusions, trauma, beak trimming, etc.), the clinician must give careful consideration to the extent of these dermal papillae, as damage to them can result in pain, increased sensitivity, or loss of object and food manipulation abilities, feeding, preening, and social behaviours.

The cere, the fleshy area around the nares, is found only in owls, parrots, and pigeons.

Malformation

Definition/overview

Malformation of the beak may be congenital or acquired. Congenital conditions include lateral deviations of the maxilla (wry/scissor beak) (**Fig. 12.3**) and mandibular prognathism (**Fig. 12.4**). Acquired conditions include lateral deviations of the maxilla (wry/scissor beak), compression deformities of the mandible or maxilla (**Fig. 12.5**) and other malformations, often associated with a 'softening' of the beak due to malnutrition.

DOI: 10.1201/9781003272519-12

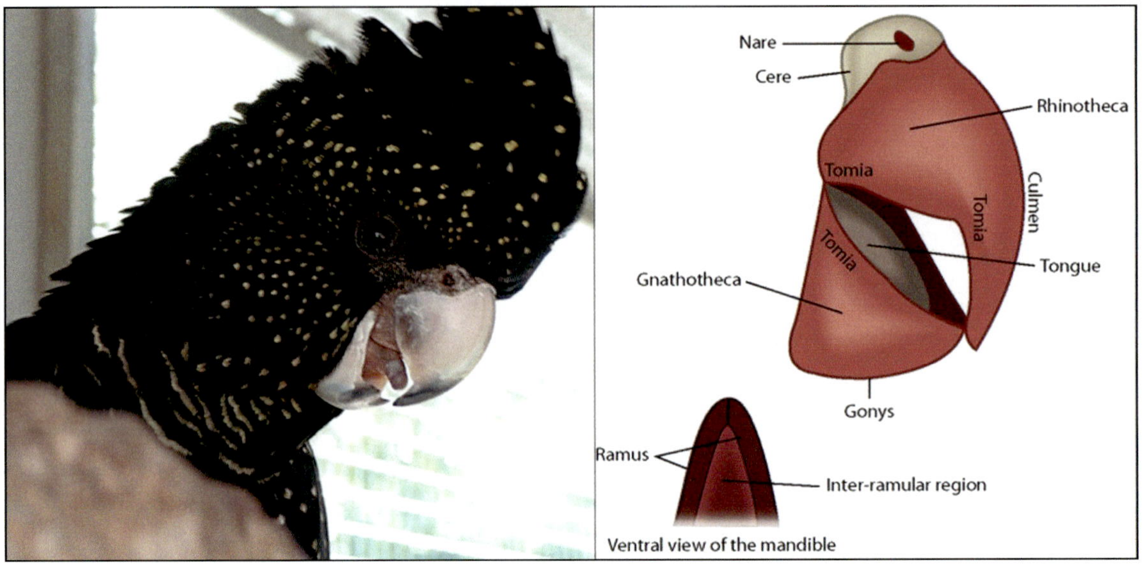

Figure 12.2 **The keratinised structures of the beak.**

Figure 12.3 **Scissor beak in a juvenile macaw.**

Figure 12.4 **Prognathism in an Eastern rosella.**

Aetiology

Congenital malformations are thought to be associated with disruptions of the growth centres of the premaxilla or other facial bones. The causes of this growth disturbance have not been determined but are considered to be the result of genetic, nutritional, or incubation factors (temperature, humidity, ventilation, or turning of the egg), or as the result of hand-feeding practices.

Acquired malformations are thought to be due to damage to all or part of the germinal epithelium of the beak or to the poorly calcified beaks of neonates. It has been suggested that feeding from one side of the beak only, compressive trauma to the rictus (the caudal oral opening between the angle of the jaw and

Figure 12.5 **Lateral compression of the rhinotheca in a juvenile macaw.**

Figure 12.6 A variant of scissor beak where the prefrontal bone of the skull has collapsed on one side, turning the whole beak to that side.

the rhinotheca) during restraint of the bird's head during feeding, or direct trauma to the rictus and underlying tissues from a syringe tip may be causative factors.

A variant of scissor beak is sometimes seen when the prefrontal bone of the skull has collapsed on one side, turning the whole beak to that side (**Fig. 12.6**).

Clinical presentation

In lateral deviations of the maxilla (wry/scissor beak), the maxillary beak is deviated laterally from the level of the cere or beak tip. The side of the mandible is no longer in wear, and the maxilla often overgrows. This condition is most common in macaws up to 2 months of age, but acquired scissor beak can occur in any species at any age.

Mandibular prognathism is most common in cockatoos aged 1–2 weeks. The maxilla sits inside the mandible.

Compression deformities of the mandible are most commonly seen in macaws up to 5 weeks of age. Other malformations are often associated with sinusitis, *Cnemidocoptes* infection (**Fig. 12.7**), trauma, and psittacine beak and feather disease (PBFD).

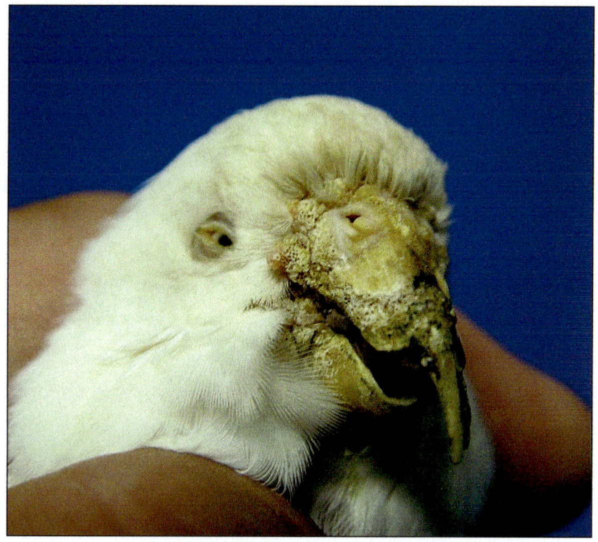

Figure 12.7 Beak deformity in a budgerigar with chronic *Cnemidocoptes* mite infection.

Management

Conservative treatment may be effective in very young chicks. Applying gentle digital pressure for 10 minutes two to three times daily may straighten a deviated maxilla or lift the maxilla up and forward in prognathic chicks. Cases that do not respond to this treatment, or older birds with calcified beaks, will need more aggressive therapy.

Deviated maxilla

Beak trimming, involving grinding the overgrown gnathotheca on the side contralateral to the maxillary deviation and the overgrown tomia on the maxilla, allows the maxilla to move back into a normal position. At the same time, the occlusal ledge inside the maxilla has to be reshaped so that it is perpendicular to the lateral walls of the rhinotheca. The bird should then be encouraged to chew hard objects (e.g., branches) frequently, and the beak reshaped every 2 weeks until normal.

Acrylic ramps or prostheses can be used to augment the trimming. An acrylic ramp is built on the gnathotheca on the same side as the maxillary deviation, forcing the maxilla into a normal position. The ramp needs to be high enough so that the bird cannot open its mouth wide enough to get its maxilla over the top of the ramp. This technique frequently fails due to damage to the gnathotheca during the attachment of the ramp.

Trans-sinus pinning is an 'orthodontic' procedure, designed to provide constant lateral tension on the deviated maxilla to guide it into a more anatomically normal position. A K-wire or small Steinmann pin is placed perpendicular to the skull, through the frontal sinuses just caudal to the craniofacial hinge joint on a line between the lateral canthus of the eye and the cere. Care must be taken not to position this pin too caudally or too low. In macaws, a small bony protuberance marks the point where the pin is introduced on one side and exits on the other. On the side of the beak contralateral to the direction of the deviation, the pin is bent 90° as it leaves the skin. The pin should run the length of the maxilla, and the end (level with the beak tip) is curled over on itself. The other end of the pin is trimmed and curled over on itself so that it is flush against the skin and cannot be pulled through the skull.

The tip of the beak is then placed under tension by means of a rubber band around the beak and the distal end of the wire. In older birds, this rubber band is replaced by a piece of wire running through a hole in the distal rhinotheca. The tension is maintained until the beak has straightened. In young birds, this can be as soon as 10–14 days; older birds may require 10–14 months. The tension can be adjusted by loosening or tightening the wire/rubber band as required (**Fig. 12.8**).

Once the beak has straightened, the wire/rubber band should be removed. If, after several days, the beak has remained in a normal position, the pin can be removed.

Scissor beak associated with prefrontal bone collapse is not, at this time, amenable to treatment. These birds usually adapt to their disability and do well. Regular beak trimming is often required due to the lack of occlusal wear.

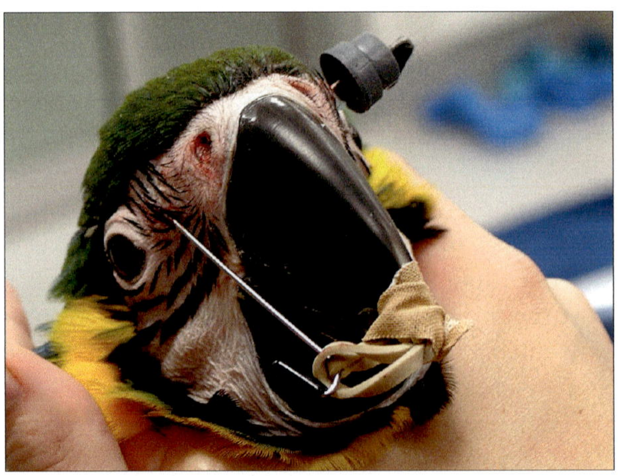

Figure 12.8 **Trans-sinus pinning to correct 'scissor beak'.**

Prognathism

An acrylic prosthesis can be placed over the rhinotheca. One or two small pins can be placed through the rhinotheca to provide 'anchor points' to prevent the prosthesis from falling off. The goal is to force hyperextension of the craniofacial hinge joint into a more normal range of motion, allowing for a 'firming up' of the joint so that the upper mandible cannot be overflexed to the point of mandibular prognathism. There is no correction of 'growth' of the beak, but rather guidance and reduction of the abnormal range of motion of the craniofacial hinge joint. The prosthesis must be long enough to prevent the bird from opening its mouth wide enough to place the prosthesis inside the gnathotheca.

A trans-sinus pinning technique can be used in older birds. It requires a pin placed through the frontal sinuses as described earlier. The ends of the pin are then bent forward to meet just rostral to the beak tip and are linked to form a triangle. A second pin is placed through the rhinotheca midway along its length, with the triangular trans-sinus pin resting on it. Cerclage wire is then used to link the tip of the beak with the apex of the triangle. This arrangement also extends the craniofacial hinge joint into a more normal position.

Prognosis

Many birds with untreated malformations will adapt to the deformity and learn to eat well. However, they will usually do better if the malformation can be corrected. With appropriate techniques, follow-up, and care, the prognosis for most beak malformations is good.

Trauma

Definition/overview

Common injuries include puncture wounds of the rhinotheca or gnathotheca (**Fig. 12.9**), avulsion of the rhinotheca or gnathotheca or both (**Fig. 12.10**), mandibular or maxillary fracture, and mandibular symphyseal fracture (**Fig. 12.11**).

Aetiology

Causes include bite wounds from other birds and occasionally from other animals, entrapment in wire or metal objects, particularly in cages, and blunt

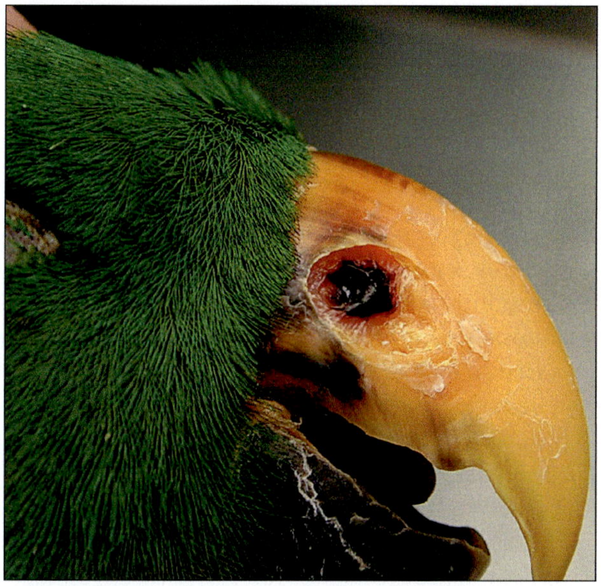

Figure 12.9 Puncture wound in the rhinotheca of an eclectus parrot.

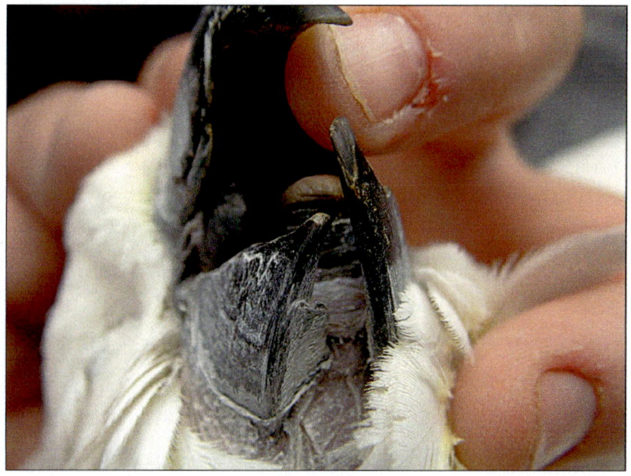

Figure 12.11 Mandibular symphyseal fracture in a cockatoo.

force trauma resulting from striking objects while in flight (e.g., ceiling fans, walls, windows).

Management

Initial treatment should be aimed at relieving pain, stopping bleeding, and controlling infection. Once this initial assessment and treatment are finished, attention needs to be given to providing nutritional and fluid support to the patient. Tube feeding or placing an oesophagostomy tube may be required until the patient is able to eat by itself.

Once the patient is stable, consideration must be given to how to repair the injury, if indeed repair is feasible.

The dermis that produces the rhamphotheca is present at the base of both the upper and lower beaks, and also entirely covers the dorsal and lateral surfaces of the maxilla and mandible. The rhamphotheca, therefore, grows from deep to superficial as well as from caudal to rostral. Any injury to the beak that damages this dermis can result in a loss of keratin. This will usually grow out with time. With severe injuries, where the bone underlying the dermis is crushed or fractured, the loss of bone and dermis usually results in very limited regrowth.

Puncture wounds, if superficial, can be treated conservatively by keeping them clean and administering antibiotics and analgesics as appropriate. Deeper wounds that may affect the structural integrity of the beak should be carefully debrided, dried, and then 'plugged' with dental acrylic. As granulation tissue fills in the wound, the acrylic plug will be forced out. This may be a matter of days or perhaps weeks. The plug can be replaced until the clinician is assured the beak's integrity is no longer at risk.

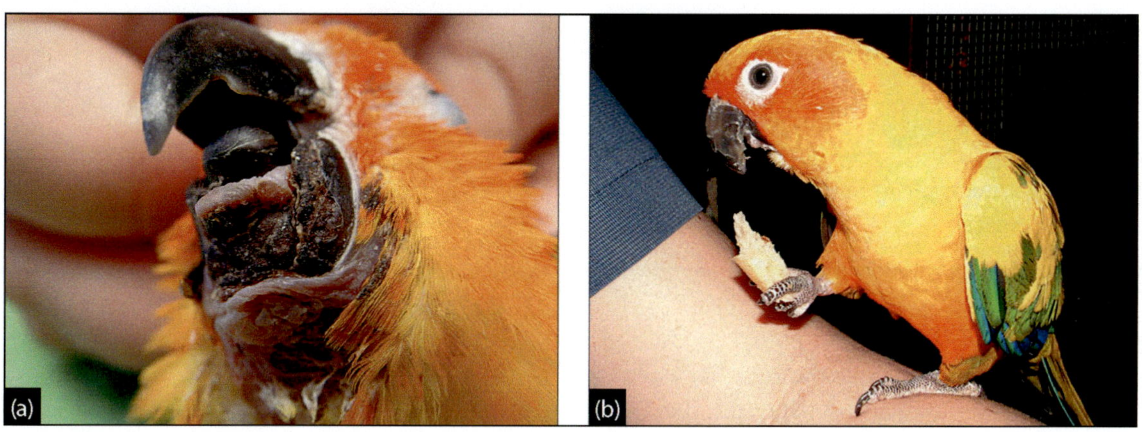

Figure 12.10 (a) Avulsion of the lower beak in a sun conure and (b) subsequent recovery.

Avulsion of all or part of the mandible or maxilla is a common injury. As a general rule, avulsions of less than the rostral third of the maxilla will usually result in regrowth of the avulsed portion. Loss of more than this amount will usually result in permanent deformity. Surprisingly, many birds will adapt to this deformity and do well, as long as infection and pain are controlled, and the bird is nutritionally supported until the wounds have healed. Overgrowth of the opposing rhinotheca or gnathotheca is common, requiring regular trimming. Beak prostheses rarely work in birds, as the kinetic forces transmitted into the prosthesis by normal beak actions (e.g., eating, gripping objects) rapidly lead to implant failure.

Fractures of the mandible or maxilla are more difficult to handle. With fresh injuries, repair with cerclage wire may be effective in some cases. Splinting the fractured area by taping the beak shut may also be of benefit. Nutritional support via an oesophagostomy tube is obviously essential.

Mandibular symphyseal fractures, if fresh, can be treated by wiring with cerclage wire and sealing the fracture line with tissue glue. Care must be taken not to allow the tissue glue to seep into the fracture line. If non-union results, or if the injury is already several days old when first presented, many of these birds will adapt and be able to eat well. Again, nutritional support may be required via tube feeding or an oesophagostomy tube until the bird is able to eat by itself. Long-term regular beak trimming is usually required.

Prognosis
Most beak injuries have a good prognosis. Despite severe injuries, many birds adapt and survive quite well.

Hyperextension of the maxilla
Definition/overview
This is predominantly a problem in macaws. The maxilla is hyperextended, leaving the beak slightly open. Simple reduction, even under anaesthetic, is usually unsuccessful.

Aetiology
Trauma, particularly biting hard onto a solid object and forcefully hyperextending the maxilla, leads to the palatine bone moving rostrally and then luxating dorsally beyond its normal end-point on the mesethmoidal bony ridge at the ventral base of the braincase. This then engages the infraorbital septum, locking the maxilla into a hyperextended position.

Management
Under anaesthesia, a Steinmann pin or closed artery forceps is introduced through a lateral approach via the infraorbital sinus, just caudal to the beak commissure and dorsal to the palatine bones. It is inserted to a depth half the width of the skull. This can be done blindly or with the guidance of an endoscope. Several large blood vessels are in this area.

The beak is hyperextended further to 'disengage' the palatine bone from the septum. With the beak still hyperextended, the pin or forceps are used to lever the palatine bone ventrally, disengaging it from the septum and allowing the beak to close. Additional stabilisation can be provided by passing absorbable suture around the suborbital arch and jugal bones bilaterally or around a trans-maxillary pin placed in the proximal central maxilla. The bird is treated with analgesics for 5–7 days and prevented from climbing or biting hard foods for 2 weeks.

Prognosis
The prognosis is usually good.

Overgrown maxilla
Aetiology
An overgrown maxilla is often associated with chronic liver disease (**Fig. 12.12**), although the

Figure 12.12 **Chronic liver disease, especially accompanied by rhinothecal bruising, is associated with liver disease.**

Figure 12.13 **Chronic malnutrition can cause overgrowth of the beak.**

exact pathogenesis is unclear. Other possible causes include damage to the germinal epithelial layer, malnutrition (**Fig. 12.13**), lack of occlusal wear (e.g., scissor beak), lack of opportunity to chew, *Cnemidocoptes*, and neoplasia. In some species, such as the long-billed corella, it may be normal (**Fig. 12.14**).

Diagnosis

A thorough review of the patient's history, including diet and history of head trauma, is required.

Haematology and biochemistry are carried out to determine health status, with a particular emphasis on assessment of the liver. Radiography or CT imaging of the head may be necessary in some cases.

Figure 12.14 **A long beak is normal in this long-billed corella.**

Treatment

The diet should be corrected and appropriate treatment given for liver disease if present (see Chapter 20). Corrective trimming is carried out as required, and the opportunity for increased chewing behaviour should be provided. Any other causative factors should be identified and remedied.

Cnemidocoptes (scaly face mite)

Aetiology

The species of mites are *Cnemidocoptes pilae* (parrots and passerine birds) and *Cnemidocoptes mutans* (poultry), also known as scaly face (or leg) mites. It appears that the mite spends its entire life cycle on its host, although it may be transmitted between birds via dead skin and scales. It is probable that immunosuppression and genetic factors play a role in the development of clinical signs.

Clinical presentation

There is a slight variation between affected species:

- Budgerigars, kakarikis, neophemas, and *Polytelis* spp. develop proliferative crusty, exuberant growths on the beak, cere, face, legs, the margin of the vent, and on the wing tips. A characteristic 'honey-combed' appearance is due to the mite tunnelling through the keratin. With severe infection, marked beak deformities can occur through damage to the dermis under the rhamphotheca. These include thickening of the beak, overgrowth of all of the beak, or overgrowth of part of the beak. Weakening of the keratin can cause part of the beak to crumble and break off (**Fig. 12.15**).

Figure 12.15 ***Cnemidocoptes* in a budgerigar.**

- Canaries develop 'tassel foot', keratin tags arising from the caudal edge of the podotheca along the tarsometatarsus, giving the appearance of 'tassels'. The scales on the podotheca may also become hyperkeratotic.
- Poultry also develop hyperkeratosis on their legs, which can become swollen and painful, causing the birds to limp.
- Owls are rarely affected, but if so, they develop bilateral proliferative papillary hyperkeratosis on the feet and legs.

Diagnosis

The classical 'honeycomb' appearance is typical. Scraping the lesion is usually rewarding. The mite can be identified by its dorsal shield and scapular setae.

Management

Ivermectin or moxidectin is used at 0.2 mg/kg topically, orally, or parenterally, repeated every two weeks until the crusts disappear (usually three treatments are required). Topical creams and liquids (e.g., mineral oil) are often not effective, as the whole bird must be treated.

Prognosis

The prognosis is good, although beak deformities may become permanent if treatment is initiated too late.

INFECTION

Bacterial and fungal infections

Aetiology

Bacterial and fungal infections may result from the extension of nasal and sinus infections into the bone and keratin of the beak, or they may be the result of trauma (see earlier). Causative organisms include *Candida*, *Aspergillus*, *Cryptococcus*, *Mycobacterium*, and *Pseudomonas* spp. Other bacteria can also be isolated.

Clinical presentation

The beak and nares become distorted. Underrunning of the keratin occurs, which could lead (in severe cases) to sloughing of all, or part of, the rhamphotheca (**Fig. 12.16**).

Figure 12.16 **Bacterial pyogranulomatous dermatitis in a sun conure.**

Diagnosis

Diagnosis is based on clinical signs, transillumination of the beak to determine the depth and extent of the lesion, and biopsy and culture.

Management

Surgical debridement is carried out as required, and antimicrobial therapy is administered as indicated by culture.

Viral infections

Aetiology

Viruses affecting the beak include beak and feather disease virus in cockatoo species (**Fig. 12.17**), avipox virus in all species, and polyomavirus in finches.

Clinical presentation, diagnosis, and management

See Chapter 11.

Neoplasia

Definition/overview

Tumour types involved in the beak include fibrosarcoma, squamous cell carcinoma, liposarcoma, and keratoacanthoma (**Fig. 12.18**).

Figure 12.17 **Overgrowth of the rhinotheca and necrotic underrunning of the keratin and dermis seen with PBFD in cockatoos but not other species.**

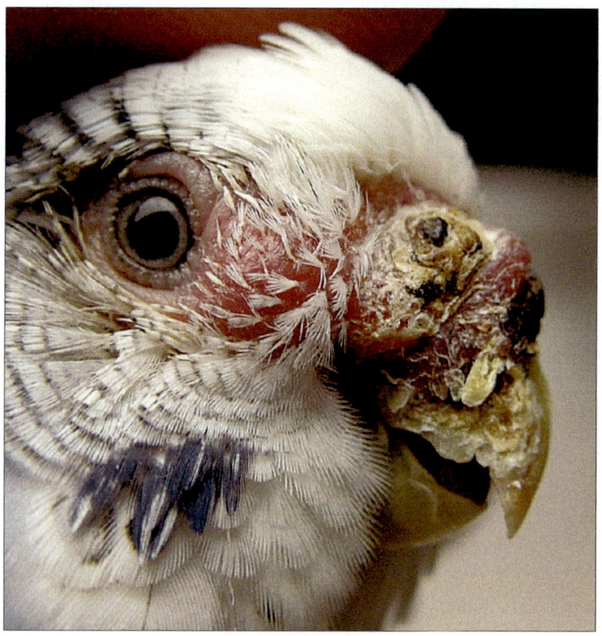

Figure 12.18 **Fibrosarcoma in the beak and cere of a budgerigar.**

Diagnosis

Diagnosis is by biopsy.

Management

The use of chemotherapy and radiation in the treatment of avian neoplasia continues to be explored. See Chapter 34 for more details.

Prognosis

The prognosis is guarded, although newer treatment protocols, including radiation, offer more hope.

KERATIN FLAKES ON THE BEAK

The continued growth of keratin culminates in flakes or sheets of keratin on the surface of the beak that are normally exfoliated by the bird rubbing its beak on abrasive substrates (e.g., branches, grooming perches). If these abrasive surfaces are not available, the beak may take on a flaky appearance or become thickened (particularly rostrally). Often, these flakes can be easily removed by simply peeling them off with a fingernail. Occasionally, trimming with a rotary hobby grinder (e.g., Dremel®) is necessary to reshape the beak into a more normal shape.

DISORDERS OF THE CERE

The cere is the naked skin situated at the base of the maxilla found in parrots, pigeons, and owls. The nares are located within the cere.

Hypertrophy

This may be caused by *Cnemidocoptes* (see earlier) or, more commonly, hyperoestrogenism in budgerigar hens (**Fig. 12.19**). The latter usually presents as a smooth (or slightly roughened) hypertrophic growth

Figure 12.19 **Cere hypertrophy in a female budgerigar.**

Figure 12.20 **Cere abscess and rhinoliths.**

on the cere. It is easily peeled off with a fingernail, with no underlying haemorrhage. Malnutrition, especially hypovitaminosis A, may result in cere hypertrophy.

Colour change

Increasing brown colouration of the cere in female budgerigars is indicative of oestrogenic activity associated with the onset of the breeding season.

Sertoli cell tumours in male budgerigars may present as a blue cere turning brown. Treatment with surgery may be successful in selected cases. Some birds will respond (in the short term) to GnRH agonists such as leuprolide or deslorelin.

Pallor of the cere is often seen in budgerigars with chronic illnesses of any type.

Distortion

Chronic respiratory infections may present as 'cere abscesses', swelling and distortion of the cere due to an underlying accumulation of caseated pus in the frontal sinus and nasal cavity (**Fig. 12.20**). Rhinoliths, large plugs of caseated debris in the nasal cavity under the nares, are commonly seen in African grey parrots but can be seen in any species. Affected birds commonly present with sneezing, swelling of the cere, and occlusion of the nare (usually unilateral). Rhinoliths must be differentiated from tumours and fungal granulomas. Many can be gently debrided under general anaesthesia, with the area thoroughly flushed after completion. In some cases, surgical debridement may be necessary. Samples should be submitted for histopathology and culture to identify the underlying cause.

Permanent asymmetry of the size and shape of the nares often results after cere abscesses or untreated chronic infections erode the supporting bone and soft tissues.

FURTHER READING

Altman B (1997) Beak repair, acrylics. In: *Avian Medicine and Surgery*. RB Altman, SL Clubb, GM Dorrestein, K Quesenberry (eds). WB Saunders, Philadelphia, pp. 787–799.

Clipsham R (1997) Beak repair, rhamphorthotics. In: *Avian Medicine and Surgery*. RB Altman, SL Clubb, GM Dorrestein, K Quesenberry (eds). WB Saunders, Philadelphia, pp. 773–786.

Doneley RJT (2021) Transsinus pinning to correct lateral deviation of the upper beak in Juvenile Macaws. *Journal of Avian Medicine and Surgery* 35(1): 68–74.

Gelis S (2006) Evaluating and treating the gastrointestinal system. In: *Clinical Avian Medicine, Vol 1*. GJ Harrison, TL Lightfoot (eds). Spix Publishing Inc, Palm Beach, pp. 411–440.

Lessner EJ, Echols MS, Paul-Murphy JR, Speer BL, Holliday CM (2023) Grey parrot (Psittacus erithacus) beak papillae and nerves identified using novel 2-D and 3-D imaging modalities. *American Journal of Veterinary Research* 84(7):ajvr.23.03.005.

Olsen GH (2003) Oral biology and beak disorders of birds. *Veterinary Clinics of North America: Exotic Animal Practice* 6(3):505–521.

Schmidt RE, Lightfoot TL (2006) Integument. In: *Clinical Avian Medicine, Vol 1.* GJ Harrison, TL Lightfoot (eds). Spix Publishing Inc, Palm Beach, pp. 395–410.

Speer BL (2002) Trans-sinus pinning to address scissor-beak deformities in psittacine species. In: *Proceedings of the Annual Conference of the Association of Avian Veterinarians Australian Committee*, pp. 283–290.

Speer BL (2012) Surgical procedures of the psittacine skull. In: *Proceedings of the Annual Conference of the Association of Avian Veterinarians.*

CHAPTER 13

DISORDERS OF THE EYE

ANATOMY

The size of the eye is extremely large in relation to the head, particularly when compared with that of mammals; in many birds, the two eyes together outweigh the brain. Large eyes equal a large image projected on the retina, which contributes to visual acuity. The globe can be one of three basic shapes (**Fig. 13.1**).

- Flat globes are found in most diurnal birds with narrow heads. The short distance between the cornea and the retina means that the image projected onto the retina is relatively small, with correspondingly low visual acuity.
- Globular globes are found in diurnal birds with wider heads, such as insectivorous wing-feeders, crows, and diurnal birds of prey. The cone-shaped eyeball results in greater visual acuity.
- Tubular globes are found in nocturnal birds of prey; the elongated shape provides the greatest visual acuity.

The lower eyelid is thinner, more extensive, and more mobile than the upper lid. The lower lid usually contains a fibroelastic tarsal plate. Filoplume feathers are present near the eyelid margin and have a protective and tactile function. In most species, the eyelids only close when sleeping; therefore, the nictitating membrane, which is well-developed, actively mobile, and lying beneath the eyelids on the nasal side of the eye, is responsible for blinking. There is a nictitating membrane gland.

The conjunctiva is morphologically similar to that of mammals. Tears are produced by the Harderian gland and the lacrimal gland, which are present inferior and lateral to the globe. The tears drain into the conjunctival sac on the bulbar surface of the lower lid and then exit via the inferior and superior nasolacrimal puncta at the medial canthus. Meibomian glands are absent in birds. In budgerigars and some other species, a nasal or salt gland lies in the orbit dorsomedial to the globe, and the duct of this gland pierces the frontal bone and enters the nasal cavity.

The oblique and rectus extraocular muscles are thin and poorly developed, and no true retrobulbar muscle group is present. Head movements generally compensate for the reduced mobility of the globe.

The cornea is small compared to the rest of the eyeball and is made up of five layers: an anterior (outer) stratified squamous epithelium; an anterior (outer) limiting lamina (Bowman's membrane), not always differentiated in birds and not found in mammals; the substantia propria, consisting of bundles of collagen fibres, which forms the bulk of the corneal wall; a posterior limiting lamina (Descemet's membrane); and a posterior (inner) layer of simple cuboidal epithelium.

The sclera is reinforced by a continuous layer of hyaline cartilage, which, in the zone nearest the cornea, is modified into a ring of 10–18 small, roughly quadrilateral, overlapping bones called the scleral ossicles (**Fig. 13.2**). The ossicles strengthen the eyeball and provide attachments for the ciliary muscles. The scleral venous sinus (canal of Schlemm) is conspicuous in some species, but small or almost invisible in others; it lies at the limbus (junction between the cornea and sclera). The trabecular reticulum, or pectinate ligament, in this region (a wide-meshed plexus of connective tissue fibres) joins the limbus to the iris and to the ciliary body. The spaces between these fibres form the spaces of the iridocorneal angle (spaces of Fontana) through which the aqueous humour drains into the scleral venous sinus (**Fig. 13.3**).

The uvea, the vascular part of the eye wall, consists of the choroid, the ciliary body, and the iris. The choroid is a thick, highly vascular, darkly pigmented layer coating the retina. A tapetum lucidum is only found in a few nocturnal species. The choroid

DOI: 10.1201/9781003272519-13

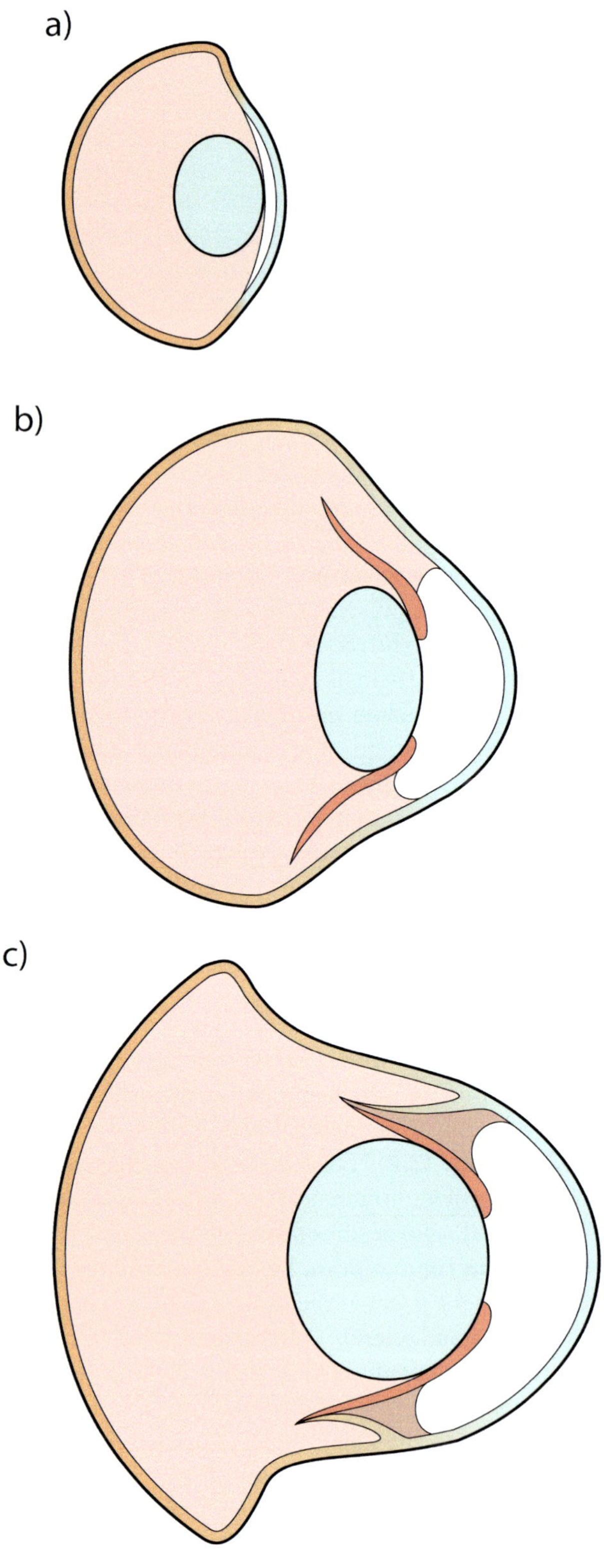

Figure 13.1 **Avian eye globe shapes.**

continues as the ciliary body and the iris. The ciliary body suspends the lens by the zonular fibres; it also forms small folds (ciliary processes), which are pressed against the rim of the lens by the ciliary

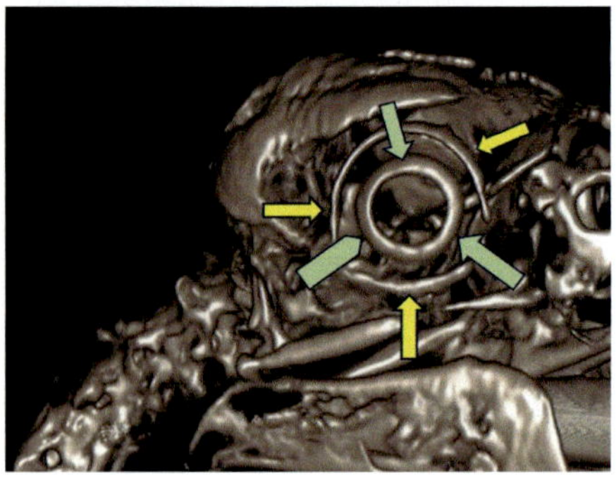

Figure 13.2 **3D reconstruction of a skull CT (long-billed corella) showing scleral ossicles (green arrows) and orbital rim (yellow arrows).**

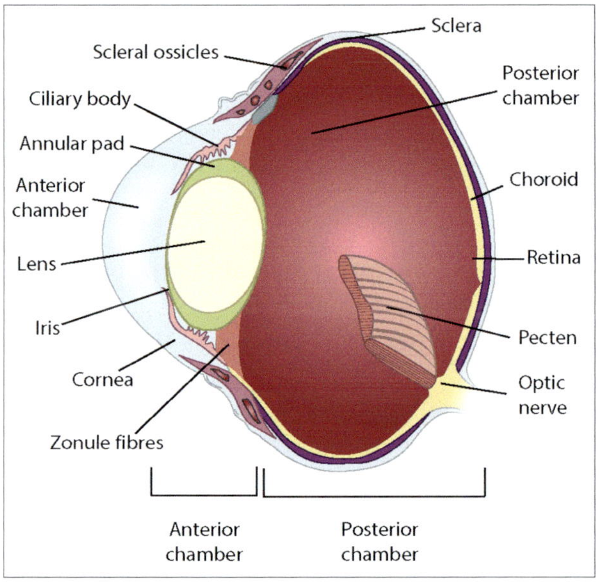

Figure 13.3 **The anatomy of the avian eye.**

muscles. The iris is dark in most species, but highly coloured in some. It forms a round aperture in most species. The ciliary muscles and the sphincter and dilator muscles of the iris are striated muscles, in contrast to the smooth muscles of mammals.

The retina arises as a direct continuation of the brain. It consists of an external, non-sensory single layer of cuboidal epithelium containing pigment (pigment epithelium) and an internal transparent and thicker neuroepithelium (sensory retina) containing several types of neurones and glial cells. Rods and cones are present in birds, serving similar functions as in mammals. The retina is thick compared to other vertebrates and contains an array of photoreceptors and several possible combinations of areas and foveas specialised for more acute (and often stereoscopic) vision. It is completely devoid of blood vessels and derives its nutrients from both capillaries within the choroid, external to the pigment epithelium, and the well-vascularised pecten with the vitreous body.

Areas are circumscribed thickenings of the sensory retina involving thinner and longer visual cells that improve the resolving power and are, therefore, associated with improved visual acuity in birds. There is almost always one central area, but often two or even three distinct areas (a circular lateral area and a horizontal linear area) are present. Foveas are depressions within either a central or lateral area or both. Not all areas have foveas, but foveas are only found within an area. Visual cell density is greater in the fovea than elsewhere in an area, and its shape acts to magnify the retinal image and increase its resolution. Although some species have no fovea, most have one or two. The location, depth, and relative position of these foveas exhibit considerable variation depending on the species. The presence of three distinct retinal areas (central, lateral, and linear), two of which (central and lateral) possess a fovea, is a unique avian adaptation that permits the formation of three separate and distinct visual fields (visual tridents), two lateral monocular fields (one for each eye), and a central binocular field.

Birds can detect a spatial frequency much higher than that of mammals: 160 frames/second compared with 60 frames/second in humans. This higher spatial frequency can cause problems under artificial light, which has a frequency of 100–120 frames/second and can therefore produce a stroboscopic effect, which may contribute to some behavioural disorders. High-frequency lights are therefore recommended for indoor birds.

The pecten, unique to birds, arises from the site of exit of the optic nerve and projects a variable distance into the vitreous body in the ventral and temporal quadrant of the fundus. It consists almost exclusively of capillaries and extravascular pigmented stromal cells; there are no muscular or neurological tissues. It can be conical, vaned, or pleated. The pleated pecten is the most common structure, arising as a single accordion-pleated lamina held in place at its free (apical) end by a more heavily pigmented crest or bridge of tissue. The mass, shape, number and arrangement of pleats, their extent of pigmentation, and their relationship to the ventral ciliary body vary considerably in different birds. The size of the pecten and number of pleats do not coincide with the size of the eye but seem to be related to the behaviour of the bird towards illumination and its general level of activity. Active diurnal birds with high visual acuity and monocular vision have a larger and more pleated pecten, while those nocturnal species with poor vision usually have a smaller pecten of simpler morphology. It may play a number of roles: nutrition of the retina; secretion of glycosaminoglycans and other products; regulation of intraocular pressure through the secretion of fluids; light absorption, which would reduce internal reflections that may possibly interfere with the production of a clear image; the perception of movement; and perhaps orientating birds in space by acting as a sextant, casting a shadow upon the retina to permit the latter to estimate the angular position and movement of the sun.

The lens of birds is much softer than that of mammals due to a fluid-filled lens vesicle between the annular pad and the body of the lens. This softness allows rapid accommodation. The lens is optically clear, rendering ultraviolet radiation visible. An annular pad runs around the equator of the lens, adjacent to the ciliary processes. This is well developed in diurnal predators but reduced in nocturnal species and flightless species. It has extensive equatorial thickening for contact with the ciliary body. Contraction of the ciliary muscles thrusts the ciliary body inward against the annular pad, transmitting the stress directly to the softer lenticular centre. This occurs evenly along the entire extent of the equatorial lens and allows for flexible accommodation and focuses light directly onto the retina. The transparency of the avian lens allows the passage of a broader spectrum of light than mammals, allowing birds to see not only blue, green, and red, but also ultraviolet and fluorescent.

Both the anterior and posterior chambers of the eye are filled with aqueous humour, which is responsible for regulating intraocular pressure and maintaining the proper shape and rigidity of the eyeball. It is continually secreted into the posterior chamber by the ciliary body and then circulates through the pupil into the anterior chamber. It then drains into the scleral venous sinus through the spaces of Fontana at the iridocorneal angle.

The attainment of exceptional avian visual abilities is the result of both the structure and function of the eye. The positioning of the eyes allows, in many birds, a wide field of vision. The large size of the eye permits the formation of a large retinal image. The actions of the cornea, lens, and aqueous humour provide a superb accommodative apparatus, increasing the depth of focus. The retina acts as a parabolic reflector, allowing all-around visual acuity, unlike mammals, which only have central vision acuity. Finally, the presence of elaborate and diverse retinal areas and foveas provides a high degree of acute monocular and binocular vision.

ASSESSING THE EYE

Distant examination

Initially, the bird should be watched from a distance (flying, moving around the cage or room). If there are eye problems, the bird may have difficulty finding food or show reluctance to fly or move around. There may be abnormal head posture; many birds with unilateral blindness will tilt their heads so that the 'good' eye is uppermost. The bird may have an exaggerated startled reaction.

Physical examination

Symmetry, position, and mobility of globes should be assessed. The eyes of a normal bird are clear, symmetrical, and centred in the socket. The normal conjunctiva is pale pink and moist.

Hydration can be evaluated by tenting the skin over the eyelids.

The periorbital area is assessed for swelling, discolouration, pustules, scars, scabs, or abnormal growths. Conjunctivitis, sunken eyes, and ocular discharges are abnormal. Pale or white colouration of the conjunctiva can indicate anaemia. Periorbital feather loss can indicate the bird is rubbing its eyes against perches and the cage sides, suggestive of pain or irritation.

The iris colour should be noted. The colour is an indication of age or sex in many species of birds. Young Amazon parrots have brown irises that become red-orange as they age. Young African grey parrots have brown irises that change to grey, then white, by 15 months of age. Young macaws have brown irises that change to grey within the first year, and then between 1 and 3 years of age, the colour changes from grey to yellow. Young cockatoos have brown irises. In some species, the adult female cockatoo has red-brown irises, and adult males have dark brown to black irises. (Note that not all hens have red-brown irises.)

Examination of the globe

The cornea and anterior chamber should be examined with focal illumination and transillumination.

Fundic examination can be difficult, as the striated muscle in the iris makes pupillary dilation difficult; atropine and tropicamide are ineffectual. Therefore, neuromuscular blocking agents are often required. Techniques that have been used include:

- Intracameral injection of d-tubocurarine. Topical application of this drug does not work well, as it does not penetrate the cornea sufficiently. The drug, therefore, has to be administered directly into the anterior chamber using a 27–30 gauge needle, carrying with it a substantial risk for injuries to intraocular structures causing hyphaema, increasing intraocular pressure (IOP), transmission of conjunctival flora with consecutive uveitis, and systemic side effects if larger doses than recommended are used. This technique may best be used therapeutically for the prevention of posterior or anterior synechia.
- Topical vecuronium bromide (4 mg/ml). Vecuronium is a synthetic curariform non-depolarising neuromuscular blocking agent. It competes for nicotinic acetylcholine receptors at the motor end plate. It does not work topically in South African black-footed penguins but has worked in citron-crested and sulphur-crested cockatoos and African grey and blue-fronted Amazon parrots with minimal systemic effects.
- Similarly, rocuronium bromide (20 μL/eye; 10 mg/mL) applied topically produces mydriasis after 20 minutes, with the optimum effect at approximately 90 minutes.
- General anaesthesia, especially with ketamine or air sac perfusion, utilises an air sac catheter placed in the caudal thoracic air sac to administer and maintain general anaesthesia using isoflurane and oxygen (0.3 L/kg/min). This technique frees the head for examination and treatment. However, while it has been reported to be successful in inducing mydriasis in pigeons, it does not appear to work as well in cockatoos.

Once the pupil is dilated, the fundus can be examined using a focused light beam (e.g., Finoff Transilluminator, Welch Allyn®) and a double aspherical ophthalmoscopic lens (at 30, 40, 60, 78, and 90 diopters refractive power). A 30D lens is used in birds with larger pupil diameters (nocturnal and diurnal raptors); a 78D lens is used for pigeons and larger parrots; and a 90D lens is used in smaller parrots. Alternatively, monocular indirect ophthalmoscopy can be performed in all birds, even the smaller species. The retina, fovea, and pecten can all be visualised. All birds with traumatic head injuries should have an ocular examination, as the incidence of retinal haemorrhage and detachment is relatively high in these patients.

Responses and reflexes

- Menace reflex (assesses cranial nerves II and VII). Eye blink, pulling away of the head, or aggressive action of the beak can be provoked by bringing the hand towards each eye.
- Pupillary light response (assesses cranial nerves II and III). The avian pupil's movements are poor in response to light but rapid in response to accommodation or voluntary control. Although there is no consensual pupillary light response (as there is complete decussation of the optic nerves at the chiasm), the orbital bone is so thin that a response can often be detected in one eye after light is shone in the other.
- Corneal reflex. A symmetrical eye-blink normally occurs when the cornea is gently touched with a moist cotton swab. Failure to do so indicates a lesion in cranial nerves V and VII.
- Palpebral reflex and facial sensation. A symmetrical eye-blink should be elicited by touching each side of the face or the lateral canthus of the eye. Failure to do so suggests a lesion in cranial nerves V and VII.
- There should be symmetrical deviations of the eyes when the head is moved into different positions, with the eyes always returning to the centre. Failure to do so suggests either a bilateral lesion in cranial nerve VIII or a brainstem lesion (especially in the pons and midbrain).

Ancillary testing

- Cytology and culture of the conjunctival flora should be performed in cases of conjunctivitis or ocular discharge.
- A Schirmer tear test is performed as with other species; in clinically normal birds, the result is 3–12 mm, but there is a wide variation between species. Standard reference intervals, using filter strips of various widths, for 42 species from 7 orders show a wide range of interspecific variations. Strigiformes show conspicuously low values. Local anaesthesia in the cornea significantly reduces tear movement along the strip.
- The Phenol Red Thread (PRT) test (Zone-Quick, San Mateo, California) was developed as an alternative to the Schirmer tear test for the evaluation of tear production in animals. It is a yellow cotton thread, 75 mm long, impregnated with the pH indicator phenol red (phenolsulphonephthalein). One end of the thread is folded to a length of 3 mm and, at the time of use, opened to an appropriate angle and placed within the ventral conjunctival fornix. The thread remains in this position for 15 seconds. The colour of the thread changes to red when wet and, once the thread is removed, the length of this change is measured in millimetres similar to the Schirmer tear test. The PRT test was developed to overcome the disadvantages of Schirmer's test, including variable results, poor repeatability, and low sensitivity in detecting dry eyes. A test time of only 15 seconds is required, in comparison to 5 minutes per eye needed for the Schirmer test. It causes minimal discomfort and therefore is less likely to induce reflex tear production. Normal results appear to lie in the range of 8–23 mm, with no effect from local anaesthesia discerned.
- Fluorescent staining of the cornea can detect small ulcerations and lacerations.
- Tonometry is an indirect method of measuring intraocular pressure (IOP). Various types of tonometers are used in veterinary medicine; applanation tonometers have been used to measure the IOP of domestic animals, but the small size of the avian globe restricts their use in these species. The rebound tonometer, with a small probe, can be used to measure IOP in birds. The IOP values of Columbiformes, Psittaciformes, and Passeriformes have been reported as 5.8 ± 1.96 mm Hg, 5.14 ± 0.06 mm Hg, and 5.79 ± 1.94 mm Hg, respectively.
- Ultrasonography allows detection of retinal detachment or sequestration of the pecten bridge.
- More advanced tests include:

1. Electroretinography is the recording of electrical potentials after light stimulation. It gives no information about vision but does assess retinal function.

2. Fluorescein angiography is the observation of the distribution of fluorescein within ocular blood vessels following intravenous injection. This enables the diagnosis of subtle haemorrhages of the pecten and choroid, atrophy of vessels, and retinal disease.

DISORDERS OF THE EYE

Lids and periorbital region

Poxvirus (in mild cases) may cause unilateral blepharitis with eyelid oedema, starting about 10–14 days after infection. It may also cause keratitis and, less commonly, anterior uveitis. Prophylactic vitamin A injections and antibiotic therapy may decrease the severity of the infection. Most cases are self-limiting and resolve over a few weeks. In more severe cases, scabby sores can develop on the eyelids, resulting in distortion and scarring of the eyelids.

Chlamydial infections and other causes of sinusitis may lead to blepharitis, thickening of the eyelids, and matting or loss of periorbital feathering (due to lacrimation and rubbing).

Proliferative lesions on the eyelids and periorbital region may be due to *Cnemidocoptes pilae*, papillomas, or vitamin A deficiencies.

Feather loss around the eye without any evidence of inflammation or conjunctivitis can be due to excessive grooming by a cage mate. This is often seen in finches when they are overcrowded or in parrots during the breeding season.

Eyelid abnormalities include:

- Cryptophthalmos (failure of eyelid formation, resulting in the fusion of the eyelid margins) is occasionally seen in cockatiels and other parrots. It can be acquired or congenital. Surgical correction is usually unrewarding, but affected birds appear to have some degree of vision through their eyelids (**Figs. 13.4, 13.5**).
- Blepharophimosis (narrowing of the palpebral fissures without fusion of the eyelid margins) is occasionally seen in all species. Again, surgical correction is usually unrewarding.
- Symblepharon (adhesion of the eyelid to the globe) can be seen as a sequel to severe conjunctivitis. Surgical correction is possible in some cases.
- Scarring and deformity are occasionally seen after avian poxvirus infections and traumatic injuries.

Figure 13.4 **Cockatiel with acquired cryptophthalmos following a severe chlamydial infection.**

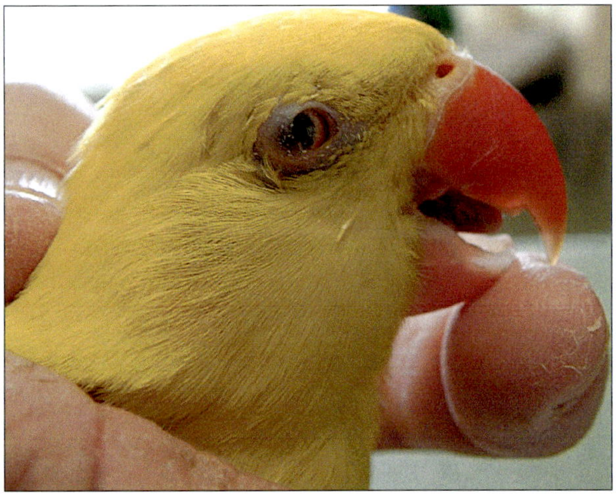

Figure 13.5 **Indian ringneck parrot with congenital cryptophthalmos.**

Globe and orbit

Microphthalmia, a congenitally abnormally small eye, is rarely seen. In some cases, the eye is structurally and functionally normal, but in other cases, cataracts, retinal dysplasia, or retinal detachment result in complete blindness. Chronic conjunctivitis, due to conjunctival exposure, can be treated with good hygiene and topical antibiotics as indicated.

Anophthalmia, the actual total absence of a globe, is rare. Many cases may contain a rudiment of ocular tissue, which can result in a weeping sinus.

Exophthalmia, the protrusion of the globe from the orbit, may be due to malformation of the orbit associated with metabolic bone disease, or it may be secondary to infection (bacterial, fungal, or trichomonads) in the infraorbital sinuses. Other reported causes include trauma, inflammation of the Harderian gland, or neoplasia of the orbit or globe. Sinusitis is the most common cause and may require surgical removal of caseated material to achieve resolution. Retrobulbar sinusitis is difficult and frustrating to treat because of the caseous nature of the exudate and the lack of surgical access to the sinus. Surgical trephination of the sinus followed by flushing and mechanical debridement may assist in some cases.

Panophthalmitis, inflammation of the globe, may be caused by infection or trauma. Early clinical signs are referable to the extension of the inflammatory process throughout the eye, and include pain, blepharospasm, hypopyon, corneal oedema and ulceration, and secondary glaucoma. Chronically, the eye may develop phthisis bulbi, a shrunken, fibrotic globe. Treatment requires aggressive antimicrobial therapy, NSAIDs, and temporary tarsorrhaphy in early cases, but in more chronic or non-responsive cases, enucleation is often the preferred treatment.

Glaucoma, increasing pressure within the globe, may be primary or secondary to trauma. The globe is enlarged and painful. Glaucoma must be differentiated from exophthalmos. Diagnosis is based on clinical signs and intraocular pressure, as measured by a Tonopen® or Tonovet® (see earlier). Intraocular pressures are between 9.2 and 22 mmHg in normal bird eyes. One suggested treatment is topical dorzolamide applied three times daily initially, then titrated to effect.

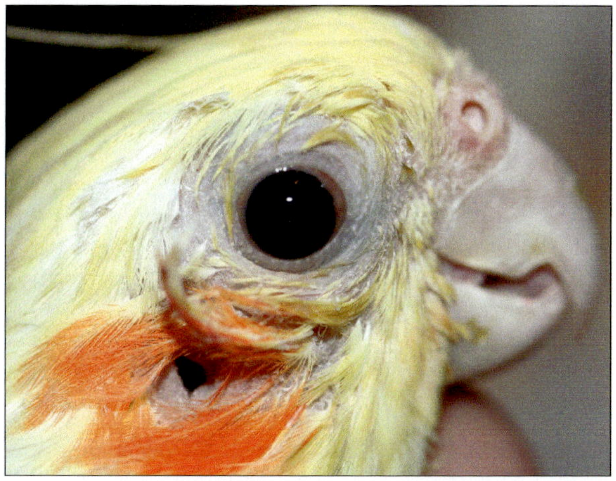

Figure 13.6 **Cockatiel with conjunctivitis.**

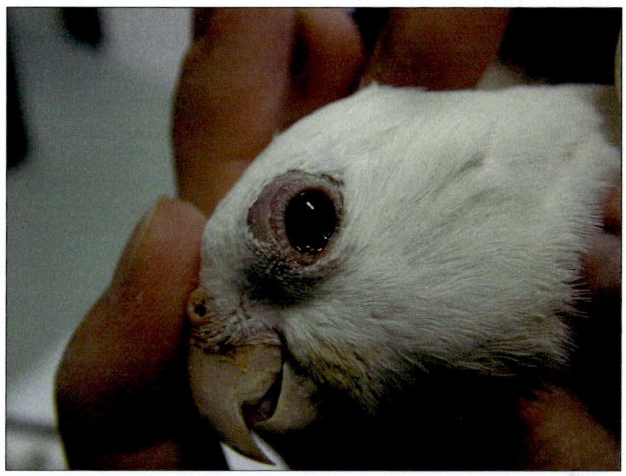

Figure 13.7 **Proliferative conjunctivitis in a galah.**

Conjunctiva

Overview

Thickening and hyperaemia of the conjunctiva, often associated with epiphora or ocular discharge, may be seen with conjunctivitis, sinusitis (see Chapter 23), or neoplasia (**Figs. 13.6, 13.7**).

Aetiology

Conjunctivitis may be due to:

- Bacterial infections.
- Chlamydiosis.
- Mycoplasma.
- Mycobacteria.
- Viral infections (e.g., adenovirus, poxvirus, cytomegalovirus [in Gouldian finches]).
- Fungal infections (*Aspergillus*, *Candida*, *Cryptococcus*).
- Parasites: 'eye' worms (*Oxyspirura mansoni*, *Ceratospira* spp., and *Thelazia* spp.); cryptosporidial infections. *Plasmodium* spp. can cause eyelid swelling in canaries and domestic poultry. Microsporidiosis has been reported to cause conjunctivitis. Trichomoniasis is an uncommon cause of conjunctivitis.
- Foreign bodies and physical irritants such as smoke or chemical fumes.
- Allergies, especially secondary to nicotine.

Neoplasia involving the eyelids includes conjunctival papillomas, haemangiomas, squamous cell carcinomas, and melanomas (**Fig. 13.8**).

Diagnosis

Diagnosis involves cytology and culture of the conjunctiva. Conjunctival biopsy or PCR of a conjunctival swab are also useful diagnostic tools.

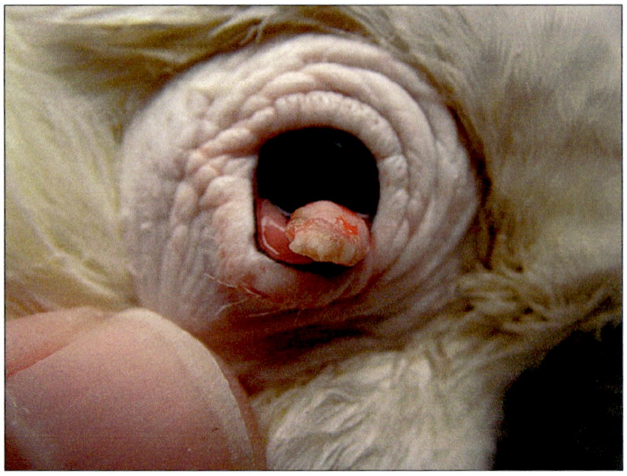

Figure 13.8 **Cockatoo with a conjunctival tumour (haemangioma).**

Management

This is determined by the aetiological agent. Topical therapy, if indicated, should be in drop form. Eye ointments should be avoided, or at least used very sparingly, because of the likelihood that grooming will cause the ointment spread throughout the feathering. Ointments containing cortisone should not be used at all, as systemic absorption and subsequent immunosuppression and hepatopathies can be significant.

In many cases, especially chlamydiosis, systemic as well as topical therapy are indicated.

Cornea

Overview

Xerophthalmia, a lack of tear production, is occasionally seen with vitamin A deficiency. A long-term deficiency may result in irreversible keratoconjunctivitis sicca.

Cholesterol deposits are seen frequently in parrots, especially Amazons, usually on a high-fat diet or with other metabolic issues. They appear as painless, multifocal, white, non-inflammatory corneal lesions. Even with the resolution of the dietary problems, these corneal precipitates do not resolve but rarely seem to cause a problem.

Infectious keratitis may be bacterial or fungal or caused by microsporidia. Non-infectious keratitis may be caused by trauma or a foreign body. Corneal ulceration is often associated with foreign bodies such as dust or seed husks.

Management

Antimicrobial therapy, topically and/or systemically, is given as indicated. Flurbiprofen topically or meloxicam systemically may be used for analgesia and anti-inflammatory therapy.

The cornea can be protected by a temporary tarsorrhaphy. Third eyelid flaps have been used, but they are difficult to perform, and experience in other species indicates they are not as useful as originally thought.

In severe cases, enucleation may be required. If possible, a specialist's opinion should be sought before taking this step.

Uvea

Acute uveitis

Aetiology

Acute uveitis may be caused by trauma, infection (e.g., reovirus, West Nile virus, and toxoplasmosis), septicaemia, or lens rupture.

Clinical presentation

Signs include photophobia, blepharospasm, corneal oedema, aqueous flare, hyphaema, hypopyon, miosis, dyscoria, iris thickening or discolouration, and anterior/posterior synechiae (**Fig. 13.9**).

Figure 13.9 **Hawk with acute uveitis.**

Management

In general, birds respond to treatment for uveitis better than mammals. Anti-inflammatory therapy is required using topical (e.g., flubiprofen) and systemic (e.g., meloxicam) NSAIDs.

Chronic uveitis

Aetiology

This is a sequela to acute uveitis.

Clinical presentation

Signs include diffuse corneal oedema, posterior synechiae causing pupillary seclusion and iris bombé, anterior synechiae, secondary glaucoma, cataract, retinal atrophy and chronic detachment, and blindness.

Management

Anti-inflammatory therapy is given as for acute uveitis: topical (e.g., flubiprofen) and systemic (e.g., meloxicam) NSAIDs.

Lens

Cataracts

Aetiology

Familial cataracts are reported in Yorkshire and Norwich canaries. A fully penetrant autosomal recessive gene is responsible. Acquired cataracts can be due to uveitis (see earlier), toxins, or ageing. Idiopathic cataracts have been noted in a number of species (**Fig. 13.10**).

Figure 13.10 **Mature kakariki hen with acquired cataracts due to ageing.**

Treatment

In many cases, no treatment is required or feasible due to other disease processes in the eye affecting vision. Phacoemulsification has been used successfully in many species and may be used if the retina is believed to be intact.

Luxation

This may occur secondary to trauma or uveitis. Lens luxation also leads to further uveitis. If this lens-induced uveitis does not respond to NSAID therapy, phacoemulsification is necessary to remove the lens. If this is not available, enucleation should be considered.

Retina

Retinitis may occur due to infectious agents (e.g., *Toxoplasma*), trauma, or diseases of the CNS. Retinal detachment may occur following trauma or retinitis. Retinal dysplasia has been noted sporadically. Retinal degeneration may be due to malnutrition, toxins, or hypoxia, but no specific cause has yet been identified.

FURTHER READING

Gancz AY, Malka S, Sandmeyer L, Cannon M, Smith DA, Taylor M (2005) Horner's syndrome in a red-bellied parrot (*Poicephalus rufiventris*). *Journal of Avian Medicine and Surgery* 19(1):30–34.

Karai V, et al. (2020) Intraocular pressure measurements using rebound tonometry in eight different species of companion birds. *Journal of Avian Medicine and Surgery* 34(4):338–342.

Kern TJ (1997) Disorders of the special senses. In: *Avian Medicine and Surgery*. RB Altman, SL Clubb, GM Dorrestein, K Quesenberry (eds). WB Saunders, Philadelphia, pp. 563–589.

Korbel RT (2000) Avian ophthalmology: A clinically oriented approach. In: *Proceedings of the Annual Conference of the Association of Avian Veterinarians*, pp. 439–456.

Korbel RT (2007) Avian ophthalmology: Principles and application. In: *Proceedings of the Annual Conference of the Association of Avian Veterinarians*, pp. 191–200.

CHAPTER 14

DISORDERS OF THE EAR

ANATOMY AND PHYSIOLOGY

Birds have a keen sense of hearing and a high degree of equilibration. Their excellent voice production and remarkable ability to imitate sounds have inferred an exceptional degree of sound analysis (pitch discrimination) within a wide range of auditory frequencies. Their aerial mode of life demands sonic acuity, in addition to a well-coordinated balance and position sense. In many ways, the avian ear closely resembles the mammalian ear, but it is simpler in structure and reptilian in design. It includes three separate but contiguous anatomical segments (the external, middle, and inner ears). These develop completely independently and from different embryonic primordia, and then combine to form a synchronised functional unit.

The external ear collects sound waves from the outside air and conducts them to the middle ear. It consists of a collecting device, the external acoustic meatus (a small aperture, nearly always circular, which opens externally on the side of the head), and a simple conducting tube (**Fig. 14.1**). It terminates medially as a partition, the tympanic membrane (eardrum), completely separating the external ear from the middle ear. It is covered by specialised contour feathers (the ear coverts) in most birds; the width of the feathers is related to sound localisation, and their fine structure varies in accordance with auditory efficiency. The coverts lying on the rostral aspect of the meatus reduce the drag caused by turbulence in flight and thus diminish the masking of sound by noise generated from turbulence in the external ear; the barbs of these ear coverts lack barbules, so the sound waves are not obstructed. On the caudal aspect of the meatus, these specialised feathers combine into a tight funnel, which is particularly enlarged in songbirds, parrots, and falcons.

The middle ear is an air-filled, ossicle-containing space (tympanic cavity) that receives sound waves as mechanical vibrations of the tympanic membrane in its lateral wall and transfers them in an amplified form to the inner ear at its medial wall. It is directly continuous with the pharynx via the pharyngotympanic tube (Eustachian tube), which enters the pharynx at the infundibular cleft. The middle ear also communicates with a large group of accessory air cavities occupying the surrounding skull bones and then extending into the mandible and beak in some species. The right and left tympanic cavities communicate with each other via interconnecting air sinuses; this has been implicated in the transference of pressure fluctuations emanating at the round window.

Figure 14.1 **Normal external auditory meatus in a macaw.**

DOI: 10.1201/9781003272519-14

In contrast to mammals, birds transfer sound with a single skeletal element, the columella, which extends medially across the tympanic cavity to form a direct connection between the tympanic membrane and the fluid within the inner ear. The columella is homologous to the mammalian stapes; the mammalian incus and malleus are homologous to the avian quadrate and articular bones, respectively. The tympanic membrane and columella function as a mechanical transformer that matches the impedance of air and inner ear fluid, facilitating the transfer of sound energy. The tension in the tympanic membrane is altered by the columellar muscle, which attaches to the columella and to the tympanic membrane itself. Vibrations of the tympanic membrane are carried to the perilymph of the inner ear by the extracolumella cartilage in contact with the tympanic membrane and the rod-like bony columella, which is implanted medially in the vestibular window. The cochlear (round) window lies near the vestibular window; it is in contact with the scala tympani of the inner ear.

The inner ear is responsible for the initial analysis and characterisation of the sound vibrations and for maintaining equilibrium. It consists of two very complex, fluid-filled components or labyrinths: one membranous, the other bony. The membranous labyrinth includes several nonauditory receptive areas composing the vestibular labyrinth and a single organ of hearing, the cochlear labyrinth. The vestibular labyrinth is a series of canals and ducts filled with endolymph and surrounded by perilymph. It is encased within the bony labyrinth. The cochlea is a relatively short and slightly curved tube containing the cochlear duct (scala media) filled with endolymph. It ends at the lagena, which contains a group of sensory cells; afferent nerve fibres from this area appear to end in the auditory centres of the medulla.

Movement of endolymph within the vestibular labyrinth provides the bird with its proprioception. Vibrations carried through the cochlear duct are detected in the brain as sound.

OTITIS EXTERNA

Aetiology

Otitis externa is an uncommon disorder. It may result from infection with bacteria (*Escherichia coli*, *Staphylococcus aureus*, *Pasteurella multocida*, *Proteus mirabilis*, *Pseudomonas aeruginosa*, *Klebsiella oxytoca*, *Enterococcus* spp.), fungi (*Candida* spp., *Microsporum gallinae*), or mycobacteria, or infestation with arthropod parasites (*Cnemidocoptes*). It may occur as an extension of generalised skin disease or may be due to neoplasia or trauma.

Figure 14.2 **Otitis externa in a budgerigar.**

Clinical presentation

Affected birds often demonstrate head shaking, yawning, head tilt, and ataxia. Closer examination may reveal an aural exudate (serous, purulent, or haemorrhagic) (**Fig. 14.2**). There may be caseated material in the ear canal (**Fig. 14.3**). Swelling and erythema, auricular feather loss, and self-inflicted excoriation are often seen. There may be hyperkeratosis of the ear canal and surrounding skin. In some chronic cases, the ear canal may become so stenotic as to close over.

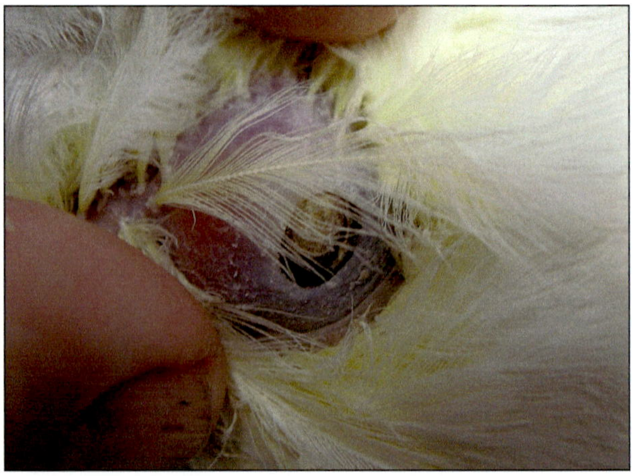

Figure 14.3 **Otitis externa in a cockatoo.**

Diagnosis

Diagnosis is based on clinical signs, cytology, and culture. The narrow ear canal makes visualisation with anything larger than a 1.9 mm rigid endoscope difficult, if not impossible.

In cases that fail to respond to therapy, a judicious biopsy may be required to better evaluate the disease process.

Management

Systemic and topical antibiotics are given as indicated by culture. Meloxicam or another NSAID can be given if the ear appears to be particularly painful or swollen. Careful flushing with saline may assist in the removal of exudate; if this is unsuccessful, debridement with a loop curette may be required.

OTITIS MEDIA AND OTITIS INTERNA

Aetiology

Otitis media and otitis interna are uncommon. They can occur as a result of developmental abnormalities seen in some canaries. Other causes include an extension of otitis externa or sinusitis, PMV-3 infection (especially in Australian *Neophema* spp., canaries, and finches), poxvirus, herpesvirus, iatrogenic damage following treatment with aminoglycoside antibiotics (e.g., gentamicin), and neoplasia.

Clinical presentation

Signs include head tilt, 'star gazing', torticollis, and vestibular disorders (loss of proprioception).

Diagnosis

Diagnosis is based on clinical signs, PMV-3 serology, or virus isolation (from tissue samples), if appropriate. In larger birds, CT or MRI may be of value in confirming a diagnosis. As birds do not possess a tympanic bulla (the structure most associated with otitis media in dogs and cats), changes in the quadrate bone and loss of air-filled spaces in the middle ear and associated bones are associated with otitis media. Endoscopic evaluation of the tympanum and clinical signs of vestibular disease complete the diagnosis.

Management

Antimicrobial and anti-inflammatory therapy may be of some value, although the prognosis is guarded, as treatment usually requires the removal of caseated pus and debris. A myringotomy, perhaps combined with an ear canal ablation, may allow gentle irrigation and removal of caseated pus in the middle ear. The skin is incised around the external meatus, and careful dissection is used to mobilise and excise the external ear canal. Care must be taken in the dorsomedial region, where the external ear connects to the inner ear and vestibular system. Complications of this surgery include recurrence of the otitis, draining tracts, and facial palsy.

FURTHER READING

Delk KW, et al. Diagnostic imaging of peripheral vestibular disease in a Chinese goose (*Anser cygnoides*). *Journal of Avian Medicine and Surgery*, vol. 28, no. 1, 2014, pp. 31–37.

Kern TJ (1997) Disorders of the special senses. In: *Avian Medicine and Surgery*. RB Altman, et al. (eds). WB Saunders, Philadelphia, pp. 563–589.

Schmidt RE, Reavill DR, Phalen DN (2003) Special sense organs. In: *Pathology of Pet and Aviary Birds*. RE Schmidt, et al. (eds). Iowa State Press, Ames, pp. 197–212.

Shivaprasad HL (2007) The avian ear: Anatomy and diseases. In: *Proceedings of the Annual Conference of the Association of Avian Veterinarians*, pp. 127–133.

CHAPTER 15

DISORDERS OF THE PELVIC LIMB

ANATOMY AND PHYSIOLOGY

Pelvis

The pelvis is made up of the ilium, ischium, and pubis. The pelvis is incomplete ventrally, allowing the passage of an egg during oviposition (note that some flightless birds, e.g., ratites, have a fused ventral symphysis). The arched shape of the pelvic bones enables birds to carry the weight of their body in a bipedal standing position. The ilium is also fused to the synsacrum, 10–23 vertebrae (notarial, sacral, and caudal) fused to each other, completing a rigid framework to support the pelvic limbs.

The acetabulum is deep, with an articular facet (the antitrochanter) on the dorsal aspect of the acetabulum, articulating with the antitrochanteric articular facet of the femur. The socket itself is perforated.

The movements of the pelvic joint are mainly extension and flexion; abduction and adduction movements of the joint are virtually non-existent.

Femur

The femur is a stout and comparatively short bone that slopes cranially to bring the legs forward towards the centre of gravity (**Figs. 15.1, 15.2, 15.3**). A patella is present in most birds. It is associated with the tendon of the femorotibial muscle and articulates with the patellar sulcus of the femur.

Movements of the knee (stifle) joint are extension and flexion, although medial rotation of the tibiotarsus is possible.

Tibiotarsus

The tibiotarsus is formed by the fusion of the tibia and the proximal row of tarsal bones; the hock joint, therefore, is an intertarsal joint between the tibiotarsus and the tarsometatarsus. The proximal end of

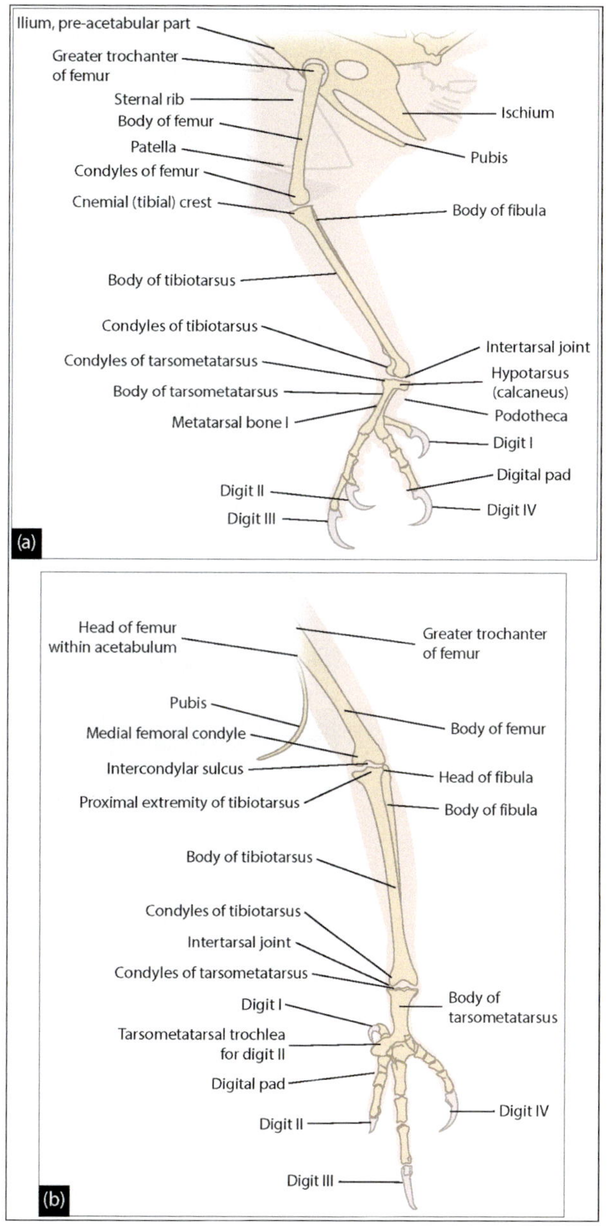

Figure 15.1 **Anatomy of the pelvic limb.**

the tibiotarsus carries the cranial cnemial crest, to which are attached the muscles of the thigh and leg. The trochlea on the caudal surface of the distal end

DOI: 10.1201/9781003272519-15

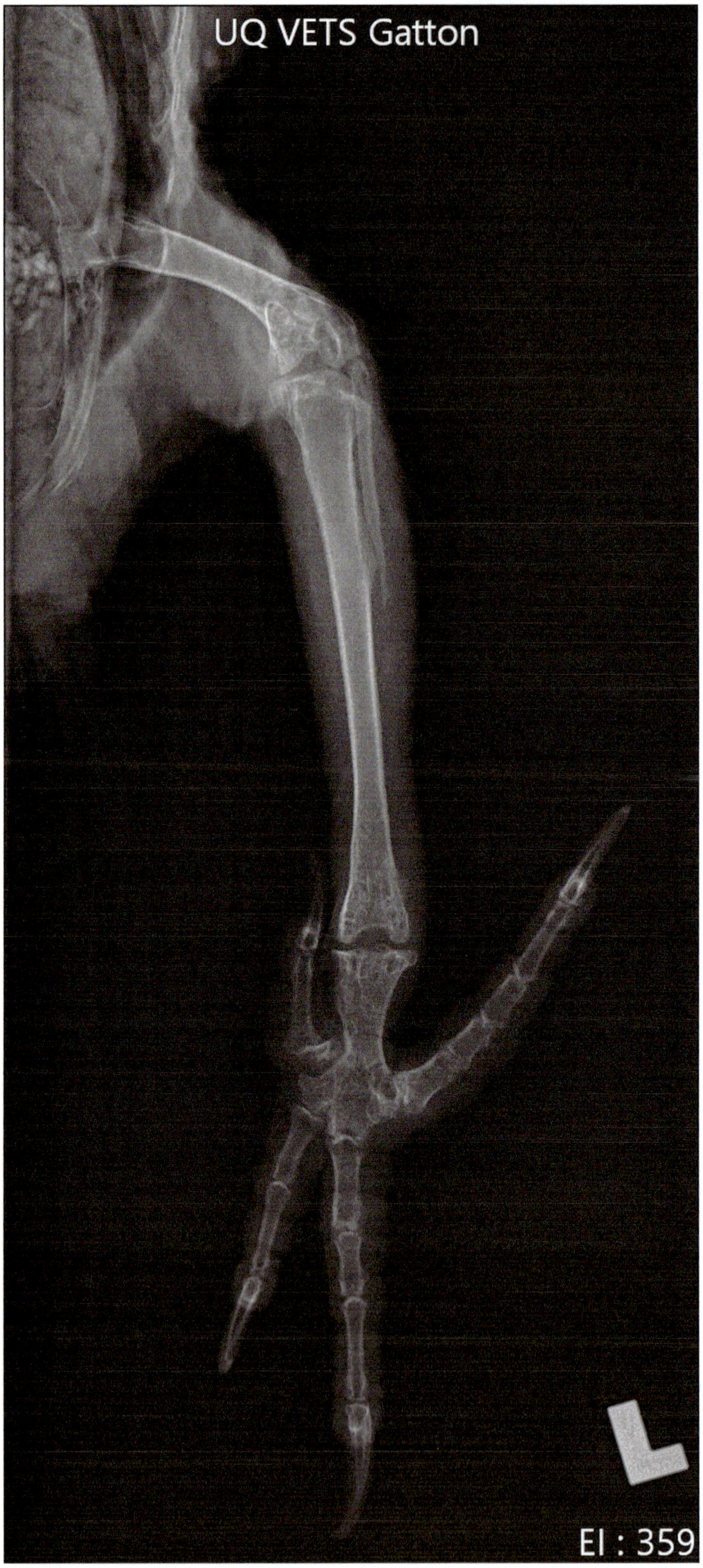

Figure 15.2 **Anatomy of the pelvic limb.**

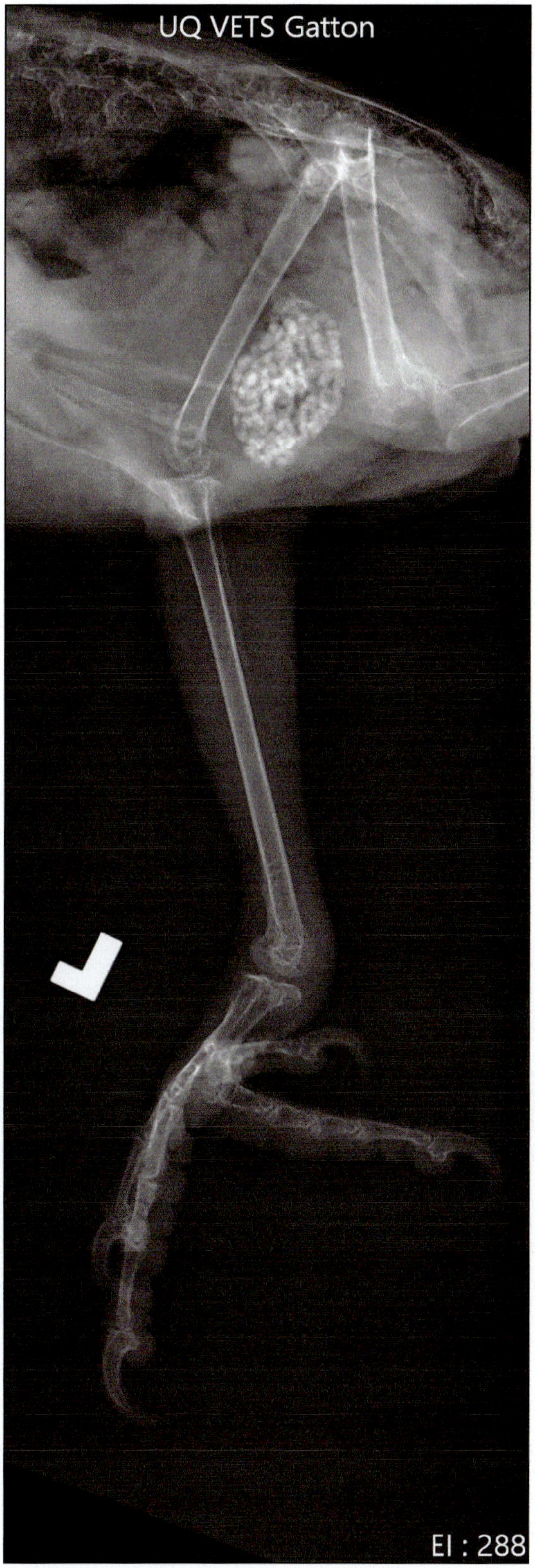

Figure 15.3 **Anatomy of the pelvic limb.**

of the tibiotarsus articulates with a block of fibrocartilage known as the tibial cartilage; the tendons of the gastrocnemius and superficial flexors pass over this cartilage, and the tendons of the deep digital flexor muscles pass through it.

The fibula extends two-thirds of the way down the tibiotarsus, to which it is fused. The reduction in its size limits the rotation of the leg.

Movements at the intertarsal joint (hock) are flexion and extension.

Tarsometatarsus

The tarsometatarsus, formed by the fusion of the distal row of tarsal bones to the three main metatarsal bones of digits II, III, and IV, is usually shorter than the tibiotarsus except in long-legged birds. On the plantar aspect of the proximal extremity of the tibiotarsus is a process, the hypotarsus, which in most birds has grooves and high crests and is perforated by canals. Within the grooves and canals lie the digital flexor tendons.

Digits

There are four digits in parrots: I has two phalanges and is usually directed backwards; II has three phalanges; III has four phalanges; and IV has five phalanges. The fourth digit is directed caudally in parrots. The main movements of the digits are flexion and extension, and there is a certain amount of abduction and adduction (**Fig. 15.4**).

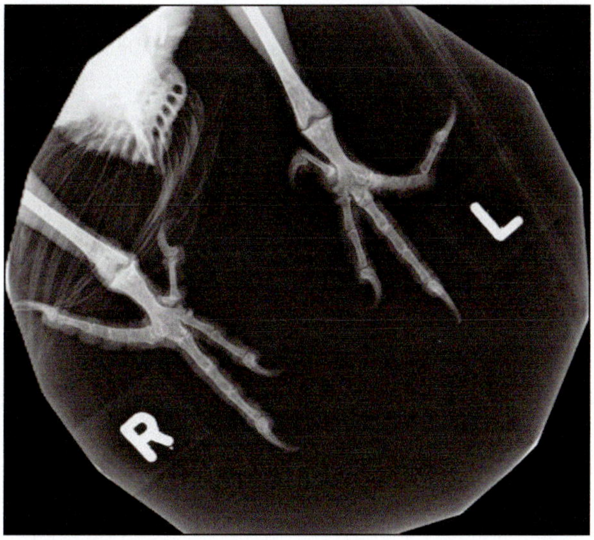

Figure 15.4 **The phalanges. Note the subluxation in P4 of the left leg.**

Joints

The anatomy of the joints is discussed in more detail in Chapter 33.

Podotheca and nails

Scales, raised areas of highly keratinised epidermis separated by folds of less keratinised skin, cover the non-feathered part of the leg, known as the podotheca. The claws, which enclose the terminal phalanx of each digit, are made up of two plates: the strongly keratinised dorsal plate enclosing the dorsal and lateral aspects of the phalanx, and the softer ventral plate forming the sole of the claw. The dorsal plate grows faster than the ventral plate; therefore, the nails curve downwards.

Pneumaticity

The pelvic limb bones vary in their degree of pneumaticity, occupied by diverticula of the air sacs. In the pelvic limb, the femur is pneumatic, and to a lesser extent, other bones may be as well.

Hyperostotic polyostosis

Medullary bone is labile bone, found in the femur, ulna, radius, pectoral girdle, and vertebrae in hyperoestrogenic birds (e.g., reproductively active hens, hens with ovarian cysts, or cocks with Sertoli cell neoplasia). It consists of interconnected bone spicules, resembling ordinary embryonic bone. About 2–3 weeks before laying, under the influence of oestrogen, the spicules grow out from the endosteal surface of the cortical bone to fill (or partially fill) the medullary canal. The spaces between the spicules are occupied by blood sinuses and red marrow. The spicules themselves contain osteocytes and are covered with variable numbers of osteoblasts and osteoclasts. There are no Haversian canals. Neither the collagen fibres nor the hydroxyapatite crystals are orientated, indicating this bone has no structural role – rather, this medullary bone acts as a calcium reservoir for egg laying.

A cycle of formation and destruction of medullary bone alternates during the breeding season as calcium demand waxes and wanes between eggs. At the end of the breeding season, when oestrogen levels decline, much of the medullary bone is reabsorbed, and the bony trabeculae become shorter, narrower, and much less continuous.

Clinically, it is seen as increased radiopacity of these bones. Fractures of long bones that have a solid medullary cavity can be difficult to repair as, without a medullary and endosteal blood supply, there may be a poor blood supply to the fracture site. This frequently results in a non-union of the fracture. Repair of these fractures should include drilling out part of the medullary bone and utilising bone grafts. In medical and small animal orthopaedics, the use of bone morphogenetic protein (BMP) shows great promise in stimulating healing of non-union fractures; it is possible that may be an avenue to be explored in avian medicine (**Fig. 15.5**).

Perching

Perching in birds is assisted by the digital flexor tendons, which cross the caudal surface of the intertarsal joint. When the bird crouches, flexion of the joint

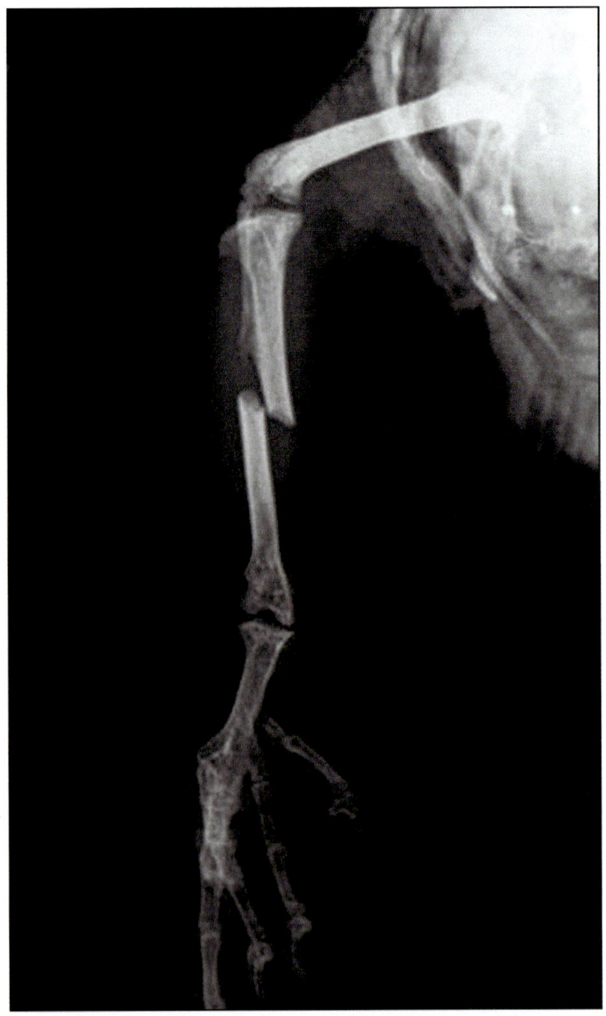

Figure 15.5 **Mid-shaft tibiotarsal fracture of a reproductively active hen with hyperostotic polyostosis.**

passively tenses the tendons and thus automatically clamps the digits around the perch. Further locking is provided by means of minute projections on the undersurface of the digital flexor tendons at the metatarsophalangeal joints. These projections are forced, by the weight of the crouching bird, to interdigitate like a ratchet with ridges on the inside of the tendon sheaths. When the bird rises, the intertarsal joints are extended; this removes the tension from the flexor tendons and allows the digits to be extended.

CLINICAL SIGNS OF PELVIC LIMB DISEASES

Birds with leg disorders may hold the limb raised at all times, or they may 'toe touch', i.e., place the affected limb on the ground/perch for only a brief period of time before again taking their weight with the other limb. Birds with loss or damage to the nerve supply will usually exhibit paresis and be unable to raise the leg or weight-bear on it.

Limbs may be rotated medially or laterally, or they may be bowed.

Some birds with leg injuries will extend their wing(s), touching the primary feathers to the ground to act as a crutch to take weight off the injured leg. If examination of the wing fails to demonstrate a lesion, ensure the legs are carefully examined.

A lameness that shifts from foot to foot, especially on a perch, suggests that there may be bilateral lameness. This is often associated with articular gout (**Fig. 15.6**).

Figure 15.6 Toe-touching lameness of the right leg. This bird has articular gout.

EXAMINATION OF THE PELVIC LIMB

Before restraining the bird, perform a distant examination of the whole bird at rest. Then restrain the bird using a towel (see Chapter 3). It is vital to avoid further injury to the leg during this, and the use of sedation (e.g., midazolam intranasally) may be of benefit. Where possible, have an assistant (not the owner) restrain the bird – it is a two-person job to examine or anaesthetise a conscious bird. Most parrots can be held for a brief examination, but usually a general anaesthetic is needed for a more detailed examination.

The leg examination begins by extending both legs together by pulling on the claw of the third digit, and then releasing them. Either resistance or failure to return to normal position is significant. Flex and extend the hock joint; the centre of the foot should align with the cnemial crest. Lateral or medial positioning may indicate an angular limb deformity (see later) (**Fig. 15.7**).

Palpate the long bones and the joints for pain, swelling, crepitus, or abnormal movement. Examine the digital joints for white deposits and swelling consistent with articular gout (**Fig. 15.8**).

Examine the plantar aspect of the feet. Long-term pressure causes the skin to lose its texture, the

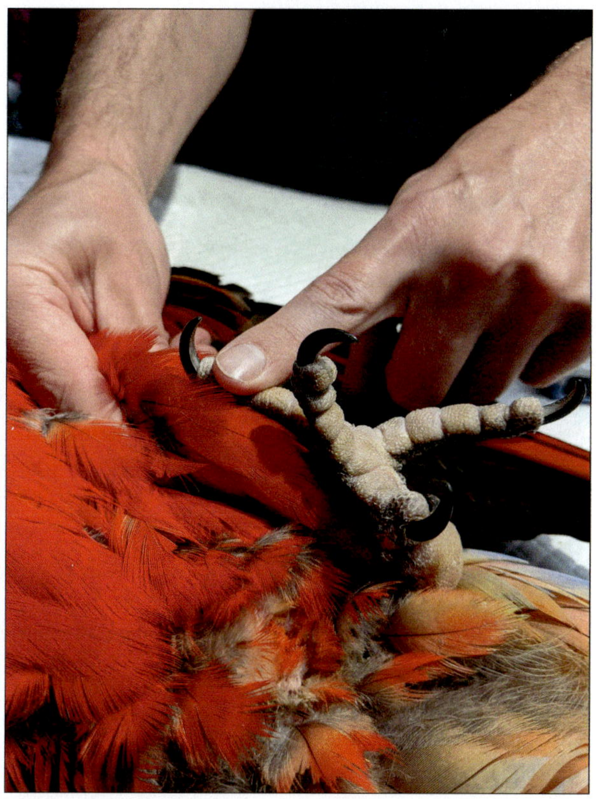

Figure 15.7 Flexing and extending the hock joint to check for an angular limb deformity.

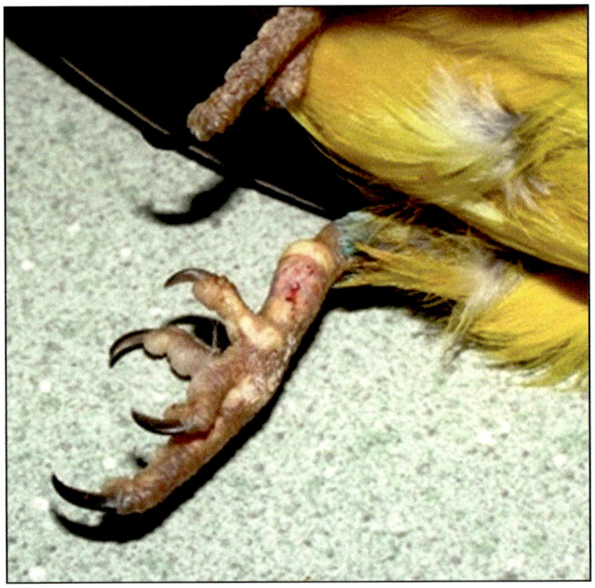

Figure 15.8 **Articular gout (same bird as Fig. 15.6).**

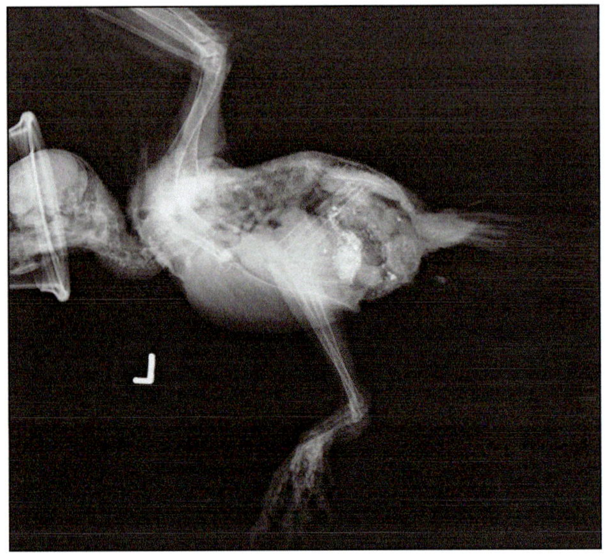

Figure 15.10 **Kyphosis at the synsacral vertebral junction.**

digital pads to flatten, and the scales to lose their form (**Fig. 15.9**). Unilateral changes suggest that the bird is lame on one leg, usually the leg that has the least affected plantar surface. The longer that the bird has been affected by lameness and/or illness, the more obvious the changes.

Examine the integument over the synsacrum and palpate the vertebral column, especially at the synsacral vertebral junction (**Fig. 15.10**). If there is any concern in this area, wet the feathers with a small amount of alcohol to check for bruising in this area.

RADIOGRAPHY

The next step is radiography: whole body ventrodorsal and lateral views, then focused limb radiographs. Orthogonal views of the leg should be taken, i.e., ventrodorsal and craniocaudal. Comparable radiographs of the contralateral limb (if unaffected) are often useful as a guide to normal (see Chapter 8).

The radiographs should be examined to assess:

- The bones
- Fractures
- Bowing or other conformational abnormalities
- Inadequate mineralisation
- Osteolysis
- Sclerosis
- Increased medullary bone (polyostotic hyperostosis)
- Periosteal and/or endosteal reaction
- Bone cysts
- The joints
- Luxations and subluxations
- Fractures

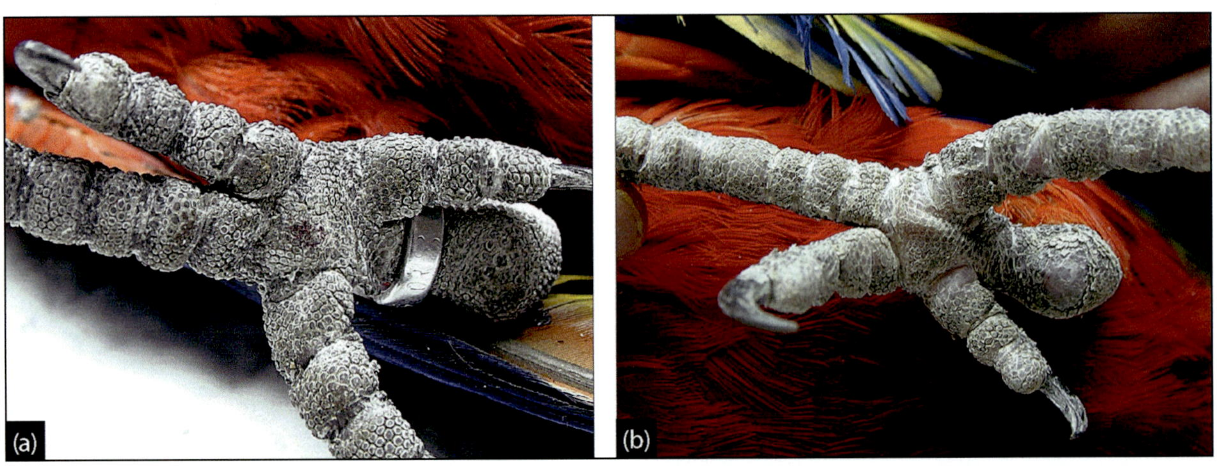

Figure 15.9 **Examine the plantar aspects of the feet for pododermatitis. (a) Normal plantar surface of the foot in a Greenwing macaw; (b) Grade 1 pododermatitis with smooth calluses forming on the heel and pads of the foot and digits.**

- Sclerosis of the subchondral bone
- Narrowing of the joint space
- Osteophyte formation
- Soft tissue
- Swollen joints
- Muscle atrophy

DISEASES OF THE PELVIC LIMB

Malformations

It can be difficult to distinguish between congenital and acquired malformations, especially in neonates and juveniles. Factors that can affect normal skeletal conformation include:

- The bird's parents:
  - Genetics. Hereditary malformations have been identified in some birds.
  - Health. In particular, subclinical heavy metal toxicosis has been associated with congenital limb malformation in the offspring of affected birds.
  - Diet. Poor quality diets fed to parent birds are often reflected in the health and conformation of the offspring. In particular, calcium–phosphorous imbalances in the parental diet will often result in chicks with limb malformations.
  - Incubation procedures and parameters. Artificial incubation is more likely to produce chicks with limb malformations than natural incubation. Incorrect incubator temperature, humidity, ventilation, and rate and degree of egg turning can have deleterious results on the development of the embryo and result in malformations.
  - The nesting environment. Nest box design can have an effect on neonates in several ways: slippery floors are believed to contribute to coxofemoral subluxation ('splay leg') in budgerigars; and inexperienced parents who enter the nest box quickly and land heavily on the chicks can cause injuries resulting in malformations.
  - Hand-rearing.
  - Although many hand rearers now use good quality commercial hand-rearing diets, there are still occasions where homemade diets are used, commercial diets are not used according to the manufacturer's directions, or commercial diets are supplemented with other ingredients. These actions may result in skeletal malformations.
  - Birds that are hand-reared in a brightly lit, roomy container are often encouraged to move around more than they would in a dark and cramped nest box. This can lead to premature and excessive weight-bearing on the legs, with subsequent bowing and malformations.
  - Many chicks reared on a good quality diet are then weaned onto a seed-based diet, often prematurely. This decrease in the nutritional value of the diet fed to a still-growing chick frequently results in acquired malformations.

Nutritional causes of malformations

Nutritional secondary hyperparathyroidism and rickets (collectively referred to as metabolic bone disease) are underlying factors behind many of these problems. Rickets is a metabolic bone disease in growing animals caused by a vitamin D3 deficiency. This can be due to inadequate nutrition, lack of exposure to sunlight, defective vitamin D activation, defective vitamin D receptors, hypoparathyroidism, renal failure, renal phosphate loss, or gastrointestinal malabsorption. The result is impaired mineralisation of osteoid tissue or epiphyseal cartilage, leading to thinning and weakening of the bones and excessive growth of cartilaginous structures. This, in turn, causes deformity at the ends of the bones, particularly in the proximal tibiotarsus, the beak (rubber beak), the head of the ribs, and sometimes the costochondral junction (rachitic rosary). Radiographically, there may be widening and distortion of the growth plates.

Note that osteomalacia, or osteoporosis, is a similar condition to rickets, but occurs in adult birds housed indoors and suffering from a vitamin D3 deficiency. There is a generalised thinning of the bone, with pathological micro- or gross fractures and resultant pain.

Nutritional secondary hyperparathyroidism is commonly seen in both juvenile and adult birds fed a diet that is either calcium deficient or has an excess of phosphorus (or both). For example, seed-only diets may have a calcium:phosphorus ratio as low as 1:10. Fruits, nuts, and most vegetables are also calcium deficient and/or have an excess of phosphorous. When these diets are fed to birds, especially parents rearing chicks or to recently fledged juveniles, the parathyroid gland releases parathyroid hormone, which, amongst other effects, withdraws calcium from the bone to maintain normal serum calcium levels. The result is thin bones that bend or break easily, leading to malformations.

Coxofemoral subluxation (splay leg)

Aetiology

Splay leg may be the result of a lack of nonslip substrate in the nest box. This allows the legs to splay out laterally away from the body, subluxating the coxofemoral joint and leading to laxity or damage of the medial collateral ligaments of the stifle, angular limb deformities of the femur, tibiotarsus, and tarsometatarsus, or slipped tendons (**Fig. 15.11**). Sometimes, if the chicks have metabolic bone disease as well, folding fractures of the tibiotarsus will result in secondary 'splay' deformities.

Management

Early recognition and correction are important; once skeletal ossification occurs, conservative techniques such as splinting or hobbling are unlikely to be effective. Very young birds can have their legs hobbled into a normal position to guide the leg into a normal position during growth. Various techniques have been used to achieve this goal, including tying the feet together with bandages or placing the chick's legs into a foam block. It is important to adjust the hobbles every 1–2 days (to allow for the chick's growth) until the problem is corrected. However, if the chick is not presented until after ossification is complete, surgical derotational osteotomies may be required.

Figure 15.11 **Coxofemoral luxation (splay leg) in a budgerigar.**

ANGULAR LIMB DEFORMITIES

Leg rotation

Aetiology

Rotation of the femur or tibiotarsus occurs when more weight is borne on one side of the growth plate than the other. Growth on this side of the physis is inhibited, while on the other side growth continues normally. Common causes include incubation problems, nutritional imbalances, and trauma. In chicks (e.g., ratites), leg rotation may occur for a range of reasons including insufficient exercise, high-energy diets, trauma, and heated flooring. In altricial chicks, such as parrots, the inciting cause is rarely determined.

Clinical presentation

The result of a leg rotation is a deviation of the growth of the limb from the midline: if the growth is directed laterally, it is termed a valgus deformity; if medial to the midline, it is termed a varus deformity (**Figs. 15.12, 15.13**). The deformity may be a bowing or rotation of the affected bone (**Figs. 15.14a, 15.14b**). This then has a cascading effect on the muscles, tendons, and joints of the affected leg. As the chick grows, this rotation often worsens as tendons contract and even mineralise.

The direction and severity of the rotation can be assessed by flexing the tibiotarsal–metatarsal joint so that the tarsus meets the cranial proximal tibiotarsus (**Fig. 15.7**). In the normal chick, the centre of the tarsus should be in alignment with the cnemial crest.

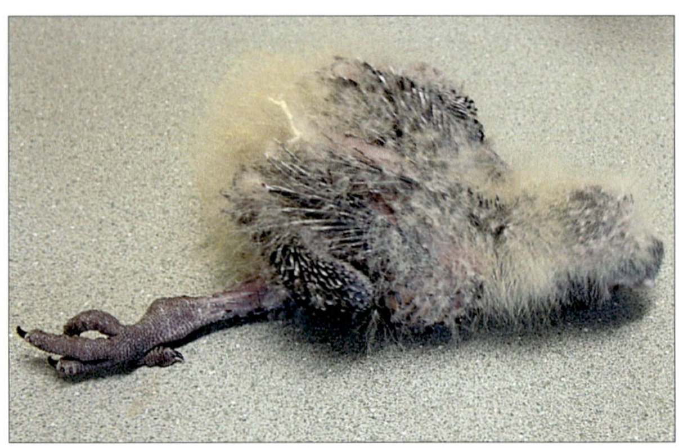

Figure 15.12 **Angular limb deformity in a black cockatoo nestling.**

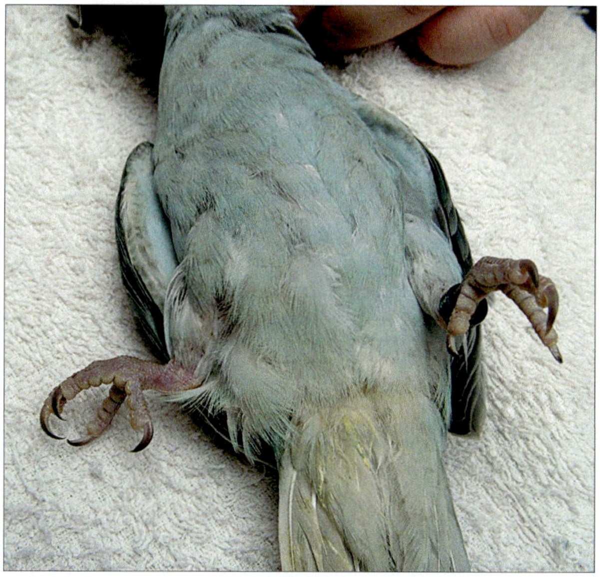

Figure 15.13 **Angular limb deformity in a juvenile Indian ringneck.**

Deviation medial or lateral to this midline indicates an angular limb deformity.

Management

Corrective surgery, in the form of derotational osteotomies, is the treatment of choice, but it must be done sooner rather than later. Timing is essential when addressing affected joints; waiting until the bird has finished growing may result in irreversible tendon and joint contracture (**Fig. 15.15**).

Figure 15.15 **This angular limb deformity in a macaw chick will require corrective surgery – but when?**

Bowed legs

Aetiology

Malnutrition, especially when combined with excessive and premature weight-bearing, causes one or both tibiotarsi to bow outwards, cranially, or caudally.

Clinical presentation

The bowing of the tibiotarsii shortens the legs and changes the placement of the phalanges. The result is that the chick weight-bears on the lateral phalanges (P3 and P4), giving the chick a 'pigeon-toed' appearance. The gait may be shortened. Radiology can determine the severity of the malformation.

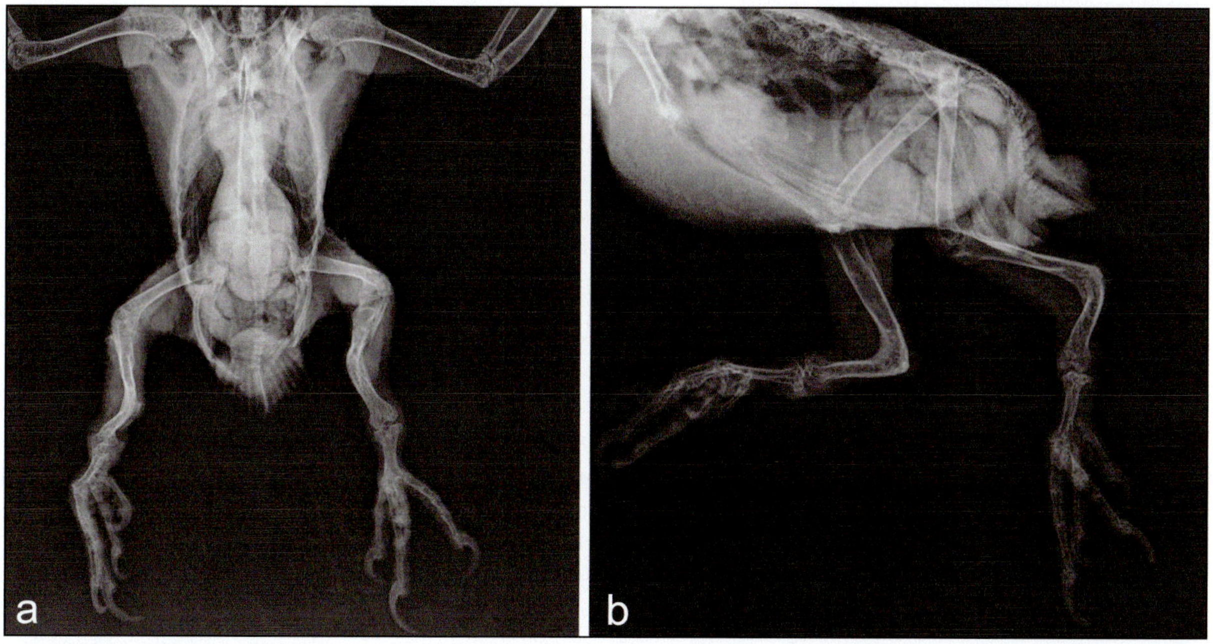

Figure 15.14 **(a) Severe angular limb deformity in a juvenile eclectus parrot. (b) Severe angular limb deformity in a juvenile eclectus parrot.**

Management

If a severe deformity is present, a derotational osteotomy may be required to correct this condition.

Osteomyelitis

Aetiology

Osteomyelitis may result from infection (localised or systemic disease) by bacteria (both aerobic and anaerobic implicated), fungi (*Aspergillus* and *Candida*), or mycobacteria. It may also be the result of trauma or neoplasia.

Clinical presentation

There is soft tissue swelling around the lesion. The lesion is painful, sometimes shown as lameness or an inability to fly. There may be a fracture (and subsequent instability) associated with the infection.

Diagnosis

Radiographically, there is usually clearly defined lysis of bone, as opposed to the more diffuse changes associated with neoplasia (**Fig. 15.16**). (Note that mycobacterial infections can also be diffuse.) The pus is caseous and does not drain; if an abscess forms, it enlarges to form an expanded outline within the bone, especially if it has formed within the medullary cavity. There is often a marked proliferative periosteal reaction.

Haematologically, there may be a leucocytosis, with heterophilia, monocytosis, and basophilia.

Management

Surgical debridement of the lesion is usually required to remove the caseated pus and necrotic bone.

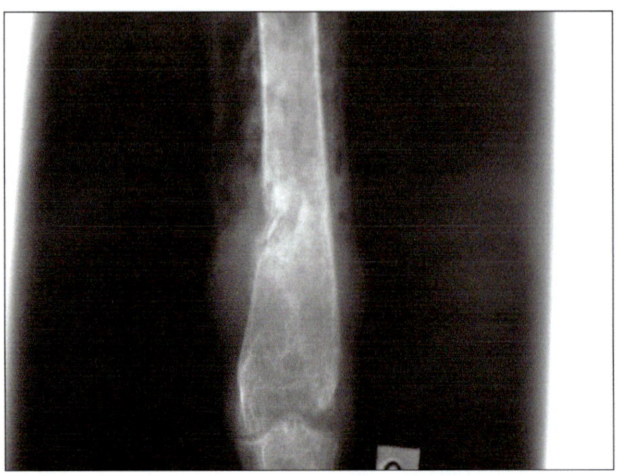

Figure 15.16 **Osteomyelitis following a dog bite.**

Antibiotic-impregnated polymethylmethacrylate beads, placed into the cavity left by the abscess after debridement and lavage, allow antimicrobials to be delivered locally in high concentrations. Antibiotics used include gentamicin, amikacin, tobramycin, lincomycin, clindamycin, neomycin, ceftiofur, and cephalothin. Bacterial and fungal culture and antibiotic sensitivity should be performed, ideally before beginning treatment, to guide the choice of antibiotic therapy. With the exception of intra-articular placement, implanted beads can be left in place indefinitely.

Systemic antibiotic therapy should be directed by culture and continued long term (2–3 months).

PATHOLOGICAL FRACTURES

Aetiology

Pathological fractures commonly occur in chicks with nutritional secondary hyperparathyroidism or rickets. The fractures most commonly occur in the long bones – the humerus, the radius and ulna, the femur, and the tibiotarsus. They are often first noticed when the bird fledges and attempts to fly. Flapping of the wings may be enough to fracture the humerus, radius, or ulna, while a heavy landing can break the tibiotarsus or femur. It is common for multiple fractures to be present in one bird. Other birds will suffer fractures while still in the nest; these fractures have healed by the time the bird fledges, but deformities become obvious when the bird is on the perch or tries to fly.

Diagnosis

Physical examination and radiographic assessment are usually sufficient to make the diagnosis. Measuring total and ionised calcium can assist in diagnosing hypocalcaemia and assessing its severity.

Management

Some of these fractures will heal with minimal intervention, but this invariably results in some degree of bone deformity. Attempts to repair these fractures surgically can be frustrated when the contralateral leg is required to support the bird's weight and lacks the cortical strength to do so, resulting in another fracture or bowing of that leg. Slinging the chick to minimise this problem can cause other issues such as compression of the sternum and ribs. These factors need to be considered before undertaking such a procedure, making the prognosis guarded in many cases.

TOE ANTEROFLEXION

Definition/overview

Anteroflexion of P4 (and occasionally P1) is a developmental issue in psittacine paediatrics, more often encountered where there has been subclinical stunting or substrate mismanagement (**Fig. 15.17**). Once the first phalangeal bone has ossified normally, it is fused to the tarsometatarsus and becomes less flexible. If the anteroflexion has not been corrected (by splinting) by that time, the anteroflexion may be permanent.

Management

The affected digit can be splinted into a normal position, but the abnormal angulation of the joint means that once the splint has been removed, the digit will return to an abnormal position. This can mean the nail is not in wear on a perch and will overgrow, requiring regular trimming or even amputation of the last phalanx and nail bed.

Figure 15.17 **Anteroflexion of P1 in a scarlet macaw.**

NEOPLASIA

Aetiology

Neoplasia of skeletal structures can be either primary (osteomas, osteosarcomas, chondromas, chondrosarcomas, osteochondromas, fibrosarcomas, haemangiomas) or metastatic (air sac carcinomas, other carcinomas).

Clinical presentation

The tumour usually presents as a firm swelling, with varying degrees of pain.

Diagnosis

Radiographically, there are varying degrees of bone changes, including proliferative lesions, osteolysis, and pathological fractures (**Fig. 15.18**).

Neoplasia must be distinguished from osteomyelitis. This may require a deep bone biopsy using a fine needle aspirate or a bone biopsy needle (e.g., Jamshidi bone biopsy needle). In small birds, an 18–22 gauge needle can be used to obtain a core sample: the needle is guided through the centre of the lesion from one side to the other, and then a stylet of sterile stainless-steel wire is pushed through the needle to extrude the biopsy specimen. Care must be taken to sample the centre of the lesion, as proliferative bony reactions on the periphery of the lesion may mask the origin of the lesion.

Management

Treatment usually requires amputation, followed by adjuvant chemotherapy and/or radiation.

JOINTS

Swollen joints

Swollen joints, often associated with pain and lameness, are a common presentation at any age. The swelling can be associated with articular gout, arthritis (septic or degenerative), or luxation of the joint.

Articular gout

Definition/overview

Chronic renal insufficiency, with a gradual elevation of plasma uric acid levels, results in the deposition of uric acid crystals in the synovial capsules and

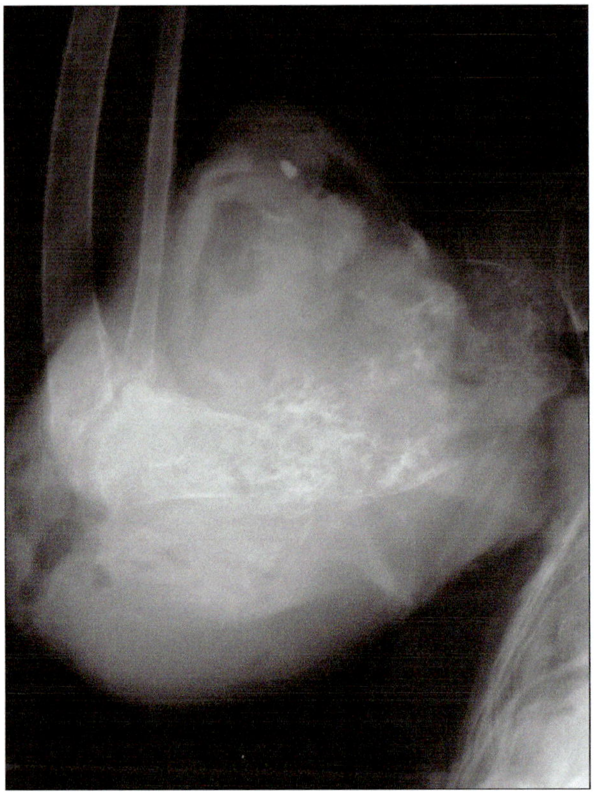

Figure 15.18 **Osteosarcoma in the humerus of a galah. A lesion such as this would require amputation and adjuvant therapy (radiation, chemotherapy).**

tendon sheaths of the joints. These uric acid crystals provoke an inflammatory response that causes the swelling and pain associated with articular gout.

Clinical presentation

Gross lesions typically consist of soft, painful swellings on the feet at the metatarsophalangeal and interphalangeal joints. White urate tophi may be visible within these swellings (**Fig. 15.8**). The bird may have a shifting lameness, and the toes are often held in an extended position rather than curled around the perch.

The diagnosis and management of this condition are discussed in more detail in Chapter 24.

Joint luxations

Definition/overview

The term 'joint luxation' refers to the complete displacement of a bone from the joint, whereas in a subluxation the bone is only partially displaced and some contact between the two articular surfaces still occurs. In most avian species, luxations occur infrequently compared with other orthopaedic conditions. All joints can be affected, but the femorotibiotarsal joint and elbow joints are most commonly affected.

Aetiology

Luxation may be the result of trauma or a congenital condition.

Clinical presentation

Localised swelling, pain, and reluctance to use the affected limb are observed.

Diagnosis

Diagnosis is based on the clinical signs referable to luxation and radiography.

Management

Conservative therapy is often unsuccessful because of the joint anatomy and the powerful contraction of flight or leg muscles.

Surgical repair is often required. This is further described in Chapter 33.

Prognosis

The prognosis for a full return to normal use of the joint is fair to guarded; however, a pain-free leg can usually be achieved even if it is not fully functional.

Arthritis

Aetiology

Swollen, painful joints (arthritis) can be due to traumatic injuries; septic inflammation (*Streptococcus* spp., *Pseudomonas* spp., *Chlamydia psittaci*, *Mycoplasma* spp., reovirus); microfilaria, especially in wild-caught Indonesian cockatoo species; articular gout (see earlier); degenerative changes; or neoplasia.

Diagnosis

Diagnosis is based on the patient's history, physical examination, haematology and biochemistries, radiography, and fine needle aspirate with cytology and culture where appropriate (**Fig. 15.19**).

Management

The cause should be identified and removed if possible (e.g., weight reduction in obese birds); septic joints may need to be opened and lavaged; caseated debris should be removed if possible. Long-term antibiotic therapy (as dictated by culture and

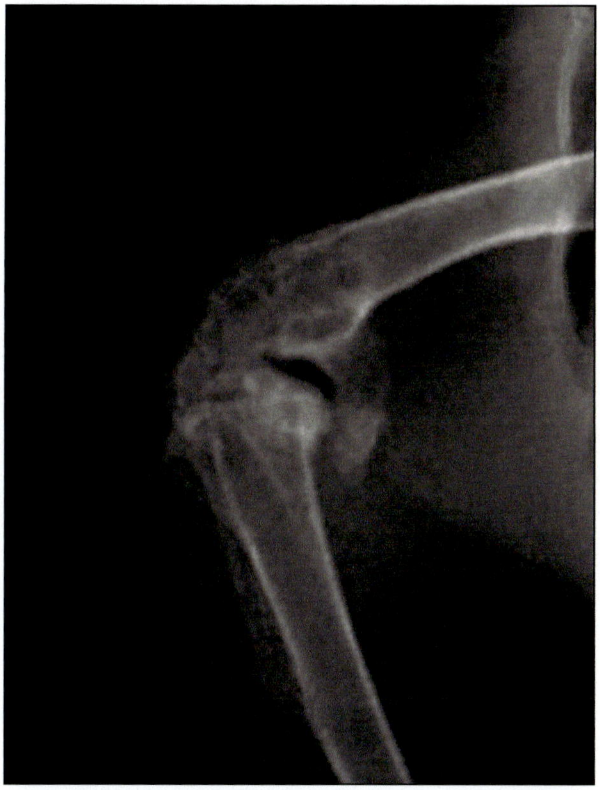

Figure 15.19 Arthritis in the knee of an aged galah.

sensitivity testing) will be required in cases of bacterial infection.

Pain relief may be achieved with NSAIDs, oral tramadol, gabapentin, or a combination of these drugs.

Attempts have been made to improve joint physiology using pentosan polysulphate and glucosamine. These are unproven therapies in birds at this time. Anecdotal reports are encouraging, but adverse effects of pentosan polysulphate (including death) have been recorded at high doses, so these therapies should be used with caution.

Husbandry should be adapted to cater to disabilities. Arthritic birds may have mobility problems and may have difficulties grasping small-diameter perches. Wide, flat perches and providing easier access to food and water may be beneficial.

PODODERMATITIS

Definition/overview

Pododermatitis is also known as bumblefoot. It refers to an inflamed and often infected lesion on the plantar surface of the foot. It is most commonly seen in raptors but can be seen in other species as well, especially gallinaceous birds, passerines, and parrots. It is a disease of captivity, being rare or non-existent in wild birds.

Aetiology

The plantar surface of the foot is protected by a thick layer of stratified squamous epithelium, which in turn is covered by a layer of keratin. Over the surface of this keratin is a layer of papillae, which are thought to evenly spread the weight-bearing requirement of the foot.

When this barrier is eroded or breached, an inflammatory response is often provoked, and infection can be introduced. These infections are usually associated with *Staphylococcus aureus*, but other bacteria including *E. coli*, *Pasteurella* spp., *Klebsiella* spp., *Clostridium* spp., *Corynebacterium* spp., *Bacillus* spp., *Diplococcus* spp., *Nocardia* spp., *Actinobacillus* spp., *Actinomyces* spp., *Aeromonas* spp., *Proteus* spp., and *Pseudomonas* spp. have been implicated. *Candida* spp. and *Aspergillus* spp. may also be involved in some cases.

The inflammation and/or infection may extend into the joints, tendons, and bones of the foot.

Contributing factors to the erosion or breaching of the keratin barrier include trauma, hypovitaminosis A, obesity, perches (smooth, regular surfaces or fine sandpaper-covered perches), long periods of inactivity leading to excessive weight-bearing without relief, and excessive weight-bearing on one leg due to a problem with the other leg (e.g., unilateral lameness may lead to pododermatitis in the contralateral leg). In many cases, pododermatitis in one foot will lead to some degree of pododermatitis in the other foot.

Clinical presentation

Raptor veterinarians have developed a classification scheme for pododermatitis that provides clinicians with a treatment plan and guide to prognosis. This classification scheme can be extended to other species with satisfactory results.

Class I

There is early devitalisation of a prominent plantar area without disruption of the epithelial barrier. It is subdivided into:

- Hyperaemia (bruise) or early ischaemia (a blanched area with compromised capillary perfusion)
- Hyperkeratotic reaction (an early callus)

Class II

There is localised inflammation/infection of underlying tissues in direct contact with the devitalised area, with no gross swelling. It is subdivided into:

- Puncture wound
- Ischaemic necrosis of the epithelium (a penetrating callus or scab)

Class III

More generalised infection: gross inflammatory swelling of underlying tissues is present. The origin may be puncture wounds or ischaemic necrosis; however, by this stage the initial cause is of minor significance in comparison with the ongoing pathology. It is subdivided into:

- Serous (acute) – oedema and hyperaemia of the tissues
- Fibrotic (chronic) – attempts at encapsulation and confinement
- Caseous – accumulation of necrotic debris

Class IV

There is established infection with gross swelling and involvement of deeper vital structures. Radiology and surgical exploration will often be required to differentiate class III from class IV. Class IV is a chronic condition causing tenosynovitis and occasionally arthritis and osteomyelitis.

Class V

This is an extension of class IV, characterised by crippling deformities.

Diagnosis

A thorough history is taken to investigate the bird's husbandry, diet, and previous medical problems. A thorough physical examination is conducted to classify the pododermatitis and identify any concurrent or predisposing problems (e.g., leg or spinal injuries). Ancillary diagnostic steps include culture and sensitivity testing of infected lesions and radiography of any cases worse than class II.

Management

This is usually determined by the classification of the condition:

- Class I carries a favourable prognosis, as there is no evidence of infection. The changes generally respond to conservative husbandry changes, including changing perching surfaces and applying topical emollients.
- Class II carries a good prognosis, as infection is localised. Such lesions respond well to surgery, as the total affected area is easily resected and epidermal defects are characteristically small. Hence, the architecture of the weight-bearing structures of the plantar aspect is maintained intact. This class will generally not respond to conservative treatment.
- Class III traditionally carries a good to guarded prognosis, as infection is well established, and structural changes have affected the foot. Some can be treated as for class II; however, the majority should be treated by complete surgical removal of all affected tissue, followed by first-intention healing.
- Class IV carries a guarded to poor prognosis, as infection is harboured in and affects deeper vital structures, making surgical debridement difficult or impossible. In view of the chronicity, pockets of encapsulated infective tissues are often present, which, if not cleared will result in later recurrence.
- Class V carries a poor to hopeless prognosis and may require euthanasia.

Treatment regimens include improving the bird's health and nutritional status, usually by converting to a formulated diet, and improving the bird's husbandry by supplying natural branches for perches. Some cases of pododermatitis may benefit from padded perches until the condition has resolved.

If surgery is contemplated, culture and sensitivity testing should be performed in advance so that the results are available at the time of surgery. This allows an appropriate antibiotic to be used at the time of surgery rather than several days later when postsurgical fibrotic encapsulation of pathogenic bacteria is underway. While waiting for the results, broad-spectrum antibiotic coverage and analgesia can be commenced.

Surgery is aimed at debulking the infection, removing all caseous debris and infected and fibrotic

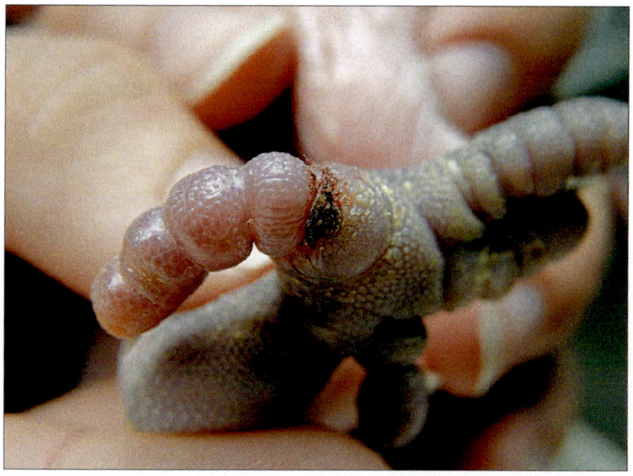

Figure 15.20 **Toe constriction in a macaw nestling.**

tissue, and closing the site so that primary intention healing can occur. Placing antibiotic-impregnated methylmethacrylate beads into the area before closure can assist with the long-term delivery of antibiotics into the affected area.

After surgery, it is important that pressure is relieved from the surgery site. This is achieved by padded bandages or gauze/rubber 'doughnuts' to distribute pressure evenly while keeping the surgery site clean and away from potentially contaminated surfaces.

Close attention must be given to the contralateral foot to ensure it does not develop pododermatitis as well.

CONSTRICTED TOE SYNDROME

Definition/overview

Constricted toe syndrome is a circumferential constriction caused by foreign material or fibrous bands, which may result in avascular necrosis of the digit distal to the constriction (**Fig. 15.20**).

Aetiology

Two forms are seen:

- The formation of constricting fibrous bands on the toes of neonatal chicks is possibly due to low humidity in the nest box, incubation issues, or ergot-like intoxication.
- Entanglement of the toes with artificial or natural fibres in birds of any age.

Clinical presentation

In the early stages, the toe distal to the constriction may appear swollen and oedematous. If the bird is perching, lameness may be noted. As the condition progresses, circulation to the toe is lost. Cyanosis of the toe is followed by a desiccated appearance; finally, the affected area drops off, sometimes leaving a nub of necrotic bone protruding.

Management

Early recognition and prompt treatment increase the chances of a successful outcome. After 24 hours, the prognosis for saving the toe becomes poor. If the patient is presented after circulation loss is severe and necrosis is apparent, amputation is usually required.

Artificial or natural fibres need to be carefully removed. The use of general anaesthesia and magnification is strongly recommended.

Mild or early neonatal cases may respond to increasing the humidity around the toe (by using moisturising creams) and massage.

More advanced or severe cases will require surgery. Two techniques have been advocated:

- A deep longitudinal incision on either side of the toe to sever the band and relieve the constriction.
- Complete excision of the band should be followed by a circumferential skin anastomosis using only two or three sutures to establish skin apposition without tension. Following the anastomosis, a release incision should be made on the medial and lateral aspects of the digit longitudinally across the anastomosis to allow swelling to occur without compromising circulation.

After surgery, a hydroactive dressing should be used to keep the area moist and prevent scar formation while the incisions heal. The toe must be monitored during healing for any reformation of the constricting band.

HYPERKERATOSIS

Definition/overview

Hyperkeratosis is seen in most species, but canaries and poultry are the most commonly reported.

Aetiology

Usually associated with *Cnemidocoptes* infection (see Chapter 12), there is some suspicion that a nutritional deficiency (possibly zinc and/or biotin) associated with all-seed diets may be involved in some cases. Secondary bacterial and fungal infections may contribute to the pain sometimes seen with this condition.

Clinical presentation

The scales and skin on the podotheca become hyperkeratotic and may be painful.

Management

Treatment with ivermectin or moxidectin, combined with correction of nutritional deficiencies, is usually curative. In refractory cases, biopsies and/or cultures may be needed to identify secondary pathogens.

SELF-MUTILATION OF THE FEET AND TOES

Definition/overview

Self-mutilation is a complex problem, with no simple answers or solutions. Affected birds will bite and chew at their feet and toes, sometimes to the point of amputating their own toes.

Aetiology

Some of the suggested/proven causes include dermatitis; allergic dermatitis; underlying pain due to conditions such as osteomyelitis, healing fractures, soft tissue infection or inflammation, tendonitis, and arthritis; chemical irritants (e.g., nicotine); vasculitis (e.g., immune-mediated, frostbite); obsessive-compulsive disorders; or neuritis and neuralgia.

Neuritis is the inflammation of a nerve or group of nerves characterised by pain, loss of reflexes, and atrophy of the affected muscles. It may be viral (e.g., proventricular dilatation disease [PDD]), traumatic (e.g., entrapment of the superficial nerves by a fracture or external skeletal fixator), or toxic (e.g., lead). Other aetiologies undoubtedly exist but are, at this time, obscure. In humans, the causes of neuralgia are varied and can include chemical irritation, inflammation, trauma (including surgery), compression of nerves by nearby structures (e.g., tumours), and infections. In many cases, however, the cause is unknown.

Neuralgia is pain that follows the path of a specific nerve. It is suspected, but not yet proven, in some birds that self-mutilate their feet and toes.

Management

Wherever possible, causative agents and secondary infections should be identified and eliminated. The bird should be prevented from mutilating the area through the use of bandages or Elizabethan collars. NSAIDs can be given to reduce inflammation and provide analgesia. Gabapentin has been used to treat neuralgia in humans and small mammals. Limited use in birds suggests that it may be a useful medication for self-mutilation disorders.

TOE TAPPING IN ECLECTUS PARROTS

Definition/overview

Toe tapping is the bilaterally symmetrical rhythmic extension and contraction of the digits of both feet, manifested when the bird is at full rest. Wing tip 'flipping' may also be seen in some individuals – an apparently involuntary flicking movement of the wing tips. When stimulated, these 'toe tappers' are able to stop the behaviour. Self-inflicted trauma is not a feature of this condition.

Aetiology

Numerous aetiologies have been suggested, including post-polyomavirus neuritis, nutritional deficiencies, and lead toxicosis. Oversupplementation with vitamins appears to be emerging as the 'favourite' aetiology. A dietary history usually reveals the excessive use of vitamin supplements, often in conjunction with a formulated diet. This theory is not yet proven.

Management

Discontinuing the supplementation often resolves the problem. In some cases, reducing the amount of formulated diet or simply changing brands is also required. As more research is done into this condition, it is likely that more appropriate therapies will be developed.

LEG BAND CONSTRICTION

Definition/overview

Swelling and vascular necrosis of the distal foot occur due to the tourniquet effect of a constricted leg band (**Fig. 15.21**).

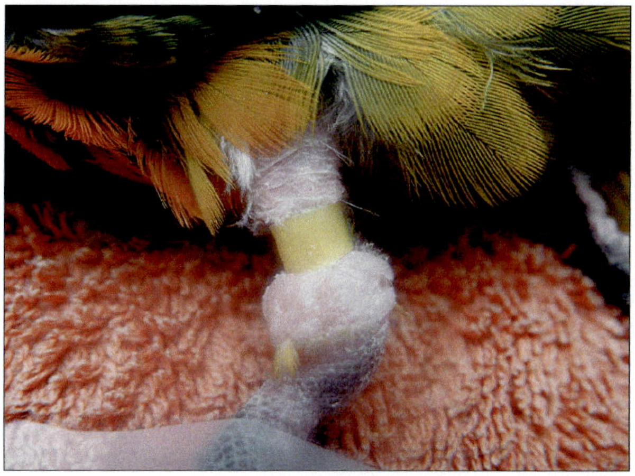

Figure 15.21 **Leg band constriction.**

Aetiology

This constriction develops as a result of applying too small a band to a growing bird; applying a soft 'split band' (usually made of aluminium) that is chewed on by the bird, causing it to overlap and constrict (**Fig. 15.22**); accumulation of scale and debris under the band (this is sometimes seen in budgerigars with *Cnemidocoptes* infection); or trauma and resultant swelling of the leg, with the (once loose) band becoming too tight.

Clinical presentation

Initially, there is pain (shifting lameness, chewing at foot), followed by swelling and more pain. Eventually, the foot may necrose and 'fall off'.

Management

Treatment in the early stages requires the removal of the band, followed by a supportive bandage and NSAIDs (e.g., meloxicam). However, owners need to be advised that constriction of the periosteum of the tarsometatarsus often leads to avascular necrosis of the bone below the band. In these cases, and where there is significant superficial necrosis, amputation of the leg may be required. In some cases, the ring is all that is holding the foot onto the leg, removing the band results in the foot coming away with it. In rare cases, removal of the ring and reperfusion of the distal limb may result in acute hyperkalaemia, with subsequent cardiac arrhythmia and death.

Figure 15.22 **A leg band chewed on by the bird, causing it to overlap.**

TOE NECROSIS

Aetiology

Any condition that compromises circulation within the toe can result in necrosis and loss of the toe. These conditions include ergotism; fibre constriction; deep infections of the skin, tendons, or bone (bacterial or fungal); frostbite (see later); trauma; neuritis leading to self-trauma; vasculitis; and contact dermatitis.

Clinical presentation

Initially, affected toes will often be swollen and appear discoloured. The bird may exhibit lameness, 'foot-shaking' behaviour, or chew at its foot (**Fig. 15.23**). As the condition progresses, the toe will become dark and dry, eventually falling off if not chewed off by the bird or amputated.

Diagnosis

In the early stages, the bird should be anaesthetised to allow detailed examination of the foot for fibrous constrictions. Radiology may demonstrate underlying bone lesions. Biopsy (sometimes via amputation of an affected toe if multiple toes are affected) is often rewarding. If circulation is thought to have been lost, cutting the nail through the nail bed, or pricking to distal toe with a needle can establish if circulation is intact or lost.

Treatment

In the early stages, the restoration of circulation is vital if the toe is to be saved. If the circulation appears to be intact, isoxsuprine, a beta-adrenergic agonist, can be used as a vasodilator. It appears to be more functional on tissue that is relatively healthy than tissue that is significantly damaged. Another drug that may be of benefit is pentoxifylline; this

Figure 15.23 **Toe necrosis and self-mutilation in a cockatoo.**

drug improves blood supply through inflamed vessels by reducing leukocyte adhesion to vascular endothelium and by making red cells more 'pliable', allowing them to pass through damaged vessels. It is more effective in damaged tissue, probably doing relatively little for the blood supply to healthy tissue.

Other treatments are determined by the results of other diagnostic testing.

If the circulation is lost, amputation is the best option.

FROSTBITE

Aetiology and pathogenesis

Frostbite is a localised tissue injury following the freezing and thawing of soft tissue. It can be a frequent problem in birds in colder climates, often when the onset of cold weather is unexpected, when the birds are on metal surfaces (e.g., wire), or when there is pre-existing circulatory compromise to the toes and feet. Non-native birds housed in cold climates may also be at greater risk.

The pathophysiology involves both the initial freezing injury and secondary vascular impairment. The initial cooling of the tissue results initially in vasodilation (often pulsatile in nature) followed by vasoconstriction as the cooling develops into freezing. Damage to the affected area therefore results from both direct cellular damage and from ischaemia; the latter is often more significant than the first.

The damaged tissue releases inflammatory mediators, which trigger further vasoconstriction, platelet aggregation, and thrombosis, leading to a cycle of further microvascular damage, hypoxia, tissue damage, and more inflammatory mediator release. This cascade may not be just limited to the damaged extremity; death associated with aseptic vegetative valvular endocarditis is well-recorded in birds with frostbite.

Diagnosis

A history of exposure to cold, combined with evidence of pain (lameness, self-mutilation) and a proprioceptive deficit in the affected limbs, can lead to a presumptive diagnosis before physical changes become obvious. After 24 hours, oedema and swelling of the toes may become obvious, but the line of demarcation of necrosis may take several weeks to become evident.

Treatment

It is essential that treatment be initiated as soon as possible before necrosis is evident. There are three goals in treatment: maintaining optimal blood supply to affected tissue, avoiding secondary infections, and providing analgesia.

If the tissue is still frozen, rapid rewarming in a warm water bath is recommended over gradual rewarming. Avoid massaging the affected area, as this is likely to cause further tissue damage. Nonsteroidal anti-inflammatory drugs may reduce inflammation and inhibit the release of inflammatory mediators. Pentoxifylline improves blood supply through inflamed vessels by reducing leukocyte adhesion to vascular endothelium and by making red cells more 'pliable', allowing them to pass through damaged vessels. Isoxsuprine, a beta-adrenergic agonist, can be used as a vasodilator but it appears to be more functional on tissue that is relatively healthy than tissue that is significantly damaged. As such, its use may be limited to the first few days after injury.

Topical aloe vera cream may be of benefit, although its mode of action remains obscure.

Secondary infections with either *Clostridia* spp. or normal skin flora can be quite common. The use of antibiotics directed at these bacteria, such as amoxicillin–clavulanic acid, for several weeks is recommended.

Frostbite can be an extremely painful condition and, as such, it is appropriate to provide analgesia through the use of opioids and tramadol in addition to non-steroidal anti-inflammatories.

Surgical debridement or amputation should be delayed until the line of demarcation becomes obvious. This may take 3 to 6 weeks.

OVERGROWN NAILS

Aetiology

Nails may overgrow when they fail to wear normally (inappropriate perch size or failing to grip properly due to toe or foot abnormalities), when the bird is malnourished or has chronic liver disease, or with diseases such as psittacine beak and feather disease (PBFD).

Clinical signs

The nails become overgrown to an extent that the bird is unable to grip a perch or walk normally. In some cases, the nails may become twisted and curled (**Fig. 15.24**).

Treatment

Affected nails should be trimmed regularly using nail clippers (for small birds) or a rotary hobby grinder (e.g., Dremel®). Predisposing factors should be identified and corrected or removed.

Figure 15.24 **Overgrown nails in an aged macaw.**

MISSING NAILS AND TOES

Aetiology

Missing nails and toes can be due to aggressive cage mates; unsafe caging or cage furniture; ergotism; toe constriction due to fibrous band formation (constricted toe syndrome) or foreign bodies (e.g., cotton thread); or frost damage.

Management

This is usually diagnosed after the event, and treatment is rarely required. Birds that are missing toes or nails rarely appear to be handicapped.

FURTHER READING

Echols MS (2007) Avian kidney disease, part I: types of renal disease. In: *Proceedings of the Annual Conference of the Association of Avian Veterinarians Australian Committee*, pp. 101–116.

Echols MS (2007) Avian kidney disease, part II: diagnosis of renal disease. In: *Proceedings of the Annual Conference of the Association of Avian Veterinarians Australian Committee*, pp. 117–128.

Echols MS (2007) Avian kidney disease, part III: treatment of renal disease. In: *Proceedings of the Annual Conference of the Association of Avian Veterinarians Australian Committee*, pp. 129–137.

Harcourt-Brown NH (2002) Orthopedic conditions that affect the avian pelvic limb. The Veterinary Clinics of North America. *Exotic Animal Practice*, 5(1), 49–81. https://doi.org/10.1016/S1094-9194(03)00046-X.

Quesenberry K (1997) Disorders of the musculoskeletal system. In: *Avian Medicine and Surgery*. RB Altman et al. (eds). WB Saunders, Philadelphia, pp. 523–539.

Remple JD, Forbes NA (2000) Antibiotic-impregnated polymethyl methacrylate beads in the treatment of bumblefoot in raptors. In: *Raptor Biomedicine III*. JT Lumeij, et al. (eds). Zoological Education Network, Lake Worth, pp. 255–266.

Stanford M (2006) Calcium metabolism. In: *Clinical Avian Medicine, Vol 1*. GJ Harrison, TL Lightfoot (eds). Spix Publishing Inc, Palm Beach, pp. 141–151.

Wellehan JFX (2003) Frostbite in birds: Pathophysiology and treatment. *Compendium on Continuing Education for the Practicing Veterinarian*, 25(10), 776–781.

CHAPTER 16

DISORDERS OF THE PECTORAL LIMB

ANATOMY

Pectoral girdle

The pectoral girdle is made up of three bones: the scapulae, the coracoids, and the clavicles (**Fig. 16.1**).

- The scapula is strongly attached to the dorsal ribs and, in some species, reaches to the ilium. It is proportionally much smaller and flatter than that of most mammals. This means that the space for muscle attachments is limited, as is the leverage that these muscles can produce. The scapulae have considerably less mobility compared with mammals, and this is practically restricted to the vertical plane.
- The coracoid is massive in most birds, functioning to hold the wing away from the sternum during flight. It extends between the sternum and shoulder joint and forms an acute angle with the scapula.
- The clavicles serve as a transverse spacer, bracing the wings apart, and provide attachment for attachment of muscles that produce the downstroke of the wings. They fuse ventrally to form the furcula (in many parrots, they are united only by cartilage or fibrous tissue). The ventral part of the furcula is attached to the apex of the sternal keel by ligaments.

These three bones come together at the triosseal foramen (canalis triosseus), through which the tendon of the supracoracoid muscle passes. This muscle lifts the humerus for the upstroke of the wing. The glenoid cavity, formed by the scapula and coracoid, is directed laterally and allows abduction and adduction of the wing.

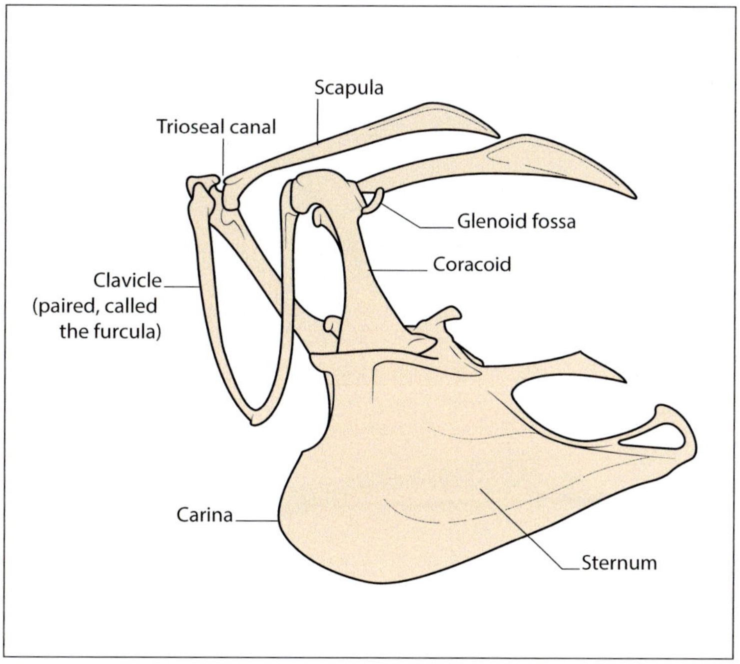

Figure 16.1 The pectoral girdle, including its interaction with the sternum. The humerus (not shown) articulates with the girdle at the triosseal canal.

DOI: 10.1201/9781003272519-16

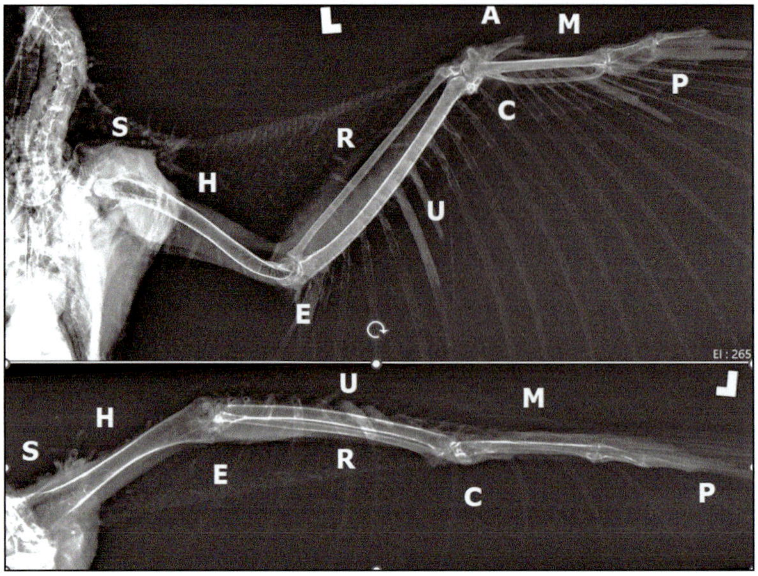

Figure 16.2 **Conventional radiograph demonstrating the anatomy of the wing. Shoulder (S); humerus (H); elbow (E); radius (R); ulna (U); carpus (C); alula (A); metacarpals (minor and major) (M); phalanges (P).**

Pectoral limb

Bones

The structure of the bones in the avian wing is similar in many respects to those in the forelimb of mammals (**Figs. 16.2, 16.3**). Evolutionary adaptations have been made to allow flight. The bones and joints are as follows:

- The humerus, in most pet species, is a pneumatic bone; the lateral diverticulum of the clavicular air sac enters through the pneumatic foramen on the medial side of the greater tuberosity. Attempts have been made to relate this to the power of flight, but there seems to be little or no correlation. When the wing is folded, the humerus lies close against the rib cage. In flight, the humerus rotates outwards such that the dorsal edge of the humerus becomes the trailing edge of the wing, demonstrating the range of movement that the humerus is capable of, a range of movement that includes elevation, depression, protraction, retraction, and dorsal and ventral rotation.
- The elbow joint is a shallow synovial joint consisting of three joints enclosed by a single capsule: between the humerus and the radius,

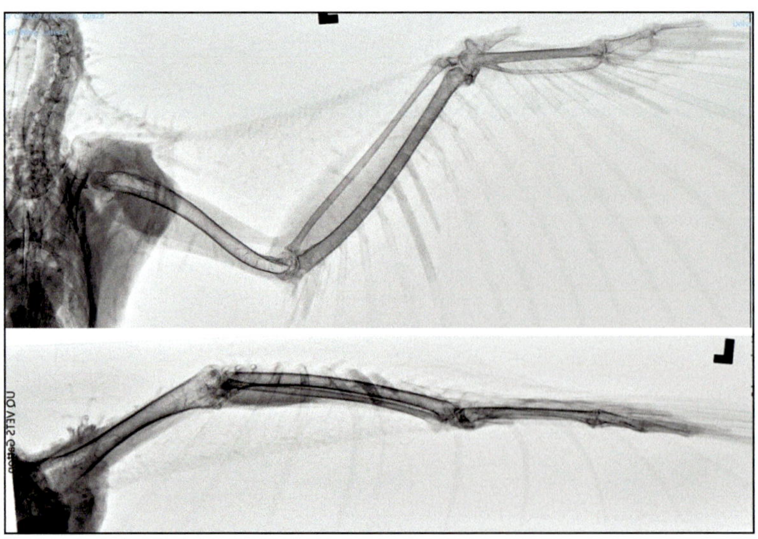

Figure 16.3 **Greyscale inversion image demonstrating the same anatomy as Figure 16.2.**

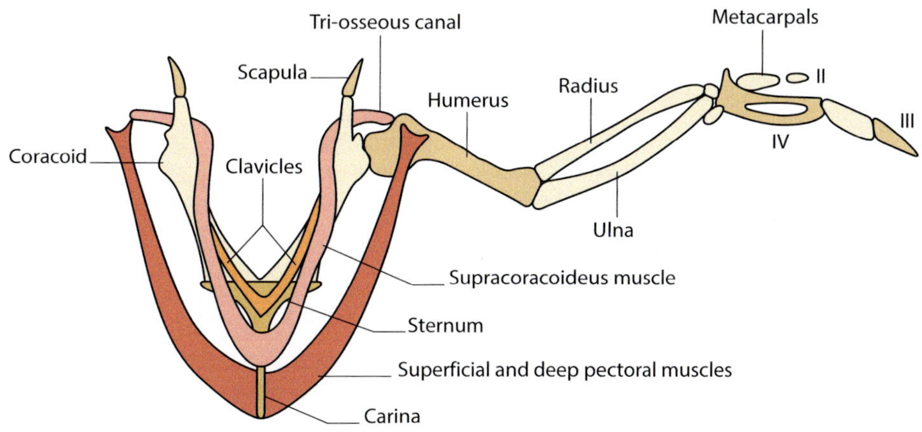

Figure 16.4 **The muscles of flight (transverse view).**

the humerus and the ulna, and the radius and ulna. Movement of the joint is mainly flexion and extension.

- The ulna is larger than the radius. The two bones are separated, and both are slightly bowed along their length. The extension of the elbow results in the longitudinal displacement of the radius with respect to the ulna, which provokes the concordant extension of the carpus and, if fully extended, the rotation of the bones of the antebrachium about the longitudinal axis. The secondary flight feathers are anchored to the ulna by ligaments.
- The carpal ('wrist') joint is formed by the radial carpal bone (cranial) and the ulnar carpal bone (caudal) articulating with the carpometacarpus. Again, movement is predominantly limited to flexion and extension.
- The manus consists of the carpometacarpal bones and the digits. The major and minor metacarpals are fused proximally and distally with an interosseous space. There are three digits: the alular digit with one phalanx; the minor digit, also with one phalanx; and the major digit with two phalanges. The rigid primary flight feathers are embedded in the carpometacarpus.

Muscles

Movements of the humerus are elevation (upstroke of the wing), depression (downstroke of the wing), protraction, retraction, dorsal rotation (when the leading edge of the wing is rotated dorsally), and ventral rotation (when the leading edge of the wing is rotated ventrally) (**Fig. 16.4**). The following is a brief discussion of the major muscles only.

- The main elevator of the humerus is the supracoracoideus muscle, which extends upwards from the keel, through the triosseal canal, and inserts on the dorsal side of the humerus.
- The main depressor muscles of the humerus are the pectoral muscles, extending from the clavicle and part of the sternum to the pectoral crest on the proximal humerus.
- The biceps and brachialis muscles flex the elbow joint, while the triceps extend the joint.
- The extensor carpi radialis is the main extensor of the carpal joint, while flexion uses the flexor carpi ulnaris.

Other structures

Struts

Struts are isolated bony rods that stretch across the medullary cavity of pneumatic bones in flighted species (they are absent in non-flighted species). They are mainly found on the proximal and distal ventral sides of wing bones in flighted species, developing at locations that have a higher risk of local buckling due to combined bending and torsion loading associated with flight (**Fig. 16.5**).

Since bones are in a constant state of growing and regrowing (remodelling), the reinforcing struts form on an 'as-needed' basis.

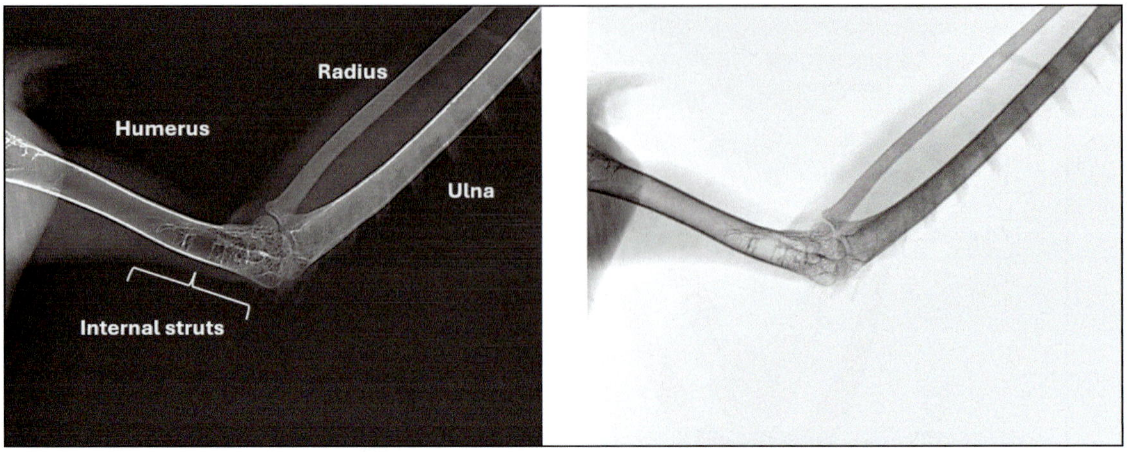

Figure 16.5 **Intramedullary struts, often prominent in juvenile birds with metabolic bone disease.**

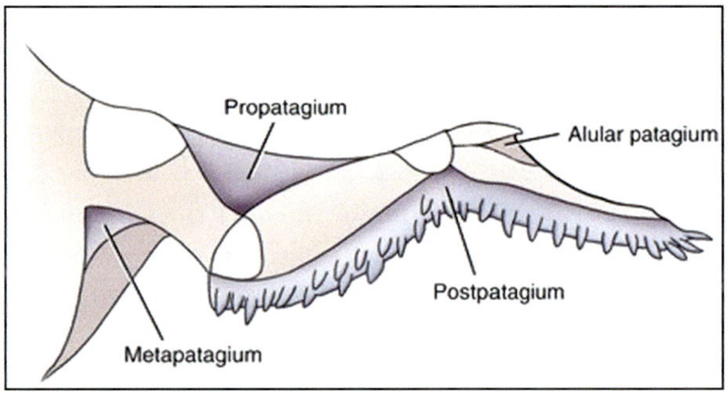

Figure 16.6 **The patagial membranes of the wing.**

Wing web

The propatagium is a triangular fold of feathered skin on the leading edge of the brachium and antebrachium, which increases the surface area of the wing. Its cranial free edge is the elastic propatagial ligament. The propatagium is supported internally by a number of muscles and tendons that alter the shape of the propatagium according to the requirements for flight.

A smaller fold of skin on the trailing edge of the wing between the body and the brachium is the metapatagium. The postpatagium can be found on the caudal margin of the antebrachium and the manus (**Fig. 16.6**).

CLINICAL SIGNS OF WING DISEASE

Most birds with a wing problem will be presented for a 'dropped' (or drooped) wing, i.e., the wing is 'hanging' or being held in the wrong position. While a dropped wing may be due to a neuropathy, it is more commonly due to injury to the bones or muscles. Less commonly, it is due to the increased weight of a mass on the wing or in the bone.

Observing how the wing is held may indicate the location of the problem. The lower the wing is held, the lower the injury is likely to be.

- The primary feathers are trailing on the ground – the injury usually affects the manus or distal antebrachium.
- The wing is lower than that on the normal side, but the primary feathers are not touching the ground – the injury likely affects the elbow or humerus.
- The whole wing is dropped, but the primaries are above the level of those on the opposing side – the injury likely affects the coracoid or shoulder joint (causing the wing to rotate).

Note, though, that this is a general rule, not an absolute rule. Small birds with a fractured wing, for example, may look quite normal. An accurate diagnosis requires a detailed physical inspection and radiography.

Some birds with leg injuries will extend their wing(s), touching the primary feathers to the ground and acting as a crutch to take weight off the injured leg. If an examination of the wing fails to demonstrate a lesion, ensure the legs are carefully examined to rule out this behaviour.

Many injuries or other problems (e.g., ulcerating neoplasms) can be associated with bleeding. Because the wing is usually held against the body wall, a lesion on the ventral aspect of the wing often results in blood being transferred to the feathers on the flank. This dried blood is often mistaken for a wound on the body, and time is lost looking for a wound there rather than on the wing.

Note that a bilateral wing droop may indicate a loss of pectoral muscle mass associated with chronic disease and weight loss – the wings themselves may be completely normal.

EXAMINATION OF THE WING

Before restraining the bird, perform a distant examination of the whole bird at rest. Then restrain the bird using a towel (see Chapter 3). It is vital to avoid further injury to the wing during this, and the use of sedation (e.g., midazolam intranasally) may be of benefit. Where possible, have an assistant (not the owner) restrain the bird – it is a two-person job to examine or anaesthetise a conscious bird. Most parrots can be held for a brief examination, but usually a general anaesthetic is needed for a more detailed examination.

Gently extend both wings using the tips of the longest primary feathers of each wing to gently pull both wings simultaneously into full extension. Watch for asymmetry in the degree of extension that can be achieved and the 'force' required to do this.

Visually and by palpation, examine each long bone and each joint on each wing separately, and repeat the exam, looking at both wings simultaneously for comparison.

Look and feel for disability, deformity, crepitus, and swelling of the bones and joints.

The joints will often have a good degree of laxity when flexed but are usually unable to be moved dorsoventrally when extended.

Examine the integument covering the ventral aspect of the wings and body, especially in the axilla. Examine any scabs or areas where there is exudation: old and dried, or fresh. Remove small feathers with gentle plucking and wet large feathers with chlorhexidine. Pluck the small ones and wet the large ones; use dilute benzalkonium chloride or a similar antiseptic. If the skin is torn, be very careful while plucking as it is easy to make the tear worse. When plucking, hold the skin and pluck against the tension. Do not attempt to pluck too many feathers at once. Assess whether any damage to the skin is traumatic or self-inflicted, and whether it is associated with any underlying pathology.

Investigate hard, sharp, raised scabs, and irrigate with saline to soften the area for examination. Old, dried-out bone has a very black appearance. Beware of pneumatised bones (such as the humerus) when flushing, as the flush may enter the respiratory tract.

RADIOGRAPHY

The next step is radiography: whole body ventrodorsal and lateral views, then focused wing radiographs. Orthogonal views of the wing should be taken, i.e., ventrodorsal and craniocaudal. If the pectoral girdle is to be examined, an H-view may give greater detail, particularly of the coracoids and the clavicles. Comparable radiographs of the contralateral limb (if unaffected) are often useful as a guide to normal (see Chapter 8).

The radiographs should be examined to assess:

- The bones
  - Fractures
  - Bowing or other conformational abnormalities
  - Inadequate mineralisation
  - Osteolysis
  - Increased medullary bone (polyostotic hyperostosis)
  - Increased cortical struts
  - Periosteal and/or endosteal reaction
  - Bone cysts
- The joints
  - Luxations and subluxations
  - Fractures
  - Sclerosis of the subchondral bone
  - Narrowing of the joint space
  - Osteophyte formation
- Soft tissue
  - Swollen joints
  - Muscles

SKELETAL ABNORMALITIES

Fractures, luxations, and subluxations

Fractures, luxations, and subluxations are commonly seen in practice. Companion and aviary birds usually present soon after accidental injury (flying into aviary wire, window strikes, ceiling fan injuries, etc.). Treatment of these injuries is discussed in Chapter 33.

Metabolic bone disease

Definition/overview

See Chapter 15 for a discussion on the development of skeletal malformations in young parrot chicks. Many chicks, reared on poor-quality diets (or by parents on poor-quality diets), suffer from nutritional secondary hyperparathyroidism. These chicks may develop bowing of the long bones, or even pathological fractures, under the weight of the flight feathers or as the chick attempts to fly when fledging.

Clinical signs

Clinical signs include an inability to fly normally or at all. Many chicks have bilaterally dropped wings.

Diagnosis

Radiography (**Fig. 16.7**) often reveals:

- Bowing of the long bones
- Pathological (folding) fractures
- Increased medullary struts in the humerus
- Thin cortices
- Loss of bone

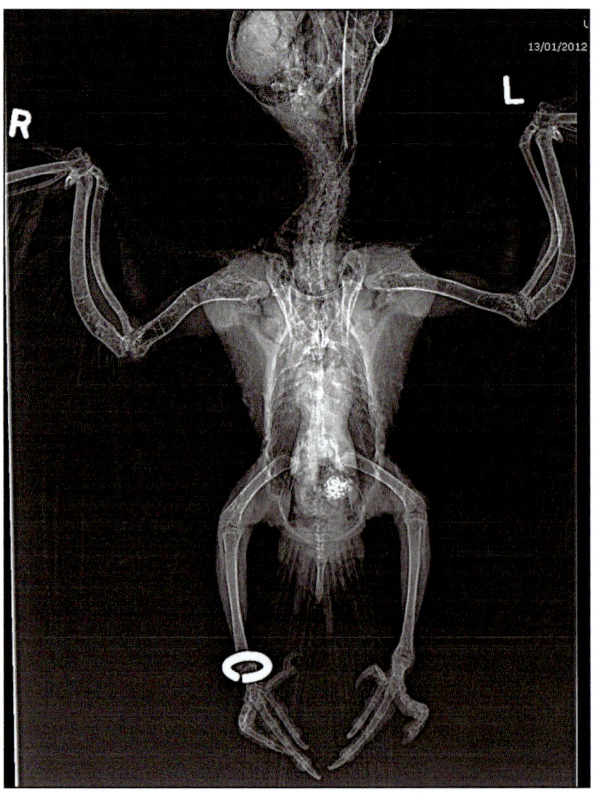

Figure 16.7 **Severe metabolic bone disease in a juvenile glossy black cockatoo. Note the bowing of the radius, ulna, and tibiotarsii, malformation of the humerus, and increased intramedullary struts, bowing of the legs, and the varus deformity of both the legs and feet.**

Treatment

This condition is difficult to treat, as malformations of the bones have already developed before the condition becomes clinically apparent. Corrective osteotomy is difficult as the bone cortices are often too thin or soft to hold implants. If multiple limbs are affected, euthanasia may be the best course on welfare grounds.

Angel wing

Definition/overview

'Angel wing' is a colloquial term for a valgus rotation of the distal wing resulting in dorsolateral rotation of the primary flight feathers, seen in young birds. It can be either unilateral or bilateral. It arises because of a valgus deformity of the growing metacarpal bones or a rotation of the carpal joints. Although most commonly described in larger waterfowl (e.g., geese and swans), it has also been reported in parrots (macaws, conures, and budgerigars), bustards, raptors, and herons. Both captive and wild birds have been reported to develop this condition.

Aetiology

The condition is likely multifactorial, presumably involving incubation and hatching problems, genetic predisposition, the environment (e.g., exposure to toxins), management factors, and malnutrition (a diet high in protein, with a calcium–phosphorous imbalance, and hypovitaminosis D3). As a result of these factors, during periods of fast growth and feather production, the body size increases at a faster rate than bone ossification develops, and the weight of the growing flight feathers (relative to the growth and strength of the wing muscles) places excessive force on the muscles and ligaments of the carpal joint and the inadequately mineralised metacarpal bones. This causes the manus to hang down, and then twist and rotate laterally. The result is a dorsolateral rotation of the primary flight feathers, which protrude when the wing is folded to the body during rest (angel wing) (**Fig. 16.8**).

Diagnosis

The diagnosis of angel wing is normally a presumptive one based on physical examination. Flexion and extension of all joints in the wing can be either normal or abnormal; sometimes joints may only exhibit abnormal laxity. The clinical presentation, however,

Figure 16.8 **Angel wing in a juvenile macaw. Note the size (and weight) of the primary feathers.**

is quite typical and diagnostic imaging is not considered necessary in most cases.

Treatment

1. If recognised early in the disease process (1–2 days), the use of figure-of-eight bandages to realign the primary flight feathers with the angle of the wing, in combination with the control of the growth rate by use of restrictive feeding and calcium and vitamin D3 supplementation, will often correct the problem in 1 week.
2. Juvenile birds not presented until after several days can be treated similarly, but the correction time may be extended to several weeks. Twice-weekly reassessment, passive range of motion physiotherapy, and bandage changes are recommended until corrected.
3. Mature birds will usually require a derotational osteotomy, transecting the metacarpal bones, rotating the distal fragments into a normal position, and transfixing the fragments surgically (see Chapter 33).

NEUROLOGICAL DISORDERS

Definition/overview

The brachial plexus consists of nerve roots from four or five spinal nerves, usually involving distal cervical to proximal thoracic spinal nerves. There are two major branches: the ventral fascicle, which gives rise to multiple nerves that innervate the ventral aspect of the wing, including both the supracoracoideus and pectoralis muscles, via the pectoral, bicipital, median–ulnar, median, and ulnar nerves; and the dorsal fascicle, which supplies the extensor musculature of the wing and integument on the dorsal aspect via the radial, axillary, and anconeal nerves. The ventral branch is the most important for flight muscle innervation.

Avulsion/dysfunction of this plexus or the associated nerves results in wing droop and an inability to fly.

Aetiology

Nerve damage that presents as wing droop can be caused by:

- Brachial plexus avulsion, which may result from trauma (severe abduction of the scapula and thoracic limb, or displacement of the shoulder and scapula caudally).
- Radial nerve dysfunction can be associated with either traumatic injuries to the nerve where it crosses the dorsal aspect of the humerus or iatrogenic damage during surgery.
- Viral disease (e.g., paramyxovirus).
- Lead toxicosis.

Clinical signs

A complete wing droop or paralysis, inability to fly, failure to regain function, and shoulder muscle asymmetry attributed to denervation (disuse atrophy) are all consistent with neuropathy of the wing.

Diagnosis

Neurological disease is often a diagnosis of exclusion, although it should be considered in all cases of traumatic wing injuries.

Treatment

If neurological innervation does not return within several weeks or if the wing is becoming traumatised (through self-mutilation or lack of sensation), amputation should be considered.

SOFT TISSUE MASSES

Definition/overview

The relatively thin envelope of soft tissue covering the thoracic bones means that swellings are readily palpable, but their origin is often not readily determined nor is the involvement of surrounding tissues.

Clinical signs

Clinical signs can include wing droop, fresh/dried blood on the feathers of the wing or flank, displacement of the wing, feather-damaging behaviour, or self-mutilation.

Aetiology

The differential diagnosis for soft tissue masses includes:

- Recent trauma with swelling and bruising (often green in colour).
- Bony callus or non-aligned bone healing.
- Nodular granulomas of an infectious nature (bacterial, mycobacterial, parasitic, or fungal).
- Neoplasia, the origin of which can be:
  - Underlying bone or cartilage origin (osteosarcoma, chondrosarcoma).
  - Connective tissue origin (e.g., fibrosarcoma, fibroma)
  - Muscular origin (e.g., rhabdomyoma, leiomyosarcoma)
- Xanthomas, commonly observed as diffuse, yellow, friable swellings, or discrete nodular masses, consisting of multinucleated giant cells, macrophages, and cholesterol clefts. They are not neoplastic but can be locally invasive. Hypercholesterolaemia is thought to contribute to their formation. It is thought they may originate as a bruise resulting from wing trauma. As the blood clot is left, an inflammatory reaction to the cholesterol left *in situ* results in the formation of a xanthoma.

Diagnosis

The diagnosis of soft tissue swelling will rely on a combination (or all) of the following: the patient's history, physical examination findings, diagnostic imaging, and biopsy (fine needle or excisional).

Treatment

Treatment will be determined by the final diagnosis. Many neoplastic conditions, if not already metastasised, will respond well to amputation. Small, localised lesions may respond well to electrochemotherapy using cisplatin or bleomycin.

Xanthomas may respond to dietary modifications and perhaps statin therapy, but if bleeding occurs, amputation may be required.

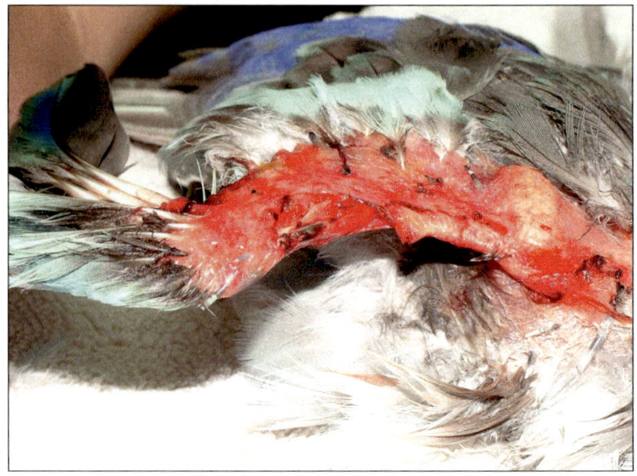

Figure 16.9 **Propatagial trauma in a princess parrot.**

DISEASES OF THE PROPATAGIUM

Propatagial injuries

The propatagium is a vital and complex structure. Injury to this structure may result in a reduction or inability to fly. The propatagium is frequently injured by the bird flying into wires or electric cables, and the injury itself may at first be missed (**Fig. 16.9**).

The most important injured structure is usually the ligament of insertion of the M. tensor propatagialis pars longus, forming the leading edge of the wing and keeping this edge tensed even when the wing is half-opened. The rest of the propatagium is supported by the M. tensor propatagialis pars brevis, which can be more complicated in its anatomical structure, varying between genera and even species; it is, however, less likely to be injured.

This elasticity often results in the propatagium cut like a piece of pie: the largest wound at the leading edge and the pointed tip of the pie nearest to the elbow. The cut skin edge is frequently desiccated and may be infected. It is unusual for the tendons from the supporting muscles to be obvious.

Treatment

Under anaesthesia, the propatagium is plucked clean, taking care not to tear the propatagium further. Firstly, identify and isolate the tendon from the pars longus and then debride the area around it of dead tissue. Dissect out the tendon ends, remove dead tendon, and suture the tendon together using a 4/0 polydioxanone (PDS Ethicon) suture. Failure to suture the tendon leaves the wing with a substantial loss of ability to stretch to full extension. Next, cover at least one surface, or preferably both, with skin. The tendon appears to run in a tunnel in the leading edge of the wing, and this tunnel should be reapposed if possible on all surfaces to surround the tendon. Initially, it is more important that the tissue heals than that the propatagium is kept at its original length.

Finally, attempt to repair as much skin as possible. Areas with a skin deficit must be covered with a sheet of hydrophilic dressing, which may be sutured into place. (This will not make a complete hole through the propatagium heal; there must be one surface of skin.)

The propatagium may be so shortened that it will only allow the wing to open to half its normal distance. This will need to be stretched over a period of time – this is best achieved by leaving the wing unstrapped but keeping the bird cage restrained for 10 days. Once the wing has healed, the amount of exercise for the wing can be increased until it is allowed to fly in a reduced manner. This is usually started about a month to 6 weeks after the initial surgery. The wing should be checked every few weeks and a comparison made under general anaesthesia.

Prognosis

Even after a substantial loss of skin and tendon, the propatagium will eventually appear to return to the same length as the other wing, and the bird may be able to fly normally.

Propatagial/axillary dermatitis

Definition/overview

This is a moist, exudative dermatitis of the ventral aspect of the ventral propatagium and axillary areas (**Fig. 16.10**). It appears to be very painful, and self-mutilation adds to the clinical appearance. This condition has been reported in grey parrots (*Psittacus erithacus*), hyacinth macaws (*Anodorhynchus hyacinthinus*), cockatiels (*Nymphicus hollandicus*), cockatoos (*Cacatua* spp.), and lovebirds (*Agapornis* spp.). It is sometimes referred to as superficial chronic ulcerative dermatitis (SCUD).

Aetiology

The aetiology of this condition is unclear but is likely to be multifactorial. Feather loss on the patagial membranes (behavioural or hormonal) may lead to an increased surface humidity of the patagia, especially when the wing is folded. This in turn

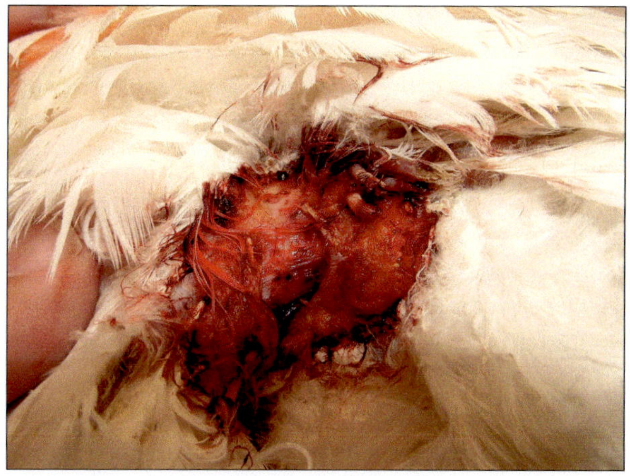

Figure 16.10 **Axillary dermatitis, or superficial chronic ulcerative dermatitis (SCUD), in a corella.**

may lend itself to an increased surface bacterial and yeast burden. Malnutrition may lower the innate immune system in this region, allowing colonisation of these microbes in the superficial and deeper layers of the skin. Cutaneous hypersensitivity to airborne or topical chemicals may play a role. In some cases, feather cysts or cutaneous neoplasia may be initiating factors.

Bacterial and yeast infections are likely to be secondary to initiating primary causes. A large number of bacterial species have been isolated from these lesions, as have a number of yeasts and fungi (*Malassezia*, *Candida*, and *Trichosporon* spp).

Clinical signs

Because the lesions are often concealed under the wing or feathers, affected birds may not be presented until the condition is advanced. The client may not be aware of the severity, bringing the bird to the clinic for focal feather damage, and perhaps blood on the beak.

On physical examination, there is generally a loss of downy and contour feathers along the ventral aspect of the propatagium or in the axilla. A focal area of skin is typically ulcerated, and a crust or scab is usually found on the surface or margins of the affected area. This area can be quite extensive, involving the axilla and extending onto the ventral aspects of the brachium. One or both wings can be affected, and it is not uncommon to have both the propatagium and the postpatagium involved.

The lesions are often very painful, as evidenced by the patient's response when attempts are made to extend the wing. This extension may be limited by the contraction of the wing webs.

Diagnosis

Given the often advanced and aggressive nature of this condition at the time of presentation, a reasonable active investigation is recommended from the start. This work-up should include:

- Haematology and plasma biochemistry analysis. This often is non-specific, with leucocytosis and elevated creatine kinase (CK) being common findings associated with tissue trauma.
- Radiology, focusing on the detection of underlying skeletal lesions.
- Cytology (impression smears or tape preps) of the lesions, and microbiologic culture and sensitivity testing taken from beneath any crusts (see Chapter 11).
- Skin biopsies and histopathology of skin and feather follicles can provide important clues into the initial events if biopsy samples are collected at the time of the initial presentation. If left until after a treatment trial, valuable information may be lost.

Treatment

Treatment of this condition should be based on several considerations:

1. The lesions are often quite advanced and aggressive by the time the individual is aware of the problem. Treatment has to be aggressive and broad-based and is likely to take some time (months) to reach an effective resolution. The client must be educated as to the severity of the problem.
2. It is a painful condition, and the bird is likely to continue self-trauma when it can reach the lesions. Bandaging of the wing is generally not advisable, as this may trap microorganisms and moisture, and prevent regular assessment and treatment of the wounds. Therefore, the use of Elizabethan collars may be necessary.
3. Constriction of the patagium may cause permanent contracture of the elastic tensor muscle tendons but, conversely, stretching the wing to minimise contracture may tear a healing wound open again.

Treatment can be directed at:

1. Analgesia. Multimodal therapy with drugs such as meloxicam, tramadol, and gabapentin can provide effective pain control.
2. Cleaning the wounds. Gently clean and irrigate the wound with povidone iodine (1% or less) and chlorhexidine (1:40). Note that povidone iodine has good antibacterial properties but may disrupt fibroblasts and leukocytes and has the potential to be absorbed systemically.
3. Topical therapy. Oil-based ointments should be avoided, as they may retard healing and have the potential for systemic absorption. Silver sulfadiazine cream has broad-spectrum antibacterial properties and is effective against *C. albicans*. It can be applied once or twice daily after cleaning and gently drying the wounds. Aloe vera gel has also been used. Topical corticosteroids must be avoided because of their potent immunosuppressive effects.
4. Antibiotics. The choice of systemic antibiotic should be based upon culture and sensitivity results. Empirical antibiotic therapy might include cephalexin, clavulanic acid–amoxicillin, trimethoprim–sulfamethoxazole, and fluoroquinolones. Systemic antifungal therapy might include fluconazole and itraconazole.
5. Adjunctive therapies: The following may assist in wound resolution:
    a. Long-term use of an Elizabethan collar.
    b. Dietary modifications, if indicated.
    c. Essential fatty acid supplementation.
    d. Hormonal therapy, e.g., deslorelin or leuprolide, if it is thought there may be an underlying hormonal basis for the condition.

Prognosis

This is a particularly challenging condition to resolve. Constant extension and flexion of the wings disrupt and slows the wound healing. The wounds will often remain susceptible to recurrent opportunistic infection – this must be monitored and treated accordingly. The normally elastic patagial tissue may be replaced with scar tissue, restricting the normal range of motion of the wing. Good owner compliance is essential, as is a realistic expectation of the treatment length and outcome. The owners should be warned that, once the problem becomes chronic, the odds of eventual complete resolution are lower than if initial therapy is aggressive.

FURTHER READING

Orosz SE (2002) Clinical considerations of the thoracic limb. The Veterinary Clinics of North America. *Exotic Animal Practice*, 5(1), 31–48. https://doi.org/10.1016/S1094-9194(03)00045-8.

Powers LV, Van Sant F (2006) Axillary and Patagial Dermatitis in African Grey Parrots (*Psittacus erithacus*), in *Proceedings of the Annual Conference of the Association of Avian Veterinarians*, 2006.

Zsivanovits P, Monks DJ, Forbes NA (2006) Bilateral Valgus deformity of the distal wings (Angel Wing) in a Northern Goshawk (*Accipiter gentilis*). *Journal of Avian Medicine and Surgery*, 20(1), 21–26. http://www.jstor.org/stable/40236516.

CHAPTER 17

DISORDERS OF THE SPINE

ANATOMY

Vertebrae

The requirements of flight have limited the flexibility of the avian spinal column, as the rigidity of the spine decreases the need for extensive musculature to maintain a streamlined and rigid posture during flight. The most mobile part is the cervical spine, made up of 11–12 vertebrae (in parrots). They provide sufficient flexibility for a bird to reach its tail and uropygial gland. The notarial vertebrae, carrying the ribs, are somewhat flexible in parrots but are fused in many other species to form the notarium. The notarial joint spaces are minimally mobile but have the potential for movement and probably absorb a small amount of compression with each wing stroke. A similarly slightly flexible notarial–synsacral joint connects the notarium to the synsacrum, made up of 10–23 fused notarial, lumbar, sacral, and coccygeal vertebrae. It is followed by 5–8 free coccygeal vertebrae and then the pygostyle (4–10 fused coccygeal vertebrae), which supports the tail retrices (**Fig. 17.1**).

The intervertebral disc in birds, unlike mammals, lacks the jelly-like nucleus pulposus centre. The cartilaginous annulus fibrosus and the anterior and posterior end plates (attaching the discs to vertebrae) remain. A histological analysis of avian intervertebral discs shows they are composed of collagen protein. The lack of flexibility and mobility of much of the avian spine may reflect this disc structure.

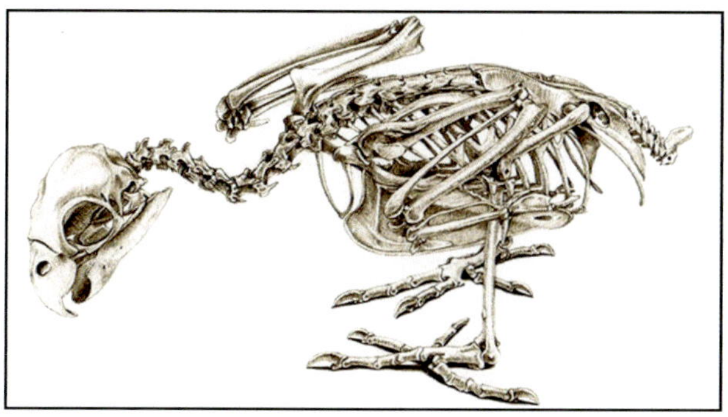

Figure 17.1 **Spinal anatomy. (Adapted from van Grouw K, *The Unfeathered Bird*, Princeton University Press, 2013.)**

DOI: 10.1201/9781003272519-17

Spinal cord

Unlike mammals, the avian spinal cord is the same length as the neural canal, with no cauda equina at its termination. The dorsal and ventral roots of the spinal nerves pass separately through the dura, coming together in the intervertebral foramina and exiting laterally between the vertebrae, with the total number varying between species (due to the species variance in the number of vertebrae). The size of the spinal roots and nerves is determined by the relative importance of the nerves; for example, in flying birds, the nerves in the brachial plexus region are larger than those in the lumbosacral plexus.

The cord diameter increases at the cervical enlargement (associated with the brachial plexus supplying the wings) and the lumbosacral enlargement (associated with the lumbosacral plexus supplying the legs) (**Figs. 17.2, 17.3, 17.4**).

The spinal cord is enclosed, protected, and supported by three meninges: the dura, the arachnoid mater, and the pia mater (**Fig. 17.5**). The dura mater is the thick outer meninx; initially fused with the periosteal lining of the vertebral canal at the foramen magnum, it is then separated from the periosteum by an epidural space extending along the cervical and thoracic regions. This space is filled with a gelatinous material and, on the dorsal side, the vertebral venous sinus. The dura then fuses with the periosteum from the caudal thoracic region to the caudal end of the canal, although the venous sinus intervenes irregularly between the dura and the periosteum. The arachnoid mater is a thin, delicate layer lying near the dura, while the pia mater is attached to the spinal cord. The space between these two meninges is the subarachnoid space, occupied by fine denticulate

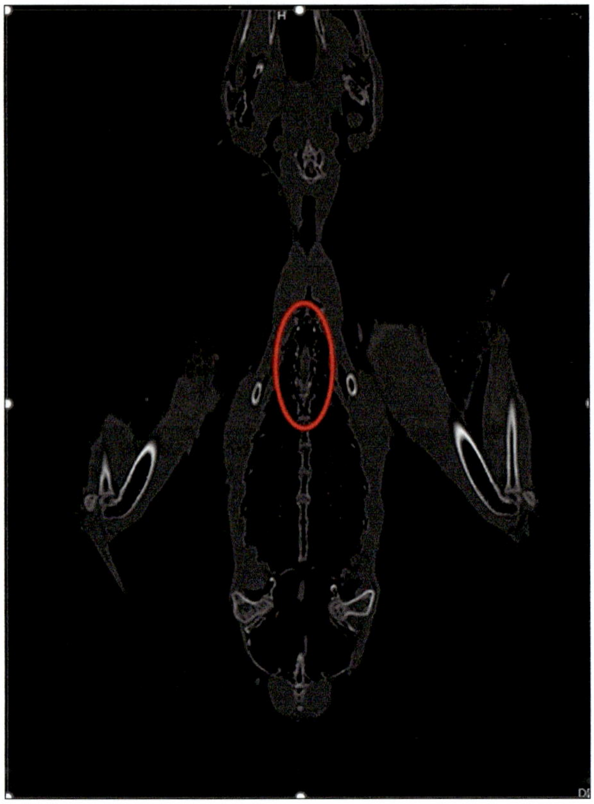

Figure 17.2 Cervical enlargement (brachial plexus) (CT coronal section).

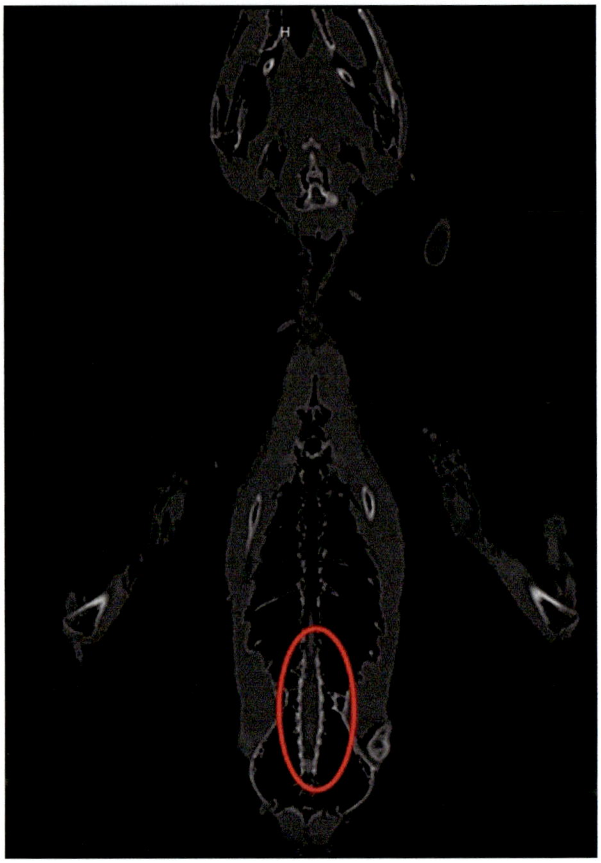

Figure 17.3 Lumbosacral enlargement (lumbosacral plexus) (CT coronal section).

ligaments suspending the spinal cord from the dura and a small amount of cerebrospinal fluid (CSF).

The spinal cord derives its blood supply from the vertebral arteries (cervical and cranial thoracic regions) and intersegmental branches of the descending aorta (from the level of the heart caudally). Branching from the intersegmental arteries are the vertebromedullary arteries, which divide and penetrate the dura. These arteries also supply three longitudinal arteries (the ventral spinal artery and two dorsolateral spinal arteries). The dorsolateral spinal arteries provide blood flow to the white matter on the dorsal cord, while the ventral spinal artery provides vascularity through smaller vessels (the sulcal arteries) to the bulk of the cord. Each spinal cord segment receives six to ten sulcal arteries.

DISEASES OF THE VERTEBRAL COLUMN

Hereditary, congenital, and developmental abnormalities

Definition/overview

These may be caused by genetics; teratogens (e.g., organophosphates); malnutrition (either parental or juvenile); malpositioning within the egg; or artificial incubation (problems with temperature, humidity, ventilation, and turning frequency).

These abnormalities are not uncommon in companion and aviary parrots; they have been reported in Indian ringnecks (*Psittacula krameri*), Alexandrine parrots (*Psittacula eupatria*), cockatoos (genus *Cacatua*), cockatiels (*Nymphicus hollandicus*), eclectus (*Eclectus roratus*), macaws (*Ara* spp.), Amazons (*Amazona* spp.), grey parrots (*Psittacus erithacus*), and conures (*Aratinga* spp. and *Pyrrhura* spp.).

Although often described only in terms of the physical abnormality itself, they may have more wide-reaching consequences than just a bend in the spine. Impingement on spinal nerves, compression of the air sacs, and dystocia are just some of the physical effects of spinal abnormalities.

Aetiology

Abnormal flexion of the spine occurs due to malformation or collapse of a vertebra. In dogs, these malformations take the form of butterfly vertebrae, wedge-shaped vertebrae, and hemivertebrae:

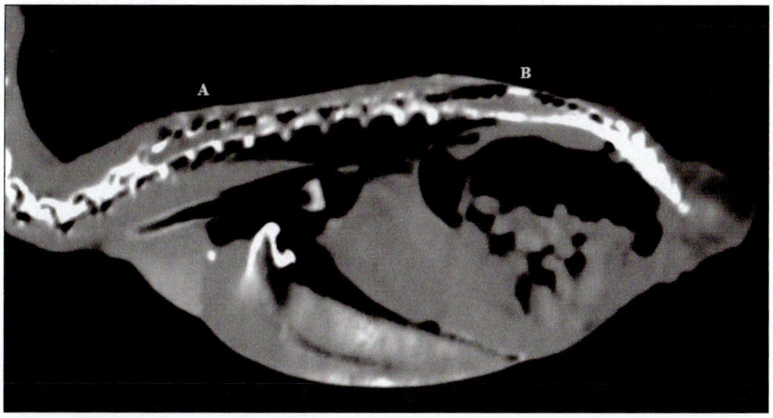

Figure 17.4 **Cervical enlargement (A) and the lumbosacral enlargement (B) (CT sagittal section).**

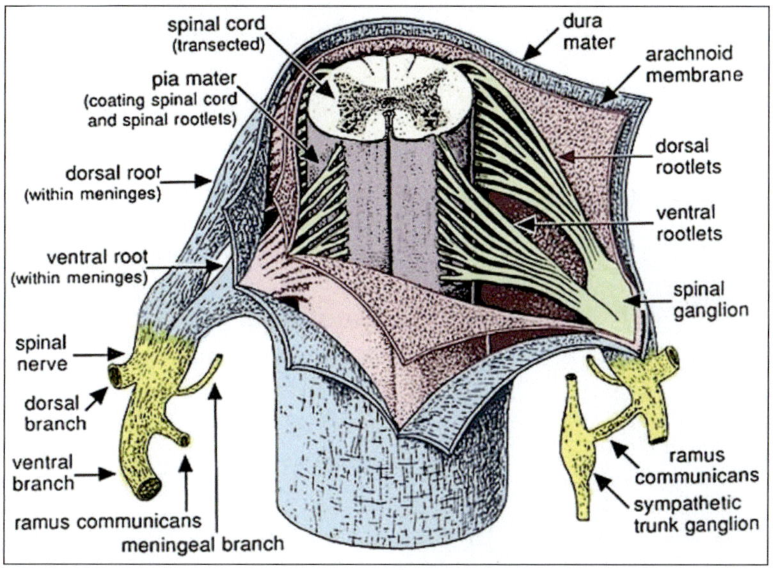

Figure 17.5 **Spinal cord anatomy.**

- Butterfly vertebrae have a sagittal cleft through the body of the vertebra and a funnel shape at the ends. This gives the appearance of a butterfly on an X-ray. It is caused by the persistence of the notochord (which usually only remains as the centre of the intervertebral disc) during vertebra formation. There is a widening of the affected vertebra. The bodies above and below the butterfly vertebra adapt to the altered intervertebral discs on either side by showing concavities along the adjacent end plates.
- Hemivertebrae are wedge-shaped vertebrae, where the wedge is due to part of the vertebra not forming, probably due to a lack of blood supply.
- Wedge-shaped vertebrae are complete vertebrae but are wedge-shaped rather than block-shaped.

No such classification has been attempted in birds yet, probably because of the practical difficulties in imaging these cases. These abnormalities may be congenital (e.g., genetics, embryonic malposition, parental malnutrition).

Collapse of the vertebrae can be associated with trauma, metabolic bone disease, or osteomyelitis.

These malformations usually occur in an area of some mobility, i.e., in the notarial–synsacral and synsacral–coccygeal regions.

Spinal bifida

Spinal bifida occurs sporadically in pet birds. Incomplete closure of the embryonic neural tube results in an incompletely formed spinal cord. In addition, the vertebrae overlying the open portion of the spinal cord do not fully form and remain unfused and open. This allows the abnormal portion of the spinal cord to protrude through the opening in the bones. There might be a fluid-filled sac surrounding the open spinal cord.

Kyphosis

Kyphosis (from Greek *kyphos*, 'a hump') is an abnormally excessive convex kyphotic curvature of the spine in the notarial, synsacral, and coccygeal regions (**Fig. 17.6**). In dogs, it is associated with dorsal hemivertebrae, ventral wedge-shaped vertebrae, and butterfly vertebrae. While most cases of kyphosis are mild and only require routine monitoring, serious cases can be debilitating (**Fig. 17.7**).

Figure 17.6 **Kyphosis in an African grey parrot.**

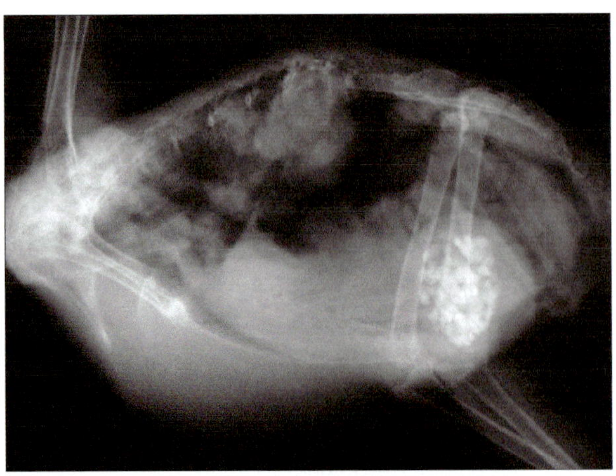

Figure 17.7 **Radiograph of the bird in Fig. 17.6, showing a fungal granuloma invading and collapsing the spine, causing kyphosis.**

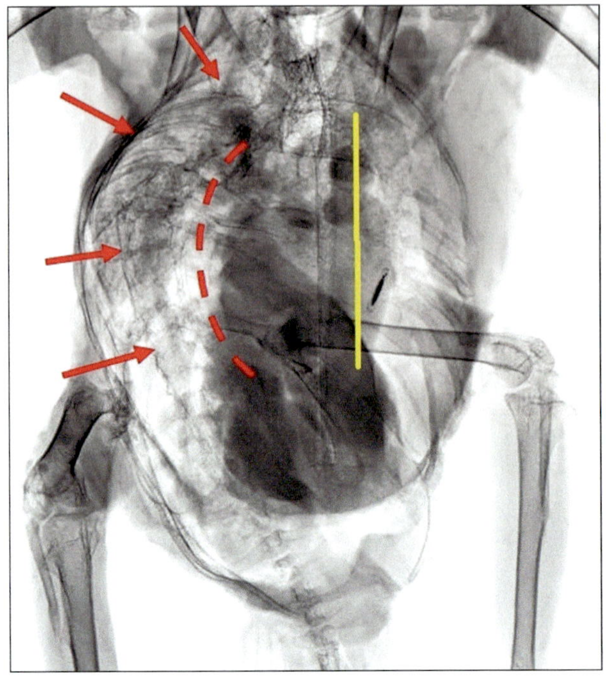

Figure 17.8 **Scoliosis in a juvenile macaw. The radiograph shows lateral curvature of the spine (red arrows and line) away from the sternum (yellow line).**

Scoliosis

Lateral deviation and curvature of the spine occurs due to malformations of vertebrae similar to kyphosis (**Fig. 17.8**). In poultry, it has been associated with genetics, melatonin deficiency through pinealectomy, or continuous light exposure. The indication is that there is a threshold level of serum melatonin below which scoliosis may develop, probably in conjunction with other factors that have yet to be identified. Deficiencies of copper, manganese, or vitamin B6 have been shown to aggravate the condition. How this correlates with the occasional scoliosis seen in companion birds remains unclear, but it raises issues about parental nutrition and hand-rearing practices.

Wry neck (torticollis)

The term 'wry neck' describes a condition where there is either vertebral malformation or tendon contracture in the neck of a newly hatched chick. The result is the neck is twisted to varying degrees (**Fig. 17.9**). The cause appears to be multifactorial, with genetics, incubation, and infectious diseases implicated. In turkey poults, this condition has been associated with mycoplasma infection, extending into the cervical vertebrae from the cervical air sacs, and leading to osteoarthritis, misshapen articular facets, and ultimately cervical deformities. In other species,

Figure 17.9 **Wry neck in a 6-week-old grey parrot chick.**

torticollis has been associated with viral infections (e.g., avian polyomavirus) and cerebrospinal nematodiasis (e.g., *Baylisascaris* spp.).

Clinical signs

Kyphosis and scoliosis can present as a series of physical effects resulting in clinical signs.

- *Dyspnoea and tachypnoea*: The coelom of patients with severe kyphosis can be so compressed that it causes dyspnoea and tachypnoea as the abdominal and thoracic air sac space is lost. These patients present with increased respiratory rate and effort and decreased exercise tolerance. Radiographic assessment may be needed to assess the severity of the compression, but it is unlikely that medical therapy, e.g., bronchodilators will have any positive effects in these birds.
- *Dystocia*: Reproductively active female birds with moderate to severe spinal abnormalities are likely to experience dystocia. For this reason, the owners of such birds should be counseled about controlling reproductive activity, including the use of hormonal therapies as indicated.
- *Referred pain and sensation*: It is likely that spinal abnormalities will have other, perhaps less immediately obvious, effects associated with altered biomechanical loading along the entire vertebral column. The biomechanical consequences of spinal curvature abnormalities in dogs remain to be fully evaluated, but there is a strong correlation between these lesions and intervertebral disc disease (IVDD). It is not a far reach to suggest that, while IVDD has not been diagnosed in birds, impingement of the spinal nerves as they exit the intervertebral spaces could result in neuralgia and referred pain. This may be a contributing cause to clinical syndromes such as foot/toe chewing, axillary mutilation, and feather damaging behaviour.
- *Pododermatitis*: It is likely that the altered biomechanical loading along the vertebral column may result in reduced movement of limbs, causing an inability to fly and perhaps uneven weight-bearing, leading to pododermatitis. Many patients with spinal abnormalities associated with malnutrition have long bone abnormalities as well (e.g., bowing or rotation), and it may be difficult to distinguish between those abnormalities and spinal abnormalities as the cause of reduced locomotory abilities.

Birds affected with wry neck may show a multitude of clinical signs, including ataxia, opisthotonos (the head is turned backwards so the bird appears to be 'star gazing'), and torticollis (the head is twisted to the side). The neck can usually be gently turned to a normal position but reverts to its twisted position when released. It does not appear to be painful.

Treatment

No treatment is currently available for kyphosis and scoliosis, but with supportive care, many of these birds can live long and apparently happy lives.

Wry neck offers a better prognosis for treatment if treated early in the disease process. The application of a foam neck brace to hold the chick's head and neck in a more normal position may give good results. The neck brace will have to be left in place for several weeks until the bone is more developed (**Fig. 17.10**).

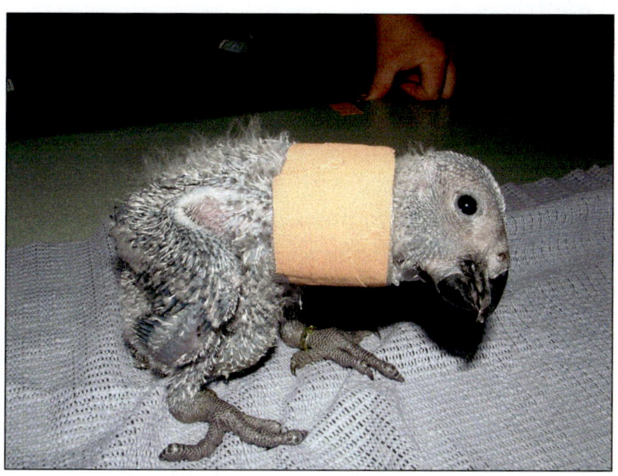

Figure 17.10 **A foam brace used to correct wry neck in the chick in Fig. 17.9.**

TRAUMA

Spinal fractures

Definition/overview

Spinal trauma, while common in wildlife patients following collisions with vehicles, etc., is a less common injury in companion and aviary birds. The rigidity in the avian spine needed for stability for flight also provides protection to the spinal cord. Further stability is provided by the ribs articulating with the sternum. Fusion of the caudal thoracic, lumbar, sacral, and caudal vertebrae forms the synsacrum. Because of this high degree of stability, injury to the spinal cord sufficient to cause clinical signs is generally severe and often involves the cord between fused segments, notably between the notarium and synsacrum (the only mobile section of the notarial–synsacral spine).

Clinical signs

The degree and persistence of clinical signs are directly related to cord compression. As the spinal nerves radiate laterally from the cord through the ventral foramina at the same level as the vertebral column, lesion localisation is more straightforward than in mammals (where the spinal nerves pass caudally from the cord).

Birds with spinal trauma generally show varying degrees of paresis or plegia in one or more limbs. Hind limb paraplegia or paraparesis, decreased pain perception, and the presence of upper motor neurone signs (hyperreflexia and weakness) can be observed. The cloacal tone may be normal or increased. If the lumbosacral plexus is affected, hyporeflexia may be apparent. Cloacal tone may be reduced.

Diagnosis

Lesion localisation is inferred through palpation, neurologic assessment (see Chapter 26), and ancillary testing.

Most fractures are non-displaced, with neurological deficits associated with direct stretching injury to the cord and compression secondary to soft tissue swelling around the spine and cord. Plain film radiography performed on initial presentation may therefore be less sensitive in detecting the injury than CT or MRI. After a few days/weeks, bone remodelling and periosteal response increase the likelihood of detecting the injury using radiographic evaluation.

Treatment

Symptomatic and supportive care is indicated, including non-steroidal anti-inflammatory drugs and rest. Splinting/bandaging may not be feasible in many cases. Surgical stabilisation has been tried in several cases, but with minimal success.

Prognosis

If a return to normal function, especially deep pain, is not evident after 3–5 days, it is unlikely that the bird will recover.

Pygostyle injury

Definition/overview

Of the 15 coccygeal vertebrae, the first five are incorporated into the synsacrum, the following six are free and mobile, and the last four are fused together to form the pygostyle. The tail feathers (usually six pairs) attach to the pygostyle and are used for steering and braking during flight, especially during take-off and landing. The uropygial gland lies dorsal to the levator muscles of the tail and pygostyle.

Poorly performed wing trims, especially in young birds or heavy-bodied birds, often result in repeated heavy falls to the ground. This repetitive trauma can cause broken tail feathers, soft tissue swelling, infection, and fractures of the pygostyle. Pain and irritation can lead to self-trauma in the area.

Occasionally, the ventral skin, between the pygostyle and the vent, will tear after a heavy impact. These 'tail split' injuries have to be distinguished from a cloacal prolapse.

Clinical signs

Clinical signs reflect the nature and extent of the problem. They include:

- Tail flicking
- Feather-damaging behaviour, which may lead to folliculitis and dystrophic/missing tail feathers
- Self-mutilation of the area
- Bleeding around the base of the tail

Diagnosis

Under general anaesthesia, blood and crusts should be gently removed from the tail. Badly damaged feathers may need to be removed. Radiography of the pygostyle may reveal fractures, osteomyelitis, or osteolysis.

Treatment

In early to mild cases, analgesia and antibiotic therapy may resolve the lesions. Tail split injuries should be cleaned and sutured.

Moderate to severe cases may require amputation of the tail (pygostylectomy). An incision is made just proximal to the openings for the rectrices, both dorsally and ventrally. The skin is reflected proximally by blunt and sharp dissection, revealing the proximal base of the tail feather follicles. The attachments of the follicles are then severed from lateral to medial by radiosurgery. The incision is made as distally as possible to maximise the available skin for closure. The pygostyle is disarticulated from the tail and removed with the tail feathers. Bleeding is usually minimal and controllable with direct pressure. The skin edges are trimmed and then closed in a simple interrupted pattern.

Regardless of the treatment option chosen, attention must be paid to restoring normal flight, e.g., by imping wing primary feathers, or to soften landings through confinement, low perches and soft padding on the cage floor.

Vertebral osteomyelitis

Definition/overview

Vertebral osteomyelitis may be a result of infection by bacteria or fungi (*Aspergillus* and *Candida*). Infection can be spread haematogenously (trauma or neoplasia may provide an entry portal to the bone or spinal cord), or by local dissemination from neighbouring tissue. If the dura or its blood supply is breached, infection may spread to the spinal cord itself.

Clinical signs

Clinical signs of a localised lesion in the spinal cord are dependent on the location and severity of the lesion. Cranial to the lesion, neurological function remains normal; caudal to the lesion, neurological function is reduced or absent. This demarcation of signs allows for the localisation of the lesion on examination and the detection of often subtle radiographic changes. See the earlier section 'Spinal fractures' for more detail.

Diagnosis

Radiographically, there is usually clearly defined lysis of the bone. There may be a marked proliferative periosteal reaction.

Haematologically, it may demonstrate leucocytosis, with heterophilia, monocytosis, and basophilia.

Treatment

Often, by the time a diagnosis is reached, the disease may be too advanced for a good response to therapy.

FURTHER READING

Bruggeman BJ, Maier JA, Mohiuddin YS, Powers R, Lo Y, Guimara˜es-Camboa N, Evans SM, Harfe, BD (2012) Avian intervertebral disc arises from rostral sclerotome and lacks a nucleus pulposus: Implications for evolution of the vertebrate disc. *Developmental Dynamics Wiley Periodicals, Inc.* 241: 675–683.

Doneley B (2018) Spinal abnormalities in parrots. *Proceedings of the Annual Conference of the Association of Avian Veterinarians Australasian Committee.*

Welle KR (2019) Rectricectomy with pygostylectomy for resolution of feather trauma in an Umbrella Cockatoo (Cacatua alba). *Journal of Avian Medicine and Surgery* 33(2): 193–197.

CHAPTER 18

DISORDERS OF THE MUSCLES AND TENDONS

ANATOMY

The musculature of birds is well adapted for flight or terrestrial movement (or both). The pectoral, wing muscles, pelvic, tail, and neck muscles are well developed, while those of the notarium and synsacrum are less so, as the need for strength along the back is catered for by the semi-fusion of this part of the spine. The well-differentiated neck muscles contribute to balance during flight, while the powerful tail musculature controls the steering feathers (rectrices).

Compared to mammals, the skeletal muscles of birds contain less connective and adipose tissue, with a higher muscle fibre density. The connective tissue encloses the muscle belly, defining it from its surroundings and allowing smooth movement. It also provides channels for vessels and nerves.

The muscle bellies are found as close to the body as possible, moving the centre of gravity to the middle of the body for flight. Consequently, the tendons are usually quite long, and those in the limb and back may partially ossify for additional strength (**Fig. 18.1**). For most of their length, the tendons carry no apparent blood supply – nutrition is obtained via the tendon sheath fluid.

As birds do not have a smooth muscle diaphragm, their breathing is reliant on the contraction and relaxation of the skeletal intercostal and sternal muscles, drawing the ribs out and sternum down on inspiration, and in reverse on expiration.

Tendons are energy modulators that can store, conserve, amplify, and transmit energy. They transmit the biomechanical forces generated by muscle contraction, playing a major role as both force amplifiers and force attenuators. Their functionality

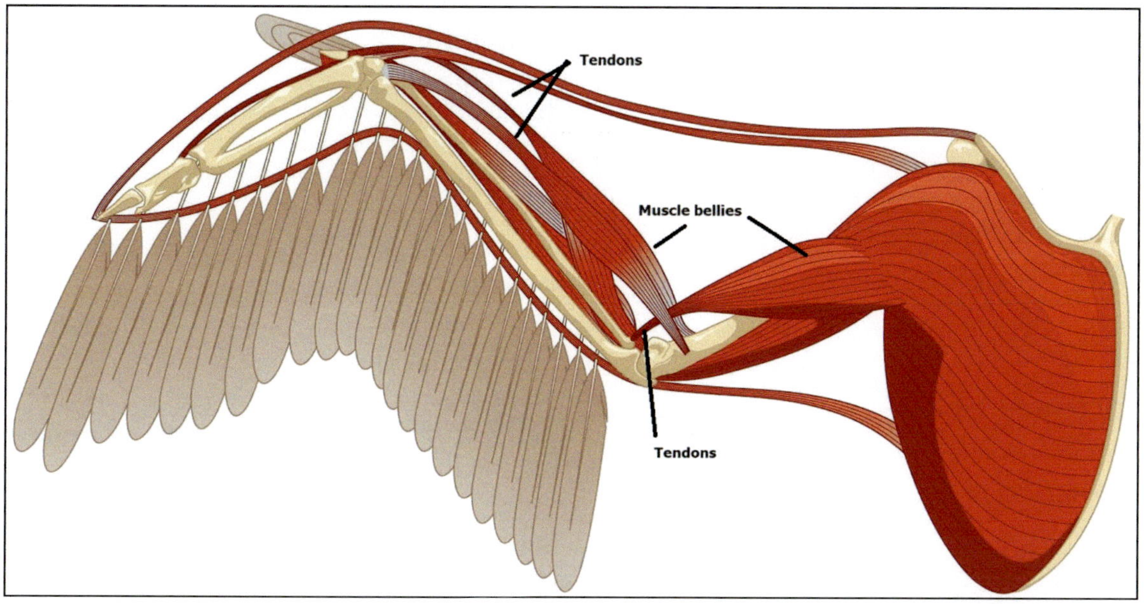

Figure 18.1 **Muscles of the avian wing, showing the proximal muscle bodies and the distal tendons.**

DOI: 10.1201/9781003272519-18

is derived from a combination of two biomechanical properties: high tensile strength and elasticity.

DISEASES OF TENDONS

Tendon contracture ('joint disease')

Aetiology

Enforced immobility of a joint in a flexed condition (e.g., when a wing or leg is splinted) can lead to contracture of the tendons controlling that joint. This can result in a permanent reduction in the joint's range of movement.

Management

Treatment requires a programme of gradual extension and flexion of the joint, performed under general anaesthesia, to break down the fibrous adhesions causing the contracture. The prognosis for return to normal joint flexibility is guarded.

To prevent this problem, joints should be immobilised for no more than 2 weeks, if possible. If this is not possible, weekly physiotherapy should be performed (e.g., passive flexion and extension of the joint through its normal range of movement). This usually requires a short general anaesthetic to remove the splint, perform the physiotherapy, and then replace the splint.

Tendon tears

Aetiology

Traumatic injury to the tendon occurs either directly by a blow or indirectly by a sudden, excessive movement of the limb that overtaxes the tendon.

Clinical signs

Tendon injuries, either tears or avulsions, can lead to an immediate loss of function of the muscle and, if healing is inadequate, result in disuse atrophy of the muscle.

Diagnosis

Diagnosis can be difficult and relies on clinical signs, palpation of the tendon (and comparison with the other limb), advanced imaging (CT or MRI), and perhaps ultrasound.

Treatment

When a minor tendon is severed, the effect on the bird may be trivial because its action is duplicated by another muscle and tendon, and there is no loss of function. However, if a major tendon is severed, with no duplication of action, normal function in that area is lost and repair is warranted.

Restoration of tensile strength is the prime objective of tendon repair, but at the same time, gliding function must be maintained if normal limb function is to be regained. The tensile strength of a tendon post-repair can be compromised by excessive scar tissue formation. This, however, may be minimised by controlled exercise during the latter stages of healing.

The lack of a continuous blood supply to the tendons makes surgical repair (e.g., end-to-end anastomosis using a Bunnell tendon suture pattern) difficult; this usually fails because of avascular necrosis at the surgical site. If the tendon does heal, the muscle and tendon usually are restricted in their movement because of a mass of adhesions.

Therefore, waiting several weeks after the initial injury (to allow the tendon to adhere to the tendon sheath and develop a blood supply) is often necessary. Alternatively, the severed ends can be stitched to the tendon sheath soon after the initial injury and then joined later. After 3–4 weeks, the now-vascularised tendon ends can be joined.

Avulsion lesions can be more difficult to treat. Possible surgical repair options include reattachment with a lag screw placed through the bone fragment, or a staple or a screw and spiked washer to reattach the tendon to the bone. Alternatively, the tendon can be sutured to the adjacent soft tissues.

Tendonitis

Aetiology

Tendonitis may be infectious (may start in a joint and extend into the surrounding tendons) or non-infectious. Infectious causes include bacteria, *Mycoplasma* spp., and reovirus. Non-infectious causes include trauma, articular gout, and tendon contracture (see earlier).

Clinical presentation

There is lameness or reluctance to use the limb, and localised swelling. Radiology may be required to rule out fractures or luxations.

Management

Inflammation is reduced with NSAIDs such as meloxicam. Antimicrobial therapy is given when appropriate.

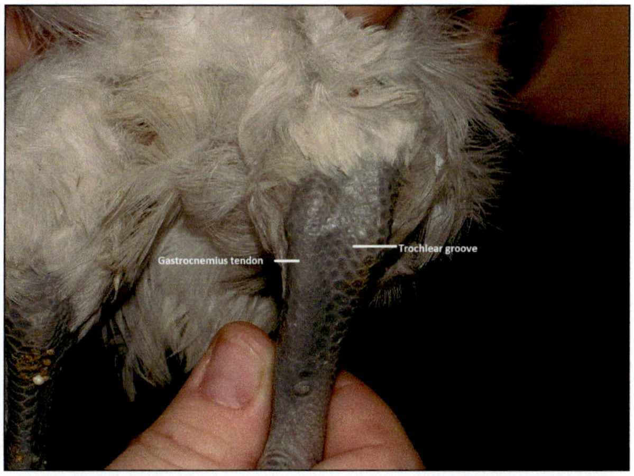

Figure 18.2 **Medial luxation of the gastrocnemius tendon in a chicken.**

Slipped tendon

Definition/overview

Medial or lateral luxation of the gastrocnemius tendon occurs over the tibiotarsal condyles (**Fig. 18.2**). It is most commonly seen in juvenile poultry, waterfowl, and ratites (where it is associated with nutritional deficiencies) but has been seen in parrots with leg deformities.

Diagnosis

Diagnosis is made based on clinical signs and palpation of the joint and tendon.

Management

This condition often requires arthrodesis of the hock joint.

Surgical replacement of the tendon and repair of the tendon retinaculum are invariably unsuccessful, the repair often breaking down within hours or days when the bird attempts to use the affected leg.

A technique has been described that involves placing a K-wire into the tarsometatarsus on the medial side of the ligament, with the proximal end of the pin protruding from the bone and effectively preventing the ligament from slipping medially. This technique appears promising, although a degree of joint stiffness often results. Any bone deformities predisposing to the tendon luxation must be corrected.

MUSCLES

Congenital

Muscular dystrophy

Muscular dystrophy is seen in chickens and turkeys, but not in companion birds. Muscle fibres are lost and are replaced by fat.

Arthrogryposis

Congenital flexure or contracture of joints occurs secondary to atrophy of skeletal muscle and subsequent fibrosis of the muscle. This atrophy is due to congenital neurological defects, either anatomical deficits or following parental alkaloid plant toxicosis (e.g., lupines).

Non-inflammatory conditions

Muscle atrophy

Aetiology

Atrophy may be caused by disuse (e.g., painful conditions of a limb, inability to fly, joint contraction following excessively long splinting of an injured limb, or pressure on the sciatic nerve from a renal tumour). It may also result from ageing and denervation caused by trauma or neoplasia (**Fig. 18.3**).

Generalised chronic disease can cause muscle atrophy. Acute illness in birds sees a demand for energy greater than dietary intake. This triggers glycogenolysis, which quickly exhausts the body stores of glycogen. Fat oxidation cannot meet the demands for energy and, after a few days, the process of gluconeogenesis using muscle protein (catabolism) begins. This results in muscle atrophy. The pectoral muscles are most noticeably affected, with the keel

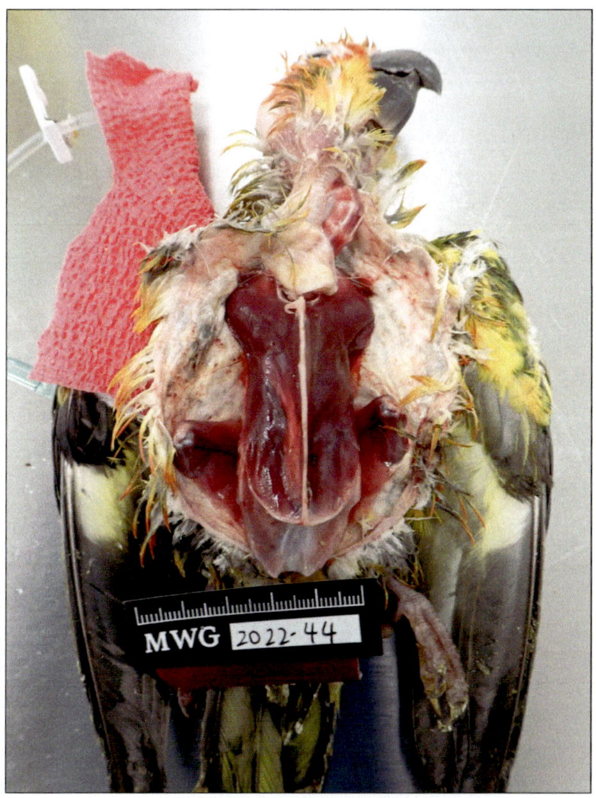

Figure 18.3 **Pectoral muscle atrophy in a sun conure.**

bone becoming more prominent. This condition is colloquially known as 'going light'. It is not a disease syndrome in itself; it is simply an indication of a chronic disease process.

Management

Disuse atrophy and denervation require identification and correction of the underlying problem, if possible.

Muscle atrophy due to chronic disease requires correction of the underlying disease problem, supplemental feeding (e.g., tube feeding for debilitated birds that cannot eat normally), and good-quality food (e.g., formulated diets to assist in the recovery and place the bird into an anabolic state). Recovery of the lost muscle mass is invariably slow, taking several days or weeks to return to normal.

Vitamin E and selenium deficiency

Aetiology

Vitamin E and selenium deficiency is seen in piscivorous species fed a diet of improperly frozen and thawed fish. Any species fed a diet containing rancid polyunsaturated fats can develop the same condition.

Clinical presentation

Vitamin E and selenium deficiency can result in cardiomyopathy and muscle lesions, showing as generalised weakness.

Management

Recovery is usually rapid once nutritional deficiencies are corrected.

Ionophore toxicosis (monensin, lasalocid, salinomycin, narasin)

Aetiology

Ionophore toxicosis is seen in non-gallinaceous birds fed ionophore-medicated poultry feeds. Ionophores are used in poultry as a coccidiostat.

Clinical presentation

Toxicosis is seen as generalised weakness and paresis due to damage to skeletal muscle.

Management

No specific treatment is available. The drug should be removed from the animal's diet, and supportive care should be provided until the patient recovers.

Inflammatory conditions: Non-infectious

Trauma

Aetiology

Muscle inflammation may be caused by blunt trauma (e.g., flying into objects such as ceiling fans or aviary wire) or iatrogenic trauma (e.g., intramuscular injections, especially enrofloxacin, doxycycline, and oxytetracycline).

Clinical presentation

Bruising and swelling can be seen (**Fig. 18.4**), and this may develop into necrosis or, with chronicity, fibrosis can replace all or part of the muscle.

Management

Severe or painful muscle injuries (**Figs. 18.5, 18.6**) should be treated with an analgesic (e.g., meloxicam, tramadol, or butorphanol).

Exertional/capture myopathy

Definition/overview

This is usually seen in flamingos, crowned cranes, Canada geese, sandhill cranes, and wild turkeys (USA) that have been recently captured or handled.

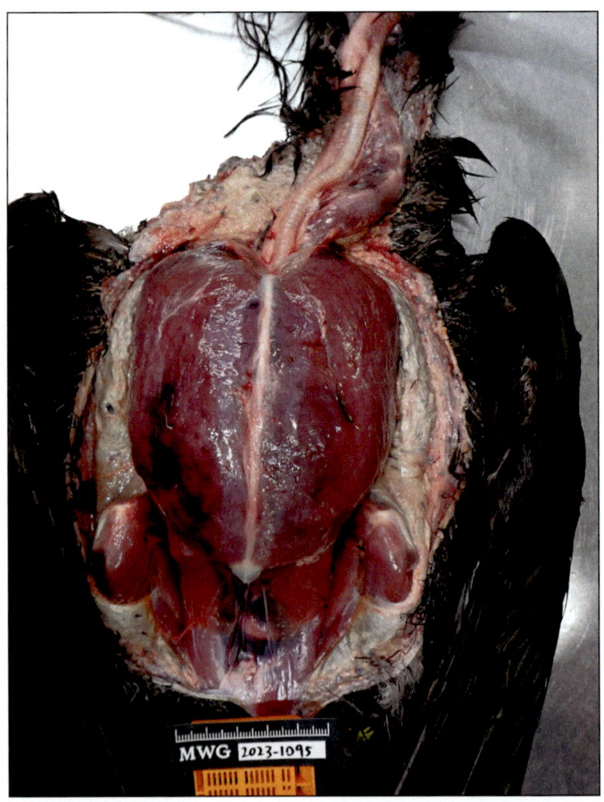

Figure 18.4 **Muscle bruising in a northern bald ibis.**

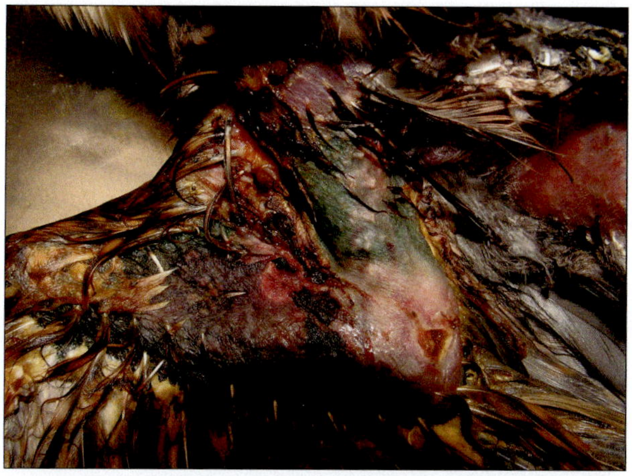

Figure 18.5 **Muscle trauma and bruising in a peregrine falcon following a collision with a vehicle. Note how, after a few days, bruising turns green due to biliverdin.**

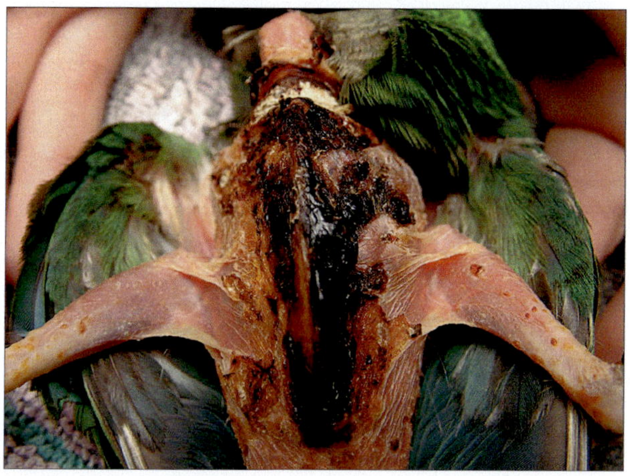

Figure 18.6 **Adult Quaker parrot with severe necrosis and mutilation of the pectoral muscles. This bird was diagnosed with 'Quaker mutilation syndrome', but the cause of this problem remains unclear.**

Clinical presentation

Muscle necrosis and intramuscular haemorrhage are commonly noted, followed by fibrosis with chronicity.

Diagnosis

Diagnosis of exertional myopathy is based on a history of recent capture or trauma, clinical signs, and elevation of AST, ALT, CK, LDH, and serum potassium.

Management

Treatment includes fluid therapy, vitamin E, and physiotherapy. The use of corticosteroids has been advocated by some authors; their benefits must be weighed against possible adverse effects on the liver and immune system.

Prognosis

Prognosis in severe cases is poor.

Inflammatory conditions: Infectious

Viral

Viral myositis is uncommon. Acute polyomavirus and adenovirus infections may cause muscle pallor and haemorrhage (**Fig. 18.7**).

Bacterial

Bacterial myositis is usually an extension of an infection in the skin or underlying bone; however, foreign bodies (e.g., microchips implanted without attention to sterility) can introduce an infection. The infection may be aerobic or anaerobic.

Fungal

Fungal myositis usually occurs as an extension of infection from air sacs or by haematogenous spread. It is similar in appearance to bacterial myositis and granulomas and can only be distinguished histologically.

Figure 18.7 **Muscle haemorrhage associated with adenovirus infection in an Amazon parrot.**

Parasites

Sarcocystis

Aetiology

Caused by *Sarcocystis* spp., a parasite of many species of birds, with a worldwide distribution, there are believed to be six species of *Sarcocystis* that infect birds. It has an obligate two-host life cycle that alternates between a sexual intestinal phase in the definitive host and an asexual multi-organ (especially the lungs) and ultimately muscular cyst phase in the intermediate host. The definitive host is usually a carnivore that eats an animal (the intermediate host) with mature cysts in its muscles. The intermediate host is usually an herbivorous or insectivorous bird. Transmission to the intermediate host is usually mechanical, in the form of insects such as flies and cockroaches. There is no direct spread from bird to bird, although several birds in one collection can be infected simultaneously.

Clinical presentation

While some intermediate hosts have evolved with both the parasite and the definitive host (and therefore do not develop clinical illness), naïve species exposed to *Sarcocystis* often develop clinical signs.

Acute pulmonary disease, due to asexual reproduction in the lung, causes severe dyspnoea and often acute death. Encephalitis can develop in parrots and raptors (posterior or unilateral paresis, intention tremors, and head tilt). Cardiac disease can develop in some birds if cysts localise in the myocardium. Severe myositis can occur in some birds (**Fig. 18.8**); other birds may have the cysts present but display no clinical signs.

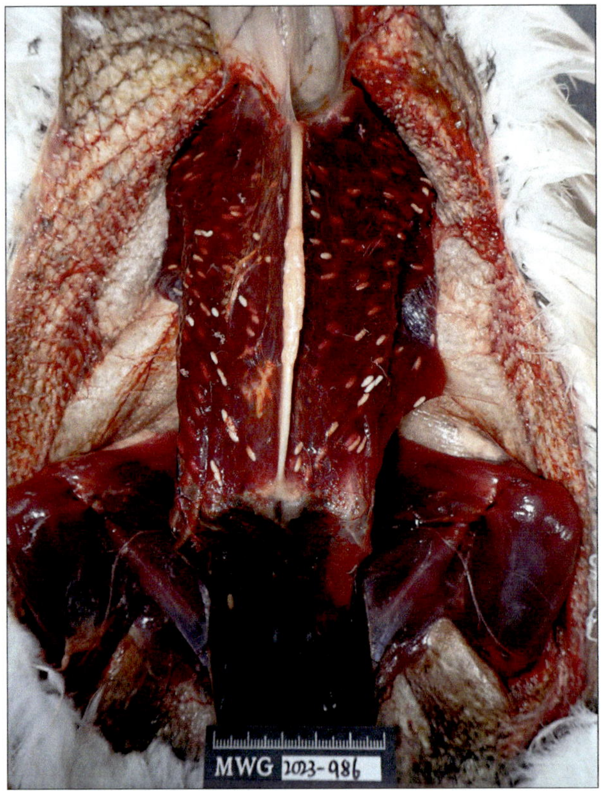

Figure 18.8 **Myositis in a lesser flamingo associated with *Sarcocystis* infection.**

Diagnosis

Diagnosis in the live bird is difficult. Blood tests show elevated CK and AST, and leucocytosis. An immunofluorescent antibody (IFA) test is available in some countries. Muscle biopsy of the quadriceps muscles gives better results than the pectoral muscles because the muscular cysts there are less likely to degenerate before reaching maturity.

Management

For the surviving mate, pyrimethamine is given at 0.5–1 mg/kg orally q12h for 2–4 days followed by 0.25 mg/kg for 30 days. Treatment for affected birds includes the triazinone anti-protozoal drugs (toltrazuril, diclazuril, and ponazuril), amprolium, primaquine phosphate, and sulfadoxine–pyrimethamine.

Control involves keeping definitive hosts out of the aviary or immediate surroundings and utilising effective insect control.

Leukocytozoon

Aetiology

Leukocytozoon is a protozoan parasite of erythrocytes and leucocytes, but with schizogony occurring in tissues other than the blood. Usually, the parasite is non-pathogenic, but in some cases, it can cause disruption of skeletal and cardiac muscles.

Clinical presentation

Clinical signs can include anorexia, weakness, haemoglobinuria, depression, and dehydration.

Management

Chloroquine, primaquine, or pyrimethamine may be used for treatment. Exposure to biting insects should be prevented.

Neoplastic

Muscle neoplasia is uncommon in companion birds. It can be primary (rhabdomyoma, rhabdomyosarcoma) or secondary (lymphosarcoma, melanoma). Metastatic disease is rare.

FURTHER READING

Harcourt-Brown NH (2002) Orthopedic conditions that affect the avian pelvic limb. The Veterinary Clinics of North America. *Exotic Animal Practice* 5(1): 49–81. https://doi.org/10.1016/S1094-9194(03)00046-X.

Koenig HE, et al. (2016) *Avian Anatomy: Textbook and Colour Atlas* (Second Edition), 5m Publishing.

Quesenbery K (1997) Disorders of the musculoskeletal system. In: *Avian Medicine and Surgery.* RB Altman, SL Clubb, GM Dorrestein, K Quesenberry (eds). WB Saunders, Philadelphia, pp. 523–539.

Villar D, Kramer M, Howard L, Hammond E, Cray C, Latimer K (2008) Clinical presentation and pathology of sarcocystosis in psittaciform birds: 11 cases. *Avian Diseases* 52(1): 187–194.

CHAPTER 19

DISORDERS OF THE GASTROINTESTINAL TRACT

ANATOMY AND PHYSIOLOGY

Oropharynx

The choana is a median fissure in the palate connecting the oropharynx to the nasal cavity. The palate is usually ridged laterally and rostrally to the choana and is associated with the dehusking of seeds and other foods. Caudal to the choana and palate is the infundibular cleft, a slit-like opening in the midline that is common to the right and left pharyngotympanic (Eustachian) tubes. This cleft cannot be closed by atmospheric pressure, making changes in altitude while flying possible. There is also lymphatic tissue (the pharyngeal tonsil) abundant in the wall of the cleft.

The tongue is supported by the hypobranchial (hyoid) apparatus. It has many adaptations for collecting and manipulating food and for swallowing. Only parrots have intrinsic muscles within the tongue. Just caudal to the tongue is the laryngeal mound, which carries the glottis. It has several rows of backward-pointing papillae to aid in swallowing.

The salivary glands are found in the roof of the oropharynx (maxillary, palatine, and sphenopterygoid glands), the angle of the mouth and cheeks, and the floor of the oropharynx (mandibular, lingual, and cricoarytenoid glands). They are best developed in birds that evolved on a dry diet.

Oesophagus and crop

The oesophagus is thin-walled and distensible, with a relatively greater diameter than that found in mammals. Longitudinal folds on the internal surface allow for this distension; the size and degree of these folds are proportional to the size of the food particles swallowed. The oesophagus is lined with incompletely keratinised squamous epithelial cells and mucus glands (especially in the distal oesophagus).

The crop is an enlargement of the oesophagus found in many (but not all) birds (e.g., it is very prominent in parrots but very small in most passerines). Its lining is similar to that of the oesophagus, except that it has no mucus glands. It serves as a food storage area for birds that eat rapidly and then move to a safer area to give the food time to pass farther down the digestive tract. The exit from the crop into the distal oesophagus is at the level of the thoracic inlet, on the right side of the midline of the neck. A fold of crop tissue lies over the exit, which then makes an S-shaped turn into the distal oesophagus.

Proventriculus and ventriculus

There is no obvious boundary between the distal oesophagus and the proventriculus, other than a lack of internal folds. The proventriculus is lined with mucus-secreting columnar epithelial cells. Within the lamina propria are the gastric glands, multi- or unilobular glands lined with tall columnar mucus cells. They discharge into an alveolus, which then drains into the central cavity of the lobule. Secondary ducts collect from different glands, which then empty into a primary duct, and this empties into the proventricular lumen. These glands may be present throughout the proventriculus or contained in defined tracts or areas. They produce hydrochloric acid and pepsin.

Between the proventriculus and the gizzard is the intermediate zone, a variably developed region with a microscopic structure somewhere between the two. It may narrow to form an isthmus between the proventriculus and the gizzard.

The gizzard (or ventriculus) varies in size and shape between species. Those species that eat soft food (e.g., lorikeets) have smaller, rounder gizzards, which can be somewhat difficult to distinguish from the proventriculus. Other species have a thickened, biconvex gizzard, with a wall consisting of smooth

DOI: 10.1201/9781003272519-19

muscle bands, rich in myoglobin. Asymmetrical arrangement of these muscle bands results in both rotatory and crushing movements when the gizzard contracts. The gizzard is lined with simple columnar epithelium, with crypts containing exits for tubular glands in the lamina propria. These tubular glands secrete hard vertical rods that interconnect laterally for greater strength. Between them is a softer horizontal matrix of a carbohydrate–protein complex secreted by the cells of the epithelium and crypts. This matrix hardens with the effect of hydrochloric acid. The vertical rods project slightly out from the horizontal matrix. This layer of rods and matrix is known as the cuticle or koilin layer. When combined with the asymmetrical rotatory grinding of the ventricular muscles, these rods and the matrix are very effective at crushing and grinding food into a soft pulp.

There is also a pyloric region connecting the gizzard to the duodenum. Its lining is microscopically intermediate between the gizzard and duodenum. Its function is unclear.

Intestinal tract

The duodenum forms a narrow U-shape on the right side of the gizzard, with the descending duodenum proximal to the ascending duodenum. It is lined with mucus-secreting goblet cells. The bile and pancreatic ducts often open near each other at the distal end of the ascending duodenum. There may be two bile ducts (the common hepatoenteric duct and the cystoenteric duct) and two to three pancreatic ducts.

The jejunum and ileum are usually arranged in a number of narrow U-shaped loops at the edge of the dorsal mesentery on the right side of the coelomic cavity. The vitelline (Meckel's) diverticulum is the short blind remnant of the yolk sac; it can be used to differentiate the jejunum from the ileum.

The large intestine is very short, separated from the ileum by the ileorectal sphincter. In some species, caeca arise from the rectum at the junction of the rectum with the ileum. Their form and size vary, and they are reduced or absent in parrots, swifts, and pigeons.

Cloaca

The cloaca is the common exit for the gastrointestinal, urinary, and reproductive tracts. As a point of terminology, the cloaca is the chamber; its opening to the skin is the vent. The cloaca is divided internally by two mucosal folds into three compartments: the coprodeum, the urodeum, and the proctodeum (**Fig. 19.1**). This structure is similar in all birds; the main variation is the presence or absence of the phallic structures of the proctodeum.

The coprodeum is the most cranial and largest compartment. There is no distinction between the rectum and coprodeum (except in the ostrich, which has a rectoperineal fold, and Anatidae, where there is an abrupt change in the gross appearance of the mucosa). Some species have villi and folds on the mucosa; others have none.

The urodeum is the middle and smallest compartment, separated from the other two compartments by two circular mucosal folds. The coprourodeal fold is a cranial annular fold that stretches to become a thin diaphragm if the coprodeum is full of faeces; it may close during egg laying to prevent defecation while the bird is laying an egg. The uroproctodeal fold is a caudal, semicircular dorsolateral fold that fades out ventrally. The urogenital ducts open into the urodeum on the dorsolateral mucosa (the ureters dorsally, the genital ducts laterally). The ureter opens via a simple opening; the ductus deferens opens via a conical papilla. In immature hens a homologue of the ductus deferens papilla may be present, but it disappears with maturity. In the mature hen the left oviduct opens ventrally and laterally relative to the left ureter. There may be a small mound at its opening. In immature birds it is covered with a membrane that disappears at maturity.

The proctodeum is the short caudal compartment between the lips of the vent and the uroproctodeal

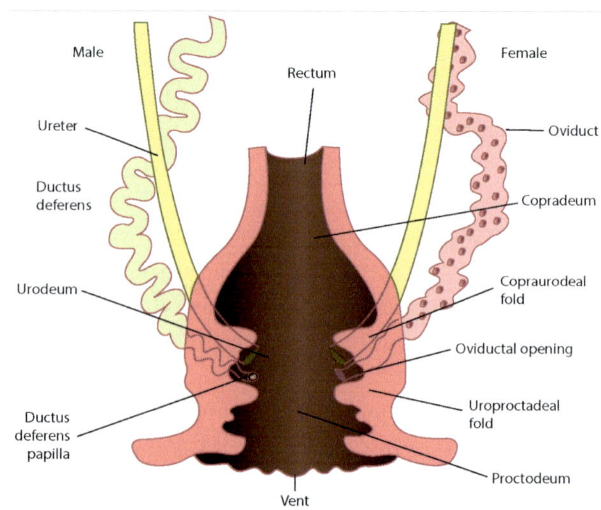

Figure 19.1 Schematic diagram illustrating both the male and female components of the cloaca.

fold. In immature birds an opening in the dorsal wall leads into the cloacal bursa.

The vent, the exit from the cloaca, is a transverse slit guarded by dorsal and ventral lips. This horizontal arrangement is the reason why purse-string sutures are unsuitable for closing the vent in birds.

DISORDERS OF THE GASTROINTESTINAL TRACT

The oropharynx and crop

Candidiasis

Definition/overview

Candidiasis, also known as thrush, moniliasis, sour crop, and crop mycosis, is most commonly due to an overgrowth of the yeast *Candida albicans*, which can be part of the normal intestinal flora and is frequently isolated from droppings of normal birds. It is primarily a disease of the upper gastrointestinal tract (oropharynx, oesophagus, and crop) but can be found lower in the tract on occasion.

Predisposing factors to this overgrowth include young birds that are not fully immunocompetent; prolonged antibiotic use; concurrent immunosuppressive conditions (e.g., debilitation, psittacine beak and feather disease [PBFD], malnutrition); poor hygiene in the bird's environment and food preparation; failure to clean excess formula from the skin or mouth of hand-reared chicks; high concentrations of sugar in fruit and hand-rearing formulae providing an optimal medium for the growth of yeast; and alkaline crop contents, seen when crop stasis occurs for other reasons, encouraging yeast overgrowth.

Clinical presentation

Affected birds show general malaise, weight loss, and reduced growth rates. There may be regurgitation/vomiting. Crop emptying times are increased, the crop may be distended with fluid and mucus, and the crop wall becomes thickened. Diphtheritic membranes may be present in the oropharynx, and crusty lesions can be seen on the commissures of the mouth. Infection can extend through the choana into the sinuses, causing signs of upper respiratory tract infection. It can also present as dermatitis (see Chapter 11).

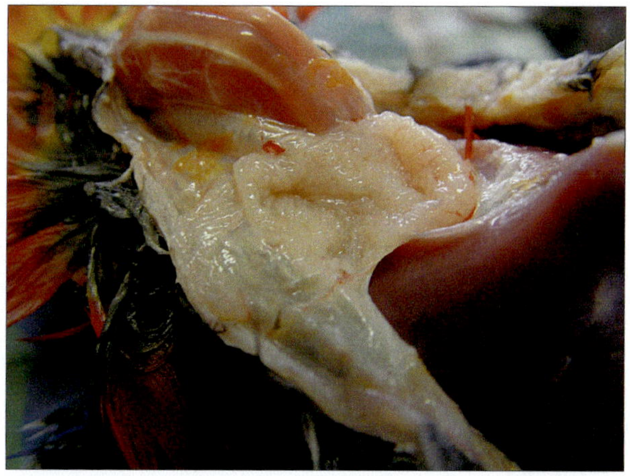

Figure 19.2 **Lorikeet with candidiasis. Note the thickening of the ingluvial wall.**

Diagnosis

Diagnosis is based on history, clinical signs, and lesions, as culture alone may not distinguish between normal flora, dietary yeast, and pathogenic overgrowth. Cytology (Gram stain) of throat swabs, crop washes, and faeces may reveal the yeast in three different forms: oval non-budding yeast, budding yeast, and pseudohyphae formation:

- Oval non-budding yeast is frequently of dietary origin.
- Large numbers (>5 per high-powered field) of budding yeast are significant.
- The presence of pseudohyphae may indicate tissue invasion.

Gross lesions (**Fig. 19.2**) must be distinguished from pox virus, trichomoniasis, hypovitaminosis A, and internal papilloma disease (IPD).

Management

Underlying factors must be addressed (e.g., environment, hygiene, husbandry, food handling, and concomitant antibiotic therapy). Antifungal therapy includes nystatin, ketoconazole, fluconazole, or itraconazole. Nystatin is not absorbed systemically and must, therefore, be given into the mouth (rather than by crop needle) as it requires direct contact with the yeast to be effective. Systemically absorbed antifungals may be preferred if the yeast infection includes the lower gastrointestinal tract. Acidification of crop contents by adding apple cider vinegar to drinking water or hand-rearing formula (10 ml/litre of water) may decrease crop pH and discourage the growth of yeast.

Trichomoniasis

Definition/overview

Trichomoniasis is also known as canker (pigeons and budgerigars) or frounce (raptors). It is most commonly seen in budgerigars, lorikeets, pigeons, and raptors, but it has been reported in most species. It has a worldwide distribution.

Aetiology

The causative organism is *Trichomonas* spp. (*T. gallinae* and *T. columbae*), a protozoan parasite with an undulating membrane and four anterior flagella. It has a direct life cycle, with the protozoa being passed from bird to bird via feeding (courtship behaviour, feeding young) or via contaminated drinking water (faeces, crop secretions). The organism dies quickly outside the host.

Pathophysiology

Following ingestion of the protozoa, they remain free-swimming in the bird's saliva and crop fluids. Although they do not attach to the mucosa, they will cause inflammation, followed by necrotic stomatitis and ingluvitis, often with diphtheritic membranes. This is often complicated by secondary bacterial infections.

Clinical presentation

Signs include unthriftiness and high mortality in young birds; gagging, regurgitation, vomiting; diarrhoea; ptyalism; weight loss; and diphtheritic membranes in the oropharynx, oesophagus, and crop (**Fig. 19.3**).

In raptors it may extend into the sinuses (causing a caseous sinusitis) or into the Eustachian tubes (causing a head tilt). Young pigeon squabs develop a visceral form of the disease involving the liver, gastrointestinal tract, and navel (often referred to as 'wet canker').

Diagnosis

A crop wash or throat swab usually (but not always) shows motile protozoa. Rarely, there is a single lesion in the thoracic oesophagus causing clinical signs, but the protozoa are not readily detectable. These cases are difficult to diagnose antemortem. Motile protozoa are occasionally found in the faeces.

Figure 19.3 **Budgerigar with ingluvial ulceration due to trichomoniasis. Note that seed is present from the mouth to the lesion but not beyond that point. Starvation is a common cause of death.**

Management

Treatment with a nitroimidazole (metronidazole, dimetridazole, ronidazole, carnidazole) is usually effective. Treatment with antibiotics for secondary bacterial infection is often warranted. Some cases with severe ulceration and diphtheritic membranes will die despite treatment. In some raptors, surgical debridement of lesions may become necessary.

In pigeons, strategic treatment of the flock may be necessary (i.e., during the breeding season, at weaning, immediately prior to and after the racing season). The extent of treatment required in pigeons is the subject of some debate, as there is evidence that infection with lentogenic strains of *Trichomonas* spp. provides some protective immunity against velogenic strains.

Hypovitaminosis A

Definition/overview

Vitamin A is formed in the liver through the conversion of beta-carotenes (from vegetables and fruit) and retinol (from liver and fish oils). Deficiencies in vitamin A (or its precursors) result in keratinisation of epithelial cells, causing squamous metaplasia of the mucous membranes of the oropharynx. The metaplasia blocks the ducts of the salivary and mucus glands, causing abscesses in these glands.

The authors' experience is that birds that have evolved with high levels of beta-carotenes in their natural diet (i.e., fruit and vegetables) appear to be more susceptible to hypovitaminosis A than birds that have evolved a much poorer quality diet (e.g., seeding grasses). Birds such as eclectus parrots, Amazon parrots, African grey parrots, and lorikeets are more likely to show clinical signs of deficiency than cockatiels and budgerigars.

Aetiology

Hypovitaminosis A is caused by a nutritional deficiency of vitamin A, e.g., all-seed diets.

Clinical presentation

White or pale swellings are seen in the inter-ramular region, beneath the tongue, or elsewhere in the oropharynx. There is blunting of the choanal papillae. Ptyalism occurs, and sinusitis is common.

Diagnosis

History and clinical signs are very suggestive. Differential diagnoses include trichomoniasis, pox virus, candidiasis, and IPD.

An aspirate of the swellings shows keratinised epithelial cells rather than inflammatory cells.

Management

Treatment includes parenteral vitamin A and dietary correction. Surgical removal of masses may be required in extreme cases, but severe haemorrhage is likely to ensue.

Poxvirus infection

Birds with avian poxvirus infection may have diphtheritic membranes in the oropharynx. See Chapter 11 for more detail.

Internal papilloma disease

Birds with IPD may have papillomas in the oropharynx. This condition will be discussed in more detail later in this chapter.

Crop stasis

Definition/overview

Crop stasis, where the crop fails to empty, is commonly seen in pre-weaning chicks, but it can also be seen in adult birds.

Aetiology

There are several possible causes. Crop stasis may occur because of generalised ileus caused by systemic illness (e.g., severe renal disease), stunting, foreign bodies below the level of the crop, chilling, heavy metal toxicosis, or dehydration. Conditions affecting the crop, such as foreign bodies, an overstretched/atonic crop, infectious ingluvitis (e.g., *Candida albicans*, bacteria), fibrous food impaction, or crop burns, may also cause stasis. Dietary factors may also be involved (e.g., cold food, watery food, food that settles out, chronic overfeeding, or overly dry food).

Clinical presentation

The crop fails to empty within 6 hours. It is palpably distended, often filled with fermenting food and water ('sour crop'). Regurgitation may occur, often upon palpation of the crop. There are clinical signs of dehydration (sunken eyes, weight loss, wrinkled skin, thick saliva).

Management

If possible, the aetiological agent should be identified and eliminated. The crop should be emptied (by surgery if necessary) and lavaged with warm saline. Parenteral fluids are given (to correct dehydration) until crop motility is restored, and antimicrobials are given as appropriate. Motility modifiers (metoclopramide, cisapride, and fennel or cumin tea) may assist, but their efficacy is not universal.

Once the bird is rehydrated, small, watery meals should be given often. Predigesting hand-rearing formulae with pancreatic enzymes can liquefy the diet without diluting it.

Chronically distended, nonresponsive crops may require support with an elastic sling ('crop bra') or, as a last resort, crop reduction surgery.

Thermal injuries

Definition/overview

Thermal injuries are less common than in past years, as aviculturists have become aware of the problem. Novice hand-rearers, however, still present chicks with this problem.

Aetiology

The most common aetiology is a hand-rearing formula that has been overheated in a microwave oven and then fed before it has cooled sufficiently. 'Hot spots' (small foci of super-heated food) may be present in a mix and be overlooked when (if) the food temperature is checked. Some cases may be due to chicks coming into contact with incandescent light bulbs or heating pads, particularly while the crop is distended after a recent feed.

Clinical presentation

In the early stages (1–2 days), affected chicks may be lethargic and refuse feeding. Physical examination may reveal mild to moderate crop stasis and erythema of the most prominent part of the ventral crop. After the initial stage of erythema, blanching of the affected tissue develops. A crust then forms over the area; when it lifts off, a fistula is usually revealed, often with food leaking from it (**Fig. 19.4**).

Diagnosis

Thermal injuries must be differentiated from a crop perforation (see later).

Management

Surgical resection of the burn and repair of the crop and skin are necessary to effect a cure. However, surgery must be delayed until the fistula has formed, and all devitalised tissue has become obvious (usually 4–7 days after the incident; see Chapter 32). The crop has an incredible ability to stretch, and even large crop resections seem to be well tolerated by most young birds. Subsequent feedings will obviously need to be reduced in volume depending on the postoperative size of the crop.

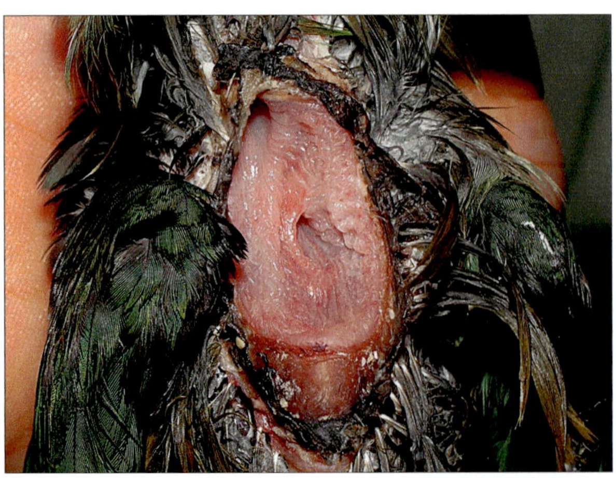

Figure 19.4 Juvenile parrot whose outer crop wall has sloughed off following a crop burn.

While waiting for the burn to become clearly demarcated, the chick must be given supportive care, including analgesia and antibiotic coverage, and small feeds given frequently to avoid distending the crop. Placement of an oesophagostomy tube may be necessary in some cases (see Chapter 6).

Crop perforations

Aetiology

Crop perforation usually occurs when using a metal feeding tube. The tube perforates the crop because either the chick has an unrestrained feeding response and thrusts up against the tube, or the person feeding the chick does so roughly or impatiently. Food can be deposited outside the crop and under the skin, in some cases without being noticed.

Clinical presentation

Early cases may be presented because of blood on the feeding tube when it is withdrawn. These birds are usually asymptomatic. More advanced cases may be presented because of apparent crop stasis, often with severe systemic illness. Distension is palpable in the crop region, but food cannot be aspirated with a feeding tube. The skin overlying the crop is often discoloured and gas may be palpated under the skin.

Management

Surgical removal of the food deposited under the skin, followed by debridement and flushing of the subcutaneous tissues and repair of the crop injury, is essential. Prognosis is determined by the time lapse since the initial injury and the degree of sepsis present in the patient.

Ingluvoliths and other foreign bodies in the crop

Aetiology

Ingluvoliths are chronic concretions of starch and other materials that form in the crop. Other foreign bodies can be ingested by the bird or fed to it by its parents. They can include fibrous material, seed and nut husks, large pieces of food, and pieces of wood.

Clinical presentation

Signs include regurgitation, gagging, vomiting, weight loss, crop stasis, and ptyalism. Some birds are presented extremely emaciated and dehydrated; these birds have a guarded prognosis.

Diagnosis

Diagnosis is based on palpation of the crop and plain and contrast radiography.

Management

Endoscopic retrieval through the mouth is feasible in some cases. Ingluviotomy may be required to remove some objects (see Chapter 32). Small ingluvoliths can be lubricated with liquid paraffin (mineral oil) and massaged until they break down into sludge or small particles that can pass through the digestive tract or be aspirated via a large-bore feeding tube.

Parasites

Capillaria contorta (hair worm, thread worm)

Definition/overview

Capillariasis occurs in quail and pheasants. The worms burrow into the mucosa of the oesophagus and crop, creating tracts that fill with blood, producing hyperaemic streaks and some diphtheritic lesions. The life cycle can be direct or indirect.

Clinical presentation

Clinical signs include anorexia, dysphagia, diarrhoea, and weight loss.

Diagnosis

Diagnosis is made by finding double-operculated eggs on faecal examination.

Management

Treatment is with an appropriate anthelmintic. Levamisole, ivermectin, and moxidectin have all been used. It should be noted that resistance is common, and it is important that follow-up screening is performed to assess the efficacy of the treatment.

Spiruroid worms (Spiruroidea)

Definition/overview

Spiruroid worms occur in wild corvids and Strigiformes and are also reported in passerines and parrots. They produce granulomatous lesions in the oropharynx, crop, and proventriculus, with a worm protruding. The life cycle requires an arthropod vector.

Management

Treatment is with ivermectin or moxidectin, followed by the manual removal of the worms if possible.

PROVENTRICULUS AND VENTRICULUS

Fungal infections

Macrorhabdus

Definition/overview

Macrorhabdus ornithogaster, also known as avian gastric yeast (AGY) or megabacteria, is an anamorphic ascomycetous yeast, a large, gram-positive, periodic acid–Schiff (PAS)-positive, highly pleomorphic yeast.

Disease is often associated with concurrent illnesses (e.g., circoviral infections, stress, chlamydiosis). It colonises the gastric isthmus of many species of birds; in large numbers, it can cause a maldigestion disorder and proventricular ulceration. It is probably transmitted by the ingestion of faecal material, although parent birds may pass it to their chicks when feeding them.

Clinical presentation

Macrorhabdus is most recognised in budgerigars and lovebirds, but it has been diagnosed in many species including parrots, canaries, Gouldian finches, goldfinches, ostriches, and poultry (including quail and partridges).

In wild Australian birds, the organism is commonly found in recently fledged galahs (*Eolophus roseicapilla*) and short-billed (little) corellas (*Cacatua sanguinea*) with chronic catarrhal enteritis and weight loss. These birds often have comorbidities such as PBFD and motile protozoal parasites. The full host range of *M. ornithogaster* in parrots is unknown, and infection should be considered in any species of psittacid with gastrointestinal disease.

Clinical signs include weight loss and weakness, polyphagia (although the bird is often grinding its food and then letting it fall from its mouth), diarrhoea, and vomiting. In acute cases, especially in parrotlets, the birds may regurgitate food, and there may be blood staining around the beak. Undigested seed may be passed in the droppings, and melena may be seen in some cases.

Diagnosis

Faecal examination is used to diagnose *M. ornithogaster* in the live bird. As it often washes off a slide during staining, it is important that a wet mount be examined under 40× magnification with the stage condenser nearly closed. A rapid way of concentrating *M. ornithogaster* and separating it from other solid matter in the faeces is to homogenise a dropping with approximately 20 times its volume of physiologic saline in a small tube, let it sit for 10 seconds, and then examine a small drop of the suspension collected from the meniscus. Because *M. ornithogaster* takes longer to settle than most other material in the faeces, they are more easily seen in wet preparations after this treatment (**Fig. 19.5**).

Gram staining and other cytology stains show variably staining large 'cigar-shaped' gram-positive organisms (**Fig. 19.6**). It is a long, straight, narrow rod with rounded ends, 2 µm to 4 µm wide and 20 µm to 80 µm long. On occasion, a single Y-shaped branch is found at one end. Most sick birds will pass large amounts of the organism, but asymptomatic birds may only shed the organism inconsistently. If doubt remains, either a Calcofluor stain or a polymerase chain reaction (PCR) test can be used to demonstrate the organism.

A PCR assay to detect *M. ornithogaster* in the faeces is also available in North America (Veterinary Molecular Diagnostics, Milford, Ohio).

Necropsy findings include a dilated proventriculus (**Fig. 19.7a**), ulcerations of the proventricular–ventricular isthmus (**Fig. 19.7b**) and large numbers of the organism in mucosal scrapings (**Fig. 19.7c**).

Management

The recommended treatment is amphotericin B, given at 100 mg/kg twice daily by crop gavage for 30 days. Resistance may develop, and some birds will relapse after treatment. Iodine preparations, lupheneuron, ketoconazole, terbinafine, and itraconazole have been shown to be ineffective; fluconazole is toxic to budgerigars at the doses needed to be effective against *Macrorhabdus*; nystatin may be effective in some birds, but resistance is likely.

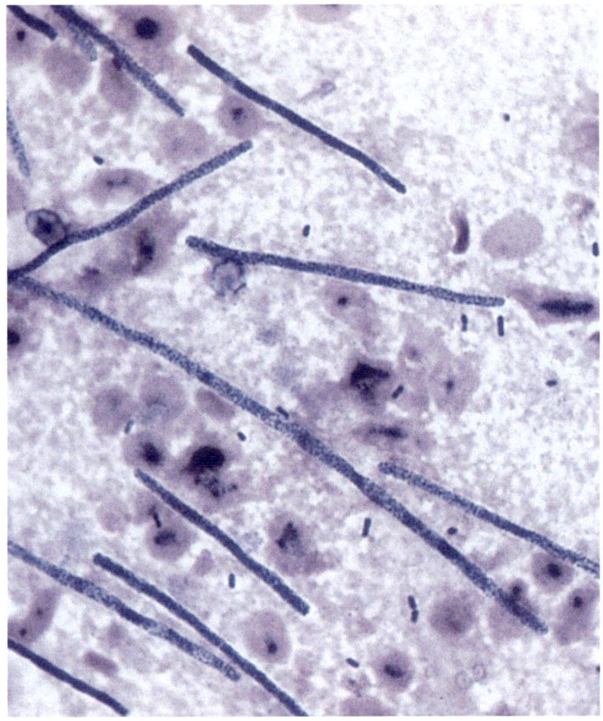

Figure 19.6 ***Macrorhabdus* organisms on a Gram stain.**

Figure 19.5 ***Macrorhabdus* organisms on a wet prep.**

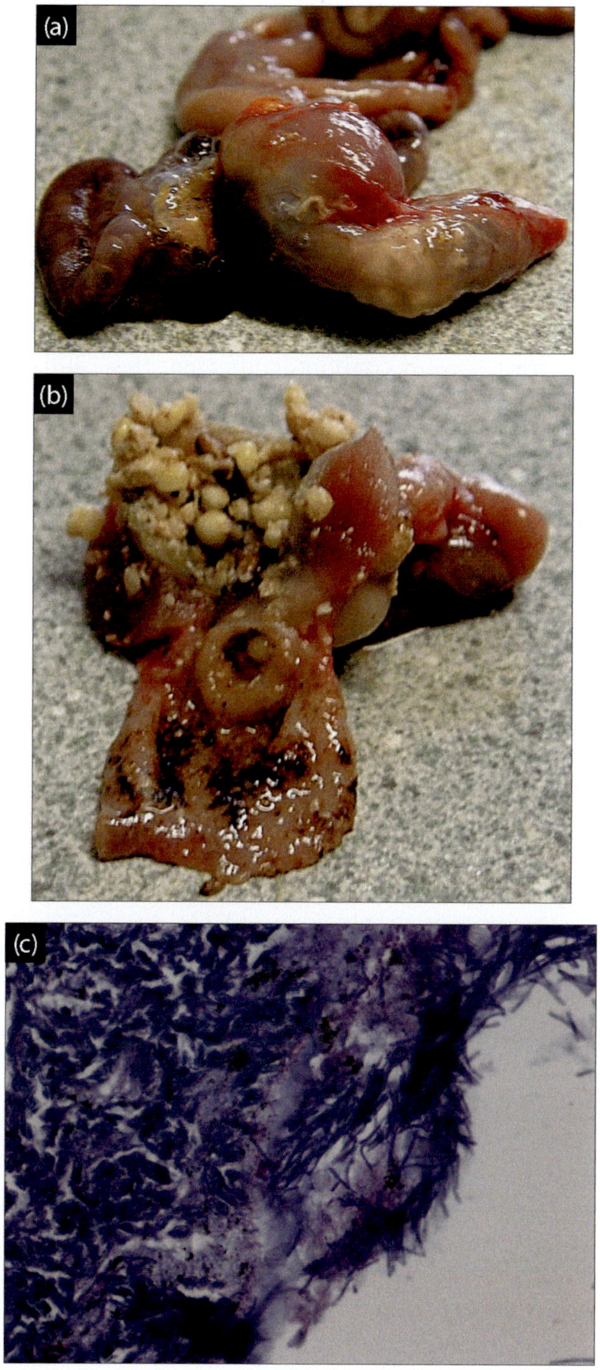

Figure 19.7 **Dissection of the gastrointestinal tract of a cockatiel reveals a dilated proventriculus (a) and ulcerations in the proventricular–ventricular isthmus (b), both suggestive of *Macrorhabdus* infection. A stained mucosal scrape confirmed *Macrorhabdus* presence (c). Note the 'cigar-shaped' organisms present on the right-hand edge of this image.**

Flock treatment is problematic; a water-based formulation of amphotericin B (Megabac-S, Vetafarm, Wagga Wagga, Australia) may not clear a flock and may lead to the development of resistant strains. Sodium benzoate has been used in some flocks with variable results – palatability is poor, and it appears to be toxic to many birds, causing salt poisoning.

As *Macrorhabdus* does not appear to be egg-transmitted, artificial incubation and rearing of chicks in isolation can produce 'clear' birds as a basis for a *Macrorhabdus*-free flock.

Candidiasis

Candidiasis occasionally causes ventriculitis in small passerines (see earlier).

Zygomycetes and Aspergillus spp.

Definition/overview

These fungi (*Rhizopus*, *Mucor*, and *Absidia* spp., and *Aspergillus* spp.) are ubiquitous, usually fast-growing fungi. They will occasionally invade the mucosa and muscular layers of the proventriculus and ventriculus (**Fig. 19.8**). They may be confined to the gastrointestinal tract or disseminated widely through the body, possibly haematogenously.

Clinical signs

Affected birds are usually presented with vomiting/regurgitation and weight loss.

Diagnosis

Antemortem diagnosis is difficult, although endoscopic biopsy and brush cytology may be rewarding. Radiology using contrast may demonstrate thickening of the proventricular wall and fungal granulomas elsewhere in the body.

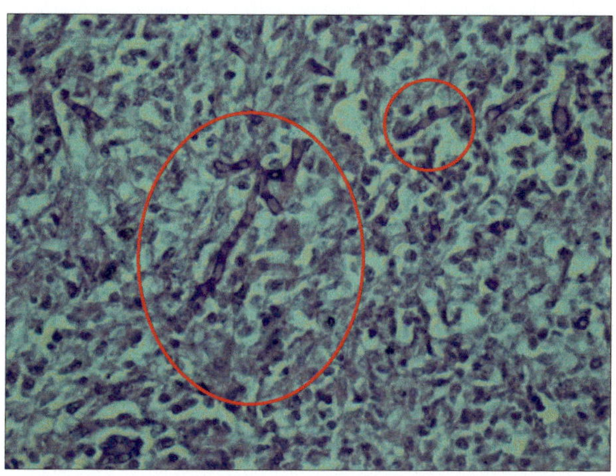

Figure 19.8 **Stained tissue sample of proventricular tissue from a green-cheeked conure shows fungal hyphae (circled) invading mucosal cells.**

Management

Long-term treatment (1–2 months) with systemic antifungal drugs (fluconazole, itraconazole, ketoconazole) is required.

Viral infections

Avian ganglioneuritis

Definition/overview

Avian ganglioneuritis, also referred to as proventricular dilatation disease, macaw fading syndrome, myenteric ganglioneuritis, infiltrative splanchnic neuropathy, and neuropathic gastric dilatation, is a progressive fatal neuropathy. It has a worldwide distribution and has been reported in most psittacine species. Suggestive lesions have also been described in toucans, honeycreepers, canaries, weaver finches, Canada geese, and roseate spoonbills. It usually affects adult birds, but it has been reported in chicks as young as 2 weeks.

Aetiology

Research since 2008 has confirmed that the disease is commonly associated with the previously unknown avian bornavirus (ABV), a negative-encoded, single-stranded, non-segmented RNA virus of the order Mononegavirales. There are several genotypes, of which two, ABV4 and ABV2, are the most common in parrots around the world. ABV-C in canaries and ABV-CG in Canada geese in North America have been described.

It must be noted that not all birds infected with ABV go on to develop avian ganglioneuritis, as many apparently healthy birds carry ABV for prolonged periods of time.

The virus appears to be shed predominantly in urine, but the route of transmission remains unclear. Several routes have been disproven, including faecal–oral and aerosolisation. Vertical transmission has also been shown to be possible in psittacine birds, but it is uncommon. Although the virus has been shown to be present in the urine of some birds, sometimes independently of faeces, the urine–oral or urine–nasal route is yet to be proven. Regardless of the route of transmission of the virus, only a few individuals of a flock may be infected, even when there is close contact among the birds (Lierz, 2016). Cockatiels may be asymptomatic carriers of ABV.

Pathogenesis

The incubation period of ABV is extremely variable, with periods as short as 11–14 days in unweaned chicks and months to years for older birds having been reported. It appears that, in most adult birds, the incubation period is 1–3 months.

Avian ganglioneuritis is characterised by the presence of lymphoplasmacytic inflammation of the brain and nerves, especially in the enteric ganglia and nerve plexuses. Although the anterior gastrointestinal tract is most commonly affected, these inflammatory infiltrates can be found in other nerves, including the brachial, sciatic, and vagus nerves. It is believed that the lesions seen are the result of the action of cytopathic T-cells rather than a direct cytopathic effect of the virus. The role of antiganglioside antibodies has not yet been fully determined.

Clinical presentation

Avian ganglioneuritis is a segmental neuropathy, with clinical signs dependent on the nerves (and therefore the innervated organs) affected. Most commonly, clinical signs are related to gastrointestinal dysfunction: passing whole food in droppings, regurgitation, weight loss, anorexia, lethargy, and depression.

Occasionally, CNS and peripheral nervous system signs are seen, including ataxia, abnormal head movements, progressive paresis leading to paralysis, seizures, and proprioceptive or motor deficits. Lesions in the eye, when present, are characterised by moderate to severe perivascular cuffing in the optic nerves and in the choroid, ciliary body, and occasionally in the iris and pecten. This can result in severe retinal lesions and blindness. On rare occasions, avian ganglioneuritis causes sudden death due to an effect on the conduction pathways of the heart.

Secondary infections, including endoparasitism, in the dysfunctional gastrointestinal tract are common.

Diagnosis

Although, in classical cases, the clinical signs of avian ganglioneuritis are highly suggestive of the disease, clinicians should strive to obtain a definitive diagnosis in order to advise their clients on prognosis and risk to other birds.

Haematology and biochemistry are often unrewarding, with no consistent changes noted in either.

Some birds may have a non-regenerative anaemia of chronic disease.

In classical cases, radiology will demonstrate a dilated, thin-walled proventriculus. This can be better defined through the use of contrast media. This finding is, however, not pathognomonic for avian ganglioneuritis and can also be seen with gastrointestinal neoplasia, heavy metal toxicosis, and gastrointestinal foreign bodies.

Contrast fluoroscopy is a useful tool to demonstrate impaired movement of contrast through the tract. In the normal parrot, boluses of ingesta leave the crop and travel unidirectionally along the distal oesophagus to the proventriculus at an approximate rate of 1 bolus per minute. Little or no barium remains in the oesophagus between boluses. The proventriculus is relatively immobile, but a large contraction followed by partial emptying into the ventriculus should be seen every few minutes. Ventricular movement, often described as 'washing machine-like', is the constant grinding and mixing of ingesta prior to emptying into the duodenum. Normal intestinal peristalsis is bi-directional and wavelike. Birds with avian ganglioneuritis, by contrast, show little (if any) movement of the proventriculus, and the normal movement of the ventriculus is replaced by a shallow and irregular flutter of the ventricular wall. Peristalsis of the small intestine may become fast and erratic or slower than usual.

Accurate diagnosis of ABV infection in a bird requires a combination of PCR-based and serological tests. The presence of both virus (PCR-positive) and antibody (serology-positive) may be considered to indicate an ABV-positive status. Birds that are PCR-positive should be isolated and retested 4 to 6 weeks later to rule out a false positive due to sample contamination. Birds that are serology-positive may be retained in the same flock/enclosure, but should also be retested 4 to 6 weeks later. In general, tests for both PCR and serology need to be negative for two consecutive tests before a bird may be considered negative for ABV.

Since many birds that are infected with ABV do not show or develop clinical signs, and may not develop signs, the presence of ABV infection is not proof that a sick parrot has avian ganglioneuritis. Therefore, a clinically unwell bird with an enlarged proventriculus should be diagnostically worked up to exclude other differential diagnoses such as gastrointestinal neoplasia, heavy metal toxicosis, and gastrointestinal foreign bodies, regardless of its ABV infection status.

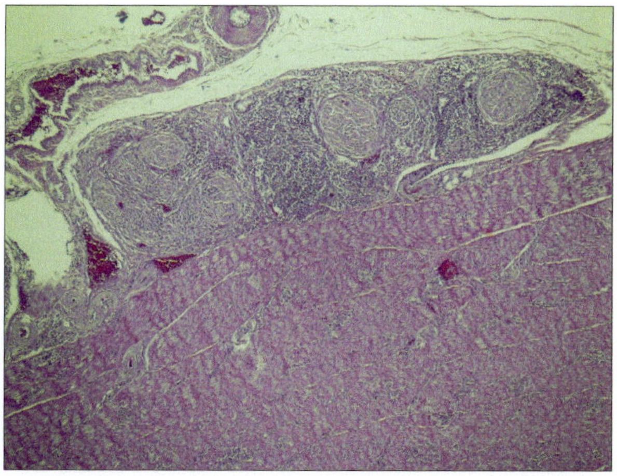

Figure 19.9 **Stained tissue sample of proventricular tissue from a bird with avian ganglioneuritis showing lymphoplasmacytic infiltrate ganglioneuritis.**

Definitive findings on histopathology include non-suppurative lymphoplasmacytic ganglioneuritis of central and peripheral nerve tissue, with associated myositis and atrophy of the gastrointestinal tract, leading to dilatation and impaction (**Fig. 19.9**). Adrenal glands may also show a lymphoplasmacytic infiltrate. The biopsy specimen must include a blood vessel and associated nerves. The myenteric plexus in the tunica muscularis of the proventriculus and ventriculus contains more lesions. The segmental nature of the disease means that the pathologist must examine stepped sections of the specimen to ensure lesions are not overlooked. Antemortem crop biopsies have been used as a diagnostic tool for avian ganglioneuritis, but only 50% of birds showing signs may be positive. Lesions of the splanchnic nerves of the ventriculus have been found in 100% of birds exhibiting signs of avian ganglioneuritis, and these enlarged nerves may be visible grossly through the endoscope.

Necropsy usually reveals the characteristic dilated, thin-walled proventriculus, often distended with ingesta (**Fig. 19.10**). The vitreous of the eye, the brain, spinal cord, peripheral nerves, adrenal glands, and major organs are suitable samples to submit for histopathology and, as mentioned earlier, for reverse transcription PCR. Immunohistochemistry and viral isolation are alternate means of diagnosis.

Treatment

There is 100% mortality in untreated birds diagnosed with avian ganglioneuritis. Most birds that

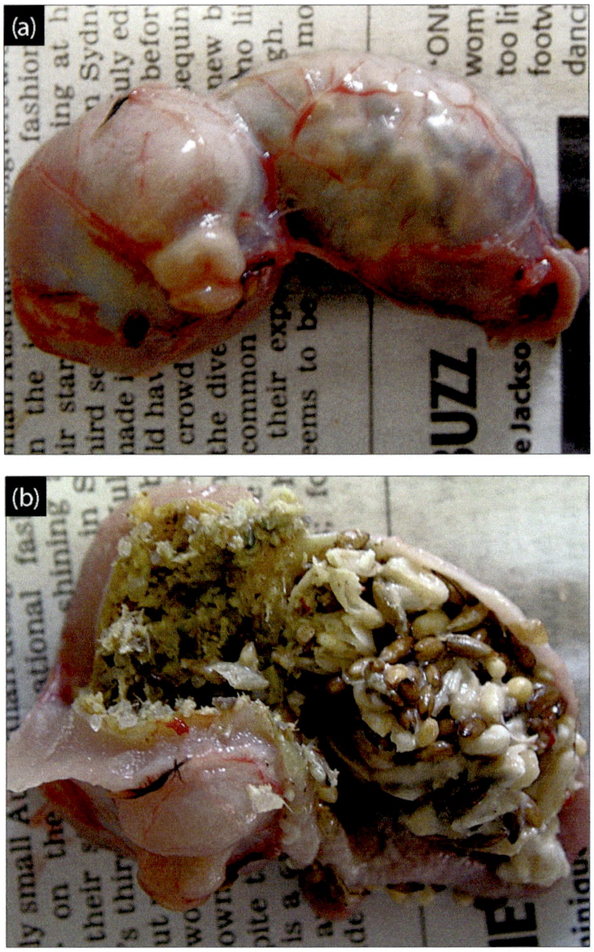

Figure 19.10 **An excised proventriculus from an eclectus parrot with avian ganglioneuritis is clearly enlarged and dilated (a), and when cut open, has a thin wall stretched over the contents (b).**

are treated also die, albeit months or years after their diagnosis.

Based on the inflammatory nature of this disease, it makes sense to seek to reduce this inflammation. The first treatment that was reported to be successful involved the use of the non-steroidal anti-inflammatory drug, celecoxib, a cyclooxygenase (COX)-2 inhibitor. Complete recovery in some birds, with return to normal function and biopsy showing resolution of the neuritis, was reported. Birds are given 10 mg/kg once daily for 6–24 weeks, with treatment cessation based on the bird regaining weight and resolution of lesions on biopsy. Similar success has also been reported with robenacoxib given intramuscularly at 2 to 10 mg/kg weekly, decreasing to monthly subsequently.

There is one report that suggests meloxicam may worsen the onset and severity of clinical signs associated with avian ganglioneuritis, so its use may not be encouraged until further evidence is available.

Cisapride, metoclopramide, and high-fibre diets have all been used to assist gastrointestinal motility, with varying degrees of success.

Control

Basic biosecurity principles are important in the control of ABV transmission, including:

- Testing of all birds in the flock
- Quarantine and testing of all new and sick birds
- Control of traffic
- Good sanitation and thorough cleaning of all areas using detergents and suitable disinfectants

Elimination of ABV is possible in large psittacine collections if a rigorous testing and exclusion protocol is used. All birds in the collection are tested using both PCR and concurrent serological tests. After the first round of testing, birds are separated into positive (if both PCR and serology tests are positive), questionable (only one test is positive), and negative (both tests negative) groups. Strict biosecurity measures are required to prevent disease transmission between groups. Decisions will also need to be made regarding the positive group. Euthanasia of positive birds is one option, but it may not be acceptable because they are generally healthy and may have significant genetic value. Alternatives to euthanasia would include rehoming or managing these birds as breeders in a separate facility,

The questionable and negative groups are retested in 4 to 6 weeks. Transfer of birds between these two groups will be based on the results of the second round of testing. The negative group is considered cleared of ABV when two consecutive combined tests are negative for all the birds in the group. Birds from the questionable group should be tested until two consecutive combined tests are negative, upon which they may be housed with the negative group. The process needs to be restarted whenever a positive bird is found, or if a clinical case of avian ganglioneuritis/ABV is found in the group.

Efforts to eradicate ABV from a zoological or private avicultural collection require appropriate facilities and are costly in terms of time and money. The facility must possess the spatial and manpower capacity to manage flocks separately. Furthermore, the rehoming of positive birds should be coupled with strong biosecurity, ethical, and legal considerations.

Parasitic infections

Finches, pheasants, quail, and poultry are most commonly affected.

Acuaria spp. (gizzard worm)

Aetiology

Acuaria spp. (gizzard worm) are the most common worms involved. They include *A. skrjabini*, *A. humulosa*, and *Dispharynx nasuata*. In canaries and finches, these worms inhabit the proventriculus (causing swelling of the proventriculus mucosa) and under the koilin lining of the ventriculus, where they cause thickening of the ventricular wall. The intermediate hosts include weevils, grasshoppers, slaters, and other insects.

Clinical presentation

Clinical signs are usually ill-thrift and death.

Diagnosis

Diagnosis can be made on necropsy or by demonstrating the typical double-operculated eggs in the faeces.

Management

Levamisole, moxidectin, and ivermectin can be used, but resistance is common, so treatment efficacy must be monitored. A recently released benzimidazole, flubendazole, registered for poultry, offers promise for resistant infestations, but its use in non-poultry species is off-label. Control of intermediate hosts is important.

Foreign bodies

Definition/overview

A bird's curiosity and exploring nature will often lead it to ingest foreign bodies. The normal functioning of the ventriculus – the tougher koilin lining and the presence of grit, when combined with the grinding effect of the powerful ventricular muscles – effectively grinds up most of these foreign bodies within a few days. While the great majority of these foreign bodies cause little or no harm, others can cause gastrointestinal blockages, damage the gastrointestinal mucosa, and release toxins such as lead and zinc, which are then absorbed and become a systemic problem.

Unless the bird is displaying clinical signs consistent with foreign bodies causing a problem, most can be left and monitored to see if the body is removing them unaided. The addition of a small amount of grit to the bird's diet may assist this process. The administration of paraffin oil or psyllium may help to move smaller particles through the tract.

Clinical presentation

Clinical signs include anorexia, weight loss, regurgitation, decreased faecal output, melena, and signs consistent with heavy metal toxicosis.

Diagnosis

Radiography, both plain and contrast, is useful. (The entire tract must be evaluated; don't identify a foreign body in the crop and overlook the one in the proventriculus!)

Haematology and biochemistry can be used to evaluate systemic changes.

Management

If a foreign body needs to be removed, it is easier to remove it from the crop or proventriculus rather than waiting until it is in the ventriculus or intestinal tract. Endoscopy, either through the mouth or via an ingluviotomy, can be used to remove foreign bodies in the crop or proventriculus.

A proventriculotomy or ventriculotomy can be used to remove foreign bodies in the proventriculus and ventriculus. These surgeries have a high morbidity associated with them and should not be undertaken lightly.

Smooth, non-penetrating foreign bodies in the crop, proventriculus, or ventriculus may be removed by flushing:

- Anaesthetise the patient, intubate, and pack the choanal slit with gauze. Place the patient in either ventral or dorsal recumbency.
- Pass a flexible, large-bore feeding tube into the thoracic oesophagus through the exit from the crop in the right lateral dorsal corner of the thoracic inlet. This can be done blindly or with guidance from endoscopy.
- Tilt the bird's head down at an angle of 30°.
- Flush the proventriculus gently and steadily with warmed saline. The flush can be allowed to run out or it can be aspirated with gentle suction.

- Collect the flush to identify and count the number of foreign bodies retrieved.
- Repeat the flush until all the foreign bodies have been retrieved.

Neoplasia

Definition/overview

The most common gastrointestinal neoplasias are papillomas associated with IPD (see later "Disorders of the cloaca") and gastric carcinomas, found at the proventricular–ventricular junction. Death from gastric neoplasia may be due to haemorrhage, gastric perforation and sepsis or endotoxic shock, or anorexia and subsequent catabolism.

Diagnosis

Diagnosis is made by radiography, endoscopic biopsy, or necropsy.

DISORDERS OF THE INTESTINAL TRACT

Ileus

Definition/overview

Ileus is decreased motility or complete stasis of the intestinal tract.

Aetiology

Causes include foreign body obstruction, decreased or absent neurological function (e.g., lead toxicosis, avian ganglioneuritis), enteritis, peritonitis, pancreatic disease (e.g., zinc toxicosis), and intussusception or torsion of the intestine.

Clinical presentation

Clinical signs include regurgitation, dehydration, lethargy, and decreased faecal output.

Diagnosis

Radiography shows fluid- or gas-filled loops of the intestine, often with proventricular dilation.

Management

Dehydration should be corrected, and causative factors identified and rectified. Motility enhancers such as ranitidine, cisapride, or metoclopramide may be of assistance.

Parasites

Ascarids (roundworms)

Definition/overview

Species identified include *Ascaridia hermaphrodita*, *A. columbae* (shared between pigeons and parrots), *A. galli* (shared between gallinaceous birds and parrots), and *A. platycercii* (restricted to parrots). They are most commonly seen in birds with access to the ground and are particularly common in budgerigars, cockatiels, Quaker parrots, and princess parrots.

Pathophysiology

The worms are found in the small intestine of the affected bird, particularly in the duodenal loop (**Fig. 19.11**). They have a direct life cycle.

Clinical presentation

Clinical signs include lethargy, poor condition, diarrhoea, and death.

Diagnosis

Diagnosis is by faecal examination (**Fig. 19.12**).

Management

Most anthelmintics are effective. Ascarids are resistant to disinfectants but susceptible to desiccation, steam, and flames.

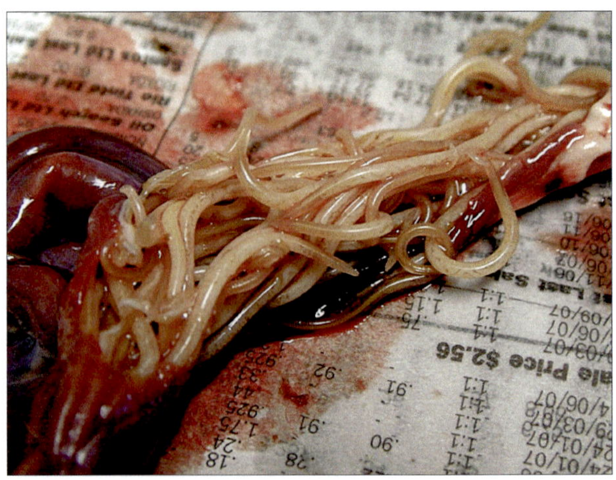

Figure 19.11 An excised duodenal loop from a princess parrot is found to contain large numbers of roundworms (*Ascaridia* sp.).

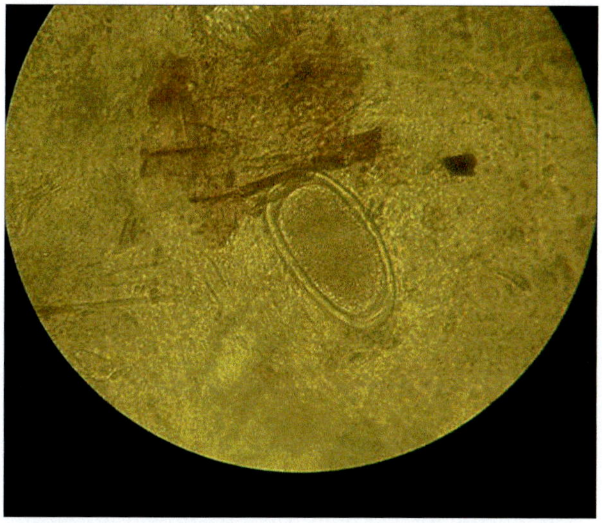

Figure 19.12 Avian faecal smear containing a roundworm (*Ascaridia* sp.) egg.

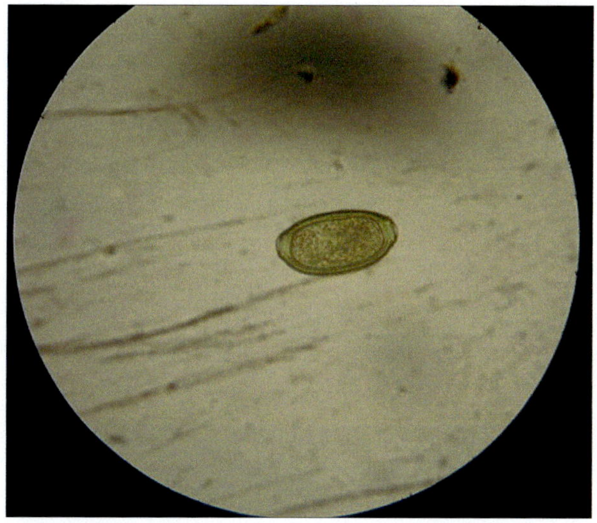

Figure 19.13 Avian faecal smear containing a threadworm (*Capillaria* sp.) egg. Note the eggs are double-operculated.

Capillaria (hair worm, thread worm)

Definition/overview

Species identified include *Capillaria annulata*, *C. contorta* (which causes pseudomembrane formation in the oesophagus and crop [see earlier]) and *C. obsignata* (found in the small intestine).

Pathophysiology

Capillaria embed in mucosa, causing a marked thickening and inflammation of either the crop or the intestinal mucosa. The worms have a direct or indirect life cycle, using insect vectors. Eggs are infective in two weeks. They can remain infectious in the environment for several months.

Clinical presentation

Signs include anorexia, dysphagia, diarrhoea, and weight loss.

Diagnosis

Faecal examination reveals a double-operculated egg (**Fig. 19.13**).

Management

Ivermectin, moxidectin, levamisole, or flubendazole (off-label use) can be trialled. Anthelmintic response needs to be monitored, as resistance is common. Access to insect vectors should be controlled.

Heterakis (caecal worm)

Definition/overview

Species identified include *Heterakis isolonche* in quail and *H. gallinarum*, which affect gallinaceous birds.

Pathophysiology

The life cycle is indirect, through earthworms.

Clinical presentation

There are usually no clinical signs, but the worms are significant because they transmit *Histomonas* spp. (see later).

Management

Most anthelmintics can be used. If possible, access to intermediate hosts should be controlled.

Cestodes (tapeworms)

Definition/overview

Species identified include *Raillietaenia*, *Choanotaenia*, *Gastronemia*, *Idiogenes*, and *Amoebotaenia*. Tapeworms are commonly found in finches and Old World parrots (African grey parrots, cockatoos, lorikeets, and eclectus parrots). The life cycle is indirect and may be complete in 3–4 weeks. Intermediate hosts include grasshoppers, beetles, ants, horseflies, earthworms, slugs, snails, and crayfish.

Clinical presentation

Signs include catarrhal enteritis, emaciation, and anorexia.

Diagnosis

Diagnosis is based on the finding, on faecal examination, of oncospheres (tapeworm embryos with six hooks).

Management

Praziquantel and fenbendazole may be used. Exposure to intermediate hosts should be controlled.

Coccidia

Definition/overview

Species identified are numerous but include *Eimeria*, *Isospora*, and *Caryospora* spp. This parasite tends to be host-specific.

- *Isospora* oocysts have two sporocysts, each with four sporozoites. They are most common in parrots, passerines, and Piciformes.
- *Eimeria* oocysts have four sporocysts, each with two sporozoites. They are most common in Galliformes and Columbiformes.
- *Caryospora* oocysts have one sporocyst and eight sporozoites. They are most common in raptors.

Pathophysiology

The life cycle is direct. The bird ingests the sporulated oocysts lying in its environment. Oocysts undergo schizogony in the intestine (asexual reproduction, producing multiple parts). If the schizonts produced are deep in the intestinal mucosa, they may damage the mucosa when they divide, causing enteritis. After schizogony, they differentiate into male (microgametocytes) and female (macrogametocytes) forms. They then undergo sexual reproduction (gametocytic), which is non-pathogenic. This produces non-sporulated oocysts, which are shed in faeces. Oocysts sporulate in a warm, moist environment to become infective.

Coccidia require 6–8 days to complete the life cycle, with clinical signs seen 4–6 days after infection.

Clinical presentation

Affected birds may not show clinical signs until stressed. Signs include lethargy, weight loss, diarrhoea (sometimes with blood and mucus), dirty vents, and death.

Diagnosis

Oocysts can be found in faeces on smear and/or flotation. Note that clinical signs can be present before oocysts are detectable in the faeces.

Management

Anti-coccidial treatments include amprolium, toltrazuril, ponzuril, and sulphadimethoxine. Birds are treated daily for 2–3 days, then treatment is repeated after 5 days to treat those organisms that were in the prepatent period when the first treatment was given.

Access to infected faecal material is prevented by using concrete floors that are cleaned regularly or by having suspended wire floors. Food and water dishes should be kept off the ground.

Cryptosporidia

Definition/overview

Cryptosporidia are enteric coccidians that usually lie just inside the limiting membrane of the enterocytes. They can also attach to the proventricular epithelium, the respiratory epithelium, the conjunctival sac, the urinary tract, and the bursa. The oocysts are the smallest of any coccidia, with four naked sporozoites and no sporocysts in oocysts.

Pathophysiology

Cryptosporidia develop intracellularly at an extracytoplasmic location on the apical surface of epithelial cells. They are often secondary pathogens. Although not very host-specific, they appear to be infective only to other birds, not to mammals. The life cycle is direct, usually via water contamination. The oocysts are sporulated when passed and, as such, are fully infective immediately.

Clinical presentation

Cryptosporidiosis is often asymptomatic. Clinical signs may include depression, dehydration, anaemia, anorexia, persistent diarrhoea, malabsorption problems, coelomic pain, vomiting, coughing, sneezing, and nasal discharge.

Diagnosis
Diagnosis is based on histopathology using modified (Kinyoun's) acid-fast stains. Infective cysts may be seen on faecal flotation and can be detected via PCR.

Treatment
Treatment is with oral paromomycin sulphate 100 mg/kg, q12h for 7 days.

Management
Cryptosporidium oocysts are extremely resistant in the environment. Outside the body, they can remain infectious for 6–8 months.

- They are highly resistant to chlorine disinfection levels used in water systems. The required chlorine levels generally preclude the use of chlorine as a reliable control method.
- Boiling for at least 1 minute or filtration with a 1 micrometre filter will also decontaminate drinking water.
- Ammonia can inactivate *Cryptosporidium* but requires an 18-hour exposure time.
- Hydrogen peroxide is more effective than chlorine.
- Alcohol-based hand sanitisers are ineffective.

In clinical cases, owners should be advised that *Cryptosporidium* can be a zoonotic disease and aggressive disinfection is required. Porous items such as perches and toys should be discarded or baked in the oven. Cages should be scrubbed to remove all organic material.

Giardia

Definition/overview
Giardia has a wide geographic and host distribution. Its zoonotic potential is unclear. It has been reported in most avian species, although it has not been reported in finches and canaries. It has a direct life cycle, although cysts are shed only intermittently.

Clinical presentation
Signs include weight loss, depression, ruffled feathers, and chronic diarrhoea. Persistent feather picking and pruritus may be seen in cockatiels (see Chapter 11). Infected chicks may have poor growth rates; many die.

Diagnosis
Cysts or trophozoites can be detected in the faeces. As the trophozoites are not stable outside the bird, faecal examination should be performed on droppings within 10 minutes of defecation.

Management
Nitroimidazoles are generally effective.

Treatment
Good hygiene is essential to break the life cycle. *Giardia* cysts may survive in chlorinated water but are inactivated by quaternary ammonium compounds.

Cochlosoma

Definition/overview
Cochlosoma is a motile protozoan that is common in waterfowl and finches. In Australia it is common in cockatiels. Bengalese (society) finches can carry the parasite without clinical disease, but when used as foster parents for Gouldian finch chicks, there is often severe loss of the chicks. It has a direct life cycle, with transmission due to the ingestion of infective trophozoites in faecal material and contaminated food and water.

Clinical presentation
Clinical signs include diarrhoea, weight loss, and death.

Diagnosis
Diagnosis is made on examination of a fresh faecal smear.

Management
Drug resistance is common. Nitroimidazoles are used, but the treatment must be monitored for efficacy. Care must be taken when fostering Gouldian finches with Bengalese finches, as Gouldians are very susceptible to this parasite.

Hexamita (Spironucleus spp.)

Definition/overview
Hexamita is a motile protozoon with eight flagella (six anterior and two trailing). It has been reported in pigeons (*Hexamita columbae*), game birds (*Hexamita*

meleagridis), cockatiels, lorikeets, grass parrots, galahs, and Australian king parrots. It has a direct life cycle. Concurrent disease and parasitism are common.

Clinical presentation
In pheasants and pigeons, it is most commonly seen at 6–12 weeks of age. It causes decreased appetite, lethargy, emaciation, and intractable diarrhoea. Similar signs are seen in parrots.

Diagnosis
Diagnosis is made on examination of a fresh faecal smear; the parasite swims in a smooth, linear fashion.

Management
Nitroimidazoles are used, but the treatment must be monitored for efficacy. Supportive therapy is essential, as many birds are emaciated, dehydrated, and hypoglycaemic on presentation.

Histomonas meleagridis

Definition/overview
Histomonas meleagridis is a motile protozoan that affects turkeys, quail, peacocks, and pheasants. Chickens and guinea fowl may harbour latent infections. The disease in turkeys and peacocks is known as 'blackhead'. The life cycle is indirect, with transmission occurring through the ingestion of infected eggs of the caecal worm *Heterakis gallinarum.*

Clinical presentation
Clinical signs include weight loss, increased thirst, decreased appetite, and brown-black to mustard-coloured diarrhoea. In chickens there may be blood in the droppings. There can be a high mortality rate, especially among juveniles.

Diagnosis
Diagnosis is based on clinical signs, identification of *Heterakis* eggs in droppings, and, on necropsy, caecal cores (caecal necrosis) and hepatomegaly with multifocal yellow necrotic lesions.

Management
Treatment is with nitroimidazoles. Control of *Heterakis* is required (see earlier).

Microsporidiosis (encephalitozoonosis)

Definition/overview
Microsporidia sp. is a gram-positive, acid-fast, obligate intracellular fungus, measuring 1–2 μm. It appears to have a direct life cycle, probably through oral transmission although the aerosol route is also thought to be feasible. *Encephalitozoon hellum* has been isolated from eclectus parrots, Amazons, lovebirds, budgerigars, and finches. It has also been isolated from humans, where it can cause keratoconjunctivitis (sometimes severe), sinusitis, and, uncommonly, systemic disease.

Pathophysiology
Increasing evidence strongly suggests that birds are the primary host for E. hellem and that most bird infections are self-limiting and do not result in disease. When disease does occur, there may be some element of immunosuppression (immaturity, stress, overcrowding, or concurrent disease, e.g., pasteurellosis, PBFD). Latent infections may exist.

Clinical presentation
Signs include anorexia, lethargy, weakness, diarrhoea, stunting, ruffled feathers, weight loss, and neonatal mortality. Chronic conjunctivitis and sinusitis, as well as exophthalmos, corneal oedema, and blepharospasm, were reported in a double yellow-headed Amazon parrot.

Diagnosis
PCR may be available. Histopathology using silver stain or acid-fast staining is necessary, as standard stains may not reveal the organism.

Management
Microsporidiosis should be considered zoonotic. Because shedding is so common in lovebirds, practitioners should advise clients not to keep lovebirds if there is an immunosuppressed person in the household. Likewise, lovebirds should not be kept in retirement homes and facilities caring for the chronically ill.

Treatment is with albendazole (50 mg/kg orally q24h for 5 days).

Mycobacteria
Definition/overview
Most avian infections are caused by *Mycobacterium avium avium* and *M. genavense. M. tuberculosis* and

M. bovis are occasionally reported but are considered rare.

Mycobacteriosis is most commonly an enteric infection, but it may be respiratory or disseminated. It is considered to be a leading infectious cause of avian deaths in many zoological parks with mixed-species aviaries, where up to 14% of annual mortality can be attributed to mycobacteria.

Although not conclusively proven, it should be considered a potential zoonotic disease.

Pathophysiology

Transmission is by ingestion of infected faeces, either directly or via contaminated food and water. Insects may act as mechanical vectors. Pulmonary disease by the aerogenous route is rare. The incubation period following infection can be months or years. Following ingestion, the mycobacteria penetrate the mucosa and colonise under the serosa. It enters the blood supply and is removed from circulation by the liver, spleen, and bone marrow. The lack of well-developed regional lymph nodes in birds may be one reason for the ease with which this disease becomes systemic. Lesions therefore develop in the intestinal wall, liver, spleen, and bone.

Clinical presentation

Weight loss and severe muscle atrophy are the most consistent clinical signs. Coelomic distension, hepatomegaly, and dilated, fluid-filled, thickened intestines may be palpable. Lameness, due to arthritis or osteomyelitis, may be present. Nodular or diffuse keratinous skin lesions at the mucocutaneous junctions of the eyes and beak have been reported in parrots.

Diagnosis

Antemortem diagnosis in birds is difficult and often inconclusive, particularly in the early stages. Clinical signs combined with marked leucocytosis with monocytosis and elevations in AST and bile acids increase the suspicion of mycobacteriosis as the cause of illness. Radiographs may show an enlarged liver, kidney, and spleen, as well as focal, cloudy densities in the medullary cavities of long bones with osteolysis. Endoscopy and biopsy are used to confirm the diagnosis.

Faecal acid-fast stains may reveal the organism, but as *Mycobacterium* is only shed intermittently, and other non-pathogenic acid-fast staining can also sometimes be found in avian faeces, this is not an effective screening tool. An intradermal test can indicate the presence of mycobacteria in a flock, but it is unreliable and difficult to interpret in an individual, non-poultry avian species. PCR performed on cloacal swabs or faecal samples may have value, but similar problems as faecal acid-fast stains exist. Pooling samples for PCR may partly overcome the unreliability associated with intermittent shedding. Selection of the appropriate PCR test is also important. For example, performing a pan-*Mycobacterium* PCR on faecal samples is not an effective screening tool for the same reasons as faecal acid-fast stains. However, performing PCRs specific for *M. avium avium* or *M. genavense* may be a useful non-invasive test for ruling in mycobacteriosis. Pan-*Mycobacterium* PCRs are more useful when performed on organ, e.g., hepatic, biopsy samples, as there should be no mycobacteria of any species present in these organs. Sequencing of the positive PCR sample can subsequently be performed to identify the *Mycobacterium* species present.

Necropsy usually reveals granulomas in the liver (**Fig. 19.14a**), spleen, and subserosa of the intestinal

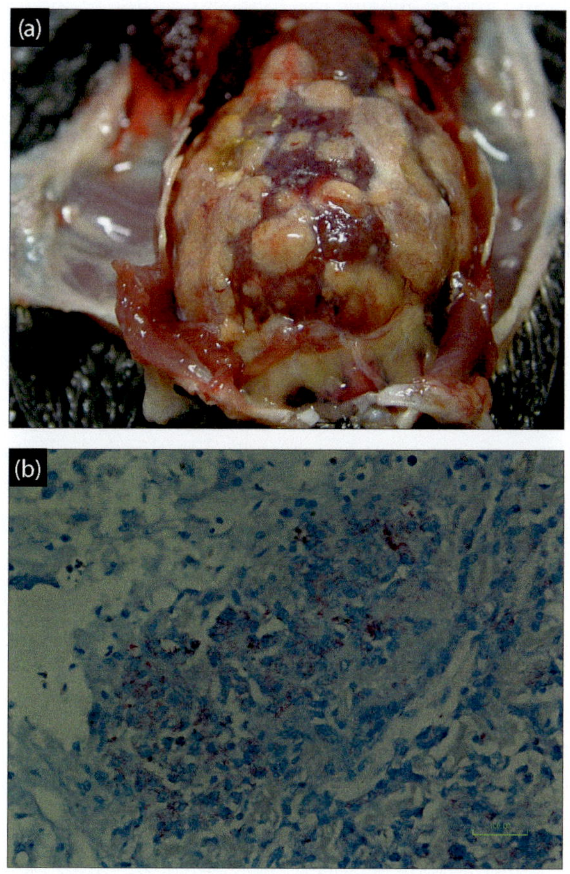

Figure 19.14 **Post-mortem examination of a red cap parrot with a mycobacterial infection is found to have disseminated granulomas throughout its liver (a), and a stained liver tissue smear confirms the presence of acid-fast bacteria (stained red) (b).**

tract. There may be hepatomegaly, splenomegaly, and enlarged, thickened intestines.

Histopathology shows a spectrum of lesions from nodules of large foamy histiocytes packed with acid-fast bacilli (**Fig. 19.14b**) to giant cell-containing granulomas that are often caseous but not cavitated or calcified. PCR can be performed on samples from these lesions for confirmation. Amyloidosis is seen in approximately 20% of cases.

Management

Husbandry improvements to eliminate exposure to a high environmental load of the pathogen and stressors that may lead to immunosuppression are an important aspect of disease management. Consideration must be given to whether to treat because of the possible zoonotic implications.

Resistance develops rapidly; therefore, combinations of drugs must be used. Treatment must be given for 9 months or more, as the organism is only susceptible when replicating, which occurs every 16–20 hours, and even less in caseated lesions. One should aim to achieve the highest possible blood levels, not consistent levels; therefore, once-daily dosing is used. Commonly used combinations include isoniazid, ethambutol, rifampin, and enrofloxacin.

Mycobacteria are resistant to typical physical and chemical destruction. They can survive for months to years in the environment.

BACTERIAL ENTERITIS

Several bacteria are commonly implicated in bacterial enteritis in birds. Most are secondary pathogens, although some can be primary. They include *E. coli*, *Clostridium* spp., *Salmonella* spp., and *Chlamydia psittaci* (see Chapter 20). Many other species of bacteria can be involved, usually as secondary invaders.

Treatment is based on bacterial culture and sensitivity, and the identification and elimination of predisposing factors and concurrent diseases.

Escherichia coli

Definition/overview

E. coli is a gram-negative, non-spore-forming bacillus. It has a worldwide distribution and is a normal inhabitant of mammalian gastrointestinal tracts. Most *E. coli* strains produce endotoxins.

E. coli can be isolated from clinically normal parrots. It is debatable whether the isolation of *E. coli* from the cloaca of clinically normal parrots indicates a definitive pathogen, a potential pathogen, or a commensal organism, or is merely transient. It is more common in cockatoos than in other parrots.

Aetiology

Infection with *E. coli* is primarily the result of inadequate hygiene and faecal contamination of water sources, food, perches, floors, and the general environment.

Clinical presentation

Signs vary from chronic diarrhoea to severe generalised septicaemia (lethargy, anorexia, weakness, ruffled feathers, diarrhoea, and death). *E. coli* can also cause respiratory infections, infertility, ingluvitis, and arthritis.

Clostridium spp.

Definition/overview

The clostridial organisms found in birds include *Clostridium perfringens* types A and C, *C. colinum*, and *C. tertium*. They are anaerobic gram-positive bacilli.

Pathophysiology

An abrupt change in diet may disrupt the intestinal microflora, allowing Clostridia to proliferate in the upper intestine. They can cause necrotic enteritis, 'gangrenous dermatitis', and malignant oedema.

Clinical presentation

Clostridial infection is often seen in juvenile cockatoos with concurrent enteric problems; it produces a characteristic foul-smelling diarrhoea.

In lorikeets, it may present as acute haemorrhagic diarrhoea, followed by death.

Diagnosis

Spore-forming bacteria are seen on faecal Gram stains. Severe necrosis and petechial haemorrhages are found in the mucosa of the duodenum and jejunum.

Management

Addition of apple cider vinegar (10 ml/litre of water) has been effective in treating juvenile cockatoos with simple enteritis. Penicillin derivatives and metronidazole are usually effective in more severe cases.

Salmonella spp.

Definition/overview

Salmonellae are (usually) motile, gram-negative bacteria. *Salmonella* spp. are members of the family Enterobacteriaceae. There are over 2,100 serotypes, including the Arizona group. There are five subgenera; subgenus I is the most important to birds. Subgenus III (*S. arizonae*, *S. hinshawii*) is occasionally reported in birds, especially those in contact with reptiles. Two organisms are host-specific and non-motile: *S. pullorum* (pullorum disease) and *S. gallinarum* (fowl typhoid). *S. typhimurium* is the most common parrot isolate. It is also the most common isolate in human beings with salmonellosis, making it important as a zoonosis.

Pathophysiology

Faeces of chronically infected carrier birds (often asymptomatic) are one of the most common sources of infection. Salmonellae can survive for extended periods in organic matter, faeces, and dirt (up to 2 years). Ingestion of contaminated food or water or direct contact by aerosolisation of faecal or feather dust is the most common mode of transmission.

Improper husbandry and sanitation can spread the infection through the contamination of seeds, fruits, vegetables, drinking water, and food containers.

Salmonella spp. can be a primary pathogen; some serotypes can penetrate the mucosal barrier. Virulent strains can penetrate intact intestinal mucosa, while non-virulent strains require a mucosal lesion to enter the host. The incubation period can be as short as 3–5 days in acute cases.

Clinical presentation

The disease can be peracute, acute, chronic, or subclinical depending on the number of organisms, their serotype, and the age, species, and condition of the host. Clinical signs include depression, lethargy, anorexia, weight loss, diarrhoea, a soiled vent, dehydration, and crop stasis. Lameness and wing droop are associated with septic arthritis (swollen joints). In breeding birds with subclinical infections, there is often poor hatching or excessive fledgling mortality. Neurological signs may be seen, including convulsions.

Diagnosis

Diagnosis is based on necropsy findings: hepatomegaly, splenomegaly, air sacculitis and pulmonary congestion, caecal cores (caecal necrosis) in game birds, ulcerative enteritis, orchitis, and oophoritis.

Bacteria are present on histological sections, and there is focal necrosis in the affected organ, with organisms found within the necrotic areas. Culture of the digestive tract or liver sections, and faecal culture can be used for organism identification.

Antemortem diagnosis can be difficult because of intermittent shedding. A serological test is available for poultry.

Management

Antibiotics are given based on culture and sensitivity. Treatment must be given for 3–8 weeks.

Control of vectors (wild birds, rodents) and prevention of contamination of food and water are required. Good hygiene practices must be implemented. A vaccine is available for pigeons, but it may not be fully protective. Birds should be vaccinated 4 weeks prior to a show or race season. Autogenous vaccines may be produced from faecal cultures.

DISORDERS OF THE CLOACA

Prolapse

Aetiology

Cloacal prolapse can be caused by excessive straining: masturbatory behaviour in male cockatoos, sexual overwork in waterfowl and ratites, oviductal disease (egg binding, salpingitis), increased intracoelomic pressure (fluid accumulation, organomegaly, or a coelomic mass), cloacal disease (cloacoliths, cloacitis, IPD), or constipation or diarrhoea.

It may also be related to the loss of cloacal tone due to spinal cord disease or chronic trauma.

Most commonly, the prolapse may only involve the cloaca itself, but prolapses of the oviduct or rectum can also occur.

Clinical presentation

There is a protrusion of the cloacal mucosa, with or without the rectum or oviduct, through the lips of the vent (**Fig. 19.15**). Phallic prolapse may be seen in waterfowl and ratites.

Cloacal prolapse is usually accompanied by straining and grunting while defecating, and blood around the vent or in the droppings. If the prolapse

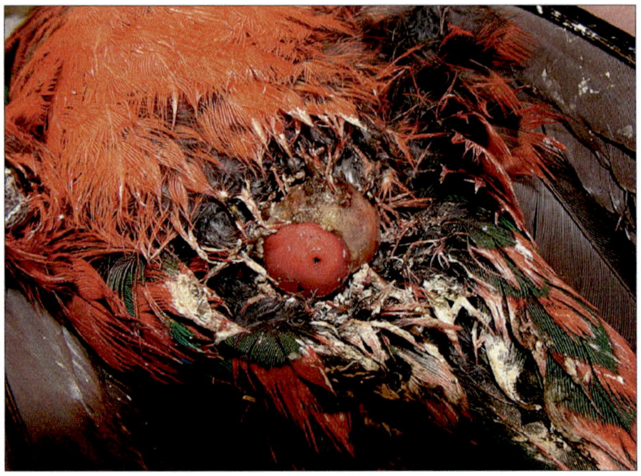

Figure 19.15 **King parrot hen with a prolapsed cloaca.**

has been long-standing, the prolapsed tissue may be oedematous and superficially necrotic. The peri-cloacal feathers are often soiled with foul-smelling faecal material as a consequence of the bird's inability to defecate normally.

Diagnosis

Small prolapses need to be differentiated from papillomas (see later).

Management

Initial treatment is to stabilise the patient (if required), reduce the prolapse, hold it in place, and treat any swelling and damage to the cloacal mucosa until the aetiology can be determined.

Under general anaesthesia, the prolapsed tissue is gently cleaned and, if possible, the prolapsed tissue is identified (cloaca, rectum, or oviduct). Endoscopy may be of value, following the centre of the prolapse back into the body. If the prolapse is oedematous, the swelling can be reduced with the osmotic effect of sugar or glucose poured onto the mucosa. Once the cloaca is clean and normal-sized, the prolapse should be gently reduced. Moistened cotton buds can be used to return the tissue to its normal anatomical position, and it can be flushed with saline to inflate and clean it. Instilling 1–2 ml of silver sulphadiazine cream has both a soothing and antibacterial effect. As the vent is elongated rather than rounded, a purse-string suture is not recommended. Instead, one to two vertical mattress or simple interrupted sutures should be placed on both sides of the vent in order to reduce the size of the vent opening. This opening must be sufficiently large to allow the bird to defecate and urinate (**Fig. 19.16**).

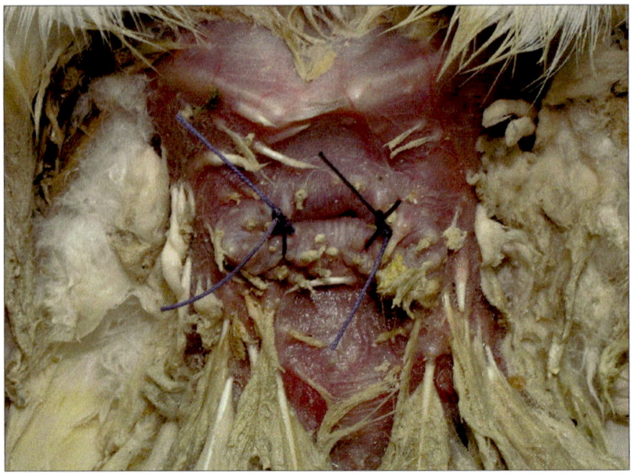

Figure 19.16 **Cockatoo with a prolapsed cloaca managed using temporary ventoplasty.**

The cause of the prolapse is identified through a thorough history and physical examination, haematology and biochemistry, cloacal culture (caution must be taken in interpreting this, as the possibility of contamination is high), whole-body radiographs, or cloacal endoscopy with/without biopsy.

Follow-up treatment is aimed at minimising the possibility of recurrent prolapses by eliminating or minimising the impact of the inciting cause and performing surgery to maintain the cloaca in place (cloacopexy or ventoplasty) (see Chapter 32).

Long-term follow-up on cloacal prolapses indicates a reasonably high degree of recurrence, regardless of the techniques used. This must be communicated to the client before commencing any treatment.

Internal papilloma disease (IPD)

Aetiology

The causative virus is psittacid herpesvirus 1 (PsHV-1) genotypes 1, 2, and 3. This is the same virus that causes Pacheco's disease (see Chapter 20). It is thought that birds with IPD have survived an initial episode of Pacheco's disease. Recently, a novel herpesvirus, PsHV-2, was identified in grey parrots with papillomas.

Pathophysiology

IPD has been reported mainly in macaws (green-winged, blue and gold, great green, scarlet, severe, and military), Amazon parrots, conures, and hawk head parrots. It has also been reported in cockatiels, budgerigars, and African grey parrots. Lesions are rarely seen in eclectus parrots and cockatoos.

The mode of transmission is through ingestion of material contaminated by oral and faecal secretions from infected birds. There is no evidence of egg transmission, although parent-reared chicks with IPD-positive parents have developed the disease. Infected birds may develop a necrotising hepatitis (Pacheco's disease); untreated, these birds will die. Birds that survive, or those that remain subclinical, often go on to develop IPD.

Clinical presentation

The clinical signs depend on the location of the papillomas:

- Cloacal. Most birds with papillomas will have lesions in their cloaca. Often the first signs noticed are straining to defecate, pasting of the vent, and blood on droppings. The bird may develop enteritis and cloacoliths. There may be a red mass protruding from the cloaca (**Fig. 19.17**). Cloacal examination reveals papillomatous lesions ('cobblestone' appearance of the mucosa). These may blanch with the application of 5% acetic acid (vinegar), but this test is neither specific nor sensitive. The appearance of the lesions is characteristic; if the lesions appear inflamed, a deep biopsy may be necessary to identify lesions underlying the inflammation.
- Oropharynx. Lesions of the choanal slit and glottis are more common in macaws (**Fig. 19.18**). Lesions may also be seen in the crop, proventriculus, and ventriculus in some of the larger macaws. Clinical signs of papillomas are rare, but may include dysphagia, dyspnoea, wheezing, anorexia, and weight loss.

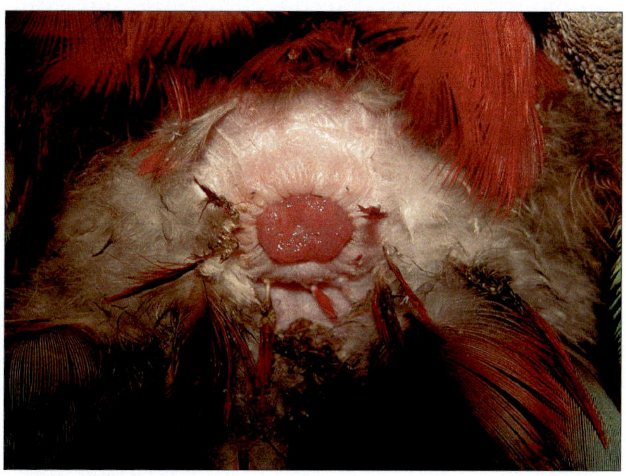

Figure 19.17 **Green-winged macaw presents with a red mass protruding from its cloaca suggestive of internal papilloma disease.**

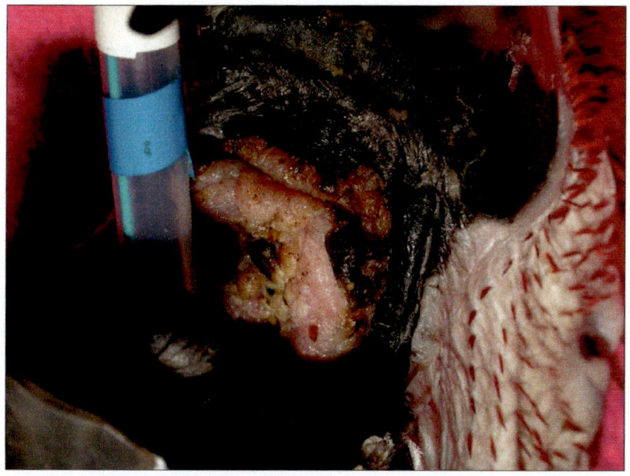

Figure 19.18 **Scarlet macaw with oropharyngeal lesions as a consequence of internal papilloma disease.**

- Extension into the intestinal tract, nasal mucosa, and nasolacrimal duct is extremely rare.
- Some birds may ultimately develop bile duct or pancreatic carcinomas.

Diagnosis

Gross findings are usually sufficient for diagnosis. Blanching the lesions with acetic acid (vinegar) is often suggested as a means of gross diagnosis, but this recommendation is falling out of favour because of its low specificity and sensitivity.

Cloacal or choanal biopsy shows a characteristic histological picture. There is a proliferation of undifferentiated columnar epithelial cells arranged in a pseudo-stratified fashion, forming fronds of villi and papillae on a thin fibrovascular stalk.

A PCR on cloaca swabs or organ biopsy samples can confirm the diagnosis or be used when screening asymptomatic or in-contact birds.

Elevations in GGT may indicate the development of bile duct carcinomas. (This may not be reliable in Amazon parrots.) AST and ALT remain unchanged. Ultrasonography and CT scanning may detect bile duct carcinomas before endoscopy.

Treatment

Lesions may wax and wane over a period of time, and there is anecdotal evidence that they may be worse when the affected bird is reproductively active.

A variety of techniques have been tried for the removal of papillomas: cauterisation with silver nitrate; cryosurgery; surgical debridement with radiosurgery or laser; cloacal mucosal stripping;

and imiquimod (Aldara) application (an interferon-immune response modifier). Mixed results have been obtained.

Most birds initially respond to treatment but later regress. Aggressive surgical debridement may result in strictures and should be reserved for severe cases only. Debridement of a small area may stimulate a (temporary) immune response and regression of the lesions.

Cloacoliths

Definition/overview

These are dried concretions of faecal material and urates in the urodeum or proctodeum. They are often adhered to the cloacal wall and, in extreme cases, can cause obstipation and intermittent ureter obstruction.

Clinical presentation

Signs include straining to defecate, decreased faecal output, and fresh blood in droppings and around the vent. A flatulent sound is occasionally heard.

Diagnosis

Diagnosis is based on clinical signs and palpation of the cloaca. Radiography often reveals the cloacolith. Cloacal endoscopy may be used to confirm the presence of the cloacolith and the severity of the associated cloacitis.

Management

Mild cases may respond to lubrication and gentle pressure to break the cloacolith into small pieces. Severe cases may require a cloacotomy to relieve the obstruction (see Chapter 32). Supportive care, especially fluid therapy and anti-inflammatory treatment, is essential.

Cloacal atony

Aetiology

Cloacal atony can be due to the loss of neurological innervation to the cloaca (spinal cord dysfunction or proventricular dilatation disease [PDD]) or loss of muscular strength of the cloacal sphincter. The latter may be a result of soft tissue trauma and swelling associated with egg laying, systemic illness and weakness, or chronic cloacal disease (e.g., recurrent prolapses or cloacitis).

Clinical presentation

Signs include a flaccid-appearing vent, faecal accumulation around the vent, and loss of cloacal tone (when a normal cloaca is stimulated lightly with a cotton bud, the sphincter should close in a manner similar to anal tone in a mammal). Faecal material, usually in large amounts, is easily expressed from the cloaca on palpation.

Management

Treatment includes NSAIDs and regular expression of the cloaca. Ventoplasty may be necessary to reduce the size of the vent and prevent prolapse.

Cloacitis

Aetiology

Inflammation of the cloaca can be due to cloacal atony (see earlier), cloacoliths, a retained egg in the distal oviduct, bacterial infection, or trauma. It is generally considered to be a secondary disease.

Clinical presentation

Signs include tenesmus, fresh blood in droppings, a malodorous smell from the area, and faecal pasting around the vent. The odour is often from an overgrowth of anaerobes, especially *Clostridium* species, but *Escherichia coli* and budding yeasts have also been identified.

Diagnosis

Cloacitis must be differentiated from IPD, cloacoliths, and cloacal prolapse. Endoscopy, with or without biopsy, can help determine the cause and nature of the cloacitis.

Management

Treatment includes NSAIDs and the topical application of silver sulphadiazine cream to the cloaca. Topical flurbiprofen eye drops may assist when the peri-cloacal area is very inflamed.

Neoplasia

Definition/overview

Carcinomas, smooth muscle tumours, and haemangiomas have been reported in the cloaca of birds.

Clinical presentation

Neoplasms produce similar signs to IPD and cloacitis.

Diagnosis

Biopsy is necessary to achieve an accurate diagnosis.

Management

A cloacotomy can be performed to resect a mass if necessary. Many carcinomas will metastasise, and staging via radiology, ultrasound, or CT may be indicated before undertaking any therapy.

FURTHER READING

Doneley RJT, Miller RI, Fanning TE (2007) Proventricular dilatation disease: An emerging exotic disease of parrots in Australia. *Australian Veterinary Journal* 85: 119–123.

Gancz AY, Clubb S, Shivaprasad HL (2010) Advanced diagnostic approaches and current management of Proventricular Dilatation Disease. *Veterinary Clinics of North America: Exotic Animal Practice* 13(3): 471–494.

Gelis S (2006) Evaluating and treating the gastrointestinal system. In: *Clinical Avian Medicine, Vol 1.* GJ Harrison, TL Lightfoot (eds). Spix Publishing Inc, Palm Beach, pp. 411–440.

Hadley TL (2005) Disorders of the psittacine gastrointestinal tract. In: *Gastroenterology.* TK Ritzman (ed.) *Veterinary Clinics of North America: Exotic Animal Practice* 8(2): 329–350.

Hoefer HL (1997) Diseases of the gastrointestinal tract. In: *Avian Medicine and Surgery.* RB Altman, et al. (eds). WB Saunders, Philadelphia, pp. 419–453.

Hoppes ST, Tizard I, Shivaprasad HL (2013) Avian Bornavirus and Proventricular Dilatation Disease: Diagnostics, pathology, prevalence, and control. *Veterinary Clinics of North America; Exotic Animal Practice* 16(2): 339–355.

Hoppes SM, Shivaprasad HL (2020) Update on avian bornavirus and proventricular dilatation disease: Diagnostics, pathology, prevalence, and control. *Veterinary Clinics: Exotic Animal Practice* 23(2): 337–351.

Hoppes ST, Heatley JJ, Guo J, Turner MS, Shivaprasad HL, Tizard I (2013) Meloxicam treatment in cockatiels (Nymphicus hollandicus) infected with Avian Bornavirus. *Journal of Exotic Pet Medicine* 22(3): 275–279.

Lierz M (2016) *Avian Bornavirus and Proventricular Dilation Disease: Current Therapy in Avian Medicine and Surgery.* B Elsevier Speer (ed.), St Louis, pp. 28–46.

Phalen D (2005) Diagnosis and management of Macrorhabdus ornithogaster (Formerly Megabacteria). *Veterinary Clinics of North America; Exotic Animal Practice* 8(2): 299–306.

Phalen D (2014) Update on the diagnosis and management of Macrorhabdus ornithogaster (formerly Megabacteria) in avian patients. *Veterinary Clinics of North America; Exotic Animal Practice* 17(2): 203–210.

Reavill D (2007) Lesions of the proventriculus/ventriculus of pet birds: 1640 cases. In: *Proceedings of the Annual Conference of the Association of Avian Veterinarians*, pp. 89–94.

Rossi G, Dahlhausen RD, Galosi L, Orosz SE (2018) Avian ganglioneuritis in clinical practice. *Veterinary Clinics: Exotic Animal Practice* 21(1): 33–67.

CHAPTER 20

DISORDERS OF THE LIVER

LIVER ANATOMY AND PHYSIOLOGY

The avian liver consists of the right and left lobes joined cranially in the midline. The right lobe is larger than the left, with each lobe having several small processes. The liver is enclosed in a thin and slightly elastic capsule of connective tissue, allowing its expansion. Blood is supplied to the liver by the right and left hepatic arteries and hepatic portal veins. The hepatic arteries arise from the coeliac artery, while the portal veins drain blood from the proventriculus, ventriculus, duodenum, pancreas, intestines, and cloaca. Two hepatic veins join the caudal vena cava cranial to the liver, draining blood away from the liver.

Terminal portal venules and arterioles empty into sinusoids between plates of hepatocytes. The low pressure in these sinusoids allows the hepatocytes to absorb molecules from the blood. Phagocytic Kupffer cells are also present in the sinusoids, collecting particulate matter and microorganisms. The now 'filtered' blood drains into the hepatic veins and into the heart. The oxygenated arterial blood maintains the viability of the hepatocytes. Bile canaliculi form between three to five hepatocytes and drain into a bile ductule.

A portal triad of arteriole, portal venule, and bile ductule, along with associated hepatocytes, bile canaliculi, and sinusoids, forms the basic functional unit of the liver: the hepatic acinus (**Fig. 20.1**). Hepatocytes close to these portal triads are said to be 'periportal'. Those farther away, near the hepatic venules, are called 'periacinar'. The intermediate area is termed the 'midzone'. The hepatocytes in these different areas, although morphologically the same, are biochemically different and react differently to incoming chemicals and metabolites.

Bile is produced by hepatocytes and enters the bile canaliculi and then the ductules in the portal triad, which then empty into the interlobular ducts. These in turn form the right and left hepatic ducts, which join to become the common hepatoenteric duct emptying into the duodenum. A branch of the

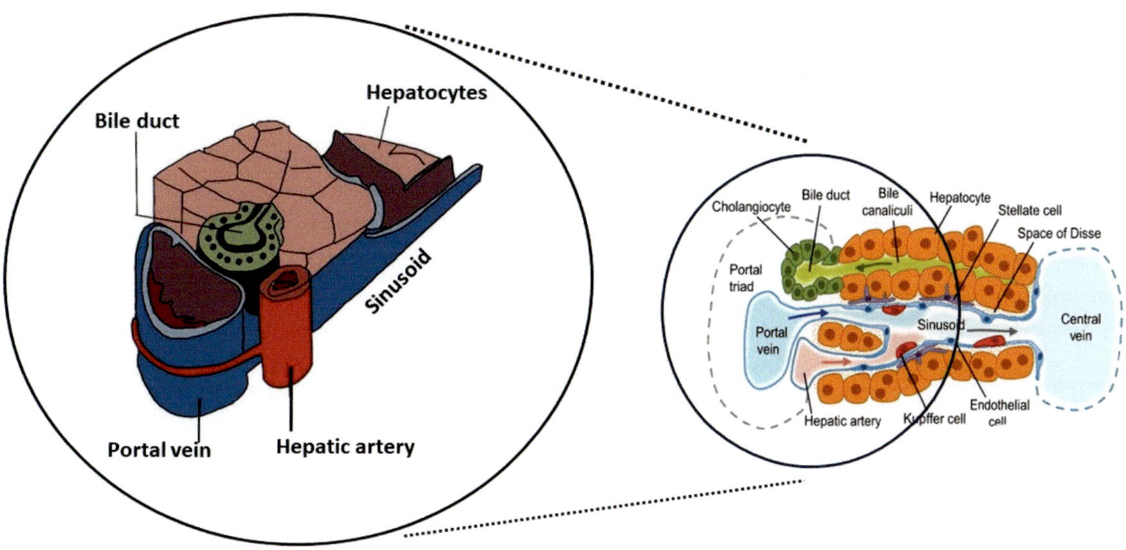

Figure 20.1 **Normal liver architecture. (Adapted from Zaefarian F, Abdollahi MR, Cowieson A, Ravindran V, Avian liver: The forgotten organ, *Animals*, 2019; 9(2):63.)**

DOI: 10.1201/9781003272519-20

right hepatic duct either forms the right hepatoenteric duct (emptying into the duodenum) or, in those birds with a gall bladder, the hepatocystic duct entering the gall bladder. (Pigeons, most parrots, and ostriches do not have gall bladders.) From there, the cystoenteric duct runs to the duodenum. Birds thus have two bile ducts emptying into the duodenum.

The liver has several functions in the body:

- Digestion. Bile contains bile acids, synthesised in the liver from cholesterol. (In birds, the primary bile acid is chenodeoxycholic acid.) In the distal duodenum, these bile acids emulsify fat, facilitating its digestion. Bile acids are then resorbed in the jejunum and ileum and recirculated through the liver. Bile also plays a role in the digestion of carbohydrates and proteins. It contains amylase and helps to activate pancreatic amylase and lipase in the duodenum. Because of the lack of biliverdin reductase and glucuronyl transferase in birds, the primary bile pigment is biliverdin, giving avian bile its characteristic green colour.
- Carbohydrate metabolism. The portal blood supply, carrying nutrient-rich blood from the gastrointestinal tract, supplies the liver with these nutrients before any other major organs. Hepatic enzymes carry out glycogenesis, protein synthesis, and lipogenesis in the well-fed bird. The glycogen, protein, and triglycerides produced in the liver enter the circulation and are used (or stored) throughout the body. If a bird is fasted (for any reason), the resultant hypoglycaemia stimulates glucagon production, which in turn activates liver enzymatic pathways to produce glucose through glycogenolysis, gluconeogenesis, and lipolysis. The liver, therefore, plays a major role in carbohydrate metabolism.
- Metabolism of metabolites, drugs, and chemicals. The liver, through its microsomal drug-metabolising enzyme system in the periacinar hepatocytes, processes both endogenous metabolites and exogenous chemicals. Hydrophobic, lipid-soluble molecules (which are difficult to eliminate) are converted by the liver to hydrophilic, water-soluble molecules and excreted in bile and urine. This is done in two phases: in the first, enzymes modify the molecules by oxidation or reduction; in the second, they are enzymatically conjugated with other molecules to become sufficiently water soluble. The best example of this is the synthesis of urea and uric acid from protein in the liver.
- Protein synthesis. The liver is the primary site of synthesis of a range of essential proteins:
  - Albumin.
  - Fibrinogen, prothrombin, and clotting factors I, II, V, VII, and IX.
  - Molecules involved in the transport of metals, hormones, and lipids (e.g., ceruloplasmin and macroglobulins).
- Antimicrobial effect. The Kupffer cells in the sinusoids are important in the clearance of microorganisms entering the portal circulation in cases of intestinal infections or surgery. They also play a role in the detoxification of bacterial endotoxins.

DISEASES OF THE LIVER

Overview of liver disease

Because of its unique metabolism and relationship to the gastrointestinal tract, the liver is an important target for the damaging effects of infectious organisms, endotoxins, drugs, and metabolic diseases. Initial damage from any of these aetiologic agents stimulates the secretion of pro-inflammatory cytokines and chemokines by the Kupffer cells and heterophils, leading to the accumulation and activation of heterophils in the hepatic vasculature. Ongoing damage and further release of cytokines and chemokines from the hepatic parenchyma result in extravasation of these heterophils and an attack on the hepatocytes, resulting in cellular death or damage (**Fig. 20.2**).

Fortunately, the liver has enormous regenerative capacity, and as long as the acinar architecture of the liver is maintained, regeneration is frequently rapid. If, however, the acinar structure collapses, regeneration either does not occur or is very haphazard; fibrosis usually results. The same cytokines and chemokines that activate heterophils will also activate stellate cells. Normally quiescent, these stellate cells (also known as Ito cells) are found in the peri-sinusoidal space (a small area between the sinusoids and hepatocytes), making up 5–8% of the total number of liver cells. When activated, these stellate cells increase in number and contractility and secrete collagen, leading to the fibrosis of damaged areas of the liver. Early fibrosis may respond to treatment or the removal of the source of injury, but more advanced

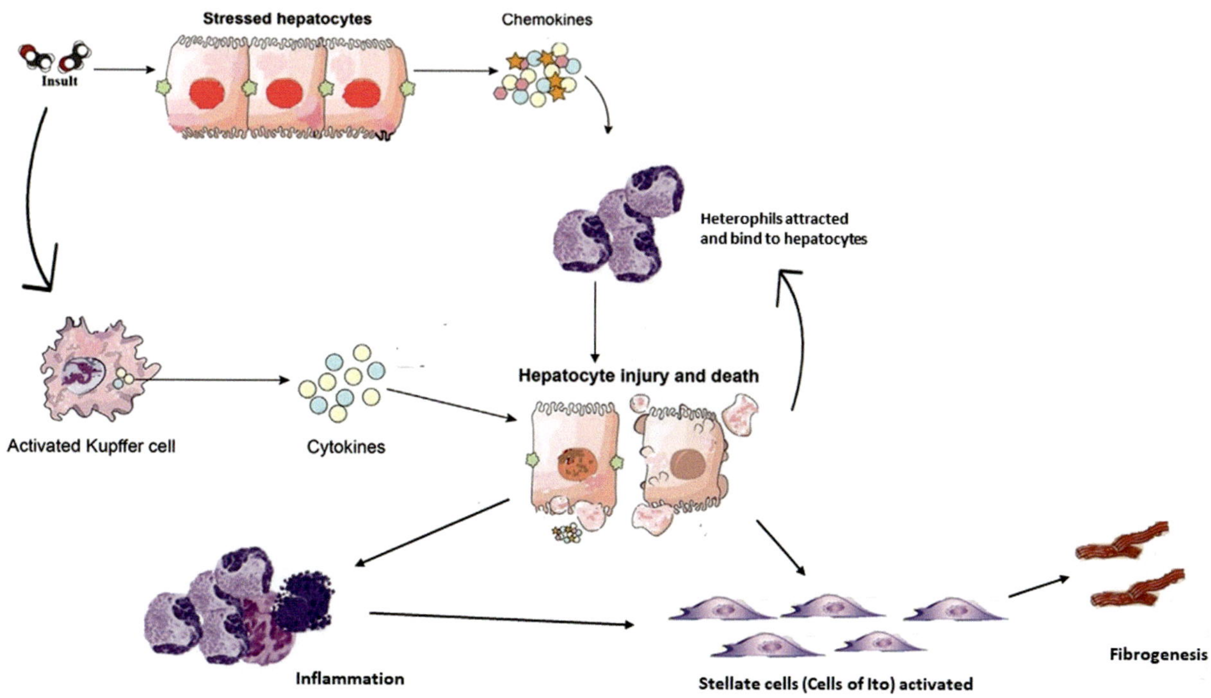

Figure 20.2 **The pathogenesis of liver disease.**

fibrosis (seen in chronic liver disease) is generally irreversible.

It is obviously important that the clinician understands this disease process and the aetiological agents that provoke this damage in order to diagnose, treat, and manage liver disease in their patients.

AETIOLOGY OF LIVER DISEASE

Liver disease can result from congenital abnormalities, trauma, metabolic and nutritional disorders, toxins, parasites, infectious agents (bacterial, fungal, viral, chlamydial, mycobacterial, and rickettsial), and neoplasia. Often the exact aetiology cannot be determined, lost in the fibrotic changes accompanying the disease.

Congenital

Extra-hepatic biliary cysts have been reported in an African grey parrot. A congenital coelomic hernia involving the liver has been reported in Japanese quail.

Trauma

Trauma can cause tearing of the liver parenchyma and/or the capsule. This occurs when the bird is subjected to massive force (e.g., a moving motor vehicle) or when the liver is friable due to other problems (e.g., hepatic lipidosis). The severity of the liver rupture and resultant haemorrhage will determine whether the bird survives the initial trauma.

Metabolic/nutritional

Visceral gout

Visceral gout is the deposition of uric acid crystals on and in organs. It is typically associated with renal disease, although dehydration and high-protein diets may be involved. In the liver, most of this deposition occurs on the capsule, but it can occur within the parenchyma where it is associated with necrosis and heterophilic inflammation.

Amyloidosis

Amyloidosis is most commonly seen in raptors, waterfowl, and passerine birds. Amyloid A is a degradation product of inflammatory proteins, and its deposition is commonly seen in birds with chronic disease. Affected livers are usually enlarged and friable and can be confused with hepatic lipidosis. Histologically, the amyloid is seen as a pale eosinophilic or amphophilic deposit between the cells, compressing them.

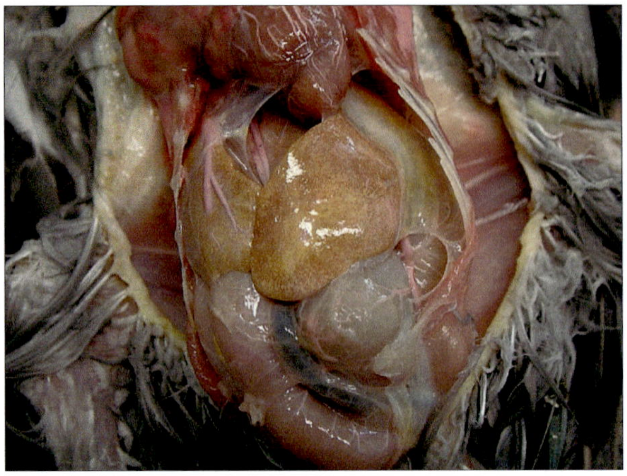

Figure 20.3 **Hepatic lipidosis. Note the lipaemic blood in the vasculature.**

Hepatic lipidosis (fatty liver disease)

Hepatic lipidosis occurs when excessive fatty acids are consumed (in the form of dietary fat), when there is increased lipolysis (e.g., diabetes mellitus or egg-laying activity), decreased fatty acid oxidation in the liver, or when there is a decreased ability of the liver to secrete processed fatty acids back into the circulation (due to dietary deficiencies of lipotropic factors such as choline, biotin, and methionine). Affected livers are enlarged, pale yellow, and friable (**Fig. 20.3**). Histologically, there is vacuolation of the hepatocytes.

Iron storage disease (haemosiderosis)

Haemosiderosis occurs when there is more iron in the circulation than is needed for erythrogenesis, allowing the iron to accumulate in the liver. It results from excess iron intake, either dietary or via excessive blood transfusion. This condition is commonly seen in Sturnidae (mynahs, starlings), Paradisaeidae (birds of paradise), Ptyonorhynchidae (bowerbirds), Bucerotidae (hornbills), and Ramphastidae (toucans and toucanettes). It has also been reported in parrots, especially lories and lorikeets. Affected livers are enlarged and are usually golden-brown in colour, often with scattered dark foci. Iron can be seen histologically in the hepatocytes and Kupffer cells. There may be an associated inflammatory process with lymphocytes and occasional heterophils.

Lipofuscinosis

Lipofuscin pigment accumulates in hepatocytes secondary to a range of diseases. It is due to excessive biological oxidation at the cellular level. Vitamin E deficiency has been suggested as one possible cause.

Toxic

Toxic causes of liver disease include:

- Drugs: alcohol, dimetridazole, medroxyprogesterone
- Plants: pyrrolizidine alkaloids, oleander, gossypol, avocado fruit
- Aflatoxins, ergot
- Heavy metals: lead, copper, iron
- Pesticides: metaldehyde, phosphorous, vitamin D3 analogues

Parasitic

Protozoa

Apicomplexa

Cryptosporidia may occasionally be found attached to biliary epithelium, causing the proliferation of the epithelium and a mild chronic mononuclear reaction.

Systemic *Isospora* is primarily reported in passerines, especially canaries. It undergoes schizogony in the liver, causing a generalised inflammatory reaction composed of macrophages, plasma cells and lymphocytes. The organism may be found in macrophages and lymphocytes.

Birds infected with *Sarcocystis* and *Toxoplasma* are usually intermediate hosts. They produce a similar inflammatory reaction to systemic *Isospora*. Organisms can be difficult to locate.

Haemoprotozoa

Plasmodium (avian malaria) is a widespread blood parasite. Some species, especially penguins, appear to be more susceptible. Schizogony occurs in the reticuloendothelial cells of many organs, with merozoites released to infect erythrocytes. Affected livers are enlarged, and in falcons, they often appear grey-black. There is infiltration of the liver with macrophages, plasma cells, and lymphocytes, with the organism found in some of the inflammatory cells.

Haemoproteus is usually non-pathogenic, with schizonts occasionally found in the endothelial cells of the liver.

Leukocytozoon affects many species, with mortalities reported in ducks and geese. Where liver damage does occur, it is usually acute and severe, with haemorrhage and necrosis, but minimal inflammation. Gametocytes can be found in peripheral blood smears.

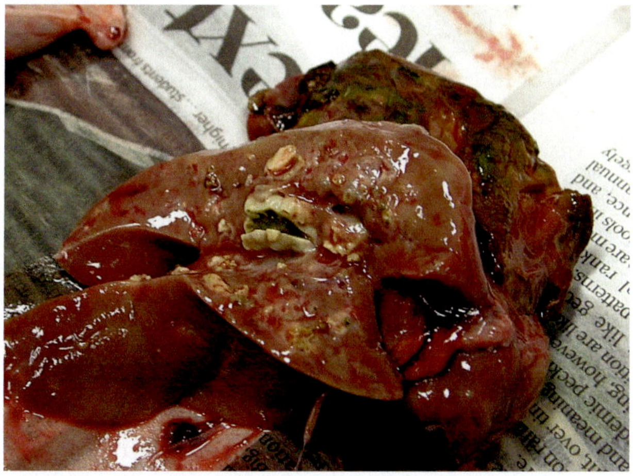

Figure 20.4 **Granulomatous hepatitis in a peacock due to histomoniasis (blackhead).**

Flagellates

Histomonas causes the disease known as blackhead in domestic poultry, especially peafowl and turkeys. It produces classical white-yellow granulomas throughout the liver parenchyma (**Fig. 20.4**). The organism can be found in the lesions, although it may resemble macrophages and be difficult to detect.

Trichomonas is usually a gastrointestinal parasite, but heavy infections can spread to the liver. Necrotic lesions with the parasite found at the periphery are diagnostic.

Trematodes

Flukes of the family Dicrocoelidae inhabit the bile ducts of many avian species, including parrots, Anseriformes, Ramphastidae, and ratites. Wandering flukes may be found in dilated bile ducts, and schistosomes can be found in dilated sinusoids. There is usually minimal inflammation unless degenerating eggs are present. *Fasciola* flukes in emus provoke an eosinophilic response, with macrophages and lymphocytes present. Granulomas may form with giant cells and fibrosis.

Nematodes

Intestinal nematode larvae can migrate through the liver, causing extensive fibrosis, bile duct hyperplasia, and inflammatory cell infiltrates.

Infectious

Bacteria

Both gram-positive and gram-negative bacteria can cause liver disease, usually secondary to septicaemia or enteritis. *Salmonella* spp., *E. coli*, *Pseudomonas* spp., *Yersinia* spp., and *Campylobacter* spp. are common isolates. Affected livers are usually swollen, with grey-white foci throughout the parenchyma. Multifocal hepatocyte necrosis with a heterophilic inflammatory response is usually seen. Bacteria are usually seen in macrophages and Kupffer cells. Additionally, endotoxins arising from bacterial enteritis can enter the portal circulation, damaging the periportal hepatocytes.

Viruses

Herpesvirus has been isolated from nearly all species of birds. Serotypes involved include Pacheco's disease virus (see later), pigeon herpesvirus, owl herpesvirus, falcon herpesvirus, crane herpesvirus, stork herpesvirus, quail herpesvirus, and finch herpesvirus. Other strains of herpesvirus, while not specifically attacking the liver, may produce some degree of hepatic damage. Affected livers are enlarged, with variable yellow-grey mottling and haemorrhage. Histologically, there is acute necrosis with variable inflammation, syncytial cell formation, and intranuclear inclusion bodies.

Polyomavirus is primarily recovered from parrots and finches but may affect a range of birds. Affected livers are enlarged and friable, with the degree of severity often varying according to the species (**Fig. 20.5**). There is multifocal or midzonal necrosis and haemorrhage, with characteristic inclusion bodies in the Kupffer cells (**Fig. 20.6**).

Adenovirus rarely causes clinical disease unless there are other immunosuppressive factors at work (e.g., psittacine beak and feather disease [PBFD], juvenile birds). Many species of birds are affected including parrots, poultry, pigeons, and ostriches. Affected livers are discoloured, with scattered yellow-grey areas present (**Fig. 20.7**). There is multifocal necrosis and haemorrhage, non-suppurative cholangitis, and large basophilic intranuclear inclusion bodies in the hepatocytes.

Paramyxovirus causes hepatomegaly in isolated cases, but its primary effects are neurological,

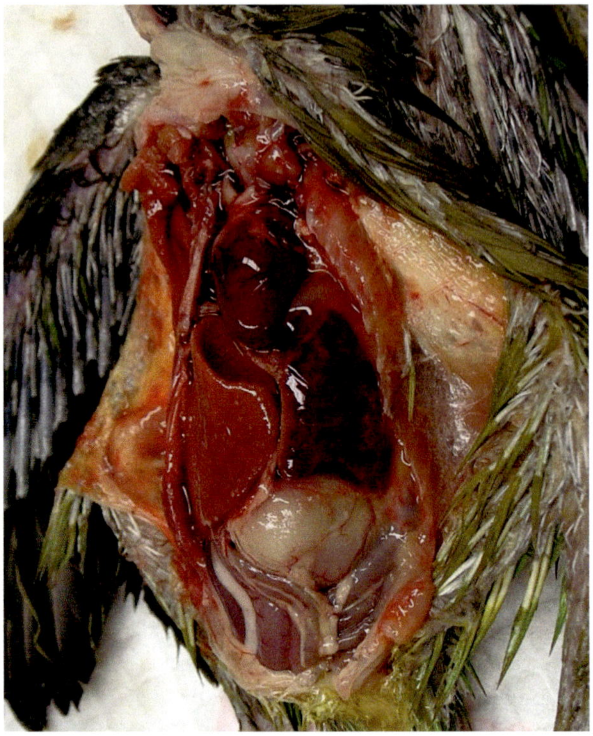

Figure 20.5 **The liver of an Alexandrine nestling that died of avian polyomavirus.**

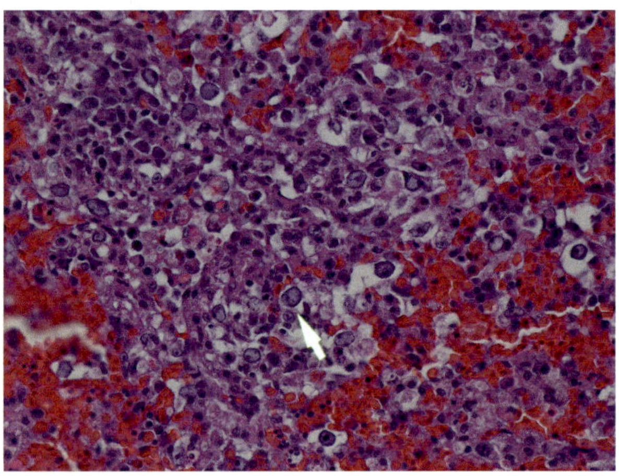

Figure 20.6 **Polyomavirus inclusion bodies (arrow), with extensive haemorrhage.**

gastrointestinal, and respiratory. Histologically, there is a lymphoplasmacytic infiltrate in the periportal area.

Circovirus is occasionally seen as an acute disease in young birds. In these birds, and in some adults with severe feather changes, there may be mild necrosis and a lymphohistiocytic inflammatory reaction in the portal areas. Inclusion bodies can be seen in the Kupffer cells.

Reovirus has been recovered from parrots, Galliformes, Anseriformes, raptors, pigeons, and chickens. Affected livers are enlarged with scattered grey-white or yellow foci. Histologically, there is hepatocellular necrosis with minimal inflammation. Hepadnavirus is the cause of duck viral hepatitis, resulting in hepatic necrosis and periportal inflammation. Togavirus (eastern equine encephalitis) causes enlarged livers with some necrosis in many species of birds.

Fungi

Fungal infections involving the liver are usually opportunistic, spreading from other sites within the body. *Aspergillus* spp. affecting nearby air sacs have been reported to invade the liver locally, causing hepatic necrosis.

Chlamydia

The obligate intracellular bacterium *Chlamydia psittaci* is a common cause of liver disease in all species of birds. Affected livers are enlarged, discoloured, and may show grey-yellow foci of necrosis. There is multifocal to confluent necrosis with a mononuclear inflammatory reaction. Organisms may be found in macrophages and hepatocytes (**Fig. 20.8**).

Mycobacteria

Mycobacterial infection in birds is primarily a gastrointestinal infection, with caseated tubercles occurring in the intestinal mucosa and the liver. Histologically, early lesions are comprised primarily of heterophils and macrophages, with only a few microorganisms. As the lesions progress, large epithelioid macrophages containing mycobacteria

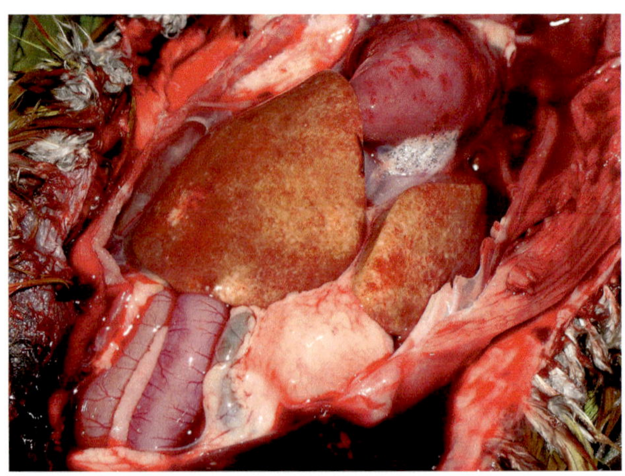

Figure 20.7 **Adenoviral hepatitis seen on necropsy of an Amazon parrot.**

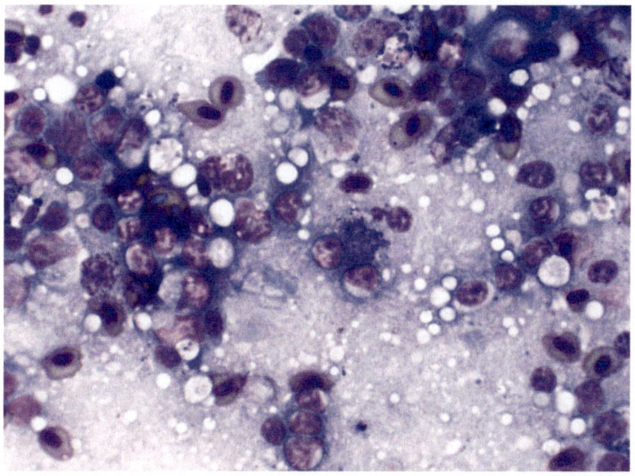

Figure 20.8 **Cytology of the liver, showing chlamydial elementary body inclusions.**

appear. These are best seen with an acid-fast stain (**Fig. 20.9**).

Rickettsia

Aegyptianella pullorum causes hepatitis in many species of birds, especially in the Mediterranean region. Intra-erythrocytic inclusion bodies can be seen with Giemsa stain.

Neoplastic

Hepatic neoplasia is usually primary, although metastatic neoplasia can occur. Primary bile duct tumours are more common than hepatocellular tumours. In parrots, these may be associated with internal papilloma disease (IPD).

Idiopathic

Sometimes referred to as 'hepatopathy', 'chronic active hepatitis', or 'hepatic cirrhosis', this is a common condition in many species of birds. No one single cause has been identified, and it is probably multifactorial, possibly involving an immune-mediated component. Affected livers are often shrunken, pale, and fibrotic. In the early stages of the disease, there is hepatic vacuolisation, a pleocellular inflammatory infiltrate in the periportal areas, bile duct proliferation, and mild fibrosis. As the condition progresses, the fibrosis worsens, and diffuse biliary hyperplasia develops (**Fig. 20.10**).

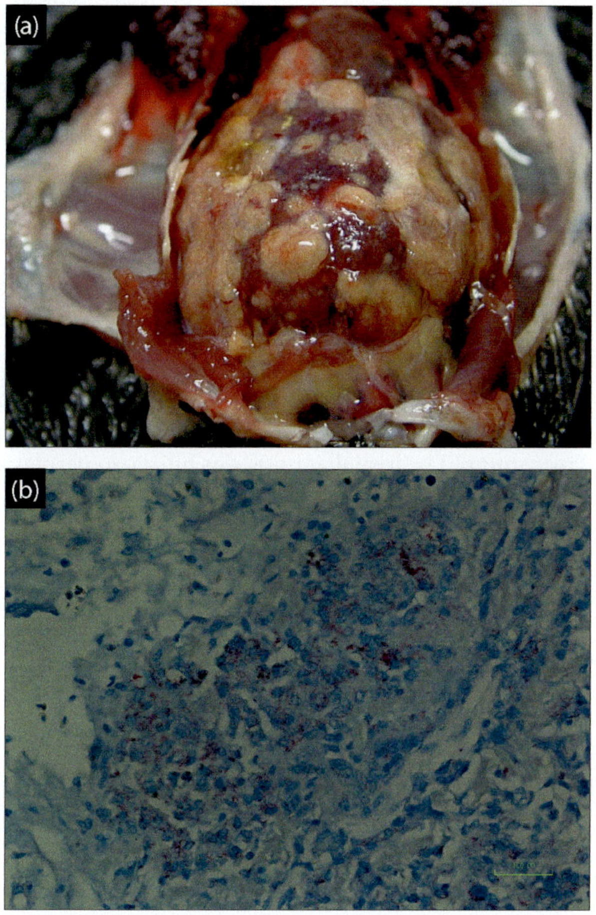

Figure 20.9 **Granulomatous hepatitis in a parrot (a). Acid-fast staining of the liver reveals numerous acid-fast bacilli consistent with mycobacteriosis (b).**

CLINICAL SIGNS

Clinical signs seen in birds with liver disease can reflect any or all of the following malfunctions.

Cholestasis

Swelling of the hepatocytes and inflammatory infiltrates can cause occlusion of the biliary system. This results in the increased retention of bile and a subsequent rise in serum bile pigment levels. Because of the lack of biliverdin reductase, it is uncommon for birds to produce bilirubin. Icterus is therefore uncommon in birds. However, increased levels of biliverdin in the blood due to cholestasis will result in biliverdin-stained urates and urine, giving these waste products a green discolouration. Decreased bile secretion into the intestine can result in maldigestion (weight loss, diarrhoea).

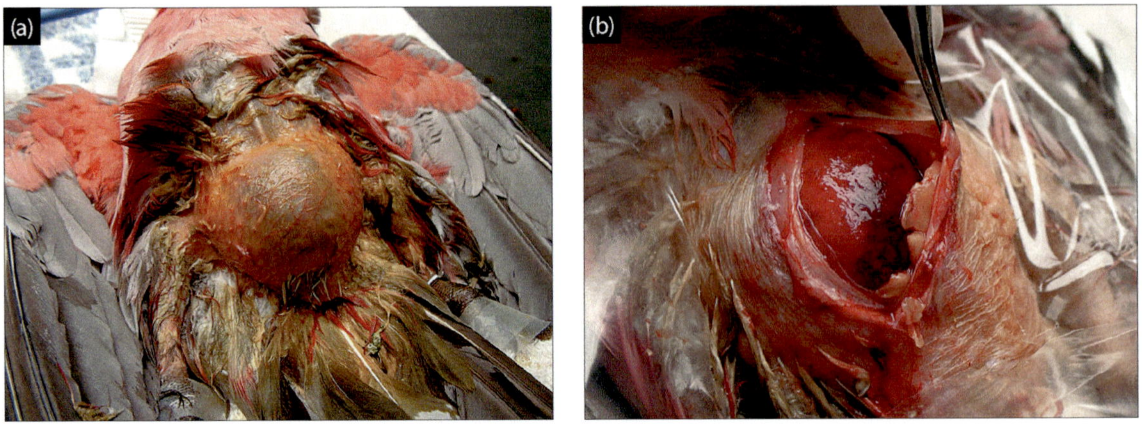

Figure 20.10 **(a) Coelomic distension due to hepatomegaly and (b) the liver visualised on an exploratory laparotomy. The final diagnosis was biliary hyperplasia of unknown origin.**

Inadequate conversion of ammonia to uric acid and urea

Failure to convert ammonia to uric acid and, to a lesser extent, urea leads to increased serum levels of ammonia and the onset of hepatic encephalopathy. Signs include weakness, depression, personality changes, behavioural disturbances, seizures, and paresis. This is uncommon in granivorous birds such as parrots because of the lack of dietary encephalopathic precursors.

Protein synthesis deficits

Inadequate production of clotting factors can lead to coagulopathies. This can be as subtle as mild haemorrhages in the rhinotheca (**Fig. 20.11**) or as dramatic as sudden death due to internal haemorrhaging.

Figure 20.11 **Overgrown rhinotheca and bruising in an eastern rosella with chronic liver disease.**

Decreased synthesis of ceruloplasmin can result in decreased iron mobilisation and subsequent anaemia. Decreased albumin can be one factor in the development of ascites.

Abnormal carbohydrate and fat metabolism

Hypoglycaemia, weight loss, and debility are frequently seen in birds with severe liver disease. Nutritional deficiencies resulting from liver disease may be the cause of the dermatological signs seen in affected birds. These include feather and skin changes, and overgrown beaks and nails (**Fig. 20.12**).

Failure of Kupffer cell activity

Endotoxaemia and bacteraemia may develop when the Kupffer cells are unable to perform their function correctly.

Portal hypertension

Increased blood pressure within the sinusoids can result in the development of ascites, as oedema in the liver results in the movement of a modified transudate out of the liver into the peritoneal cavity.

Inadequate or inappropriate metabolism of drugs and chemicals

Decreased ability of the hepatocytes to modify or metabolise drugs and chemicals can lead to unexpected aberrant responses to administered drugs.

Figure 20.12 Overgrown rhinotheca and poor plumage in a galah with chronic liver disease.

Other clinical signs

These result from other, less obvious processes:

- Bile salts deposited in the skin may be the cause of some of the apparent pruritus seen in birds with liver disease.
- Pain from the stretching of the capsule over an enlarged liver can cause feather-picking over the torso.
- Occlusion of the space normally occupied by the air sac by either the liver or ascites can result in dyspnoea.

DIAGNOSIS

Clinical pathology

Haematology will usually reflect the nature of the aetiological agent or the chronicity of the disease.

Biochemistry testing indicates either hepatocellular damage (AST, CK, LDH, GLDH), function (plasma protein, bile acids), or cholestasis (GGT). Cholesterol, triglycerides, VLDL, LDL, and HDL can be markers of fat metabolism, e.g., hepatic lipidosis. If indicated, coeliocentesis and fluid analysis can be performed (ascites arising from liver disease is usually a modified transudate). See Chapter 7.

Diagnostic imaging

Radiography

Two well-positioned views are required: lateral and ventrodorsal. Both the size of the liver and the displacement of nearby organs should be examined. See Chapter 8.

Lateral view

On the lateral view, the liver should not extend past the end of the sternum, and the proventriculus should slope down towards the ventriculus at an angle of 30°–45°. There should be little or no space between the heart and the liver.

Ventrodorsal view

On the ventrodorsal view, the liver shadow should not extend past a line joining the shoulder and acetabulum. Care must be taken in interpreting this view so that the cardiac shadow or the proventriculus is not confused with the liver (**Fig. 20.13**).

Ultrasonography

Ultrasonography offers several advantages over radiography, in that it allows examination of the internal structure of the liver, as well as examination of the nearby heart and pericardium, spleen, and gastrointestinal tract. The avian liver is homogeneous, finely granular, and contains transverse and longitudinal blood vessels throughout the parenchyma. The gall bladder, when present, lies to the right of the midline. Changes in echogenicity and size of the liver should be noted.

Endoscopy/coeliotomy and biopsy

After using clinical pathology and diagnostic imaging to focus on the liver as the source of the patient's problem, the next step is to accurately assess the aetiology and pathophysiology of the disease process. To do this requires invasive technology: endoscopy or an exploratory coeliotomy to visually assess the liver and a biopsy to examine the histological basis of the problem. Several approaches have been described: through an incision behind the caudal ribs on either side, through the caudal thoracic air sac, and through a ventral midline incision.

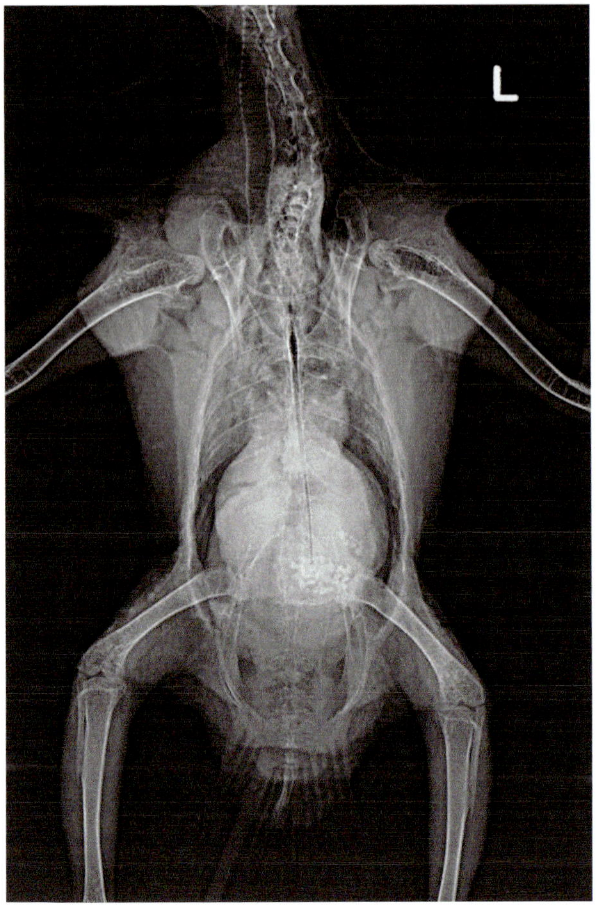

Figure 20.13 **Radiograph demonstrating hepatomegaly in a budgerigar.**

TREATMENT

Treatment of an avian patient with confirmed liver disease should have three objectives: support of the patient and correction of the abnormalities caused by the liver disease, treatment of the specific condition, and creation of an environment within the liver most favourable to regeneration of normal liver tissue.

Supportive care

Problems include dehydration, anorexia, weight loss, hypoglycaemia, hepatic encephalopathy (in raptors and other carnivorous birds), anaemia, septicaemia, dyspnoea (enlarged liver and/or ascites), bleeding disorders, and diarrhoea.

Supportive care, therefore, revolves around addressing these problems. See Chapter 5 for more details on supportive care, including fluid therapy, assisted feeding, blood transfusions, oxygen therapy, and warmth.

Retrospective studies in both people and animals have shown that protein restriction is contraindicated in liver disease and should only be used in cases where severe hepatic encephalopathy is present. Patients with liver disease are often in a catabolic state, and protein restriction may exacerbate both the catabolic state and the liver disease, making the development of hepatic encephalopathy more likely. Protein levels in the diet should therefore be increased to allow for anabolic regeneration of the liver. (In carnivorous birds, there may be the need to increase protein tolerance by feeding high-quality protein and using lactulose.) Diets high in simple and complex carbohydrates (such as vegetables, rice, and pasta) offer many advantages to patients with liver disease. Vegetable protein appears to lack many encephalopathic components; the higher fibre alters the intestinal flora to minimise ammonia production; carbohydrate metabolism is useful for hypoglycaemic patients; and insoluble fibre (e.g., psyllium) binds many endotoxins and noxious bile acids. Fat levels in the diet can be left untouched unless cholestasis causes a maldigestion problem with subsequent diarrhoea, or in cases of hepatic lipidosis. Conversion to a formulated diet achieves many of these requirements and is the cornerstone of treatment for liver disease.

Vitamin supplementation may be given, but caution is needed with vitamins A and D to prevent toxicosis. B group vitamins are needed in many animal species to improve healing, promote normal metabolism, and improve energy utilisation by cells. They are stored in the liver as co-enzymes and may become depleted with liver damage and anorexia. Supplementation with a multivitamin B complex is often indicated.

Treatment of specific conditions

Amyloidosis

Therapy for amyloidosis is aimed at correcting the underlying problem. Colchicine has been used in dogs to minimise further deposition and may be of use in birds. Other treatments used in humans and dogs have included immunosuppressive therapy, chemotherapy, and dimethyl sulfoxide (DMSO) (oral or subcutaneous). No reliable treatment has been shown to work in all cases, and success is limited.

Hepatic lipidosis

Hepatic lipidosis in birds is most commonly due to the feeding of a high-fat, low-protein diet, where the fat becomes the major source of calories. All-seed diets are a typical example of this sort of diet, but hepatic lipidosis occurs in most avian species. Treatment requires a reduction of dietary fat, an increase in dietary protein, the provision of carbohydrates to replace fat as the main source of calories, and supplementation with vitamin E and vitamin B complex.

Iron storage disease

Most cases of iron storage disease are believed to be associated with a high-iron diet. Treatment consists of feeding a low-iron diet, minimising the absorption of iron from the intestinal tract, and reducing the level of iron in the body. As the ascorbic acid found in citrus fruit reduces the ferric ion (Fe^{3+}) to the more easily absorbed ferrous ion (Fe^{2+}), citrus fruit should be eliminated from the diet. Commercial diets for susceptible species (Ramphastidae and mynahs) are now made with low iron levels (<100 ppm). Tannin added to the diet also binds iron in the intestine, reducing absorption. Iron levels in the body are reduced by weekly phlebotomy (1% body weight) and chelation therapy with deferoxamine.

Parasitic

Drugs used to treat hepatic protozoan infections are the same as those used for intestinal or systemic infections. Trematode infections are often difficult to treat. Praziquantel has been shown to reduce egg laying but may not eliminate the parasite itself. Albendazole has been shown to eliminate *Fasciola hepatica* infections in emus.

Infections

Bacterial, fungal, viral, mycobacterial, and chlamydial infections all have different treatments, according to the isolate and its sensitivity.

Neoplastic

Up until recently, hepatic neoplasia has been considered untreatable. Recent work suggests that some neoplasms may respond to chemotherapy, and drugs such as carboplatin and cisplatin are being trialled for their efficacy.

Creation of an environment favourable to regeneration

Given the liver's ability to regenerate, treatment to stabilise hepatocyte cell membranes, maintain the acinar structure, and minimise fibrosis will have a positive effect on patients with liver disease. Several drugs have been trialled in mammals to achieve these aims, and they show potential in avian medicine.

Ursodeoxycholic acid

Birds produce cholic, allocholic, and chenodeoxycholic acids. These bile acids are hydrophobic and may act as biologic detergents, causing solubilisation of cell membranes and hepatocellular lysis when cholestasis leads to elevated liver levels. Ursodeoxycholic acid (UDCA) is a hydrophilic bile acid that decreases the detergent effects of the hydrophobic bile acids, thereby reducing their reducing hepatotoxic effects. It may also increase the flow of bile from the liver, reducing the concentration of hydrophobic bile acids and therefore their toxic effects. UDCA is used as an adjunct in the treatment of cholestatic and necroinflammatory liver disorders. Its use as a single-agent therapy in mammals is relatively ineffective.

Colchicine

Colchicine is used as an antifibrotic drug. It inhibits the synthesis and secretion of collagen by stellate cells and fibroblasts and may induce the production of collagenase. It may also play a role in stabilising hepatocellular membranes and facilitating the excretion of hepatic copper. It may also block the synthesis and secretion of serum amyloid A (SAA) by hepatocytes, preventing the formation of amyloid-enhancing factor and subsequent amyloid deposition. Anecdotally, it has been used as an antifibrotic agent in birds with some success.

Silibinin

Silibinin, the extract from the milk thistle *Silybum marianum*, is frequently recommended as a therapy for avian disease. It is widely used in medical and veterinary medicine for the treatment of liver disease, even though its mechanisms of action are not fully understood. It is thought to have antioxidant and free radical scavenger effects as well as cell membrane stabilising effects that prevent toxins from entering hepatocytes. It is believed to stimulate hepatocyte regeneration while at the same time inhibiting stellate cells from depositing collagen fibres, reducing hepatic fibrosis. It may also inhibit the cytotoxic, inflammatory, and apoptotic effects of tumour necrosis factor.

However, in its original state, its water solubility and bioavailability are poor. Recent developments have improved both solubility and bioavailability, but not all available commercial products utilise these advances. Care must be exercised in the selection of a suitable formulation. If these problems can be overcome, silymarin shows great promise as a hepatoprotective agent in the treatment of liver disease.

S-adenosyl-methionine (SAMe)

Methionine is an essential amino acid requiring conversion to S-adenosyl-methionine (SAMe) to be active. SAMe is part of three major biochemical pathways in the body that serve to activate or excrete drugs, maintain functional cell membranes, and for anti-inflammatory effects and cellular regeneration. The conversion of methionine to SAMe requires SAMe synthetase, an enzyme produced in the liver and decreased or deficient in the presence of liver disease. A lack of SAMe results in a deficiency of glutathione, which is important in many metabolic processes and cell detoxification. Exogenous SAMe (rather than methionine) requires supplementation in animals with liver disease and has been shown to increase liver and red cell glutathione levels and/or prevent its depletion. It may also have a hepatoprotective effect. Its use in birds has not been trialled extensively but may prove to be a valuable adjunct to the treatment of liver disease.

CHLAMYDIOSIS

Aetiology

Chlamydia are gram-negative, non-motile, obligate intracellular bacteria of the order Chlamydiales, family Chlamydiaceae, genus *Chlamydia*. There are nine recognised species, of which eight are of veterinary interest:

- *C. psittaci*: birds, humans
- *C. abortus*: sheep, goats, cattle
- *C. pneumoniae*: humans, reptiles, amphibians, horses, koalas
- *C. pecorum*: ruminants, marsupials
- *C. felis*: cats
- *C. caviae*: guinea pigs
- *C. suis*: pigs
- *C. muridarum*: mice

Chlamydia psittaci

There are five serovars (A–E); different serovars infect specific species (e.g., A infects parrots, E infects Columbiformes). Parrots are most frequently infected with serovar A (genotype A) strains but can also be infected with serovar B strains. Serovar A strains are highly pathogenic. Chlamydiosis, also known as psittacosis and ornithosis, has been a known zoonotic infection for about 100 years.

Clinical presentation

The virulence of the strain affects the clinical course, with clinical signs ranging from inapparent to severe septicaemia and acute death. Typically, though, the disease affects either the respiratory tract, the gastrointestinal tract and liver, or both.

Respiratory signs

These include conjunctivitis, often with loss or matting of periocular feathers, dyspnoea (tail bobbing, mouth-breathing), and sneezing with a purulent nasal discharge and sinus distension.

Gastrointestinal/hepatic signs

These include diarrhoea, biliverdinuria (green urates and urine), and a 'sick bird look' (fluffed, lethargic, anorexic, and weight loss).

Other signs

There may be poor feathering and neurological signs (torticollis, tremors, and convulsions). Polyuria is frequent due to the nephrotoxic effects of elementary bodies. Affected birds may be infertile.

Pathophysiology

This can be a highly contagious disease, with transmission occurring through ingestion (faecal material, mutual feeding, and feeding chicks) or inhalation (respiratory secretions, aerosolised feather, or faecal dust). Egg transmission may occur. Shedding of the organism can start 72 hours after infection, and birds may be shedding for up to 10 days before they start to show clinical signs. The incubation period can be as short as 4 days and up to 1–2 years.

The infection can be latent and activated during stress. There are indications that shedding in pigeons is greater in hot weather, possibly as a stress response to heat. A carrier state may exist. Immunity to infection is short-lived and birds are susceptible to reinfection shortly after treatment ends.

Diagnosis

Diagnosis can be difficult, as there is no single 'best' test. Testing can rely on antigen detection or antibody detection, while ancillary testing can be used to support the presence of disease, rather than just the organism. The Centers for Disease Control and Prevention (2010) classifies the level of diagnosis as confirmed, probable, or suspect:

- A confirmed case of avian chlamydial infection is defined based on one of the following:
  - Isolation of *C. psittaci* from a clinical specimen.
  - Identification of chlamydial antigen by the use of immunofluorescence (fluorescent antibody) in the bird's tissues.
  - A fourfold or greater change in serologic titre in two specimens from the bird obtained at least 2 weeks apart and assayed simultaneously at the same laboratory.
  - Identification of *Chlamydia* within macrophages in smears or tissues (e.g., liver, conjunctival, spleen, respiratory secretions) stained with Gimenez or Macchiavello stain.
- A probable case of avian chlamydial infection is defined as a compatible illness and one of the following:
  - A single high serologic titre in a specimen obtained after the onset of clinical signs.
  - *Chlamydia* antigen (identified by the use of enzyme-linked immunosorbent assay [ELISA], PCR, or fluorescent antibody) in faeces, a cloacal swab specimen, or respiratory tract or ocular exudates.
- A suspected case of avian chlamydial infection is defined as one of the following:
  - A compatible illness that is not laboratory confirmed but is epidemiologically linked to a confirmed case in a human or bird.
  - A bird with no clinical signs and a single high serologic titre or detection of chlamydial antigen.
  - Compatible illness with positive results from a non-standardised test or a new investigational test.
  - Compatible illness that is responsive to appropriate therapy.

This classification highlights the complexity of confirming a diagnosis of chlamydial infection, and it also demonstrates the potential difficulty in stating that a bird is free of chlamydia.

Antigen detection

The respiratory epithelium (choanal/oropharynx) and the cloaca yield the most reliable samples for antigen detection, and a single combined swab is often submitted for testing.

- Isolation. Cell culture is the gold standard.
- Cytology. *Chlamydia* inclusions are most easily demonstrated in impression smears from serosal membranes, liver, spleen, and affected air sacs. Smears are stained with Giemsa or Macchiavello stain to show intracytoplasmic inclusions. Negative staining does not rule out *Chlamydia*.
- Immunofluorescence antibody testing. Commercial FA conjugate for detection of *Chlamydia* spp. may cross-react with *M. avium*, giving false-positive results.
- ELISA tests are available; however, be aware of false positives from *S. aureus* in the sample. ELISAs that detect the *Chlamydia* lipopolysaccharide (LPS) should be avoided because it is serologically related to the LPS of Enterobacteriaceae. False positives result from the presence of cross-reactive antibodies in the serum; false negatives occur when *Chlamydia* is shed intermittently.
- Latex agglutination tests (e.g., Clearview® Chlamydia MF, UiPath). A high number of false positives is seen due to bacteria.
- PCR tests. Be aware of false negatives caused by presence of inhibitors of the PCR reaction such as blood, serum, urine, faeces, sputum, hair shafts, and lab reagents specifically haematin, DMSO, sodium chloride, phosphate-buffered saline, melanin, heparin, detergents, and glove powder. DNA cannot distinguish between live organisms and dead organisms (which may be contaminants).

Antibody detection (serology)

A positive serologic test result is evidence that the bird has been infected by *Chlamydia*, but it does not indicate whether it has an active infection or a persistent titre following exposure. Most serological tests detect IgG, which does not appear until at least 2 weeks after infection. False-negative results can therefore occur in birds that have acute infections and have not yet seroconverted. IgG titres may also persist following successful treatment and should not be used to monitor treatment. In the individual bird, a serology result should be interpreted considering the bird's clinical signs, the history of the bird (or flock), ancillary testing (see later), and antigen detection. Available tests include:

- Elementary body agglutination detects IgM but may give false positives to *S. aureus*, *P. multocida*, and *Sarcina* sp.
- Direct complement fixation (CF) can be used to measure IgG, but it is best used with paired serum samples tested.
- ELISA (e.g., Immunocomb® Avian Chlamydia psittaci antibody test kit, Biogal, Galed Labs) detects IgG.
- BELISA (blocking ELISA). This inhibitory ELISA is a very sensitive test, marketed in Germany.

Ancillary testing

Haematology may demonstrate moderate anaemia, leucocytosis, absolute or relative heterophilia, and monocytosis. Blood chemistry may show elevations in CPK, AST, LDH, total protein, and bile acids if the liver is involved. Uric acid may be elevated if the kidneys are involved. Electrophoresis shows elevations in total globulins, and beta and gamma globulins, and a decrease in albumin.

Radiology may demonstrate hepatomegaly and/or splenomegaly, air saculitis, and pneumonia. A biopsy of liver, spleen, or air sac may also be helpful.

Acute lesions found at autopsy include fibrinous peritoneal exudate, air saculitis, splenomegaly, hepatomegaly, perihepatitis, pericarditis, myocarditis, bronchopneumonia, catarrhal enteritis, nephrosis, orchitis, epididymitis, and oophoritis. Chronic lesions include liver and kidney cirrhosis, pancreatic necrosis (budgerigar and pigeon), and non-purulent meningitis (**Fig. 20.14**).

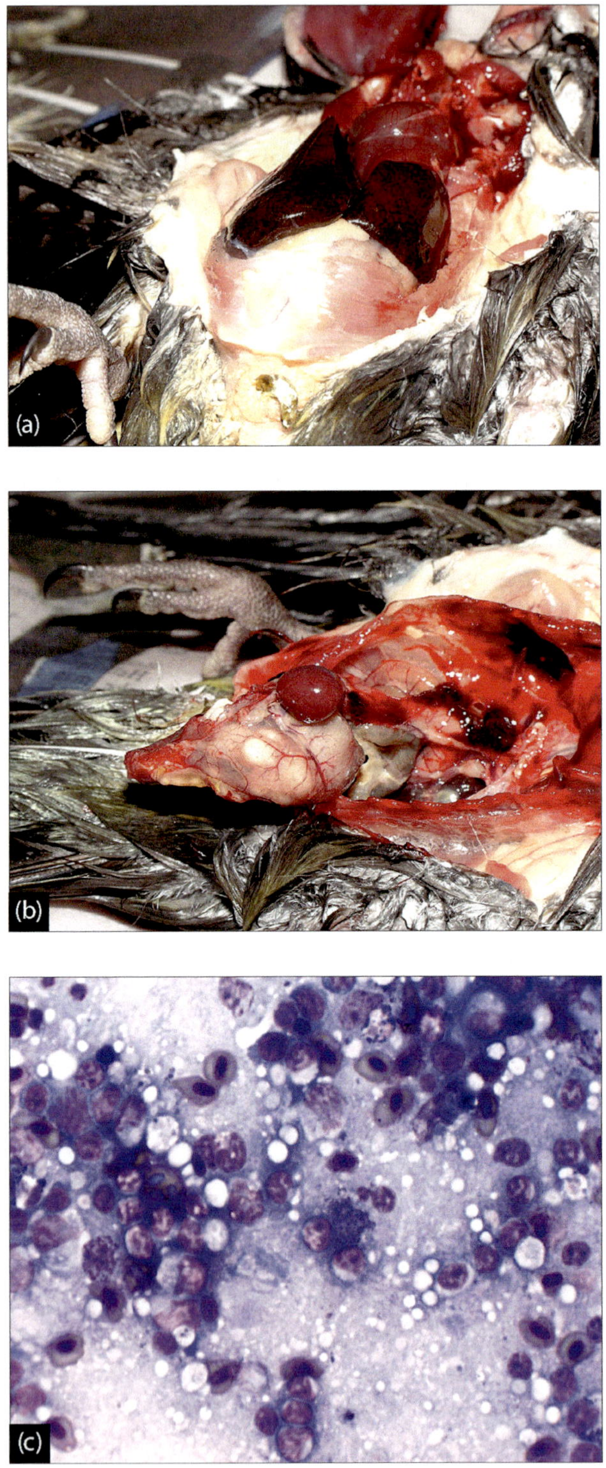

Figure 20.14 **Necropsy of a cockatiel reveals hepatomegaly (a) and splenomegaly (b) due to chlamydiosis. Subsequent cytology confirms the presence of chlamydial inclusion bodies within hepatic tissue (c).**

Differential diagnosis

Differential diagnoses include adenovirus, acute PBFD, herpesvirus, paramyxovirus, influenza A, and bacterial infections, especially Enterobacteriaceae.

Treatment

Tetracyclines

Tetracyclines inhibit *Chlamydia* protein synthesis and are active only when intracellular reticulate bodies are actively replicating. Disadvantages include low intracellular concentration, immunosuppression, chelation with calcium (dietary calcium should be reduced to 0.7% or less), and inhibition of autochthonous flora. The duration of therapy is empirically set at 45 days, the average lifespan of a macrophage. Oxytetracycline is poorly absorbed from the digestive tract of birds and should not be used. This leaves two members of the tetracycline group available for therapy:

- Chlortetracycline is renally excreted and should be used cautiously in birds with renal disease. Therapeutic tissue levels are difficult to maintain when administered through drinking water.
- Doxycycline is excreted extra-renally (faeces, bile). Adequate plasma levels are often achievable in medicated drinking water and seeds. It may cause hepatic necrosis with elevation of AST, which resolves with cessation of therapy.

Enrofloxacin

Clinical improvement and cessation of faecal shedding have been reported for 5 weeks post-treatment. This is not fully documented, and enrofloxacin may not be effective clinically. Field trials indicate that it may not be as effective as doxycycline.

Azithromycin

Azithromycin is a macrolide antibiotic, able to penetrate macrophages. There are anecdotal reports of effectiveness, but no clinical trials have been conducted yet to confirm this.

Disinfection

Free elementary bodies are unstable and can be inactivated in the environment within days. They are sensitive to heat, quaternary ammonium products, 70% ethanol, and 3% hydrogen peroxide.

Clinical signs in man

Chlamydia infection is a zoonosis, causing fever, chills, pneumonia, headache, weakness, fatigue, myalgia, chest pain, anorexia, nausea, vomiting, and diaphoresis. It is important to discuss the zoonotic implications of this disease with clients when chlamydia is diagnosed in a bird.

PACHECO'S DISEASE

Aetiology

Pacheco's disease is caused by psittacine herpesvirus 1 (PsHV-1). There are four recognised genotypes of PsHV-1, with different genotypes affecting different species of birds in different ways.

- All four PsHV-1 genotypes have been shown to cause Pacheco's disease in Amazon parrots.
- Only PsHV-1 genotypes 2, 3, and 4 have been identified in African grey parrots with Pacheco's disease.
- Only genotype 4 causes mortality in macaws and conures.
- Parrots from Australia, Indonesia, the Philippines, and Southeast Asia appear to be somewhat affected by any of the four PsHV-1 genotypes.
- PsHV-1 genotype 2 is the second leading cause of Pacheco's disease outbreaks in Europe but has not been detected in parrots with Pacheco's disease in the USA.

These genotypes appear to have co-evolved with several species of Neotropical (Central and South American) parrots. In their host species, they do not appear to cause disease. However, disease may result when other parrots are infected with a PsHV-1 genotype to which they are not evolutionarily adapted.

Infected birds may:

- Become acutely sick and die (all genotypes); this is known as Pacheco's disease.
- Become acutely sick, respond to therapy, and then become lifelong carriers.
- Develop IPD (genotypes 1, 2, or 3). See Chapter 19.
- Develop biliary adenocarcinomas (genotype 3).

Pathophysiology

Asymptomatic carriers (especially conures, Amazon parrots, and macaws) may serve as virus reservoirs, shedding the virus for life in faeces and respiratory/pharyngeal secretions. Shedding may be triggered by periods of stress. Birds dying of Pacheco's disease will shed large amounts of virus into their environment. Most infections occur when contaminated material is ingested. The incubation period may last 5 days to several weeks. Severe hepatic and splenic necrosis usually ensues, with occasional pancreatic and intestinal necrosis occurring in some birds.

Birds that survive the initial disease may develop IPD within a year of the initial infection.

Figure 20.16 **Haemorrhage from the nares and oral cavity in a macaw that died of Pacheco's disease. (Photo courtesy B Dalhausen.)**

Clinical presentation

Sudden death is a common presentation (**Fig. 20.15**). Birds exposed to virulent strains of PsHV-1 and surviving long enough to develop clinical signs will show evidence of lethargy, regurgitation, diarrhoea, biliverdinuria, and neurological signs, often dying within 3–10 days of exposure to the virus (**Fig. 20.16**).

Birds that are infected with PsHV-1 genotype 3 and survive the initial infection may develop bile duct and pancreatic duct carcinomas. Birds with bile duct carcinomas will often exhibit signs of chronic liver disease, such as weight loss, an overgrown beak, and poor feather quality, eventually dying of chronic liver failure. Pancreatic duct carcinomas are much less common, and clinical signs are vague and nonspecific. Bile duct carcinomas develop in the months and years following the onset of mucosal papillomas.

Figure 20.15 **An outbreak of Pacheco's disease can cause devastating loss in a collection. This group of *Neophema* spp. parrots died acutely as a result of such an outbreak. (Photo courtesy B Speer.)**

Diagnosis

Autopsy may show no lesions or hepatomegaly and splenomegaly, with a mottled tan discolouration of the liver (**Fig. 20.17**). Histopathology shows multifocal necrosis in the liver and spleen with intranuclear eosinophilic inclusion bodies.

Viral-specific PCR probes using oral and cloacal swabs have been developed to detect infected birds.

Treatment

Without treatment, birds that develop Pacheco's disease die. However, virus replication is rapidly

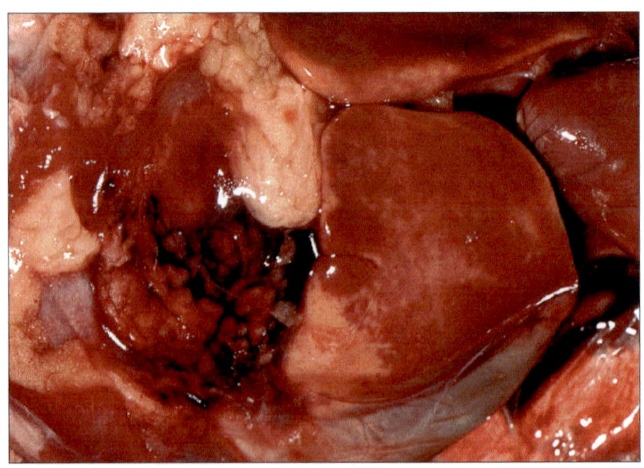

Figure 20.17 **Haemorrhagic hepatitis in a bird that died from Pacheco's disease. (Photo courtesy R Schmidt.)**

inhibited by acyclovir, an acyclic purine nucleoside, and birds in the early stage of infection can be saved with treatment. It works best in in-contact birds not showing signs of illness. All infected birds that survive the initial disease (or never develop illness) become carriers.

Management

Subclinically infected birds are readily detected by PCR, so all new birds should be screened for this disease in quarantine. Birds that have survived an outbreak should be regarded as carriers. All positive birds should be held in isolation. Although there is strong evidence of parent-to-offspring transmission after hatch, vertical transmission through the egg does not appear to occur. Therefore, positive birds can be used as breeders, as long as incubation and hand-rearing are used to isolate the chicks from their parents.

An inactivated vaccine was available in the United States but appears to have been discontinued.

FURTHER READING

Compendium of measures to control *Chlamydia psittaci* infection among humans (psittacosis) and pet birds (avian chlamydiosis) (2010). http://www.nasphv.org/Documents/Psittacosis.pdf

Doneley B (2004) Treating liver disease in the avian patient. *Seminars in Avian and Exotic Pet Medicine* 13(1): 8–15.

Hochleithner M, Hochleithner C, Harrison LD (2006) Evaluating and treating the liver. In: *Clinical Avian Medicine, Vol 1*. GJ Harrison, TL Lightfoot (eds). Spix Publishing Inc, Palm Beach, pp. 441–449.

Phalen D, Tomaszewski E, Styles D (2009) Epizootiology, diversity and pathogenicity of Psittacid Herpesviruses. Proceedings of the Annual Conference. Association of Avian Veterinarians, Milwaukee, Wisconsin: 47–51.

Styles DK, Tomaszewski EK, Phalen DN (2004) Psittacid herpesviruses associated with mucosal papillomas of neotropical parrots. *Virology* 235: 24–35.

Tully TN (2006) Update on *Chlamydia psittaci*: A short comment. In: *Clinical Avian Medicine, Vol 2*. GJ Harrison, TL Lightfoot (eds). Spix Publishing Inc, Palm Beach, pp. 679–680.

CHAPTER 21

DISORDERS OF THE PANCREAS

ANATOMY

The avian pancreas consists of three lobes. The dorsal and ventral lobes are supported and separated by the pancreatic artery within the duodenal loop, and the splenic lobe runs more laterally up to the spleen, as an extension of the ventral lobe. The pancreas is found within the duodenal loop, lying on the ventral wall of the right coelom (**Figs. 21.1, 21.2**).

The pancreas has both endocrine and exocrine functions. While the amount of endocrine tissue is proportionally greater than that of mammals, over 99% of the pancreatic mass has an exocrine function.

- The exocrine pancreas consists of compound tubuloacinar glands divided into lobules. These glands secrete amylase, lipase, proteolytic enzymes, and sodium bicarbonate into the ascending duodenum via three pancreatic ducts.
- The endocrine pancreas (1–2% of the pancreatic mass) is predominantly located in the splenic lobe of the pancreas. It consists of scattered clusters of endocrine cells (the islets). Two types of endocrine islets have been described: the larger islets containing predominantly glucagon-secreting A-cells and the smaller islets containing predominantly insulin-secreting B-cells. Other endocrine cells (somatostatin-secreting D-cells and pancreatic polypeptide-secreting PP cells) are randomly scattered throughout the islets. These cells synthesise and release their peptide products directly into the bloodstream. On a hormone-weight to tissue-weight basis, insulin is the most concentrated hormone in the entire pancreas, followed by glucagon, APP, and somatostatin.

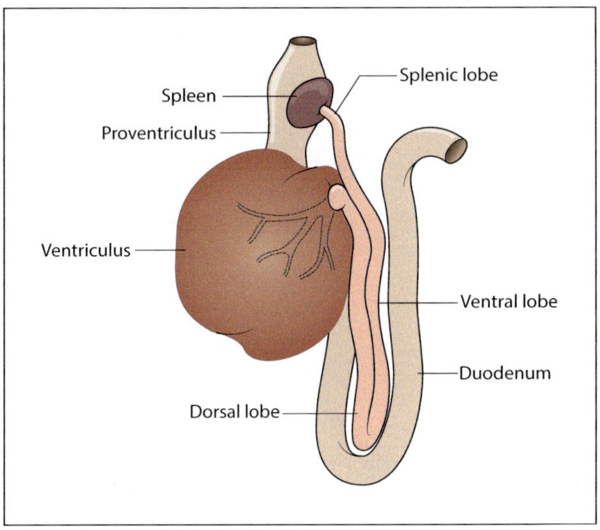

Figure 21.1 **The avian pancreas.**

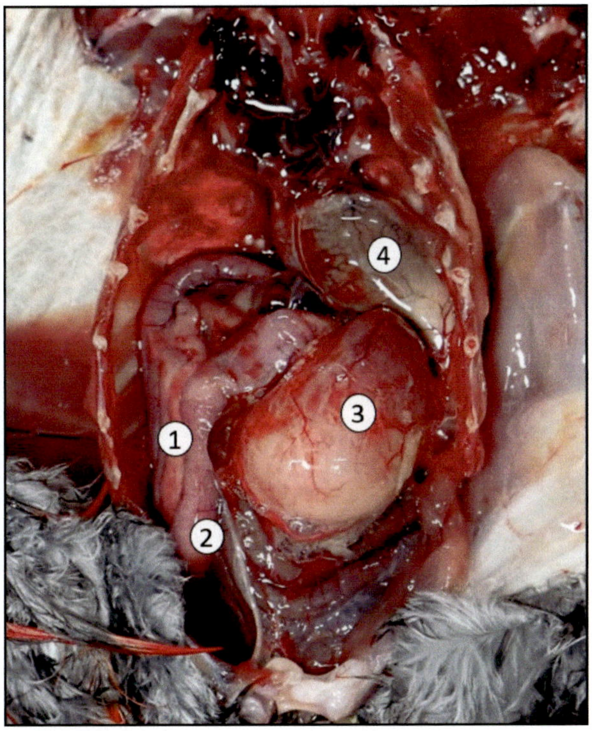

Figure 21.2 **The pancreas in a lorikeet: (1) pancreas, (2) duodenal loop, (3) ventriculus, (4) proventriculus. The pancreas is located within the duodenal loop, with the ventral lobe against the coelomic wall on the right side of the body.**

DOI: 10.1201/9781003272519-21

PHYSIOLOGY

Exocrine pancreatic secretion, which is at a higher rate than that of mammals, is controlled by both neural and hormonal mechanisms. Immediately after a bird starts eating, pancreatic secretion begins, apparently via a vagal reflex. Distension of the proventriculus stimulates a hormonal response involving a vasoactive intestinal polypeptide; this results in pancreatic secretion. Diet can also affect the rate of secretion, with diets high in fat and carbohydrates increasing the activity of amylase and lipase.

The endocrine function of the pancreas involves the balance and negative feedback of three major hormones: insulin, glucagon, and somatostatin. These are released in response to absorbed nutrients, cholinergic input, and probably hormonal stimulation (**Fig. 21.3**).

- Glucagon (A-cells) acts primarily as a catabolic agent by increasing blood glucose and free fatty acid levels through its gluconeogenic, glycogenolytic, and lipolytic effects. Thought to be the dominant hormone in glucose regulation, levels in avian plasma have been reported to be at least 10–80 times higher than in mammals, with a plasma glucagon to insulin ratio reported to be 5–10 times higher in birds compared to mammals. The catabolic function of glucagon helps birds meet the high metabolic demands of egg laying, sustained flight, and fasting. Alpha cells secrete glucagon secondary to free fatty acid and cholecystokinin stimulation, and secretion is inhibited by glucose.
- Insulin (B-cells) acts primarily as an anabolic agent by decreasing blood glucose concentrations through (1) promoting intracellular translocation of glucose (by increasing the availability of glucose transport carriers), (2) lipogenesis, and (3) anti-gluconeogenic processes. Avian plasma levels of insulin are much lower than in most mammals, but it is more potent. It is not anti-lipolytic in birds as it is in mammals. Its role in carbohydrate metabolism in birds is unclear, with apparent species differences; for example, carnivorous birds are thought to be more insulin-dependent than granivorous species. Insulin is known to decrease glucagon secretion in many species.
- Somatostatin (D-cells) inhibits and controls the release of glucagon and insulin. It inhibits islet cell hormone secretion, slows the absorption of nutrients (especially glucose and lipids), and inhibits lipolysis.
- Pancreatic polypeptide (F/PP cells) is primarily involved in lipogenesis, but it appears to stimulate the secretion of gastrin and other gastric chemicals and inhibits gastrointestinal motility.

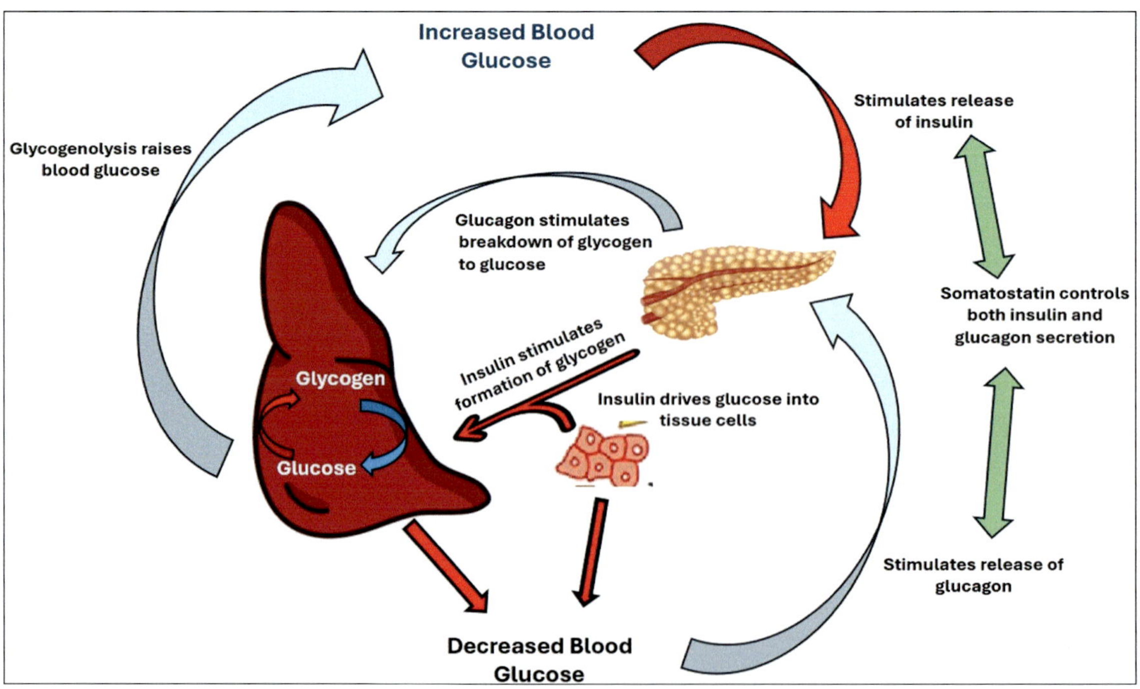

Figure 21.3 **Glycaemic control.**

It also mobilises liver glycogen but has no effects on plasma glucose levels; instead, the pancreas, duodenum, ileum, spleen, bone marrow, and probably liver and proventriculus, are its target organs.

DISEASES OF THE PANCREAS

Diseases of the pancreas can affect the exocrine pancreas, the endocrine pancreas, or both. Pancreatic disease is infrequently described in birds, but as our ability to diagnose diseases improves, the incidence of diagnosed diseases rises.

Exocrine

Pancreatic disease

Definition/overview

The incidence of primary pancreatic disease in birds is a matter for some discussion. However, pancreatic disease secondary to other disease processes is probably underdiagnosed in avian medicine. This undoubtedly reflects the lack of specific and sensitive tools to diagnose this disease. Our understanding of avian pancreatic disease is still in its formative stages.

Aetiology

Causes of pancreatic disease include obesity (often when combined with high-fat diets or fatty meals), toxicosis (particularly zinc, mycotoxins, and selenium; see **Fig. 21.4**), trauma, viral infection (including paramyxovirus-3, adenovirus, polyomavirus, avian influenza type A, infectious bronchitis, and herpesvirus), chlamydiosis, bacterial infection, egg yolk peritonitis, and neoplasia. Of particular note is an acute pancreatic necrosis seen in Quaker parrots, possibly caused by a high-fat diet, but the exact cause has not been determined.

Pathophysiology

Activation of the digestive enzymes (trypsin, protease, and phospholipase) within the gland results in pancreatic autodigestion. Damage to the pancreatic cell walls allows the release of these enzymes into the intracellular space and ducts, and this in turn causes the production of unopposed free radicals, which cause even more damage. This again releases more enzymes, and the cycle continues. The result is inflammation, cellular destruction, and finally fibrosis.

Continued bouts of pancreatitis can result in extensive fibrosis and loss of pancreatic tissue. Although the endocrine function (see later) is less commonly affected, chronic pancreatic disease can result in exocrine pancreatic insufficiency.

Clinical presentation

Gastrointestinal dysfunction causes vomiting, ileus, weight loss, polyuria, polydipsia, and coelomic distension. The condition is painful, and signs of this include anorexia, lethargy, feather picking, aggression, and obsessive chewing on the cage and other items.

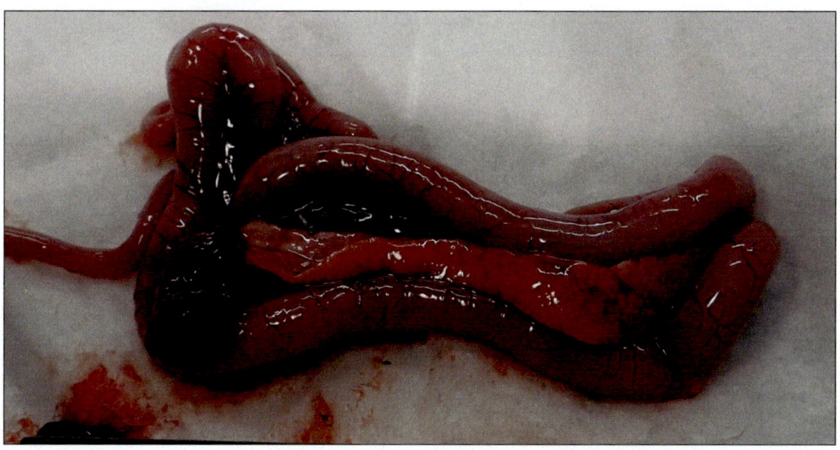

Figure 21.4 **The pancreas of a barn owl with multifocal areas of acini loss and protein accumulation within distended pancreatic ducts with flattened squamous epithelium, possibly from a toxic insult.**

Diagnosis

Clinical signs suggest a diagnosis. Significant elevations may be found in serum amylase and lipase concentrations. However, these biochemistries are neither specific nor sensitive for pancreatic disease, and a diagnosis of pancreatic disease based solely on hyperamylasaemia or elevated lipase cannot be justified. Ultrasonography can be utilised, but the pancreas is often difficult to visualise. Pancreatic biopsy can be used, either endoscopically or through a ventral midline coeliotomy (see later) (**Fig. 21.5**).

Management

If possible, the causative agent should be identified and eliminated (e.g., chelating zinc toxicosis, conversion from a seed diet to a lower-fat formulated diet, surgical treatment of yolk peritonitis). Analgesia and fluid support are required, and antibiotic therapy is given if indicated.

Exocrine pancreatic insufficiency (EPI)

Aetiology

Pancreatic insufficiency, an inability to digest food properly, may be caused by congenital pancreatic atrophy, prolonged caloric deficiency, or end-stage chronic pancreatic disease (see later) with loss of exocrine tissue.

Clinical presentation

The loss of exocrine pancreatic tissue results in maldigestion, particularly of fatty foods and starch. Undigested fats and starch account for steatorrhoea (seen as pale, bulky, and often malodourous faeces) and weight loss. In mammals, depletion of body stores of cobalamin occurs rapidly, but this has not been looked at in birds at this time. Clinical signs seen in affected birds include lethargy, generalised malaise, weight loss, and steatorrhoea, often described as 'popcorn' droppings (**Fig. 21.6**). Polyphagia is variable.

Diagnosis

Diagnosis is based on clinical signs and pancreatic biopsy. Plasma biochemistry analysis is rarely conclusive but may indicate pancreatic involvement.

A ventral midline coeliotomy is performed and the duodenal loop exteriorised. The dorsal and ventral lobes of the pancreas can be seen within the duodenal loop. Overall, the degree of pancreatic atrophy

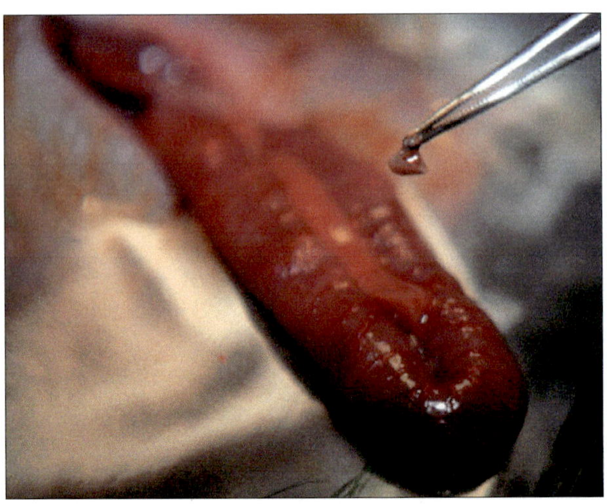

Figure 21.5 **Pancreatic biopsy: after exteriorising the duodenal loop, the tip of the dorsal lobe is elevated and removed.**

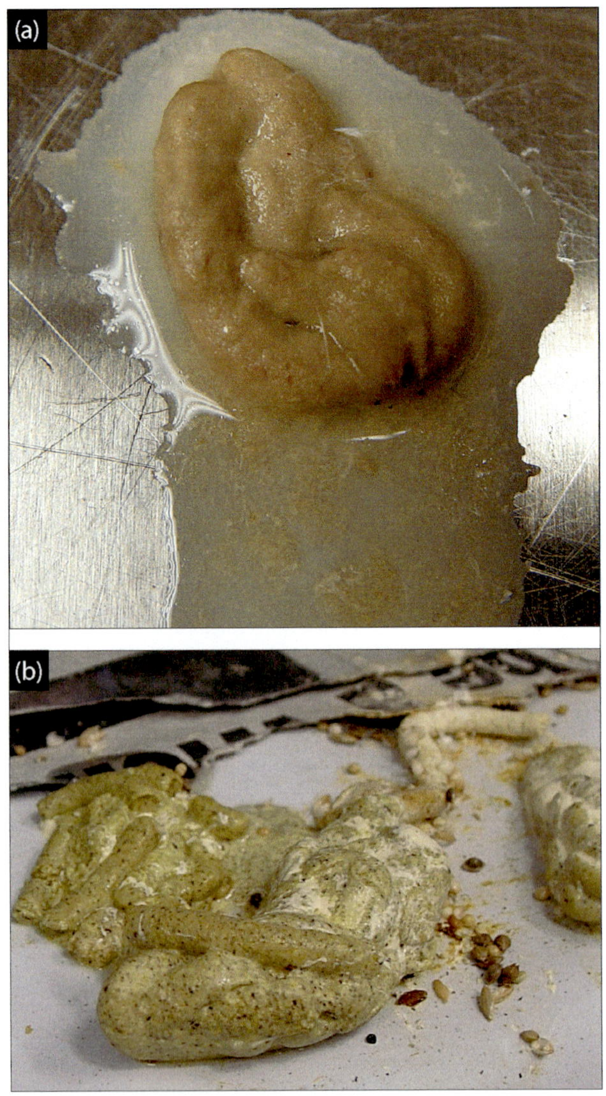

Figure 21.6 **Undigested fat in the droppings of a cockatoo with pancreatic atrophy.**

can vary; typically, an affected pancreas appears small and irregular in shape. Lifting the tip of the ventral lobe reveals the pancreatic blood vessels. These vessels are avoided while the tip of the pancreas is biopsied. Other areas of the pancreas can be biopsied, if necessary, again avoiding the pancreatic vasculature. The splenic lobe of the pancreas is difficult to visualise and care must be taken not to place excessive tension on the duodenal loop when trying to access the splenic lobe.

An alternative technique is an endoscopic biopsy through the right abdominal air sac.

No evidence of iatrogenic post-biopsy pancreatitis has been demonstrated; recovery is usually uneventful.

Histologically, there is a variable loss of zymogen granules and shrinkage of acinar cells.

Management

If possible, the causative agents should be identified and eliminated. The affected bird should be converted to a formulated diet and vegetables, which are supplemented with pancreatic enzymes in a powdered form. Palatability can be an issue in some birds, and intake must be monitored. Response to therapy is based on weight gain and a return to normal faeces.

Pancreatic neoplasia

Definition/overview

Primary pancreatic neoplasia is usually either carcinomas or adenocarcinomas. Metastatic spread of other neoplasias to the pancreas is uncommon. Bile and pancreatic duct hyperplasia and papilloma formation are seen in some birds with internal papilloma disease associated with PsHV 1 genotype 3.

Clinical presentation

Signs are like those seen with pancreatic disease or exocrine pancreatic insufficiency. Coelomic distension may be appreciable if the mass is large enough.

Diagnosis

Antemortem diagnosis is like that of other pancreatic diseases and requires ultrasonography and biopsy.

Management

Chemotherapy (e.g., carboplatin) has been used in a few cases with variable degrees of success. The prognosis is generally guarded to poor.

Endocrine

Diabetes mellitus

Overview

Hyperglycaemia is a hallmark of diabetes mellitus. Glucose may travel freely dissolved in the circulation or it can be taken up by erythrocytes via glucose transport proteins (GLUTs). The ability of avian erythrocytes to take up glucose declines as the cells mature, such that 95% of circulating glucose is present in the plasma instead of the erythrocytes (unlike humans, where it is nearly evenly distributed, and dogs and cats where 87–93% is carried in the plasma). The result is that birds naturally have normal blood glucose concentrations that are nearly double the levels measured for mammals of similar body size and are resistant to insulin-mediated glucose uptake into tissues. Glucose levels often vary among and within individual birds of the same species in response to dietary, pharmacological, and other factors such as stress and diurnal activity. (Glucocorticoids are released in response to stress to provide energy to deal with stressors and to increase glucose bioavailability for the brain.)

Diabetes mellitus in mammals, characterised by chronic hyperglycaemia and other metabolic abnormalities, is due to a lack of effect by insulin on glucose metabolism due to either an absolute lack of insulin or an inability of the insulin to lower blood glucose levels. Birds differ from mammals because of the effect of glucagon; diabetes mellitus in birds could theoretically occur due to a lack of insulin effect or an excess of glucagon effect.

Diabetes mellitus is most commonly diagnosed in budgerigars, cockatiels, and galahs. It has also been diagnosed in larger parrots, toucans, mynahs, and other species.

Aetiology

Understanding the pathogenesis of diabetes mellitus in birds is complicated by the role of glucagon and somatostatin in glucose metabolism in birds. In granivorous birds the pancreatic islets have a substantially higher percentage of glucagon-secreting cells compared with the mammalian pancreas. Correspondingly, the plasma insulin:glucagon ratio in these birds appears to be two to five times higher than that of mammals; it, therefore, appears that glucagon is the dominant hormone in glucose metabolism.

Glucagon is a powerful catabolic hormone that stimulates gluconeogenesis, lipolysis, and glycogenolysis, thus increasing blood glucose. This process begins within hours of a meal, allowing the continual production of glucose to meet the body's demands for energy. Insulin, on the other hand, is a powerful anabolic hormone, increasing the availability of glucose transport carriers, which allows easier transfer of glucose into the cell. It is anti-gluconeogenic. The effect of these hormones is modulated by somatostatin, pancreatic polypeptide, prolactin, thyroxine, and growth hormone. It is, therefore, unclear whether diabetes mellitus in granivorous birds is due to an excess of glucagon or to a lack of insulin effect. The insulin:glucagon ratio and its regulation by somatostatin may be the critical factor in regulating glucose concentration rather than the actual concentration of the individual hormones.

Diabetes mellitus in mammals is classified as:

- Type I (insulin dependent) is characterised by a lack of insulin production by B-cells in the pancreas due to genetic makeup, autoimmune disease, or other causes targeting the B-cells.
- Type II (non-insulin dependent) is characterised by decreased insulin sensitivity (insulin resistance), without any overt pancreatic disease.
- Other specific types of diabetes, characterised by nonspecific destruction of pancreatic tissue associated with other pathological conditions or diseases or drugs that cause insulin resistance.

No similar classification system has been applied to diabetes mellitus in birds, and it may be that attempts to do so will be thwarted by the differences in glucose metabolism in birds compared with mammals. However, it would appear that the equivalents to these diabetes classifications do occur in birds.

- Type I diabetes, with selective absence or destruction of pancreatic islets, may be associated with pancreatitis (either primary or secondary to female reproductive pathology such as yolk coelomitis), iron storage disease, viral infection (Herpesvirus, Paramyxovirus, Adenovirus, Polyomavirus), toxins (zinc, fungal toxins, selenium), neoplasia (pancreatic islet carcinoma, pancreatic adenocarcinoma), or other unknown causes.
- Type II diabetes, with peripheral resistance to insulin, is commonly associated with obesity but can also be due to high levels of corticosteroids (including iatrogenic dosing), megestrol acetate, medroxyprogesterone acetate, growth hormone, glucagon, and high iron. A link between iron storage disease and diabetes mellitus is being increasingly recognised.

Pathogenesis

In both conditions, chronic hyperglycaemia damages cells and tissues in the body and leads to polyuria, glucosuria, polydipsia, lethargy, and a switch to the utilisation of fatty acids for energy, resulting in increased production and disposal of ketones in the urine (ketonuria).

Clinical presentation

Diabetes mellitus has been reported infrequently across multiple avian orders including Accipitriformes, Anseriformes, Columbiformes, Galliformes, Passeriformes, Piciformes, Sphenisciformes, and Psittaciformes. It has been reported in cockatoos, macaws, grey parrots, Amazon parrots, cockatiels, budgerigars, conures, and eclectus parrots. Cockatiels, budgerigars, macaws, and toucans appear to be over-represented in the literature. Although there is no reported age, sex, or hereditary predisposition, it may be that middle-aged females are more at risk.

The principal signs are polyuria, polydipsia, polyphagia, and weight loss (although this is not consistent) (**Fig. 21.7**). In some cases, especially those with concurrent disease, nonspecific signs such as depression, lethargy, vomiting, or anorexia may be seen.

Figure 21.7 **Extreme polyuria in an obese galah with type II (insulin-resistant) diabetes mellitus.**

Diagnosis

Diagnosis is based on the patient's history, clinical signs, and the demonstration of persistent hyperglycaemia and glucosuria.

- Normal avian blood glucose levels (12–25 mmol/l) are 150% to 300% above those of mammals, and this can go even higher when the bird is stressed (e.g., with capture and restraint). Birds readily exhibit significant hyperglycaemia with stress, including handling and restraint for venepuncture. However, blood glucose levels persistently above 38–44 mmol/l are more consistent with a diagnosis of diabetes mellitus.
- As the renal threshold for glucose in birds is 33 mmol/l, stressed birds, as well as diabetic birds, can demonstrate glucosuria, making it a non-specific indicator of diabetes.
- Fructosamine levels evaluate proteins resulting from an irreversible, non-enzymatic, and insulin-independent binding of glucose to serum proteins. It has not been fully evaluated in birds yet; it therefore remains an unproven but potentially useful tool for monitoring diabetic patients. While fructosamine may indicate the diabetogenic metabolic status in some birds, it appears to undergo short-term changes (resulting in a shorter half-life for serum proteins) and, thus, cannot be used as a long-term parameter of blood glucose concentration.
- Despite glucagon's lipolytic function, which would be expected to result in the production of beta-hydroxybutyrate and ketones, ketonuria is not consistently reported in confirmed or suspected cases of diabetes mellitus.

It is worth repeating that a single-point finding of hyperglycaemia and/or glucosuria does not equate to a diagnosis of diabetes mellitus. The diagnosis of diabetes mellitus is therefore somewhat complicated and relies on input from all aspects of the bird's presentation.

Treatment

The goal of treatment is to lower the patient's blood glucose levels and restore a more normal glucose metabolism. This is achieved by determining and correcting the primary disease process that is inducing diabetes mellitus, stabilising the unwell patient, and correcting the hyperglycaemia. The latter may be achieved through the use of insulin or oral hypoglycaemic agents.

If insulin is used, the patient should be hospitalised until an optimal dose of insulin is determined (via blood glucose curves) and the bird is stable. The goal of insulin therapy in birds should be to keep the patient mildly hyperglycaemic to avoid life-threatening hypoglycaemia.

The immediate use of short-acting insulin (0.1–0.2 U/kg) can stabilise the patient initially, but long-term control at home usually requires longer-acting insulin (NPH or ultralente). Again, a glucose curve should be performed, and dose rates adjusted accordingly. The dose rates vary considerably and should be based on the observed effects (reduction in PU/PD). They range from 0.067 to 3.3 U/kg q12–24h.

Compounded and diluted insulins both have inherent risks. Insulin molecules do not mix well in water, leading to uneven suspensions and irregular dosing of patients. Some insulins, including glargine and detemir insulins, should not be diluted at all. The shelf life of diluted insulin has not been well studied.

Oral hypoglycaemic agents (sulphonylureas, e.g. glipizide) may assist some patients, although reports of their efficacy vary. Sulphonylureas lower blood glucose concentrations in both diabetics and non-diabetics. The exact mechanism of action is not known, but these agents are thought to exert their effect primarily by stimulating the beta cells in the pancreas to secrete additional endogenous insulin. Extra-pancreatic effects include enhanced tissue sensitivity to circulating insulin and reduction of the production of hepatic basal glucose. The mechanisms causing these effects are yet to be fully explained. For these medications to be effective, pancreatic beta cells need to be functional.

Monitoring response

Monitoring the effects of either treatment can be difficult because, as in mammals, blood glucose measurements in the hospital may be affected by stress and be higher than values at home. Other complicating factors include body size and the need for repeated venepuncture.

Depending on the size of the patient and the frequency of sampling, the use of a laboratory analyser may not be possible. Instead, a handheld glucometer can be utilised, but it must be recognised that they have not been validated in parrots. Therefore, values should be interpreted with caution. Furthermore, dose adjustments are more difficult to make in avian patients based on the glucose curve results without dilution or compounding insulin, each of which has risks.

For smaller patients, where glucose curves with a handheld glucometer may not be practical, monitoring using urine dipsticks may be more feasible. Monitoring of clinical signs (e.g., reduced urine output, reduced thirst) can also help guide therapy and should be performed in all patients. Observing water intake and urinary output may be more practical measures for owners to take at home.

Conversion to a formulated diet is an important part of treatment. Not only does it correct many of the causative factors (e.g., obesity), but it also provides a high-fibre diet that assists with regulating blood glucose. In many (but not all) cases, dietary conversion and treatment of primary disease processes can result in a clinical resolution of the signs of diabetes mellitus, removing the need for ongoing insulin therapy.

Prognosis

Management of a diabetic bird requires a dedicated owner.

- For smaller birds, management with oral antiglycaemic medications may require less input from the owner; however, these medications are often less effective compared to insulin therapy, and clinical symptoms may persist or worsen, often leading to death or euthanasia within months.
- Larger parrots can be more easily treated and monitored with insulin. Survival times for larger parrots receiving insulin therapy vary from days to years depending on the time of diagnosis and the underlying cause(s).

It must be emphasised that the owner must be committed to giving once or twice daily injections of insulin, and regular monitoring both at home and in the hospital. If the owner cannot commit to this level of care and oral antiglyacemic medications are not controlling symptoms, leading to a poor quality of life, euthanasia should be considered.

FURTHER READING

Denbow DM (2000) Gastrointestinal anatomy and physiology. In: *Sturkie's Avian Physiology*. E Whittow (ed). 5th ed. San Diego: Academic Press: 299–325.

Doneley B (2000) Acute pancreatitis in psittacine birds. In: *Proceedings of the Annual Conference of the Association of Avian Veterinarians Australian Committee*, pp. 261–267.

Gancz AY, et al. (2005) Diabetes mellitus in large psittacines: A possible relationship with excessive iron storage. In: *Proceedings of the Annual Conference of the Association of Avian Veterinarians Australian Committee*, pp. 267–269.

Gelis S (2006) Evaluating and treating the gastrointestinal system. In: *Clinical Avian Medicine, Vol 1.* GJ Harrison, TL Lightfoot (eds). Spix Publishing Inc, Palm Beach, pp. 411–440.

Hudelson KS, Hudelson PM (2006) Endocrine considerations. In: *Clinical Avian Medicine, Vol 2.* GJ Harrison, TL Lightfoot (eds). Spix Publishing Inc, Palm Beach, pp. 541–557.

McCleery B (2022) Diabetes Mellitus in Psittacine birds. ExoticsCon Conference proceedings.

Pilney AA (2008) The avian pancreas in health and disease. *Veterinary Clinics of North America: Exotic Animal Practice* 11: 25–34.

Schmidt RE, Reavill DR (2014) Lesions of the avian pancreas. *Veterinary Clinics of North America: Exotic Animal Practice* 17(1): 1–11.

Speer B (1998) A clinical look at the avian pancreas in health and disease. In: *Proceedings of the Annual Conference of the Association of Avian Veterinarians Australian Committee*, pp. 57–64.

Sweazea KL (2022) Revisiting glucose regulation in birds – A negative model of diabetes complications. *Comparative Biochemistry and Physiology Part B: Biochemistry and Molecular Biology* 262: 110778–110778.

CHAPTER 22

DISORDERS OF THE CARDIOVASCULAR SYSTEM

ANATOMY

The avian heart lies in the cranial midline of the body, within the rib cage and enclosed by the liver rather than the lungs. The basic anatomy is like that of the mammalian heart, with two atria and two ventricles. The right cranial vena cava and the caudal vena cava form a sinus venosus with left and right sinoatrial valves. A sinus septum separates the opening of the left cranial vena cava from the openings of the other two veins. The right atrioventricular valve is muscular with no chordae tendineae, while the left atrioventricular valve is membranous (similar to mammals) but is a continuous sheet without defined cusps. The resistance to blood flow is, as with mammals, lower in the pulmonary circulation than the systemic circulation, leading to more pressure required in the left ventricle. Accordingly, the left ventricular wall is thicker than the right. The aortic and pulmonary artery valves are similar to those in mammals, while the left and right pulmonary veins merge in the atrium to form a single vessel entering the left atrium. The aorta curves to the right, as it is derived from the right fourth arterial arch and right aorta.

The heart is enclosed in the pericardium, a semi-rigid fibrous sac that contains only a small amount of serous fluid to allow lubrication for the heart within the sac. The pericardium is adherent to the sternum, the cranial and caudal thoracic air sacs, the liver, and the hepatic peritoneum (which is, in turn, adherent to the ventral vertebral column). These multiple connections provide mechanical stability to the apex of the heart and to several large, central blood vessels that are adherent to the pericardium. The relative rigidity of the pericardial sac acts to mechanically couple the ventricles (i.e., limit acute increases in cardiac size).

The ascending aorta gives off the coronary arteries, supplying the myocardium, and then two brachiocephalic trunks, supplying the head, wings, and flight muscles. These brachiocephalic trunks give off the carotid arteries before carrying on to the wings and flight muscles. The descending aorta runs caudally and dorsally, just ventral to the vertebral column. As it does, it gives off arteries supplying the internal organs and the legs (**Fig. 22.1a**).

The arteries eventually give rise to arterioles and then capillary beds within tissue. These perfuse the tissues before entering the venous system. The avian venous system cranial to the heart differs from the mammalian system in that there are two cranial vena cavae; the right jugular vein is larger than the left, and there is an anastomosis between the left and right jugular veins at the base of the head, allowing some of the venous blood to enter the larger right jugular vein (**Fig. 22.1b**).

Caudal to the heart, the major features of the venous system are the renal and hepatic portal systems. Blood returning from the legs and lower intestine enters the kidneys via the external iliac veins. If the renal portal valves (in the common iliac veins) are partially closed due to parasympathetic stimulation, blood is diverted to the cranial and caudal renal portal veins and perfuses the kidneys. The cranial renal portal veins empty into the caudal vena cava via the common iliac veins. Blood flowing through the caudal portal veins eventually returns to the heart via the internal vertebral venous sinuses and the caudal mesenteric vein. On the other hand, under sympathetic stimulation, the renal portal valves open, and blood from the external iliac veins enters the common iliac vein and then the caudal vena cava, bypassing the kidneys completely (**Fig. 22.1b**).

Blood returning from the legs and caudal intestinal tract in a resting bird can therefore enter either the cranial renal portal vein or the caudal renal portal vein, but not both at the same time. The control of this flow is not yet understood. Blood that enters the caudal renal portal veins will make its way into

DOI: 10.1201/9781003272519-22

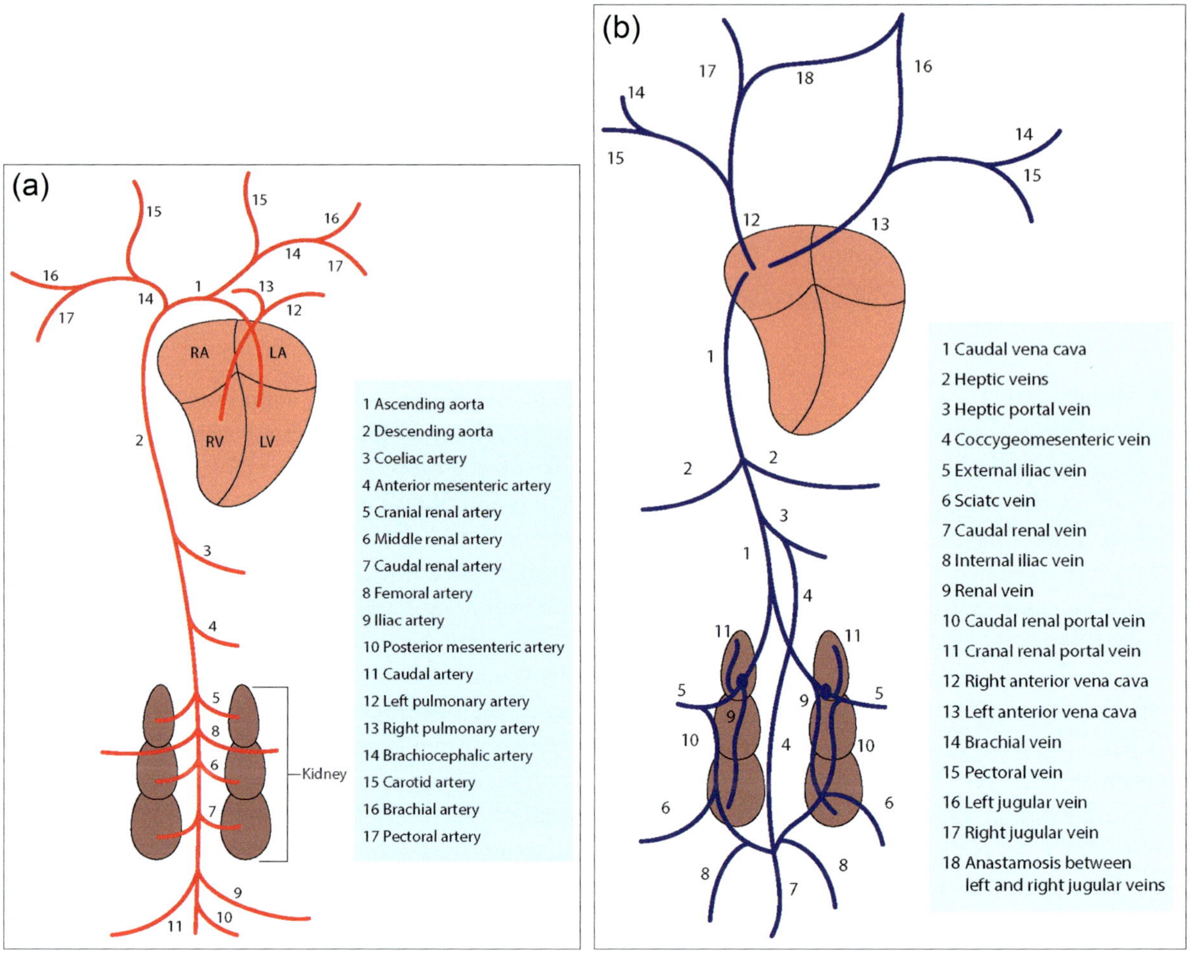

Figure 22.1 **Labelled diagrams of the major arteries (a) and minor veins (b) within the avian cardiovascular system.**

the caudal mesenteric vein and from there to the hepatic portal veins and the liver. The hepatic portal veins drain into the hepatic sinusoids, and from there to the hepatic veins. The right and left hepatic veins enter the caudal vena cava in the liver, which exits through the dorsal part of the right liver lobe to the heart.

PHYSIOLOGY

The cardiovascular system of any animal plays several major roles in the body: delivering oxygen and metabolites throughout the tissues of the body, removing metabolic waste products from these tissues, and assisting in thermoregulation. Birds have evolved a high-performance cardiovascular system to meet the demands imposed by flying, swimming, and running.

The heart is relatively larger than that of mammals and beats rapidly (usually greater than 200 beats/minute but capable in some species, e.g., hummingbirds, of reaching 1,000 beats/minute). This is achieved by having a larger numbers of myocytes with a small diameter allowing for more rapid depolarisation.

This combination of a larger heart with a fast rate gives a relatively greater cardiac output (mls/kg/minute) than mammals. When exercising (e.g., flying), birds increase their cardiac output by increasing their heart rate rather than their stroke volume (mls/beat). For example, a flying pigeon can increase its heart rate from 115 beats per minute to over 600 beats per minute. During this exercise, the stroke volume decreases marginally, but the overall result is that cardiac output is increased more than fivefold – an output up to seven times greater than that of a man or dog when all are at full exercise.

In turn, this high cardiac output requires a high arterial blood pressure (140–250 mm Hg) to adequately perfuse the capillary beds under a wide range of physiological conditions. This blood pressure is maintained by the baroreflex, driven by mechanoreceptors in the arterial walls of the aorta. The primary cardiovascular response to changes in blood pressure is to alter the cardiac output by increasing or decreasing the heart rate.

Much of the heat loss in many birds is through the distal leg and feet, regulated by the rate of blood flow to these areas. In cold conditions, intermittent brief periods of very great flow help to counter the effects of frostbite. In very hot conditions, the blood flow increases as well but allows radiant heat loss. This can increase during exercise by up to 10–15 times the amount of heat transfer compared to when at rest.

CARDIAC DISEASE

Overview

It is thought that somewhere between 10% and 40% of companion birds have some form of cardiovascular disease based on retrospective pathology studies. It is relatively rarely diagnosed antemortem, but as diagnostic techniques improve, an increase in the incidence of diagnosed cardiovascular disease can be expected.

Predisposing factors for cardiac disease include restricted exercise, poor nutrition, and keeping birds in a climate in which they did not evolve (e.g., keeping South American parrots in European winters).

Congestive heart failure (CHF)

Birds with cardiac disease often present with signs of congestive heart failure (CHF), the compensated condition associated with fluid retention that results from a sustained inadequacy of the cardiac output to meet the demands of the body. CHF is not a primary disease but the result of any structural or functional change in the heart. The causes are numerous and include endocardial, epicardial, myocardial, and combined diseases.

The pathophysiology of congestive heart failure involves both backward failure and forward failure. Backward failure involves increased atrial and venous pressure due to a failing ventricle, while forward failure involves decreased renal blood flow resulting in sodium and fluid retention. In response to low blood volume, renin is released from the juxtaglomerular cells of the kidney. Renin acts on circulating angiotensinogen to form angiotensin I, which is converted to angiotensin II. Angiotensin II stimulates aldosterone synthesis. Aldosterone causes sodium and fluid retention. Both mechanisms ultimately result in increased venous and capillary pressure, causing more fluid to escape by transudation into the interstitial spaces.

Clinical presentation

Pulmonary oedema predominates in isolated left-sided CHF, while right-sided CHF results in peripheral venous congestion, hepatic congestion, ascites, and pericardial effusion. A combination of all these signs may be present when both sides of the heart are affected.

Early-stage signs include periodic weakness, syncope, lethargy, and exercise intolerance. As the disease progresses, dyspnoea, coughing, and coelomic distension due to ascites and hepatomegaly are seen. Neurological signs, both central and peripheral, can be observed with poor circulation and perfusion.

Underlying cardiac diseases

Aetiology

Congenital heart disease

The diagnosis of intraventricular septal defects is becoming increasingly common, especially in umbrella cockatoos. Most are haemodynamically insignificant, but 2% will be associated with congestive heart failure. Blood is shunted from left to right, which leads to right ventricular failure and ascites secondary to valvular insufficiency. This is usually seen in birds aged 1–3 years. Duplicitas cordis, multiplicatis cordis, and ectopia cordis have all also been reported.

Endocardial disease

Vegetative endocarditis of the aortic and mitral valves may cause vascular insufficiency, lethargy, and dyspnoea. It is most common in birds with chronic infections (e.g., salpingitis, hepatitis, and bumblefoot). Commonly implicated bacteria include *Streptococcus* spp., *Staphylococcus* spp., *E. coli*, *Pasteurella* spp., *Pseudomonas aeruginosa*, and *Erysipelothrix rhusiopathiae*. Lesions consist of yellow irregular masses on any of the heart valves.

Myocardial disease

Myocarditis can occur secondary to viral, bacterial, mycotic, and protozoan infections (**Fig. 22.2**). Myocarditis has been reported with haemosiderosis, polyomavirus infection, chlamydiosis, PDD,

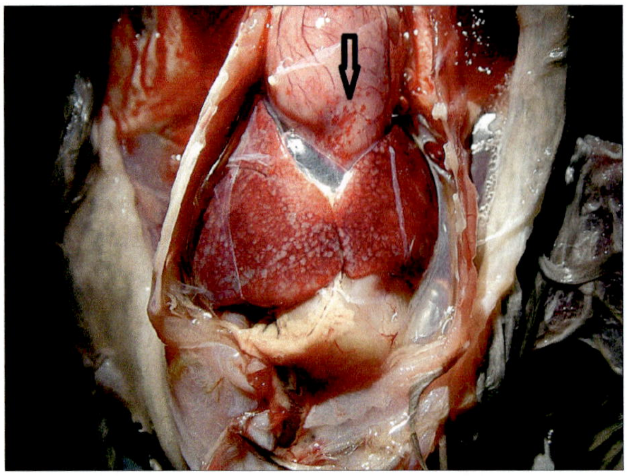

Figure 22.2 **Necropsy of an African grey parrot reveals myocardial haemorrhages (as indicated) due to septicaemia following a bacterial infection.**

Sarcocystis infections, pasteurellosis, and mycobacterial and *E. coli* infections.

Cardiomyopathies arise from similar mechanisms to those that occur in mammals (**Fig. 22.3**). Myocardial ventricular hypertrophy (**Fig. 22.4**) can occur with any condition that results in an elevated preload on the heart, including pulmonary hypertension and atherosclerosis. As an example, it has been noted in young broilers (4–6 weeks old) selected for high meat yield. The oxygen demand by the large muscle exceeds what the heart/lungs can provide. This results in polycythaemia, right ventricular dilatation/hypertrophy, valvular hypertrophy/insufficiency, and right-sided failure. Vitamin E and selenium deficiencies cause cardiomyopathy in gallinaceous birds. It may also cause myocardial and skeletal muscle degeneration in ratites less than 6 months old and myocardial degeneration in cockatiels.

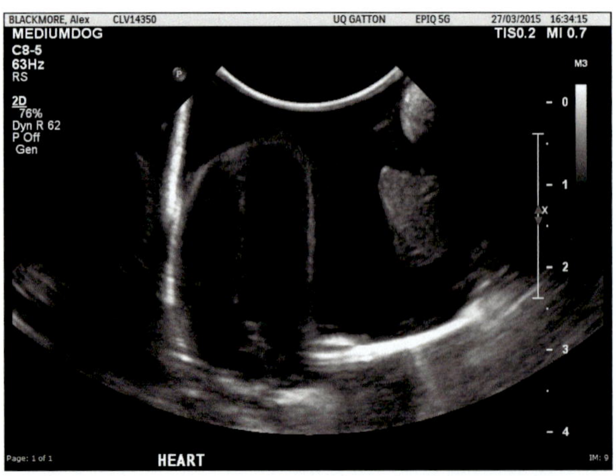

Figure 22.3 **Ultrasound of dilated cardiomyopathy in an Alexandrine parrot.**

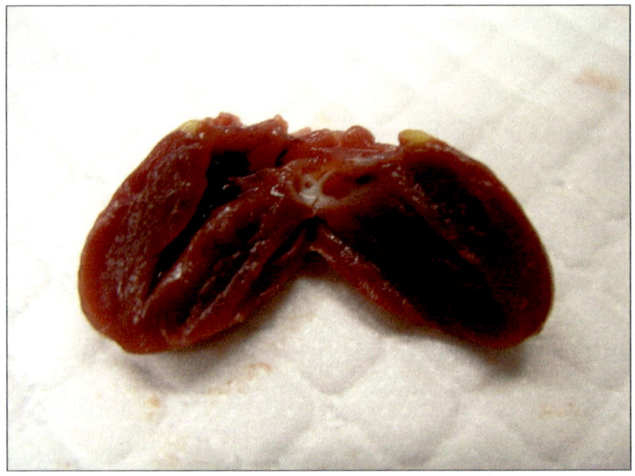

Figure 22.4 **Hypertrophic cardiomyopathy in a galah.**

Ruptures of the myocardium may occur secondary to degenerative, inflammatory, or neoplastic conditions of the myocardium or aneurysms of the myocardial vessels.

Myocardial infarctions are occasionally seen when an embolism originating from damaged muscle or valvular endocarditis lodges in a coronary blood vessel.

Myocardial damage has been reported with avocado poisoning and with visceral gout.

Pericardial disease

Pericardial effusion is a common finding in birds. It may result from cardiac or systemic disease and may be inflammatory (due to pericarditis or myocarditis) or non-inflammatory. Transudative effusions may be due to congestive heart failure or systemic conditions causing hypoalbuminaemia.

Fibrinous pericarditis may lead to adhesions of the epicardium to the pericardium and subsequently to constrictive pericarditis. Serofibrinous pericarditis is most common with bacterial (*E. coli*, *Streptococcus* spp., *Listeria*, *Salmonella*, *Chlamydophila*, *Mycoplasma*, *Mycobacterium*) and viral (reovirus in Galliformes and polyomavirus in parrots) infections (**Fig. 22.5**).

Visceral gout may lead to a non-infectious pericarditis (**Fig. 22.6**).

Severe pericardial effusion or restrictive pericarditis can compress the heart, resulting in impaired ventricular filling (diastolic dysfunction) and subsequent decreases in stroke volume and cardiac output. This condition (cardiac tamponade) can become life-threatening (**Figs. 22.7a, 22.7b**). Because the intramural pressure of the thinner-walled right ventricle is overcome more rapidly than that of the left,

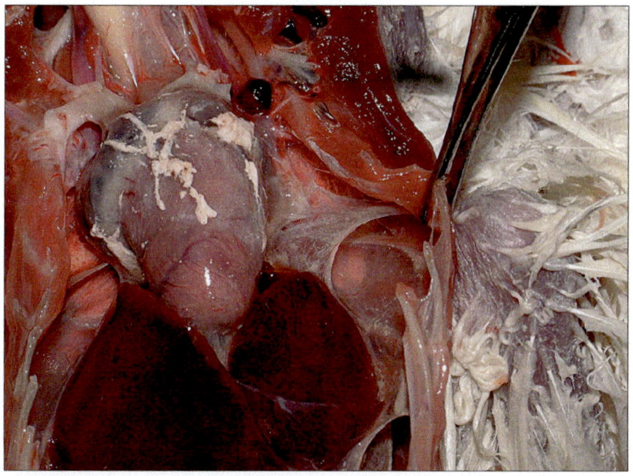

Figure 22.5 **Necropsy of a cockatoo reveals pericarditis and air sacculitis.**

Figure 22.6 **Necropsy of a scarlet macaw reveals visceral gout. Note that uric acid crystals have precipitated out over the pericardium, giving it a white appearance.**

cardiac tamponade results more quickly in right-sided CHF than in left-sided failure.

Arrhythmias

Arrhythmias can be normal (see later) but can range from clinically insignificant to life-threatening and/or terminal events.

Sinus bradycardia can be induced by vagal stimulation and can be converted by the administration of atropine. Various anaesthetics and sedatives (e.g., halothane, methoxyflurane, xylazine, and acepromazine) have been reported to cause sinus bradycardia when atropine is not given simultaneously. Hypothermia, which may accompany long-term anaesthesia, may potentiate this arrhythmia. Pathological conditions that may induce sinus bradycardia and sinus arrest include hypokalaemia, hyperkalaemia, thiamine deficiency, and vitamin E deficiency. Several toxins have been reported to induce bradycardia, including organophosphorous compounds and polychlorinated biphenyls. Reflex vagal bradycardia may occur when pressure is exerted on the vagus nerve by neoplasms, and space-occupying lesions impinging on the vagal nerve should be considered when unexplained atropine-responsive bradycardia is seen. Atrioventricular (AV) nodal escape rhythm has been noted in ducks with sinus bradycardia from hyperkalaemia.

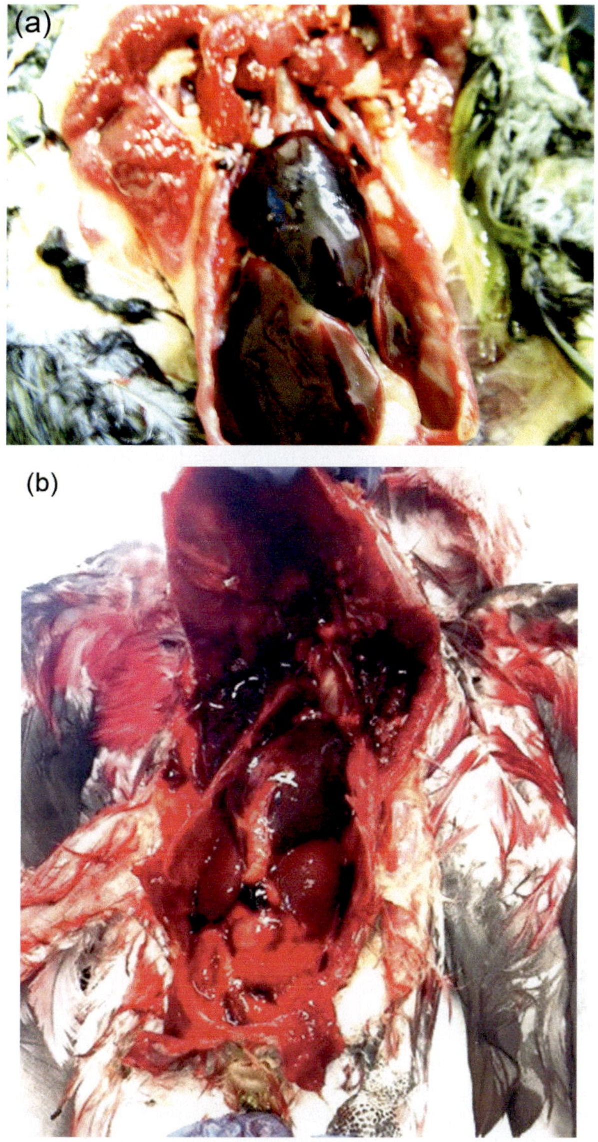

Figure 22.7 **(a and b) Traumatic haemopericardium causing pericardial tamponade.**

Atrial arrhythmias

- Sinus tachycardia is considered non-pathologic. The most common cause of sinus tachycardia is nervousness, but stress and pain may also precipitate the condition.

- Atrial tachycardias may be seen as a result of pathological conditions of the atrium. Sinus tachycardia has been reported in chickens infected with avian influenza virus. When the heart rate is rapid, the P-wave may be superimposed on the T-wave (P on T phenomenon). This has been recorded in 16% of normal Amazon parrots and in 6% of African grey parrots.
- Atrial fibrillation occurs when electrical impulses are generated in the atrium in a rapid and irregular way, and the atrium is in a state of permanent diastole. Impulses reach the AV node at a high frequency and at irregular intervals, and hence the ventricular rhythm is irregular. The ECG is characterised by the absence of P-waves, normal QRS complexes (which may have increased amplitude and duration because of ventricular hypertrophy), and irregular SS intervals.
- Supraventricular tachycardias may originate from the sinoatrial node (sinus tachycardia) or the junctional area (junctional tachycardia). Differentiation between the two can be accomplished by measuring the PP interval. This interval is perfectly equidistant in atrial tachycardia but may be irregular in sinus tachycardia due to vagal effects. Junctional tachycardias can be diagnosed by the presence of inverted P-waves in lead II.

Ventricular arrhythmias

- Ventricular premature contractions (VPCs) are characterised by QRS complexes that are unrelated to the P-waves. Bigeminy is a rhythm characterised by alternating normal beats and VPCs, while in trigeminy, two normal beats are followed by one VPC. VPCs in birds have been associated with hypokalaemia, thiamine deficiency, vitamin E deficiency, Newcastle disease and avian influenza viruses, myocardial infarction due to lead poisoning, and digoxin toxicity.
- Capture beats (normal P-QRS complexes in between VPCs) and fusion beats (a QRS complex intermediate between a normal P-QRS complex and a bizarre QRS complex that is formed by the simultaneous discharge of the ectopic ventricular focus and the normal sinoatrial node) are common with ventricular tachycardia.
- AV dissociation is a special form of ventricular tachycardia. The atrial and ventricular rhythms are independent of each other, whereby the atrial rate is lower than the junctional or idioventricular rate.
- VPCs, ventricular tachycardia, and ventricular fibrillation may occur during periods of hypoxia and with the use of halothane. Changes in the configuration of the T-wave and ST segment should alert the clinician that myocardial hypoxia is present and more severe ECG abnormalities are imminent.

AV node arrhythmias

- First-degree heart block occurs when the impulse through the AV node is delayed, resulting in an increased PR interval. This may be caused by halothane and xylazine, and the PR interval may increase to three to four times its normal value. This causes severe bradycardia that is reversible with atropine.
- Second-degree heart block occurs when some impulses do not reach the ventricles, but the majority of P-waves are followed by a QRS complex. Second-degree heart block Mobitz type 1 (Wenckebach phenomenon) has been reported as a physiological phenomenon in 5% of trained racing pigeons and is seen occasionally in asymptomatic parrots and raptors. In this form of AV block, the PR interval lengthens progressively until a ventricular beat is dropped.
- Third-degree AV block is characterised by independent activity of atria and ventricles, where the frequency of the atrial depolarisations is higher than that of the ventricular depolarisations. The ventricular complexes may have a normal configuration or may be wide and bizarre, depending on the site of ventricular impulse formation. Complete AV block has been seen in chickens with hypokalaemia.

Diagnosis

Physical examination

Auscultation of the heart is best performed over the left and right ventral sternum. Pleural or pulmonary fluid accumulation may cause muffled lung sounds or rales when a bird is auscultated over the back between the shoulder blades. Unique to birds, hepatomegaly can cause muffled heart sounds as the enlarged liver partly envelops the heart. Subtle murmurs may best be heard when a bird is anaesthetised. Arrhythmias may be detected.

Peripheral pulse palpation can be performed over the femoral and brachial arteries, while capillary refill can be assessed by pressing the basilic vein on the medial aspect of the elbow. The presence of jugular vein distension or pulsation is suggestive of poor venous return. Coelomic palpation should be performed to detect ascites and/or hepatomegaly. Finally, poor peripheral perfusion is suggested by cyanosis or pallor of the feet, mucous membranes, and facial skin.

Cardiac biomarkers

Cardiac assessment in human and small animal medicine now frequently uses biomarkers for diagnostic and prognostic purposes.

- Troponins are regulatory proteins that are part of the contractile apparatus of skeletal and cardiac muscle tissue. They are not present in smooth muscle tissue. The troponin complex consists of three interacting and functionally distinct proteins (troponin I, T, and C). Troponins are considered leakage markers. Damage to cardiac myocytes resulting in loss of membrane integrity causes the release of cardiac troponin I (cTnI) into the circulation. The severity of cardiac disease has been correlated with cTnI levels, while persistent elevations suggest irreversible myocardial damage. In dogs, cTnI levels begin to rise after 2 hours and peak in 12–24 hours.
- Brain natriuretic peptide (BNP) is a neuroendocrine hormone with a primary role in fluid homeostasis by increasing the glomerular filtration rate and inhibiting the renin–angiotensin–aldosterone system. It is found in myocardial tissue and released in response to atrial muscle stretch or injury. NT-proBNP is the metabolically inactive precursor to BNP and appears to be more suitable for analysis due to a longer half-life in plasma. NT-proBNP relies on renal excretion and may be elevated in non-cardiac diseases, such as renal failure.

The use of these cardiac biomarkers is an area of avian medicine waiting for further evaluation. Cardiac troponin I measurements appear to be potentially useful in birds, but until reference intervals in different species have been determined, caution is warranted in their interpretation.

Radiography

Radiographic detection of cardiovascular abnormalities may be difficult, although an enlarged cardiac silhouette or microcardia can often be visualised. The maximum width of the cardiac silhouette in a ventrodorsal radiograph should be (in most species) 35–45% of the length of the sternum, 51–61% of the width of the rib cage, and 545–672% of the width of the coracoid. See Chapter 8 for more detail.

Radiographic detection of an enlarged cardiac silhouette with muffled heart sounds is suggestive of pericardial effusion. An increased cardiac silhouette with normal heart sounds is suggestive of an enlarged heart. Microcardia is indicative of severe dehydration or blood loss that has resulted in hypovolaemia.

Other radiographic changes that suggest cardiac disease include congestion of pulmonary vessels, pulmonary oedema, pleural effusion, hepatomegaly, and ascites.

Echocardiography

Echocardiography is the imaging technique that generally provides the most diagnostic information. Probes with small coupling surfaces and frequencies of at least 7.5 MHz are necessary, and the ultrasound devices should be able to produce at least 100 frames/second. Bipolar echocardiography is used to examine the heart in the end-systolic and end-diastolic stages.

In small birds, the echocardiographic image of the heart is best obtained by sweeping through the liver. B mode provides slices of the heart and allows anatomical evaluation of cardiac structures. M mode makes perpendicular slices of the B mode, allowing for quantitation of cardiac dimensions and timing of cardiac events. Colour Doppler facilitates identification of turbulence associated with valvular insufficiencies, stenosis, and shunts.

Electrocardiography

This provides a graphic record of sequential electrical depolarisation–repolarisation patterns of the heart, detected on the surface of the body by changes in electrical fields (**Fig. 22.8**). Electrocardiography may be useful for the detection of cardiac enlargement from hypertrophy of any of the four cardiac chambers. It is indispensable for the diagnosis and treatment of cardiac arrhythmias and is also useful in monitoring changes in electrolyte concentrations during the treatment of metabolic diseases that alter electrolyte balance. Because the myocardium is very

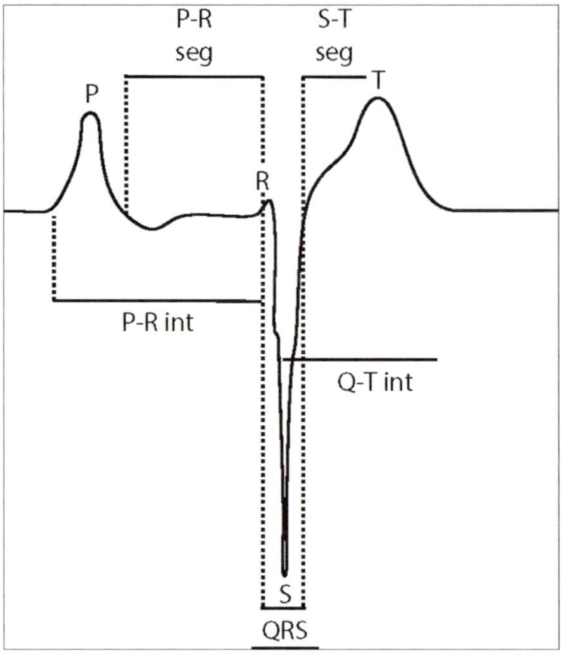

Figure 22.8 This lead II electrocardiogram provides a graphic record of sequential electrical depolarisation–repolarisation patterns of the heart, detected on the surface of the body by changes in electrical fields: • The P-wave signifies that the atria have depolarised, causing contraction and ejection of their complement of blood into the ventricles. • The PR segment indicates the short delay in the atrioventricular node that occurs after the atria contract, which allows complete filling of the ventricles before ventricular contraction occurs. • The QRS complex represents ventricular depolarisation and contraction with the ejection of blood into the aorta and pulmonary artery. The Q-wave is the first negative deflection, the R-wave is the first positive deflection, and the S-wave is the first negative deflection following the R-wave. • The ST segment and T-wave depict the repolarisation of the ventricles. In clinically asymptomatic parrots, the ST segment may be very short or even absent, the S rising directly into the T-wave ('ST-slurring').

sensitive to hypoxia, electrocardiography can be a reliable indicator of oxygenation.

Leads are attached to both wings and the left leg. In large or well-muscled birds, one may need to place the V electrode to the right or left of the sternum for greater amplitudes of all complexes. Needle electrodes placed subcutaneously are superior to alligator clips for use in avian patients. The Biolog® hand-held ECG machine has been used with one foot (lead) on the chest, one on the right wing, and one on the right thigh. QRS complexes were larger than with standard machines, but they were comparable.

Different lead systems are used to record differences between various limbs:

- Standard lead system leads I, II, III.
- Unipolar or augmented lead system: avR, avL, avF.
- Unipolar chest lead: V1–V10.

The sequence of depolarisation and repolarisation is recorded:

- Atrial depolarisation: P-wave (the P-wave is positive on I, II, III, and avF, and is negative on avR).
- Ventricular depolarisation: QRS complex. This is negative in I, II, III, and avF; there is no Q-wave in I, II, and III (the negative mean electrical axis of ventricular depolarisation in birds occurs because the depolarisation wave begins sub-epicardially and then spreads through the myocardium towards the endocardium. Ventricular depolarisation of the canine ventricles starts sub-endocardially).
- Ventricular repolarisation: T-wave (positive on II, III, and avF; negative on avR and avL; ± or absent on I).
- T-P frequently overlap.

The paper needs to run at 50 mm/second or greater, as heart rates can vary from 100 to 1,000 beats per minute (bpm) (e.g., budgerigars: 600–750 bpm; cockatiels: 496–604 bpm). Some authors state paper speed needs to be at least 100 mm/s. For smaller species, paper speed may need to be set at 200 mm/second.

Determination of heart rate: The top of the ECG paper has a series of marks. The marks are 75 mm apart. The number of complexes that occur in three seconds is counted and multiplied by 20 to get beats/minute.

Determination of heart rhythm:

- Is the heart rate normal or abnormal for the species?
- Is the heart rhythm regular or irregular?
- Is there a P-wave for every QRS complex, and is there a QRS complex for every P-wave?
- Are the P-waves related to the QRS complexes?
- Do all the P-waves and all the QRS complexes look alike?

Determination of mean electrical axis:

- Find an isoelectric lead.
- Use the six-axis reference system chart and find which lead is perpendicular to the isoelectric lead.
- Determine if the perpendicular lead is positive or negative on the tracing and examine the angle value on the six-axis reference system. Compare these values with reference values.
- The average mean electrical cardiac axis is between 83° and –162°.

Measurements are all made on the lead II rhythm strip:

- P-wave. This signifies that the atria have depolarised, causing contraction and ejection of their complement of blood into the ventricles. With right atrial hypertrophy, the P-wave becomes tall and peaked (P pulmonale), and with left atrial hypertrophy, the P-wave becomes too wide (P mitrale). P pulmonale has been associated with dyspnoea induced by aspergillosis or tracheal obstruction. A tall, wide P-wave is suggestive of biatrial enlargement and is common with the influenza virus in gallinaceous birds.
- PR interval. This indicates the short delay in the AV node that occurs after the atria contract, which allows complete filling of the ventricles before ventricular contraction. In the normal pigeon ECG, a small T_a-wave can be seen in the PR segment, indicating repolarisation of the atria. This may also occur in some asymptomatic parrots and gallinaceous birds. This finding is considered normal.
- QRS complex. This represents ventricular depolarisation and contraction with the ejection of blood into the aorta and pulmonary artery. The Q-wave is the first negative deflection, the R-wave is the first positive deflection, and the S-wave is the first negative deflection following the R-wave. When there is no R-wave, the negative deflection is called a QS-wave. The largest wave in the QRS complex is depicted with a capital letter (i.e., Rs or rS). Two measurements are made on the QRS complex. The duration is measured from the beginning of the R-wave to the end of the S-wave. The second is the amplitude of the S-wave, measured from the baseline downwards. Low voltage ECGs often occur in birds with pericardial effusion. A QRS complex that is too wide or too tall indicates left ventricular hypertrophy or a bundle branch block. Prominent R-waves are suggestive of right ventricular hypertrophy.
- ST segment. Along with the T-wave, this depicts the repolarisation of the ventricles. In clinically asymptomatic racing pigeons and parrots, the ST segment is often very short or even absent. When present, it may be elevated above the baseline, which should not be interpreted as a sign of myocardial hypoxia, as it is in the dog.
- T-wave. In the normal avian ECG, this is always in the opposite direction to the main vector of the ventricular depolarisation complex, and always positive in lead II. All measurements are made in lead II. When the T-wave changes its polarity, it suggests that myocardial hypoxia is occurring. The same is true for a T-wave that progressively increases in size (e.g., during anaesthesia). T-wave changes may also occur in association with electrolyte changes (e.g., increased T-wave amplitude with hyperkalaemia).
- QT interval. Prolongation might be associated with electrolyte disturbances (hypokalaemia, hypocalcaemia) and sedation, possibly due to a decreased heart rate. The QT interval may be prolonged in African grey and Amazon parrots.

NORMAL ARRHYTHMIAS

Sinus arrhythmia, sinus arrest, sinoatrial block, and wandering pacemaker have been reported in association with normal respiratory cycles and are considered physiological in birds.

The normal rhythm of the heart is established by the sinoatrial node. A normal sinus rhythm does not vary in rate from beat to beat. An increase in vagal activity may decrease the heart rate, while a decrease in vagal activity may increase the heart rate. The heart rate may increase during inspiration and decrease during expiration, and hence the SS interval may not be equidistant. The associated rhythm is called sinus arrhythmia.

Sinus arrest is an exaggerated form of sinus arrhythmia and can be diagnosed if the pause is greater than twice the normal SS interval. Sinoatrial block occurs when an electrical impulse from the sinoatrial node fails to activate the atria. The pauses are exactly twice the SS interval. Wandering pacemaker refers to a continuous shifting of the pacemaker site in the sinoatrial node or the atrium, and an alteration in P wave morphology.

Blood pressure

The determination and monitoring of blood pressure has recently been recognised as a valuable diagnostic tool in the avian patient. Systolic blood pressure is the pressure exerted against the blood vessel wall during systole.

Direct arterial pressure measurement requires specific skills and is both invasive and expensive. Consequently, indirect blood pressure measurement techniques, based on the detection of blood flow beneath an inflated cuff, are more commonly used but have not been found to correlate well with direct blood pressure measurements. The ultrasonic Doppler flow detector (Parks Medical Electronics Inc.) uses ultrasonic waves to detect arterial blood flow distal to a blood pressure cuff and produce an audible signal.

An appropriately sized cuff (depending on the size of the bird) is placed around the distal humerus or femur and taped in place. The Doppler probe is then placed above the radial carpal bone or tibiotarsal bone in order to detect arterial blood flow and secured in place. The cuff is inflated until the Doppler signal is lost, and then slowly deflated. The pressure on the cuff when the first sound is heard is marked and recorded as the systolic pressure (see Chapter 31).

Management

Congestive heart failure

As mentioned earlier, congestive heart failure is a combination of inadequate cardiac output and an increased preload. In left heart failure, an increase in the pressure in the left atrium and the pulmonary veins leads to pulmonary oedema. In birds with right heart failure, ascites, liver congestion, and hydropericardium are commonly seen. Treatment, therefore, revolves around improving the contractility of the heart, reducing afterload, and removing excessive retained fluid in the lungs or liver/coelom.

Contractility is improved by the use of ionotropes. The most commonly used drug is digoxin (0.02–0.05 mg/kg q24h). This drug is indicated for myocardial dysfunction, chronic mitral insufficiency and chronic volume overloads. However, it is contraindicated for hypertrophic cardiomyopathy, ventricular tachycardia, and sinus or AV node disease. Adverse effects relate to myocardial toxicity; therefore, patients should be monitored for clinical improvement and via ECG for prolonged PR time. Serum concentrations should be measured after 1 week of therapy or 1 week after the dose is changed. The dose can be increased if the serum concentration is <0.8 ng/l 8–10 hours after dosing.

Another drug that is showing promise in the treatment of congestive heart failure in birds is pimobendan. This drug acts as a positive inotrope (increased cardiac contractility), a positive lusitrope (improved ventricular relaxation), and a peripheral vasodilator. In small animal medicine, it has a wider safety margin than digoxin but has yet to be extensively trialled in birds. Recent research has suggested that the dose rates used in small animals (0.25 mg/kg q24h) may not be adequate, with suggested doses for birds of 6–20mg/kg q12–24h being recommended.

Excessive neurohormonal activation is blocked by angiotensin-converting enzyme (ACE) inhibitors (e.g., enalapril 1–5 mg/kg q12h). ACE inhibitors block the formation of angiotensin II, thereby blocking the renin–angiotensin–aldosterone system. They also have a diuretic effect. If the bird is on concurrent diuretic therapy, it may be best to start the ACE inhibitor at a low dose while reducing the other diuretic therapy and then gradually increase the dose of the ACE inhibitor. Side effects include hypotension, reflex tachycardia, gastrointestinal disorders, renal dysfunction, and hyperkalaemia.

Signs of fluid retention are treated with diuretics such as furosemide (0.15–2 mg/kg q12–24h) or spironolactone (1–13 mg/kg PO q8–12h). Overuse or high doses of diuretics can reduce the blood volume and trigger the renin–angiotensin–aldosterone system. Therefore, it is necessary to use the lowest effective dose.

Cardiac arrhythmias

Treatment is not required unless the patient is clinically affected by the arrhythmia. The underlying cause should be treated.

- Supraventricular tachycardia:
  - Digoxin
  - Beta blockers (e.g., propranolol 0.2 mg/kg q24h)
  - Calcium channel blockers
- Ventricular tachycardia:
  - Beta blockers

  - Lignocaine (2.5–5.0 mg/kg IV)
- Bradycardia and AV blocks; limited therapy is available:
  - Atropine (2.5 mg/kg IV) may be of some benefit
  - Glycopyrrolate (0.01–0.02 mg/kg IM)
  - Ephedrine 0.5 mg/kg q8h

Pericardial effusions

Diuretics are usually contraindicated because of the severe hypotension seen in many patients with reduced cardiac preload. If used, the bird should be monitored carefully for signs of hypotension (including weakness, ataxia, and collapse). Furosemide, with or without an ACEI, may be useful to reduce low volumes of pericardial fluid. Endoscopic pericardial fenestration or ultrasound-guided pericardiocentesis may give temporary relief, but the effusion may return when the fenestration heals.

ATHEROSCLEROSIS

Definition/overview

Atherosclerosis is a diffuse or local degenerative condition of the internal and medial tunics of the wall of arteries, causing chronic inflammatory fibroproliferative vascular disease. The degenerative changes include the proliferation of smooth muscle cells and the deposition of collagen, proteoglycans, and cholesterol. As tissue damage progresses, dystrophic mineralisation occurs in the wall of the blood vessels. The resultant plaque protrudes into the vessel lumen and retards blood flow (**Fig. 22.9**), increases surface tension, and may result in embolism formation (although rare in birds), aneurysm, and possible vessel rupture (**Fig. 22.10**). These changes are most commonly seen in the aorta, the brachiocephalic trunks, and the pectoral and carotid arteries, but any blood vessel can be affected. The resulting clinical signs are caused by ischaemia, arterial aneurysm, stenosis, and cardiac consequences (e.g., congestive heart failure and arrhythmias). Atherosclerosis is likely an underlying factor in most non-infectious cardiovascular diseases in pet birds.

Risk factors are an individual's measurable characteristics that predict the probability of developing clinical disease. With atherosclerosis, the risk factors appear to be species, age, sex, diet, activity, and plasma cholesterol. Many avian orders may be affected, but Psittaciformes and Anseriformes appear to be particularly susceptible. Within the parrot family, Amazon parrots, Grey parrots, and cockatiels seem to be particularly prone, whereas cockatoos and macaws appear to be uncommonly affected. Other risk factors appear to include age (most affected birds are more than 8–15 years old; birds aged 20–30 years have a dramatically higher incidence), sex (females are more prone to severe lesions than males, possibly because of vitellogenesis and the mobilisation of lipids from the liver under the influence of oestrogen), high plasma cholesterol and triglyceride levels, lack of exercise, and high-fat diets (i.e., seed). It is unclear if chlamydial infections create an additional risk factor.

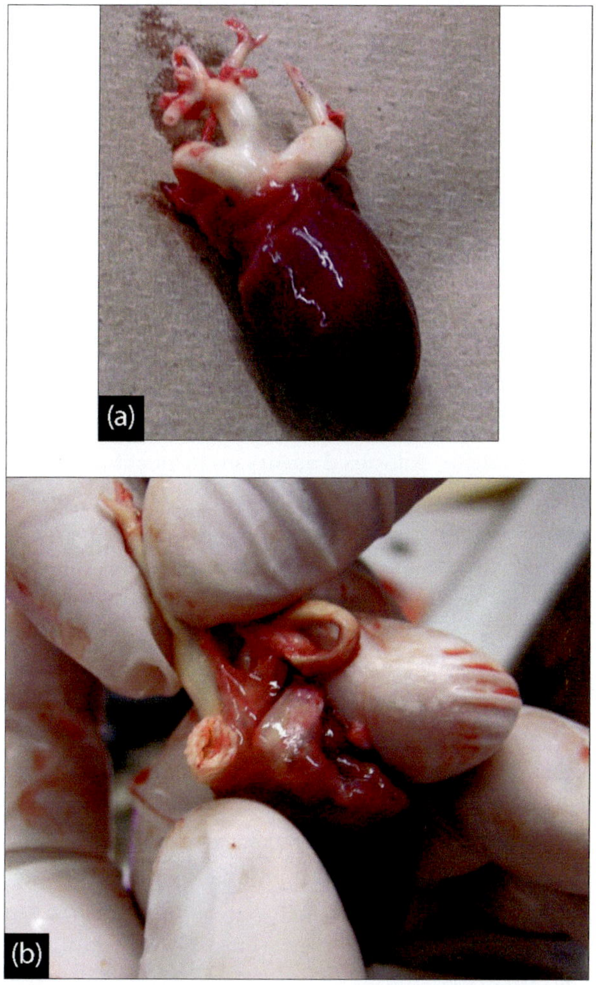

Figure 22.9 **An excised atherosclerotic heart from an African grey parrot shows a proliferation of smooth muscle cells and the deposition of collagen, proteoglycans, and cholesterol in the aorta.**

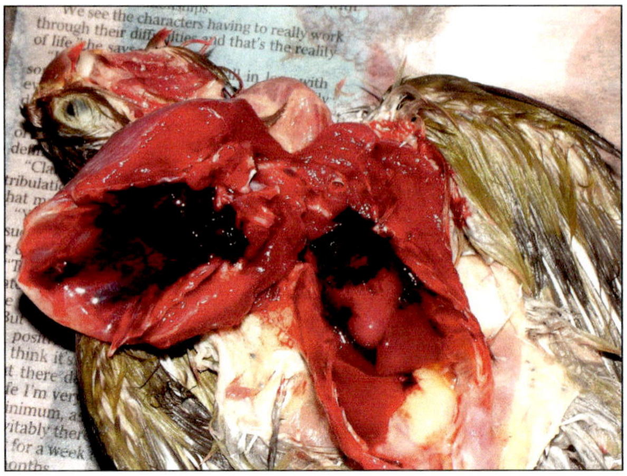

Figure 22.10 **Ruptured aneurysm in an elderly macaw.**

Aetiology

The accumulation of pathogenic material in the arterial wall has been explained by the insudative theory. Normally, a transfer of plasma proteins occurs through the arterial wall, with subsequent removal from the outer coats by lymphatic vessels. During this process of permeation, fibrinogen and very low-density lipoproteins are selectively entrapped in the connective tissue of the arterial wall. Their presence stimulates reactive changes that give rise to the production of atherosclerotic lesions. Variations in vascular permeability and arterial blood pressure can explain the preference for atherosclerotic lesions in certain areas.

What triggers these changes is still the subject of debate, but it may include genetic predisposition in some birds, hyperlipaemia, endothelial inflammation, toxins, immune complexes, and hypertension. The role of cholesterol and lipoproteins remains unclear.

Clinical presentation

Clinical signs associated with atherosclerosis in avian species can vary depending on the severity of lesions, location of lesions, and concurrent disease. Clinical signs are attributed to advanced lesions, whereas early and intermediate lesions are generally subclinical. Atherosclerosis is rarely diagnosed antemortem in birds, and the condition is often associated with sudden death. Subtle and intermittent signs include dyspnoea, weakness, signs of congestive heart failure (see earlier), and neurological signs (e.g., ataxia, seizures, tremors, and paresis of one or both pelvic limbs). Some birds show reduced mentation, blindness, anisocoria, and vestibular signs.

Diagnosis

Atherosclerosis is difficult to diagnose antemortem in birds. Clinical signs, knowledge of known risk factors, and the application of currently available diagnostic tests may be of some value.

There may be elevated plasma cholesterol, triglycerides, and low-density lipoprotein levels, but this finding is not consistent and may be confounded by the effects of female reproductive physiology. High total blood cholesterol and triglyceride levels are considered major risk factors for atherosclerosis. Lipoproteins – low-density lipoprotein (LDL) and high-density lipoprotein (HDL) – may not be as useful in birds as they are in mammals. Most birds transport cholesterol mainly in the form of HDL (unlike humans, where LDL is predominant). Therefore, higher HDL values may not be associated with decreased risks of developing atherosclerosis as it is in humans – in fact, the converse may be true.

Cardiac biomarkers, discussed earlier, may assist in detecting cardiac disease associated with atherosclerosis, but it may be already advanced by the time this methodology detects a problem.

Radiography offers a useful but somewhat insensitive screen for atherosclerosis. Atherosclerosis targets the major arteries (brachiocephalic arteries, pulmonary arteries, and ascending and abdominal aorta) and these arteries are visible on plain radiographs. Although insensitive, radiographic detection of arterial calcification is fairly specific to advanced atherosclerotic lesions (**Fig. 22.11**).

Angiography, using fluoroscopy or CT, may detect narrowing of arterial lumens and may be the preferred means of testing for atherosclerosis.

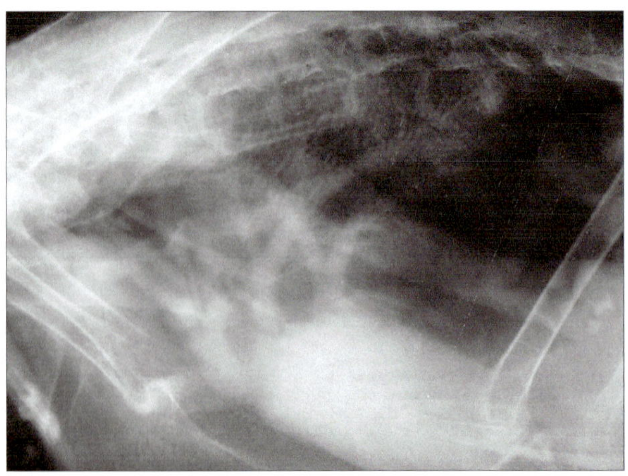

Figure 22.11 **Radiograph of an elderly cockatoo showing signs of atherosclerosis. Note the evidence of arterial calcification, which is fairly specific to advanced atherosclerotic lesions.**

Management, treatment, and prevention

Atherosclerotic lesions cannot be resolved, but diet, husbandry, and lifestyle changes may help to prevent, slow progression, or decrease the size of the lesions. Treatment is centred on managing the risk factors and medical therapy to manage/treat the clinical effects of the disease.

Managing risk factors

Identifying and minimising the known risk factors may assist in the prevention of atherosclerosis. Controlling and treating reproductive problems such as chronic egg laying and oophoritis, as well as dietary management and increasing physical activity, may have beneficial effects on a bird before atherosclerosis develops or becomes advanced.

Other risk factors, such as hypercholesterolaemia and hypertriglyceridaemia, may require medical management:

- Statins. A variety of statins (e.g., atorvastatin, rosuvastatin) have been advocated to lower cholesterol levels, but none appears to be consistent in their efficacy, nor have their pharmacodynamics been determined. Target levels of blood cholesterol and LDL that would reduce atherosclerotic risks have not been determined
- Omega-3 and omega-6 fatty acids. Omega-3 and omega-6 fatty acids show promise as a means of decreasing cholesterol levels in both mammals and birds. The ratio of omega-3 to omega-6 (3:1) is more important than absolute levels. Flax seed oil is a suitable source of omega-3 fatty acids for birds.

Treating cardiovascular disease

Medical management of associated cardiovascular disease focuses primarily on improving peripheral perfusion. Medical management of congestive heart failure, as discussed earlier, may also be appropriate if clinical signs are present.

Clinical signs of peripheral arterial disease have been anecdotally treated with isoxsuprine, ACEIs, and pentoxifylline.

- Isoxsuprine is a peripheral vasodilator that causes vascular smooth muscle relaxation predominantly through an α-adrenoceptor blockade. Benefits may be obtained when confirmed (or suspected) cases are treated at 10–15 mg/kg PO q12h.
- ACEI, including enalapril, results in vasodilation by blocking the formation of AII. (AII promotes vasoconstriction and venoconstriction by mediating the release of catecholamines, which act on the vascular smooth muscle via α-adrenergic receptors.) Enalapril has been used at 1.25–5 mg/kg PO q8–12h and may be synergistic with isoxsuprine.
- Pentoxifylline has been used for the treatment of peripheral vascular and cerebrovascular disease by improving microcirculatory blood flow. It appears to increase tissue perfusion, mitigate inflammation, and attenuate atherosclerotic plaque formation. It has been used at doses of 15–25 mg/kg PO q8–12h.

Beta blockers (e.g., propranolol, oxprenolol) may have a protective effect against the development of atherosclerotic plaques. Verapamil, nifedipine, and diltiazem (calcium channel blockers) may decrease the extent and distribution of atherosclerotic lesions. However, the lack of accurate means of measuring arterial blood pressure in clinical cases to both diagnose and monitor the response to treatment makes these therapies difficult to justify at this time.

HYPERTENSION

Definition/overview

Blood pressure is determined by cardiac output and systemic vascular resistance. Cardiac output is determined by heart rate, preload, and myocardial contractility. Tachycardia and increased myocardial contractility usually do not result in hypertension, as there is a corresponding decrease in systemic vascular resistance. Acute fluid overload can trigger a hypertensive episode.

The common denominator in the majority of hypertensive patients is an increase in systemic vascular resistance. This increase is mediated by increased levels of circulating catecholamines, increased α-adrenergic activity, and activation of the renin–angiotensin–aldosterone system.

Aetiology

The underlying factors that contribute to hypertension in parrots have not been well studied and are often not determined. Primary or essential hypertension, although common in people, has only been occasionally documented in dogs. Its occurrence in other species, including birds, is unknown. It is believed to be due to excessive catecholamine release caused by chronic stress.

Renal failure is relatively common and is associated with severe hypertension (through activation of the renin–angiotensin–aldosterone system) in dogs and cats, particularly in patients with protein-losing nephropathy (glomerular nephritis). The association between protein-losing nephropathy and hypertension in birds has not yet been documented, although glomerular nephritis is commonly found. Renal tumours may induce hypertension (through activation of the renin–angiotensin–aldosterone system) and are common in parrots. An association between renal tumours and hypertension has been documented in parrots.

Atherosclerosis, as discussed earlier, is another risk factor for the development of hypertension.

Clinical presentation

Signs include ocular lesions (retinal haemorrhages, detachments, and/or papilloedema), neurological signs (depression, confusion, ataxia, and seizures), cardiovascular signs (associated with ventricular hypertrophy and congestive heart failure), and renal system abnormalities (associated with glomerulonephritis).

Diagnosis

The difficulty in diagnosing this disease in parrots lies in the difficulty in reliably measuring arterial blood pressure, making diagnosis (and even defining normal blood pressure) difficult. Diagnosis is based on clinical signs and indirect (Doppler) blood pressure measurement. Although this measurement is not reliable in parrots, direct arterial blood pressure measurement is invasive and rarely clinically practical. Consistently elevated indirect blood pressure greater than 270 mm Hg may be suggestive of hypertension in parrots.

Treatment

In patients without hypertension-related disease, treatment should be initiated cautiously with the goal of reducing blood pressure by 25% over several weeks. In patients with acute, severe hypertension-related disease, a rapid reduction in blood pressure may be necessary.

Left untreated, hypertension can result in CHF and ventricular hypertrophy. In other species, hypertension can be treated with a variety of drugs such as beta blockers, ACE inhibitors, calcium channel blockers, and diuretics. In the small number of parrots diagnosed and treated at the time of writing, the use of amlodipine (0.1–0.4 mg/kg PO q24h) and enalapril (0.3–0.5 mg/kg q12h PO) has been shown to gradually lower the patient's blood pressure.

FURTHER READING

Beaufrère H (2013) (ed) Atherosclerosis and vascular medicine. *Journal of Exotic Pet Medicine* 22(4): 315–420.

Beaufrère H, Cray C, Tully T (2014) Association of plasma lipid levels with atherosclerosis prevalence in Psittaciformes. *Journal of Avian Medicine and Surgery* 28(3): 225–231.

Fitzgerald BC (2022) Cardiovascular diseases in pet birds. The Veterinary Clinics of North America. *Exotic Animal Practice* 25(2): 469–501.

Hanley SH, Helen GM, Torrey S (1997) Establishing cardiac measurement standards in three avian species. *Journal of Avian Medicine and Surgery* 11: 15–19.

Johnson HJ, Phalen DN, Kondik VH, Tippit T, Graham DL (1992) Atherosclerosis in psittacine birds. In: *Proceedings of the Annual Conference of the Association of Avian Veterinarians Australian Committee*, pp. 87–93.

Lichtenberger ML, Orosz S (2008) Cardiology: From anatomy to treatment. In: *Proceedings of the Annual Conference of the Association of Avian Veterinarians Australian Committee*, pp. 237–249.

Pees M, Krautwald-Junghanns ME, Straub J (2006) Evaluating and treating the cardiovascular system. In: *Clinical Avian Medicine, Vol 1.* GJ Harrison, TL Lightfoot (eds). Spix Publishing Inc, Palm Beach, pp. 379–394.

Wells SM, Sleeper M (2008) Cardiac troponins. *Journal of Veterinary Emergency and Critical Care* 18(3): 235–245.

CHAPTER 23

DISORDERS OF THE RESPIRATORY TRACT

ANATOMY AND PHYSIOLOGY

Respiratory tract

The size and shape of the external nares (nostrils) are variable between species. They are located at the top of the beak (except in the kiwi, where they are located at the tip of the beak). In parrots, pigeons, and owls) they are located within the fleshy cere. Just inside the nares is the operculum, a cornified flap of tissue.

The nasal septum is partly bony and partly cartilaginous. It is complete in parrots and many other species and separates the nasal cavity to the level of the choana. Within the nasal cavity are the nasal conchae (turbinates), usually divided into three parts: the scroll-like highly vascular rostral turbinates with their stratified squamous epithelial lining; the middle turbinates, also scroll-like but with a mucociliary lining; and the caudal turbinates with an olfactory epithelial lining innervated by the olfactory nerve. The vascularity of the turbinates assists in controlling the rate of water and heat loss from the body. The infraorbital sinus network is connected to the nasal cavity in its middle and caudal regions. However, the connections between the nasal cavity and the sinuses are such that it is difficult to sample the sinuses by a nasal flush.

The infraorbital sinuses are located around the eye, the upper beak, the mandible, and the pneumatised sections of the skull (**Fig. 23.1**). They are classified as rostral (in the maxillary rostrum or bill), periorbital/preorbital (rostral to the orbit), infraorbital (medial to the eye), mandibular (mandibular rostrum), and postorbital (surrounding the opening of the ear). The right and left sides communicate and have diverticula in Anseriformes, Psittaciformes, and insectivorous Passeriformes. They do not communicate in non-insectivorous Passeriformes.

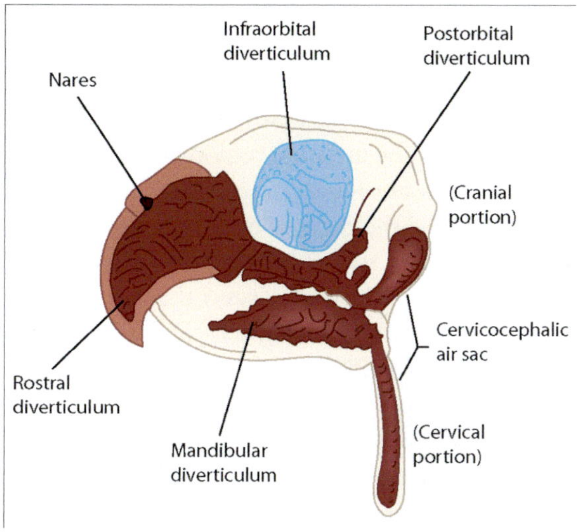

Figure 23.1 **Labelled diagram of the sinus network within a psittacine skull.**

Each infraorbital sinus has five diverticula and two chambers:

- Rostral diverticulum. Within the beak, surrounded by premaxillary bone.
- Maxillary chamber. Underneath the nares, lateral to the rostral diverticulum, extending back to the preorbital diverticulum.
- Preorbital diverticulum. In the space between the nares and the rostral aspect of the orbit. It is bordered medially by the nasal cavity and dorsolaterally by the nasal and frontal bones. It connects to the suborbital chamber.
- Suborbital chamber. Beneath the bony orbit. It is bordered medially by the caudal extent of the nasal cavity and laterally by the jugal and prefrontal bones.
- Infraorbital diverticulum. Above the suborbital chamber, extending behind the suborbital arch and medial to the eye. It is the largest diverticulum.

DOI: 10.1201/9781003272519-23

- Postorbital diverticulum. This has two parts: the caudal portion is the preauditory diverticulum, extending caudally to the bony orbit and bound by the temporal and quadrate bones laterally; and the cranial portion is immediately caudal to the orbit.
- Mandibular diverticulum. Communicating with the maxillary portion near the postorbital diverticulum. Occupies the mandibular bone.

In addition, the cervicocephalic air sac communicates with the most caudal aspect of the infraorbital sinus. This air sac does not play a role in gas exchange, nor does it communicate with the lower respiratory tract. It has two divisions: the cranial cephalic (from the occipital region to just behind the cere), which is not found in many species, including the macaw; and the cervical (from the tympanic area and extending in two columns bilaterally down the neck). It is thought to play several roles: insulation for heat retention, control of buoyancy, reducing the force of impact with water in fish-eating birds, and support of the head during sleep or flight. A jugular venepuncture may result in blood entering this air sac and appearing as epistaxis.

The upper respiratory tract serves several functions: a sense of smell, filtration of airborne debris, thermoregulation, and water conservation.

Lower respiratory tract

Birds do not have a true larynx as such but rather have a glottis. This has no role in voice production; its main role is preventing food from passing into the trachea. There are four cartilage structures in the glottis: the cricoid, a scoop-shaped cartilage with left and right lateral wings; the procricoid, a small cartilage that articulates with the cricoid wings on the dorsal midline; and the two arytenoids that form the margins of the glottis. The glottis is in the laryngeal mound behind the tongue. During inspiration, the glottis is raised to the choana and opened, allowing air to be inspired without opening the mouth.

Unlike mammals, the avian trachea is composed of complete cartilaginous rings, each shaped like a signet ring, with the broad part forming the left and right walls alternately. These rings, therefore, partially overlap each other. They may be ossified in passerines and some larger species. The tracheal lumen diameter progressively reduces caudally but is still larger than that of a comparable mammal: the typical avian trachea is 2.7 times longer and 1.29 times wider than that of comparably sized mammals. This means that the avian tracheal dead space is 4.5 times greater than that of comparably sized mammals. Birds compensate for this with a low respiratory frequency (one-third of that of mammals) and an increased tidal volume (four times that of mammals). It is important to understand this anatomy when intubating birds. Non-cuffed tubes are preferable, but if cuffed tubes are used, they should not be inflated, and smaller tubes are necessary to prevent iatrogenic trauma to the tracheal lining caudal to the glottis.

Most companion birds will have a tracheobronchial syrinx: the last of the tracheal rings fuse into a syringeal box, which joins to the first of the bronchial rings. (There are also tracheal and bronchial syrinxes in other species, e.g., storks and owls.) The syrinx consists of a number of variably ossified cartilages and vibrating soft structures.

The syringeal cartilages consist of:

- The tympanum, a direct continuation of the trachea, formed by the fusion of several tracheal rings. It is commonly ossified.
- The tracheal syringeal cartilages, C-shaped flattened cartilages attached to the pessulus at one end and free at the other.
- The pessulus, a wedge-shaped cartilage with the blade lying dorsoventrally, dividing the airway vertically.
- The bronchial syringeal cartilages, three to five paired C-shaped rings forming the divided part of the syrinx.

The vibrating structures of the syrinx include:

- The paired medial tympaniform membranes that form the medial surface of the divided part of syrinx, held between free ends of bronchial syringeal cartilages.
- The paired lateral tympaniform membranes that form the membranous areas between the cartilages on the lateral aspect of the syrinx.
- The lateral labium, a pad of elastic tissue projecting into the lumen of the syrinx from the cartilage of the lateral wall.
- The medial labium that projects into the lumen from the pessulus.

The syrinx is controlled by both intrinsic and extrinsic muscles. The number of intrinsic muscles varies between species: songbirds have five pairs, parrots have only two pairs, and some ratites and Galliformes have none at all. There are three sets of extrinsic muscles: one pair of cleidohyoid muscles, from the clavicle to the glottis, which pulls the trachea caudally and relaxes the muscles around the syrinx; the tracheolateral muscle, from the caudal trachea to the syrinx, enclosing the trachea ventrally and laterally, which tenses the syrinx; and the sternotracheal muscle, from the craniolateral sternum to the trachea, just cranial to the syrinx, which fuses with the tracheolateral muscle.

There are two theories on how birds vocalise. The first holds that the vibration of the tympaniform membranes produces sound; the second that the compression of the bronchial elements against the median parts of the syrinx forms narrow slots through which air is forced during expiration, causing whistling sounds.

Avian lungs are not lobed as are mammalian lungs. Approximately one-quarter of the lung volume is enclosed between ribs; avian lungs weigh about the same as those of mammals (on a weight basis) but are more compact and take up 50% as much space as in mammals. They extend from the first to the seventh rib in Psittaciformes but may extend to the ilia in some species (**Fig. 23.2**).

Each lung receives one of the two primary bronchi, formed by the bifurcation of the trachea at the syrinx. The bronchus enters the lung ventrally and obliquely at the junction of the cranial and middle third of the lung and then passes dorsolaterally to the lung surface and turns caudally to its opening into the abdominal air sac. The bronchi have a well-developed internal, circular, smooth muscle layer and longitudinally orientated smooth muscles. Acetylcholine, pilocarpine, and histamine induce contraction, and atropine blocks these effects. Each primary bronchus gives off four groups of secondary bronchi:

- Mediodorsal (seven to ten). Originate from the dorsal wall of the primary bronchus and are located over the costal surface of the lung.
- Lateroventral (eight). Arise from ventral wall of the primary bronchus and are located in the ventral part of the costal surface of the lung; they enter the abdominal and caudal thoracic air sacs.
- Laterodorsal (variable number). Arise from the lateral wall of the primary bronchus and extend laterally towards the costal surface.
- Medioventral (four to six). Arise from the dorsomedial wall of the cranial third of the primary bronchus and run medially on the ventral (septal) surface of the lung, servicing three-quarters of the septal surface of the lung. They are the largest of the secondary bronchi.

The secondary bronchi give rise to the parabronchi (tertiary bronchi), which anastomose with other parabronchi. They are divided into two groups: the paleopulmonic and the neopulmonic parabronchi. The paleopulmonic parabronchi come off the mediodorsal and medioventral secondary bronchi. They form the medioventral–mediodorsal system in the cranial and dorsal region of the lung, making up about two-thirds of the lung. Air flows unidirectionally, caudal to cranial, in this region of the lung, which is the major site of gas exchange and is more efficient than the neopulmonic lung. The remainder of the lung (ventrolateral) is the neopulmonic region.

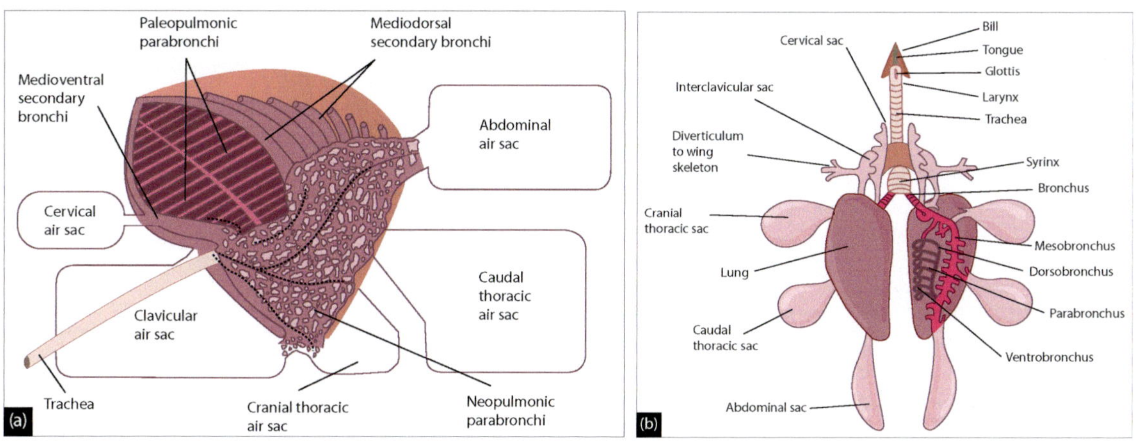

Figure 23.2 **Schematic drawings of the avian lower respiratory tract showing the key components as seen in (a) lateral view and (b) ventral view.**

It is most advanced in chickens, pigeons, and passerines; absent in emus and penguins; and minimal in storks, cormorants, cranes, ducks, gulls, owls, and buzzards. Air in this region changes direction with each phase of breathing (i.e., it is bidirectional).

The parabronchi are uniform in diameter throughout the lung and lined with simple squamous epithelium. The inner lining of these parabronchi is pierced by numerous openings into individual chambers, called atria – pocket-like polygonal cavities lined with flat or cuboidal epithelium and coated in surfactant. The openings into the atria are surrounded by smooth muscle with parasympathetic and sympathetic innervation. At the bottom of each atrium are infundibula: openings that lead to air capillaries. These air capillaries branch and freely anastomose with each other; their small diameter means that the pressure gradient for oxygen diffusion is greater than in mammals. They are intimately entwined with a network of blood capillaries, making them the site of gaseous exchange.

Extending from the lungs are nine air sacs: the unpaired clavicular and the paired cervical, anterior thoracic, posterior thoracic, and abdominal air sacs. (Chickens and some other species have fused cervical air sacs, leaving them with eight air sacs.)

The clavicular air sac is a large, unpaired, and complicated sac occupying the thoracic inlet and extending into the extrathoracic diverticula (the humerus, coracoid, scapula, and clavicle) and the intrathoracic diverticula (around the heart and along the sternum). The first, second, and third medioventral bronchi form the main connections to the clavicular air sac.

The cervical air sacs arise from the first medioventral bronchus. They form two median chambers lying between the lungs and dorsal to the oesophagus, leading into a pair of vertebral diverticula on each side of the vertebral column, one inside the neural canal and one outside. They also invade the vertebrae.

The cranial thoracic air sacs arise from medioventral secondary bronchi and lie dorsolaterally in the coelom. The caudal thoracic air sacs are found caudal to the cranial thoracic air sacs and arise from lateroventral secondary bronchi and primary bronchi.

The abdominal air sacs arise from lateroventral secondary bronchi and primary bronchi and lie between the caudal thoracic air sacs. They are the most variable in size but are often the largest air sacs. They carry air to the leg and pelvic bones through perirenal and femoral diverticula.

Although the mesenteric oblique septum separates the cranial and caudal coelomic cavities, there is no smooth muscular diaphragm to aid in respiration. Instead, birds rely on the movement of the ribs and sternum to move air through the respiratory tract. During inspiration, the striated external intercostal muscles pull the ribs cranially, laterally, and ventrally. At the same time, the sternum, coracoids, and furcula move ventrally and cranially, pivoting at the shoulder joint. Combined, these movements have the effect of increasing the coelomic volume. Expiration is simply a reversal of these movements, using the internal intercostal muscles and the coelomic muscles.

The air moves through the airways on a two-breath cycle (**Fig. 23.3**):

- First inspiration. Air moves through the trachea into the primary bronchus and neopulmonic region, and then into the caudal air sacs. Some

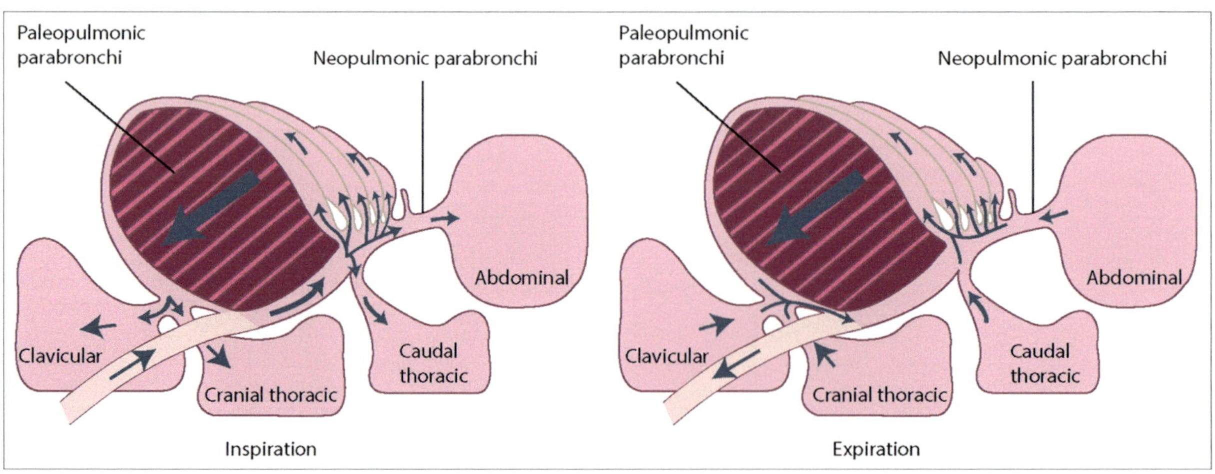

Figure 23.3 **Schematic diagrams illustrating the pattern of airflow through the respiratory tract during inspiration and expiration.**

air may enter the paleopulmonic region and start gaseous exchange.

- First expiration. Air moves from the caudal air sacs into the paleopulmonic region; a small volume of air (12%) escapes from the caudal air sacs through the bidirectional neopulmonic region to escape through the trachea.
- Second inspiration. Air moves from the paleopulmonic region into the cranial air sacs.
- Second expiration. Air moves from the cranial air sacs out through the bronchi and trachea; a small volume of air (12%) escapes from the caudal air sacs through the bidirectional neopulmonic region to escape through the trachea.

As mentioned earlier, gaseous exchange occurs in the air capillaries. The cross-current arrangement between parabronchial airflow and parabronchial blood capillaries in the paleopulmonic region provides a highly efficient system of gaseous exchange, so efficient that birds need less ventilation to achieve a higher level of oxygenation of blood than mammals.

DISEASES OF THE RESPIRATORY SYSTEM

Upper respiratory tract

Sinusitis

Aetiology

The complex anatomy of the infraorbital sinuses and diverticula lends itself to the establishment and persistence of upper respiratory tract infections. When combined with factors such as hypovitaminosis A (causing squamous metaplasia of the epithelial lining of the sinuses and decreased normal function and resistance to infection), irritation of the sinus linings (e.g., ammonia toxicosis in poorly ventilated, unhygienic cages and aviaries, cigarette smoke, or other aerosol pollutants), extremes of humidity (too dry, too moist), or choanal atresia (see later), it is little wonder that such infections are common in birds. Anatomic abnormalities such as choanal atresia, collapsed frontal bones, and traumatic injuries can also predispose to infection, as can inhaled foreign objects (e.g., millet seed, hand-rearing formula) and neoplasia (carcinoma, adenocarcinoma, fibrosarcoma, lymphosarcoma, and malignant melanoma).

Infectious agents are usually secondary invaders taking advantage of the aforementioned conditions. They include bacteria (*E. coli*, *Haemophilus*, *Klebsiella*, *Pasteurella*, *Pseudomonas*, *Mycobacterium*, spirochaetes in cockatiels), viruses (herpesvirus, poxvirus, reovirus, and avian influenza), fungi (*Aspergillus* spp., *Candida albicans*, *Cryptococcus*, and *Zygomycetes*), Chlamydia, *Mycoplasma*, and even parasites (*Trichomonas*, *Cryptosporidia*).

Clinical presentation

Clinical signs include oculonasal discharge, occlusion of the nares, exophthalmos (or enophthalmos in macaws [sunken eye syndrome]) and distension of the sinuses (**Fig. 23.4**). This sinus distension can be quite marked as the lateral wall of the sinus is skin rather than bone as in mammals. There may be matting or loss of the periocular feathering and conjunctival hyperaemia or thickening. Affected birds may sneeze, shake their heads, and scratch at their face with their feet or rub their face excessively on objects such as perches. There may be hyperinflation of the cervicocephalic air sac.

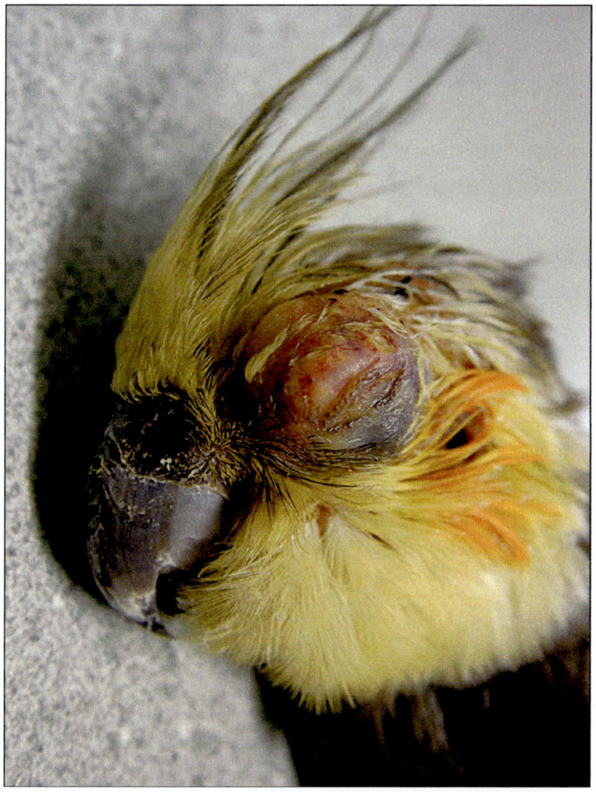

Figure 23.4 **Cockatiel with sinusitis. Note the yellow caseated pus distending the sinus, dorsal to the eye, and the periocular feather loss.**

Lockjaw syndrome may be seen in cockatiels aged 3–10 weeks. After an initial course of oculonasal discharge and sinus swelling, they are unable to open their beaks and starve to death.

Diagnosis

Cytology and culture are helpful (see Chapter 6 and Chapter 7). Diagnostic imaging, including contrast radiographs (using barium or other contrast agents), CT, and MRI, can be used to examine the anatomy and patency of the infraorbital sinuses and to localise a problem (e.g., a pocket of caseated debris). Endoscopy, via the choana, can be useful when a foreign body is suspected. Transillumination with an intense focal light can be used to examine the sinuses and rhinal cavity.

Management

The underlying cause must be identified and corrected. Nutritional modification and the use of parenteral vitamin A are particularly important in many cases.

If the nidus of infection is not found and removed, recurrence is common. Therefore, it is important to remove caseous exudate and other debris within the sinuses. Nasal flushing using normal saline or diluted acetylcysteine may remove some caseated material, but the complex anatomy of the sinuses makes it difficult to flush the infraorbital sinuses successfully. Infraorbital flushing and aspiration, using the same approaches as those used for sampling, can achieve more than nasal flushing in some cases. Surgical debridement and drainage via trephination into the affected sinus is an aggressive, but often successful, therapy. Before attempting this surgery, it is important to review the anatomy and consult surgical texts. It is extremely easy to damage underlying structures such as the eye or nerves.

Oral and/or parenteral antimicrobial therapy are given, based on cytology and culture. Topical therapy may also be helpful, using nebulisation or application of drops into the nares, if caseous pus has not accumulated in the sinuses.

Rhinitis

Aetiology

Causes of rhinitis are similar to those of sinusitis (see earlier). Occlusion of the nares can occur due to cere hypertrophy, *Cnemidocoptes* infection, or rhinoliths. See Chapter 12.

Clinical presentation

Clinical signs include sneezing, rubbing at the face, nasal discharge, and asymmetry in the size of the nares. In more chronic cases, the external nare may be occluded with a dried plug. When this is removed, a large hole may be left where the nare once was; the surrounding bone and keratin having been eroded by chronic infection (**Fig. 23.5**).

Diagnosis

Diagnosis of the causative organism may require cytology and culture.

Figure 23.5 **Princess parrot with a rhinolith before (a) and after removal (b).**

Management

Rhinoliths can be removed with a small curette. Caseated debris is removed by nasal flushing. Topical or parenteral therapy is given based on culture and sensitivity. Parenteral vitamin A and nutritional management are often indicated.

Ruptured cervicocephalic air sac

Aetiology

As the cervicocephalic air sac communicates with the infraorbital sinuses rather than the lungs, this condition may be the result of chronic sinusitis (see earlier), trauma, or localised air saculitis.

Clinical presentation

The neck becomes distended with air (**Fig. 23.6**).

Diagnosis

Physical examination, aspiration of air, and radiography of the neck will indicate the diagnosis. Endoscopy of the air sac may help to localise the site of rupture, and biopsy of the air sac may better determine the aetiological agent. This condition must be differentiated from aerophagia, where the crop is filled with air. This is done by introducing a feeding tube or crop needle into the crop to see if it can be deflated with a syringe.

Management

The underlying cause should be identified and treated. Lancing the distension with a scalpel or the bevelled edge of a needle, tearing through both the skin and air sac, will deflate the air sac. This may have to be done several times until the originating rupture has healed. Refractory cases may require the placement of an acrylic stent in the most prominent area of the distension. Purpose-made stents are available in the USA and are implanted surgically under general anaesthesia. An alternative technique has been developed that creates a shunt between the cervicocephalic air sac and the clavicular air sac, avoiding the maintenance of an external stent.

Choanal atresia

Definition/overview

Failure of the choana to properly form during development results in a persistent membrane or bony plate at the palate of the nasal cavity, occluding one or both of the internal nares and blocking the normal drainage of nasal secretions into the oral cavity. The condition is predominantly reported in African grey parrots and occasionally in cockatoos and Amazon parrots. The author has seen it in several Alexandrine parrots and Indian ringneck parrots; there is no reason why it could not occur in other species. It is present from an early age.

Clinical presentation

Clinical signs include unilateral or bilateral mucoid discharge and sinus swellings around the eyes (**Fig. 23.7**). A choanal slit may be present, although the internal nares are not patent.

Figure 23.6 **Sulphur-crested cockatoo with a ruptured cervicocephalic air sac which has led to subcutaneous emphysema.**

Figure 23.7 **An Indian ringneck presents with choanal atresia. Note the bilateral mucoid discharge and sinus swelling around the eyes.**

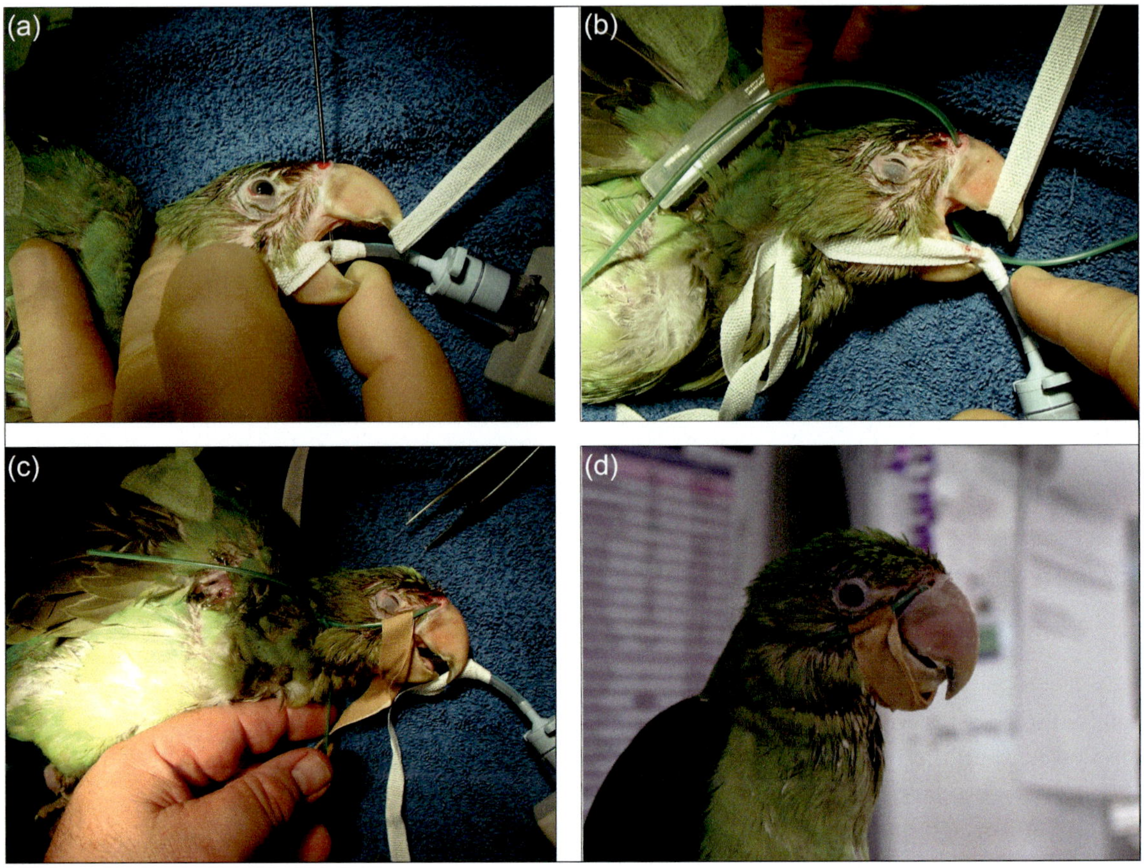

Figure 23.8 **Surgical correction of choanal atresia. (a) An appropriately sized Steinman pin is used to bore a hole through the nare on the occluded side(s) to the choanal slit. (b) An 8 Fr red rubber catheter or nasogastric tube is passed through each hole, leaving the ends long. Openings are cut into the sides of the tube at the nares to allow mucus to drain. (c) The ends of the tube are tied behind the head and a chin strap is made of tape to prevent the tubes from slipping off the top of the head. (d) The tubes are left in place for 4–6 weeks to allow the tracts to epithelialise, creating permanent openings. Once the tubes are removed, the nares are flushed twice daily for 2 weeks to help keep the openings free of debris.**

Diagnosis

Saline flushed into the nares fails to enter the mouth. Instead, the infraorbital sinuses swell. Rhinography using contrast media can be used to demonstrate the lack of communication between the sinuses and the choana. CT can be used to demonstrate the presence of mucus in the sinus and nasal cavity, and the lack of patency of the internal nares. Rigid endoscopy can be used to visualise the choanal slit and assess the internal nares.

Management

Surgical correction is required. An appropriately sized Steinman pin is used to bore a hole through the nare on the occluded side(s) to the choanal slit (**Fig. 23.8a**). The pin is positioned so that it is perpendicular to the dorsal surface of the beak and is directed from the external nare towards the midline. This positioning should have the pin exiting through the location of the internal nare. There may be some bone to pass through and, while haemorrhage is usually minimal, in some cases it may be significant and potentially fatal. An 8 Fr red rubber catheter or nasogastric tube is passed through each hole, leaving the ends long (**Fig. 23.8b**). Openings are cut into the sides of the tube at the nares to allow mucus to drain. The ends of the tube are tied behind the head and a chin strap is made of tape to prevent the tubes from slipping off the top of the head (**Fig. 23.8c**). The tubes are left in place for 4–6 weeks to allow the tracts to epithelialise, creating permanent openings (**Fig. 23.8d**). Once the tubes are removed, the nares are flushed twice daily for 2 weeks to help keep the openings free of debris. It must be emphasised that if the tract does not epithelialise, the procedure will fail.

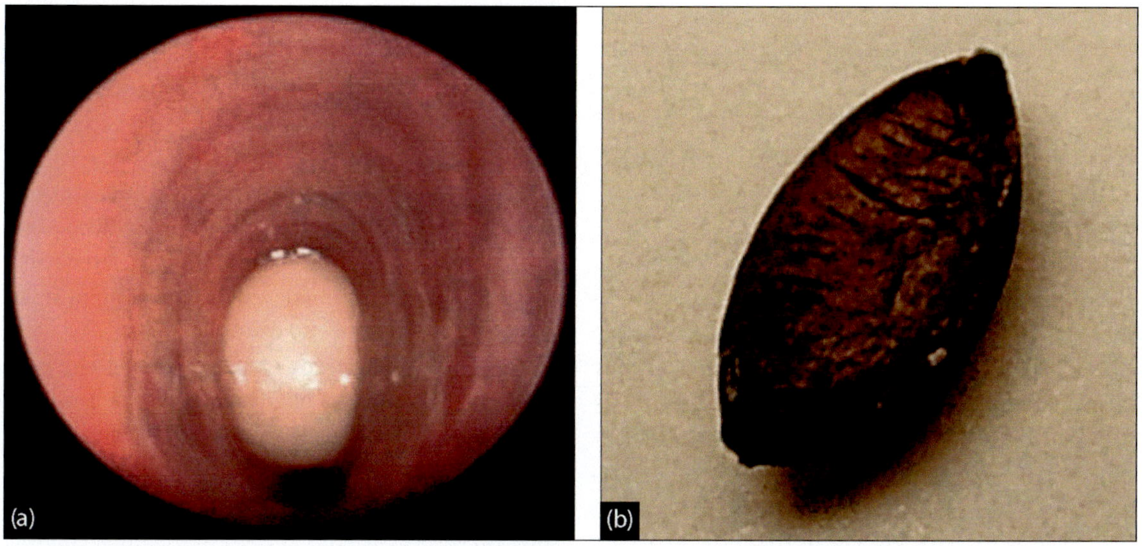

Figure 23.9 **Endoscopy of the trachea of a dyspnoeic cockatiel revealed this millet seed obstructing the trachea (a). (Photo courtesy A Lennox.) After removal (b), it can be seen that the millet seed is a perfectly accidental inhalation and subsequent tracheal obstruction.**

Lower respiratory tract

Tracheal obstruction

Aetiology

The complete rings of the avian trachea make a functional obstruction (e.g., a collapsing trachea as seen in dogs) unlikely. The same anatomical feature makes external compression equally unlikely, although a fractured coracoid or thyroid enlargement has been known to compress the trachea. This leaves a physical obstruction of the trachea as the most likely problem to be seen in practice. It is the authors' experience that these physical obstructions can result from three aetiologies: an inhaled foreign body; a diphtheritic, granulomatous, or neoplastic lesion within the trachea or syrinx; or a stricture following an insult to the tracheal mucosa.

Inhalation of millet seeds causing a complete, or near-complete, obstruction is frequently seen in cockatiels (*Nymphicus hollandicus*) (**Fig. 23.9**). Other inhaled foreign bodies have included hand-rearing formula or even insects. If the foreign body is large enough to occlude the diameter of the trachea, acute collapse and death frequently result. If the object is smaller than the tracheal diameter, acute respiratory distress is more likely to be the presenting sign.

Diphtheritic and granulomatous lesions are often slower to develop, making clinical signs of dyspnoea usually evident before acute collapse occurs. Diphtheritic lesions, in the authors' experience, are more common in cockatoos, mid-tracheal in location, and are associated with a bacterial infection often secondary to the lodgement of inhaled objects such as plant material. Granulomatous lesions are more often seen in the syrinx and are often associated with aspergillosis (**Fig. 23.10**).

Strictures are usually associated with trauma, either due to external wounds (such as those resulting from bites from another bird) or internal trauma following endotracheal intubation or other mucosal insults (e.g., inhaled foreign material). The

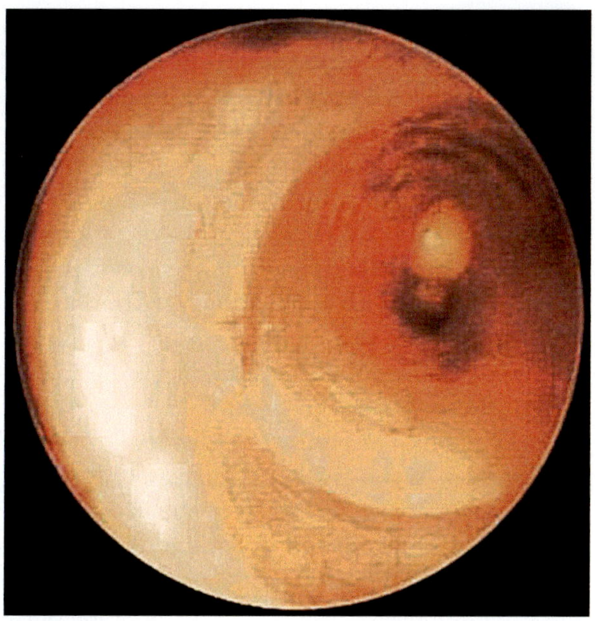

Figure 23.10 **Endoscopic examination of the syrinx of a dyspnoeic African grey parrot revealed a glistening white obstruction – an *Aspergillus* granuloma.**

development of a stricture is usually more insidious, with obvious clinical signs often only seen when the bird's respiration is exerted by stress, exercise, or handling. Careful observation of the bird at rest may reveal subtle mouth breathing and tail bobbing.

Clinical presentation

The clinical signs seen in a bird with a tracheal obstruction depend on the length of time the obstruction has been present and the degree of obstruction that is present. Birds with an acute obstruction may present weak and collapsed with the wings partially extended, the neck extended, obvious mouth breathing and sternal lift, and audible inspiratory noises. Occasionally, the neck will be stretched out.

However, birds with slowly developing obstructions may show milder signs that become more obvious only when the bird is exerted by stress or handling. These birds compensate for the obstruction by keeping still and have a slightly increased respiratory effort. Vocalisation is decreased and there may be a change in voice. As the degree of obstruction increases, the clinical signs become more obvious.

Diagnosis

While the aforementioned clinical signs are suggestive of an obstruction, they are not always pathognomonic. Lower respiratory tract disease, i.e., pulmonary or air sac disease, can appear very similar even to experienced clinicians.

Endoscopy usually offers the most rapid and accurate tool for diagnosing tracheal obstruction. The authors' approach is to mask induce the patient with isoflurane and then perform a left flank laparoscopy to evaluate the lungs and air sacs. An air sac catheter is then placed unless contraindicated (e.g., by the presence of fluid, severe airsaculitis, or pulmonary disease), and anaesthesia is maintained via this route (see Chapter 6). With practice and preparation, this can be performed in 1–2 minutes. With an alternate airway secured, tracheoscopy can then be performed to evaluate the trachea and syrinx.

Radiography can be used to identify extramural compression of the trachea. Contrast radiography may be necessary in some birds to obtain a definitive diagnosis. A small amount of barium (0.5 ml/kg) is placed in the trachea using a catheter. Barium is inert and will not induce irritation or inflammation of the respiratory mucosa.

Management

The treatment of tracheal obstruction revolves around two key principles: the obstruction must be removed and an alternate airway must be maintained until this is done.

Once an alternate airway (the air sac catheter) has been secured, the next step is to relieve the obstruction. The approach taken here will be determined by the location of the obstruction, the type of obstruction, and the size of the obstruction.

Small, non-adherent obstructions in the mid- to caudal trachea have been removed in a variety of ways. These include removal with grasping forceps, suction, tracheotomy, or the introduction of an IV catheter caudal to the obstruction, directed rostrally. The tip of the catheter is used to dislodge the foreign body, and it can then be 'blown out' using air from a syringe attached to the catheter.

Similar obstructions located in the syrinx may be approached through a distal tracheotomy and removed surgically. (Another technique, once recommended but now out of favour, was to use a small bone pin to push the obstruction through the syrinx into the supporting tissues, where it could be left to form a granuloma outside the airways.) The patient is positioned in dorsal recumbency with the shoulders elevated 45° to provide the surgeon with visual exposure to the thoracic inlet. A skin incision is made along the ventral midline, and the crop retracted if necessary (this may require some dissection of subcutaneous tissue). If the obstruction is located in the syrinx, the interclavicular air sac is entered to expose the sternotracheal muscles, which are transected with radiosurgery. This mobilises the syrinx and allows it to be brought rostrally. Great care must be taken not to tear the bronchi off the syrinx. The tracheotomy is made through the ventral half of the annular ligament between the tracheal rings, rather than through them. If approaching the syrinx, the incision is made three to five rings rostral to the syrinx. Closure of the trachea is achieved using a fine, synthetic, monofilament, absorbable material (e.g., 5-0 to 8-0 PDS) in a simple interrupted pattern. The sutures should be preplaced and positioned at 30° circumferentially so that a 50% tracheotomy is closed with two to three sutures. They should encompass at least two rings on each side of the tracheotomy, and the knots are tied external to the tracheal lumen.

Diphtheritic or granulomatous lesions present a different problem. Although they can be endoscopically debulked, care must be taken to avoid iatrogenic trauma to the tracheal mucosa. In the short term this can lead to bleeding, worsening the degree of obstruction, and in the long term it may contribute to stricture formation. Nebulisation with hypertonic saline and/or acetylcysteine can help to 'dissolve' these lesions. Great care, however, must be taken to ensure the patency of the air sac catheter, as the lesions may swell prior to dissipation and completely obstruct the airway. If the lesion is an *Aspergillus* granuloma, injections of diluted amphotericin B trans-tracheally may be of assistance as well, although it must be remembered that this drug can be irritating to mucosal surfaces (see the end of the chapter).

Tracheal strictures can be removed by resection and anastomosis. The flexibility of the avian neck confers a similar flexibility on the avian trachea, which is relatively longer than its mammalian counterpart. This inherent flexibility makes the resection of even long pieces of trachea feasible in birds – up to 10%. A skin incision is made over the stricture and careful sharp dissection is performed to mobilise the trachea from the surrounding connective tissues. The stricture is then excised by transecting the trachea on either side of the affected area (**Fig. 23.11**). The incised ends of the trachea are apposed with four single interrupted sutures (5-0 to 8-0 PDS) placed equidistantly around the tracheal circumference, each suture enclosing two tracheal rings on each side of the anastomosis and entering the tracheal lumen. The loose connective tissue around the trachea and the skin are closed routinely. Stricture at the site of an anastomosis is considered the major postoperative complication, generally developing 3–6 weeks after surgery. This, however, is an uncommon occurrence in birds.

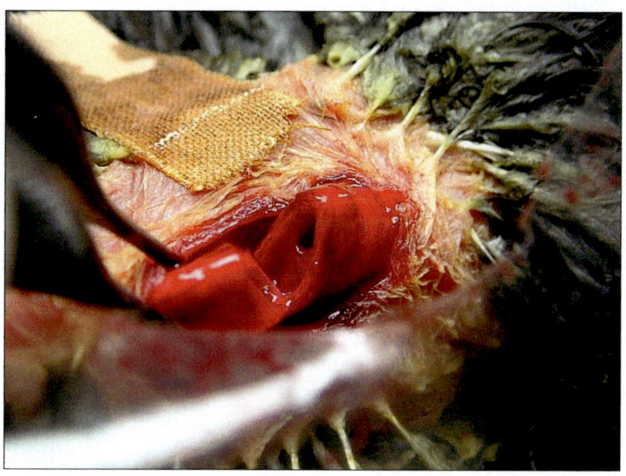

Figure 23.11 **Black cockatoo undergoing surgery to remove a tracheal stricture by resection and anastomosis. The stricture is excised by transecting the trachea on either side of the affected area.**

Tracheitis

Aetiology

Tracheitis is usually seen in conjunction with upper respiratory disease or other lower respiratory diseases, although primary tracheal disease can be seen. Vitamin A deficiencies can play a role in the development of tracheal disease. Aetiological agents include bacteria, fungi (*Aspergillus*, *Mucor*, and *Candida* spp.), viruses (poxvirus, herpesvirus), chlamydia, parasites (*Syngamus trachea*, gapeworm; *Sternostoma tracheacolum*, air sac mite; *Cryptosporidia; Trichomonas*), inhaled irritants (cigarette smoke, other pollutants), and trauma.

Clinical signs

Birds with tracheitis often present with coughing and gagging. If the syrinx is affected, there may be a loss or change of voice. In some cases, there may be audible clicking noises on both inspiration and expiration. As the disease advances, there may be increased respiratory effort and open-mouth breathing.

Diagnosis

Clinical signs are somewhat similar to those seen in a bird with a tracheal obstruction (see earlier). Differentiation between the two is usually made endoscopically. Once an obstruction has been ruled out, diagnostic efforts move towards identifying the cause of the tracheitis. Diagnostic testing includes tracheal wash (combined with cytology and culture), transillumination of the trachea, radiography, and faecal parasite testing.

A tracheal wash is performed in birds to obtain samples of the trachea, bronchi, lungs, and potentially air sacs. The patient is anaesthetised or sedated, and a sterile tube, e.g., a nasogastric tube or an IV catheter, is inserted through the glottis (being careful not to contaminate it on the oral mucosa). Sterile normal saline (0.5–2.0 mL/kg body weight) is quickly infused into the trachea and immediately reaspirated. Recovery should be 25–50% of the volume infused; this is submitted for cytology and culture.

Transillumination of the trachea is most often indicated in small birds with evidence of tracheal disease. Tracheal mites, strictures, and exudates can be visualised in small birds such as finches and canaries.

The skin over the trachea is moistened with alcohol, and a focal intense light is passed through the skin and trachea. If present, mites will be seen as small, moving black dots.

Treatment

Specific therapy is directed based on the basis of the diagnosis or laboratory testing.

If the patient is severely dyspnoeic, consideration should be given to placing and maintaining an air sac catheter until the bird is stabilised.

Supportive care is important. While nebulising is frequently recommended, care must be taken not to exacerbate bacterial or fungal infections through increasing humidity in the trachea. Various medications have been used with oxygen and sterile saline as a delivery vehicle. If used, nebulisation should be done for 15–30 minutes every 6–8 hours for a minimum of 3 days beyond the resolution of clinical signs. Mucolytic drugs should not be used if air sac involvement is suspected, as birds lack the ability to clear the exudates from the air sacs.

Pulmonary parenchyma disease

Aetiology

Infectious

Infections may be bacterial (*E. coli*, *Klebsiella pneumoniae*, *Pasteurella multocida*, and *Pseudomonas aeruginosa* [**Fig. 23.12**]), fungal (*Aspergillus*, *Cryptococcus*), chlamydial, mycobacterial, mycoplasmal, viral (paramyxovirus, polyomavirus, herpesvirus, avian influenza, canary pox, and adenovirus), or parasitic (*Sternostoma* spp., *Sarcocystis* spp., *Cryptosporidium* spp., *Toxoplasma* spp., *Atoxoplasma* spp.).

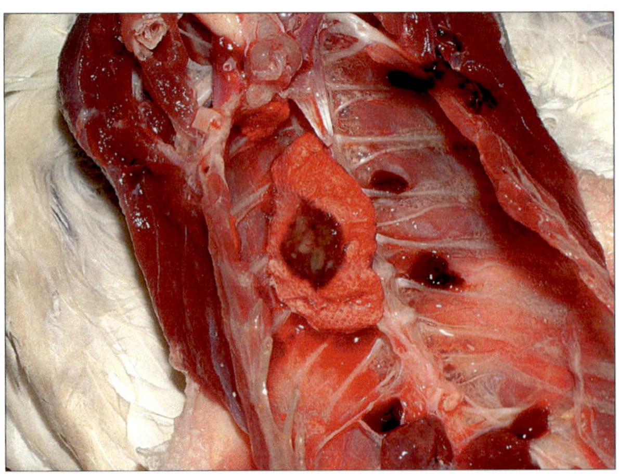

Figure 23.12 **Necropsy of a cockatoo reveals pulmonary parenchymatous abscesses. These can be bacterial, fungal, or mycobacterial in origin. Note that abscesses are often found on the dorsal aspect of the lung, making surgical access difficult.**

Pulmonary hypersensitivity

Also known as chronic obstructive pulmonary disease, this is most commonly seen in blue and gold macaws housed with birds producing feather dander (e.g., cockatoos, grey parrots).

Inhaled toxins

These include smoke, polytetrafluoroethylene (PTFE; Teflon®), cigarette smoke, aerosol sprays, carbon monoxide, burning cooking oils, foreign bodies, aspirated food material, yolk embolism, and fat embolism from bone marrow.

Pneumoconiosis

This is the accumulation of dust-laden macrophages. It may arise from an incident that occurred months or years previously.

Pulmonary congestion or oedema

Pulmonary oedema secondary to congestive heart failure occurs occasionally in pet birds.

Neoplasia

Pulmonary tumours include fibrosarcoma, carcinoma, and metastatic disease.

Clinical presentation

Signs include tachypnoea and/or dyspnoea, mouth breathing, voice change, exercise intolerance, and general signs of illness.

Diagnosis

Auscultation may detect audible respiratory noises. If these are heard on inspiration, upper respiratory tract disease should be suspected. Expiratory wheezes, rales, and rasps are more suggestive of lower respiratory tract diseases. It must be noted, however, that even severe pathology may not cause audible detectable changes. Mild exercise in the form of wing flapping will often aid in the detection or localisation of abnormal respiratory sounds. Following mild exercise, the respiratory effort of a normal bird should return to a steady state within minutes. Prolonged respiratory recovery time is an indication that further diagnostic testing is necessary.

Haematology and biochemistries

Polycythaemia (PCV 60–80%) may be present in birds with chronic hypoxia due to pulmonary parenchymatous disease, e.g., pulmonary hypersensitivity.

Radiography

Birds with pneumonia can have a variety of radiographic changes ranging from a diffuse to a uniform abnormal/thickened parabronchial pattern. Macaws with pulmonary hypersensitivity may demonstrate a more accentuated reticular pattern due to the thickening of the interatrial septa.

Endoscopy

This is carried out via the cranial and caudal thoracic air sacs. See Chapter 9.

Tracheal/lung wash

This is performed using 0.5–2.0 ml/kg of 0.9% saline (see earlier).

Lung aspirates and/or biopsy

These can be utilised for cultures, cytology, or histopathologic evaluation. Lung parenchyma biopsies can be obtained by two endoscopic approaches: the cranial thoracic air sac (caudal lung) or the intercostal approach, through the third intercostal space (dorsolateral portion of the lung). The clinician needs to be aware of the potential for haemorrhage; it should be considered only when less invasive diagnostic techniques are ineffective or when a biopsy is necessary to determine and initiate life-preserving therapy.

Management

Treatment of parenchymatous disease can be frustrating if not diagnosed early, since most birds do not show clinical signs until the disease is advanced. Treatment for 4–6 weeks is generally necessary.

Supportive care

Oxygen therapy can be beneficial, but prolonged exposure to 100% oxygen can be dangerous. Oxygen concentrations should not exceed 35–50%, and should be given for no more than 2 hours before going back to room air for a short period. Low flow intra-nasal oxygen via a nasal line can be used in suitable patients (e.g., waterfowl).

Nebulisation using 0.9% normal saline with or without antibiotics is administered for 15–30 minutes three to four times daily for a minimum of 3 days beyond the resolution of clinical signs. Nebulised particles must be smaller than 3 μm to penetrate into the lung parenchyma.

Care should be taken in the use of mucolytic drugs in birds because they lack mechanisms to clear the exudates from their air sacs.

Medications

Systemic antimicrobial therapy is given as indicated by the diagnosis and/or culture. Topical antimicrobials, delivered by nebulisation (see earlier), can be used in conjunction with systemic therapy.

Birds with acute onset of respiratory distress following exposure to feather dust, smoke, or PTFE may benefit from the administration of rapidly acting, short-duration corticosteroids (methylprednisolone sodium succinate, 15–30 mg/kg IM or IV) and oxygen supplementation. Bronchodilation with terbutaline (0.05–0.1 mg/kg IM q6–8h) may assist in some cases. Albuterol via inhaler is also helpful.

Surgery

Surgery (partial pneumonectomy) may be required to remove granulomas.

Ongoing care and prevention

The use of air filters may be necessary in poorly ventilated areas. Macaws should not be housed with cockatoos or African grey parrots. Birds should not be housed near kitchens, especially if non-stick cookware is in use.

Air sac disease

Aetiology

Infectious causes include bacteria, fungi (*Aspergillus*, *Cryptococcus*), chlamydia, mycobacteria, and parasites (*Serratospiculum* spp., *Splendidofilaria* spp.).

Aspiration pneumonia – associated with overzealous eating, vomiting/regurgitation, oral disease, hand-feeding, or in sedated/anaesthetised birds – frequently causes disease in the caudal lungs and air sacs because the first inspiration brings air directly to these areas.

Air sac carcinomas are uncommon but may be seen occasionally.

Extramural pressure on air sacs may be the result of obesity, organomegaly (liver, oviduct, ovary [physiological enlargement, ovarian cysts]), neoplasia, or fluid effusions (yolk-related peritonitis, ascites, and heart failure).

Clinical signs

Signs include increased respiratory effort, decreased exercise tolerance, and weight loss. Auscultation over the bird's back may detect friction rubs.

Diagnosis

Radiographically, severe airsaculitis can often be documented by the appearance of 'air sac lines' – thin lines of increased radio-opacity where two air sacs meet. In chronic cases, increased respiratory effort can result in hyperinflation of the air sacs, giving the appearance of extra intra-coelomic air.

Size and position of internal organs can also be assessed to detect extramural pressure on the air sacs. If the air sac space is reduced, ultrasound and/or contrast radiology can be used to better delineate internal structures.

Endoscopy can be used to both evaluate and biopsy the air sacs, as long as they are not occluded by internal organ enlargement or ascites.

A blind air sac wash can be done by inserting an intravenous catheter between the last two ribs. The needle is inserted, and the catheter advanced into the air sac. This will place the catheter into the caudal thoracic (shallow) or abdominal (deep) air sac. Sterile saline (1–2 ml/kg) is instilled and collected for evaluation.

Management

The causative agent should be identified and treated. Treatment is similar to that for lung diseases (see earlier).

The air sacs are only a few cell layers thick and normally have a minimal blood supply. However, with most disease processes, neovascularisation of the air sacs develops, and therapeutic concentrations can often be achieved using systemic antibacterial therapy. Humidification and topical treatment of the air sacs via nebulisation are usually necessary in addition to systemic antibiotic treatment.

Aspergillosis

Aetiology

Aspergillosis is a non-contagious infectious disease caused by ubiquitous soil saprophytes belonging to the genus *Aspergillus*, a filamentous fungus. The most common species involved is *Aspergillus fumigatus*, but *A. flavus*, *A. niger*, *A. glaucus*, and *A. nidulans* are occasionally isolated as well.

These fungi are opportunistic pathogens, causing clinical infections when high concentrations of spores are inhaled or when the host is immunosuppressed. Risk factors for infection include species predilection (e.g., grey parrots appear to be more represented than other species in the number of reported cases), exposure to spores (e.g., in nest boxes or poorly ventilated, dirty enclosures), immunosuppression (e.g., young birds or those with concurrent diseases such as chlamydiosis), hypovitaminosis A affecting the resilience and resistance of the respiratory mucosa, and aspiration pneumonia (see earlier). Other predisposing factors include contact with organic litter and mouldy feed, stress (e.g., hospitalisation), and the use of corticosteroids.

Pathogenesis

The type of disease induced is dependent on the source and number of spores contacted and the general condition of the bird involved. Healthy birds can generally withstand exposure to a high concentration of spores. However, in conditions as described earlier, diseases can result.

The pathogenesis of avian aspergillosis is still not known exactly. *A. fumigatus* is a thermotolerant fungus that grows within a temperature range of 20°C to 50°C. It grows well at temperatures over 40°C – the temperature normally found within the avian air sac. The anatomy and function of the avian respiratory system allow direct inhalation of fungal spores into the caudal air sacs while precluding a mechanism for their removal.

In other species, phagocytic cells (macrophages) and respiratory clearance mechanisms are normally the first-line defence against inhaled spores. However, the local immune defences of the avian respiratory tract are not sufficient to resist fungal colonisation. As mentioned, there is no effective mechanism for the removal of inhaled material. The low numbers of resident macrophages, with only a limited capacity to kill fungal spores, combined with heterophils that lack the ability to effectively phagocytise fungal spores, allow fungal colonisation. After initial inhalation and dissemination of spores, fungal plaques and necrotic debris build up in the airways and air sacs (**Fig. 23.13**), which may eventually obstruct the airways (leading to an acute presentation of a chronic problem). Haematogenous

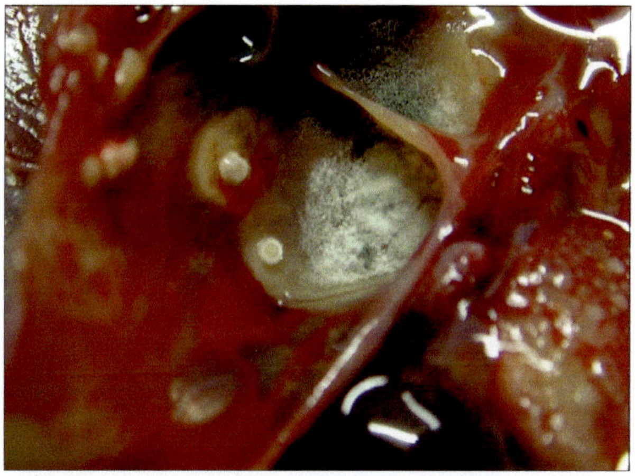

Figure 23.13 Necropsy of an Amazon parrot reveals fungal plaques due to aspergillosis.

spread, or spread into adjacent organs and body systems, may result in disseminated lesions (generally focal granulomata) involving pneumatic bone, peritoneum, internal organs, or the CNS.

Clinical signs

Clinical signs depend on the location and severity of the disease. Lesions can become established in any area of the avian respiratory tract but are especially common in the caudal thoracic and abdominal air sacs, where granulomatous plaques may develop, or as a focal granulomatous lesion in the trachea, especially at the syrinx. Less commonly, the infraorbital sinuses can be infected. As mentioned earlier, invasion of adjacent organs and body systems can cause disease outside the respiratory tract.

If the infection is localised in the sinuses, the clinical signs will be as described earlier for any sinus infection, i.e., oculonasal discharge, sneezing, sinus swelling, head shaking, etc. Tracheal infections are usually localised in the syrinx and often present acutely. Initial signs may include a voice change and decreased activity. As the trachea and bronchi become progressively blocked, the bird will exhibit open-mouth breathing, tachypnoea, increased respiratory effort, and audible respiratory noises.

Lung and air sac disease usually presents as a chronic disease, but acute fulminating disease ('brooder pneumonia') can also occur when multiple lesions in the lung germinate at the same time. This is most common in young chicks exposed to high levels of *Aspergillus* spores during incubation, hatching, or immediately after hatching, especially where there is poor hygiene in incubators, hatchers, or nest boxes. Affected chicks present in acute respiratory distress and invariably die quickly.

More commonly, the presentation is chronic, characterised by lethargy, weight loss, mild to moderate dyspnoea, and prolonged recovery time after exercise. If disseminated spread has occurred, it may lead to confusing clinical signs such as ataxia, kyphosis, seizures, lameness, and biliverdinuria.

Diagnosis

Antemortem diagnosis of aspergillosis is often difficult; either the disease presents so acutely that there is little time for diagnostic testing, or the clinical signs are so nondescript as to be of little assistance in the diagnosis.

As always, a thorough physical examination is indicated if the patient is stable. In chronic cases, auscultation of the lungs and air sacs may detect inspiratory rasps, rales, and whistles. This is, however, not definitive for aspergillosis, and a confirmed diagnosis will rely on further testing such as haematology, radiography, endoscopy, cytology, fungal culture, and serology.

Haematology often, but not always, demonstrates a marked leucocytosis, usually characterised by a monocytosis and heterophilia. This is not reliable as some birds may have a poor immune response with total white blood cell counts in the normal range or just above.

Radiographic signs consistent with aspergillosis include focal densities in air sacs or lungs, loss of definition of the air sacs, asymmetry in their size and shape, or hyperinflation. Air sac lines may be visible. If the lungs are involved, there may be an exaggerated parabronchial pattern and some consolidation of the lung pattern.

Endoscopy usually offers the best chance of confirming a diagnosis. See Chapter 9 for equipment and techniques. Tracheal endoscopy, often done after air sac cannulation has stabilised the patient by providing an alternate airway, can detect syringeal granulomas as smooth, white, glistening masses in the syrinx. Endoscopy of the air sacs may demonstrate diffuse opacity and/or white and yellow fungal plaques that may have a superficial layer of blue, grey, or aqua mould. These lesions can be sampled for cytology, PCR, or culture, debrided from the air sac surface, excised, or treated directly with an antifungal medication.

A positive PCR result or culture of *Aspergillus* spp. from a tracheal or lung wash can be strongly suggestive of the disease. However, care should be taken in the interpretation of a positive result on its own. Because the organism is ubiquitous in the environment, can be isolated from the avian respiratory tract of non-infected birds, and is a common laboratory contaminant, a positive result without supporting clinical signs and other diagnostic tests can be misleading.

Serological testing using an indirect ELISA test for antibodies for aspergillosis can be used to both detect disease and monitor response to treatment. Although the indirect ELISA can detect antibodies as early as 7 to 10 days post-infection, in subacute infections or in patients with a poor humoral response (sequestered granulomatous infections on air sac membranes or immunocompromised patients), testing for antigen may provide more information than testing for antibody response alone. On the other hand, in more chronic cases in which antigen levels may be low, ELISA for the detection of antibodies may be more helpful.

Another line of serological testing is the detection of *Aspergillus* galactomannan, a component of the fungal cell wall released during growth. This is often combined with a serum protein electrophoresis panel to detect inflammatory proteins. This test is far from receiving universal acceptance as false positive results appear to be not uncommon. These may represent exposure to *Aspergillus* spores without disease or cross-reactive antigens. Care must be taken not to overinterpret these tests in a bird not displaying clinical signs consistent with the disease.

Treatment

The treatment of patients diagnosed with aspergillosis revolves around the twin concepts of supportive care and specific therapy directed against the fungus. Supportive care measures such as oxygen therapy, air sac catheterisation, nebulisation, and antibiotic therapy (directed against secondary bacterial infections) have already been discussed in this chapter and in Chapter 5. Treatment should also be directed at the elimination of predisposing causes. In particular, the chronic form of aspergillosis is usually associated with immunosuppression, making immune support critical. Many birds become inappetent (compromising between eating and breathing), leading to further reduction in muscle mass and weight as a consequence of inadequate energy and protein intake. This often requires tube feeding to provide sufficient protein and energy, and supplemental vitamin A is often beneficial.

Syringeal granulomas

Syringeal granulomas may extend into both bronchi, making resolution difficult if not impossible. However, single granulomas not involving the bronchi may respond well to treatment as long as an air sac catheter is in place and patent during the treatment (see Chapter 6). Nebulisation and trans-tracheal injections of amphotericin B (see later) can 'dissolve' the granuloma, clearing the airway. After a few days of therapy, the patient will often become dyspnoeic as the granuloma swells and completely blocks the syrinx – hence the vital necessity for a patent air sac catheter. Success is achieved when the patient coughs up the granuloma, or it dissipates and passes down into the lung or air sac. Once the initial crisis is past, long-term systemic and topical therapy is required to ensure the infection has been cleared (see later).

Lower respiratory tract infections

Lesions in the lungs and air sacs are more difficult to diagnose and are often well advanced when first detected. Occasionally, granulomas can be surgically removed, but there can be no guarantee that all the lesions have been resected, necessitating medical therapy. Antifungal therapy must be appropriate for *Aspergillus* spp., must be delivered topically and systemically, and must be given for several months if it is to be effective. Suitable antifungal drugs include:

Amphotericin B, a polyene fungicidal drug, binds ergosterol and alters fungal cell membrane permeability. It must be given intravenously, by trans-tracheal injection, or by nebulisation. It is a tissue irritant and is poorly absorbed from the intestinal tract, making intramuscular or oral administration contraindicated.

Ketoconazole, *itraconazole*, and *voriconazole* are azole antifungal drugs. They inhibit cytochrome P450-dependent 14α-sterol demethylase, an enzyme essential for the fungal cell wall synthesis of ergosterol, an essential component of the fungal cell membrane. All are suitable for oral medication, although the effects and safety of ketoconazole are less predictable than the others. The adverse effects of itraconazole have been related to hepatic toxicity, varying between species, but are particularly common in African grey Parrots, even at lower dose rates.

Clotrimazole and *enilconazole* are also azoles but are poorly absorbed. This makes them most useful when nebulised or delivered directly to a lesion endoscopically.

Terbinafine, an allylamine, inhibits squalene epoxidase, an enzyme that is critical for fungal sterol and cell wall synthesis. It is administered orally either by itself or in combination with itraconazole.

Response to therapy for more chronic diseases is rarely rapid, and the patient must be monitored carefully for evidence of clinical deterioration or drug toxicosis. Therapy for all forms of aspergillosis should be continued for 1–3 months after the resolution of clinical signs.

Prognosis

The prognosis for any patient with aspergillosis is guarded to poor. Drug resistance is rare but not unheard of, and many patients are in an advanced disease state when first presented.

FURTHER READING

Antinoff N (2001) Understanding and treating the infraorbital sinus and respiratory system. In: *Proceedings of the Annual Conference of the Association of Avian Veterinarians Australian Committee*, pp. 245–260.

Beernaert LA, Pasmans F, Van Waeyenberghe L, Haesebrouck F, Martel A (2010) Aspergillus infections in birds: A review. *Avian Pathology* 39(5): 325–331.

Hillyer EV (1997) Clinical manifestations of respiratory disorders. In: *Avian Medicine and Surgery.* RB Altman, SL Clubb, GM Dorrestein, K Quesenberry (eds). WB Saunders, Philadelphia, pp. 394–411.

Lichtenberger M, Orosz SE (2007) Acute respiratory distress (ARD) – from anatomy through treatment. In: *Proceedings of the Annual Conference of the Association of Avian Veterinarians Australian Committee*, pp. 151–159.

Simone-Freilicher (2008) Avian sinusitis: The sniffles, the sneezes, and the silent. In: *Proceedings of the Annual Conference of the Association of Avian Veterinarians Australian Committee*, pp. 251–260.

Tell LA (2005) Aspergillosis in mammals and birds: impact on veterinary medicine. *Medical Mycology* 43 Suppl 1(0): S71–3.

Tully TN, Harrison GJ (1994) Pneumonology. In: *Avian Medicine: Principles and Application.* BW Ritchie, GJ Harrison, LR Harrison (eds). Wingers Publishing, Lake Worth, pp. 556–606.

CHAPTER 24

DISORDERS OF THE URINARY SYSTEM

ANATOMY AND PHYSIOLOGY

The kidneys lie in the renal fossae of the synsacrum, each divided into three divisions: the caudal, middle, and cranial divisions (**Fig. 24.1**). (Note: These are not lobes.) The distinctions between these divisions are not always clear. The spinal nerves and sacral plexus pass through the kidneys between the middle and caudal divisions. The surface of the kidney is covered in rounded projections, the renal lobules.

Each renal lobule is a pear-shaped, elongated piece of tissue wedged between the interlobular veins and enclosed by its perilobular collecting tubules. At the tapering end of the lobule, the collecting tubules converge to form the medullary collecting tubules (medullary region or cone of the lobule); this tapering end also contains the nephronal loops (loops of Henle) of the medullary nephrons. The wide part of the lobule is the cortical region; it contains nephrons of both cortical and medullary types (but not the medullary nephronal loops).

Several lobular medullary regions converge into a single cone-shaped assembly of collecting tubules in a connective tissue sheath (known as the medullary region of a renal lobe), draining into a single collecting duct. Several of these ducts combine to form a secondary branch of the ureter.

Although there is a lobular cortex and medulla, there are no distinct renal cortical and medullary regions because both lobes and lobules are embedded in tissue at differing depths in the kidney. There is a higher number of medullary regions in birds that conserve water and, therefore, a smaller volume of cortical regions. This implies a higher proportion of mammalian-type nephrons and, therefore, a better countercurrent concentration.

There are two types of nephrons: cortical (or reptilian) nephrons with no nephronal loop, and medullary (or mammalian) nephrons with a nephronal loop penetrating the medullary region. Both types

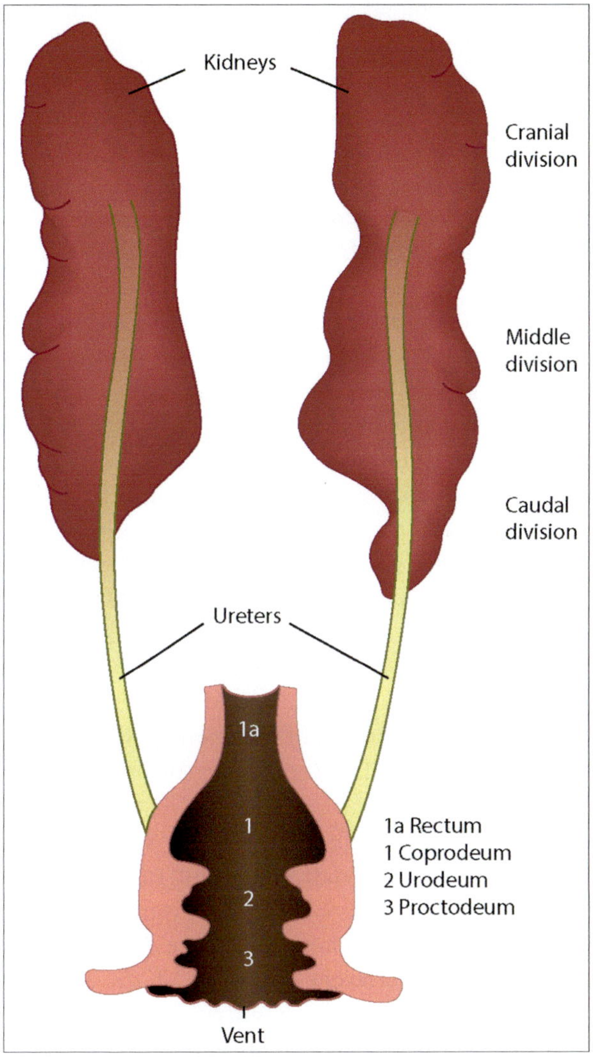

Figure 24.1 **Labelled diagram of the location of the key components of a bird's urinary system.**

start with a renal corpuscle: a glomerular capsule (Bowman's) enclosing the glomerulus (tuft of capillaries). Collecting tubules lie both superficially on the surface of the cortical region (perilobular) and within the medullary region (medullary). Several form a collecting duct, which then forms a secondary branch of the ureter.

DOI: 10.1201/9781003272519-24

The arterial blood supply to the kidneys comes from the cranial, middle, and caudal renal arteries. The kidney also receives a venous supply via the cranial and caudal renal portal veins, which form a venous ring encompassing both kidneys. Blood enters this ring from the external iliac vein, the ischiatic veins, the internal iliac veins, and the caudal mesenteric veins. Afferent renal veins come off the ring and enter the renal parenchyma to become the interlobular veins. The renal portal valve is located in the common iliac vein; when it is open (adrenergic stimulation), blood is diverted into the caudal vena cava and away from the kidney.

The ureter starts within the depth of the cranial division and continues caudally in a groove on the ventral surface of the middle and caudal divisions. It receives primary branches, which in turn receive secondary branches from the collecting ducts draining several renal lobes. The ureter opens into the urodeum.

The urine:plasma osmolar ratio in most birds can only reach 2.0–2.5, compared with 25–30 in mammals. Birds will excrete 1% of filtered water, compared with mammals that excrete less than 0.1%. The concentration ability of the avian kidney lies with the mammalian nephrons; because most birds have more reptilian nephrons than mammalian nephrons, they do not produce concentrated urine. This is believed to be, in part, due to the need for water to transport the more viscous uric acid through renal tubules. Uric acid, the end product of protein metabolism in birds, is produced in the liver and removed from the blood by a combination of filtration in the glomerulus (10%) and tubular secretion in the proximal part of the nephron (90%).

Urine production is controlled by arginine vasotocin (AVT), the avian equivalent of antidiuretic hormone (ADH). Increased plasma osmolality stimulates the hypothalamus to produce AVT. This in turn constricts afferent arterioles of the reptilian nephrons (reducing the glomerular filtration rate [GFR]) and increases the permeability of the collecting ducts of the mammalian nephrons. The result is decreased urine production and therefore decreased plasma osmolarity.

Water resorption also occurs in the rectum during retrograde flushing from the cloaca. Up to 15% of urine water can be resorbed in this manner, but this is reduced by polyuria and stress-induced defecation. If the urine is too concentrated, a concentration gradient across the rectal mucosa cannot be achieved, and so resorption would be limited. However, as urine osmolality increases, birds are able to gradually increase plasma osmolality, thus preserving the urine:plasma osmolar ratio and allowing water resorption. Birds with functional salt glands can decrease plasma osmolality by excreting salt, but this is not applicable to parrots.

Effective osmoregulation therefore requires:

- Normal plasma osmolality
- Sufficient functional nephrons
- Normal production of, and response to, AVT
- Efficient cloacal water resorption

RENAL DISEASE

Mechanism of renal disease

Avian renal disease is a frequently diagnosed but often poorly understood problem in birds. It is associated with multiple diseases and has many causes. Understanding the mechanisms of renal disease, the possible aetiologies, and the effects on other parts of the body will assist the clinician with both diagnosis and treatment.

The mechanisms of renal damage in birds have not been studied as much as in mammals, but there are sufficient similarities to extrapolate and understand how renal disease occurs. Any insult to the kidney – ischaemic, inflammatory, or toxic – results in the release of prostaglandins and thromboxanes. This in turn results in changes in renal vascular resistance, blood flow, recruitment of inflammatory cells and other physiologic effects. Thromboxane, in particular, leads to renal vasoconstriction, resulting in decreased glomerular filtration rate (GFR) and delivery of oxygen and nutrients to tubular cells, further adding to renal damage.

The flow-on effect of this renal damage is seen in conditions such as gout (discussed later in this chapter) and fluid loss leading to dehydration.

Aetiology

Renal disease may be classified broadly into two categories: inflammatory and non-inflammatory.

Inflammatory conditions

Nephritis – inflammation of the kidney – may involve the interstitial tissue, the tubules, or the glomerulus (although 'glomerulonephritis' is typically reserved

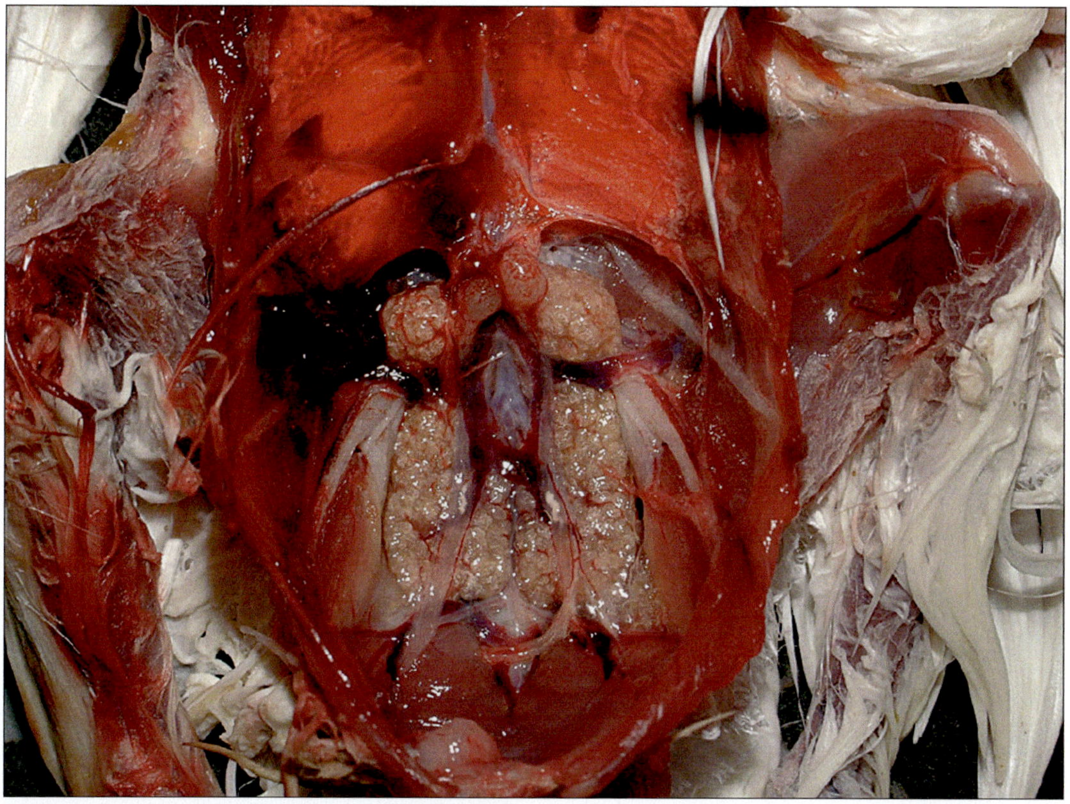

Figure 24.2 **Necropsy of a cockatiel with nephritis. Note that by the time the kidneys have reached this advanced stage of disease, identifying the aetiology is difficult.**

for glomerular lesions). The term 'pyelonephritis' is technically incorrect in avian medicine as bird kidneys lack a renal pelvis. Causes of inflammatory renal disease may be infectious or non-infectious (**Fig. 24.2**).

Infectious causes include:

- Viruses – adenovirus, circovirus, coronavirus, herpesvirus, orthomyxovirus, polyomavirus, paramyxovirus, poxvirus, and retrovirus.
- Bacteria (most bacteria, including mycobacteria).
- Chlamydiosis.
- Parasites – coccidia (waterfowl and raptors), cryptosporidia (rare in all species), microsporidia (lovebirds, budgerigars, and finches), *Toxoplasma*, *Sarcocystis* (systemic infection), trematodes (waterfowl), and schistosomiasis (waterfowl).
- Fungal infections – *Aspergillus* infections invade either from neighbouring air sac lesions or by haematogenous spread after fungal invasion of blood vessels.
- Immune-mediated – Membranous glomerulonephritis is an immune-mediated condition following chronic antigenic stimulation (e.g., following a polyomavirus infection).

Non-infectious causes include egg yolk peritonitis and trauma (generalised trauma or following endoscopy or surgery).

Non-inflammatory conditions

Non-inflammatory renal disease (nephrosis) can result from:

- Amyloidosis – Most commonly seen in waterfowl, small passerines, and raptors as an apparent immune-mediated response. Multiple organs, including the kidneys, are affected, with the liver most affected.
- Toxins – rodenticides (vitamin D3 analogues), aminoglycosides (gentamicin, amikacin), heavy metals (lead, zinc, cadmium, mercury, arsenic), mycotoxins (aflatoxin, oosporein, citrinin, ochratoxin), and others (hypernatraemia, acetone, allopurinol, ethylene glycol, glycine, oxalic acid, selenium).
- Nutritional – hypercalcaemia, hypervitaminosis D3, hypovitaminosis A, hypervitaminosis A. A syndrome is seen in colour mutation varieties of cockatiels, budgerigars, and lovebirds fed on a 100% formulated diet. Currently, it is thought

that these diets may contain excessive levels of vitamin A for these species, although this has yet to be proven.

- Metabolic – disseminated intravascular coagulation (DIC), haemochromatosis, haemoglobin deposits, lipidosis, and nephrogenic diabetes insipidus.
- Genetic/congenital – renal cysts, agenesis, and hypoplasia.
- Decreased renal perfusion associated with dehydration or hypovolaemia.
- Degenerative – renal mineralisation, tubular nephrosis.
- Obstructive dysuria due to egg binding, cloacoliths, and uroliths.
- Neoplasia, including carcinomas, adenocarcinomas, round cell tumours, sarcomas, and embryonal nephromas.

Clinical presentation

Most birds with renal disease are presented with advanced disease. Questioning the owner usually reveals a history of polyuria and polydipsia. Other historical findings may include access to toxins, previous health problems, and other birds that have been affected.

Birds with renal disease are often fluffed and lethargic, with varying degrees of muscle wastage and weight loss. Many are anorectic but still have an increased thirst. Despite this increased water intake, the bird will appear dehydrated (sunken eyes, thin toes, wrinkled skin, and thickened saliva). Regurgitation or vomiting is common, especially if the bird is severely polydipsic. Persistent polyuria (distinguished from stress or excitement-induced polyuria) is often noticeable in the cage or during the consultation (**Fig. 24.3**). Lameness or wing droop associated with articular gout may be seen in chronically affected birds (**Fig. 24.4**).

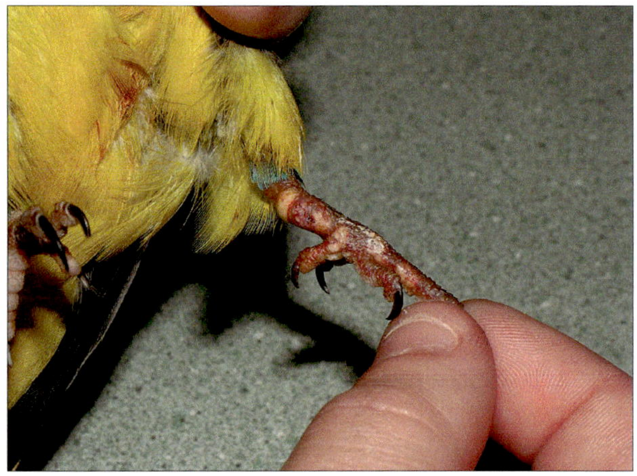

Figure 24.4 **Uric acid deposits in the tendons and joints of a Neophema species parrot.**

Figure 24.3 **Evidence of polyuria from wet newspaper in association with bird droppings. Note that polyuria must be distinguished from diarrhoea.**

Differential diagnoses

Other than renal disease, there are a number of conditions associated with similar clinical signs, including:

- Liver disease.
- Pancreatic disease.
- Gastrointestinal disease.
- Septicaemia.
- Pituitary or pineal gland neoplasia is reported as a cause of polyuria/polydipsia, especially in budgerigars.
- Diabetes mellitus is frequently reported in parrots, particularly budgerigars and cockatiels (see Chapter 21).
- Diabetes insipidus has been reported in several species of birds.
- Hyperadrenocorticism due to adrenal neoplasia has been reported as a cause of polyuria/polydypsia.
- Renal phosphate flush seen with low-calcium, high-phosphorous diets (i.e., seed diets) is believed to be a cause of polyuria/polydypsia.
- Psychogenic polydipsia is a relatively rare condition, often associated with juvenile hand-reared cockatoos. Stress, excitement, and fear may also be contributing factors.

These conditions need to be considered when working up a bird with polyuria and polydipsia.

Diagnosis

Water intake

Polydipsia should be confirmed quantitatively if possible. This can often be done by the owner at home, lessening the probability of a stress-induced polydipsia with secondary polyuria. Water intake greater than 120 ml/kg/day confirms polydipsia.

Haematology and biochemistry

Haematology and biochemistry can be helpful (see Chapter 7). Many birds with renal disease have mild anaemia and an elevated uric acid level. (Persistent hyperuricaemia in a well-hydrated bird is highly suggestive of renal disease.) Blood urea nitrogen may be elevated if the bird is dehydrated but is not a reflection of renal function. The calcium:phosphorous ratio may be low in birds with severe renal disease as calcium is excreted and phosphorous retained. This is, however, not a consistent finding.

Toxicology

Heavy metal (lead and zinc) levels can be assessed. While this is useful for lead toxicosis, it is doubtful if measuring zinc levels is of great diagnostic value. See later in this chapter for further discussion on this topic.

Urinalysis

Urinalysis is indicated when there is persistent (not transient) polyuria. Interpretation of the results must be done with caution, as faecal contamination is the norm, and this must be taken into account. Factors to be checked include:

- Colour. Biliverdinuria is suggestive of liver disease (a common cause of polyuria and polydipsia), while haematuria is strongly suggestive of lead-induced nephropathy (see Chapter 4).
- Urine specific gravity (USG) is difficult to interpret. Birds do not concentrate urine in the kidneys, but rather in the rectum. Normal birds, therefore, can have a wide variation in specific gravity. In polyuric birds, the USG has been reported as ranging from 1.005 to 1.020. The main value of USG, therefore, lies in cases with a persistently low USG and those where USG fails to increase with water deprivation.
- Dipstick evaluation:
    - pH. Normally 6.0–7.5. Lower readings in parrots may indicate acidosis. Higher values can indicate bacterial metabolism. Care must be taken to ensure that the colour of the urine does not influence the colour readings on the dipstick.
    - Protein. Normally, only trace amounts are recorded. Higher levels can indicate faecal contamination, renal disease, haemoglobinuria, haematuria, hyperproteinaemia, or sepsis. Proteinuria is not usually a feature of renal disease in birds.
    - Glucose. Normally, zero amounts are registered. Glucosuria indicates faecal contamination, stress, diabetes mellitus, or renal damage. Persistent glucosuria without hyperglycaemia, in the authors' experience, gives a guarded to poor prognosis.
    - Ketones. Ketonuria indicates severe catabolism or complicated diabetes mellitus. It also indicates a poor prognosis.
    - Blood. Haematuria/haemoglobinuria can indicate faecal or cloacal contamination, or

it can be associated with severe renal disease, and warrants a sediment examination.

- Sediment can be examined under the microscope, at first unstained, and then stained with methylene blue. The following can be observed:
  - Desquamated epithelial cells from the cloaca.
  - Leucocytes and erythrocytes are rare in normal urine. More than two to three per high power field should be considered abnormal.
  - Casts (granular or cellular) can be seen with renal disease. Haemoglobin casts may account for positive haemoglobin readings seen on dipsticks where no erythrocytes are seen.
  - Small numbers of gram-positive bacteria can originate from the faeces. Large numbers of bacteria should be considered abnormal. Comparison with a faecal Gram stain may help determine their origin.
  - Uric acid crystals (small and spherical) are normal. Other crystals are occasionally seen, but their significance is unclear.

Radiography

Plain view radiographs allow visualisation of the size and density of the kidneys (see Chapter 8). In a lateral view, with both acetabulae on the same plane, the normal kidneys lie in the sacral area and do not extend ventrally past the ventral rim of the acetabulae. Enlarged kidneys can often be seen below these limits (**Fig. 24.5**). Radiodense mineralised kidneys can occasionally be seen. Both metastatic and dystrophic calcification can occur within the kidney as a result of renal disease, excessive dietary calcium, or hypervitaminosis D3. Mineralisation is not an indication of chronicity, as it can begin within days of the original insult. Radiodense metallic-like particles seen in the gastrointestinal tract can indicate lead or zinc toxicosis as possible causes of renal disease.

Endoscopy

The avian kidney is readily accessible using endoscopy (see Chapter 9). Renal gout, mineralisation, or physical anomalies can be visualised. Endoscopy also enables the clinician to biopsy the kidney. Biopsy is perhaps the single most important tool in the accurate diagnosis of renal disease. Haemorrhage is minimal, and the resultant histology (and, if appropriate, culture) allows a more accurate diagnosis, a more meaningful prognosis, and a more concise treatment plan. It should be noted, though, that renal histologic lesions are rarely pathognomonic for a specific disease process, as many different diseases cause similar lesions.

Water deprivation test

Once other possible causes of polyuria/polydipsia have been eliminated, a water deprivation test can be used to differentiate between psychogenic polydipsia and diabetes insipidus. *This test is potentially very dangerous and must be performed with great care and caution.* The rationale behind a water deprivation test is that a bird with the ability to osmoregulate will concentrate urine in response to increased plasma osmolality, due in turn to water deprivation. A gradual water deprivation test is preferable to an abrupt water deprivation test in order to overcome the problem of renal medullary washout and cloacal concentration gradients. Water is gradually restricted by 10% per day over 3–5 days, and then deprived completely. This process gives the kidneys and the cloaca every chance to respond to gradually increasing plasma osmolality. Close monitoring is mandatory, as severe dehydration can result if the patient is unable to concentrate urine. Plasma proteins, PCV, urine specific gravity, and weight should be monitored every 3–4 hours for 12–48 hours. The test should stop when the patient loses 5–7% of its body weight, begins to show signs of distress, or is able to concentrate its

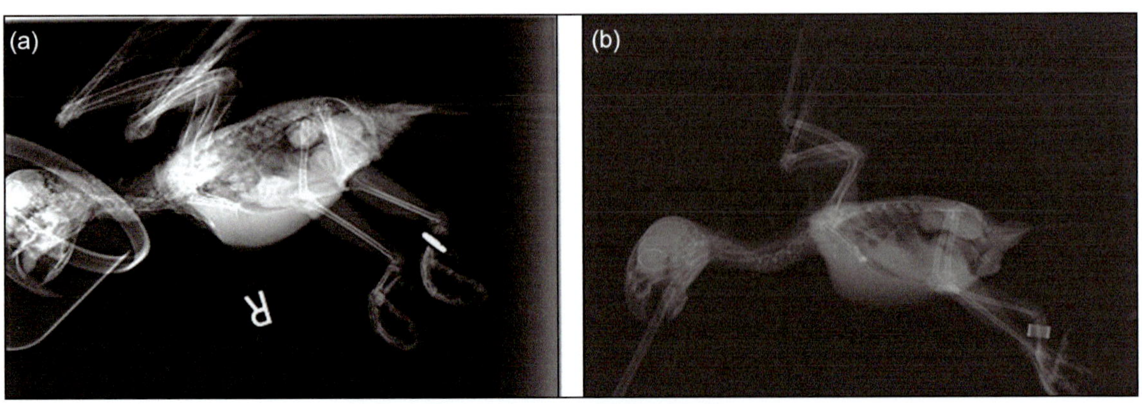

Figure 24.5 **Radiographs of renomegaly in (a) macaw and (b) lorikeet.**

urine. Birds with psychogenic polydipsia should tolerate this test well and reduce their urine output and concentration (increased specific gravity) and show an increase in PCV and plasma protein, all consistent with dehydration. Failure to concentrate urine in the absence of other possible causes indicates diabetes insipidus. A vasopressin response test can then be performed to distinguish between neurogenic and nephrogenic diabetes insipidus.

Vasopressin response test

Diabetes insipidus (a deficiency of, or failure to respond to, arginine vasotocin [AVT]) has been recorded in birds. Birds that have had other causes of polyuria/polydipsia ruled out and have failed to concentrate their urine in a gradual water deprivation test are likely to have either neurogenic or nephrogenic diabetes insipidus. Desmopressin acetate (an AVT analogue) is given orally at doses of 0.02–0.2 mg/kg. A reduction in polyuria and polydipsia within 30 minutes of administration confirms the diagnosis of neurogenic diabetes insipidus. Nephrogenic diabetes insipidus fails to respond to the desmopressin.

Treatment

Rehydration and diuresis

Initial supportive care includes diuresing the patient and maintaining good hydration. Parenteral fluids are given at 100–200 ml/kg/day (divided into 2–3 doses) for 3 days, followed by 50–100 ml/kg/day until uric acid levels return to normal and remain normal. Fluids can be given IV, IO, or SC. If the patient is severely dehydrated, peripheral circulation is often compromised, and SC fluids may not be absorbed (this is seen as subcutaneous oedema hours after fluid administration). These birds should be given IV or IO fluids until better hydrated. If the patient is mildly dehydrated and is still eating, oral administration of fluids is acceptable, as long as vomiting or regurgitation does not occur.

If fluids are not enough to diurese anuric or oliguric birds, furosemide can be given IV. Great care must be taken with this therapy, especially in lorikeets, which appear to be exceptionally sensitive to this drug. As this drug will cause a drop in plasma electrolytes, it is important to maintain the bird on balanced electrolyte fluids such as lactated Ringer's solution.

Reduce plasma uric acid levels

This fluid therapy may be sufficient to lower uric acid levels to normal (<500 μmol/l) in 1–3 days. If they do not come down, or if severe hyperuricaemia (>1200 μmol/l) is present when first seen, medical therapy may be required.

- *Allopurinol* decreases uric acid production by inhibiting xanthine oxidase, which is required to convert hypoxanthine to xanthine and subsequently to uric acid. It should be noted that allopurinol is toxic to the red-tailed hawk (*Buteo jamaicensis*) and should not be used in this species or other raptors.
- *Urate oxidase* appears to degrade excessive uric acid to allantoin, which is more easily cleared by the kidneys. It acts on uric acid alone and does not interfere with the metabolism of purines. There needs to be further evaluation to better understand the use and potential long-term effects of this drug in birds. It is currently only available as a costly injectable medication.
- *Colchicine* reversibly inhibits xanthine dehydrogenase (lowering uric acid levels), blocks the synthesis and secretion of serum amyloid A, and decreases the formation and increases the breakdown of collagen. It is therefore used to lower uric acid levels, treat amyloidosis, and reduce the fibrosis associated with renal disease.

Protein restriction

There is still controversy over the role of dietary protein in birds with renal disease. In theory, moderate protein restriction should be instituted, but this in turn may result in malnutrition and a catabolic state. A safe recommendation is that birds with hyperuricaemia and/or gout should not consume diets with protein levels greater than what is considered normal for the given species.

Vitamin A

As hypovitaminosis A is suspected to contribute to renal disease, if the patient's dietary history suggests chronic malnutrition, a single IM injection of vitamin A can be given at the beginning of the therapy, followed by appropriate dietary modification. Care must be taken to avoid oversupplementing the patient, as hypervitaminosis A has also been associated with renal disease.

Specific therapy

More specific therapy can be given when a specific diagnosis is achieved by renal biopsy or other laboratory tests.

Membranous glomerulonephritis

Membranous glomerulonephritis is commonly diagnosed in older birds by endoscopic biopsy. Therapy involves the removal of the antigen responsible for the immune-mediated reaction (if possible), aspirin at 1 mg/kg orally q24h, and omega-3 and omega-6 fatty acid supplementation, mixed in a ratio of omega-6:omega-3 of 6:1 and given at a dose of 0.1–0.2 ml/kg PO q24h. It is expected that 2–4 weeks of treatment would be needed before changes in the bird's health would be appreciated.

This therapy is not used specifically for the treatment of membranous glomerulopathies and may be used for any inflammatory form of renal disease.

Amyloidosis

Currently, no therapy is available.

Bacterial nephritis

Treatment of bacterial nephritis with appropriate antibiotics should be based on culture and sensitivity results when available. Otherwise, suspected bacterial-induced nephritis should be treated with broad-spectrum bactericidal antibiotics for a minimum of 6 weeks. Aminoglycosides, known to be nephrotoxic, should be avoided.

Dietary-induced renal disease

Dietary-induced renal disease can often be treated by discontinuing pellets and changing the diet to whole grains, seeds, and vegetables. If after 3–6 months all signs of renal disease are gone, pellets (<50% of total diet) can be cautiously added to the diet.

Renal fibrosis

Renal fibrosis can be treated with colchicine until the fibrosis resolves (and is confirmed histologically). Otherwise, it can be used for 6–12 months or until laboratory abnormalities normalise.

Zinc or lead toxicosis

Zinc or lead toxicosis can be treated with chelation therapy (calcium EDTA, d-penicillamine). Zinc toxicosis usually only requires treatment until the metallic particles in the gastrointestinal tract have been removed (approximately 4–5 days if left to pass naturally). This needs to be confirmed radiographically before discontinuing treatment. Birds diagnosed with lead toxicosis should be chelated twice daily until clinically normal, then twice weekly for 6–8 weeks. At the end of this period, blood lead levels should be retested.

UROLITHIASIS

Definition/overview

The formation of ureteroliths in birds is rare. The stones are composed of uric acid crystals and a proteinaceous matrix. They are typically lodged in the ureters and may be unilateral or bilateral. Hypovitaminosis A may be a contributing factor.

Clinical presentation

Signs include non-productive straining, depression, and anorexia. Other clinical signs will depend on the extent of renal involvement (see earlier).

Diagnosis

Radiography reveals radiopaque obstructions of the ureter. Intravenous pyelography may help to define the obstruction and ureters. Endoscopic examination of the ureter may reveal a distended ureter proximal to the obstruction. If only one kidney is involved, it is unlikely that serum biochemistries will show elevations in uric acid (assuming the other kidney is normal).

Treatment

Diuretic therapy with fluids may help to 'flush' the stone out. The stone may be able to be milked along the ureter with an endoscope or during a surgical approach. If these techniques fail to relieve the obstruction, a ureterotomy may be required. Care must be taken to minimise postsurgical stricture formation.

GOUT

Overview

In birds, protein metabolism sees ammonia initially converted to glutamate and then to glutamine, glycine, and aspartate for incorporation into the purine synthetic pathway, which results in xanthine. This in turn is metabolised in the liver by xanthine oxidase to form uric acid. This process occurs primarily in the liver and to a small extent in the kidney. This uric acid forms soluble salts (urates) with ammonia, sodium, or potassium ions and is then transported via the bloodstream to the kidneys where 80% is actively secreted by the proximal tubule (the remainder is filtered by the glomerulus). Glucose, amino acids, a small amount of urea, and electrolytes are found in the glomerular filtrate in the same concentration as the plasma, but most of this (except the urea) is resorbed in the proximal tubule to produce hypotonic urine. The filtrate then entering the collecting ducts and ureters therefore consists of uric acid as a supersaturated colloidal suspension in hypotonic urine, with a small amount of urea. Once this suspension reaches the cloaca, the uric acid and ammonia are passed in the bird's droppings (the urates), while the urine is back-flushed into the rectum where it is concentrated across the rectal mucosa.

Anything – renal disease, dehydration, or obstruction of the ureters or cloaca – that disrupts this excretion of uric acid will see the blood levels of uric acid start to rise (hyperuricaemia). Since uric acid is actively secreted by the proximal tubules, hyperuricaemia does not occur until at least 70% of renal function is lost. When the levels are high enough (>600 μmol/l) the solubility of sodium urate in plasma is exceeded, and precipitation of monosodium urate crystal begins. It is the authors' clinical impression that acute renal disease, causing a sharp rise in blood uric acid levels, is more likely to result in visceral gout, while a slower rise (seen with more chronic renal disease) gives rise to articular gout.

Visceral gout is the deposition of uric acid on the surface of the pericardium, liver, and spleen as a white coating (**Fig. 24.6**). Deposits are also found within the lamina propria of the gastrointestinal tract and within the muscles, heart, and kidney. Urate crystals cause mechanical damage to tissues, and this may account for the sudden death so often seen in birds affected with visceral gout.

Articular gout, on the other hand, results from the accumulation of urates in the synovial capsules and tendon sheaths of the joints (**Fig. 24.4**). This is the result of the cooler temperature of the extremities encouraging precipitation, combined with a slower rise in blood uric acid levels. The intense inflammatory reaction provoked by the urate crystals causes swelling and pain.

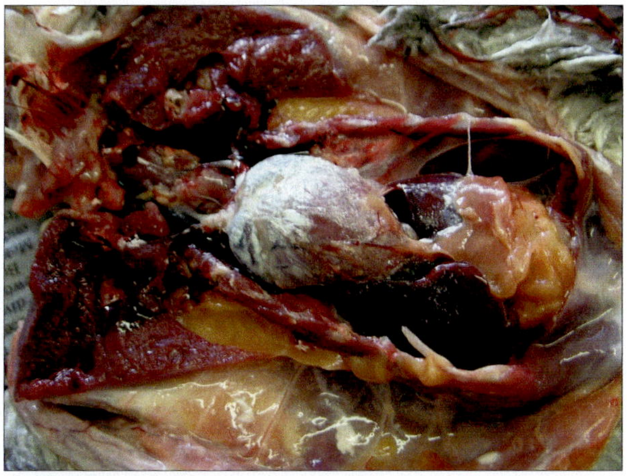

Figure 24.6 **Deposition of uric acid on the surface of the pericardium and liver.**

Clinical signs

Many birds with visceral gout are presented dead or dying within hours or days of showing signs of renal disease, as described earlier. Occasionally, birds will live long enough for deposits of uric acid to become visible in the subcuticular tissues.

Birds with articular gout are frequently presented for lameness. Polyuria, polydipsia, and weight loss are common. Close physical examination shows swelling of the joints of the toes, legs, and sometimes the wings. These swellings are often associated with grossly visible white deposits under the skin (**Fig. 24.7**). If the swellings are in the joints of the toes, the affected toe(s) is usually held extended, with the bird unable to grip the perch. If both feet are affected, the lameness will shift from foot to foot.

Diagnosis

Visceral gout is frequently a diagnosis made on necropsy (**Fig. 24.8**). Impression swabs made of the white deposits on the visceral organ surfaces will show the typical needle-shaped uric acid crystals. Due to their water-soluble nature, urates will dissolve in formalin and therefore the crystalline form will not be seen on conventionally fixed tissue. However, urates can be seen in alcohol-fixed tissue using Gomori's methenamine silver impregnation technique.

The diagnostic tests described previously for diagnosing renal disease (haematology, biochemistry, heavy metal testing, and radiology) are utilised in both live birds with suspected visceral gout and those likely to have articular gout.

Fine needle aspirates from swollen joints, examined as a wet preparation, typically reveal uric acid

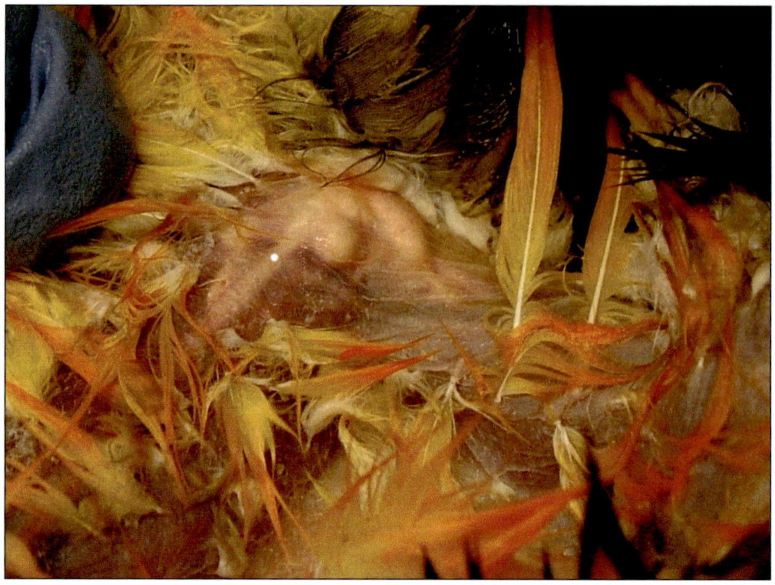

Figure 24.7 **Grossly visible white deposits under the skin associated with visceral gout.**

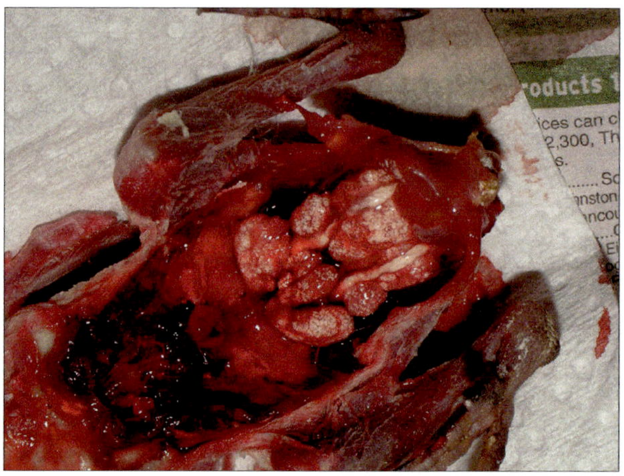

Figure 24.8 **Necropsy of a neonatal caique chick reveals visceral gout.**

crystals surrounded by a sterile pyogranulomatous infiltrate. Alternatively, the murexide test can be performed by placing a small amount of the suspect material on a slide and mixing it with nitric acid. Use a flame to evaporate and dry the mixture. Once cool, add one drop of concentrated ammonia. If urates are present, a mauve colour will appear.

Treatment

Fluid therapy, vitamin A, allopurinol, and colchicine as described previously are the first line of treatment for both visceral and articular gout. Allopurinol and colchicine should be continued until the hyperuricaemia and swelling have resolved.

In addition, analgesia should be provided for birds with articular gout. As NSAIDs are generally contraindicated in renal disease patients, opioids should be employed, e.g., butorphanol or tramadol.

If necessary, gout deposits around joints can be carefully removed by making a small incision over the swelling and expressing the urate material. As this is both a painful and bloody procedure, the patient should be anaesthetised, and the incision bandaged afterwards to maintain some pressure on it and prevent secondary infection.

Prognosis

The prognosis for birds affected with gout, either visceral or articular, is guarded to poor. Birds with advanced articular gout are often in severe pain, and quality of life decisions need to be made by the owner and the veterinarian.

ZINC TOXICOSIS

Overview

Zinc is an essential micronutrient for birds, required for a variety of purposes including many enzyme functions, skin health, and fertility. It is absorbed from the gastrointestinal tract, bound to metalloproteins, and utilised where required. Excess levels are excreted through the kidneys and pancreas. Toxicosis occurs when the amount circulating in the body exceeds the ability of the pancreas and kidneys to maintain homeostasis. The intestinal tract, kidneys, and pancreas are the organs affected by zinc toxicosis.

Aetiology

Zinc toxicosis results from the ingestion of large amounts of zinc in the form of galvanised metal particles, wire, some coins, cosmetic jewellery, etc. Although some authors suggest that this can be a chronic intoxication, it is the authors' experience that this problem is an acute intoxication as zinc is not stored in the body unless the body's homeostatic mechanisms have been overwhelmed. Clinical signs and death can follow within 1–3 days of ingestion of excessive amounts of zinc.

Clinical signs

Clinical signs relate to the organs damaged by the zinc:

- Direct irritation of the gastrointestinal tract from the presence of zinc-containing particles will result in enteritis, seen as ileus (crop stasis), vomiting, and diarrhoea.
- Renal damage will result in polyuria and polydipsia, often very pronounced.
- Pancreatic damage will result in ileus and vomiting.

Consequently, many affected birds will present with an acute history of polydipsia, polyuria, and a crop distended with fluid. Passive regurgitation frequently follows palpation of the crop.

Often, birds are presented dead; necropsy usually shows a well-nourished body with extensive visceral gout and pale, swollen kidneys. This is, however, not pathognomonic for zinc toxicosis.

Other clinical signs that have been attributed to zinc toxicosis, such as feather-damaging behaviour and neurological signs, cannot be correlated with this toxicosis and should not be regarded as clinical signs consistent with zinc poisoning.

Diagnosis

Clinical signs and a history of access to large amounts of zinc (e.g., a new aviary made with heavily galvanised wire) can lead to a tentative diagnosis of zinc toxicosis. Biochemistry findings typically include elevated uric acid and amylase levels.

As previously mentioned, it is doubtful if measuring zinc levels is of great diagnostic value. Normal plasma zinc values in parrots are <30.6 µmol/L (2.00 ppm), although cockatoos and eclectus parrots may be higher (38.2–45.9 µmol/L; 2.5–3 ppm). There is, however, a natural diurnal variation in blood zinc levels, as well as a slight elevation in unwell or stressed birds. There is also the risk of contamination of the sample with zinc (e.g., from rubber stoppers on syringes). While elevations of a magnitude of four or five times the normal level are suggestive of zinc toxicosis, slight elevations are often misinterpreted as zinc toxicosis.

Radiographs demonstrating metallic particles in the crop, proventriculus, and ventriculus of a bird displaying consistent clinical signs strengthen the diagnosis of zinc toxicosis.

Histopathologic findings from biopsy or necropsy will often demonstrate pancreatic necrosis and renal nephrosis (degenerative, non-inflammatory lesions of the kidney).

Treatment

Treatment of zinc toxicosis revolves around:

- Removal of absorbed zinc via chelation therapy, e.g., Ca EDTA.
- Removal of zinc in the gastrointestinal tract by flushing the proventriculus and ventriculus (see Chapter 19) or by allowing it to pass through. It has been shown that most metallic particles will move through the gastrointestinal tract in 3–7 days. Adding a small amount of insoluble grit to the diet can hasten this process, but tube feeding bulk laxatives such as peanut butter or mineral oil has not been shown to be advantageous.
- Correction of dehydration through IV, IO, or SC routes.

Removal of the fluid in the crop and denying access to drinking water until the dehydration is resolved will limit the possibility of regurgitation and aspiration of crop contents.

Response to treatment is usually rapid, with most birds showing improvement 24–36 hours after the start of treatment. Once the metal in the gastrointestinal tract has been removed, chelation therapy can be discontinued.

Prevention

Limiting access to zinc is obviously an important preventative measure. New aviary wire should be examined for metallic tags or flakes, which should be removed. Excessive galvanising can be removed by washing the wire with diluted acetic acid (vinegar) and then painting the wire to prevent rust.

FURTHER READING

Echols MS (2006) Evaluating and treating the kidneys. In: *Clinical Avian Medicine, Vol 2.* GJ Harrison, TL Lightfoot (eds). Spix Publishing Inc, Palm Beach, pp. 451–492.

Echols MS (2007) Avian kidney disease, part I: Types of renal disease. In: *Proceedings of the Annual Conference of the Association of Avian Veterinarians Australian Committee*, pp. 101–116.

Echols MS (2007) Avian kidney disease, part II: Diagnosis of renal disease. In: *Proceedings of the Annual Conference of the Association of Avian Veterinarians Australian Committee*, pp. 117–128.

Echols MS (2007) Avian kidney disease, part III: Treatment of renal disease. In: *Proceedings of the Annual Conference of the Association of Avian Veterinarians Australian Committee*, pp. 129–137.

Lumeij JT (1994) Nephrology. In: *Avian Medicine: Principles and Application*. BW Ritchie, GJ Harrison, LR Harrison (eds). Wingers Publishing, Lake Worth, pp. 748–804.

Orosz S, Dorrestein GM, Speer BL (1997) Urogenital disorders. In: *Avian Medicine and Surgery.* RB Altman, SL Clubb, GM Dorrestein, K Quesenberry (eds). WB Saunders, Philadelphia, pp. 614–644.

Puschner B, St Leger J, Galey F (1999) Normal and toxic zinc concentrations in serum/plasma and liver of psittacines with respect to genus differences. *Journal of Veterinary Diagnostic Investigation* 11(6): 522–527.

Raidal SR, Echols MS (2007) The advantages and disadvantages of excreting uric acid. In: *Proceedings of the Annual Conference of the Association of Avian Veterinarians Australian Committee*, pp. 87–100.

CHAPTER 25

DISORDERS OF THE LYMPHOID AND HAEMATOPOIETIC SYSTEMS

ANATOMY AND PHYSIOLOGY

Lymphoid organs

The lymphoid organs are classified as primary (thymus and bursa of Fabricius) or secondary tissue (spleen, Harderian gland, pineal gland, bone marrow, and scattered aggregates of lymphoid tissue). An important feature to note is the lack of lymph nodes in birds; rather, lymphoid tissue is diffusely disseminated in many tissues, including the liver, spleen, kidney, and gastrointestinal tract.

The thymus is found in the neck, often in multiple sites extending from the angle of the jaw to the thoracic inlet. It consists of lobules of epithelial cells, each covered by a connective tissue capsule. Each lobule has an outer dark cortex and an inner light medulla. Lymphocytes are most dense in the cortex, while thymic corpuscles (islands of reticular tissue known as Hassall's corpuscles) are present in the medulla. The thymus is at its largest in the sexually immature bird. It serves as the source of T lymphocytes, which are the circulating cells responsible for cell-mediated immune responses. Approximately 65% of the mononuclear cells in the spleen and 80% of the mononuclear cells in the blood of chickens are T cells.

The bursa of Fabricius is a dorsal median diverticulum of the proctodeal region of the cloaca. It contains a central cavity that forms a single large cavity opening into the proctodeum. The internal wall of the bursa is folded and covered by simple columnar or pseudostratified columnar epithelium. Lymphoid nodules are located between these epithelial folds. Each nodule has a cortex and a medulla, separated by a basement membrane and epithelial cells. Lymphocytes, plasma cells, and macrophages are found in the cortex, while lymphoblasts and lymphocytes are found in the medulla. In the embryo, the medulla is the first lymphoid organ to produce immunoglobulins. Mature B lymphocytes, responsible for humoral immunity, migrate from the bursa to peripheral/secondary lymphoid tissue. This occurs as early as day 17 of incubation. It also produces a hormone, bursapoietin, which stimulates the movement of B cells from the yolk sac to the bursa and induces the maturation of bone marrow cells. The bursa involutes as the bird ages; in parrots, this may take as long as 18–20 months, compared with chickens where the bursa is involuted by 2–3 months of age.

The spleen is found on the right side of the junction between the proventriculus and the ventriculus. It varies in size and shape; it may be round, elongated, or slightly triangular, depending on the species of bird. There are no well-defined trabeculae; instead, there is a basic network of reticular fibres and cells. Lymphoid tissue, known as white pulp, surrounds the arteries and is responsible for lymphopoiesis. Numerous venous sinuses are present, surrounded by lymphocytes, macrophages, and elements of circulating blood. This red pulp is responsible for the phagocytosis of aged erythrocytes. Both white and red pulp contribute towards antibody production. The spleen does not function as a significant blood reservoir.

Disseminated lymphoid tissue is found in the Harderian gland in the third eyelid, throughout the alimentary tract (caecal tonsils, oropharynx, small intestine, and caudal oesophagus), and as solitary nodules in all organs and the bone marrow. It responds to antigens similarly to the spleen.

Granulopoiesis in birds seems to follow similar stages to those seen in mammals, i.e., myeloblast, promyelocyte, myelocyte, metamyelocyte, and granulocyte. While the bone marrow is the primary site of granulopoiesis, other organs such as the spleen, liver, kidney, lungs, thymus, gonads, and pancreas may also be involved. Lymphopoiesis, occurring in the bursa, thymus, and spleen, has

DOI: 10.1201/9781003272519-25

three stages: the lymphoblast, the prolymphocyte, and the lymphocyte.

The immune response

The immune system serves two primary purposes: it clears infection from the body, and it then develops a pathogen-specific resistance to protect the bird from future infections. It does this through a combination of nonspecific defences (including barriers such as the skin and mucosa, and the innate immune system – macrophages, heterophils, thrombocytes, and complement) and specific defences, including humoral (B cells) and cell-mediated immunity (T cells).

Macrophages (in tissue) and monocytes (in blood) identify and consume damaged cells and foreign materials. They are attracted to sites of inflammation by lymphokines produced by damaged cells. Once this phagocytosis is complete, more lymphokines are released by the macrophages and monocytes, which attract B and T cells.

B cells produce antibodies (proteins that coat and neutralise an invading pathogen and 'mark' infected cells so that the immune system can identify and destroy them). An antigen, usually a protein on the pathogen, may activate a B cell directly or initially bind to an accessory cell that leads to the activation of T and then B cells. Either way, the B cell then differentiates into the antibody-producing plasma cell. Antibody response can be either a primary response, stimulated the first time the body is infected with a specific pathogen, or a secondary response to subsequent reinfections. A primary response is characterised by a lengthy (1–2 weeks) latent phase during which immunocompetent cells are activated, leading to a progressive increase in a circulating antibody. This initial antibody is usually IgM; it declines after reaching a peak. A secondary response is characterised by a shortened latent period, with a peak of antibody production, usually IgG (although some species utilise IgY), occurring earlier and at higher levels. Young birds originally passively derive both IgM and IgG (IgY) from the egg yolk and albumen, but an active humoral system begins to develop approximately two weeks after hatching, reaching maturity at 4–6 weeks of age.

Following the humoral response is the cell-mediated response. T cells originate in the thymus, and therefore cell-mediated immunity is dependent on the normal development of the thymus. Lymphokines (chemotactic factor, thrombocyte migratory inhibitory factor, interleukins 1 and 2, several types of interferons, and lymphocytotoxin) attract both B and T cells to an area of inflammation. There are three types of T cells: T helper cells, which release chemical signals controlling other cells involved in the immune response; T cytotoxic cells, which destroy infected and damaged cells; and T suppressor cells, which modulate the effect on the immune system, preventing overstimulation and autoimmune damage.

Erythrocytes

The avian erythrocyte is elliptical in shape with a centrally located, oval nucleus. It is usually 10–15 μm in length (smaller cells are found in smaller species), compared with the typical biconcave mammalian erythrocyte, which is 6–7 μm in diameter. Typically, healthy erythrocytes have a uniform colour and size when examined in a stained blood smear. They have a relatively short half-life (28–45 days) compared with that of mammals (120 days in humans), which results in a greater number of immature erythrocytes in the peripheral circulation. The cytoplasm of these immature erythrocytes stains more basophilic than that of mature erythrocytes (polychromasia), they are non-uniform in size (anisocytosis), and the nuclear chromatin is less condensed than in mature erythrocytes. In most healthy parrots, they represent approximately 1–5% of the total erythrocyte count.

The major function of erythrocytes is oxygen transport, but recently it has been reported that they may participate in some immune responses that contribute to host defence. Erythropoiesis occurs in the bone marrow (and, in the embryonic chick, the yolk), although ectopic erythropoiesis can occasionally be found in the spleen and liver. There are seven stages of development:

- Rubriblasts are large cells with large central and round nuclei, coarse chromatin, and large nucleoli. The cytoplasm is deeply basophilic with clear spaces (mitochondrial spaces).
- Prorubricytes resemble rubriblasts but lack nucleoli and have mitochondrial spaces in the cytoplasm.
- Basophilic rubricytes have a homogeneous basophilic cytoplasm and round nuclei with clumped chromatin.

- Initial polychromatic rubrycites are smaller than basophilic rubrycites. The cytoplasm of these cells is basophilic to slightly eosinophilic, an indication of the beginning of haemoglobin synthesis.
- Late polychromatic rubrycites have a round to oval shape and a more eosinophilic cytoplasm.
- Polychromatic erythrocytes resemble mature erythrocytes but are more oval in shape; their cytoplasm is slightly basophilic, and chromatin appears less condensed.
- Mature erythrocytes.

Erythropoiesis is controlled by several factors such as oxygen concentration in tissues and hormones. Hypoxia stimulates the production and release of erythropoietin from the kidneys, which then stimulates erythropoiesis. Other hormones, such as androgens and adrenocorticoids, also stimulate this process. Polycythaemia, either secondary or primary, may suppress erythropoiesis.

THE EFFECT OF DISEASE ON THE BONE MARROW

The bone marrow is the main site of granulopoiesis (granulocyte production) and erythropoiesis (erythrocyte production) (see earlier). It is also a source of some long-lived lymphocytes. Various disorders can affect the bone marrow, producing lesions that can be hypocellular or hypercellular, granulocytic or erythroid. It is important that the evaluation of bone marrow aspirates requires comparison with peripheral blood.

Infectious diseases (bacteria, fungi, chlamydia, mycobacteria) usually trigger a hypercellular granulocytic response, leading to a leucocytosis. Some viral infections (herpesvirus, poxvirus, reovirus, polyomavirus), however, may have the opposite effect, inducing leucopaenia. All infectious diseases have the capacity to cause a hypocellular erythroid response, leading to non-regenerative anaemia.

Various toxins can affect the bone marrow, causing a broad hypocellular response. These include ochratoxin A (produced by *Aspergillus ochraceus*), sulphaquinoxaline, cisplatin, and fenbendazole. Lead toxicosis induces a hypocellular erythroid response, leading to anaemia.

Neoplasia, including myeloproliferative disorders (neoplastic proliferation of non-lymphoid haematopoietic cells in the bone marrow that subsequently invade the spleen and liver) and lymphoid leukaemia (a tumour of the peripheral blood and bone marrow), is uncommon. Haemangiosarcoma can be a primary neoplasm of the bone marrow. They can cause anaemia, but histologically, it may be difficult to tell which cell lines are involved, particularly if the cells are immature.

DISEASES OF THE LYMPHOID SYSTEM

Thymic cysts

Aetiology

The aetiology of thymic cysts is unknown, but they may be congenital abnormalities associated with persistent thymopharyngeal ducts or acquired during thymic involution. They are similar in appearance and location to ultimobranchial cysts.

Clinical presentation

The cysts may fill with a colloidal fluid or gel and present as swellings on the neck. If large enough, they may cause dysphagia and dyspnoea.

Management

Surgical removal of the cysts is recommended but is often technically difficult because of extensive attachments to surrounding tissues.

Premature thymic and/or bursal atrophy

The thymus normally involutes as a bird approaches maturity, while the bursa normally involutes at 18–20 months of age. Premature thymic and bursal atrophy results in the loss of the lymphocyte population (T- cells) and humoral immunity (B cells). This is particularly important in young birds, making them more susceptible to many bacterial diseases and some fungal infections.

Causes include acute or chronic disease (e.g., viral [circovirus, polyomavirus, avian pox virus, adenovirus, and herpesvirus], bacterial, yeast, and parasitic infections [*Cryptosporidium*]); toxins (oil, mycotoxins, pesticides); malnutrition (hypervitaminosis D3, hypovitaminosis A, calorie-deficient diets); and poor management leading to environmental stress (excessive temperatures, humidity, and noise).

Splenic atrophy

Hypoplasia or atrophy of the spleen may occur secondary to bursal disease. Lymphoid necrosis can be due to chronic disease or other severe stress. Atrophy also occurs naturally with age.

Splenomegaly

Definition/overview

Unlike mammals, the spleen is a lymphoid organ, not a blood reservoir. Splenic enlargement, therefore, reflects lymphoid stimulation or lymphoid hyperplasia/neoplasia.

Aetiology

Infectious causes include:

- Acute viral diseases, including polyomavirus, adenovirus, and herpesvirus.
- Bacterial infections (especially *Salmonella* spp. and *Yersinia pseudotuberculosis*) or septicaemia from gram-negative bacterial infections.
- Chlamydiosis.
- Mycobacterial infections.
- Mycotic infections.
- Parasites, including *Atoxoplasma*, haemoparasites (e.g., *Plasmodium* spp.).

Non-infectious causes include amyloidosis or neoplasia (lymphosarcoma, fibrosarcoma, haemangiomas, haemangiosarcomas, and metastatic carcinomas).

Diagnosis

Diagnostic imaging (radiology [**Fig. 25.1**], CT [**Fig. 25.2**], and ultrasound) can demonstrate splenomegaly but not determine the cause. Haematology, biochemistries, and other diagnostic modalities for infectious disease (e.g., serology, PCR) should be pursued. Endoscopic biopsy of the spleen via left flank coelioscopy has proven to be safe (mild to moderate haemorrhage can be observed but does not require intervention) and effective (good quality biopsy samples can be obtained).

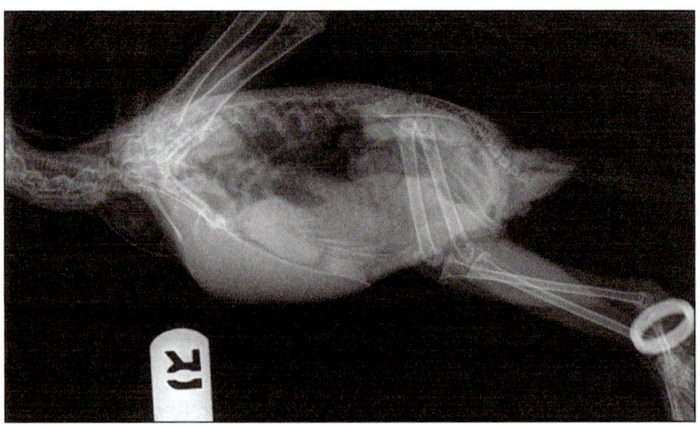

Figure 25.1 **Lateral radiograph demonstrating splenomegaly.**

Treatment

Treatment is directed by clinical pathology and biopsy results.

Lymphosarcoma

Definition/overview

Lymphosarcomas (**Fig. 25.3**) arise from lymphocytes, either of thymic origin (T cells) or bursal origin (B cells), found in lymphoid tissue. They subsequently metastasise to other tissues. It is the most common lymphoid neoplasia reported in parrots and passerine birds, and is most commonly multicentric, with tumours developing anywhere in the body where lymphoid tissue can be found, in particular the intestine, liver, spleen, and kidney. They may be solid, cystic, or haemorrhagic. Less common forms of lymphosarcoma include:

- Epitheliotropic lymphosarcoma, where neoplastic lymphocytes with an affinity for the epidermis and adnexal structures diffusely infiltrate the skin, presents as a pruritic hyperkeratotic dermatitis associated with alopecia.
- Non-epitheliotropic cutaneous lymphosarcoma has been reported in several birds.
- Several cases of exophthalmos in parrots have been diagnosed as retrobulbar lymphosarcoma.
- Intestinal lymphosarcoma has been reported in several birds where the presenting complaint was weight loss, accompanied by inappetence and lethargy.

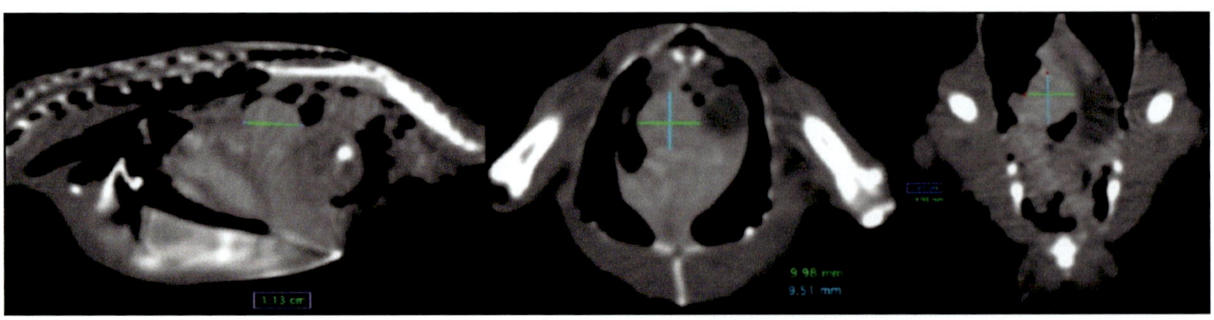

Figure 25.2 **CT images demonstrating splenomegaly.**

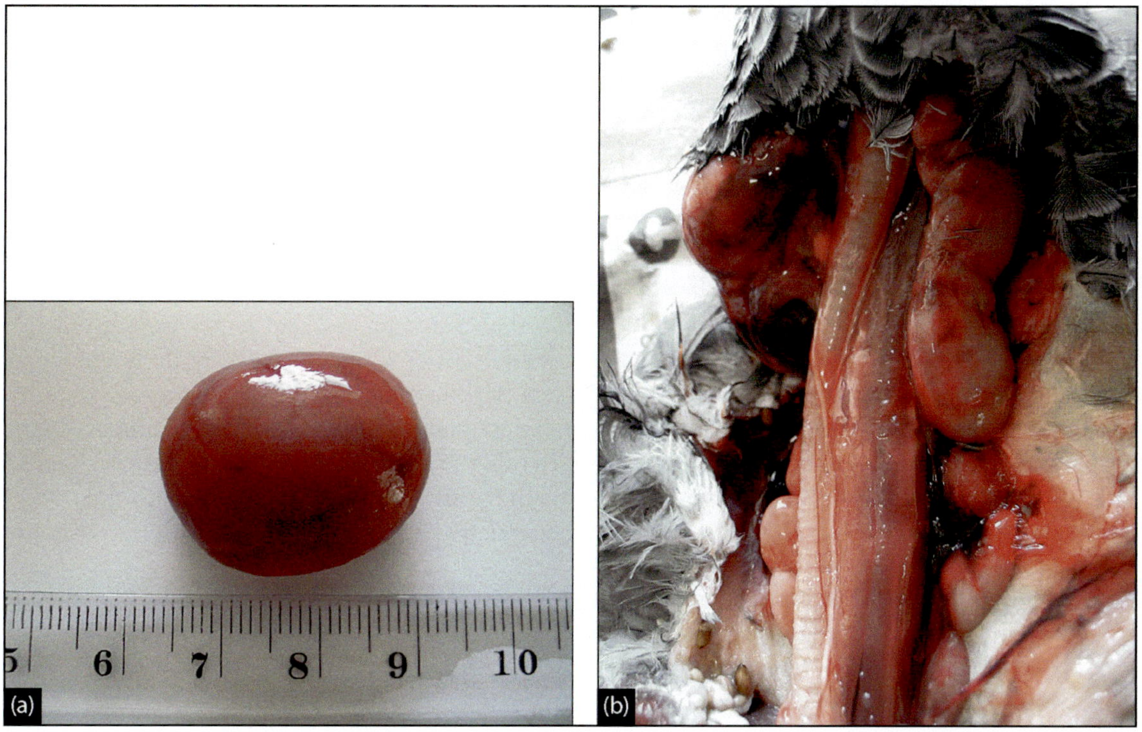

Figure 25.3 **Necropsy of an African grey parrot reveals massive splenomegaly that was due to neoplastic infiltration (a) and lobulated masses running down the neck; the thymic gland infiltrated with neoplastic cells (b).**

Aetiology

Lymphosarcoma has been associated with several viruses in poultry. A viral aetiology, while long suspected, has not yet been demonstrated in parrots, and so the aetiology of lymphosarcoma is unknown in this family of birds.

Clinical signs

Clinical signs are non-specific, reflecting the location and extent of the lesions. Signs can include:

- 'Sick bird look' – lethargy, depression, and anorexia.
- There may be weight loss or weight gain.
- Respiratory changes may include dyspnoea (open-mouth breathing, tail bobbing), and dysphonia.
- Gastrointestinal signs may include dysphagia, regurgitation, decreased faecal output, or diarrhoea.
- Cutaneous signs include pruritus, feather loss, and self-mutilation.
- Neurologically, there may be lameness, decreased grip strength/inability to perch, blindness, ataxia, and inability to fly.
- Some birds may die suddenly without premonitory signs.

Diagnosis

Physical examination findings depend on the location of the neoplasm and the organs involved. Possible additional findings include dehydration, cachexia, cataracts, uveitis, hyphema, exophthalmos, periocular swelling, nystagmus, proprioceptive deficits, wing droop, coelomic distension or mass, cutaneous mass, self-excoriation, cutaneous ulcers, cutaneous nodules, periarticular or bony swelling, oral cavity swelling or mass, and sinus swelling.

Haematology may demonstrate a leukaemic profile, although the combination of leukaemia and lymphosarcoma is uncommon in birds. Hypercalcaemia, commonly seen in dogs with lymphosarcoma, is rare in birds. Other biochemical changes may reflect hepatic or renal dysfunction.

Plasma biochemistry analysis may indicate organs that are affected. Hypercalcaemia of malignancy is rarely reported in birds.

Monoclonal hyperglobulinaemia is suggestive of a possible lymphoproliferative disorder and has been reported in several birds with lymphosarcoma. The monoclonal bands or spikes found using protein electrophoresis are considered to be representative of a single clone producing an abnormal amount of a single immunoglobulin and may reflect the production of acute phase proteins.

Diagnostic imaging (radiology, CT, and ultrasound) may demonstrate enlargement of the liver, spleen, and kidney.

Although fine-needle aspiration of palpable masses is technically simple, the results may not be diagnostic for lymphosarcoma. Surgical biopsy (incisional or excisional) allows the pathologist to examine not only the cell type but also the tissue architecture, giving a more accurate diagnosis. Bone marrow aspirates appear hypercellular and densely packed with neoplastic lymphocytes. Immunohistochemistry staining of the biopsy then allows differentiation of T-cell and B-cell lymphosarcoma, an important guide to both treatment and prognosis (B cell has a *b*ad prognosis, while T cell has a *t*errible prognosis).

Management

Chemotherapy protocols for avian lymphosarcoma have been developed and have moderate degrees of success. Most are extrapolated from small animal medicine, as only limited reports of treatment in birds are available. Some protocols that have been suggested include:

1. Oral medications
   i. Prednisolone and cyclophosphamide
   ii. Lomustine (CCNU)
2. Injectable protocols
   i. COP-L protocol
      a. L-asparaginase injection
      b. Oral prednisolone (tapering dose)
      c. Oral weekly cyclophosphamide
      d. Vincristine injections weekly IV
   ii. Single-agent adriamycin injections q 3 weeks IV
   iii. UW-Madison Cyclic Combination
      a. L-asparaginase
      b. Vincristine
      c. Cyclophosphamide
      d. Adriamycin
      e. Prednisolone

Dose rates for these protocols have not been fully evaluated in birds and need to be used with caution. The doses used are extrapolated from treatment regimens in small animals. Owners should be made aware of the limitations of current knowledge and the possibility of unexpected side effects.

See Chapter 34 for more discussion on the use of chemotherapy in birds.

Leukaemia

Definition/overview

Lymphosarcoma and leukaemia are both neoplastic proliferations of leucocytes. While lymphosarcoma is a proliferation of genetically altered lymphocytes that originates in lymphoid tissue and disseminates to other organs (including bone marrow), leukaemia is the proliferation of haematopoietic stem cells that originates in bone marrow and disseminates to other organs.

Diagnosis

Affected birds may show signs of lethargy and weakness, while haematology demonstrates dramatic lymphocytosis. The advanced stage of lymphoid leukaemia at the time of diagnosis may make it difficult to identify the origin of the neoplasia in birds, but immunohistochemistry can determine if the predominant cell type is a T cell or B cell.

Treatment

Treatment is warranted when the bird shows clinical signs of disease, a dramatic increase in the peripheral lymphocyte count, evidence of organ involvement, associated haematologic abnormalities such as anaemia, and a rapid lymphocyte doubling time in less than 6 months. Complete cure is rare; the aim is to reduce the number of circulating lymphocytes while sparing the bone marrow. Most treatment protocols are based on an alkylating agent (e.g., chlorambucil or cyclophosphamide) paired with prednisone.

PRIMARY DISORDERS OF THE IMMUNE SYSTEM

Immune suppression

Immune suppression can occur either directly or indirectly.

Direct suppression is the inhibition of plasma cells, lymphocytes, and macrophages, preventing antibody production, cell-mediated immunity, or antigen processing. It occurs with:

- Drugs. Tetracycline, tylosin, and gentamicin can decrease antibody production.
- Aflatoxicosis. Depresses complement activity and decreases phagocytic activity; also impairs cell-mediated immunity through inhibition of thymus-associated lymphocytes.
- Some viruses (e.g., influenza A virus) undergo antigenic variance wherein the surface proteins of the virus change and prevent immediate detection.
- Bursal and thymic atrophy (see earlier).

Indirect suppression occurs through the activation of the adrenal gland (by environmental stress) and the subsequent production of corticosterone. This, in turn, inhibits antibody-forming cells and macrophage activity.

Immune-mediated disorders

Allergies

Allergic dermatitis has been reported as a cause of feather picking and dermatitis in parrots. Confirmation can be difficult, as histological lesions are not definitive. Intradermal skin testing in birds is still to be completely validated, but it remains promising.

Allergic pneumonitis (respiratory hypersensitivity) is seen occasionally in macaws exposed to powder down/feather particles from other parrots and organisms such as *Aspergillus* spp. See Chapter 23.

Immune-mediated haemolytic anaemia

Definition/overview

Immune-mediated haemolytic anaemia (IMHA) occurs when antibodies produced by the bird's immune system target the bird's own erythrocytes. It is seen infrequently in poultry and is rarely documented in parrots (only two documented cases at the time of writing).

Clinical presentation

Signs include lethargy, weakness, weight loss, pallor, polyuria, and biliverdinuria.

Diagnosis

Haematology demonstrates a strongly regenerative anaemia with a predominance of round, small erythrocytes (spherocytes), leucopaenia and/or leucocytosis, and elevated plasma protein. Serum biochemistries are non-specific.

Saline autoagglutination may be positive. Blood is diluted with sterile saline (3:1 dilution) in an EDTA tube. This is then centrifuged and the supernatant removed. The remaining cells are washed several times with saline solution, and then examined grossly and microscopically. If IMHA is present, autoagglutination (clumping) of the erythrocytes is visible grossly; microscopic examination is used to confirm true autoagglutination and not rouleaux formation.

Radiographic changes include marked splenomegaly and possibly hepatomegaly.

Liver, spleen, and bone marrow biopsies can be evaluated both for evidence of other haemolytic diseases (e.g., malarial parasites) and evidence of regenerative changes (in the liver, spleen, and bone marrow) and haemosiderosis (in the liver). Bone marrow cytology may not demonstrate the expected low myeloid-to-erythroid cell ratio as seen in mammals. This may be because, in birds, erythropoiesis occurs in the bone marrow primarily, but maturation of the erythroid cell series occurs intravascularly, within bone marrow, and in other haematopoietic organs such as the liver and spleen. These cells may therefore be seen in blood smears and liver and splenic biopsies.

Management

Treatment options for parrots with IMHA involve the administration of immunosuppressive agents such as corticosteroids, cyclosporine, and cyclophosphamide. These drugs have not been widely used in birds, and their use must be extrapolated from small animal medicine.

Membranous glomerulonephropathy

Definition/overview

See Chapter 24. Glomerular disease is an important cause of end-stage renal disease in birds.

Aetiology

The cause of glomerulopathies is generally assumed to be immune-mediated, but the inciting aetiology is often unknown. PCR has been used to identify polyomavirus particles in the kidneys of many affected birds.

Diagnosis

Although proteinuria is the hallmark sign of glomerulonephritis in mammals prior to the onset of clinical renal insufficiency, avian leucocytes lack the proteolytic enzymes that would potentially damage the glomerular basement membrane (and allow protein leakage). Therefore, birds may not develop pathological proteinuria with glomerulopathies.

Renal biopsy is the best way to definitively diagnose glomerular (and other) kidney diseases in birds.

Management

Regardless of the possible causes, identifying and correcting underlying inflammation and/or systemic disease(s) are the first steps in managing renal disease. Omega-3 and omega-6 fatty acids and low-dose aspirin have shown promise in treating this condition.

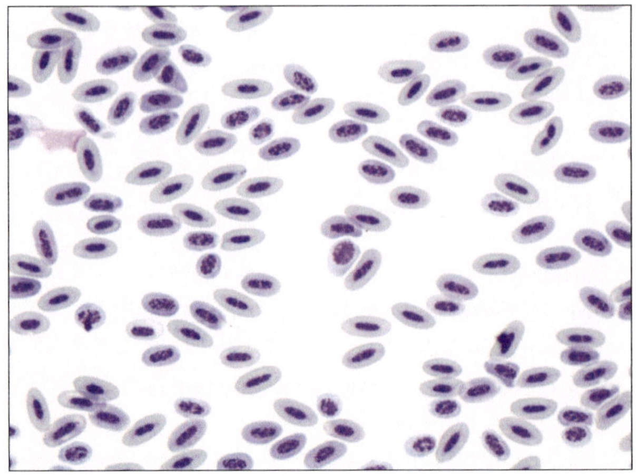

Figure 25.4 **Blood cell sample with polychromasia (variability in erythrocyte cytoplasmic colour) and anisocytosis (variability in erythrocyte size) indicating mild regenerative anaemia.**

Transfusion reactions

Transfusion reactions occur due to the incompatibility of donor red cells with host plasma. They have been reported in birds given multiple heterologous transfusions. Transfusion reactions may result in haemolysis of the donor red cells. Death may be the only sign associated with a transfusion reaction in a bird.

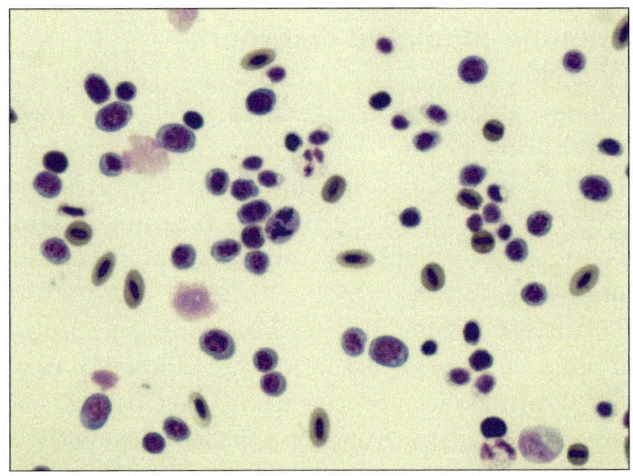

Figure 25.5 **Blood cell sample with mitotic figures, poikilocytosis (variability in erythrocyte shape), and binucleate erythrocytes indicating pronounced regenerative anaemia.**

DISEASES OF THE HAEMOPOIETIC SYSTEM

Anaemia

Definition/overview

Anaemia is usually the result of increased loss, increased destruction, and/or decreased production of erythrocytes. It is classified as regenerative, non-regenerative, haemolytic, or haemorrhagic.

- *Regenerative anaemia.* This is characterised by the presence of polychromasia, reticulocytosis, macrocytosis, and anisocytosis (**Figs. 25.4, 25.5**). It is an indication of an appropriate bone marrow response. In order to distinguish between polychromasia and reticulocytosis, the blood smear must be vitally stained with new methylene blue.
- *Non-regenerative anaemia.* This is the most common type of anaemia described in birds. It arises due to a lack of appropriate bone marrow response combined with the relatively short life of avian erythrocytes. It is characterised by a lack of a regenerative response (i.e., little or no reticulocytosis, polychromasia, or anisocytosis). Possible aetiologies include:
  - Acute, overwhelming infections: bacterial, viral (e.g., PBFD), or fungal.
  - Chronic diseases include chronic inflammatory conditions, chronic infectious diseases (tuberculosis, colibacillosis, salmonellosis, aspergillosis, chlamydiosis), some viral diseases, hypothyroidism, and neoplasia.
  - Toxicosis.
  - Nutritional deficiencies.

- *Haemolytic anaemia*. This is often regenerative in nature and is indicated by increased polychromasia, macrocytosis, anisocytosis, and reticulocytosis. Possible aetiologies include:
- Haemoparasites (*Plasmodium* spp. and *Aegyptianella* spp.).
- Bacterial septicaemia.
- Acute toxicosis (oil ingestion, lead, aflatoxicosis).
- Immune-mediated conditions.
- *Haemorrhagic anaemia*. This results from blood loss due to trauma, gastrointestinal parasitism, coagulation disorders (as in conure bleeding syndrome), organ rupture or ulceration, aneurysms, and some viral diseases. Following acute haemorrhage, there is often a rapid response during which erythropoiesis increases dramatically. Healthy birds are remarkably adept at dealing with acute blood loss, and haemorrhagic shock rarely occurs. Studies have shown that after the loss of even 50% of the blood volume, the packed cell volume (PCV) returns to normal within 3–7 days.

Clinical signs

Clinical signs in birds reflect the oxygen-carrying role of erythrocytes. Anaemic birds are not well oxygenated, and so clinical signs include:

- Weakness/collapse
- Tachypnoea
- Pale tongue and oral mucous membranes
- Heat conservation responses – immobile, fluffed feathers, eyes closed

Diagnosis

Diagnosis involves a review of the bird's haematology, in particular the PCV, total plasma solids, and erythrocyte morphology. The latter is the tool most likely to assist in diagnosing the type of anaemia.

Diagnostically important morphologic characteristics of erythrocytes include the colour of the erythrocytes, their stage of development, their size, shape, and the presence of inclusion bodies.

- Variations in erythrocyte colour include polychromasia and hypochromasia. Polychromatophilic erythrocytes occur in low numbers (usually <5% of erythrocytes) in the peripheral blood of most normal birds due to the higher turnover of nucleated cells. The cytoplasm of polychromatophilic erythrocytes is weakly basophilic, and the nucleus is less condensed than in mature erythrocytes. Polychromatophilic erythrocytes are similar in size to mature erythrocytes, and they appear as reticulocytes when stained with vital stains such as new methylene blue.
  - Hypochromatic erythrocytes are abnormally pale in colour compared with mature erythrocytes, and they have an area of cytoplasmic pallor that is greater than half the cytoplasmic volume. A significant hypochromasia is reflected as a decrease in the mean corpuscular haemoglobin concentration (MCHC) and mean corpuscular haemoglobin (MCH) values.
  - Increased polychromasia occurs with blood loss and haemolytic anaemias.
  - Non-regenerative anaemia is associated with hypochromasia.

- Immature erythrocytes are uncommon in the normal blood films of birds. Immature erythrocytes are round to irregular cells with large, round nuclei, and basophilic cytoplasm. The nucleus lacks the dense chromatin clumping of the mature cell. Immature erythrocytes often appear to be larger than mature erythrocytes.
- Changes in the size of erythrocytes include microcytosis, macrocytosis, and anisocytosis. A significant change in the mean size of erythrocytes is reflected in the mean corpuscular volume (MCV). The presence of macrocytes or microcytes should be noted during the assessment of the blood film. The degree of variation in the size of avian erythrocytes (anisocytosis) can be scored from 1+ to 4+ based on the number of variable-sized erythrocytes in a monolayer field. Macrocytic erythrocytes are usually the polychromatic erythrocytes associated with erythroid regeneration.
- In most species of birds, the shape of the erythrocytes is relatively uniform.
- Inclusions within erythrocytes include:
  - Mitotic figures and binucleate erythrocytes suggest a marked regenerative response or erythrocytic dyscrasia.
  - Perinuclear rings are common artefacts of improper slide preparation (e.g., exposure to solvent or formalin fumes, or allowing the slide to dry too slowly), and they represent nuclear shrinkage.

- Clear, irregular, refractile spaces in the cytoplasm occur when blood films are allowed to dry too slowly. This artifact, which is a form of erythrocyte crenation, should not be confused with blood parasites.
- Blood parasites, such as gametocytes of Hemoproteus and Plasmodium.

Treatment

Anaemic birds with greater than 10% polychromasia (3+ and 4+ polychromasia) are exhibiting an appropriate regenerative response to their anaemia. Those with a smaller response, however, are not. The number of reticulocytes also indicates a bird's current response to anaemia. Therefore, the reticulocyte count can be used in conjunction with the assessment of the degree of polychromasia to determine the bird's current erythropoietic response.

Many anaemic birds that are responding appropriately to their condition may not require more than general supportive care until their PCV has returned to normal. Other birds may require a blood transfusion and oxygen therapy. See Chapter 5 for a discussion on this treatment modality.

Polycythaemia

Definition/overview

Polycythaemia, defined as a high PCV (often greater than 60%), is uncommon in birds. It can be:

- Primary (or absolute) polycythaemia, a myeloproliferative disorder.
- Secondary polycythaemia is associated with chronic hypoxia due to:
  - Pulmonary disease.
  - Cardiac insufficiency.
  - Chronic hypoxia associated with high altitudes.
  - Increased erythropoietin production associated with renal disease.
  - Dehydration may also cause temporary polycythaemia.

Clinical signs

Clinical signs are often non-specific and may predominantly reflect the underlying condition.

Diagnosis

The history may suggest underlying clinical conditions. For example, exposure to respiratory irritants such as smoke, strong odours or fumes, sprays and aerosols, dust, powder down, and low environmental humidity may suggest a primary respiratory condition.

Necropsy/biopsy may reveal extramedullary haematopoiesis in the spleen, liver, and kidneys.

Treatment

Treatment revolves around identifying and eliminating causative factors. Phlebotomy (5–10 ml/kg every 6 weeks) to reduce the PCV has been suggested as an intermediate step while eliminating the cause of the condition.

Myeloproliferative disorders

Myeloproliferative disorders are neoplastic proliferation of non-lymphoid haematopoietic cells. These neoplastic cells generally infiltrate the spleen, liver, and bone marrow. The infiltration can result in massive enlargement of both the liver and spleen. This condition is rare in birds.

FURTHER READING

Gerlach H (1994) Defence mechanisms of the avian host. In: *Avian Medicine: Principles and Application.* BW Ritchie, GJ Harrison, LR Harrison (eds). Wingers Publishing, Lake Worth, pp. 109–120.

Johnston MS, Son TS, Rosenthal KL (2007) Immune-mediated hemolytic anemia in an Eclectus parrot. *Journal American Veterinary Medical Association* 23(7): 1028–1031.

Jones JS, Thomas JS, Bahr A, Phalen DN (2002) Presumed immune-mediated hemolytic anemia in a blue-crowned conure (Aratinga acuticaudata). *Journal Avian Medicine and Surgery* 16(3): 223–229.

Lightfoot TM (2006) Clinical avian neoplasia and oncology. In: *Clinical Avian Medicine.* Harrison G, Lightfoot TM (eds). Spix Publishing, Fort Worth Florida.

Ritchie BW (1995) Viral attack and avian response. In: *Avian Viruses: Function and Control.* BW Ritchie (ed.). Wingers Publishing, Lake Worth, pp. 47–82.

Schmidt RE (1997) Immune system. In: *Avian Medicine and Surgery.* RB Altman, SL Clubb, GM Dorrestein, K Quesenberry (eds). WB Saunders, Philadelphia, pp. 645–652.

CHAPTER 26

DISORDERS OF THE NERVOUS SYSTEM

ANATOMY

The brain

In some ways, the avian brain is like that of the mammalian brain, particularly in the hindbrain (rhombencephalon) and the midbrain (mesencephalon). Evolutionary pressures have seen the greatest differences develop in the forebrain (telencephalon and diencephalon), especially in the size and position of the neocortex. Mammals have evolved a large superficial neocortex, whereas in birds it is much smaller and deep within the cerebral hemisphere.

The hindbrain, a continuation of the spinal cord, is composed of a large medulla oblongata, a small pons, and the cerebellum. The medulla oblongata contains the cardiac, respiratory, vomiting, and vasomotor centres and so deals with the autonomic (involuntary) functions of breathing, heart rate, and blood pressure. Cranial nerves XII to V (inclusive) arise from the ventral and lateral aspects of the medulla. The cerebellum is attached to the dorsal aspect of the medulla and the optic lobe (below) by two cerebellar peduncles. It consists of a large median tri-lobed body (the vermis) flanked by two smaller cerebellar hemispheres. The cerebellum in many species is relatively much larger than that of mammals, reflecting the need for precise coordination when flying and when using a beak and tongue to manipulate objects.

The midbrain consists largely of the mesencephalic (optic) tectum, or optic lobe. This optic lobe is relatively much larger than that of mammals, reflecting the importance of vision to birds. Two cranial nerves (III and IV) arise from the ventral side of the midbrain, and the cerebellum is attached to the midbrain by the cranial cerebellar peduncle.

The diencephalon is an extension of the midbrain. On the dorsal aspect, the pineal gland arises from the epithalamus and sits in the triangle formed by the two cerebral hemispheres and the cerebellum. The pineal gland responds to light via the eyes and the brain and plays a role in diurnal and seasonal rhythms in birds. Ventrally, the optic chiasma leads to the left and right optic tracts; immediately caudal to this is the ventral surface of the hypothalamus and the hypophysis. The thalamus lies within the diencephalon and acts as a relay for afferent pathways ascending into the cerebral hemispheres. As such, it plays a role in vision and hearing, as well as other functions. The hypothalamus lies within the ventral portion of the diencephalon, continuous with the neurohypophysis and then the hypophysis. It dominates all autonomic functions such as thermoregulation, respiration, circulation, thirst, appetite, reproduction, and behaviour.

The telencephalon consists of the cerebral hemispheres and the olfactory bulb. The olfactory bulb is relatively small and protrudes from the rostral forebrain. The left and right cerebral hemispheres, separated dorsally by the median fissure, are almost smooth (particularly when compared to the mammalian brain). The thin cortex overlies a large amount of grey matter, which forms the bulk of the hemisphere. (In contrast, the mammalian cortex is much thicker.) Caudally, the cerebral hemispheres overlap the optic lobe of the midbrain. The cerebral hemispheres play a key role in the development of complex behaviours in birds, especially those involving learning.

The brain is enclosed by the same three meninges that enclose the spinal cord, i.e., the dura, arachnoid, and pia maters. The dura is attached to the periosteum except where the dural venous sinuses separate the dura and periosteum. Dural folds separate the forebrain from the optic lobes and the optic lobes from the cerebellum. The subarachnoid space expands slightly at the base of the cerebellum, dorsal to the medulla oblongata, forming the cisterna magna. Although cerebrospinal fluid (CSF) can be obtained from this cistern at the foramen magnum,

DOI: 10.1201/9781003272519-26

this procedure is extremely hazardous due to the proximity of the vertebral venous sinus.

Cranial nerves

As with mammals, birds possess 12 pairs of cranial nerves.

I. *Olfactory nerve.* A sensory nerve connecting the nasal epithelium to the olfactory bulb of the forebrain via the olfactory foramen (there is no cribriform plate as there is in mammals). Both nasal cavities are innervated by separate branches, which combine into a single nerve at the nasal bone; this then runs along the dorsal border of the interorbital septum to the olfactory foramen and then the olfactory lobe.

II. *Optic nerve.* A sensory nerve from the retina to the optic chiasma. The optic nerve is the largest of the cranial nerves, reflecting the visual nature of birds. It penetrates the skull at the optic foramen, decussates at the optic chiasma, and continues as the optic tract to the thalamus. The optic nerves are relatively short, and care must be taken when enucleating one eye that the optic nerve of the other eye is not avulsed accidentally.

III. *Oculomotor nerve.* A motor nerve arising from the midbrain and exiting the cranium either via the optic foramen or the oculomotor foramen. After exiting, the nerve divides into a dorsal branch (upper eyelid levator muscle and the dorsal rectus muscle) and a ventral branch (ventral rectus, medial rectus, and ventral oblique muscles). The ventral branch also gives off the iridociliary nerve, which, after receiving a branch of the trigeminal nerve (V), provides parasympathetic innervation to the iris and ciliary body.

IV. *Trochlear nerve.* A motor nerve arising from the dorsal midbrain, exiting through the trochlear foramen into the orbit, and innervates the dorsal oblique muscle.

V. *Trigeminal nerve.* A nerve arising from the brainstem at the caudal edge of the optic lobe. It divides almost immediately into the ophthalmic nerve and the combined maxillary and mandibular nerves. The *ophthalmic nerve* is the main sensory nerve of the nasal cavity and the wall of the eyeball. It supplies the iris and ciliary body (along with the oculomotor nerve III), the nasal mucosa, the edge and tip of the upper beak, the palate, the upper eyelid and the adjacent skin, and the rostral nasal cavity. The combined maxillary and mandibular nerves separate either before or immediately after exiting the skull through the maxillomandibular foramen and sometimes the mandibular foramen. The *maxillary nerve*, also a sensory nerve, divides into three branches: the supraorbital nerve supplying the upper eyelids and crest; the infraorbital nerve supplying the lower eyelids; and the nasopalatine nerve supplying the lacrimal glands and the glands of the nasal mucosa. The *mandibular nerve*, primarily a sensory nerve, innervates the muscles of mastication, the skin along the mandible, and the floor of the oral cavity.

VI. *Abducent nerve.* This motor nerve arises from the rostral medulla oblongata and exits the skull rostrally through the abducent foramen. It innervates the lateral rectus muscle and the nictitating membrane muscles.

VII. *Facial nerve.* This primarily motor nerve arises from the ventrolateral medulla oblongata, enters the internal acoustic meatus, and then passes along the facial canal. As it exits the facial nerve foramen, it divides into the palatine nerve and the hyomandibular nerve. The *palatine nerve* divides into a dorsal branch (ethmoidal ganglion, innervating the nictitating membrane gland, nasal gland, and nasal glands) and a ventral branch (sphenopalatine ganglion, innervating nasal glands and caudal palate). The *hyomandibular nerve* innervates depressor muscles of the mandible and the stylohyoid muscle.

VIII. *Vestibulocochlear nerve.* A sensory nerve arising from the medulla oblongata. This nerve is divided into the *vestibular part* and *the cochlear part*. It receives sensation from the ear and vestibular apparatus.

IX. *Glossopharyngeal nerve.* This nerve arises from the ventrolateral medulla oblongata with the vagus and accessory nerves. It divides into three branches: the *lingual nerve* (sensory from the epithelium of the tongue and the taste buds, motor to the sublingual salivary gland); the *laryngopharyngeal nerve* innervates the muscles of the larynx and the pharynx;

and the *descending oesophageal nerve* innervates the cervical oesophagus and trachea.

X. *Vagus nerve.* The vagus nerve arises from the medulla oblongata and exits the skull through the foramen of the vagus nerve. It runs down the neck alongside the jugular vein and, after passing through the thoracic inlet, divides into the *nerve to the carotid body*; *glandular filaments* to the thymus, thyroid, ultimobranchial gland, and the parathyroid gland; an *aortic nerve*; the *cranial cardiac nerve*; the *recurrent nerve*, which innervates the crop, oesophagus, tracheal muscles, and syringeal muscles; the *pulmoesophageal nerve* (oesophagus and lung); *pulmonary rami* (lungs); and the *caudal cardiac nerves*. The left and right vagal trunks then unite to innervate the proventriculus, ventriculus, duodenum, and the liver.

XI. *Accessory nerve.* The accessory nerve arises from both the medulla oblongata and the cranial spine. The spinal part passes rostrally through the foramen magnum and anastomoses with the rest of the nerve, which then exits the skull with the vagus nerve. It innervates the cucillaris capitus muscle (the avian equivalent to the trapezius muscle).

XII. *Hypoglossal nerve.* Arising from the ventral medulla oblongata and exiting via the hypoglossal foramen, the hypoglossal nerve anastomoses with the first cervical spinal nerve to form the *hypoglossocervical nerve*, and then with the vagus and glossopharyngeal nerves. It then divides into the descending cervical nerve (tracheal muscles), the laryngolingual ramus (tongue muscles), and the tracheal ramus (syringeal muscles). This is a motor nerve.

The spinal cord

The internal anatomy of the spinal cord in birds is like that of mammals, with a central butterfly of grey matter surrounded by white matter. Unlike mammals, the avian spinal cord is the same length as the neural canal, with no *cauda equina* at its termination. The dorsal and ventral roots of the spinal nerves pass separately through the dura, coming together in the intervertebral foramina and exiting laterally rather than caudally. The cord diameter increases at the cervical enlargement (associated with the brachial plexus) and the lumbosacral enlargement (associated with the lumbosacral plexus). The cervical enlargement is greater than the lumbosacral enlargement in flying birds such as parrots but smaller in non-flighted birds such as poultry and ratites. The dorsal column of white matter in the lumbosacral enlargement is divided sagittally by the rhomboidal sinus, a cleft occupied by the gelatinous body (glial cells rich in glycogen); its function remains unclear. There may be a similar glycogen body in the brachial region.

The spinal cord is enclosed, protected, and supported by three meninges: dura mater, arachnoid mater, and pia mater. The *dura mater* is the thick outer meninx; initially fused with the periosteal lining of the vertebral canal at the foramen magnum, it is then separated from the periosteum by an epidural space extending along the cervical and notarial regions. This space is filled with a gelatinous material and, on the dorsal side, the vertebral venous sinus. The dura mater then fuses with the periosteum from the caudal notarium to the caudal end of the canal, although the venous sinus intervenes irregularly between the dura and the periosteum. The *arachnoid mater* is a thin, delicate layer lying near the dura mater, while the *pia mater* is attached to the spinal cord. The space between these two meninges is the subarachnoid space, occupied by fine denticulate ligaments suspending the spinal cord from the dura mater and a small amount of CSF.

The spinal cord derives its blood supply from the vertebral arteries (cervical and cranial notarial regions) and intersegmental branches of the descending aorta (from the level of the heart caudally). Coming off the intersegmental arteries are the vertebromedullary arteries, which divide and penetrate the dura. These arteries also feed into three longitudinal arteries (the ventral spinal artery and two dorsolateral spinal arteries). The dorsolateral spinal arteries supply the white matter on the dorsal cord, while the ventral spinal artery gives off branches (the sulcal arteries) supplying the bulk of the cord. Venous drainage is via the vertebral venous sinus.

Spinal nerves

The spinal nerves emerge laterally between the vertebrae, with the total number varying between species (due to the species-variance in the number of vertebrae). Each nerve has a dorsal (afferent) and a ventral (efferent) root, which combine to form a mixed spinal nerve. The size of the spinal roots and nerves is determined by the relative importance of the nerves; e.g., in flying birds, the nerves in the brachial plexus region are larger than those in the lumbosacral plexus. Each spinal nerve divides into three branches: the meningeal branch (innervating the spinal meninges), the dorsal ramus (innervating the epaxial muscles and dorsal skin), and the ventral ramus (innervating the hypaxial muscles and ventral skin, as well as forming the brachial and lumbosacral plexuses).

The nerves of the wing arise from the brachial plexus, formed by the ventral rami of four or five notarial spinal nerves, known as the roots of the plexus. These roots combine to form two or three short trunks of the plexus, which then divide into a dorsal and ventral division. The dorsal divisions unite to form a dorsal fascicle, and the ventral divisions form a ventral fascicle. The dorsal fascicle forms the peripheral nerves that innervate the dorsal (extensor) muscles of the wing; these nerves include the radial nerve and the axillary nerve and serve to elevate the wing. The ventral fascicle innervates the ventral (flexor) muscles and includes the large pectoral nerves, the medioulnar nerve, the ulnar nerve, and the median nerve; these nerves serve to lower the wing. The ventral fascicle also gives rise to the supracoracoid nerve, which activates the upstroke of the wing through its innervation of the supracoracoid muscle.

The lumbosacral plexus is formed by the ventral rami of eight spinal nerves and is divided into the lumbar plexus (three roots, one of which is shared with the sacral plexus) and the sacral plexus (six roots, including one from the lumbar plexus). These plexuses then give rise to the peripheral nerves of the leg – the femoral nerve and its branches come off the lumbar plexus, and the sciatic nerve comes off the sacral plexus. The roots of the lumbar plexus are in contact with the dorsal surface of the cranial division of the kidney, while the roots of the sacral plexus and the start of the sciatic nerve are embedded in the medial division. Compression of these nerves by renal enlargement will result in paresis or paralysis of the affected leg.

The pudendal and caudal plexuses supply the nerves which innervate the tail, vent sphincter, and cloaca. The pudendal plexus is located in the region of, and within, the caudal renal division.

Autonomic nervous system

The autonomic nervous system acts as a control system functioning largely below the level of consciousness. It affects the heart rate, digestion, respiration rate, salivation, pupillary size, and other functions. Whereas most of its actions are involuntary, some, such as breathing and pupillary size, work in tandem with the brain and peripheral nerves. The autonomic nervous system in birds, as with mammals, has two divisions: the craniosacral (parasympathetic) system, which serves to conserve body resources; and the thoracolumbar (sympathetic) system, which induces physiological responses appropriate to flight. Both work together to preserve homeostasis of the internal organs via dual innervation, which is both afferent and efferent, eliciting a specific organ function or response, e.g., cardiac function, reproductive activity, gastrointestinal integrity, renal function, pain response, and pancreatic and hepatobiliary functions.

Parasympathetic system

The cranial division of this system is formed by cranial nerves III, VII, IX, and X. They distribute efferent pathways to the eye, the glands of the orbit, the nasal cavity, the salivary glands, the heart, the lungs, and the digestive tract as far as the duodenum. The sacral division is formed by the pudendal spinal plexus and pudendal nerve, and supplies the rest of the digestive tract, the urogenital tract, and the cloaca. The parasympathetic system is mediated by the neurotransmitter acetylcholine and is primarily concerned with conserving and restoring a steady state in the body.

Sympathetic system

This system is based on a chain of paravertebral ganglia (the sympathetic chain) running from the base of the skull to the tail. The first (and largest) ganglia distribute sympathetic fibres to the eye, the glands of the head, and many of the large vessels in the head. These ganglia give off the cardiac nerves, splanchnic nerves, and intestinal nerves. The sympathetic system is mediated by catecholamines (epinephrine, norepinephrine, and dopamine) and is concerned with the fight-or-flight response.

DISEASES OF THE NERVOUS SYSTEM

Introduction

Birds showing neurological disorders, including loss of consciousness, seizures, abnormal mentation, ataxia, paresis or paralysis of legs and/or wings, cloacal atony, head tilt, and nystagmus, are frequently presented to veterinarians.

Diagnosing neurological disorders requires a thorough examination in order to determine if the neuropathy is focal or diffuse. If the examination finds that dysfunction is present at more than one level, the lesion can be assumed to be either located at the highest location or multifocal. If the dysfunction is focal, an attempt is then made to localise the lesion to the head, cervical, notaral, or synsacral spinal cord. If the examination fails to localise the lesion, a metabolic or generalised neuromuscular lesion should be considered.

As with any examination, a neurological assessment begins with a detailed history and involves a thorough physical examination before moving on to diagnostic testing to confirm a diagnosis or further localise the lesion.

History

The bird

The age of the bird is important; hereditary and inflammatory conditions are more common in the young bird, while degenerative and neoplastic conditions are more frequent in the older bird. Questions should be asked about diet, as malnutrition can be a primary cause of some neurological conditions. The bird's lifestyle should be determined with regard to access to toxins and the likelihood of trauma. Reproductive activity may indicate the possibility of a yolk embolism. It is also important to ask the owner for details of any previous medical problems.

The problem

If the condition is of acute onset, one should consider toxicosis, trauma, vascular accident, or generalised severe inflammation. If onset is more chronic, degenerative changes, neoplastic disease, or low-grade inflammation are more likely. If the disorder is progressive, a clear description of the chronological development of the clinical signs must be obtained.

The owner should be asked about the bird's mental state and demeanour, including changes to personality and behaviour. A precise description of the clinical signs is important (e.g., seizures may be difficult to differentiate from syncope). Generally, seizures are characterised by ataxia; disorientation; and falling off the perch, remaining rigid, or having some form of motor activity for varying lengths of time.

Distant examination

Mental status

If the bird is showing uncharacteristic aggression, a psychomotor-like epilepsy or a space-occupying mass in the forebrain should be considered. A depressed response to stimuli may result from lesions in the brainstem or forebrain. Semi-consciousness may be caused by acute encephalitis, forebrain trauma, or neoplasia.

Posture

If there is a head tilt, lesions may be present in cranial nerve VIII. Tremors may be caused by lesions in the cerebellum or vestibular system, and falling or loss of balance may also be because of lesions in the vestibular system.

Flight

Uncoordinated flight or non-rhythmic fanning of wings may be related to a poor wing clip or lesions in the cerebellum, vestibular system, or corpora striata. If the bird has poor obstacle avoidance, lesions in the eyes, cranial nerve II, or the visual centre in the brain (neoplasia, abscess, granulomas) should be considered.

If the bird has poor take-off or landing, one should consider a poor wing clip or lesions in the cerebellum, vestibular system, visual system, spinal cord, or peripheral nerves. Paresis (flight not sustainable for lengthy periods) may be due to systemic disorders or lesions in the spinal cord or peripheral nerves.

Gait

Ataxia may be caused by lesions in the cerebellum or vestibular system. Dysmetria (inability to control or limit movement) may be due to lesions in the cerebellum. Circling can result from lesions in the vestibular system (unilateral) or corpora striata.

Poor obstacle avoidance may be due to lesions in the eyes, cranial nerve II, or the visual centre; and poor righting reflex may be a result of lesions in the vestibular system, corpora striata, visual or proprioceptive systems, cerebellum, spinal cord, or peripheral nerves.

Other

Difficulties with prehension (abnormal tongue movement, reduced beak strength) may be a result of lesions in cranial nerves V and IX–XII. Reduced perching ability (not able to grasp a perch and support weight) can be due to lesions in the peripheral receptors, corpora striata, vestibular or visual system, cerebellum, spinal cord, or peripheral nerves.

Physical examination

A physical examination can be conducted after obtaining a thorough history and observing the patient from a distance. The examination should be performed methodically and logically. However, the order of the neurological examination depends on the clinical condition and cooperation of the patient. It is important to take into account the degree of cooperation of the patient when interpreting responses to neurological tests. Examination of the neurological system is detailed in Chapter 4.

Diagnostic tests

Tests used in the diagnosis of neurological disorders include haematology and biochemistry, toxin analysis (especially blood lead levels), diagnostic imaging (radiography, CT, MRI), PCR and serology for infectious diseases such as avian bornavirus and chlamydia, electroencephalograms (continuous recordings of the electrical activity of the cerebral cortex), and electromyograms (can be used to detect diseases of motor neurone cell bodies, ventral nerve roots, nerve plexuses, peripheral nerves, neuromuscular junctions, and muscle fibres).

CENTRAL NERVOUS SYSTEM DISORDERS

Congenital/hereditary conditions

- Hydrocephalus has been seen in older parrots, although it is possibly acquired rather than hereditary.
- Cerebellar hypoplasia has been reported in two juvenile African grey parrots.
- Lafora body neuropathy has been reported mainly in cockatiels. Glycoprotein-containing cytoplasmic inclusion bodies form within neurones, probably as a result of a defect in intracellular metabolism.
- Lysosomal storage disease has been reported in emus, but not in companion birds. Teratogenic toxins such as dioxins can lead to asymmetrical brain development.

Infectious diseases

Viral diseases

Paramyxovirus 1, 2, 3, and 5

Definition/overview

Paramyxovirus 1 (PMV-1, Newcastle disease) is seen in poultry, parrots, and pigeons. PMV-2 is most commonly isolated from passerines. PMV-3 is the main paramyxovirus affecting parrots. Transmission of PMV-1 is horizontal through respiratory, faecal and oral secretions. Mechanical vectors (wind, insects, equipment, and humans) may also spread the virus. Vertical transmission is possible. The incubation period may be between 3 and 28 days, depending on the strain. PMV-1 is a zoonotic disease. In humans (the only mammal to have clinical signs), there is mild acute granular conjunctivitis, general malaise, and sinusitis, with recovery in 7–20 days.

Clinical presentation

PMV-1

PMV-1 may result in sudden death. Clinical signs can involve the CNS, respiratory, and gastrointestinal systems, including depression, tremors, paralysis, a twisted neck, ataxia, opisthotonos, torticollis, conjunctivitis, nasal discharge, polyuria, and diarrhoea.

PMV-3

PMV-3 may cause pancreatitis, lymphoplasmacytic myocarditis, otitis media, and otitis interna. In Neophemas, there is pulmonary oedema/congestion and hepatomegaly (high flock morbidity and low mortality). In nestling cockatiels, there may be opisthotonos, tremors, leg paralysis, dyspnoea, and a high mortality. In finches, there may be initial conjunctivitis, anorexia, yellowish diarrhoea, and dyspnoea.

Diagnosis

Diagnosis is made through serological testing (PMV-3 can cross-react with PMV-1) or viral isolation and culture.

Management

Vaccines (live or inactivated) are available but may not be suitable for all species. The virus is inactivated by high temperatures (>56°C), sunlight, detergents, 1% chloramine, sodium hypochlorite, lysol, phenol, and 2% formalin.

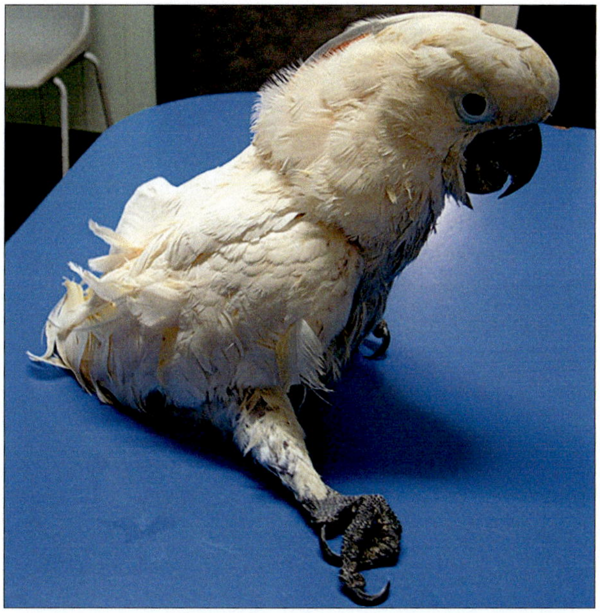

Figure 26.1 Neurological signs in a juvenile Moluccan cockatoo, subsequently confirmed as avian bornavirus.

Avian ganglioneuritis (avian bornavirus; PDD)

Avian ganglioneuritis can present as a primarily neurological disease (**Fig. 26.1**) (see Chapter 19).

West Nile virus

Definition/overview

West Nile virus is a flavivirus that is transmissible to birds, people, and horses. Birds act as a reservoir for the virus, while mosquitoes act as vectors. The disease is seen in Africa, the Middle East, Europe, Asia, and North America. In birds, death is caused by myocarditis and encephalitis.

Clinical presentation

Clinical signs include sudden death or depression, anorexia, weakness, weight loss, and recumbency. Some birds show neurological signs, including abnormal head/neck posture, ataxia, tremors, circling, disorientation, paresis, and impaired vision.

Management

Prevention revolves around vaccination and mosquito control.

Avian polyomavirus

Polyomavirus in budgerigars can cause cerebellar disease with marked intention tremors. It has also been observed in cockatoos with concurrent psittacine beak and feather disease (PBFD).

Herpesvirus

Pigeon herpesvirus can cause encephalomyelitis in pigeons. Parakeet herpesvirus affects *Neophema* and *Psittacula* parrots. They present with stargazing, stumbling, tremors, torticollis, and dyspnoea. Marek's disease is seen in poultry, turkeys, and quail.

Adenovirus

Adenovirus rarely acts as a primary pathogen; one case the authors saw was secondary to PBFD. Neurological signs have been noted in cockatiels and budgerigars. Non-suppurative encephalitis is seen on pathology.

Togavirus

Togavirus causes eastern equine encephalitis and western equine encephalitis in emus, with clinical signs including ataxia, depression, anorexia, haemorrhagic diarrhoea and emesis, and ending with death. Togavirus is also considered the cause of avian viral serositis, which can result in non-suppurative meningitis and encephalitis in parrots.

Other viruses

- A suspected viral encephalomyelitis is seen in lorikeets and *Polytelis* parrots presenting with progressive bilateral paralysis of the legs, with clenched feet.
- Avian influenza virus may cause loss of balance, ataxia, and torticollis in some birds, although neurological signs are relatively rare.

Bacterial infections

Bacteria identified as causing meningitis, encephalitis, and myelitis include *Salmonella*, *Listeria*, *Staphylococcus*, *E. coli*, *Pseudomonas*, *Klebsiella*, and *Mycobacterium*.

Fungal diseases

Aspergillus spp. can invade the brain, spinal cord, and meninges via haematogenous spread or extension from the nasal cavity, sinuses, or air sacs (see Chapter 23).

Chlamydia

Chlamydia can produce a non-suppurative meningitis (see Chapter 20). This can result in seizures, tremors, torticollis, and opisthotonos. In cockatiels, weakness or paralysis of the legs is occasionally seen.

Protozoal diseases

Sarcocystis falcatula

Sarcocystis falcatula can cause encephalitis in parrots and passerines (see Chapter 18).

Toxoplasma gondii

Definition/overview

Toxoplasma gondii infection is noted primarily in Galliformes and passerines but has been diagnosed in parrots, raptors, and owls. Cats shed the infective oocysts, which are then consumed by an intermediate host or by a bird.

Clinical presentation

Clinical signs include anorexia, diarrhoea, blindness, conjunctivitis, head tilt, circling, and ataxia.

Diagnosis

Toxoplasmosis is diagnosed by serology or histopathology.

Management

Treatment is with pyrimethamine, toltrazuril, diclazuril, or trimethoprim/sulphadiazine.

Leukocytozoon

Leukocytozoon may form megaloschizonts in the brain (see Chapter 18).

Nematodes

For the clinician and pathologist alike, the diagnosis of cerebrospinal nematode infections in birds can be problematic; the clinical signs are non-specific, and there are no suitable antemortem tests specific for these parasites. Diagnosis is often made by necropsy; fortunately, this condition is rare.

Baylisascaris procyonis (the raccoon roundworm) causes ataxia, recumbency, and other locomotor abnormalities in birds eating the faeces of infected raccoons. It has been reported in cockatoos, cockatiels, and emus in the USA.

Chandlerella quiscali is a parasite of grackles in the USA, transmitted by gnats. When emus are infected, adult worms develop in the lateral ventricles of the cerebrum, causing torticollis, ataxia, recumbency, and death.

Wild-caught cockatoos from Indonesia are occasionally found to have microfilaria plugging the small vessels in the brain.

Migrating larvae of the rat lungworm (*Angiostrongylus cantonensis*) have been reported in the brains of yellow-tailed black cockatoos (*Zanda funereus*) displaying neurological signs in Australia.

Trauma

Aetiology

Concussive head injuries are common in fully flighted birds that fly into ceiling fans, windows, or aviary wire. They may also occur when a bird is attacked by another bird, either a cage mate or a wild predatory bird.

Clinical presentation

Clinical signs include depression, head tilt, circling, and unilateral or bilateral paresis of the wings or legs. Radiographically, skeletal trauma may be visible. In many birds, an air-filled space is visible between the heart and the lung, presumably because of air sac rupture.

Diagnosis

Diagnosis is based on history and neurological examination or on necropsy. Traumatic injuries usually result in haemorrhage in the meninges, extending into the brain parenchyma. There is often associated bruising and trauma to the skin and skull.

Post-mortem pooling of blood in the venous sinuses of the calvarium is common with any cause of death and must not be confused with traumatic injuries.

Treatment

Shock, pain, and infection should be treated as appropriate. Some birds may require assisted feeding until neurologically stable.

Prognosis

Prognosis improves with neurological improvement. In some cases, this may take several weeks; improvement after 3–4 weeks is unlikely.

Cerebrovascular accidents

Aetiology

Thrombosis and/or aneurysm rupture can lead to acute ischaemia and/or haemorrhage within the brain. Common causes of these emboli include atherosclerosis in older birds and yolk emboli in reproductively active females.

Clinical presentation

Affected birds present with neurological signs that can be localised to the CNS (head tilt, nystagmus, seizures, etc.). Their signalment may suggest either cardiovascular disease or reproductive activity.

Diagnosis

Serum biochemistries may indicate either potential cardiovascular disease or reproductive activity.

Advanced diagnostic imaging, e.g., MRI, may assist in diagnosing and localising a lesion.

Treatment

See Chapter 22 for a discussion on the treatment of atherosclerosis in birds. If yolk embolism is suspected, reproductive rest is recommended. Meloxicam or other NSAIDs may be helpful.

Toxins

Lead toxicosis

Definition/overview

Lead is both an acute toxin and a cumulative toxin. If a large amount is ingested quickly, acute toxicosis develops; however, small amounts taken in gradually will accumulate within the body until tissue levels are high enough to cause disease. There are three target organs of toxic lead intake: the nervous system, the gastrointestinal system, and erythropoietic bone marrow. The source is often not determined. It may be lead shot, leaded petrol and oil, galvanised wire, lead-based paints, lead putty, solder, foil from some champagne and wine bottles, some welds on wrought-iron cages, lead weights (curtain and fishing), bells with lead clappers, lead-lighted glass, improperly glazed ceramics, batteries, bird toys with lead weights, costume jewellery, or mirror backs.

Pathophysiology

Lead is solubilised in the ventriculus by the combination of its grinding action and the low pH (2–3.5). Once solubilised, lead is absorbed across the intestinal mucosa and is bound to metalloprotein for distribution around the body. Lead competes for calcium at the myoneurological junction, resulting in neuromuscular blockage, and it affects neuronal cell adhesion molecules and glucocorticoid receptors in neurones and glial cells. It also suppresses aminoleuvulinic acid dehydrase and heme synthetase, leading to an accumulation of aminoleuvulinic acid and protoporphyrin IX in erythrocytes, which inhibits the heme synthesis cascade. This leads to decreased erythrocyte production and anaemia.

Clinical presentation

Multifocal neurological signs include hyperexcitability, ataxia, paresis and paralysis (**Fig. 26.2**), convulsions, wing droop, and head tilt. Gastrointestinal signs include ileus, vomiting/regurgitation, and abnormally coloured diarrhoea (green to black). There may be renal signs of polyuria and polydipsia, and haematuria in some parrots, particularly Amazons and galahs. Anaemia causes lethargy, depression, and weakness.

Diagnosis

Clinical pathology

Haematology reveals a hypochromic, regenerative anaemia. Erythrocytic ballooning is common but not pathognomonic.

Biochemistry shows elevated LDH, AST, CK, and uric acid concentrations, reflecting increased protein catabolism and renal dysfunction. In waterfowl,

Figure 26.2 **Lead toxicosis in a rainbow lorikeet.**

inhibition of delta-aminoleuvulinic acid dehydrogenase to levels <86 IU/l is suggestive of lead toxicosis.

Blood lead determination is a reliable indicator of toxicosis, as 90% of circulating lead is contained in erythrocytes. There is still debate as to what constitutes a 'normal' blood lead level. It might be better to say that below certain levels, clinical signs are unlikely to be seen. What effect these 'normal' levels have on the patient is difficult to assess.

- 'Normal' blood level is <0.7 μmol/l (14.5 μg/dl, 0.145 ppm).
- 0.7–1.7 μmol/l (35.2 μg/dl, 0.352 ppm) suggests exposure.
- >1.7 μmol/l confirms lead toxicosis.

Diagnostic imaging

Plain and contrast radiographs often show a dilated proventriculus and other evidence of ileus. Radiodense particles may be visible in the crop, proventriculus, and ventriculus, but the absence of these particles does not exclude a diagnosis of lead toxicosis.

Histopathology

Acid-fast intranuclear inclusions are found in epithelial cells of proximal convoluted tubules and hepatocytes. There may also be neuronal degeneration (cerebral cortex, optic lobes, medulla oblongata) and Zenker's degeneration of the myocardium, the ventriculus, and pectoral muscles (hyaline degeneration and loss of muscle striations).

Management

Chelation

Calcium EDTA is given at a dose of 20–40 mg/kg IM q12h. There are reports of doses up to 100 mg/kg q12h being administered without any evidence of nephrotoxicity. Metal ions displace the calcium in CaEDTA to form a water-soluble complex that is excreted in the urine. The authors prefer to treat twice daily until the patient is asymptomatic (usually 3–5 days), then twice weekly until blood lead levels are in the 'normal' range (often up to 6 weeks). Response is usually rapid, with clinical improvement usually noticeable within 24–36 hours.

D-penicillamine is an effective oral chelator, but it may cause nausea and vomiting in some birds.

Removal of particles from the gastrointestinal tract

Most small particles will pass through the gastrointestinal tract in 4–5 days. Larger pieces may have to be retrieved endoscopically or by concurrent flush and suction with warmed water. As a last resort, a proventriculotomy can be performed.

Supportive care

See Chapter 5. Patients will require fluid support, nutritional support, and thermal support. Blood transfusion may be required in very anaemic patients.

Botulism

Definition/overview

Botulism is most commonly a problem with game birds, ducks, and pheasants, although it occurs sporadically in other wild or exotic species.

Aetiology

Botulism is due to the ingestion of bacterial exotoxins produced by *Clostridium botulinum*, a widely dispersed saprophytic bacterium found in soil. This toxin (type C predominantly, although types A and E are sometimes identified) develops when *C. botulinum* grows in anaerobic conditions. It is present in vertebrate carcasses, decaying vegetation and dead invertebrates found in stagnant ponds, wetlands, and marshes. Fly maggots and aquatic insects accumulate the toxin and are then ingested by birds. It is most often seen in late summer and early autumn when many ponds are drying out and water levels drop, exposing dead animals and rotting vegetation.

Pathophysiology

The toxin interferes with the release of acetylcholine (ACH) at motor end plates.

Clinical presentation

Affected birds show ataxia and flaccid paralysis of the legs, wings, and neck ('limber neck'). There may be diarrhoea, loose feathers, dyspnoea, gasping, conjunctivitis, or sudden death.

Diagnosis

Diagnosis can be difficult, as there are few lesions found on necropsy. Botulinum toxin can be detected by a variety of techniques including ELISA, electrochemiluminescent (ECL) tests, and mouse inoculation or feeding trials. The toxins can be typed with neutralisation tests in mice.

Management

Treatment involves the administration of antitoxin, if it is available, and supportive care (fluids, tube feeding, warmth, etc.). Anecdotally, meloxicam may be of some value, presumably by reducing myositis in paretic/paralysed birds. The prognosis is usually guarded to poor.

A vaccine is available in some countries. Prevention revolves around preventing access to decaying vegetation and dead animals by birds.

Organophosphates and carbamates

Definition/overview

Birds are 10–20 times more susceptible to acetylcholinesterase inhibitors than mammals. These chemicals are used as pesticides. Exposure is usually through inhalation of aerial spraying or cutaneous absorption through washing or dipping. Occasionally, wild birds are poisoned by ingesting grain laced with organophosphates or carbamates in an attempt to control wild bird numbers around grain silos.

Pathophysiology

Organophosphates and carbamates are two different classes of pesticides that both inhibit acetylcholinesterase, the enzyme that breaks down the neurotransmitter ACH. This results in ACH accumulating at nerve synapses and subsequent continued stimulation at cholinergic nerve endings and myoneural junctions.

Clinical presentation

Clinical signs include anorexia, diarrhoea, crop stasis, ptyalism, ataxia, tremors, seizures, and paralysis. Death is usually due to respiratory failure.

Prolonged exposure to organophosphates may result in 'delayed neuropathy syndrome'. The organophosphate molecules bind to neuropathy target esterase; this binding then initiates degeneration of long axons in the peripheral and central nervous systems. The effects of this are seen 7–10 days after exposure as weakness, ataxia, decreased proprioception, and paralysis. This syndrome is not associated with the inhibition of acetylcholinesterase.

Management

Atropine 0.2 mg/kg IM is given as needed until the clinical signs stop. Atropine has little or no effect on nicotinic receptors and, used alone will not counteract neuromuscular paralysis. Pralidoxime 10–20 mg/kg IM is administered every 8–12 hours as needed. It is only effective if used within the first 24 hours of exposure. It is contraindicated for carbamate toxicosis.

Chlorinated hydrocarbon (DDT)/ organochlorines

Definition/overview

These chemicals have been used widely as pesticides since 1945. Exposure occurs through aerial spraying and ingestion of poisoned insects.

Pathophysiology

The mechanism of action is unclear, but it is believed to interfere with axonal transmission of nerve impulses.

Clinical presentation

Clinical signs include thin-shelled eggs, tremors, and convulsions.

Diagnosis

Diagnosis is made by measuring levels of the chemicals in fat, brain, and liver.

Management

Treatment is unrewarding.

Dimetridazole

Dimetridazole causes convulsions, wing-flapping, and opisthotonos in budgerigars, goslings, pigeons, and ducks.

Levamisole

Definition/overview

Levamisole causes toxicity by stimulating cholinergic receptors in the autonomic ganglia, neuromuscular junctions, and CNS. However, in most studies, the reported cause of death is asphyxia and respiratory failure. In poultry, toxic reactions only occur with doses in excess of 160 mg/kg and deaths at >480 mg/kg. This is why it has gained wide acceptance as a safe anthelmintic for poultry. It is considered to be contraindicated in finches and other small passerines, and administration should be carefully monitored in other species when water intake is likely to increase, e.g., during hot weather or when young are in the nest.

Clinical presentation

Levamisole causes regurgitation, ataxia, recumbency, catatonia, dyspnoea, and death in parrots and poultry. Pulmonary oedema is a feature of levamisole toxicosis in a range of species, whether given by injection or orally.

Nitrofurazone

Nitrofurazone causes screaming, convulsions, aimless running or flying, and opisthotonos. There may be depression and growth retardation.

Nutritional deficiencies

Hypovitaminosis B1 (thiamine)

Hypovitaminosis B1 causes anorexia, ascending paralysis and opisthotonos, polyneuritis with myelin degeneration, adrenal hypertrophy, and skin oedema. This is most commonly seen in wild birds that are fed meat without supplementation. Cases respond within hours to parenteral vitamin B1.

Hypovitaminosis B2 (riboflavin)

Hypovitaminosis B2 causes curled toe paralysis in poultry and nestling budgerigars, weakness, emaciation (despite a good appetite), diarrhoea, walking on hocks with toes curled inward, leg muscle atrophy, and demyelinating peripheral neuritis with nerve oedema. Chronic cases have irreversible damage.

Hypovitaminosis E and selenium deficiency

Vitamin E deficiency in young birds causes encephalomalacia, exudative diathesis, and muscular dystrophy. It is particularly common in hatchling budgerigars.

Hypovitaminosis B6 (pyridoxine)

Hypovitaminosis B6 causes jerky, nervous walking, progressing to running and flapping the wings. The bird then falls with rapid, clonic-tonic head and leg movements, which usually lead to death.

Calcium

Definition/overview

Hypocalcaemic tetany is most common in adult grey parrots.

Aetiology

It is usually associated with the long-term feeding of a diet deficient in calcium, combined with hypovitaminosis D3 (due to lack of UVB radiation or dietary supplementation). Initially, the bird will maintain normal blood calcium concentrations by utilising the calcium reservoir present in the bones, but eventually, the homeostatic system fails and the blood calcium level will fall. In a normal adult grey parrot, the blood ionised calcium concentration is kept within a very narrow range (0.96–1.22 mmol/l). Once the homeostatic system fails and the blood ionised calcium concentration falls below the normal level, the bird will demonstrate a variety of neurological clinical signs.

Clinical presentation

Initially, the bird may just twitch or flick its head. Ataxia develops and progresses to fits or convulsions.

Diagnosis

Blood calcium levels (preferably ionised calcium) should be measured and interpreted as noted earlier.

Caution must be taken not to assume that all seizuring grey parrots are hypocalcaemic.

Treatment

The clinical signs respond rapidly to parenteral calcium and vitamin D3 supplementation. Long-term therapy should involve improvement of the diet with respect to calcium and vitamin D3 content.

UVB radiation should be provided, either from natural sunlight or artificial full spectrum lights, particularly in birds exposed to low light levels kept indoors. Care must be taken when using artificial UVB lights so that UVB-induced keratitis does not develop.

Neoplasia

Pituitary adenomas

Definition/overview

These are either adenomas or adenocarcinomas arising from chromophobe cells in the pituitary gland. They most commonly occur in young to middle-aged male budgerigars but have been reported in other parrot species including cockatiels and one case in an Amazon. A recent study (Langohr et al., 2012) concluded that the high incidence of this disease in budgerigars suggests a possible genetic predisposition of budgerigars to develop somatotroph pituitary tumours with a high incidence of local invasion and with metastatic potential.

Clinical presentation

Clinical signs are related to the compression of the brain and cranial nerves from the space-occupying mass. These signs include uncoordinated wing flapping, clonic leg twitches, seizures, circling, depression, somnolence, unconsciousness, exophthalmos, mydriasis, blindness, and inappropriate behavioural responses.

If the tumour is functional, other clinical signs can include polydipsia and polyuria, feather abnormalities, cere colour changes, and obesity.

Primary brain tumours

Primary brain tumours are rare. They include astrocytoma, glioblastoma multiforme, schwannomas, ganglioneuroma, neurofibrosarcomas, pineal body tumours, undifferentiated sarcomas, and haemangiomas. Clinical signs are similar to those seen with pituitary tumours.

Epilepsy/seizures

Overview

A seizure is the clinical manifestation of a sudden surge of excessive electrical activity in the brain. There are several characteristic stages of a seizure incident: the prodromal stage, the seizure itself, and postictal stage. The prodromal stage is characterised by restlessness or anxiety lasting several hours or even days before the seizure. The seizure starts with the bird hiding, seeking the owner, or other behavioural changes minutes to seconds before the seizure itself starts, usually lasting a few minutes. Postictally, there are temporary neurologic abnormalities such as disorientation, blindness, and ataxia for a few minutes or hours.

Seizures themselves can be focal (petit mal) or generalised (grand mal).

A bird exhibiting a focal seizure generally does not lose consciousness but exhibits localised abnormal movement of a body part (e.g., wing twitching) or unusual behaviour (fear, aggression, etc.). It may progress to a generalised seizure.

Generalised seizures generally begin with a tonic phase – all of the bird's muscles contract and the patient usually falls to one side with its limbs extended. Defecation is common during this phase. After a few minutes, the clonic phase begins, in which there is rhythmic contraction of muscles (i.e., paddling, limb twitching, or chewing). This again lasts for a few minutes before the bird regains consciousness and moves into the post-ictal stage.

Status epilepticus is defined as a seizure lasting 5 minutes or more, or two or more seizures without full recovery of consciousness between seizures.

Aetiology

Epilepsy/seizures can be due to a number of causes, either intra- or extracranial.

Intracranial

- Infectious (see earlier in this chapter)
- Non-infectious
  - Toxins (lead, organophosphates, organochlorines, chocolate, caffeine, some mycotoxins)
  - Neoplasia
  - Cerebrovascular accidents
    - Ischaemic
      - Yolk embolism
      - Atherosclerosis
    - Haemorrhagic
      - Ruptured aneurysm
      - Trauma
  - Heat stress
  - Storage diseases, such as Lafora's disease
  - Idiopathic epilepsy, reported in peach-faced lovebirds, red-lored Amazons, and greater Indian hill mynahs; it is a diagnosis of exclusion, reached after ruling out other causes of seizures

Extracranial

- Metabolic disorders (e.g., hypoglycaemia, hypocalcaemia)
- Hepatopathy/hepatic encephalopathy (rare in parrots)
- Renal disease
- Respiratory insufficiency leading to hypoxia
- Cardiac insufficiency
- Nutritional deficiencies (vitamin E, vitamin B1)
- Iatrogenic, e.g., insulin overdose

Management

Initial stabilisation of the seizuring bird may require a benzodiazepine, e.g., diazepam, clonazepam, or midazolam. Blood glucose should be assessed and, if low, a dextrose bolus (1 mL/kg of 50% dextrose IV) should be given. Similarly, if ionised calcium is low, calcium gluconate should be given by slow IV injection or IM.

If possible, other causes of the seizures should be identified and eliminated. A detailed history, a thorough physical examination, haematology and biochemistry, blood lead measurement, and radiographs are appropriate starting points.

Longer-term stabilisation requires anticonvulsant therapy such as phenobarbital, levetiracetam, or gabapentin. It is important to realise that there is relatively little data about the use of anticonvulsant therapy in birds; speed of absorption and drug metabolism are unknown and dose rates are largely extrapolated from small animal medicine. Monitoring the response to treatment is usually based on the patient's clinical signs. Serial assessments of liver function should be performed.

Prognosis

Unless the inciting cause can be identified and removed, the long-term prognosis of the seizuring bird remains guarded. However, short- to mid-term control can often be achieved with good owner compliance.

PERIPHERAL NERVOUS SYSTEM DISORDERS

Many of the aforementioned disorders may present as peripheral neuropathies. There are, however, some that are confined to the peripheral nervous system.

Paresis or paralysis

Bilateral

Spinal trauma, infection, or neoplasia

Definition/overview

If trauma, infection, or neoplasia severs or compresses the spinal cord, it can result in paresis or paralysis (**Fig. 26.3**). Depending on the level and extent of the injury or neoplasm, cloacal and vent tone may also be lost. Most spinal trauma will occur at the notarial–synsacral junction, where there is some spinal flexibility.

Management

Conservative or surgical therapy is rarely successful, and a guarded prognosis for return to normal function should be given. Non-steroidal anti-inflammatory drugs and, if indicated, tramadol may provide symptomatic therapy while awaiting any response or recovery.

Obturator paralysis

Clinical presentation

A form of 'obturator paralysis' is sometimes seen in egg-laying hens. After laying an unusually large egg, or after a difficult oviposition, the hen is paretic or paralysed in the legs (**Fig. 26.4**). This is due to compression of the pelvic plexuses causing neurapraxia.

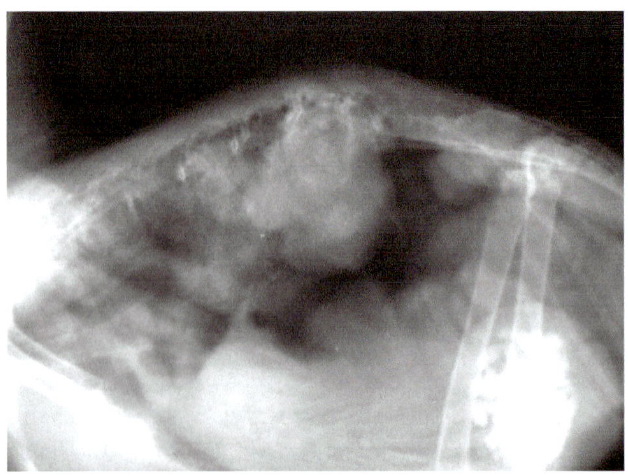

Figure 26.3 **A granuloma, later confirmed as *Aspergillus* spp., extending from a pulmonary infection, invading the spinal cord.**

Figure 26.4 **'Obturator' paralysis in an Alexandrine parrot after an episode of egg binding.**

Treatment

Rest, NSAIDs, and calcium supplementation usually result in recovery, although this may take several weeks in severe cases.

Lorikeet paralysis syndrome

Clinical presentation

Lorikeet paralysis syndrome (clenched foot syndrome) is seen in both wild and captive lorikeets. Affected birds, which can be of any age and either sex, are presented throughout the year with leg paralysis and clenched feet (**Fig. 26.5**). They are otherwise bright and alert and usually have a good appetite. Occasional individuals will become progressively more extensively paralysed.

Diagnosis

Histopathology of the cerebellum, brain stem, spinal cord, and brain shows non-suppurative leptomeningitis, perhaps consistent with viral encephalomyelitis. However, no virus has yet been identified. Lead toxicosis has been implicated in some cases, and so blood lead levels should be assessed.

Management

Other than lead toxicosis, there is no specific therapy, but some birds that are given supportive care may recover over a period of weeks or months.

Barraband (or Polytelis) paralysis syndrome

Barraband (or *Polytelis*) paralysis syndrome, seen most commonly in the Barraband or superb parrot (*Polytelis swainsonii*), is an uncommon condition similar to

Figure 26.5 **Lorikeet paralysis syndrome.**

lorikeet paralysis syndrome. No known cause has been identified. With supportive care, some birds will recover, but recurrence is seen occasionally.

Bilateral leg trauma

Definition/overview

Fledgling parrots with nutritional secondary hyperparathyroidism may present with bilateral pathological fractures of the legs.

Management
Surgical therapy may be warranted, but the prognosis is guarded.

Lead toxicosis

This may present as a bilateral paresis.

Unilateral

Unilateral trauma

Unilateral trauma, either skeletal or soft tissue, may result in unilateral paresis. Affected birds often lie in sternal recumbency with the wing on the affected side extended as a support to that side of the body. This is sometimes mistaken for wing trauma.

Sciatic nerve compression

The sciatic nerve can be compressed by pressure arising from an adjacent neoplastic enlargement of the testicle or kidney.

Pathogenesis

The roots of the lumbar plexus, giving rise to the femoral nerve, are in contact with the dorsal surface of the cranial division of the kidney, while the roots of the sacral plexus and the start of the sciatic nerve are embedded in the medial division. Compression of these nerves by renal or testicular enlargement will result in paresis or paralysis of the affected leg. This is most common in budgerigars but has been reported in other species. The high incidence of this disease in budgerigars has led to a suspicion of viral aetiology, but this remains unproven.

Clinical presentation

This is nearly always a progressive, unilateral disease with paresis on one side associated with pressure exerted on the sciatic nerve by a renal tumour (carcinoma, adenocarcinoma, and embryonal nephromas). There is normal flexion and extension of all the joints of the affected limb (**Fig. 26.6**). Withdrawal reflexes below the stifle joint typically are absent or reduced when compared to the other leg. Reflexes are normally present proximal to the stifle joint and, quite often, the bird can still move its leg from front to back but will not weight-bear. The affected leg is held loosely on the perch, with the toes partially closed.

Figure 26.6 **Unilateral leg paresis in a budgerigar with renal neoplasia.**

Diagnosis

The clinical picture is very suggestive of this disease, but confirmation can be obtained by demonstrating elevated uric acid and evidence of nephromegaly or gonadal enlargement (by either radiology, CT, or ultrasound).

Endoscopic biopsy can be used to identify the type of renal tumour present, which may, with more research, indicate the best treatment and likely prognosis.

Management

Chemotherapy (carboplatin) has been tried, but with minimal success at this stage. The prognosis is poor – once a budgerigar starts displaying obvious clinical signs of leg paresis, death usually occurs in 2–8 weeks.

Spinal neoplasia

Spinal neoplasia may produce a unilateral paresis/paralysis, although bilateral effects are more common.

Trauma

Overview

Traumatic injuries, especially to the long bones, can result in peripheral nerve damage. This damage can range along a spectrum of nerve injury: neurapraxia (temporary loss of motor and sensory function due to blockage of nerve conduction), axontmesis (an interruption of the axons, but a preservation of the surrounding connective tissues around the axon; recovery will take longer than neurapraxia), neurotmesis (both the nerve and the nerve sheath are disrupted; while partial recovery may occur, complete recovery is impossible), or complete transection. The extent of the injury determines the prognosis. In many cases, it is transient, but occasionally the damage can be permanent.

Clinical signs

Clinical signs include loss of withdrawal and pain reflexes, muscle atrophy, and paresis/paralysis. It is often difficult to assess the degree of nerve injury associated with a fracture. If the extent of the injury is confined to neurapraxia, recovery should be complete in 2–4 weeks.

Treatment

Supportive care, analgesia, and passive range of motion exercises should be provided until the extent of injury can be assessed.

Horner's syndrome

Overview

Horner's syndrome is a loss of sympathetic innervation of the eye, resulting in ptosis and occasionally miosis. The cause is usually trauma, causing interruption of the sympathetic pathway from the cranial cervical spine or the thoracic vertebrae. It is rarely reported in birds (only two cases have been documented at the time of writing), and information is therefore lacking on treatment and prognosis. However, the prognosis appears to be favourable.

Neuralgia

Overview

Foot and toe mutilation is a common presentation in psittacine patients of all species. It is perhaps more frequently observed in cockatoos, although an 'Amazon foot mutilation' syndrome has been reported. It has been proposed that this self-mutilation may be the result of neuralgia – pain that follows the path of a specific nerve.

Aetiology

The causes of neuralgia are varied. Chemical irritation, inflammation, trauma (including surgery), compression of nerves by nearby structures (e.g., tumours), and infections may all lead to neuralgia. Diabetes mellitus is another common cause of neuralgia in people. (Diabetes damages the arterial supply to the nerves, resulting in nerve fibre malfunction and sometimes nerve loss.) In many cases, however, the cause is unknown.

Clinical signs

Affected patients present with clinical signs of intense irritation of the lower extremities – foot shaking, shifting lameness, feather-damaging behaviour on the lower leg, and self-mutilation (up to and including, chewing toes off).

Diagnosis

There are no specific tests for neuralgia, and the diagnosis is usually one of exclusion. Haematology, biochemistry profiles, and skin/toe biopsies are used to rule out other conditions. Nerve conduction studies with electromyography, examining the electrical activity of nerves, may confirm the diagnosis.

Treatment

Non-steroidal anti-inflammatory drugs, such as meloxicam, may be helpful in mild cases. Opioid analgesics may be needed for a short time to control severe pain. However, these analgesics often have disappointing results. Other treatments may include nerve blocks or surgical ablation of the affected nerve using different methods, such as local radiofrequency, heat, balloon compression, and injection of chemicals.

Anti-seizure medications such as gabapentin have been found to be helpful for pain associated with neuralgia, sometimes combined with analgesics.

FURTHER READING

Antinoff N (2007) Stop the shakes! Diagnosing and treating neurological disorders in birds. In: *Proceedings of the Annual Conference of the Association of Avian Veterinarians Australian Committee*, pp. 201–215.

Delk K (2012) Clinical management of seizures in avian patients. *Journal of Exotic Pet Medicine* 21(2): 132–139.

Gancz AY, Malka S, Sandmeyer L, Cannon M, Smith D, Taylor M (2005) Horner's syndrome in a red-bellied Parrot (Poicephalus rufiventris). *Journal of Avian Medicine and Surgery* 19(1): 30–34.

Langohr IM, Garner MM, Kiupel M (2012) Somatotroph pituitary tumors in budgerigars (Melopsittacus undulatus). *Veterinary Pathology* 49(3): 503–507.

Platt SR (2006) Evaluating and treating the nervous system. In: *Clinical Avian Medicine, Vol 2*. GJ Harrison, TL Lightfoot (eds). Spix Publishing Inc, Palm Beach, pp. 493–518.

Rosenthal K, Orosz S, Dorrestein GM (1997) Nervous system. In: *Avian Medicine and Surgery*. RB Altman, SL Clubb, GM Dorrestein, K Quesenberry (eds). WB Saunders, Philadelphia, pp. 454–474.

Simova-Curd S, Nitzl D, Mayer J, Hatt JM (2006) Clinical approach to renal neoplasia in budgerigars (Melopsittacus undulatus). *Journal of Small Animal Practice* 47(9): 504–511.

Taggers A, Kelly-Bosma M, Mastakov A, Allavena R, Doneley RJT (2022) Cerebellar hypoplasia in two juvenile African grey parrots (Psittacus erithacus). *Journal of Avian Medicine and Surgery* 36(3): 308–315.

CHAPTER 27

DISORDERS OF THE REPRODUCTIVE TRACT

ANATOMY AND PHYSIOLOGY

Male reproductive tract

In the male embryo, the left gonadal region receives more germ cells than the right, leading to asymmetrical development and the development of a larger left testicle (**Fig. 27.1**). Unlike the right ovary (see later), the right testicle does not regress, so while the left testis is often larger than the right in the immature bird, this changes after maturity so that both are similarly sized. Suspended by the mesorchium, the testicles are surrounded, but not cooled, by the abdominal air sacs. The bulk of the testis is made up of thousands of convoluted seminiferous tubules with numerous anastomoses. There is no lobulation as is seen in mammals, as there are no septa present. The size of the testicle increases with sexual activity due to the increased length and diameter of the seminiferous tubules and a greater number of interstitial cells. The testicle is covered in the tunica albuginea, but there is no pampiniform plexus. The seminiferous tubules are lined by spermatogenic epithelium made up of germ cells and sustentacular cells (Sertoli cells), which provide mechanical support for the germ cells, produce steroid hormones, and may have a phagocytic role. Between the tubules are the interstitial cells (cells of Leydig), which produce androgenic hormones, especially testosterone. There may also be melanocytes present in the interstitial spaces of some species, giving the testis a black colouration.

There are three phases of spermatogenesis: the multiplication of spermatogonia; their growth into primary spermatocytes; and then the maturation of primary spermatocytes into secondary spermatocytes and then spermatids, which develop into spermatozoa. These mature spermatozoa detach and pass through a short, straight tubule into the rete testis, a thin-walled, irregular channel on the dorsomedial aspect of the testis, adjacent to the epididymis. (The rete testis is not present in all species.)

The epididymis lies on the dorsomedial side of the testis and is relatively small compared with that in mammals. It enlarges during sexual activity but has no distinct head, body, and tail because the efferent ductules, arising from the rete testis, enter along its entire length. They lead into connecting ductules and finally into the epididymal duct.

The ductus deferens runs from the epididymis to the cloaca, entering the cloaca at the urodeum. At the urodeum, it enters the receptacle of the ductus deferens, a spindle-shaped dilation embedded in the cloacal muscle. In passerines and some parrots, the caudal end of the ductus deferens forms a mass of convolutions called the seminal glomus. This enlarges in the breeding season to push into the cloaca, forming the cloacal promontory, which pushes the vent caudally. This is the main site of spermatozoal storage in these birds.

Female reproductive tract

The avian embryo has two ovaries and two oviducts (**Fig. 27.2**). During incubation, the left gonadal region receives more germ cells than the right, leading to asymmetrical development. The right ovary and oviduct usually regress, so most birds have only a left ovary and oviduct (with the exception of the kiwi and some raptors).

The ovary is located beside the cranial division of the left kidney, adjacent to the adrenal gland. The ovarian blood supply enters the ovarian hilus where it is in close contact with the dorsal coelomic wall. The arterial supply comes from the ovario-oviductal branch of the left cranial renal artery, while venous drainage is via two ovarian veins directly into the caudal vena cava. This vascular anatomy makes ovariectomy a difficult and dangerous procedure to undertake, requiring optical magnification and specialised ligating instruments.

DOI: 10.1201/9781003272519-27

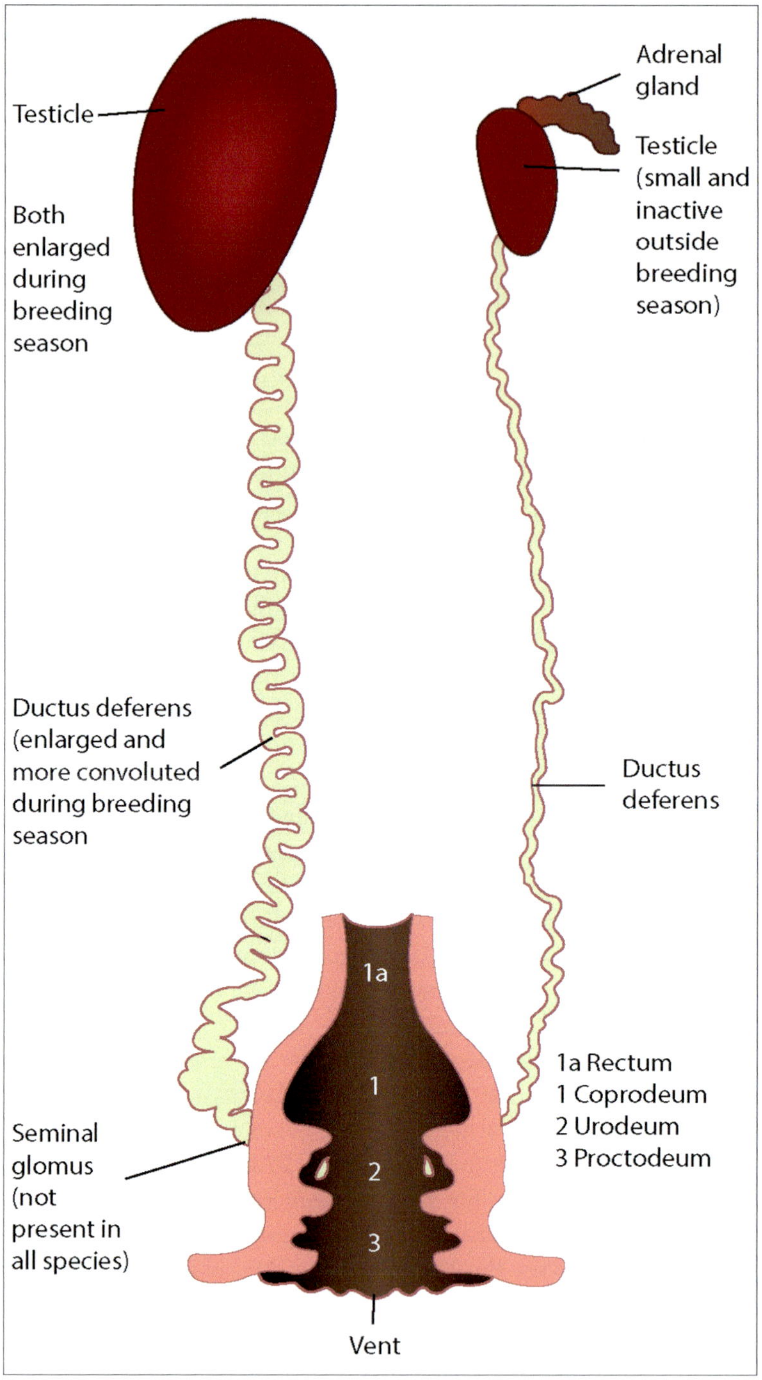

Figure 27.1 **Anatomy of the male reproductive tract.**

In seasonal laying birds (e.g., parrots), three phases of ovarian growth can be recognised:

- Prenuptial acceleration. At the beginning of the breeding season, the ovary and oviduct begin to enlarge.
- Culmination phase. Ovulation and egg laying commence.
- Refractory phase. With egg laying completed, the ovary and oviduct reduce in size.

Each follicle, containing an oocyte surrounded by the wall of the follicle, is suspended by a stalk, which possesses smooth muscle, blood vessels, and nerves. The wall of the follicle is very vascular and innervated with cholinergic and adrenergic fibres. Running across the surface is the stigma, a meridional band,

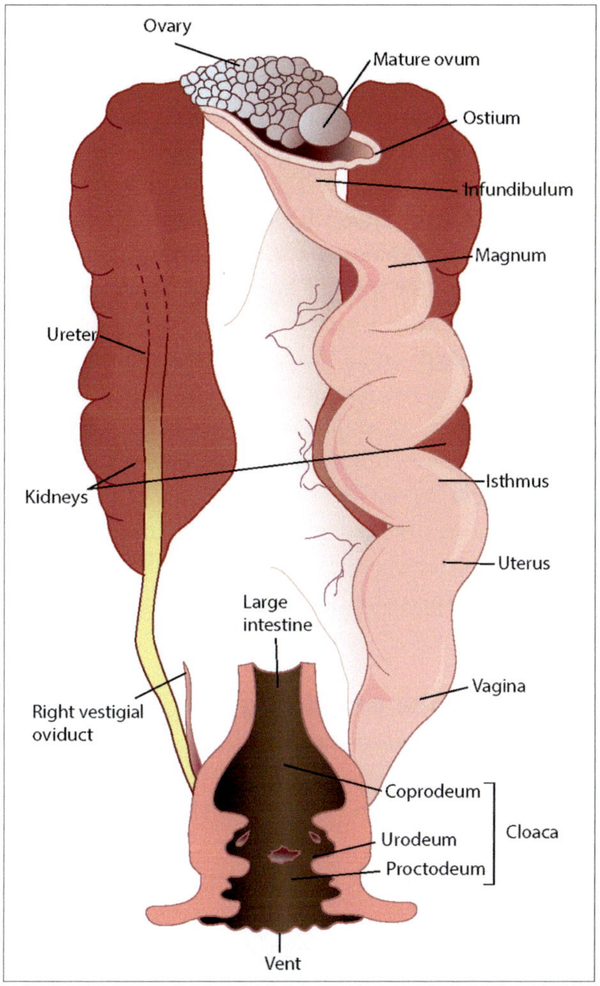

Figure 27.2 **Anatomy of the female reproductive tract.**

which is less vascular and has no connective tissue or smooth muscle. Vitellogenin or yolk, consisting of protein and lipid, is synthesised in the liver (vitellogenesis) and enters the follicle as it develops.

During ovulation, the wall of the follicle splits along the stigma, releasing the yolk and oocyte. The follicle then shrinks to a thin-walled sac, which quickly regresses and is absorbed. No corpus luteum is formed, although the regressing follicle may secrete progesterone for the first 24 hours, affecting oviposition and nesting behaviour.

The oviduct can be divided into five regions: the infundibulum, the magnum, the isthmus, the uterus (shell gland), and the vagina.

The infundibulum resembles a funnel with a thin wall; its opening, an elongated slit, faces into the ovarian pocket formed by the left abdominal air sac. The infundibulum tapers rapidly into a tubular part (the chalaziferous region) with a thickened wall and higher mucosal folds. The infundibulum is reasonably motile, moving to envelop the developing follicle and capturing it as ovulation occurs. Fertilisation occurs in the infundibulum before albumen (thick albumen immediately around the yolk, and the chalaza at each end of the yolk) is laid down by the chalaziferous glands in the tubular region. The egg passes through the infundibulum in 15 minutes.

From the infundibulum, the developing egg passes into the magnum, the longest and most coiled part of the oviduct. This transition is marked by a sudden great enlargement of the mucosal folds. There are numerous tubular glands in these folds, which secrete albumen. Passage through the magnum takes 3 hours, during which time the egg acquires albumen, sodium, magnesium, and calcium.

The next part of the oviduct, the isthmus, is short and reduced in calibre, with folds less prominent than those in the magnum. After a short band of demarcating tissue without glands, the wall of the isthmus has tubular glands similar to those of the magnum. Passage through the isthmus is slow, taking 75 minutes, during which protein is added to the albumen and the shell membranes (inner and outer) are added.

There is no distinct separation between the isthmus and the uterus (shell gland). This part is relatively short but divided into two areas: the initial short, narrow 'red region' and a larger pouch-like region. In the uterus, the longitudinal folds are transected by transverse furrows, forming leaf-like lamellae. The egg stays in the uterus for 20 hours: plumping (the addition of watery solutions) occurs in the first 8 hours, and then the eggshell is formed and calcified over another 15 hours.

The vagina, S-shaped due to smooth muscle and connective tissue, is separated from the uterus by a sphincter. The mucosal folds of the vagina are thin and low, and it has a thick muscle wall. There are no secretory glands; however, near the sphincter are the spermatic fossulae, crypts that act as a storage site for sperm for up to several weeks. Immature birds may have a membrane covering the entrance of the vagina into the cloaca; tearing of this membrane can account for the presence of blood on the shell of the first egg laid.

The oviduct is suspended from the dorsal wall of the coelom by the dorsal mesosalpinx. A ventral mesosalpinx extends ventrally from the oviduct but has a free margin. Smooth muscle in both ligaments is continuous with the smooth muscle layers of the oviductal wall, and caudally the smooth muscle in the ventral ligament condenses into a muscle cord

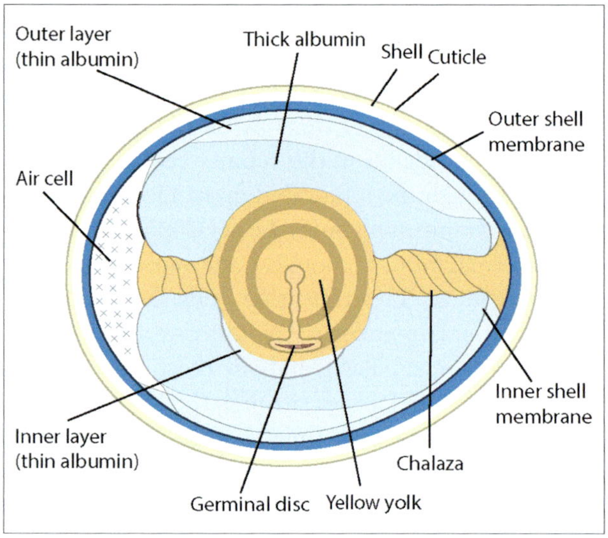

Figure 27.3 **The egg.**

fused with the ventral surface of the uterus and vagina. These ligaments may help to move the egg along the oviduct, especially in the magnum.

The egg

Lying on the surface of the yolk is the germinal disc containing the blastoderm (if fertilised) or the blastodisc (if unfertilised). Underlying the germinal disc is the 'white yolk' or latebra, which is less dense than the yellow yolk and therefore will always be uppermost regardless of the orientation of the egg. It is made up of protein (two-thirds) and fat (one-third). The yellow yolk (two-thirds fat and one-third protein) is encased in four layers of yolk membrane, which, while mechanically strong, forms a water- and salt-permeable membrane between the yolk and albumen (**Fig. 27.3**).

Albumen is less viscous than yolk, and it contains protein (ovomucin). The amount of ovomucin determines whether it is dense or thin albumen. There is a dense chalaziferous layer around the yolk, which is continuous with the chalazae at each end of the egg that merge with the shell membranes. The chalazae, therefore, suspend yolk in the middle of the egg. Beyond this chalaziferous layer are three more layers of albumen: thin inner and outer layers, and a dense layer between them. Albumen contributes to the aqueous environment of the embryo, has antibacterial components, and is a source of nutrition for the embryo.

The eggshell has three layers: the shell membranes, the testa, and the cuticle. There are two shell membranes, each composed of several layers of fibre. The inner layer is fused to the chalazae as described earlier. The outer layer is fused to the testa. They separate at the blunt end to form the air cell. The testa is made up of an organic matrix of fine fibres and an inorganic solid component of calcite (crystalline calcium carbonate). The organic matrix is made up of a thin inner mamillary layer, containing conical knobs embedded in the outer shell membrane, and a thick outer spongy layer. The inorganic component has a thin inner layer corresponding to the mamillary layer and a thick outer palisade layer corresponding to the spongy layer. Pores run through all the layers, through which gaseous and water exchange occurs. Surrounding the entire shell is the outer cuticle, a continuous organic layer that reduces water loss and is somewhat resistant to bacteria. It also has a water-repellent effect. Not all species have a cuticle on their shells.

Reproductive physiology

Most birds are seasonal breeders, with substantial variation between species in their reproductive strategies. This variation is based on the environmental cues used to trigger reproduction, the developmental stage of the chick at hatching, and the extent of parental care.

Mechanisms controlling this breeding seasonality are both endogenous and exogenous. Endogenous factors are poorly understood but are reflected in the fact that many captive birds held in constant environmental conditions still show seasonality, as do migratory birds going through a wide range of environmental changes and tropical birds with little variation in photoperiod. Exogenous factors are better understood. They can be either ultimate factors, which select for individuals that will breed when there are optimum conditions for offspring survival (e.g., food availability), or proximate factors that vary from year to year. These proximate factors are further broken down into:

- Initial predictive factors that initiate gonadal development in anticipation of breeding (e.g., photoperiod).
- Essential supplementary factors that support the initial predictive factors and initiate the final stages of gonadal development, which include social cues (e.g., breeding plumage, mate availability, courtship behaviour), territorial behaviour, climate (e.g., rainfall), and nutrition (an increase in fat and sugars in the diet).

- Synchronising and integrating factors that regulate the sequence of breeding behaviour (e.g., social interaction between a pair).
- Modifying factors that can disrupt the breeding cycle (e.g., loss of a mate or disturbance of the nest site).

All of these factors have hormonal modulators. Their input into the hypothalamus has an effect on gonadotrophin-releasing hormone (GnRH), which in turn leads the pituitary to release follicle-stimulating hormone (FSH) and luteinising hormone (LH). In the hen, FSH supports ovarian and oviductal growth, gametogenesis, and steroidogenesis, while LH also supports steroidogenesis. This steroidogenesis leads to the release of oestrogen, which affects follicular and oviductal growth, calcium metabolism, and vitellogenesis. Some female secondary behaviours are also influenced by oestrogen (e.g., courtship and nesting behaviours). In males, FSH initiates growth of seminiferous tubules and results in increased spermatogenesis, while LH promotes the development of the testosterone-producing cells of Leydig. This in turn gives rise to secondary male characteristics such as plumage changes, nesting activity, courtship behaviour, and territorial behaviour.

Progesterone is produced by granulosa cells in the large follicles as they develop under the influence of LH. This in turn causes a surge in LH production from the pituitary just before ovulation. This surge of LH stimulates the production of prostaglandin (PG) F2α from ovarian follicles, causing the follicular stigma to rupture and allowing ovulation. Progesterone inhibits further ovulation and induces behavioural and physical changes associated with incubation and brood care.

PGF2α and PGE (1 and 2) are released by the F1 (first generation of follicles produced during ovulation) and postovulatory follicles. PGE2 and PGF2α bind at specific sites in the shell gland and vagina; PGF2α binds preferentially at the shell gland, allowing PGE2 to potentiate its effects and allow relaxation of the vagina during oviposition. Therefore, PGE (1 and 2) allows relaxation of the uterovaginal sphincter, while PGF2α stimulates shell gland contractions. Uterine contractility stimulates AVT release from the pituitary, which stimulates further contractility and release of uterine PGs.

Eggs are successively laid until a clutch is formed: indeterminate layers continue to lay if eggs are removed, while determinate layers will only lay a set number of eggs. Incubation is performed by the hen only (in 25% of species), shared (54%), by the males only (6%), or by mixed strategies. Plasma prolactin levels are elevated in both sexes during incubation, which then has inhibitory feedback on GnRH release.

DISORDERS OF THE MALE

Congenital

Abnormalities include abnormally shaped testes, fusion of the cranial poles, hypoplasia, and agenesis. Usually there are no clinical signs, although aviary birds may be presented for infertility investigation. Diagnosis is by endoscopy, and no treatment is required.

Non-inflammatory

Degeneration of the testes is associated with various drugs and toxins (e.g., furazolidone, copper fungicides, and mercury), as well as age. It is diagnosed by endoscopic biopsy or needle aspiration cytology.

Atrophy may occur as the end stage of degenerative changes.

Orchitis

Aetiology

Orchitis may be bacterial (*E. coli*, *Salmonella* spp., *Pasteurella* spp.) (**Fig. 27.4**), fungal (extension from fungal air saculitis), mycobacterial (extension from systemic disease), or chlamydial (extension from

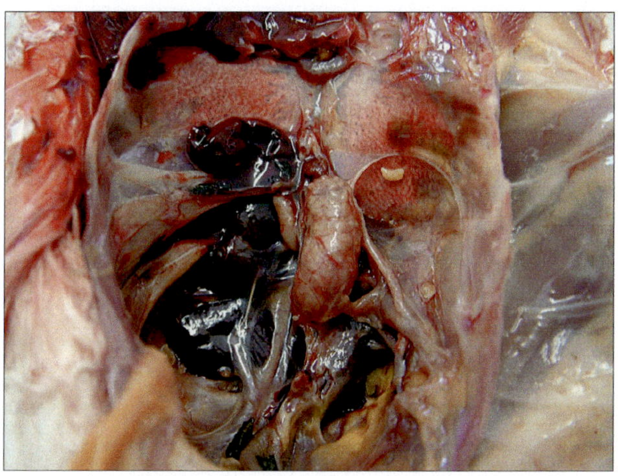

Figure 27.4 **Necropsy of a pigeon reveals orchitis due to *Salmonella* infection.**

systemic disease). Infection occurs by haematogenous spread or via an ascending infection from the cloaca/phallus.

Clinical presentation

No clinical signs are seen unless the bird is septicaemic. The presenting complaint may be infertility or, if a flock is infected, there may have been deaths. In the case of salmonellosis, there may be other clinical signs, such as polyarthritis, unrelated to orchitis.

Diagnosis

Haematology may reflect an inflammatory process. Diagnosis is by aspiration cytology or endoscopic biopsy.

Management

Antimicrobial therapy is given as indicated. Surgical orchidectomy may be required.

Neoplasia

Definition/overview

Neoplasia of the male reproductive tract is reported in numerous species. Tumours seen include Sertoli cell tumours (most common), interstitial cell tumours, seminomas, teratomas, lymphosarcoma, leiomyosarcoma, and carcinoma of the ductus deferens and epididymis.

Clinical presentation

Clinical signs include chronic weight loss, coelomic distension (often accompanied by dyspnoea due to air sac compression), unilateral paresis of the leg (due to sciatic nerve compression), and cere colour change in budgerigars (blue to brown). (Note: This colour change is not pathognomonic for Sertoli cell tumours.)

Metastatic disease is rare but may appear in the liver, kidney, and contralateral testicle. Ataxia and blindness have been reported with brain metastasis in some birds.

Diagnosis

Radiology may demonstrate an enlarged testicle and, if a Sertoli cell tumour is present, increased radio-opacity of the bones (polyostotic hyperostosis, in response to hyperoestrogenism) may be seen. On endoscopic examination, the testicles may appear cystic and can be confused with an immature/inactive ovary. Diagnosis is by endoscopic biopsy or ultrasound-guided fine-needle aspirate biopsy.

Management

Sertoli cell tumours may respond, in the short term, to GnRH agonists such as deslorelin implants or leuprolide injections. Otherwise, orchidectomy is indicated. There is a good prognosis if there is no metastatic disease.

Phallic prolapse

Aetiology

Phallic prolapse is seen in ratites and waterfowl and is associated with sexual overwork. Inflammatory changes at the base of the phallus, occasionally associated with chronic bacterial infection, can maintain the prolapse even though the bird is sexually inactive.

Clinical presentation

The prolapsed phallus protrudes from the vent. The end of the phallus may become traumatised and/or necrotic from chronic exposure and contact with the ground.

Management

Sexual rest is required. The bird should be isolated from the hens and daylight hours reduced. A change in diet may be needed, and one may consider GnRH agonists such as leuprolide acetate, human chorionic gonadotropin (HCG), and deslorelin. Anti-inflammatory treatment (e.g., meloxicam) and broad-spectrum antibiotics are indicated.

If the phallus is damaged, one may have to consider reducing the prolapse and placing lateral vent sutures to keep the phallus in the cloaca until it heals. Phallic amputation can be considered, usually as a last resort.

DISORDERS OF THE FEMALE

The ovary

Congenital abnormalities

Congenital abnormalities include congenital atresia, persistent right ovary, and congenital ovarian cysts.

All early embryos have bilateral ovaries and oviducts, but in most birds, the growth of the right ovary and oviduct is inhibited after the first trimester of incubation. It is normal for kiwis and many raptors to have two ovaries. (Kiwis have two ovaries and one oviduct; raptors have two ovaries and two oviducts.) Persistent right ovaries and oviducts are occasionally seen in species (such as parrots) that would normally have only a left ovary and oviduct.

Congenital ovarian cysts are common in budgerigars and canaries.

Oophoritis

Aetiology

Infectious oophoritis can result from either haematogenous spread of a bacterial (e.g., *Salmonella* spp., mycobacteria), fungal or viral infection (e.g., herpesvirus), or spread from adjacent air sacculitis or peritonitis. Non-infectious oophoritis may be caused by rupture of follicles and extrusion of yolk into the ovarian stroma, or by the effect of some toxins (e.g., aflatoxicosis).

Clinical presentation

Infectious oophoritis is often part of a systemic illness, and many birds will present with generalised weakness, lethargy, and weight loss. In milder cases, there may be an infertility problem or increased embryonic deaths.

Diagnosis

Haematology will usually demonstrate a leucocytosis. Free coelomic fluid may be detected by ultrasound and should be submitted for cytology and culture. Endoscopy may reveal an abnormally coloured or shaped ovary, often haemorrhagic and with abnormal follicles. The ovary of birds with non-infectious oophoritis may appear enlarged, with multifocal yellow spots of varying sizes or obvious scarring on the ovarian parenchyma.

Treatment

Antimicrobial therapy is given as indicated. If possible, the abscessed follicle(s) should be drained completely, being careful not to contaminate the coelom. Partial or complete ovariectomy may be required for chronically infected and caseated follicles.

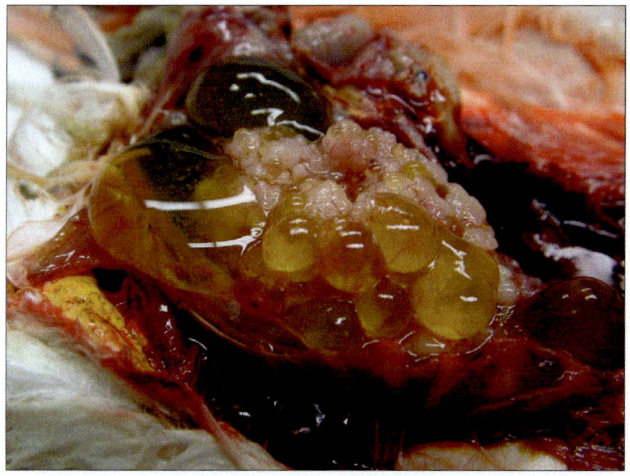

Figure 27.5 **Necropsy of a cockatoo hen that finds ovarian cysts secondary to an ovarian neoplasia.**

Ovarian cysts

Aetiology

Ovarian cysts may be congenital (see earlier) or acquired, i.e., secondary to neoplasia or oophoritis (**Fig. 27.5**). They are rarely functional, i.e., secreting hormones.

Clinical presentation

If the cysts are small, there may be no clinical signs. Large cysts (**Fig. 27.6**) may cause coelomic enlargement and dyspnoea (through compression of the air sacs). Some birds will strain, occasionally resulting in a cloacal prolapse.

Diagnosis

Diagnosis is based on ultrasonography, endoscopy, and exploratory coeliotomy. Ultrasound is

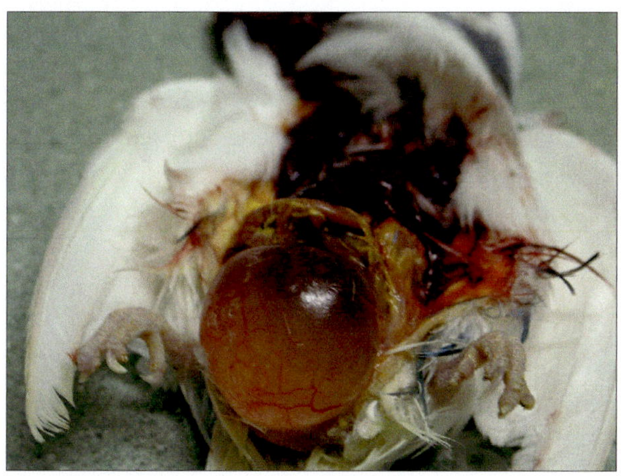

Figure 27.6 **Necropsy of a budgerigar reveals a large ovarian cyst.**

particularly useful in distinguishing free fluid (as in yolk peritonitis or ascites) from the 'contained' fluid within a cyst or cysts. Aspiration of the cyst usually gives clear, straw-coloured fluid with low cellularity.

Management

Aspiration of the cyst may be carried out by ultrasound-guided trans-coelomic aspiration, endoscopic aspiration, or surgical aspiration.

Ovariectomy (partial or complete) is difficult to achieve successfully because of the complex vascularity of the ovary. It should not be attempted without good magnification (e.g., an operating microscope).

Hormonal therapy (e.g., deslorelin, leuprolide acetate, HCG) has been suggested to reduce or resolve ovarian cysts in birds and offer a non-invasive treatment option. This, however, assumes that the cysts are functional and are not secondary to other disease processes (neoplasia, oophoritis). These GnRH agonists, however, may play a role in reducing ovarian activity and the formation of new cysts.

As long-term resolution of this condition is difficult to achieve, it is important to regularly monitor patients for recurrence.

Neoplasia

Definition/overview

Tumours of the ovaries include granulosa cell tumours, ovarian carcinomas, carcinomatosis, dysgerminoma, arrhenoblastoma, and teratoma.

Clinical presentation

If the tumour is small, there may be no clinical signs. Large tumours may cause coelomic enlargement and dyspnoea (through compression of the air sacs). There may be egg retention and oviductal impaction.

Granulosa cell tumours (and possibly other reproductive tract tumours) may be functional and cause increased plasma hormone levels. This can result in hormonal changes (e.g., polyostotic [medullary] hyperostosis).

Diagnosis

Diagnostic imaging (ultrasonography and radiography) helps to identify the presence of a tumour. Confirmation is by endoscopic or surgical biopsy.

Management

Surgical debulking or complete ovariectomy is an appropriate therapy but technically very challenging and comes with a high likelihood of recurrence if complete removal is not achieved. Adjuvant chemotherapy may assist in controlling this recurrence. The use of GnRH agonists such as deslorelin or leuprolide has been suggested and shows promise in controlling some ovarian neoplasms. The prognosis, however, remains guarded.

The oviduct

Congenital abnormalities

Congenital abnormalities include atresia, segmental aplasia, congenital cysts, and persistent right oviducts (see later).

Cystic hyperplasia

Grossly visible cysts occur, containing clear or cloudy fluid. They are probably endocrinal in origin, but this is not determined yet. Birds are usually presented for either infertility investigation, yolk-related peritonitis, or salpingitis.

Neoplasia

Adenomas, adenocarcinomas (**Fig. 27.7**), and leiomyomas have been reported. Birds are usually presented for coelomic distension, yolk-related peritonitis, or salpingitis.

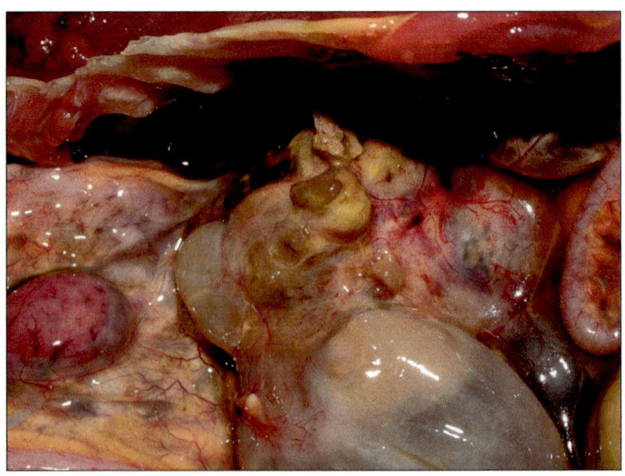

Figure 27.7 **Necropsy of a cockatoo finds a uterine adenocarcinoma.**

Salpingitis and metritis

Definition/overview

Salpingitis is inflammation of the oviduct and mesosalpinx, and metritis is inflammation of the uterus (or shell gland).

Aetiology

Salpingitis and metritis are relatively common conditions in parrots and backyard poultry. Predisposing factors include age, malnutrition, excessive coelomic fat, excessive egg laying, egg binding, and other reproductive disorders.

Primary infections are uncommon in domestic situations; they may be caused by the Newcastle disease virus or the infectious bronchitis virus.

Secondary infection may follow yolk retention or prolonged/excessive egg laying. Haematogenous or ascending infections may occur. Bacterial species involved include *E. coli*, *Klebsiella* spp., and *Pseudomonas* spp.

Clinical presentation

Affected birds often have a history of extremely good egg production. They are often on a predominantly seed diet with inadequate vitamin and mineral supplementation. There may be a history of infertility or embryonic or neonatal mortality.

Clinical signs include weight loss, ruffled plumage, anorexia, and lethargy. If the birds are still laying eggs, these eggs may be malformed (soft-shelled [**Fig. 27.8**], stress lines [**Fig. 27.9**], abnormal shape) or have streaking of blood on the shell. There may be chronic egg binding, infertility, a distended coelom, a flaccid vent, and cloacal discharge.

Figure 27.9 **Egg that has visible stress lines from a chicken with metritis.**

Diagnosis

There may be a leucocytosis, either heterophilic or monocytic, depending on the chronicity of the problem. Clinical biochemistries may show hypercalcaemia (if the bird is still reproductively active) and hyperamylasaemia (if there is concurrent pancreatic disease associated with yolk peritonitis).

Radiography may reveal retained eggs, an enlarged oviduct, or the presence of coelomic fluid. Ultrasonography can distinguish between fluid enlargement and organ enlargement and may reveal retained eggs or fluid in the oviduct. Endoscopy (if there is no free fluid in the coelom) can demonstrate

Figure 27.8 **Soft-shelled egg from a cockatiel.**

a swollen and inflamed oviduct. If coelomic fluid is present, coeliocentesis can help to differentiate the inciting causes.

Management

Conservative treatment may be attempted but is usually unsuccessful. Recommendations include reproductive rest through environmental, nutritional, and hormonal manipulation; NSAIDs (e.g., meloxicam); and antibiotics. If there is material in the oviduct (e.g., caseous pus, eggs, or fluid), the use of prostaglandins may be indicated. Using PGE2 to relax the uterovaginal sphincter and stimulate oviductal contractility may assist in this regard. Caution must be exercised with this therapy, as chronic cases may have developed adhesions to the oviductal wall, and strong contractions may lead to rupture of the oviduct.

In valuable breeding birds, an attempt to combine the aforementioned therapy with a coeliotomy and retrograde flushing of the oviduct can be made. Definitive treatment may require salpingohysterectomy (see Chapter 32).

In all probability, a diagnosis of salpingitis or metritis indicates that the bird's reproductive life is almost certainly complete. If the patient is a breeder or egg layer, this must be communicated to the owner before commencing any treatment. Conservative treatment may give some short-term success, but the problem often recurs. Salpingohysterectomy often gives good long-term results, although some birds will continue to ovulate and develop yolk-related peritonitis (see later). These birds require environmental, nutritional, and hormonal manipulation to minimise ovulation.

Yolk-related peritonitis

Aetiology

Yolk-related peritonitis is the result of an intense inflammatory reaction to ectopic yolk and ova within the coelom. This ectopic ovulation is caused either by a failure of the infundibulum to 'capture' ovulating yolk (because of fat, trauma, or disease) or by retropulsion of yolk from the oviduct into the coelom, possibly associated with metritis/salpingitis, oviductal cystic hyperplasia, or oviductal impaction, or with exuberant reverse peristalsis.

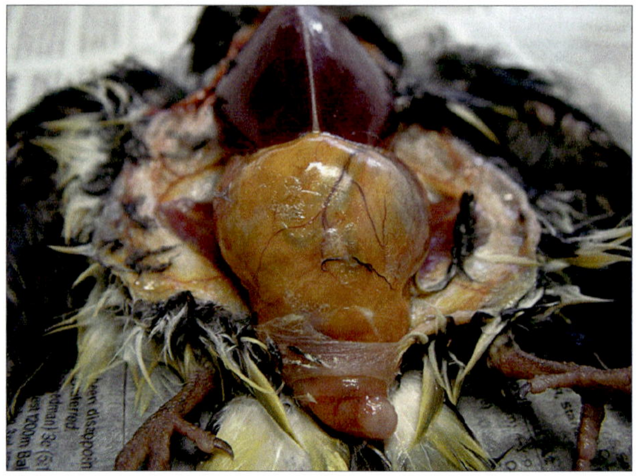

Figure 27.10 **Scarlet-chested parrot with yolk-related peritonitis has coelomic distension due to yolk and fluid.**

Clinical presentation

The condition is usually seen in high-producing hens, especially cockatiels. It is usually sterile; if septic, signs are consistent with severe septicaemia.

Clinical signs are related to a fluid-producing inflammatory reaction in the coelom: dyspnoea, coelomic distension (**Fig. 27.10**), and weakness. The bird may stop laying or may lay malformed eggs (often elongated).

Secondary diseases may develop as a result of yolk-related peritonitis, including pancreatic disease (such as diabetes mellitus), hepatitis, nephritis, splenitis, and coelomic adhesions.

Diagnosis

Haematology generally demonstrates a marked leucocytosis. Hypercalcaemia and lipaemia may be seen in reproductively active hens.

Coelomic ultrasonography confirms fluid distension of the coelom rather than organomegaly. Free fluid must be distinguished from ovarian cysts. Caseous material (inspissated yolks) may be detected. Coeliocentesis reveals variably coloured fluid (brown to yellow-pink), and cytology shows mesothelial cells, leucocytes, and pink yolk globules.

Management

Short-term therapy includes coelomic drainage, NSAIDs, antibiotics, and hormonal manipulation to stop ovarian activity (leuprolide acetate or deslorelin).

Most cases will require surgery to lavage the coelom and perform a salpingohysterectomy (see

Chapter 32). Inspissated yolks should be removed, and adhesions broken down if possible. Persistent ovulation post-surgery may occur and incite another bout of peritonitis, so the concurrent use of GnRH agonists is often recommended. Ovariectomy may be feasible, but it is fraught with danger due to the nature of the blood supply.

Egg binding (dystocia)

Overview

Egg binding is a slowed passage of the egg through the oviduct. If it is located in the uterine portion, it will have a shell deposited on it; proximal to this level, it will be encased in shell membranes only. It must be distinguished from egg retention (see later).

The lodged egg may compress the local vessels and nerves, causing lameness, paresis, paralysis, and poor organ perfusion. Pressure necrosis of the oviduct wall can lead to oviductal rupture, and obstruction of the rectum and ureters can cause metabolic disturbances.

Aetiology

Predisposing factors include age (very young and very old birds are more frequently affected), malnutrition and obesity (particularly seen in birds on all-seed diets), excessive egg production (especially in cockatiels, budgerigars, and backyard poultry), and lack of physical fitness (in caged birds).

Causes include oviductal muscle dysfunction (due to a calcium deficiency, myositis due to excessive egg production, or scarring from previous dystocia episodes), concurrent salpingitis or metritis, excessively sized or malformed eggs, and systemic problems (concurrent illness, hypothermia, environmental stress).

Clinical presentation

Signs include depression, excessive straining, persistent tail wagging, a 'penguin-like' wide stance, leg weakness, dyspnoea, collapse, and coelomic distension (**Fig. 27.11**). Death may occur suddenly.

Diagnosis

Diagnosis is based on the history of egg laying and clinical signs (see earlier). Coelomic palpation usually reveals an egg, but soft-shelled eggs can be difficult to detect. Coelomic radiography or ultrasonography may be required.

Figure 27.11 **Sun conure with coelomic distension due to egg binding.**

Treatment

If the bird shows no or only mild to moderate signs of discomfort and distress, confirm the time the last egg was laid; eggs are usually laid 23–26 hours apart and the patient may not be ready to lay.

The bird should be placed in a heated hospital cage with adequate humidity and given calcium gluconate by intramuscular injections every 3–6 hours. Oxygen may be required if the bird is dyspnoeic. Tube-feeding highly digestible, high-sugar supplements may be given to provide a rapid source of energy. Stress and handling should be minimised, and the bird should be kept in a dark, quiet environment.

If the bird fails to respond to this treatment, oxytocin may be given, but there is controversy over its efficacy in birds, as it is not an avian hormone. (The avian equivalent is arginine vasotocin, not available commercially.) Intra-cloacal PGE2 gel (**Fig. 27.12a, b**) will usually produce uterovaginal sphincter dilation and straining within 5–10 minutes (**Fig. 27.12c**). PGF2α should be avoided, as it will induce oviductal contractions without relaxing the uterovaginal sphincter. If necessary, PGE2 administration will allow manual manipulation of the egg into the cloaca and its delivery. Caution must be taken not to push the egg up against the kidneys and spine.

If the bird is distressed or dyspnoeic, an emergency exists and ovocentesis and egg collapse may be necessary

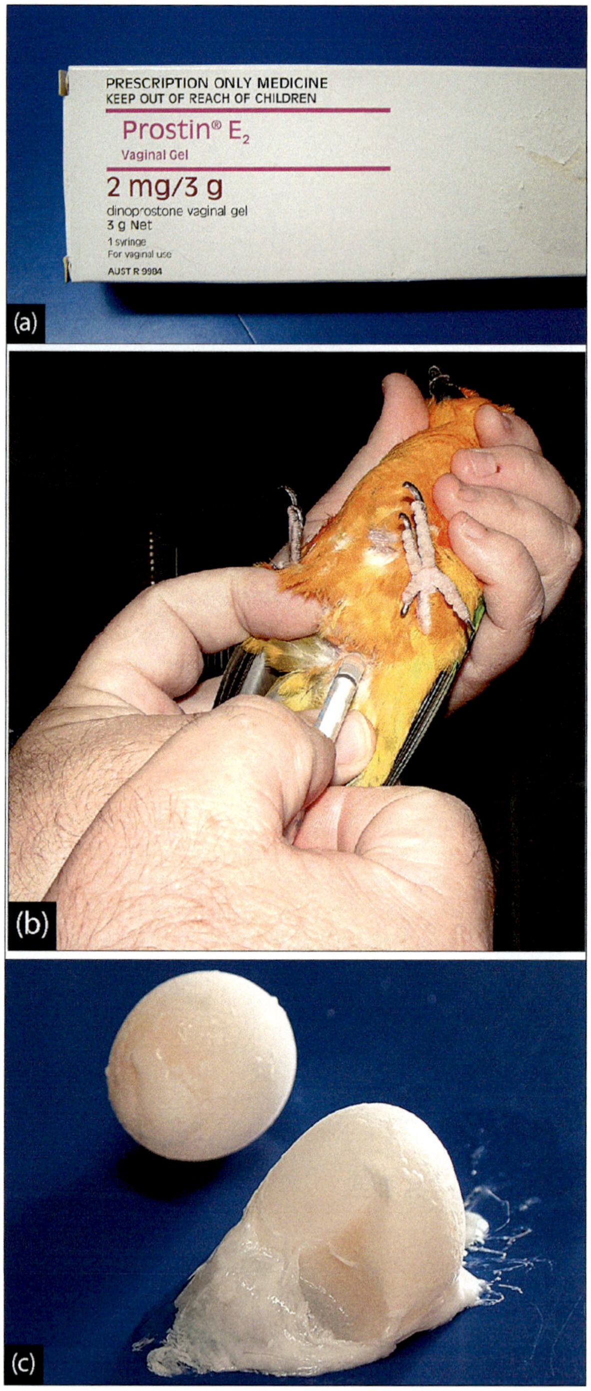

Figure 27.12 **Prostaglandin E2, available as Prostin® E2 vaginal gel (a), is the preferred treatment for egg binding in birds. It is administered directly into the cloaca using a syringe (b) and absorbed rapidly across the cloacal mucosa. Treatment is rapid and effective; two soft-shelled eggs were laid almost simultaneously by this sun conure following prostaglandin E2 treatment (c).**

(**Fig. 27.13**). A large gauge needle is introduced into the egg through the cloaca (preferably) or the coelomic wall and the contents are aspirated, while the egg is simultaneously collapsed with digital pressure. The eggshell is usually passed within 48 hours of this procedure. It must be noted that while this procedure may be lifesaving, damage to the oviduct usually requires a salpingohysterectomy at a later date.

In some cases, it may be necessary to anaesthetise the bird (mask induction with isoflurane), intubate it, and apply intermittent positive pressure ventilation (IPPV) while the egg is being manipulated through the cloaca. Coeliotomy and caesarean section may be necessary in some cases (**Figs. 27.14, 27.15**).

Prognosis

The earlier the case is presented, the better the prognosis. Simple cases have an excellent prognosis, while cases that have reached the stage where the bird is collapsed, dyspnoeic, and unable to use its legs properly have a guarded prognosis.

Ectopic eggs

Definition/overview

Ectopic egg cases present clinically very similar to egg-bound birds, with the same aetiology and clinical signs. The hen fails to pass the egg, regardless of the treatment given. This is due to the rupture of the oviduct at the level of the shell gland, leaving a fully shelled egg loose in the coelom.

Management

Treatment requires a coeliotomy to remove the egg and repair the oviduct. The prognosis is good, and some of these birds return to egg laying uneventfully, provided they are given several months without egg laying to recuperate.

Retained eggs

Definition/overview

The egg may be retained in the oviduct, but it may have collapsed, leaving only the shell. Others may have the egg still intact but are not straining to pass it. The egg is often located in the anterior coelom and usually does not cause dyspnoea.

Clinical presentation

These birds may present with coelomic distension, or they may be asymptomatic.

Diagnosis

Diagnosis is made radiographically. Eggs may have an excessively thickened shell or it may have collapsed within the oviduct.

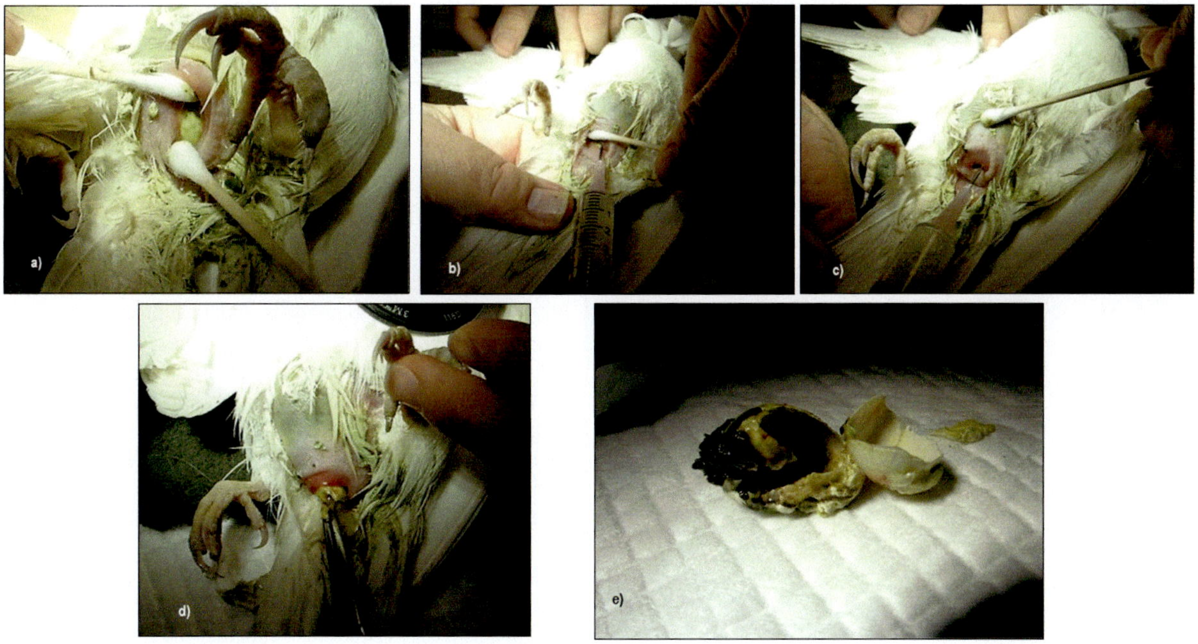

Figure 27.13 **Ovocentesis used as an emergency procedure to collapse an egg: (a) the egg is confirmed to be in the cloaca with a portion of the shell visible; (b) an 18 gauge needle is passed through the eggshell and the egg contents evacuated; (c) gentle pressure is applied to the egg with a Q-tip until it collapses; (d) the eggshell is gently removed; (e) the shell is checked to ensure removal is complete.**

Figure 27.14 **Princess parrot with coelomic distension due to a retained egg.**

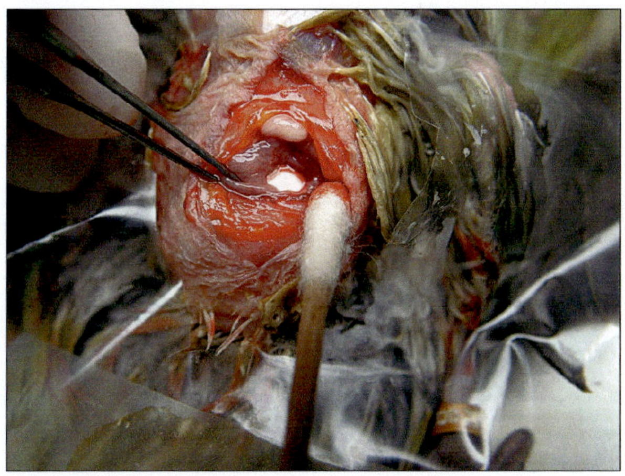

Figure 27.15 **Princess parrot undergoing a caesarean section required to deliver an egg. This procedure should be considered a last resort.**

Treatment

Conservative treatment (as described for egg binding) may work in some cases but should not be relied upon. Surgical therapy (caesarean section or salpingohysterectomy) is usually required.

Persistent right oviduct

As mentioned earlier, the right ovary and oviduct in parrots (and many other species) usually fail to develop and then regress before the chick hatches. A persistent right oviduct is an abnormality commonly recognised in chickens, with its anatomical appearance ranging from rudimentary, discontinuous, and atretic structures to full-size, and rarely, functional oviducts. Reports of the presence of a right ovary or right oviduct, functional or not, in psittacine species are limited, with most observed as necropsy findings. However, anecdotal reports suggest it is more common than often thought, with an incidence of up to 40% in birds with reproductive disease.

When present, a persistent right oviduct is rarely complete. It originates distally from the cloaca along the right caudal dorsolateral coelomic wall, then extends cranially for a variable distance along the ventral surface of the right kidney, lateral to the right ureter. It may develop cysts, endometritis, or hyperplasia. Although it rarely presents as neoplastic disease, this should be a consideration.

Clinical signs

The bird is usually presented for reproductive issues:

- Infertility investigation
- Ovarian cysts
- Yolk-related peritonitis
- Salpingitis
- Cloacal prolapse
- Hyperoestrogenism

Some of these clinical signs may develop even after a 'traditional' left-sided salpingohysterectomy, especially if the reason for surgery was to close the proposed feedback link between the oviduct and ovary. It stands to reason that the presence of a persistent right oviduct would continue this hormonal feedback loop and maintain the clinical condition.

Cystic right oviducts can be very large and produce clinical signs related to a space-occupying coelomic mass, including dyspnoea, coelomic distension, and cloacal prolapse.

Diagnosis

Diagnosis can be made based on the patient's history (especially a history of a previous salpingohysterectomy) and clinical signs consistent with reproductive activity or associated renal, gastrointestinal, or neuromuscular symptoms. Ultrasonography can be useful in detecting large oviductal cysts, while ultrasound-guided centesis of these cysts may obtain a 'watery-milk' coloured fluid.

Treatment

Surgical removal of oviductal remnants and cysts is usually indicated as a curative measure. The left flank approach, usually used for surgery of the reproductive tract, may not be appropriate for a persistent right oviduct, as the rectum and mesorectum make visualisation of the dorsocaudal lateral region of the coelomic cavity, where a persistent right oviduct would be located, difficult if not impossible from a left flank approach. Instead, a ventral midline or trans-coelomic approach is indicated.

GONADOTROPIN-RELEASING HORMONE AGONIST USE FOR MANAGEMENT OF REPRODUCTIVE TRACT DISEASES

GnRH binds to specific receptors on pituitary gonadotrophs, leading to gonadotropin secretion and synthesis. Administration of GnRH agonists results in prolonged activation of GnRH receptors, leading to desensitisation and consequently suppressed gonadotropin secretion, e.g., follicle-stimulating hormone (FSH) and luteinising hormone (LH). Birds, like many other vertebrates, possess at least three forms of GnRH (GnRH-I, GnRH-II, GnRH-III). Their ability to stimulate LH, and possibly FSH, release from the anterior pituitary is variable depending on the species, sex, and reproductive status of the bird. In addition, multiple types of GnRH receptors have been identified, which are subdivided into mammalian and non-mammalian receptors.

The use of GnRH agonists to suppress the secretion of gonadotropins and downstream hormones such as androgens, oestrogens, progestogens, and inhibin in birds is helpful in clinical practice because of the risks and difficulties of surgically desexing. However, the incongruities between avian and mammalian GnRH peptides and their receptors may explain the reduced efficacy of synthetic GnRH agonists, including leuprolide acetate and deslorelin acetate in avian species.

Research and anecdotal reports suggest that there are significant differences in implant efficacy between birds and mammals, as well as between avian species. Reports also indicate that egg laying can be suppressed for approximately 3 months in parrots with a single 4.7 mg deslorelin implant, and even longer if the proximate factors are addressed conscientiously by the owner. Similarly, experience suggests that, in backyard poultry, a 4.7 mg implant suppresses egg laying for 3–4 months, while a 9.4 mg implant extends this time by only 1–2 months. In Japanese quail, a study showed that one 9.4 mg implant stopped egg laying after 5–12 weeks, and this effect lasted for more than 6 months. However, the authors' experience has been that the duration of implants in ducks can only be measured in weeks.

Leuprolide acetate (in the depot injection formulation) works in a similar manner. It produces a steady, non-pulsatile release and thus suppresses the reproductive system. Although leuprolide acetate has been used extensively in many avian species, few controlled clinical trials exist that examine the efficacy of this drug. In addition to unknown efficacy in most avian species, the published dose range is wide, from 100 to 1200 μg/kg IM. The recommended dose range and treatment interval for most parrots is 400 to 1000 μg/kg IM every 2–3 weeks for three injections, repeated as required.

Diseases of the reproductive tract, such as chronic egg laying, oviductal disease, ovarian disease, and testicular tumours, can be managed using deslorelin implants or leuprolide, which are sometimes more feasible and readily accepted by both veterinarians and clients compared to surgical treatment.

Breeding disorders

Chronic or excessive egg laying

Definition/overview

Chronic or excessive egg laying is most commonly seen in cockatiels (**Fig. 27.16)**, but any species can be affected. Cockatiels in the wild or in the aviary will usually lay two to three clutches, each of three to five eggs, over a 12-month period. There is usually a period of several months between clutches. As companion birds, pet cockatiels will often lay this many eggs in a single month and do it month after month. Left untreated, many of these birds deplete their calcium reserves and develop problems such as egg binding and pathological fractures. Many will develop salpingitis/metritis and subsequent yolk-related peritonitis.

Aetiology

Cockatiels evolved in an arid environment and are consequently opportunistic breeders, laying eggs when environmental conditions are ideal for the survival of their offspring. In captivity, these environmental conditions are provided all year round: large amounts of readily available food and water, especially high-fat and sweet foods (e.g., seed and fruit); constant light (these birds are often housed indoors and do not have an appropriate diurnal rhythm); an 'appropriate' mate (this may not be another bird; in many cases it is the owner who has allowed an 'unnatural' bond to develop between themselves and the bird); and a secure nest site, which may be anywhere in the house or cage where the bird has established territoriality. The result is a constant drive to lay eggs.

This is often exacerbated by the owner removing the eggs as they are laid. Cockatiels are indeterminate layers and will simply lay another egg to replace the missing one.

Diagnosis

The history of excessive egg laying makes diagnosis relatively straightforward. The bird's physiological and health status should be evaluated by physical examination, haematology, biochemistry (including ionised calcium), and diagnostic imaging.

Management

Management of chronic egg laying requires identification and elimination of the environmental, dietary, and social factors triggering the bird's reproductive drive and then, if necessary, the use of hormonal manipulation or surgery to minimise or prevent further egg laying. It should be noted that complete cessation of egg laying is difficult to achieve, but, if the bird can physiologically handle the number of eggs laid each year, treatment can be said to have been successful.

Environmental modification

Normal diurnal rhythms need to be re-established. In some cases, it may be necessary to drastically reduce daylight hours artificially (to 8 hours or less) until the egg-laying behaviour has stopped, and then

Figure 27.16 **These nine eggs were laid by one cockatiel hen in one month.**

bring the bird up to normal rhythms. Note that in species where day length has little effect on breeding (e.g., zebra finches and budgerigars), changing day length will have little effect on reproductive behaviour.

Nest sites should be removed (if possible). If the bird chooses an unconventional nest site (e.g., on the floor of the cage), rearranging the area may effectively remove its attractiveness as a nest site. The bird's environment can be changed (e.g., new or rearranged cage furniture, repositioning the cage on an irregular basis). This induces a certain amount of environmental stress and reduces territoriality.

Social modification

If appropriate (and possible), the companion bird can be removed. A normal relationship between human and bird should be established through the introduction of basic behavioural training and client education. The client must be taught what behaviours sexually stimulate the bird (e.g., stroking the bird's back, kissing the bird, or allowing it to take food from the owner's mouth, etc.).

Removing the eggs of an indeterminate layer, such as a cockatiel, will induce further egg laying. Leaving the eggs in the 'nest' or replacing them with artificial eggs will often result in a hen laying a normal-sized clutch and then brooding the eggs. This temporarily shuts down (temporarily) the production of eggs.

Nutritional modification

Fat and sugar in the diet should be reduced by conversion to a formulated diet and removal of fruit from the diet. Foraging behaviour can be introduced. Typically, wild birds spend 80% of their day foraging for food, leaving 20% of the day to groom, socialise, and 'nap'. In captivity, this time allocation is nearly reversed, with birds only having to spend a small amount of time looking for food. By introducing foraging activities, there is less time for pair bonding and other reproductive activities.

Hormonal manipulation

Hormonal therapy does not work in isolation. Unless modifications in the environment, behaviour, and diet are instituted beforehand or concurrently, there will be minimal or no response to the use of hormones.

Drugs that are commonly used include:

- Leuprolide acetate, 100–700 µg/kg every 2–4 weeks.
- Deslorelin, a GnRH agonist available as an implant for dogs, is also commonly used in birds.

Surgery

Salpingohysterectomy has been proposed as a means of controlling egg production. Unfortunately, it appears that, by itself, it does not stop reproductive behaviour and subsequent ovulation, which then results in yolk peritonitis. Surgery should therefore be reserved as a 'last resort' therapy only.

LOW PRODUCTION IN THE AVIARY

Introduction

Veterinarians are occasionally asked to investigate an apparent fertility problem in an avicultural collection. The problem is generally one of three types: lack of egg production, infertile eggs, or poor hatchability.

Any investigation should start with a comprehensive review of the aviculturist's records, including the aviculturist's objectives and intentions; aviary design and construction; origin, age, and species of birds; husbandry practices, traffic flow, and nutrition; previous medical histories; production records; and description of the current problem.

In many cases, it is appropriate to do this review before visiting the avicultural complex, utilising written records, a map drawn by the aviculturist of the complex's layout and traffic flow, and photographs.

Lack of egg production

Failure of birds in private collections or zoos to reproduce due to a lack of egg laying can be costly, frustrating, and time-consuming. For aviculturists, breeding the birds may be more than just a hobby; breeding success and financial success often go hand in hand. A failure to breed a species in captivity may result in a loss of interest in that species and its loss to aviculture. In the case of breeding programs for endangered species, where the goal is to preserve the species' genetics while gaining time to eliminate an environmental threat, such failure may lead to the loss of that species.

Breeding programmes for some species, valuable or endangered, are not always successful. However, by revisiting the principles of reproduction, solutions to previously insurmountable problems may be found.

The first step is to review the client's breeding records (if available) and discuss the problem with the client. After this, the clinician should be able to determine if the problem is generalised or is limited to a particular pair, species, or part of the aviary. Generalised problems suggest an overall suppressive effect through diet or husbandry, while more localised problems may indicate a problem with individual birds or pairs.

The second step is to look at the birds.

- General body condition and health of the birds: Check representative birds and screen for flock diseases, e.g., chlamydiosis and parasites.
- Sex: Same-sexed monomorphic species are sometimes inadvertently paired, and breeding is expected. If this pair are both males, no eggs will be laid.
- Age: Juvenile birds are sometimes unrealistically expected to mature early and breed shortly after reaching puberty. Conversely, aged birds (especially those that mature early) often show a gradual decline in egg laying with age.
- Synchrony: In seasonally reproducing wild birds, older females with previous breeding experience often initially lay earlier in the season and lay larger clutches than inexperienced females. This has been attributed to females having prior experience responding to photostimulation, resulting in more robust photo-induced reproductive development, including having higher circulating levels of reproductive hormones and greater seasonal increases in follicle size than in photo-naive birds.
- Individual and species compatibility: Some birds will not breed if they are paired with an incompatible partner or if they can see or hear other birds of the same or different species during the breeding season.

The third step is to review the husbandry in the collection or for the affected birds.

- The diet: The provision of an inappropriate diet, e.g., an all-seed diet, may lead to chronic health problems and loss of libido.
- Aviary design: This should reflect the height and space requirements for the species housed there. The design must reflect species-specific behaviour, e.g., full-flight aviaries are needed for ground feeders, while canopy dwellers may do well in a suspended aviary.
- Predator control: The presence of snakes, vermin, and insects may discourage the birds from nesting.
- Nest boxes: There should be a selection of nest boxes available – different designs, heights, sizes, and orientations in the aviary. Birds should be able to feel secure when entering and exiting the nest box. The provision of a privacy screen in front of the nest opening will often make the birds more confident as they come and go from the nest box.
- Privacy provision and stress reduction: Some birds will not feel secure enough to breed if they cannot access 'private' areas in the aviary where they can indulge in courtship and mating behaviours without feeling threatened. Stress from the surrounding environment, e.g., birds of prey, dogs, children, and construction activities, must also be minimised.
- Environmental enrichment encourages birds to engage in more natural, species-specific behaviour, which, in turn, can lead to an increased sense of wellbeing and libido. Environmental enrichment is not just providing toys; it includes foraging activities, sensory enrichment, physical enrichment, occupational enrichment, and social enrichment.

Then look at the effects of the proximate factors that stimulate breeding.

- Local weather patterns, and if 'artificial weather' in the form of artificial lighting or water sprinkling systems, can be used to stimulate reproductive behaviour. Prolonged heat (>30°C) has been shown to decrease egg production in many species.
- Diet: The provision of high-fat foods at the start of the breeding season can stimulate reproductive activity. If these foods are fed year round, that stimulus can be lost. Alternatively, the provision of a low-fat formulated diet will also often fail to stimulate activity. Feeding a formulated diet as the base for aviary nutrition and then using seed and nuts to stimulate reproductive behaviour when desired can be a powerful tool in reproductive management.

- Trial different designs, locations, and orientations of nest boxes. Partially closing the entrance hole can stimulate reproductive behaviour as the birds work on the nest to prepare it for breeding. This may also help to synchronise reproductive cycles between individual birds.
- Increase the bird's feelings of privacy and security:
  - Step up vermin and snake control.
  - Consider erecting visual screens between pairs to cut down on territorial aggression that may be inhibiting reproduction.
  - Fully roof the aviary to prevent disruption by wild birds (especially raptors).
- Ensure the birds are true pairs, old enough to breed, and are compatible. Splitting up a previously unsuccessful pair and re-pairing them with other birds may be an effective breeding stimulus.

Hormonal manipulation

Just as slow-release GnRH agonists can downregulate the avian HPG axis, suppressing the release of FSH and LH, short-acting formulations can stimulate their release and increase steroidogenesis. Such short-acting formulations include buserelin acetate and lecirelin.

INFERTILITY

Successful fertilisation occurs when a male pronucleus and a female pronucleus fuse to form a zygote (i.e., syngamy). The overall success of fertilisation relies on healthy ova being produced and ovulated normally, and on good quality spermatozoa being present in the infundibulum at the precise time of ovulation.

Infertility is the failure of this to occur, and any process, male or female, contributing to it can be a cause of infertility.

Is this really an infertility problem?

The first step when investigating an infertility problem is to ascertain if eggs are truly infertile or are failing to develop. About 10% of embryos can be expected to die during incubation. Typically, there are two mortality peaks: the first and third trimester (with the larger proportion occurring near hatching). Candling an egg is often not sensitive enough to distinguish between infertility and early embryonic death. Necropsies on eggs in early incubation that are failing to develop (as determined by candling) are required to distinguish between infertile eggs and early embryonic death (see section "Poor hatchability"). If eggs that are infertile or have experienced early embryonic mortality are left to incubate to full term, their contents often decompose, making diagnosis impossible. Therefore, fresh eggs may have to be sacrificed to determine what is happening.

Once it is determined that the egg is infertile, further investigation is warranted.

Causes of infertility

There are multiple causes of infertile eggs. While some of these are sex-specific, it should be noted that there is considerable overlap between the two sexes in factors that result in infertility. Female fertility traits in birds are still understudied, and the role of the female in determining reproductive success is therefore underappreciated.

Beyond the practical advantages of studying sperm relative to ova, the over-representation of studies on male fertility may also be a consequence of the historical view that sperm are the 'active' participants in fertilisation: seeking, binding to, and penetrating the somewhat 'passive' egg. Infertility must be investigated from the perspective of both sexes.

Broken down into its parts, infertility can be due to:

- Failure of the male to transfer sperm to the female
- Failure of the sperm to be stored in the female's sperm glands
- Failure of stored sperm to be transported to the infundibulum (the site of fertilisation)
- Failure of sperm in the infundibulum to interact with the perivitelline layer (PVL) of the ovum and penetrate it
- Failure of the sperm pronucleus to locate the female pronucleus or to fuse (syngamy)

Factors that can contribute to any or all of these failures include:

- Diet
- Disease
- Failure during egg formation
- Fertility senescence (ageing of reproductive physiology)

- Pollution

Failure of the male to transfer sperm to the female

Unsuitable birds:

- Incorrect sex identification, resulting in the pairing of two hens, with one or both laying eggs, is common in low-value, sexually monomorphic species, e.g., lovebirds (*Agapornis*), where the owners have not invested in sex identification. However, it can be seen when technical mistakes have been made in sex identification.
- Two birds that are incompatible, immature, or inexperienced.
- Two birds that are not reproductively synchronised. Seasonality may affect the production of semen in some birds. For example, macaws and amazons are seasonal breeders and may not produce sufficient semen to fertilise an egg at the start of the breeding season. However, eclectus parrots, and perhaps cockatoos, produce semen year round, reducing the effect of synchronisation.

Lack of successful mating due to:

- Physical impediments to successful mating, e.g., old leg injuries, excessively heavy feathering around the vent, obesity, ill-health, impaired vision, or balance.
- Inappropriate perches preventing normal mating (too small, too large, not fixed in place).
- Lack of privacy or external stresses (other birds, animals, humans) is causing the birds to feel insecure.

Failure of the sperm to be stored in the female's sperm glands

Once semen has been deposited into the cloaca and enters the vagina, the spermatozoa begin their ascent to the uterovaginal sphincter, aided by ciliary cells lining the vagina, smooth muscles in the wall, and the inherent sperm motility. Due to intense selection, less than 1% of spermatozoa reach this area and then enter the sperm storage glands where they can survive for several weeks. It is thought that the passage of an egg through the vagina stimulates the release of the spermatozoa. This system compensates for the lack of an oestrus period to synchronise ovulation and copulation.

Failure of the sperm to reach the vaginal glands and enter them may be due to:

- Low numbers of spermatozoa.
  - Age of the bird (immaturity vs. senescence).
  - Species or individual variation
  - High ambient temperature (<40°C), especially heat waves versus consistently high (heat adaptation), will result in a decreased number of spermatozoa as well as increased morphological abnormalities.
- Abnormal or poor-quality spermatozoa may not be able to enter the vaginal sperm glands.
- Sperm competition results from genetic selection of males with healthy spermatozoa. If a cock with genetically poor-quality spermatozoa is bred with a high-quality hen, the spermatozoa may not be able to enter the glands.
- Diet affects the sperm midpiece length through unknown mechanisms.
- Female selection of sperm. It is possible that the female's sperm gland will not only store incompatible or abnormal spermatozoa but may select which male's spermatozoa will be stored.

Failure of sperm to be transported to the infundibulum

Factors affecting the passage of spermatozoa include:

- The number, morphology, and motility of the spermatozoa released from the sperm storage glands.
- Vaginal and uterine fluid is known to both assist and decrease spermatozoan motility. The composition of this fluid, including its pH, may select for certain spermatozoa and, in some infertile females, antibody-producing plasma cells in their reproductive tract may reject spermatozoa.
- Low-grade cloacitis or metritis may inhibit the passage of semen from the vent to the infundibulum. Multiple copulations can reduce fertility if they damage the female reproductive tract. Copulation also provides an opportunity for bacterial transfer, which can cause local inflammation in the vaginal wall, impairing sperm transport and reducing fertility.

Failure of sperm in the infundibulum to interact with the perivitelline layer (PVL) of the ovum and penetrate it

Following successful ovulation, the avian ovum is captured by the infundibulum where it encounters sperm. Successful fertilisation involves the initiation of multiple events in sequence: sperm–egg binding, acrosomal exocytosis, sperm penetration through the perivitelline layer, and fusion of the male and female pronuclei in the germinal disc.

The inner perivitelline layer (IPVL) is composed of a mesh of fibres that forms a three-dimensional extracellular matrix. Unlike mammals, the IPVL of birds does not inhibit polyspermy, and, in fact, a degree of physiological polyspermy is required for normal development in birds.

Sperm are known to bind to the germinal disc region in higher concentrations than elsewhere on the ovum, suggesting there must be an underlying mechanism by which sperm locate and/or preferentially bind to this region. Such a mechanism has yet to be discovered, but the egg plasma membrane is reported to differ in morphology around the germinal disc region compared with other regions of the ovum.

The germinal disc region is also known to show subtle intra- and interspecific variation in terms of morphology, and in terms of the location, size, and number of sperm penetration holes. It is currently unknown how variation in PVL structure affects fertility, but fertilisation rates are positively correlated with the number of sperm that penetrate the PVL.

Failure of the sperm pronucleus to locate the female pronucleus or to fuse (syngamy)

While multiple sperm can penetrate the PVL in birds, only one male pronucleus typically fuses with the female pronucleus in the germinal disc during syngamy. This provides additional potential for the female pronucleus itself to be selective, e.g., based on the male genotype or on the integrity of the male's DNA. The exact mechanisms of avian syngamy remain unknown.

Elucidating the full molecular mechanisms involved in avian syngamy is challenging but will be an important step in understanding how fertility can be compromised at the gametic level.

Other factors

- Failure during egg formation. The process of egg formation – from follicular development through to the release of the ovum during ovulation – is a metabolically demanding process, and problems occurring during these early stages of reproduction are likely to be an important cause of infertility. In birds, the mechanisms underpinning follicular development, maintenance, and selection have yet to be well defined, although the implications of these processes are likely significant for fertility.
- Diet. Yolk formation is energetically demanding and requires substantial changes in the body's metabolism of lipids. Lipogenesis is responsive to both hormonal control as well as dietary changes, and overfed hens exhibit symptoms of lipotoxicosis, including ovarian abnormalities and follicular atresia. Nutrients thought to be important for egg production and fertility in birds include manganese, selenium, iodine, fluoride, sodium, zinc, copper, vitamin A, vitamin E, vitamin B12, protein, and linoleic acid. An excess or deficiency in any of these can be disruptive.
- Disease. Infectious diseases such as chlamydiosis and non-infectious diseases such as cystic ovaries or ovarian neoplasia may affect fertility. The number of immunocompetent cells present in the follicles of the ovary also decreases with age, suggesting a reduction in infection resistance in older hens that could have an associated impact on egg production.
- Fertility senescence. Rates of reproductive senescence are higher in birds that invest more heavily in early-life reproduction (e.g., chickens). Initial increases in reproductive output are often followed by gradual declines in fertility with age across species.
- Environmental pollutants. Over 90,000 anthropogenic chemicals are estimated to have been released into the environment, and several hundred of these pollutants are confirmed to act as endocrine-disrupting compounds (EDCs). If passed on to developing embryos, EDCs can disrupt reproductive development and cause sterility.

Investigation and treatment

Investigating a proven infertility problem begins with a review of the birds, their health and husbandry, and the male's fertility (via semen collection and assessment).

Treatment plans may include re-pairing the birds with different mates, or the use of artificial insemination.

POOR HATCHABILITY

Definition/overview

It is normal for up to 10% of all eggs laid to fail to develop normally. Embryonic mortality above these levels should be investigated.

See "Incubation" in Chapter 28 for more details.

OTHER REPRODUCTIVE-RELATED PROBLEMS

Aggressive males

Overview

Aggression in male birds is predominantly associated with captive-bred cockatoos, but has been seen in macaws, amazons, and conures. Aggression may be directed towards a mate, neighbouring birds, or a human companion. It may be triggered by territorial behaviour or sexual frustration. Owners are often surprised by this behaviour, whether it is directed at another bird or at people.

Clinical signs

In the case of territorial aggression, intrusion into what the bird perceives as its territory gives rise to early warning signs displayed by the bird. These include outspread wings, increased vocalisations, pupil dilation and constriction ('pinning'), and fanning of the tail. If these signs are ignored, an attack may follow.

Aggression related to sexual frustration generally occurs early in the breeding season. Lack of reproductive synchrony between even closely bonded pairs leads to the male's sexual advances being ignored by the female. In the wild, the hen simply flies away, but this option is rarely present in a captive situation. The cock bird attacks the hen, inflicting physical trauma ranging from feather chewing on the face to beak avulsion or limb fractures. On occasion, the aggression only ends with the death of the hen or the separation of the pair.

Treatment

Territorial aggression in an aviary situation can often be treated by screening the aviary to minimise interaction between neighbouring birds. Where possible, owners should avoid entering the aviary, feeding, and watering the birds through a small door. Nest boxes can be designed and positioned so that they can be accessed from outside the aviary for inspection. (A safety walkway must be utilised, as some birds will chew through the nest box and escape.)

In a companion bird, aggression can be minimised or eliminated by altering the bird's environment (social and physical) to reduce or eliminate reproductive triggers that drive this behaviour. If this is not sufficient, GnRH agonists (e.g., deslorelin and leuprolide) can be utilised to 'shut down' the bird's reproductive drive.

Sexual frustration can be more difficult to manage, particularly if reproductive success is the owner's main goal in keeping the birds. In many cases, the male's aggression can be redirected towards nest preparation rather than the hen. This can be achieved by:

- Covering the nest entrance with cardboard or thin plywood, leaving a small entrance hole. The cock bird will 'work' this entrance hole to make the nesting site accessible and more suitable.
- Filling the nest box with small blocks of soft untreated wood. The cock bird will chew these into shavings, again to prepare the nest site.

Hopefully, by the time the cock has prepared the nest box, the hen will have synchronised her reproductive drive with him and will be more receptive to his courtship.

Other avenues include the use of large aviaries (giving the hen the opportunity to escape) or special two-entrance boxes with baffled interiors that allow the hen to avoid being trapped by the cock in the next box.

In some cases, trimming the cock's wings to reduce his ability to fly and chase the hen may be necessary. This can be augmented by providing elevated escape areas for the flighted hen that are inaccessible to the now-clipped cock.

Temporary pair separation is always necessary once trauma has been inflicted on the hen; this may break the aggression cycle and give the pair time to synchronise their reproductive drive. In these cases, pairs may be slowly and cautiously reintroduced. If the aggression continues, there is little choice other than permanent separation of the pair.

Aggressive beak alteration procedures have been recommended in the past but are not considered to be ethical by most veterinarians. These procedures included attaching acrylic or rubber balls to the tip of the maxilla (offering limited temporary protection) and surgically splitting the mandible at the symphysis. Birds with split mandibles suffer numerous complications, including difficulties eating, drinking, playing, and mating.

FURTHER READING

Bowles HL (2006) Evaluating and treating the reproductive tract. In: *Clinical Avian Medicine, Vol 2.* GJ Harrison, TL Lightfoot (eds). Spix Publishing Inc, Palm Beach, pp. 519–540.

Joyner KL (1994) Theriogenology. In: *Avian Medicine: Principles and Application.* BW Ritchie, GJ Harrison, LR Harrison (eds). Wingers Publishing, Lake Worth, pp. 748–804.

LaBonde J (2006) Avian reproductive and pediatric disorders. In: *Proceedings of the Annual Conference of the Association of Avian Veterinarians Australian Committee,* pp. 229–238.

Mans C, Pilny A (2014) Use of GnRH-agonists for medical management of reproductive disorders in birds. *Veterinary Clinics: Exotic Animal Practice* 17(1): 23–33.

Molter CM, Fontenot DK, Terrell SP (2015) Use of deslorelin acetate implants to mitigate aggression in two adult male domestic turkeys (Meleagris gallopavo) and correlating plasma testosterone concentrations. *Journal of Avian Medicine and Surgery*: 224–230.

Nemetz L (2010) Clinical pathology of a persistent right oviduct in psittacine birds. Proc AAV: 73–77.

Orosz S, Dorrestein GM, Speer BL (1997) Urogenital disorders. In: *Avian Medicine and Surgery.* RB Altman, SL Clubb, GM Dorrestein, K Quesenberry (eds). WB Saunders, Philadelphia, pp. 614–644.

Petritz OA, Guzman DSM, Hawkins MG, Kass PH, Conley AJ, Paul-Murphy J (2015) Comparison of two 4.7-milligram to one 9.4-milligram deslorelin acetate implants on egg production and plasma progesterone concentrations in Japanese quail (Coturnix coturnix japonica). *Journal of Zoo and Wildlife Medicine* 46(4): 789–797.

Schubot RM, Clubb KJ, Clubb SL (1992) *Psittacine Aviculture: Perspectives, Techniques and Research.* Avicultural Breeding and Research Center, Loxahatchee, FL.

CHAPTER 28

REPRODUCTIVE TECHNOLOGY

SEMEN COLLECTION, STORAGE, ARTIFICIAL INSEMINATION, AND INCUBATION

In recent years there has been increased interest in artificial insemination (AI) in aviary birds (especially endangered species held in zoos) and in backyard poultry. The use of artificial breeding methods (semen collection, storage, and artificial insemination) is intended to increase the production of valuable or rare/endangered species.

Semen collection

Semen collection in birds is complicated by the absence of a penis and the potential for faecal–urine contamination. However, it has been used in different species since the 1930s, although no standardised techniques have been developed for use across all species.

Several methods have been described for semen collection:

- Training birds to voluntarily ejaculate onto a collecting surface. This technique is very well established for birds of prey.
- Cloacal and abdominal massage has been used in birds of prey, waterfowl, ratites, and small parrots (Samour, 2002) (**Fig. 28.1**).
- Electrical stimulation (electroejaculation) of the cloaca (Lierz et al., 2013).

Successful collection of semen allows the evaluation of potential breeding males based on their semen quality, resulting in either their exclusion from the breeding programme or inclusion as a genetically valuable reserve. Furthermore, this semen can be used for artificial insemination.

The amount of semen that can be collected is governed by:

- The size and maturity of the bird.
- The time of the year, especially in seasonally breeding birds. Most birds reproduce during a rather species-specific, short time of the year.
- The technique used. Cloacal massage in cockatiels has been shown to have a success rate of 75% over a 12-month period, while electrostimulation has a success rate of 55–100% during the shorter species-specific breeding season.
- Contamination with faeces and urine.

Assessing semen quality

Assessment of the collected semen is comparable to that of humans and domestic mammals. Volume, colour, consistency, and pH value of the ejaculate as well as their level of contamination (for example, with faeces, uric acid, and/or blood) can be evaluated. Motility, vitality (via intravital staining), and the morphology of the spermatozoa are also assessed. Differences in these parameters have been noted between species (and even subspecies); wherever possible, specific reference intervals for each species should be established. Cutting-edge examination methods, like computer-assisted semen analysis (CASA), can determine different parameters for the movement and speed of individual spermatozoa. The perivitelline membrane penetration test, which has recently been modified for various bird species, plays an important role in this process. In this test, semen is placed onto a perivitelline membrane of an egg to assess whether the sperm can penetrate the membrane, which is a basic requirement for the fertilisation of an egg.

The average semen volume collected from birds such as chickens, turkeys, and ducks (approximately 100–300 μl) is considerably less than from mammals. However, the average sperm concentration of poultry semen such as chicken ($3–5 \times 10^9$ spermatozoa/ml) is much higher than that of mammals such as

DOI: 10.1201/9781003272519-28

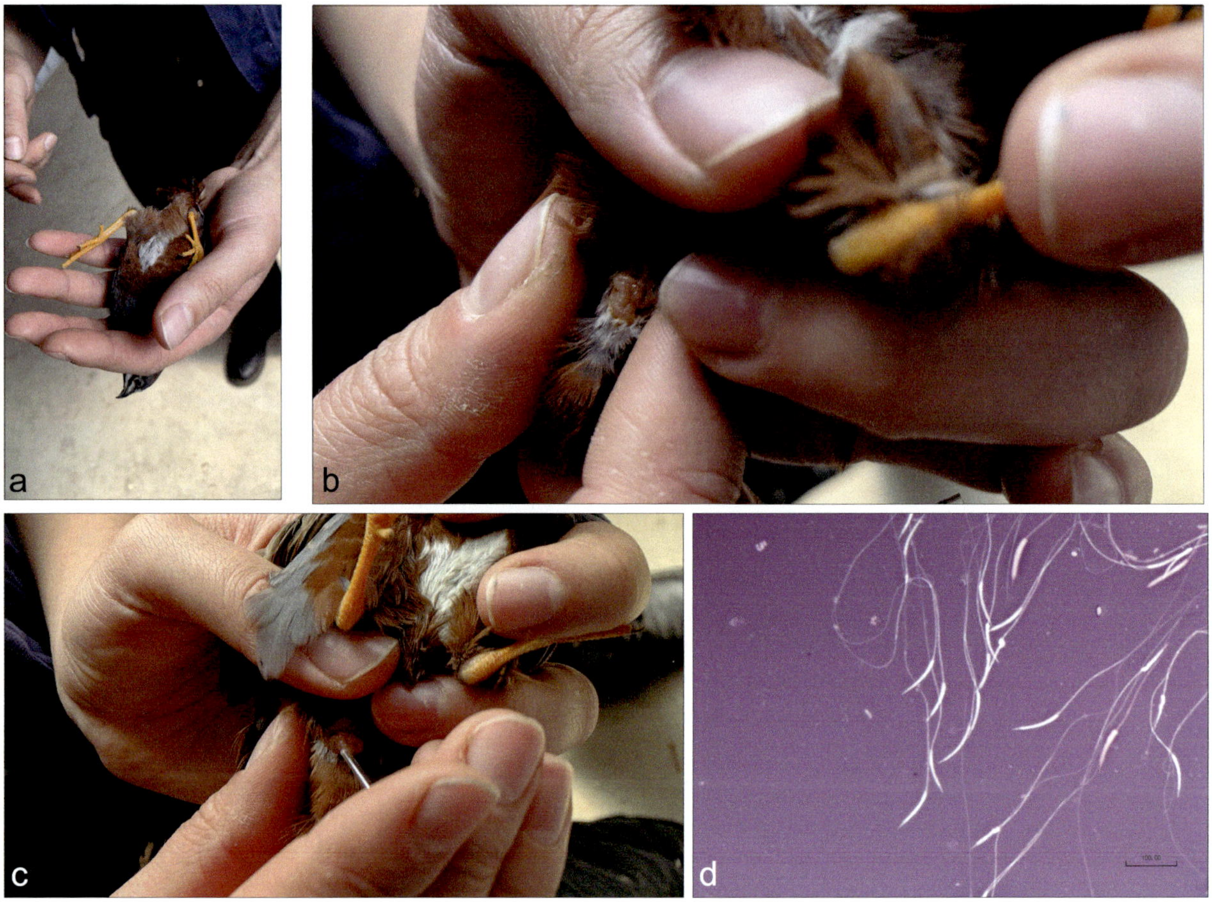

Figure 28.1 **Semen collection in a king quail. (a) Restraint of a king quail for semen collection. (b) Lateral massage of the cloaca to express semen. (c) Semen collection with a capillary tube. (d) Assessment of spermatozoa for viability. (Photos courtesy Isabella Wong.)**

bulls ($0.3–2 \times 10^9$ spermatozoa/ml) and boars ($0.3–3.5 \times 10^8$ spermatozoa/ml).

Semen storage

In chickens, insemination of a hen with approximately 100 million spermatozoa can result in a ~90% fertilisation rate. Chicken semen begins to lose fertility when stored for more than 60 minutes, although commercially available extenders can maintain acceptable fertility for up to 24 hours at 4°C.

- Ideally, insemination should use freshly collected, undiluted semen with a high concentration of spermatozoa.
- Cold storage of semen using cryoprotectants has been achieved in some – but not all – species. The cryopreservation of parrot semen has been described for only a few bird species, but most of them are anecdotal (for example, budgerigars). Due to the large physiological differences in the semen of different bird species, such protocols are not transferable to other parrot species and must be newly established every time.
- To do so, first the basic physiological data of the semen of a bird species must be determined and investigation protocols need to be established accordingly, for example, for electron microscopy. Finally, adequate diluents and cryoprotectants, as well as a freeze protocol, must be established in a series of tests to guarantee the survival of the semen during the freezing process, e.g., DNA fragmentation.
- Development of methods for semen cryopreservation would make it possible to distribute semen obtained from genetically superior birds widely, both geographically and temporally, accelerating genetic selection and preserving endangered species. At the moment, the fertility rates of frozen poultry semen are far lower than any of the domesticated mammalian species. It has been estimated that post-thawing chicken semen retains only 1.6% of the fertilising capacity of fresh semen.

Artificial insemination

Two major considerations need to be factored in when inseminating the hen: where and when.

Three sites have been described for insemination: the cloaca, the vagina, and the magnum, in increasing order of complexity. Cloacal insemination is the easiest to perform but has the risk of loss of the insemination if the bird defecates or urinates before the semen enters the vagina and oviduct. Vaginal insemination can be achieved by everting the cloaca to expose the urodeum and then placing a capillary tube of semen into the vagina. Insemination into the magnum is achieved using a 2 mm flexible endoscope and a catheter to enter the oviduct and advance the catheter into the magnum. These latter two techniques are easier to achieve when the hen has already laid eggs and the oviduct/vagina is larger and more pliable.

The timing of the insemination has more to do with the hen's seasonality rather than the precise ovulatory cycle. This is because semen is stored in vaginal crypts for days or weeks after mating/insemination and goes on to lay fertile eggs.

If ovulation proceeds normally, the ovum progresses into the infundibulum, and a second glycoprotein layer is formed around it within approximately 15 minutes, preventing additional sperm from penetrating. This short window requires precise timing of insemination and the release of sperm from female storage to ensure sufficient sperm are in the infundibulum at ovulation.

- In turkeys, sperm storage is more efficient when artificial insemination is performed just prior to (rather than just after) the onset of egg production.
- In chickens, more sperm reach the infundibulum when they are inseminated within 1–6 days of ovulation (relative to 6–8 days before ovulation).
- In parrots, insemination is recommended within 1–2 days of expected oviposition (based on the bird's behaviour and appearance) or within 1–2 hours of an egg having been laid.

The greater the time delay between insemination and ovulation, the fewer sperm will be available for fertilisation since sperm are lost passively from storage at a constant rate. The time interval between ovulation and egg laying takes approximately 24–28 hours in chickens, turkeys, and quail. Inseminations performed immediately before or after egg laying reduce fertility, possibly because egg-laying contractions impede the ability of sperm to move through the oviduct and/or the passage of sperm is blocked by the egg.

INCUBATION OF EGGS

Why incubate eggs?

Incubation allows increased production, particularly in rare or valuable species. Removing an egg shortly after it is laid often induces the hen to lay a replacement. Poor parenting by some birds means that eggs and/or chicks can be lost in the nest. Removing and artificially incubating the egg can also reduce this risk.

It was thought that hand-rearing birds made them bond closer to people; therefore, artificial incubation was (and still is) widely practised to provide 'bonded' chicks for the pet market.

In some cases, artificial incubation is used as a means of disease control. Pathogens that are not vertically transmitted through the egg, but can be horizontally transmitted, can be reduced or eliminated by artificial incubation.

When to collect eggs

Artificial incubation is rarely as efficient and effective as natural incubation. The eggs of small species (e.g., conures, cockatiels, lovebirds, budgerigars) do not do well when incubated from day 0; it is thought that vibrations from the incubator may cause decreased hatchability. In these birds, incubation should start after 2 weeks of natural incubation. Large birds also benefit from natural incubation for the first 2–2.5 weeks.

However, it may, in some circumstances, be necessary to collect eggs immediately or shortly after being laid (e.g., when the parents are destructive to the eggs or when it is desired to increase egg production; if the hen broods for 2–3 weeks, she may not lay another clutch in that breeding season).

Storing eggs prior to incubation

Some aviculturists prefer to store eggs before beginning incubation to synchronise hatching and hand-rearing. However, it is usually preferable not to store eggs, as hatchability reduces daily when stored. This loss of hatching viability is species-dependent.

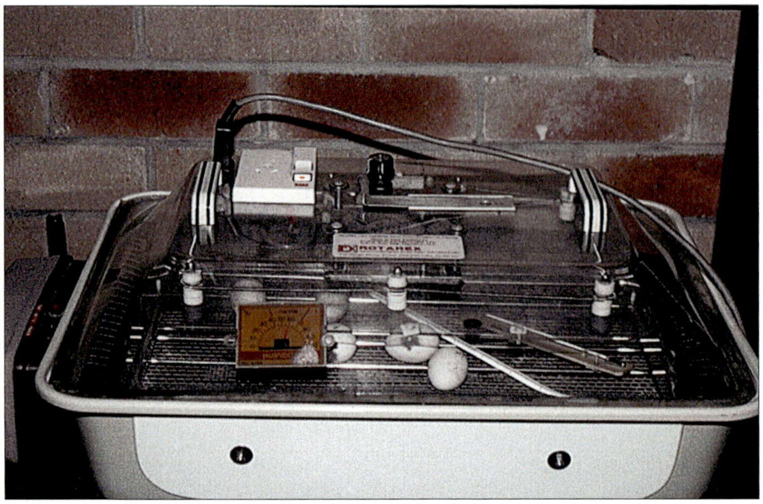

Figure 28.2 **Commercially available forced air incubator. Note that these are more effective and efficient than convection heating incubators.**

If there are no other options (and the aviculturist accepts the risk of decreased hatchability), eggs can be stored as follows:

- Temperature in the storage area should be kept between 12.8°C and 18.3°C to prevent embryonic development.
- The relative humidity should be maintained between 80% and 90%.
- The eggs should be turned 90° at least twice daily.
- Eggs should not be stored for more than 7 days.

Sanitation of the eggs

Eggs should not be cleaned if possible. Routine washing is not recommended, as this can reduce hatchability through chemicals passing through the pores in the eggshell as the egg cools. Dry cleaning with a soft brush is the preferred means of cleaning eggs. If it is necessary to wash the eggs, a 10% quaternary ammonia or 4.2% chlorine solution can be used. The water used should be warmer than the egg to prevent the egg from drawing in the washing solution through its pores. Detergents must not be used on very porous eggs.

Fumigation using formalin and potassium permanganate is sometimes recommended (20–30 minutes contact with egg), but this but may be harmful/carcinogenic to humans. Formaldehyde gas is also teratogenic in chickens.

Equipment

Incubators can be either forced air (fan-driven) or convection heated (cool air is drawn in at the bottom of the incubator, warmed, and finally exhausted past the eggs through the top of the incubator). Forced air incubators give more uniform temperature control and airflow than convection-heated incubators and are preferred by most aviculturists (**Fig. 28.2**). Incubators can only add heat to the eggs' environment; they cannot cool air.

Accurate humidity control is essential. It is important to recognise that incubators can usually only add moisture to the egg environment; they rarely are able to dehumidify it. Water reservoirs in incubators can act as a growth medium for potentially pathogenic bacteria such as *Pseudomonas*, *Aeromonas*, and *Klebsiella*, and should be cleaned regularly.

Automatic turning controls are recommended but must be checked for serviceability and vibration. Jarring and jolting of the eggs when the automatic turner is operating can reduce hatchability and increase the incidence of malpositioning.

As very few incubators can either cool air drawn in or dehumidify it, it is recommended that the incubator be kept in a clean, biosecure, air-conditioned, and dehumidified room whenever possible.

Incubation parameters

Temperature

Temperature determines the rate of embryonic growth. The correct temperature should ensure the egg hatches at the same time a naturally incubated egg does. Larger eggs need lower temperatures; smaller eggs need higher temperatures. Typically, psittacine eggs develop normally if incubated between 36.4°C and 37°C. As mentioned earlier, the temperature of the incubator room must be cooler than the incubator, allowing the incubator to warm the egg rather than trying to cool it.

Humidity

Eggs lose weight during incubation due to water loss through the shell (transpiration). The rate of water loss is dependent on shell thickness, shell porosity, and the water vapour concentration of the atmosphere. The expected weight loss is 16–22% over the incubation period. To achieve this, it is usually necessary that the relative humidity in the incubator be between 38% and 46%. The actual humidity in each incubator is determined by the weight loss of the eggs in the incubator (see later).

As incubators cannot dehumidify air, the relative humidity in the incubator room should be lower than in the incubator, allowing the incubator to add water to increase the relative humidity around the eggs.

Ventilation

Oxygen and carbon dioxide are exchanged across the shell (respiration). If the carbon dioxide builds up in the incubator, the chicks may suffocate. Therefore, there must be a continual infusion of fresh air into the incubator. Forced air incubators are more effective than convection incubators.

Turning and positioning

Turning of the egg allows proper development of the embryo and the blood vessels in the egg and allows better distribution of nutrients in the albumen. The old theory of the embryo 'sticking' to the shell is no longer considered valid. The eggs should be turned a quarter turn along the long axis every 4–5 hours. This is usually done mechanically, although older incubators rely on manual turning. The turning should be smooth, with no vibration or jolting of the eggs.

Poultry and ratite eggs are incubated vertically, with the air cell up; psittacine eggs are incubated horizontally with the air cell end slightly elevated. Eggs that are not positioned correctly within the incubator are more likely to develop malpositioned embryos or experience increased embryonic death.

Figure 28.3 Candling an egg. In this case, the egg was infertile.

Monitoring

Embryonic development can be monitored by candling (examining the egg in a darkened room with an intense focal light to transilluminate it) (**Fig. 28.3**). Candling at 7–10 days will indicate if the egg is fertile and developing properly. Clear yolks showing no signs of blood vessels or development by day 7 are infertile or died early and should be removed. Occasionally, a 'blood ring' (extravasated blood in a ring surrounding the remnants of the embryo and embryonic circle of blood vessels) is seen. This indicates early embryonic death.

Weight monitoring should also be carried out, ideally by daily weighing, although weighing every 2–3 days can be sufficient in well-developing eggs. Eggs should lose 13–22% of their weight between setting and removal to the hatcher through evaporative loss through the shell. They may lose another 3% during hatch, although excessive weight loss at this time may dehydrate the chick, making hatch difficult. Eggs with a thin or very porous shell, or those placed in a low humidity incubator, can lose an excessive amount of weight through evaporative loss. These eggs should be placed into a higher humidity. Conversely, those with thick shells or low porosity, or those placed in a high humidity incubator, can retain moisture and lose insufficient weight. These eggs should be placed into a low humidity environment to encourage more evaporative weight loss.

Hygiene

Incubators, being warm and humid and with a high biological content, can become bacterial and fungal incubators if not maintained correctly and rigid hygiene imposed. Water reservoirs should be cleaned daily, broken or dead eggs should be removed promptly, and dust filters should be cleaned every few days or more. The incubator should be emptied, aired, and disinfected as frequently as possible, using a broad-spectrum disinfectant such as F10® or Virkon®.

Record-keeping

It can be very difficult to persuade aviculturists to maintain detailed records, making it almost impossible at times to investigate incubation problems. Records that should be maintained include:

- Daily temperature and humidity readings in the incubation.
- Notes on problems such as power failures, equipment malfunctions, etc.
- For each egg, the following details should be recorded:
  - Laying date
  - Setting date
  - Setting weight
  - Candling and weighing data
  - Weight when moved to the brooder
  - Outcome of incubation and hatching (e.g., clear egg, failed to hatch, hatched without difficulty, hatched with difficulty)
  - Viability of each chick
- Overall, the number of eggs laid, fertility, and hatchability for the season and for each breeding pair should be recorded.

With these records available, a thorough investigation into a hatchability problem is feasible. The use of egg necropsies can provide more information.

Embryonic development

Embryonic growth is divided into three approximately equal phases: early, mid, and late terms.

Early term

In poultry and cockatiels, the following events occur:

- Day 1: The blastoderm develops, becoming doughnut-shaped as cellular division occurs.
- Day 2: The brain, eyes, and spinal cord develop.
- Days 2–3: The heart and blood vessels grow, and the extra-embryonic membranes form:
  - Amnion from the body wall cushions the embryo in amniotic fluid.
  - Yolk sac membrane from the gut envelops the yolk.
  - Allantois from the hindgut acts initially as the bladder.
  - Chorion, an extension of the amnion, fuses with the allantois to form the chorioallantoic membrane (CAM), the primary respiratory organ of the embryonic chick.
- Days 3–4: The body wall and viscera grow.
- Days 4–5: Limbs develop.

Mid term

There is growth with little further differentiation. Feathers begin to form, bones begin to calcify, and the beak forms.

Late term

The yolk sac is drawn up into the embryo, the umbilicus is sealed, and the albumen is actively drunk by the chick. Finally, the chick positions itself and hatches.

Hatching

Hatching is stimulated by the chick converting to pulmonic respiration from chorioallantoic membrane respiration. Increasing carbon dioxide concentration causes the neck muscles to twitch, and the chick becomes more active. The air cell expands and extends down one side of the egg 24–48 hours before internal pipping. This process is known as 'drawing down'.

The chick then penetrates the air cell (internal pipping) by rubbing its beak against the inner shell membranes until they tear. The chick is now able to breathe the air in the air cell. As CO_2 builds up in the air cell, the chick is then stimulated to use its 'egg tooth', a keratin tip on the rhinotheca, to crack the shell circumferentially until the top of the egg breaks away, a process known as the 'external pip'

Figure 28.4 **Indian ring-neck chick in an incubator shortly after hatching. Note the circumferentially cracked egg known as the 'external pip'.**

(**Fig. 28.4**). After a period of time (usually minutes) to regain its strength after the exertions of pipping, the chick crawls out of the shell, completing the hatching process.

The hatch time, the interval from internal pip to hatch, is usually 36–48 hours in most species. Hatching times less than 24 hours or more than 80 hours can indicate a problem.

Requirements for a hatcher in which an egg can be placed to hatch are:

- A lower temperature than the incubator, usually 36.4–36.9°C.
- A higher humidity, up to 70%.
- No turning.
- Adequate ventilation.

Problems with incubation

In a well-run incubation setup, 85–90% of fertile eggs should hatch without difficulty. Embryonic mortality should not exceed 10% of fertile eggs set for incubation. This normal 10% mortality is made up of:

- Early-term deaths, 3–4%
- Mid-term deaths, 1–2%
- Late-term deaths, up to 4%

Losses above these figures require investigation. The two most common causes of mortality/morbidity in the egg are malposition (the most frequent cause; see later) and inadequate moisture loss; other causes are listed in **Table 28.1**.

Egg necropsy

All eggs that fail to hatch should be necropsied. Not only does an egg necropsy allow a determination of the stage and possible cause of embryonic death, but it also provides the clinician (and aviculturist) with a true indication of fertility.

The technique is relatively simple. The egg is first weighed, and, if possible, the weight loss during incubation is calculated. The shell quality is examined (surface texture, porosity, and cleanliness). The egg is then candled. Candling allows identification of the air cell (mark it with a pencil), an indication of its size and mobility, and an idea as to whether an embryo or infection is present. Hairline cracks in the shell can also be identified. This information is recorded before moving to the next step.

The egg is placed vertically onto a support (an egg cup or a ring of PVC pipe is sufficient) with the air cell uppermost. The shell over the air cell is broken with a sharp pair of pointed scissors, revealing the air cell. The shell is then trimmed away to the level of the inner shell membrane, forming the base of the air cell. This membrane is then peeled away with fine forceps, revealing the egg contents.

At this stage, a swab for culture of bacteria and fungi can be taken from the albumen or inside the shell. If an embryo is not obviously present, examine the blastodisc on the yolk. The yolk around this disc is less dense than the rest of the yolk, so it will float uppermost. In an infertile egg, the disc will appear as a small white point. A very early embryo

Table 28.1 **Causes of embryonic mortality**

STAGE	CAUSES
Early	Breeder malnutrition
	Stored incorrectly – too hot, too long
	Infection
	Rough handling of the egg during collection
Mid	Breeder malnutrition
	Infection
	Rough handling
	Inadequate turning
Late	Incorrect incubator parameters • Temperature • Ventilation • Turning • Humidity
	Infection
	Malnutrition

(blastoderm) will appear as a small white 'doughnut' with a patch of yolk in the middle. Any blood vessel development at all is confirmation of fertility.

If an embryo is present, two determinations must be made: the positioning of the embryo (if it is late term) and the stage of development.

Embryonic position is observed before removing the embryo from the egg. The normal late-term embryonic position is head next to the air cell, right side uppermost and turned to the right, with the beak adjacent to the right wing tip and the egg tooth pointed towards the air cell; legs flexed on either side of the body and ventral to the shoulders; and the spine following the long axis of the egg. Any other position is classified as a malposition and may account for embryonic death (see **Fig. 28.5**, **Table 28.2**).

The embryo is then removed from the egg and examined more closely. An assessment of the embryo's growth and development can give an indication as to whether it is an early, mid-term, or late-term embryo (see earlier). Samples can be collected from the chorioallantois, yolk, and proventriculus for culture, and tissues from multiple organs, including the membranes, can then be placed in formalin for histopathology.

By determining the stage and cause of embryonic death, the clinician can then start to narrow the search for the underlying cause of embryonic mortality.

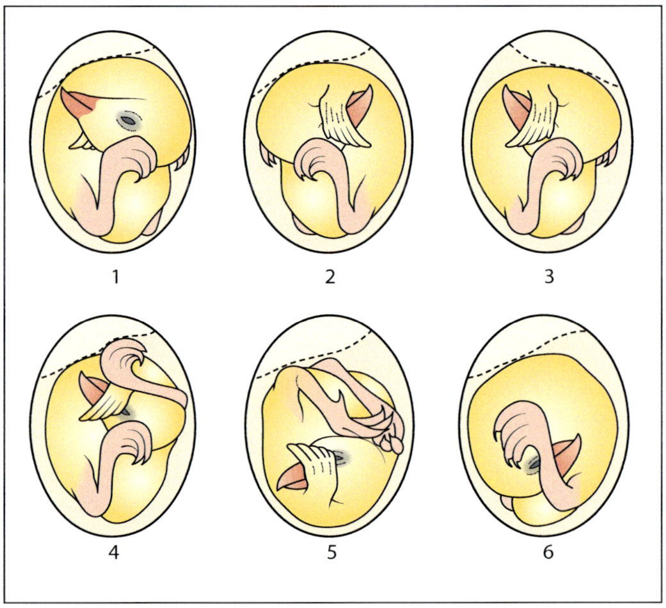

Figure 28.5 **The six classically described malpositions of the embryo in the egg. (see Table 28.2 for descriptions).**

Table 28.2 Embryonic malpositions

NUMBER	DESCRIPTION	COMMON CAUSES	OUTCOME
1	Head over right wing	• Parental malnutrition	Slightly reduced hatchability
2	Head under left wing	• Abnormal temperature • Incubator position • Parental malnutrition	Usually lethal
3	Beak away from air cell	• Abnormal incubator position; may result in pipping a blood vessel (haemorrhage)	Slightly reduced hatchability
4	Head between thigh	• High incubation temperature	Usually lethal
5	Feet over head	• Unknown	Usually lethal
6	Head in small end of egg (upside down chick)	• Low temperature • Abnormal incubator position	50% hatchability

FURTHER READING

Jordan R (1989) *Parrot Incubation Procedures.* Silvio Mattacchione and Co., Pickering.

Jordan R (2001) Incubation of psittacine eggs. *Seminars in Avian and Exotic Pet Medicine* 10(3): 112–116.

Joyner KL (1994) Theriogenology. In: *Avian Medicine: Principles and Application.* BW Ritchie, GJ Harrison, LR Harrison (eds). Wingers Publishing, Lake Worth, pp. 748–804.

Lierz M, Reinschmidt M, Muller H, Wink M, Neumann D (2013) A novel method for semen collection and artificial insemination in large parrots (Psittaciformes). *Scientific Report* 3. DOI: 10.1038/srep02066 (2013).

Olsen GH, Clubb SL (1997) Embryology, incubation and hatching. In: *Avian Medicine and Surgery.* RB Altman, SL Clubb, GM Dorrestein, K Quesenberry (eds). WB Saunders, Philadelphia, pp. 54–72.

Samour J (2002) The reproductive biology of the Budgerigar (Melopsítticas undulatus): Semen preservation techniques and artificial insemination procedures. *Journal of Avian Medicine and Surgery* 6(l): 39–49.

Schubot RM, Clubb KJ, Clubb SL (1992) *Psittacine Aviculture: Perspectives, Techniques and Research.* Avicultural Breeding and Research Center, Loxahatchee, FL.

CHAPTER 29

PAEDIATRICS

INTRODUCTION

Most companion birds (e.g., parrots and passerine birds) are altricial (i.e., when hatched they are blind, deaf, and not feathered, and therefore totally dependent on their parents or rearer) (**Fig. 29.1**). As juveniles, their health status is determined not only by their current environment but also by the input from their parents and how the eggs were incubated.

- Pre-laying factors:
  - The health of the parent birds can affect the chick via pathogens transmitted through the egg to the chick, and by the degree of transference of maternal immunity via the albumen and yolk.
  - The diet fed to the parents, particularly the levels of calcium and vitamins, will have a direct effect on the vitality, development, and health of chicks – even before hatching.
  - The maturity of the parents will affect the size of the eggs laid, the immunity passed through the egg, and the number of eggs laid.

Figure 29.1 **Parrot chicks, such as these lorikeets, are altricial and totally dependent on their parents or rearers until weaned.**

  - Parental genetics become a major issue in the breeding of mutations, where closely related birds are mated to develop certain characteristics such as colour. Unfortunately, along with desirable characteristics such as new colour, undesirable physical traits such as decreased body size or physical deformities can also occur.
- Incubation factors:
  - Artificial vs. natural incubation. As a rule, parent-incubated eggs tend to hatch stronger chicks. Advances in incubation techniques (see Chapter 28) are, by and large, resulting in better chicks. However, skill and experience are required to match the results seen in natural incubation.
  - Temperature, humidity, and hygiene in the nest box or incubator will have direct effects on the development of the chick and its hatchability and strength after hatching (See Chapter 28).
  - Care in handling the eggs by either the parents or the incubator operator, as well as the frequency and degree of rotation during artificial incubation, can affect the degree of difficulty in hatching and also influence the incidence of malpositions.
- Post-hatch factors. If the chick has been hand-reared, the following parameters in the rearing environment play important roles in the health of the chick:
  - Temperature extremes can force the chick to divert resources from its growth and health towards maintaining a constant body temperature.
  - Humidity extremes can predispose a chick to dehydration, respiratory infections, and skin problems (e.g., toe constriction).
  - Poor hygiene will substantially increase the concentration of pathogens (or potential

DOI: 10.1201/9781003272519-29

pathogens) in the chick's environment. With the chick's immune system still developing, this creates a high probability of infectious disease.

- Although the era of homemade hand-rearing diets is passing, the nutrition of the chick is still a major factor in the health of the chick. The practice of adding ingredients to a well-balanced formula (often based on anecdotal information from other aviculturists) can have major detrimental effects on the chick.
- Management of the nursery, in particular biosecurity and the quarantine of new chicks from different sources, will determine the likelihood of the introduction of infectious diseases such as avian polyomavirus.

HAND-REARING

Hand-rearing can be done for several reasons: to increase production from rare or valuable species; when the chick is orphaned; or to make a better socialised chick. Recent studies suggest that the latter is not always achieved by hand rearing, but many aviculturists maintain that hand-reared chicks make better pets.

The timing of the commencement of hand-rearing varies between the owner and the individual circumstances. Some chicks, artificially incubated, are reared from the time of hatching; others are left with their parents for 2–3 weeks and are 'pulled' for hand-rearing just before their eyes open, while others are taken just before fledging and are then reared until ready to wean. Research has shown that parrots which had been removed very early from the nest were more likely to develop stereotypic behaviours significantly more than those that had stayed longer in the nest before being removed for hand-rearing. The same research suggested that hand-reared chicks, particularly those who were incubated or removed prior to their eyes opening, were much less likely to breed successfully in later life.

The frequency of feeding is dependent on the age of the chick. A common feeding schedule is:

- Hatch to 1 week – every 2 hours
- 1 week to 3 weeks – every 4 hours
- 3 weeks to 6 weeks – every 6–8 hours
- 6 weeks to weaning – every 8–12 hours

Many breeders feed their babies on a rigid schedule of every 1 to 4 hours, depending on their ages. However, disturbing a resting or sleeping chick to feed it or forcing hungry chicks to wait for food until the next scheduled feeding is needlessly stressful for young birds. The optimal feeding schedule is sensitive to the bird's needs, including their rapid increases in weight and the concurrent increase in the volume of food required at each feeding. Breeders who no longer adhere to the rigid scheduled feeding style report neonates that sleep soundly for consecutive hours and awaken eager to eat. Rather than forcing chicks to adhere to schedules tailored to human convenience, feeding them on demand may be best for proper physical and behavioural growth.

Different techniques are used for feeding, including the use of syringes, feeding tubes, or spoons. Regardless of the technique used, it is important to stimulate a feeding response (head bobbing and thrusting) by applying slight pressure to the beak commissures before any food is introduced. When a chick is displaying a feeding response, its glottis is closed, and accidental aspiration of food is less likely. It should be noted that chicks fed with a feeding tube are often more aggressive towards people after weaning and have more health problems than those fed with a spoon or syringe. Tube feeding should therefore be reserved for sick chicks.

A variety of quality commercial diets are available today, and it is uncommon to see 'homemade' recipes. The volume fed is between 8–12% of the chick's body weight. The food should be mixed fresh for each feed and fed at approximately 41°C (normal body temperature for birds). The addition of supplements (especially calcium and vitamins) to a quality commercial diet should be discouraged.

Newly hatched chicks are not capable of endothermy and will lose body heat (or overheat) rapidly. Active warming is therefore required until the chick is feathered. As the chick grows and its feathers develop and open, the need for supplemental heat decreases, and the young chicks can be moved out of heated brooders into unheated containers that allow more movement but still resemble a hollow.

Prior to their eyes opening, neonatal parrots are responsive to light. They have evolved to begin life in the darkness of a tree hollow and may react adversely to strong light by flinching, hiding, or trembling. After their eyes open, this light stimulus may encourage premature activity and weight bearing, resulting in limb abnormalities. Restricting movement is therefore a critical factor in the prevention of limb deformities. Various containment systems employ the use of towels to provide darkness, privacy, traction, and hiding places that appear to be critical to a stress-free environment. For further comfort and safety, feeding should continue in a routine manner, and the environment must remain secure.

This better understanding of a chick's behaviour and requirements is leading to the phasing out of brightly lit, individual white containers for hand-rearing chicks. Many breeders are now rearing chicks in small, darkened containers until fledging age is reached. Hand-reared chicks reared in such a fashion appear to gain weight faster than those kept in brightly lit environments. Other breeders prefer to allow the parents to rear their young but handle the neonates briefly each day (or several times daily).

The importance of physical contact for neonates should not be ignored. Keeping multiple chicks together increases normal (but not excessive) physical stimulation. Incubator-hatched birds should have touch massage incorporated into their daily care. Placing a lightweight soft cloth on the chick's back after feeding may simulate the weight of other chicks or a parent, and often has the effect of quieting a restless chick after a feed.

A proposed alternative to hand-rearing chicks to produce tame adults is to handle the neonates briefly while they are still in the nest and being fed by their parents. The post-weaning behaviour (tameness) of hand-reared and neonatally handled parrots appears to be similar and differs substantially from that of birds which have been parent-reared (without handling). Although it has been suggested that taming parrot chicks by neonatal handling prevents the development of behavioural abnormalities associated with hand-rearing, such as abnormal sexual and social imprinting and inappropriate reproductive behaviour, this evidence is primarily anecdotal.

However, it is not known how hand-reared, neonatally handled, and parent-reared birds differ with respect to other behaviours, such as neophobia (fear of new objects or surroundings). Research has suggested that hand-reared chicks are less neophobic in their first year of life, but this does not last. Hand-reared birds may exhibit moderate levels of neophobia after weaning; later, they behave in a manner almost indistinguishable from parent-reared birds in response to a novel object. The use of environmental enrichment is recommended for all chicks, regardless of how they were reared.

Prior to fledging, babies show increasing interest in the world outside their enclosure. A partial covering of towels will enable them to see out of their container or to retreat and hide. As they become ever more inquisitive, they will spend more time looking out of the container and less time in concealment. There is often a great deal of wing flapping inside the nest as they start building up their pectoral muscles. If a perch is provided in the enclosure, the chick will often start to perch rather than stand on a flat surface.

Flight competence precedes weaning in parrot chicks – they must learn controlled flight before following their parents to various, often distant, food sources. The practice of clipping the wings of newly weaned chicks should be strongly discouraged, as this limitation to flight often ends in traumatic injuries.

Weaning is a stressful time for both the chick and the hand-rearer. The chicks will often start to refuse feeds and may regurgitate part of what has been eaten. Flight becomes common, and weight loss is normal at this stage, as they develop muscle and lose fat. Weaning requires the chick's eating skills to develop over a period of several weeks. Unfortunately, many pet stores and breeders consider a chick 'weaned' as soon as it shows interest in eating on its own; the process is gradual as the chick learns where, what, and how to eat. There are numerous motor skills necessary to accomplish this, and these take time to develop. Environmental evolutionary pressures appear to have an effect on the duration of the weaning process, with arid zone birds weaning faster than those from more tropical environments, perhaps because food availability is shorter in an arid environment.

A broad selection of a wide variety of foods should be available for the chicks at this time – formulated diets, vegetables, and fruit – and the intake of these foods should be monitored carefully. Feeding a chick by hand may encourage the chick to experiment with a range of foods.

Detailed knowledge of hand-rearing practices, including weaning ages, can be obtained from reputable aviculture literature.

EXAMINATION OF THE CHICK

History

As with any other medical case, a good history of the chick is an essential key in the diagnostic workup.

Factors to consider stem from the earlier discussion on determinants of a chick's health. They include:

- The parents – their genetics, diet, maturity, and health status.
- Incubation. Was it artificial or natural? The hatchability of fertile eggs is a key indicator of incubation performance, and hopefully, it can be found in the aviculturist's records.
- If the eggs were artificially incubated, were there any problems with hatching?
- How is the nursery managed regarding hygiene, biosecurity, and the source of eggs and/or chicks?
- What type of food has been fed, how it is it prepared, what volume is fed, and how frequently?
- If there are any siblings or other chicks, have there been any problems or deaths within the group?
- What records does the aviculturist keep? Details on hatch dates, hatch weights, growth rates, mortalities, medications, and previous medical problems are valuable sources of information but are sadly lacking in many cases.

Physical examination

Weight

The chick should be weighed, and its weight compared with the expected weight for that age, found in growth charts (if available). All chicks will gain weight each day – it is the rate of weight gain that is more important than the actual weight. This can be monitored by expressing the chick's weight as a percentage of the expected weight for that age. If a problem exists, improvement can be seen as a gain in the percentage of expected weight, rather than actual weight gain.

Posture

It is important to be aware that chicks sleep and rest in what seem to be 'awkward' positions. (For example, conure and macaw chicks often sleep on their backs.) These positions change as the chick moves; one should look for postures that do not change with movement.

Conformation

The positioning and conformation of the limbs and the spine should be checked. Common conformational abnormalities include a kinked (or 'wry') neck, scoliosis, kyphosis, tibiotarsal or femoral rotation, and anteroflexion of the toes.

Body condition

The toes and elbows in a well-nourished, healthy chick should be 'plump'. Thin toes and elbows are a good indicator in neonatal chicks of dehydration, malnourishment, or disease. Palpation of the pectoral muscles is helpful; the soft keel bone at this age should be well-fleshed with soft (but poorly developed) pectoral muscles (**Fig. 29.2**).

Behaviour

Pre-weaning chicks should be either resting, sleeping, or calling for food. Restlessness could indicate an incorrect environmental temperature or stress (e.g., excessive lighting). As they get older, the chicks still spend a lot of time sleeping but are more interested in their environment and in socialising with nursery mates.

A feeding response (vigorous extension and bobbing of the head and neck) should be easily elicited by pressing gently at the commissures of the beak. Failure to elicit a response can be an indication of disease, hypothermia, or weakness.

Figure 29.2 **A healthy sulphur-crested cockatoo chick.**

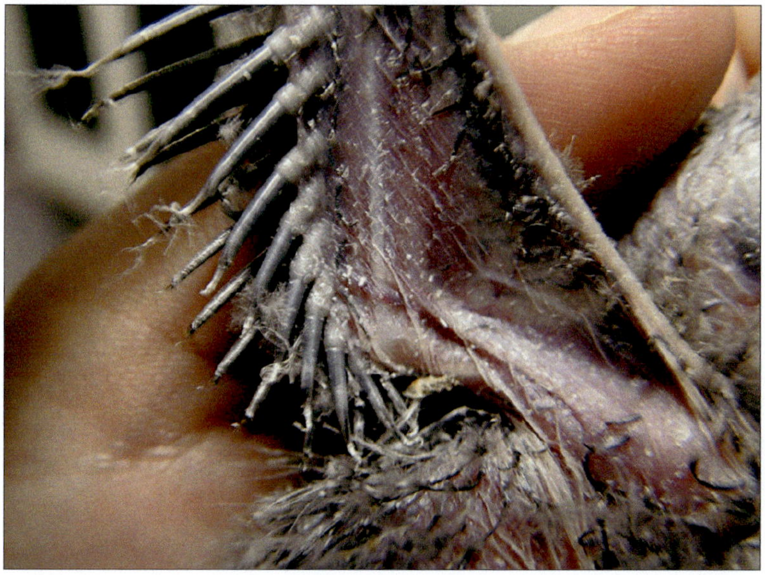

Figure 29.3 **Green-cheek conure chick with wrinkled skin and poor filling of the basilic vein, which are good indicators of dehydration.**

Skin

Prior to full feathering, the normal chick's skin should be pink or pink-yellow in colour, and soft and warm to the touch. Pallor of the skin can indicate hypothermia, anaemia, or illness. Erythematous skin can indicate hyperthermia or illness. Heavily wrinkled skin indicates dehydration (**Fig. 29.3**).

Crop

The normal crop should have some food in it at most times. It should not be overdistended, nor should it have significant amounts of air or gas in it. Rhythmic contractions of the crop should be visible in neonatal chicks, and the crop should nearly empty in 4–6 hours in all chicks.

Head

The size of the head should not be excessively large in relation to body size. The beak should have a normal conformation (see Chapter 12). There should be no sinus swellings. The nares should be open and symmetrical.

The eyes should be symmetrical and healthy in appearance. They begin to open at 10–28 days and take several days to open completely. Most Australian and African parrots hatch with their ears open. The ears of eclectus and South American species should be open within 2–3 weeks after hatching.

Oral cavity

The oral cavity should be examined for diphtheritic plaques or other abnormalities. Its colour should be noted; it is normally pale pink.

Coelom

In neonatal chicks, the abdomen should be large and convex relative to the rest of the body. The liver may be visible through the skin in very young chicks, and rhythmic contractions of the ventriculus should be visible. The duodenal loop may be visible. There should not be bruising or haemorrhage visible. The abdomen can be transilluminated with an intense focal light for closer inspection.

As the chick grows, the abdomen reduces in size relative to the rest of the body. It should be concave when palpated; a convex abdomen could indicate a degree of abdominal distension.

Feather growth

Depending on the species, some chicks are hatched naked or with wispy down feathers. Others are covered in fluffy down feathers (**Fig. 29.4**). A second wave of down feather growth begins at 1–3 weeks of age and, sometimes, even later in some species.

Pin feathers begin to emerge at 2–3 weeks of age. The body contour feathers emerge over the shoulders first; the pattern of emergence after that varies

Figure 29.4 **Yellow-tail black cockatoo chick, showing fluffy down feathers.**

between species, although usually, the body contour feathers emerge at the same time or shortly before the secondary flight feathers on the wings. Primaries may begin to develop before secondary feathers, but usually mature after them. Final feather maturity is usually not complete before the bird has weaned.

Abnormalities include:

- Feathers erupting in an unusual pattern (e.g., in a circular pattern on the crown of the head, rather than running parallel along the line of the body)
- Stress bars in the open vane
- Abnormal colouring
- Haemorrhage in the calamus
- Dystrophic development

Droppings

The faecal portion should be relatively well formed, light brown in colour, and not malodorous. A degree of polyuria is normal (especially in hand-reared chicks), but this should lessen as the chick ages. Excessive or persistent polyuria warrants further investigation.

Diagnostic testing

Microbiology is an important tool in assessing gastrointestinal flora. Gram stains and cultures are frequently used to assess crop or other gastrointestinal problems. Normal bacterial flora include *Lactobacillus*, *Streptococcus*, *Staphylococcus*, and *Bacillus* spp. Low numbers of *E. coli* are often normally cultured as well. Other gram-negative bacilli and *Candida* are rarely cultured from healthy chicks.

Clinical pathology can be used readily on chicks. It is important to note that compared to adults of the same species, chicks normally have:

- Lower PCV and higher white cell count
- Lower total protein and uric acid
- Higher CK

Radiography is an essential tool for assessing the status of a chick, but the low density of the bones and the cartilaginous growth plates in very young chicks can make the assessment of the bones difficult. The lack of fat reduces contrast in the coelom, again limiting interpretation.

MANAGING THE SICK CHICK

There are unique challenges facing the clinician when dealing with a paediatric case. These include:

- Their small body size means that there is:
  - i. A large surface area to body mass ratio, sensitising them to temperature extremes (especially hypothermia).
  - ii. Difficulty in accessing blood vessels for the administration of fluids and medications and collecting blood samples.
  - iii. Small muscle mass, making intramuscular injections difficult to administer without causing significant bruising, pain, and discomfort.
- A lack of white fat (for insulation) and brown fat (for thermogenesis) makes them reliant on their metabolism to maintain their body temperature. Although cockatiel chicks have been shown to develop endothermic responses as early as 1 week post-hatch, they are still only capable of achieving 75% of the adult ability to maintain their body temperature until they fledge. This means that:
  - i. They require the frequent ingestion of high-energy, easily digested food to maintain their metabolic rate.
  - ii. Their retention of body heat is hindered by the lack of insulation from feathers, which contributes to potential hypothermia.

- Their immature physiology means that:
    i. The ability to invoke compensatory physiologic responses (e.g., vasoconstriction and vasodilation) is limited.
    ii. The kidneys have poorly developed concentration and filtering capacity, meaning that their fluid requirements may be two to three times that of an adult bird.
    iii. The digestive tract has a rapid transit time and is easily 'stressed', leading to ileus.
    iv. The respiratory tract is immature, and the air sacs are compressed by the intestinal tract, giving a low respiratory reservoir.
    v. The still-developing immune system is often unable to cope with primary or secondary pathogens.
- Diagnostic testing can be limited by their size and age.
    i. Radiology is limited.
        a. Coelomic radiography is difficult to interpret due to the loss of contrast without coelomic fat.
        b. Evaluation of the skeletal system is difficult due to the decreased opacity of the bones and the open growth plates.
    ii. Haematologic and biochemical differences from adults can make the use of diagnostic testing difficult. Compared to an adult of the same species, chicks generally have:
        a. Lower haematocrit and higher white cell count.
        b. Lower total protein and uric acid.
        c. Higher creatine kinase.

The four H's

Regardless of the underlying aetiology, most problems with sick chicks can be associated with the 'four H's'. These have been developed by small animal emergency clinicians as an acronym for the assessment of paediatric patients. They are:

- Hypovolaemia
- Hypoglycaemia
- Hypothermia
- Hypoxaemia

Hypovolaemia

Fluid requirements for paediatric patients are higher than those of adults. This is because of increased extracellular fluid requirements, higher body surface area, greater surface area to body weight ratio, lack of body fat, higher metabolic rate, decreased ability of immature kidneys to concentrate urine, and increased respiratory rate leading to greater insensible fluid losses.

Severe dehydration is common in chicks, usually associated with gastrointestinal dysfunction (vomiting, diarrhoea, and ileus) or decreased intake (poor husbandry, inappropriate or poorly mixed diet, refusal to eat). Respiratory loss due to panting when heat-stressed can also lead to dehydration. Dehydration and hypovolaemia lead to decreased tissue perfusion and subsequent organ damage and even failure. As paediatric patients appear to lack the ability to compensate for mild hypovolaemia to increase tissue perfusion (increased heart rate and contractility, and increased vascular tone), they are extremely sensitive to the effects of dehydration.

Clinical signs include:

- Thickened mucoid saliva
- Wrinkling of the skin
- Enophthalmos
- Decreased venous return (assessed by 'blanching' the basilic vein)
- Decreased urine output and thickened urates (although polyuria can be a contributing factor to dehydration)
- Ileus (often reflected as crop stasis)

If any uncertainty is present, it is usually safe to assume the chick is dehydrated and to act accordingly.

Crop stasis, if present, should be addressed by emptying the crop and assessing its contents for bacterial or yeast overgrowth. Failure to empty the crop in a sick chick may result in regurgitation and possible aspiration pneumonia. The chick should not be fed until warmed and its hydration status is corrected.

Fluid therapy requires the administration of warmed (30–35°C) isotonic fluids. These can be given IV, IO, or SC, but if the SC route is chosen, the chick should be normothermic, as hypothermia will add to the chick's reduced ability to peripherally vasodilate, slowing fluid absorption. IO catheters can be placed in the tibiotarsus or ulna, while IV catheters can be placed in the jugular, basilic, or medial metatarsal veins. An initial fluid bolus of 3–4% of the chick's body weight can be given over 15 minutes and repeated based on the chick's response. Ongoing fluid administration at 3 ml/kg/hour can then be initiated. Care must be taken to avoid fluid overload (often seen as dyspnoea associated with non-cardiogenic pulmonary oedema). Once the chick is rehydrated and the gastrointestinal tract is functional, the oral administration of adequate volumes of fluid can be expected to maintain a suitable fluid balance.

Hypoglycaemia

Hypoglycaemia often arises due to either poor feeding practices (inappropriate diet, infrequent feeding, diluted food, or insufficient volume of food) or severe systemic illness (particularly sepsis or conditions leading to gastrointestinal ileus or other gastrointestinal dysfunction). When low dietary intake is combined with an immature gluconeogenic response and low body fat, the result can be severe hypoglycaemia. This in turn leads to neurological and cardiac disturbances (weakness, seizures, coma, and death). In the early stages of hypoglycaemia, chicks may be constantly hungry, exhibited as excessive begging behaviour. Blood glucose levels below 8 mmol/L are suggestive of hypoglycaemia.

Chicks with severe hypoglycaemia should be treated with an IV/IO bolus of dextrose 250–500 mg/kg (50% dextrose, 0.5–1 ml/kg diluted 1:4) administered over 5 minutes. Repeat boluses can be given based on response to treatment, and then once normoglycaemic, the chick's IV/IO fluids can be supplemented with 5% dextrose until the chick is eating.

Less severe hypoglycaemia may be effectively dealt with by an early return to feeding or by feeding a more appropriate diet.

Hypothermia

Neonatal parrot chicks are not capable of endothermy (controlling their body temperature) until they are nearly ready to fledge. Hypothermia in chicks arises because of the effect of the following factors:

- A large surface area to body mass ratio, making them susceptible to temperature extremes
- Lack of feathers to insulate the body
- Low reserves of white (insulating) fat
- Low reserves of brown fat (thermogenesis)
- Reduced ability to vasoconstrict or shiver
- Reduced activity (chicks are usually sleeping or eating)

When these factors are combined with inappropriate environmental temperatures or reduced food intake (decreased metabolic energy), the result can be profound hypothermia.

Hypothermic chicks are lethargic and poorly responsive to stimuli. Their limbs and bodies are cool to the touch. Peripheral vasoconstriction can be seen as a pallor to the skin. Left untreated, cardiovascular function will be compromised, resulting in organ dysfunction and finally cardiac arrest.

Hypothermic chicks should be warmed before fluid resuscitation, but care should be taken to avoid excessive peripheral vasodilation and possible hypotension. Warming can be achieved with warmed air or heat lamps, but the chick must be monitored for signs of hyperthermia (panting, hyperaemic skin, distress), especially if it is unable to move away from the source of the heat. Adequate humidity (50–60%) must be provided to prevent desiccation from the heat and worsening of the chick's hydration status.

Hypoxia

Hypoxia is common in chicks and can be associated with anaemia, the aspiration of hand-rearing formula, infectious respiratory diseases (e.g., bacterial or fungal infections), or compression of the air sacs by distended loops of the intestinal tract. Hypoxic chicks will have an increased respiratory rate and effort (mouth breathing, increased sternal lift, and tail bobbing) and may have audible respiratory noises. Cyanosis can be seen but is often difficult to appreciate. Pulse oximetry may be of benefit, but because of calibration difficulties with nucleated erythrocytes, trends rather than absolute numbers should be monitored.

Oxygen supplementation (via an anaesthetic induction chamber, flow-by oxygen, or an intranasal oxygen line) should be administered when hypoxaemia is diagnosed or suspected. Care must be taken to prevent oxidative tissue damage associated with 100% oxygen administered for prolonged periods. Antimicrobial therapy is often warranted, e.g., a combination of amoxicillin–clavulanic acid, metronidazole, and itraconazole.

Where to from here?

Once the sick chick has been stabilised, the clinician can then move on to determining the underlying reason for the chick's illness. Common causes of illness in chicks have been mentioned earlier, but the diagnosis and treatment of these diseases can only be performed once the chick is stabilised. Careful evaluation and examination are paramount in diagnosing and treating paediatric problems.

COMMON PROBLEMS

Stunting

Aetiology

Stunting is seen in the first 30 days of life. It is usually associated with improper feeding techniques (poorly balanced or incorrectly mixed diets, inadequate amounts fed, etc.), poor environmental conditions (temperature and humidity extremes), or disease (e.g., renal disease).

Clinical presentation

Signs include subnormal weight gain, reduced muscle mass (toes, wings, back should be checked), abnormal feathering (e.g., head feathers develop in a circular pattern on the crown), and oversized head relative to the size of the body (**Fig. 29.5**). Eyelids fail to open normally or when expected, and there is delayed ear opening or narrowing of the ear canal. The affected bird may suffer from chronic, recurrent infections, and constantly call and beg for food. As the chick gets older, it often develops a globose head with an elongated slender beak. The eyes may appear exophthalmic because of the misshapen skull.

Management

The predisposing cause should be identified and treated. Nutritional inadequacies should be corrected. The prognosis is good if the problem is diagnosed early and treated successfully.

Crop stasis ('sour crop')

This is a commonly seen problem in paediatric medicine, with most sick chicks having varying degrees of crop stasis. All too often, hand rearers attempt to diagnose and treat this problem without seeking to understand the underlying pathology.

Aetiology

Causes include:

- Generalised ileus (e.g., systemic illness, foreign bodies, chilling, heavy metal toxicosis, dehydration)
- Crop disorders (e.g., foreign bodies, overstretched/atonic crop, infectious ingluvitis, fibrous food impaction, crop burns)
- Dietary problems (e.g., cold food, excessively watery food, food that settles out in the crop, overfeeding, overly dry food)

Clinical presentation

Signs include the crop failing to empty in more than 6 hours, regurgitation, and loss of feeding response. Most chicks will be dehydrated on presentation (erythematous wrinkled skin, tenting of the skin, and sunken eyes).

Aspiration of the crop contents usually reveals the sour-smelling fermenting ingesta ('sour crop') or thickened ingesta (hand-rearing formula with the fluid drawn out of it).

Faecal output may be reduced, and the stool may be unformed or pasty.

Figure 29.5 **A stunted black-headed caique chick suffering from a crop burn. Note the oversized head.**

Diagnosis

Diagnosis is based on crop and faecal cytology (Gram stain) and culture, haematology and biochemistry, and radiography.

Management

The cause should be identified using the means outlined earlier and corrected where possible.

The crop should be emptied with a feeding tube and repeatedly lavaged with warm saline until a clear wash has been obtained. (In some extreme cases, e.g., foreign bodies, it may be necessary to perform an ingluviotomy.) It should always be assumed that these chicks are dehydrated, and they should be treated with parenteral fluids until crop motility has been restored.

Appropriate antimicrobials should be given as indicated by crop and faecal cytology/culture. It is overly simplistic to assume that these cases are always yeast infections; in the authors' experience, bacterial overgrowth is usually more common.

A crop 'bra' can be used if needed. This is a non-adhesive bandage placed under the crop and around the wings to 'lift and support' the atonic crop in order to allow gravity to assist with crop emptying (**Fig. 29.6**).

Once the crop has been emptied, in many cases it may be advisable to leave it empty for a few hours while dehydration is corrected. Initial feeds should be of small volumes of isotonic saline. If this moves through, solids can be added. Small, watery meals should be fed often. Pre-digesting the hand-rearing formula with a small amount of pancreatic enzymes can liquefy the diet without diluting it.

Motility modifiers (e.g., metoclopramide or cisapride) may assist in restoring motility, although their efficacy is poor if used without other supportive measures. Metoclopramide is usually ineffective unless given as a constant rate infusion. In uncomplicated cases (where the chick is otherwise bright), a strong solution of fennel tea given by crop drench may assist in restoring motility. (Fennel has been shown to promote gastrointestinal motility in cases of functional ileus.)

The prognosis is good, provided prompt and appropriate therapy is administered. However, it could also be an early indication of a more serious problem, and the underlying issue may have a different prognosis.

Thermal injuries to the crop

See **Fig. 29.7** and Chapter 19.

Crop perforations

See Chapter 19.

Orthopaedic problems

See Chapter 15.

Beak malformations

See Chapter 12.

Figure 29.6 **Juvenile black cockatoo in a crop 'bra' to assist with crop emptying.**

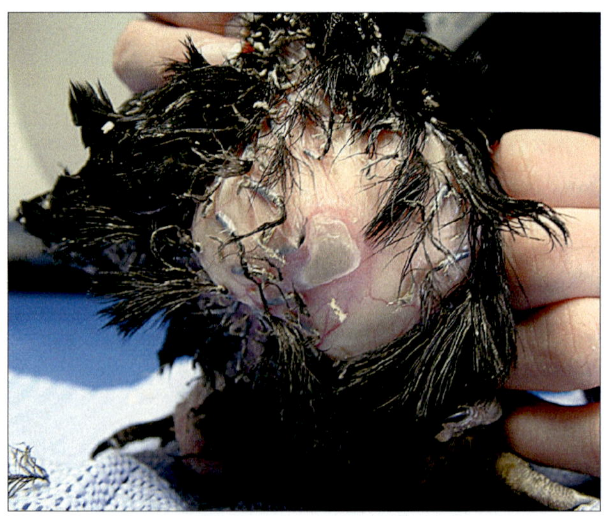

Figure 29.7 **Juvenile black cockatoo with a crop burn. Note the initial blanching of the skin prior to necrosis.**

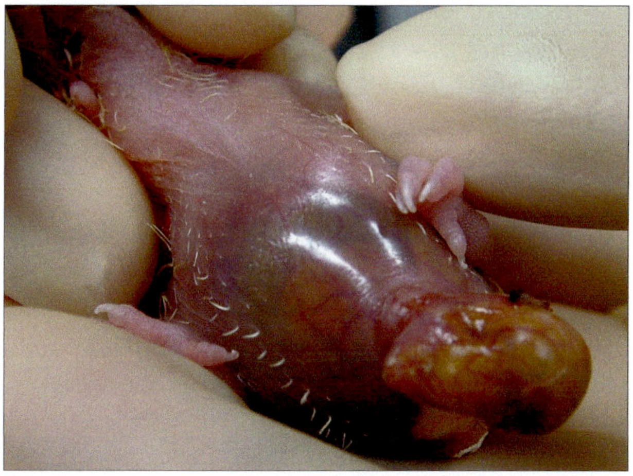

Figure 29.8 **A severe macaw chick with an externalised yolk sac.**

Omphalitis

Omphalitis (infected yolk sac) is not uncommon in incubated chicks. There are two syndromes seen: a failure to draw the yolk sac into the body prior to hatching (**Fig. 29.8**) or an infection tracking up through an umbilicus that has failed to close properly (**Fig. 29.9**).

The first condition is usually seen when the incubator temperature is too high or the humidity too low. The result is that the chick hatches before the yolk sac has been 'drawn up'. These can often be treated by ligating the yolk sac and removing it, followed by closing the umbilicus. Unfortunately, a significant number of these chicks fail to thrive and often die within days or weeks.

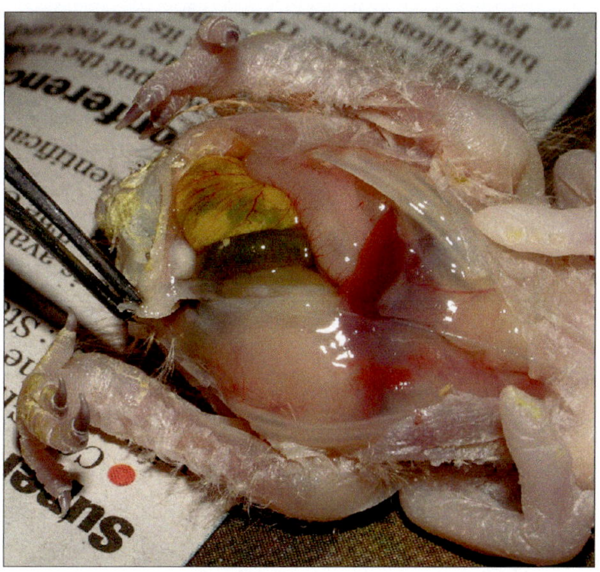

Figure 29.9 **Necropsy of a macaw chick revealing omphalitis.**

The second condition occurs when the umbilicus fails to close fully after hatch and the hygiene of the brooder is poor. The resulting infection usually kills the chick in a few days; the diagnosis is usually made by necropsy. If this becomes a problem in a particular nursery, applying an iodine-based ointment to the navel at hatch can help to prevent an ascending infection.

'Wry neck'

The term 'wry neck' describes a condition where there is either vertebral malformation or tendon contracture in the neck of a newly hatched chick. See Chapter 18 for more detail.

Foreign bodies

Chicks are by nature inquisitive animals, and it is not uncommon for them to ingest non-food items while exploring their environment. They are also vigorous feeders, and it is not uncommon for them to swallow a feeding tube that has been dislodged from a syringe. It can be surprising how quickly these foreign bodies can move from the crop into the proventriculus and ventriculus (**Fig. 29.10**).

If the foreign body is still in the crop, it can often be removed through the mouth under sedation or a

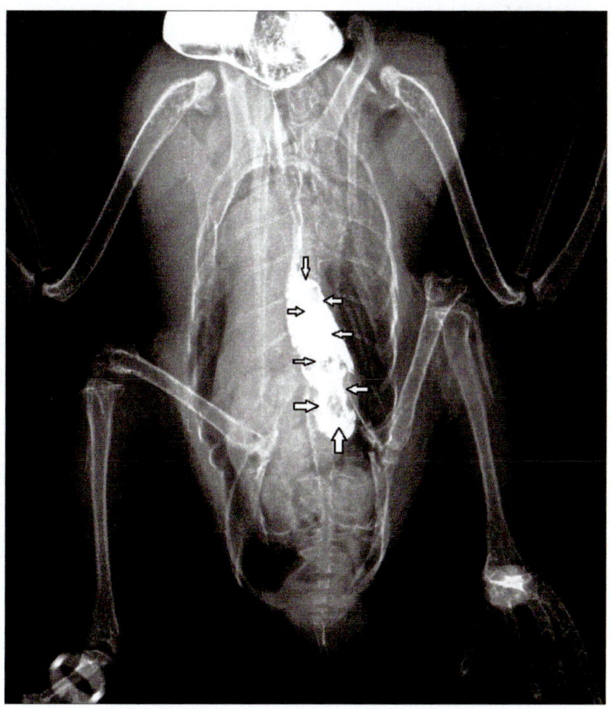

Figure 29.10 **Radiograph of an African grey chick that swallowed a feeding tube (outlined by arrows).**

short anaesthetic. If this is not possible, an ingluviotomy can be used to retrieve the object. Material that has moved into the proventriculus will require an ingluviotomy and removal by endoscopy or gentle flushing.

Infectious diseases

Definition/overview

Infectious diseases are quite common in young chicks; their low level of immunocompetence combined with often substandard rearing practices leaves them highly predisposed to infection. This same lack of immunocompetence means that the progression of an infectious disease in young birds is often rapid. Prompt and aggressive therapy is needed to save the patient.

Aetiology

Infections may be bacterial (*Pseudomonas*, *E. coli*, *Chlamydia psittaci*, other gram-negative bacteria), fungal (*Candida*, *Aspergillus*), viral (polyomavirus, adenovirus, psittacine beak and feather disease [PBFD]), parasitic, protozoal (*Cryptosporidia*, *Trichomonas*, *Cochlosoma*, Coccidia, and *Atoxoplasma* [in young canaries]), or nematodal (ascarids, Capillaria, and Acuaria).

CLINICAL PRESENTATION

Signs include lethargy, loss of feeding response, pallor or erythema of the skin, dehydration, crop stasis, vomiting/regurgitation, weight loss or failure to thrive, subcutaneous haemorrhage, feathering abnormalities, and sudden death.

Management

The aetiological agent should be identified, and the chick treated accordingly. The patient will require supportive care (see Chapter 5).

Avian polyomavirus (APV)

Aetiology

Polyomavirus is a non-enveloped virus that is relatively environmentally stable. All parrots can be affected, but it is most common in macaws, conures, eclectus parrots, and caiques. It is rare in African grey parrots, cockatoos, and cockatiels. Passeriformes, including grass finches, canaries, goldfinches, and greenfinches, can also be infected. It is unknown if the virus in these birds is antigenically related to the virus in parrots or if it can be transmitted between parrots and Passeriformes.

Pathogenesis

The virus is passed horizontally in urine, faeces, and respiratory and skin secretions. Vertical transmission is suspected, but not yet adequately proven. Infected non-budgerigar parrots shed the virus 2–7 days after being experimentally inoculated.

Infection does not always result in disease. Disease determinants include the species of the bird, the age at which it is infected, and concurrent immunosuppression (e.g., PBFD). The age of peak susceptibility varies between species:

- Budgerigars: 10–25 days old.
- Conures: <6 weeks old.
- Macaws and eclectus: < 14 weeks old.
- Lovebirds: up to 1 year (concurrent PBFD may be a factor).
- Older birds usually do not show clinical signs unless they are secondarily immunosuppressed (e.g., PBFD).

Adult birds can be infected, but they rarely show signs of disease.

Infected birds of any age that do not develop disease will still experience viral replication and shed the virus. The length of time APV is shed is determined by the age at which the bird is infected (the older the bird, the shorter the shedding period) and the species of bird:

- Budgerigars shed for up to 6 months, stopping at sexual maturity or during the first breeding cycle.
- Conures shed for 4–8 weeks, rarely up to 16 weeks.
- Macaws shed for 6–14 weeks.
- Cockatoos shed for 8–10 weeks.

Clinical presentation

In budgerigars (the primary reservoir of APV), signs include decreased hatchability and embryonic death, abdominal distension, subcutaneous haemorrhages, and feather dysplasia that may resolve after several months (primary and secondary feathers on the wings and tail are lost or fail to erupt; in the budgerigar fancy these birds are known as runners, and the condition is known as French moult). There may be neurological signs if the cerebellum is affected.

In other parrots, APV causes sudden death (especially those <15 days old) or crop stasis, weakness, yellow urates, pallor, bleeding from feather follicles, and subcutaneous bruising. Most parrots die, usually 10–14 days after exposure. In larger parrots, most deaths occur between 20 and 140 days old.

In estrilid finches (grass finches), the virus causes acute mortality in 2–3-day-old fledglings, young adults, and mature finches. Survivors may demonstrate poor feather development and misshapen lower beaks.

Infection is rare in Fringillidae finches (canaries, goldfinches, greenfinches), where three types of lesions have been described: pulmonary (firm lungs with pale areas among congestion), hepatic/splenic (hepatosplenomegaly, focal haemorrhages in the liver), and cutaneous (warts have been described on the beaks of canaries). Other concurrent infections are common (e.g., *Atoxoplasma*, poxvirus).

Diagnosis

Ante-mortem diagnosis can be achieved using polymerase chain reaction (PCR) probes. After initial infection, there is a viraemia that can be detected by blood PCR. After a short period, the virus can be detected in the cloaca by PCR. When the bird is about to stop shedding, the blood PCR becomes negative. After 6–8 weeks, the cloacal shedding stops. Therefore:

- Blood positive, cloaca negative – The early stages of infection or intermittent shedding.
- Blood and cloaca positive – Bird is actively infected and shedding.
- Blood negative, cloaca positive – The bird is about to stop shedding.
- Blood and cloaca negative – The bird has cleared the infection, no longer shedding.

Blood should be tested first. If it is negative, the bird should be quarantined for 8 weeks and can then be considered clear. If it is positive, the bird should be retested in 2–3 months.

Gross pathology findings include hydropericardium, cardiomegaly, hepatomegaly, splenomegaly, generalised pallor (muscles may appear an orange colour), and petechial haemorrhages.

Histopathological findings include:

- In larger parrots, clear to basophilic intranuclear inclusions can be found in the spleen, in mesangial cells in the kidneys, and in Kupffer cells.
- In budgerigars, inclusions can be found in most organs, including the cerebellum, renal tubules, splenocytes, and hepatocytes. The most prominent are often found in the feather follicles and growing feathers.
- Massive hepatocellular necrosis.
- Immune-mediated glomerulonephropathy.

Management

Non-budgerigar aviaries

By the time the first case is recognised, most other birds will already be infected. Therefore, vaccination or testing may not be of any benefit. Hand-rearing should stop, and chicks should be allowed to be parent-reared, as there is a lower incidence of disease in parent-reared birds (except in budgerigars). All birds to be sold should be tested by blood PCR and held for 2 weeks after a negative result before selling.

Budgerigar aviaries

Nestlings and young birds are the reservoir of infection; therefore, breeding should be stopped for 4 months (until all birds are at least 6 months old). All young birds should be removed, and the adults moved to a clean environment. Disinfection is carried out with phenolics, sodium hypochlorite, or stabilised chlorine dioxide.

Prevention

Non-budgerigar aviaries

It is best not to keep budgerigars, cockatiels, or lovebirds without testing them. Other people's birds should not be brought into the nursery. A closed aviary should be maintained (i.e., no movement in without testing first). PBFD should be tested for at the same time. If birds do have to be moved out and then back in, they must be at least 18 weeks old, and if vaccinations are being used, they should be given 4 and 2 weeks prior to movement. Traffic control is used to prevent movement from adults to chicks.

Budgerigar aviaries

A random number of birds should be tested to determine if the disease is present. Only PCR-negative birds should be brought in, including owned birds returning from shows.

Adenovirus

Aetiology

Avian adenoviruses are non-enveloped DNA viruses that are stable outside the host and resistant to many disinfectants. Although usually considered to be opportunistic pathogens, some highly virulent strains may be primary pathogens. Adenoviruses are associated with many diseases in both companion and domestic birds, including marble spleen disease (pheasants), egg drop syndrome (chickens), quail bronchitis, and pigeon adenovirus.

In parrots, it is believed the route of infection is horizontal (ingestion or inhalation), with young (immune incompetent) birds most at risk. Although vertical transmission has not been confirmed in parrots, it is suspected to occur in poultry. As an opportunistic pathogen, adenovirus has also been associated with immunosuppressive diseases such as PBFD. It is frequently isolated from persistently infected but asymptomatic birds.

Clinical presentation

In budgerigars and lovebirds (*Agapornis* spp.), adenovirus has been associated with nephritis and pancreatitis. Occasionally, neurological signs (ataxia, torticollis, and weakness) are seen as well.

In nursery-reared parrots of all species, it is occasionally seen as sporadic cases of acutely ill chicks which succumb and die within hours of presentation. Necropsy reveals a markedly enlarged, pale, mottled liver and splenomegaly (**Fig. 29.11**).

Diagnosis

Histopathology often reveals hepatic, splenic, and intestinal necrosis associated with basophilic intranuclear inclusion bodies. PCR can be used to confirm the presence of adenovirus and to detect carriers.

Management

There is no vaccine available for adenovirus. Affected birds may receive supportive care but usually die despite the care.

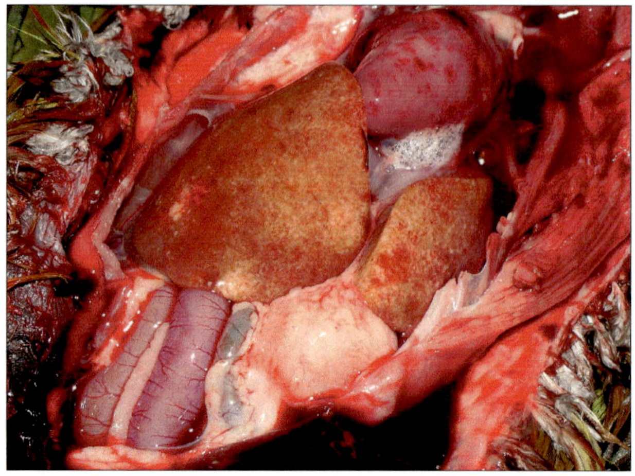

Figure 29.11 **Necropsy of an Amazon parrot chick reveals adenoviral hepatitis.**

Adenovirus is inactivated by exposure for 1 hour to formalin, aldehydes, and iodophors.

Psittacine beak and feather disease

PBFD is occasionally seen in young parrots (especially grey parrots) as an acute onset of anaemia and leucopaenia, followed by death. Necropsy reveals hepatomegaly and splenomegaly, associated with basophilic intracytoplasmic inclusion bodies.

FURTHER READING

Clubb SL (1997) Psittacine pediatric husbandry and medicine. In: *Avian Medicine and Surgery.* RB Altman, SL Clubb, GM Dorrestein, K Quesenberry (eds). WB Saunders, Philadelphia, pp. 73–95.

Flammer K, Clubb SL (1994) Neonatology. In: *Avian Medicine: Principles and Application.* BW Ritchie, GJ Harrison, LR Harrison (eds). Wingers Publishing, Lake Worth, pp. 805–841.

LaBonde J (2006) Avian reproductive and pediatric disorders. In: *Proceedings of the Annual Conference of the Association of Avian Veterinarians Australian Committee*, pp. 229–238.

Schubot RM, Clubb KJ, Clubb SL (1992) *Psittacine Aviculture: Perspectives, Techniques and Research.* Avicultural Breeding and Research Center, Loxahatchee, FL.

CHAPTER 30

BEHAVIOURAL PROBLEMS

INTRODUCTION

The behaviour of companion birds directly affects the quality of their lives, their interactions with their owners, their health, and the probability of a veterinarian maintaining an effective, long-term doctor–client–patient relationship.

As advances in avian medicine in the diagnosis, treatment, and prevention of clinical diseases grow, and as pet owners increasingly accept this new knowledge, the concept of a companion (or pet) bird is changing. Turning away from the old phrase 'caged birds' to the more acceptable 'companion birds' reflects that many birds are now seen as 'family members' and are taken to veterinary clinics as such. Clinical patterns of presenting conditions are moving away from infectious diseases and malnutrition towards age-related problems (neoplasia, heart disease, etc.) and behavioural disorders.

All bird behaviour originates from 'wild' behaviour, and while some of these are instinctive (innate) behaviours, most are acquired (learned) through experience. Companion birds need to effectively acquire learned behaviours that can make them socially acceptable as pets and companions. If they are unable to do so, this can result in behavioural issues such as attention-demanding, displacement behaviours, territoriality, aggression, feather damaging, screaming, reproductive issues, phobias, and psychotic behaviour.

Many of these problems are often chronic by the time the bird is presented to veterinarians and, like all chronic health problems, can be difficult to successfully treat and return the bird to normal. Early recognition and treatment of these problems are more likely to result in successful treatment.

PRINCIPLES

Companion birds face many challenges. These challenges are directly attributable to a lack of knowledge – and sometimes ignorance – of a bird's requirements. The two most common presentations of chronic problems seen by avian veterinarians are malnutrition and behavioural problems. Malnutrition stems from the traditional concept of birds as seed-eaters. The behavioural problems, however, stem from a lack of understanding of the unique nature of bird behaviour, and the owner's attempts to impose their own demands and expectations on their bird, expecting them to behave more like a dog or a cat or even a child.

Understanding bird behaviour is a constantly evolving subject. What was accepted as fact a few years ago has been debunked or disproved or, at best, modified. Pre-existing notions of bird behaviour and its 'modification' or 'therapy' need to be re-examined and, where appropriate, discarded. Better therapeutic success can be achieved for behavioural problems by adopting a more holistic approach.

UNDERSTANDING BIRD BEHAVIOUR

To understand bird behaviour, it is necessary to look at the behaviour of birds in the wild; after all, most pet birds are only a few generations removed from wild birds. However, to generalise and talk about 'bird behaviour' is akin to talking about 'mammalian behaviour'. There are some 9,000 species of birds; approximately 365 of these are parrots. Parrots have evolved in Australia, South America, Africa, and Asia. Different environments and environmental pressures have shaped both the physical and

DOI: 10.1201/9781003272519-30

psychological development of these birds, and trying to apply 'rules' across all 365 species is fraught with danger. However, there are some generalisations that can be made.

Avian behaviour can be categorised into two functional groups:

- Self-maintenance behaviours designed to accomplish a specific task to maintain the health of the individual. These include feeding, feather care, locomotion, and concealment.
- Social behaviours designed to communicate information to another individual. These behaviours include territoriality, concern or fear, and courtship.

Parrots are altricial; the young are hatched near-naked, blind, and perhaps deaf. They live for the first weeks of their life in a quiet, dark hollow, completely dependent on their parents (**Fig. 30.1**). Social interactions at this stage are limited, confined to their siblings and parents. After fledging, many parrots learn to socialise by contact with their parents, siblings and other birds (often their own age) in a flock (**Fig. 30.2**). An example of this is the rose-breasted cockatoo (galah). In the wild, clutches of two to five chicks usually survive. Once they fledge, the chicks are taken to a nearby 'crèche', a tree or stand of trees where other galah 'families' are interacting. Here they learn early social skills before joining a nomadic juvenile flock where they remain until reaching sexual maturity. This socialisation period teaches the juvenile birds food recognition (and how to locate it), predator recognition, sentinel duties, grooming behaviour, survival skills, and early social behaviours such as communication.

Figure 30.1 **Indian ring-neck chicks (violet colour mutation) in the nest. Note the relatively cramped and dark conditions.**

As they reach sexual maturity, they learn new skills: how to select a mate; communicate effectively with their flock; exhibit courtship behaviours and develop a pair bond; how to select, prepare, and defend a nest site; and how to reproduce and raise their young.

Some of these behaviours are believed to be innate; most are learnt. All are reinforced by the reaction the bird receives. Captive parrots, especially those hand-reared as pets, may not have the opportunity to learn these normal behaviours. Instead, their behaviours (learned as a response to a situation) – particularly as they reach maturity – may bring them into conflict with their human 'flock'.

HOW DO BEHAVIOURAL PROBLEMS DEVELOP?

Companion birds still exhibit many of the daily maintenance behaviours of their wild counterparts, but with a major difference – the time needed for foraging and consuming their food is less, as there is usually a ready and reliable source of food provided at the same time, in the same place, every day. This means more time is available each day for other activities such as socialising, grooming, playing, and even napping.

Young birds, destined for the companion bird marketplace, still retain the need for mentoring and learning of appropriate behaviours seen in their wild counterparts. These appropriate behaviours revolve around:

- Communication skills with both people and other birds
- Social skills (including conflict resolution)
- Self-care skills (such as grooming)

Problems seen in clinical practice can be attributed to one of two causes.

- Failure of the socialisation process. This is usually the result of an individual bird being hand-reared in isolation and not being taught basic social skills. Once weaned, the bird is often ignored as its novelty value wears off. The unsupervised parrot, in the absence of mentoring and guidance in its home, will often 'self-reinforce'

Figure 30.2 **Flock of galahs feeding on grass seed in Australia. Many parrots learn to socialise by contact with their parents, siblings, and other birds (often their own age) in a flock.**

behaviours based on immediate gratification. Examples of these behaviours include attention-demanding behaviour (begging calls, screaming, feather chewing) and displacement behaviours such as biting, feather-damaging behaviour, phobias, and sometimes even self-mutilating behaviour.

- Failure of people to understand normal parrot behaviour and then expecting their bird to fall into line with their expectations (**Fig. 30.3**). It was once said that there are no abnormal behaviours, just normal behaviours expressed inappropriately. Behaviours such as screaming morning and night, displaying territoriality, and certain reproductive behaviours are examples of normal behaviour that are inappropriate in a companion bird scenario. The companion bird that behaves inappropriately is less likely to have close human contact and social interaction and becomes increasingly more isolated.

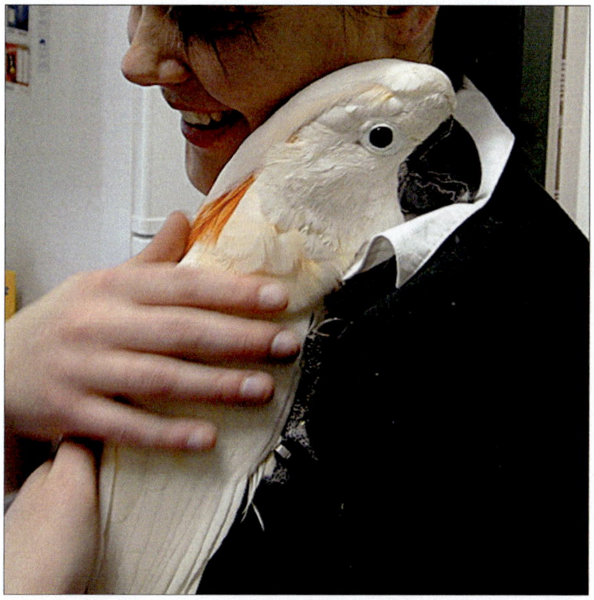

Figure 30.3 **Five-month-old Moluccan cockatoo. Many pet birds bond closely with humans at this age.**

So how do these problem behaviours develop?

The reinforcement of self-maintenance behaviours benefits companion birds, and abnormal self-maintenance behaviours are the most common behavioural disorders seen in these birds. They still have the same self-maintenance behaviours as their wild counterparts. However, they need less time for foraging and feeding behaviours (after all, the food is in a dish in front of them every morning) and therefore feather care, social communication, and displays make up more of their daily activities.

Young captive birds need continued mentoring and behavioural moulding and require guidance for the establishment of a normal bird–human flock relationship. This includes a range of normal social behaviours of flock interaction, with appropriate rules of conflict resolution and appropriate maintenance and social behaviours.

Failure to be taught or learn – these behaviours means that many young birds are not prepared for a life in captivity and may develop behavioural problems. In the absence of imposed rules, the bird will make its own rules based on immediate gratification and revolving around perceived value; however, these rules may not be socially acceptable. They eventually develop into behavioural problems, and the bird becomes unable to interact socially with people without fear or social framework. Therefore, a series of displacement or defensive behaviours develops (e.g., aggression, biting). As these behaviours develop, the bird may become even more isolated and therefore become more vocal in trying to re-establish contact with its flock.

This is often reinforced when owners respond: the bird receives a positive response (e.g., talking, feeding) and it may augment the behaviour. If, on the other hand, the bird is punished (e.g., covering the cage, time-out, water pistols), that response may augment the feeling of isolation and the problem may worsen.

Unfortunately, these problems are often chronic by the time the bird is presented to a veterinarian. Early recognition and treatment are much more likely to result in successful treatment; prevention through education of bird owners is the most preferable approach. Techniques such as incorporating behavioural training into annual wellness examinations are important steps in preventing problems, and they should be pursued vigorously by all those involved in the wellbeing of companion parrots.

HISTORICAL APPROACH TO TREATING BEHAVIOURAL PROBLEMS

Earlier attempts to treat behavioural problems in birds revolved around human concepts of 'discipline' and psychotherapy.

Humans are driven, in many cases, by fear of punishment and negative reinforcement. From early childhood, people are made well aware of the consequences of inappropriate behaviour – corporal punishment or even just the disapproval of adults. Given that early thoughts on parrot IQ were along the lines of 'as intelligent as a 3–5-year-old, but with the maturity of a 2–3-year-old', it is not surprising that people expected birds could be motivated by the same fear of negative consequences.

And, if punishment did not work, surely drugs would! The use of hormonal therapies, antidepressants, sedatives, and even narcotic antagonists was advocated as the best therapy for 'misbehaving' patients. Conventional medicine sometimes dictates that for every disease or problem there is a drug. So, studies and trials were conducted that resulted in many a companion bird sitting on a perch in a drug-induced stupefaction, and everyone (except the bird) was satisfied.

Eventually, it was realised that neither of these approaches was successful in curing behavioural problems. They might have stopped for a while, but as soon as the threat of punishment or administration of drugs was removed, the problem recurred.

NEW CONCEPTS

The first change in thinking on bird behavioural therapy came with the realisation that drugs and physical restraint devices (e.g., Elizabethan collars) just did not work. Although they may have a limited, short-term role in selected cases, these therapies may worsen a situation and do not provide a long-term benefit.

A second change in thought was in dealing with the concept of dominance. It was long thought that a strict hierarchical structure existed within bird flocks, with dominance being reflected in advantages such as the highest perch, the best food, and first choice when selecting a mate. It was further thought that in a human–bird flock situation, a similar dominance had to be maintained, with the human being the dominant member of the flock at all times. However, studies of wild birds have shown that dominance, while it does occur, is not at all strict; rather, it is a fluid arrangement with different birds being dominant at different times. A flock is inconsistent in its structure: members are lost to predators, family groups change between flocks, and new members are constantly being added. This makes a straight-line hierarchy difficult to maintain and too energy demanding to be compatible with survival in the wild. Therefore, the rigid dominance so often seen in human society does not readily transfer over to the bird world, and attempts to impose it usually involve 'flooding' a bird (overwhelming it with an imposed behaviour until it acquiesces) or using punishment as a means of instilling discipline. The problem with these techniques is that the bird learns to perform a behaviour or avoid exhibiting one, just enough to escape adverse consequences. Once these consequences are removed, the behaviour returns.

New concepts in behavioural therapy revolve around several principles:

- Birds are not people, or dogs, or cats. They are unique – they are birds.
- The best results are obtained when a bird wants to do the behaviour that the owner desires.
- A successful treatment may be a reduction in unwanted behaviour, rather than a complete cessation of it.

- Successful behavioural change in companion parrots starts with behavioural change in the parrot owner.

Birds are not people, or dogs, or cats

Imposing human values and emotions onto an animal (anthropomorphism) can hinder the behavioural modifications clinicians are trying to institute. Using terms such as angry, aggressive, jealous, and happy implies a state of mind which, realistically, we cannot truly assess. By doing so, a set of expectations is created in the mind of the owners and perhaps in the clinician as well.

Parrots are animals that can:

- Learn to feed from seasonally variable food sources.
- Learn the specific call or 'dialect' of their flock.
- Interpret subtle nuances in body language of other members of their flock.
- Perform elaborate courtship displays.
- Learn to mimic sounds in their environment in order to manipulate an environmental condition or maintain a pair bond.
- Develop monogamous pair bond relationships.
- Rigorously defend breeding territory.
- Develop flight skills advanced enough to avoid or escape predatory attacks.
- Successfully rear young and make decisions about resource availability that influence fledgling mortality.

They are not little children with feathers nor are they similar to mammals such as dogs and cats. They are unique and deserve to be treated as such. It is important to attempt to replicate the environment in which they evolved, providing them with opportunities to exhibit normal behaviours such as foraging for food, flying, interacting socially, and just having fun.

The best results are obtained when a bird wants to do the behaviour that the owner desires

The law of effect states that *behaviour is a function of its consequences*. This law of effect can be applied through reinforcement (positive or negative) or punishment. Reinforcement increases the frequency of behaviour, while punishment decreases its frequency.

Punishment involves applying a noxious consequence to a behaviour in the expectation that the bird will cease the behaviour to avoid the consequence. Examples of this include shouting at the bird, giving it a time-out, covering the cage, or squirting the bird with a water pistol. The problem with punishment is that the bird will do only just enough to avoid the consequence and no more. In fact, as soon as the consequence is removed, the behaviour usually returns. Undesired consequences of punishment can include escape/avoidance behaviour, aggression, apathy (a decreased responsiveness in behaviour), and fear, leading to phobic behaviour.

Care must be taken to avoid inadvertently reinforcing a behaviour while trying to use punishment. The classic example of this is shouting at a screaming parrot. Parrots are naturally noisy animals that seem to thrive on noise and drama; it is, after all, a natural part of their life and normal social interaction. While the owner thinks he or she is punishing an undesired behaviour, they are, in fact, giving the bird positive reinforcement and guaranteeing that the behaviour will be repeated.

Reinforcement, on the other hand, involves providing a favourable consequence (or reward) for a behaviour. It can be positive (through the addition of something to the bird's life, such as praise or a favourite food treat) or negative (removing something from the bird's life that the bird does not want or like). Either way, reinforcement increases the likelihood that the bird will repeat the behaviour in the expectation of a favourable consequence. The motivation used to stimulate a desired behaviour must be evaluated from the bird's perspective, and this can be influenced by several factors:

- The relationship between the bird and the person must be built on trust, respect, and prior positive experiences. Good communication must exist between the two: both must understand the other's body language in order to know what they want (**Fig. 30.4**).
- The bird's confidence and ability to perform the behaviour.
- The bird's past experiences in training and motivation.

Figure 30.4 **Macaws in a large free-flight aviary choosing to interact with humans.**

- Natural influences such as social interaction, breeding season, comfort, and height.
- The bird's hunger state or preference for certain feed items.

Some examples of good motivators include favourite food items, praise, petting, proximity to people, a familiar object, and even out-of-cage time.

Positive reinforcement can be used to increase or decrease the likelihood of a behaviour being repeated. It can also be used to establish and encourage other behaviours. These new behaviours can either be alternative behaviours or behaviours that are incompatible with the undesired behaviour. Either way, it can lead to the extinction of an undesired behaviour.

Successful treatment may be a reduction in an unwanted behaviour, rather than a complete cessation of it

Many behavioural problems are often chronic in nature by the time the bird is presented to a veterinarian. Feather-damaging behaviour is a classic example: by the time the owner has waited to see if it goes away, then consults the internet and tries a variety of pet shop remedies, it may be many months or even years before the bird is presented to a veterinarian. Obviously, such behaviours, now deeply ingrained and with their own set of positive motivators (perhaps only known to the bird), are going to be difficult to eradicate. This must be made clear to the owner from the beginning (i.e., that success in these cases may simply be a reduction in the behaviour, rather than its complete elimination).

Successful behavioural change in companion parrots starts with behavioural change in the parrot owner

There is no doubt that the owners of many birds are directly contributing to the problems they are experiencing and directly or indirectly reinforcing them. As mentioned earlier, people grow up in a 'punishing' environment, where transgressions are punished, and human behaviours have evolved to avoid such punishments. Birds are not like this and do not learn well in this manner. People who wish to interact and socialise with birds, preventing or overcoming behavioural problems as they do so, must change their own behaviour in order to better understand and work with their bird.

These principles call for a reassessment of how humans think of, and interact with, birds. Until we do so, our success in modifying their behaviour will be limited.

BASIC STEPS TO IMPLEMENT IN A BEHAVIOUR-MODIFICATION PROGRAMME

There are some basic steps to implement in a behaviour-modification programme when dealing with problem behaviours (Table 30.1). They include:

- Basic training
- Normalisation of social interaction
- Avoidance of unwanted behaviours
- Replacement of unwanted behaviours with acceptable behaviours

Table 30.1 **Most common behavioural problems, risk factors, possible causes, and suggested treatment**

BEHAVIOUR	RISK FACTORS	CAUSES	TREATMENT STRATEGIES

Feather damaging behaviour
• Chewing
• Stripping
• Plucking
• Snipping
	• Breed
	  • Macaws
	  • Grey parrots
	  • Eclectus
	  • Conures
	  • Cockatoos
	  • Lorikeets
	• Age
	  • Increases with age
	• Poor/inadequate socialisation
	  • Other birds
	  • People
	• Environment
	  • Sleep deprivation
	  • Lack of foraging opportunities
	  • Lack of environmental complexity
		• Anxiety/phobia, a state of apprehension, uncertainty, and fear resulting from the anticipation of a realistic or fantasised threatening event or situation, often impairing physical and psychological functioning.
		Anxiety disorders may have biological and/or environmental causes, such as:
		• Improper socialisation that many companion parrots receive during the hand-rearing process, which may be reflected in an inability to cope with otherwise normal situations. This includes separation anxiety when the owner is not present.
		• Fear, real or imagined, of a person or situation that the bird is exposed to.
		• Lack of self-confidence, due to a history of undermining events.
		• Displacement behaviours, resulting from the stress of captivity (where there is insufficient opportunity for species-typical behaviours, such as foraging, locomotion, and reproduction). This is distinct from anxiety or phobias associated with real fears, such as the household dog, garbage trucks, strangers, or caregivers leaving.
		• Attention-seeking behaviour, especially when the owner is absent. The bird uses feather-damaging behaviour to obtain attention from an owner when that owner is not engaged in an activity with the bird.
		• Impulse control disorder (or compulsive disorder), a human psychiatric construct characterised by repetitive, stereotypic motor, locomotor, grooming, ingestive, or hallucinogenic behaviours that occur out of context to their normal occurrence, or in a frequency or duration that is in excess of that required to achieve a goal, e.g., compulsive grooming in excess of that which is normally required for grooming, interferes with normal behaviour, and cannot be interrupted.
		• Prolonged reproductive behaviour, especially in cockatoos, macaws, and conures.
			• Identify and eliminate risk factors that are amenable to change.
			• Develop and introduce a species-specific environmental enrichment plan (see Chapter 2). Consider carefully whether social enrichment through introducing a new bird for companionship would help – or worsen – the problem.
			• Provide chewing opportunities, e.g., fresh browse or pinecones.
			• Interact through the training of specific behaviours between the bird and its caregivers that are unlikely to be misconstrued as courtship behaviours.
			• Implement reproductive controls, as outlined in Chapter 27.
			Treatment strategies unlikely to be successful:
			• Elizabethan collars (unless self-mutilation is occurring).
			• Psychotherapeutic drugs, unless severe anxiety is confirmed.
			• Training techniques that involve punishment or negative reinforcement e.g., water pistols and shouting.

Aggression • Lunging • Growling • Biting	• Poor socialisation with people or other birds • Overattachment (inappropriate pair bonding) to a caregiver • Previous mishandling (including painful procedures or inappropriate wing trims)	• **Fear aggression**: Occurs when a bird is unable to escape from a distressful/scary stimulus, e.g., the administration of medicine, neophobia when new items/birds/people are added to the environment. • **Conditoned/learned aggression**: Biting after learning this is an effective way of avoiding an unpleasant/unwanted situation. • **Territorial aggression**: Aggression that is directed toward people, birds, or other animals entering the bird's territory. Often worse during the breeding season and disappears when the bird is removed from its territory. • **Mate-related aggression**: Occurs when a bird that has developed a one-person bond with their caregiver; when another person approaches that caregiver, the bird may become aggressive. • **Redirected behaviours**: Occur when a bird's aggression is triggered but it is unable to reach the source of that aggression. The bird's aggression is then redirected to nearby objects or people.	**Fear aggression**: Desensitisation and counterconditioning to the source of fear, refraining from flooding the bird by forced exposure to the source of fear. **Conditoned/learned aggression**: Avoid the specific triggers for the aggression. For the unavoidable, such as medical treatments, behaviour modification, including desensitisation, counter-conditioning, and positive-reinforcement training. **Territorial aggression**: Behaviour modification and counterconditioning to people/birds in the home territory. The bird should be asked to step onto a perch to be removed from its territory before being invited to interact with people or other birds. **Mate-related aggression**: Prevent or break down one-person bonding by having non-preferred family members take over daily care for the bird, while the preferred human develops a distant relationship. **Redirected aggression**: Identify and treat/remove the underlying cause (e.g., fear, mate-related, or another motivation). This requires a thorough history to identify the actual target and motivation for the aggression.

BEHAVIOUR	RISK FACTORS	CAUSES	TREATMENT STRATEGIES
Screaming It is normal for parrots to scream; it is inappropriate for them to do it constantly or in crowded suburbia.	.	**Anxiety/phobia, including separation anxiety**: Screaming when the bird is afraid or anxious, e.g., separation anxiety when the caregiver and the bird are separated. **Inappropriate pair bonds with people**: Birds that form bonds with caregivers that mimic pair bonds seem to be more likely to vocalise when the caregiver leaves. Wild birds communicate through contact calls; companion birds may exaggerate these contact calls when they can't see their caregiver. **Attention-seeking behaviour**: This may occur when the caregiver is in the same room or to get the caregiver to return as soon as they leave. The attention the bird receives does not need to be positive to be rewarding; yelling, squirting with a water bottle, or other attempts to make the bird stop are all forms of attention, and may inadvertently reward the behaviour. **Lack of appropriate enrichment or stimulation**: Boredom through lack of stimulation may leave the bird with nothing better to do than call out.	• Client education – While the screaming may be annoying, it is normal behaviour expressed inappropriately. As such, a successful treatment may be a reduction in the noise rather than complete cessation. • Ignore all screaming other than distress calls. Responding to the screaming, either positively or negatively, may reinforce it. • If the caregiver leaves the room, they can continue to speak quietly to the bird, maintaining communication. • Develop and implement an environmental enrichment plan (see Chapter 2).

Basic training

While some feel that teaching obedience to a bird is cruel and unnatural, the opposite is true; it is cruel and unnatural to let them try to figure things out for themselves. Many pet birds are hand-reared and seem to recognise and accept people as members of their flock. This recognition, and the interaction that comes with it, is what helps to make parrots such enjoyable companion birds. Normally, other flock members would teach the juvenile bird what social behaviours are appropriate through a system of examples and trial and error. With companion birds, the role of mentor and teacher has to be filled by the owner of the bird. It is essential to have the basic requisites for training in place first in order to provide a solid foundation for acceptable social behaviours. These basics include step-up and step-down, and staying on a perch where placed.

- *Step-up.* Stepping up onto an offered hand is a foundational manoeuvre upon which most training and behavioural guidance relies. In light of what is discussed earlier, it is important that the bird performs this behaviour because it wants to, not because it has to. By using a food reward and the cue words 'step up', a bird can be taught to step up onto the owner's hand and stay there.
- *Step-down.* Stepping down allows the owner to guide the bird's movement. Once again, positive reinforcement (e.g., a food reward) and the cue words 'step down' can be used to teach the bird to step off the owner's hand and onto a suitable perch.
- *Stay.* Staying on a perch is important for a bird to experience 'normal' flock social interaction while outside of its cage with its 'flock members', without getting into mischief or requiring constant supervision. The free-roaming pet parrot is at a much greater risk of traumatic injuries and household poisonings. It is also more likely to develop pair-bonded interactions with one person and less likely to interact with other people. Portable tabletop perches are suitable for this training since they can be put anywhere. The bird is stepped down onto the perch and rewarded (verbally or with a food reward) for remaining there. If they climb down and walk around, avoid inadvertently rewarding this behaviour; simply put them back onto the perch without any verbal cue or other reward. If they stay there, offer another reward. It is usually necessary to gradually extend the time between placing the bird on the perch and offering the reward.

Normalise social interactions

Normalising social interactions has to be done on two levels:

- The interactions between the parrot and the person to whom the bird has bonded have to be modified. Interactions such as picking up the bird, scratching its head and body, and playing with it have to be stopped or reduced dramatically while the modification programme is being implemented. The bonded person can clean the cage and put in fresh food and water; however, the less interaction, the better.
- Other members of the household need the opportunity to interact positively with the bird. To avoid distracting the bird or triggering protective or territorial aggression, it is best if the bonded person is not around while this is being done. When 'out-of-cage' training is being done, the bonded person should bring the bird to a place where it does not exhibit territorial behaviours. The bird should learn to interact with other people using a combination of positive reinforcement and incremental exposure. For example, the new person can drop a preferred food reward in the bird's food bowl and then walk away. If this is done often enough, eventually the bird will look forward to seeing the new person. This person can then try offering the reinforcement to the bird through the cage bars. Eventually, the reward can be offered for stepping up on the hand and other cooperative behaviours. Alternatively, working in neutral territory, the other members of the household can cue the bird to perform simple behaviours the bird already knows how to do. This gives the bird an activity on which to focus, while at the same time receiving rewards from people other than the one the bird has bonded to. This can help build a positive relationship with the rest of the household.

The goal with these two levels of modification is to limit the positive experiences with the bonded person and, at the same time, increase the positive experiences with the new person. However, it may never be possible for the new person to interact at the same level the bonded person can with the bird, and it is unlikely positive interactions can occur while the bonded person is in the room.

Avoid unwanted behaviours

Analysis of problem behaviour should follow the ABC of behaviour:

- A = Antecedent: the situation which led to the behaviour.
- B = Behaviour: the actual behaviour displayed by the bird.
- C = Consequences: what the bird got out of the behaviour. Consequences can be positive or negative.

The consequence will determine whether a behaviour will be repeated or not. If it is not possible to alter the consequence (and therefore the bird's likelihood of repeating the behaviour), it becomes necessary to change the antecedent. For example, if a bird screams loudly each morning until it is fed, the behavioural analysis is as follows:

- The antecedent is that there is no food in the bird's bowl first thing in the morning.
- The behaviour is that the bird screams.
- The consequence is that the owner hurriedly feeds the bird, positively reinforcing its behaviour.

The bird still has to be fed each day; therefore, the consequence cannot be changed. However, if the owner places food in the bowl after the bird has gone to sleep the night before, and therefore food is available to the bird as soon as it wakes up, the antecedent has changed, and the behaviour should change (or stop).

Sometimes it is necessary for the owner to change his or her behaviour in order to change the bird's behaviour.

Replacement of unwanted behaviours with acceptable behaviours

Most birds, captive or wild, spend their time during the day performing daily maintenance behaviours that are essential for survival (**Figs. 30.5, 30.6**). These behaviours include foraging, social interaction, and feather care. Wild birds spend approximately 80% of their waking life foraging for food, with the remaining 20% of their time devoted to socialising and grooming.

Figure 30.5 **Wild cockatoo foraging for food, an activity that takes up nearly 80% of its normal day.**

Figure 30.6 **Grooming behaviour in a black-capped lory. Grooming takes up a relatively small amount of time in the wild, but captive conditions leave more time to groom, which can lead to overgrooming.**

Companion birds, with their food provided in the same dish, in the same place and at the same time each day, do not have to spend so much time foraging to survive. If 'normal' social interaction with members of the household is limited, as is often the case when owners are at work or school, then other

Figure 30.7 **A clear example of an undersized cage that doesn't provide sufficient foraging stimulation for this king parrot to alleviate boredom nor sufficient room to move.**

behaviours must be increased to fill the time (**Fig. 30.7**). This should be done using environmental enrichment (discussed in Chapter 2); otherwise, the deficit in activity may be replaced with abnormal behaviours such as stereotypical behaviours, feather-damaging behaviour (through overgrooming) (**Fig. 30.8**), or screaming. It may also heighten the bird's anticipation of the owner's presence and lead to problems with pair bonding.

By increasing the daily foraging activities available to a bird, its lifestyle can be enhanced significantly. Interestingly, several studies have confirmed that when a bird is given a choice between foraging for food or eating it out of a dish, most birds prefer to forage.

Bird behaviour is complex and, as mentioned earlier, our understanding of it is still evolving. There is no simple approach to dealing with individual cases, but by applying the principles and strategies outlined earlier, clinicians should be able to have a positive impact on the relationship between bird and owner.

Figure 30.8 **Galah that has been overgrooming, leading to feather-damaging behaviour. Galahs reared in isolation often develop behavioural problems.**

FURTHER READING

Briscoe JA, Reisner IR, Rosenthal KL (2004) Incorporating the veterinary behaviourist: A new model for the diagnosis and treatment of a feather damaging pet parrot. In: *Proceedings of the Annual Conference of the Association of Avian Veterinarians*, pp. 293–296.

Echols MS (2005) Practical use of foraging as a means of behaviour modification. In: *Proceedings of the Annual Conference of the Association of Avian Veterinarians Australian Committee*, pp. 185–194.

Friedman SG, Edling TM, Cheney CD (2006) Concepts in behaviour. Section I: The natural science of behaviour. In: *Clinical Avian Medicine, Vol 1*. GJ Harrison, TL Lightfoot (eds). Spix Publishing, Palm Beach, pp. 46–59.

Heidenreich B (2005) Addressing aggressive behaviour in birds. In: *Proceedings of the Annual Conference of the Association of Avian Veterinarians*, pp. 127–138.

Heidenreich B (2005) Solving companion parrot behaviour problems. In: *Proceedings of the Annual Conference of the Association of Avian Veterinarians*, pp. 219–228.

Heidenreich B (2006) How to train medical behaviours. In: *Proceedings of the Annual Conference of the Association of Avian Veterinarians*, pp. 135–148.

Luescher AU (2004) Rearing environment and behavioural development of psittacine birds. In: *Proceedings of the Annual Conference of the Association of Avian Veterinarians*, pp. 297–298.

Luescher AU (2006) *Manual of Parrot Behaviour*. Blackwell Publishing, Oxford.

Speer BL (2002) Practical clinical applications for the exam room. In: *Proceedings of the Association of Avian Veterinarians Australian Committee*, pp. 155–164.

Wilson L, Greene Linden P, Lightfoot TL (2006) Concepts in behaviour. Section II: Early psittacine behaviour and development. In: *Clinical Avian Medicine, Vol 1*. GJ Harrison, TL Lightfoot (eds). Spix Publishing, Palm Beach, pp. 60–72.

Wilson L, Lightfoot TL (2006) Concepts in behaviour. Section III: Pubescent and adult psittacine behaviour. In: *Clinical Avian Medicine, Vol 1*. GJ Harrison, TL Lightfoot (eds). Spix Publishing, Palm Beach, pp. 73–84.

CHAPTER 31

ANALGESIA AND ANAESTHESIA

ANALGESIA

What is pain?

Pain is a response to a noxious stimulus (one that has the potential to cause tissue damage) that serves the purpose of alerting an animal to the presence of that stimulus and thereby provoking a response that removes the animal from the stimulus and therefore prevents further injury.

It is generally accepted that birds perceive pain along neurological pathways like those in mammals. The receptors that are sensitive to pain stimuli are termed nociceptors. Within the peripheral nervous system, response can occur to either mechanical or thermal stimuli, or both. Following this peripheral detection, the stimulus is transmitted via primary afferent nerve fibres to the dorsal horn of the spinal cord. From there, the nociceptive signal is transmitted up ascending spinal tracts to the brain, where it is perceived as pain by central nociceptors.

Types of pain

Birds appear to experience different types of pain: acute, chronic, somatic, visceral, and neuropathic pain.

- Acute pain is associated with sudden tissue damage and is of short duration, but it gradually resolves as the injured tissues heal. It is relatively sharper and more severe than chronic pain.
- Chronic pain is any pain that lasts beyond the normal time expected for tissues to heal. It may become progressively worse and reoccur intermittently and may have no apparent biological cause. It includes pain such as that associated with arthritis.
- Somatic pain is associated with nociceptors in the skin, bones, joints, tendons, and muscles.
- Visceral pain is associated with the activation of nociceptors of the internal viscera. These structures are highly sensitive to distension (stretch), ischaemia, and inflammation, but relatively insensitive to other stimuli that normally evoke pain, such as cutting or burning. Visceral pain is diffuse, difficult to localise, and often referred to a distant, usually superficial, structure. In people, it is described as sickening, deep, squeezing, and dull.
- Neuropathic pain is pain caused by damage or disease affecting the somatosensory nervous system. Neuropathic pain may be associated with abnormal sensations called dysesthesia or pain from normally non-painful stimuli (allodynia). Common qualities described in people include burning or coldness, 'pins and needles' sensations, numbness, and itching.

Signs and effects of pain

Although birds perceive pain similarly to mammals, they may indicate pain in less obvious ways than mammals. They appear to respond to painful stimuli in one of three ways:

- 'Fight' responses:
  - Wing flapping
  - Aggression
- 'Flight' responses:
  - Excessive vocalisation
  - Attempts to escape
- 'Freeze' (conservation–withdrawal) responses:
  - Decreased head movement
  - Immobility
  - Closure of eyes
  - Inappetence
  - Fluffing of feathers
  - Absence or reduction of vocalisation

DOI: 10.1201/9781003272519-31

- Isolation from flock mates
- Wing droop or lameness
- Secreased grooming

It is thought that the 'fight' and 'flight' responses are more common when the bird is suffering from sudden or unexpected pain from which it is attempting to escape. In contrast, with chronic or overwhelming pain (which is perhaps more commonly seen in clinical practice), birds appear to be more likely to adopt the 'freeze (conservation–withdrawal)' responses, perhaps in an attempt to minimise the further pain that struggling would induce and to avoid attracting the attention of potential predators. Care must therefore be taken not to misinterpret lack of movement or vocalisation as an indication that the bird is not in pain (**Fig. 31.1**).

In addition to these clinical signs of generalised pain, localised pain may be reflected in feather picking or other self-mutilatory behaviours over or in the vicinity of, the painful lesion.

While we know that pain is a response to a noxious stimulus that serves the purpose of alerting an animal to the presence of that stimulus, we also now know that it has wider, more far-reaching consequences. Through stimulation of the adrenal gland and subsequent release of corticosterone, pain has an adverse effect on wound healing and the immune system. Survival and recovery rates of birds in pain are much lower than those receiving effective analgesia. As clinicians, we therefore have not only a moral obligation to provide effective analgesia but also sound medical reasons.

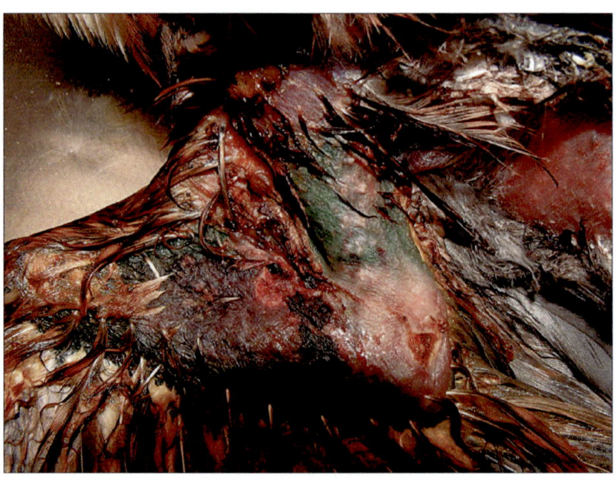

Figure 31.1 **Hawk with injuries and obvious bruising. Note that birds are not adept at displaying obvious signs of pain so the clinician should assume the patient will benefit from analgesia.**

Principles of analgesia

There are two underlying principles to consider when developing an analgesic plan for a patient.

Pre-emptive analgesia

Learning to anticipate when a patient will be in pain is helpful, as pain is easier to control if the patient is treated before it becomes painful and upset. A painful event stimulates peripheral nociceptors, which transmit the 'pain' signal to the brain via central spinal nociception pathways. Continued stimulation facilitates these pathways and evokes a painful response, a process known as 'wind up'. This facilitation may outlast the initial stimulus by a prolonged period. The underlying theory of pre-emptive analgesia is to use drugs and techniques before, during, and after a painful event to block this facilitation, preferably before it occurs. When dealing with patients who are already experiencing pain, it is not possible to apply a pre-emptive approach, which may explain why it is more difficult to get the pain under control initially, and then it requires more therapy to maintain analgesia if therapy is continued. The use of pre-emptive analgesia before performing any procedure likely to cause pain is therefore a key element in managing pain.

Multimodal therapy

The perception of pain is both a peripheral and a central event, using several types of receptors and pathways, so there is no single drug class or action that can be used as a sole analgesic agent. The concept of multimodal therapy recognises this and encourages a multilevel approach to analgesia, targeting different receptors along the pain pathway with the goal of improving analgesia while reducing individual class-related side effects.

- *Remove the source of pain.* This may involve splinting a fractured limb, removing a foreign body, or performing surgery with minimal tissue trauma and handling.
- *Reduce fear and stress.* Fear and stress may enhance the nociception pathways. Providing privacy in a warm, quiet environment to reduce apprehension and the judicious use of anxiolytics (e.g., midazolam) may serve to relieve fear and stress and therefore enhance analgesic therapy.
- *Use several classes of drugs to treat pain at different nociceptors and pathways (i.e., multimodal).* As described earlier, peripheral nociceptors detect a painful stimulus and transmit, via spinal

nociceptive pathways, a message to the brain where central nociceptors perceive it as pain. Inflammation in peripheral tissues may sensitise the peripheral nociceptors, invoking a more intense pain response at lower levels of pain. Using drugs that can reduce inflammation (and therefore decrease peripheral nociceptor sensitivity) while other drugs are used to block spinal pathways and/or central nociceptors can provide more comprehensive analgesia than one class of drug alone.

Analgesics commonly used in avian practice

The most commonly used analgesics in avian practice are opioids and NSAIDs. Local anaesthesia may also be used in certain circumstances.

Opioids

The endogenous opioidergic system is involved in setting the pain (nociceptive) threshold and controlling nociceptive processing, as well as modulating gastrointestinal, endocrine, and autonomic function. Activation of opioid receptors by endogenous neuropeptides (endorphin, enkephalin, dynorphin, and nociceptin) results in reduced neuronal cell excitability leading to a reduction in transmission of nerve impulses, along with inhibition of neurotransmitter release. The end result is the reduction or inhibition of pain.

Opioid receptors can be one of four closely related proteins. The three best known are the mu (μ), delta (δ), and kappa (κ) opioid receptor proteins. (The role of the fourth member of the opioid receptor family, the nociceptin or orphanin FQ receptor [ORL], is not clear.) In the avian brain, μ, κ, and δ receptors have been identified and mapped in the forebrain, midbrain, and spinal cord of several avian species. Differences in opioid receptor distribution, density, and functionality in birds is not yet well understood. The number, type, and distribution of these receptors appears to differ between species, leading to variability in pain sensitivity, the conscious response to pain, and the physiologic response to analgesic therapy. In pigeons, for example, the regional distribution of μ, κ, and δ receptors in the forebrain and midbrain are similar to mammals, but the κ and δ receptors were more prominent in the pigeon forebrain and midbrain than μ receptors and 76% of opiate receptors in the forebrain have been determined to be κ-type.

Opioid drugs reversibly bind to these specific receptors in the central and peripheral nervous system. These drugs are categorised as either agonists, partial agonists, mixed agonist/antagonists, or antagonists based on their ability to induce an analgesic response once bound to a specific receptor. They are used for moderate to severe pain, such as traumatic or surgical pain. Some bird species appear to have more kappa opioid receptors in the forebrain than mu opioid receptors, explaining why some birds do not respond to mu agonists such as morphine, buprenorphine, and fentanyl in the same manner as mammals. Therefore, kappa opioids such as butorphanol may be the more efficacious analgesics in some birds.

Specific drugs

Morphine is a μ receptor agonist and has not been commonly used in avian medicine. Investigations with morphine administered to chickens were confusing and clinically inconclusive.

Fentanyl is a short-acting μ receptor agonist. Due to its highly lipophilic nature, the onset of action of fentanyl is rapid as compared to other opioids due to its quick diffusion into the central nervous system. Because of its short-acting properties, fentanyl delivered via constant-rate infusion (CRI) is gaining popularity as an analgesic adjunct to inhalant anaesthesia in avian anaesthetic protocols.

Butorphanol is a mixed agonist/antagonist with low intrinsic activity at the μ receptor and strong agonist activity at the κ receptor.

Buprenorphine is a slow-onset, long-acting opiate with a unique and complex pharmacologic profile. Buprenorphine is believed to act as a partial μ agonist, but its κ receptor activities are less well defined. Several studies suggest that buprenorphine demonstrates κ receptor agonist effects, but other evidence in mammals and pigeons suggests that it also displays some κ antagonistic activities. Further work is required to determine whether clinical efficacy may be obtained using different buprenorphine doses, but at the time of writing, its use as an analgesic in parrots cannot be recommended.

Tramadol hydrochloride is a synthetic opioid that is active at opiate, alpha-adrenergic, and serotonergic receptors. It is a weak μ agonist, but its O-desmethyl metabolite (M1) is a much more potent agonist in mammals. In birds, the conversion to the M1 metabolite is variable among species but it is known that it is produced in bald eagles, red-tailed hawks, and Hispaniolan Amazon parrots.

Route of administration

Most opioid analgesics are used parenterally because of poor oral bioavailability associated with the first-pass effect. (Once absorbed, oral opioids first pass through the liver where they are metabolised, releasing a significantly lower amount of active drug into the general circulation.) For this reason, opioids are generally not given orally. An exception is tramadol, which is available in both oral and injectable formulations.

Adverse effects

The most common adverse effects reported with opioids are cardiac and/or respiratory depression. In many cases, these drugs may be reversed with antagonists, which will also terminate analgesia.

Non-steroidal anti-inflammatory drugs (NSAIDs)

NSAIDs are used to relieve musculoskeletal and visceral pain, acute pain, and chronic pain such as osteoarthritis. The pharmacologic activity of NSAIDs has been reviewed in veterinary articles and textbooks and although most reviews do not consider avian applications, it is assumed that the chemistry and mechanism of action is similar when administered to birds. The intention of recently developed NSAIDs has been to spare COX-1 and emphasise COX-2 inhibition with the goal of providing analgesia and suppressing inflammation without inhibiting physiologically important prostaglandins.

Specific drugs

Carprofen can be administered parenterally or orally and is well absorbed through the gastrointestinal tract in mammals. The mechanism of action of carprofen has not been fully elucidated. It is a weak inhibitor of COX at therapeutic doses and yet exhibits good anti-inflammatory activity. This weak inhibition of both COX isoforms may explain its wide margin of safety compared with other NSAIDs and it may achieve its therapeutic effects partially through other pathways. Much work is needed to determine appropriate doses, dosing routes, and dosing frequency of carprofen in birds.

Meloxicam is a COX-2 selective NSAID. It is available in both oral and injectable formulations and is the most widely used NSAID in avian medicine today. It appears to have poor oral bioavailability, requiring higher doses than mammals, and a short half-life requiring twice daily administration. To date it appears to have minimal effects on the kidneys and gastrointestinal tract, but its use should be avoided in patients with renal or gastrointestinal disease, or those that are dehydrated.

Celecoxib is a COX-2 selective NSAID that is relatively lipophilic in nature but poorly soluble in water. In humans, this drug is used for analgesia in conditions such as arthritis and primary dysmenorrhoea. It was historically considered standard therapy for the treatment of proventricular dilatation disease/avian ganglioneuritis in veterinary practice.

IV administration of 10 mg/kg celecoxib resulted in higher peak plasma levels than PO administration. Although the AUCs were similar, PO administration had a relative bioavailability of 110% with respect to IV administration. Based on these findings, PO administration of celecoxib may provide more effective analgesia, as well as avoidance of injection site reactions.

Mavacoxib is another COX-2 selective NSAID similar to celecoxib in properties. Although the effectiveness and safety of mavacoxib in multiple dose regimens has not yet been determined, studies in the cockatiel, revealed that Cmax values of both PO and IV administration were relatively similar, but as with celecoxib, PO administration resulted in a higher AUC than the IV administration. More importantly, mavacoxib had a long elimination half-life of 212 hours in cockatiels and 74.5 hours in the Caribbean flamingo, as a result of its low clearance in addition to a large volume of distribution.

Route of administration and frequency of dosing

NSAIDs can be given via IM, IV, or PO. Oral bioavailability is often poor, meaning higher doses are required orally than parenteral administration. Most NSAIDs appear to have short half-lives in birds, making twice daily administration advisable.

Adverse effects

The most common adverse actions of NSAIDs in mammals include effects on the gastrointestinal system, renal system, and coagulation. It is likely that these actions may apply to avian patients and care should always be taken when using these drugs, especially in dehydrated or renally compromised patients.

Local anaesthesia

Local anaesthetics block sodium channels in the nerve axon, interfering with the generation and conduction of action potentials along the nerve. When local anaesthetics are used pre-emptively, the number and frequency of impulses are reduced, thereby reducing nociceptor sensitisation, which has the beneficial effect of minimising central sensitisation.

Regional infiltration using a local line or splash block is the most common method of use. Local nerve blocks, using a nerve stimulator to locate the nerve, are showing potential in avian medicine.

Though work is lacking in birds, it appears that the duration of action for these drugs is slightly shorter than in mammals, with lignocaine lasting approximately 1–2 hours and bupivacaine lasting approximately 4–5 hours.

Adverse effects

Local anaesthetics are absorbed by the vasculature in the region being blocked. Systemic uptake of the local anaesthetics can be rapid in birds, and metabolism may be prolonged, increasing the potential for toxic reactions. Adverse effects include drowsiness and recumbency. Other toxic effects reported in birds include fine tremors, ataxia, recumbency, seizures, stupor, cardiovascular effects, and death. Toxic effects can be acute if accidentally injected intravenously.

Gabapentin

Gabapentin, a g-aminobutyric acid analogue, is similar in structure to the neurotransmitter GABA, although it is not believed to act on the same brain receptors. It was initially synthesised to mimic the structure of GABA for the treatment of epilepsy. It has been used to treat neuropathic pain in humans, dogs, and birds (although only limited reports on its use have been published). Its exact mechanism of action is unknown, but its therapeutic action on neuropathic pain is believed to involve voltage-gated N-type calcium ion channels.

Effective pain management

Pre-emptive analgesia and multimodal therapy can, in many patients, provide effective analgesia in a wide range of painful conditions. This analgesia cannot be provided simply by 'following a recipe' as found in a drug formulary. The clinician must always be guided by the patient's response to the analgesic therapy and adjust this therapy accordingly.

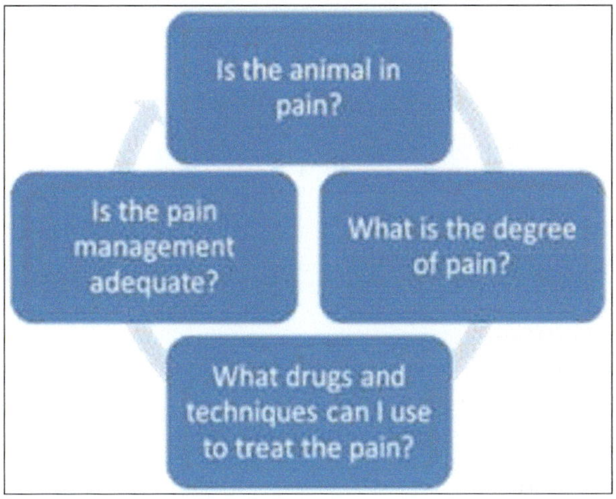

TRANQUILLISATION AND SEDATION

The terms 'tranquillisation' and 'sedation' are often used interchangeably, but they actually differ in the level of their effects. Tranquillisation reduces anxiety and induces a sense of calmness without drowsiness. The animal remains aware of its surroundings and may respond to stimuli. Sedation has a more profound effect, producing a sense of calm, lack of awareness, and drowsiness. Many drugs cannot be categorised by only one pharmacologic effect, i.e., as tranquillisers, sedatives, or analgesics. For example, butorphanol in birds produces both an analgesic and a sedative effect.

Indications

Instead of anaesthesia, and its associated effects such as hypotension and apnoea (especially in critically ill patients), sedation can be used to reduce stress and handling associated with:

- Grooming (e.g., nail, wing, and beak trims)
- Physical examination, especially in large or aggressive birds
- Diagnostic sample collection, such as venepuncture, nasal flushes, and skin biopsies (when combined with local anaesthesia)
- Minor procedures, such as wound repair, bandage changes, and small mass removal
- Diagnostic imaging

Sedation can also be used for the short-term management of seizuring activity, e.g., associated with lead toxicosis. It can also be used, again short-term, for its anxiolytic effects in birds with neophobia or other behavioural issues. Long-term use for behavioural modification is not recommended.

Commonly used drugs

CLASS	EXAMPLES	DOSE	COMMENTS
Benzodiazepines	Midazolam	2–7 mg/kg intranasally or IM	• Minimal cardiopulmonary effects • Rapid sedation, muscle relaxation, and anxiolysis • Effect is dose-dependent • Wide safety margin • Reversible with flumazenil 0.05 mg/kg
	Diazepam	6–10 mg/kg IM 6.4–16 mg/kg intranasally	
Alpha-2 agonists	Dexmedetomidine	0.1–0.2mg/kg IM	• Inconsistent effect when administered as a sole agent • Cardiopulmonary depression • Reversible with atipamezole (0.5 mg/kg IM) • Not recommended as a sole agent
Neuroactive steroid anaesthetic	Alfaxalone	10–50 mg/kg IM	• Dose-dependent sedation • Adverse effects include ataxia, hyperactivity, wing flapping, muscle rigidity, muscle fasciculations or tremors, spontaneous movement, and opisthotonos
Drug combinations	Midazolam + butorphanol	1–3 mg/kg (M) + 2–3 mg/kg (B) IM	• Lower doses may have an inconsistent effect
	Alfaxalone + midazolam	10 mg/kg (A) + 1 mg/kg (M) OM	• Smoother induction and recovery compared to alfaxalone alone
	Midazolam + ketamine	4 mg/kg (M) + 10 mg/kg (K) IM	• Heavy sedation
	Tiletamine + zolazepam	4–10 mg/kg IM	• Tiletamine–zolazepam has been used in quail and penguins • Recovery often rough

Duration of effect

Most of the sedation protocols have an onset duration of 3–5 minutes. Recovery time is drug- and dose-dependent and can often be reduced using an appropriate reversal drug (e.g., flumazenil or atipamezole). Without reversal, recovery times can be extended as long as 1–2 hours.

ANAESTHESIA

What makes avian anaesthesia different?

There are four fundamental differences between avian and mammalian anaesthesia: the anatomy of the bird, its metabolism, its tendency to lose heat quickly, and, finally, most birds requiring anaesthesia are chronically ill.

Anatomy

Trachea

- Birds have complete tracheal cartilage rings and therefore the trachea is not expandable. For this reason, only non-cuffed endotracheal tubes should be used. (If cuffed tubes are used, they must never be inflated.)
- The tracheal diameter reduces gradually from the glottis to syrinx; care must therefore be taken in selecting endotracheal tubes.
- The small diameter of the trachea in many birds makes intubation technically difficult and time-consuming. Blockage of the trachea or endotracheal tube with blood or mucus can rapidly result in hypoventilation. Even when not blocked, the small size of the required endotracheal tubes leads to increased flow resistance, making ventilation more difficult.

- Some species (notably waterfowl) have either very long, coiled tracheas or diverticula (bullae). These are anatomical adaptations allowing such birds to hold their breath underwater. It also may make it difficult to mask-induce these species, making intravenous or intramuscular induction advisable.
- Voice production arises from the syrinx, the bifurcation of the trachea. This is located within the body and is a common site for obstructions (*Aspergillus* granulomas or inhaled bird seed).

Air sacs, non-expandable lungs, and lack of a diaphragm

Respiration is achieved by movement of the sternum (up and down) and the ribs (in and out). Avian lungs do not expand and contract. Airflow through the lungs relies on the inflation and deflation of the air sacs. If the movement of the ribs or sternum is restricted (e.g., by restraint), or if enlarged organs, fat, or fluids within the body compress the air sacs, respiratory compromise can quickly develop. Prolonged positioning of the bird in dorsal recumbency can see the weight of the viscera compress the air sacs. Therefore:

- Wherever possible, place the bird in ventral or lateral recumbency. Birds in dorsal recumbency may require IPPV to properly oxygenate them.
- If any anaesthesia is likely to be prolonged more than 5–10 minutes, the patient should be intubated and ventilated, regardless of its positioning.
- Birds with ascites should be handled very carefully and may need diuretic therapy or coeliocentesis before anaesthesia.

If the coelom is opened during surgery, the bird may be able to breathe *through* the coelom, via the air sacs. This may require increasing the oxygen flow rate and deepening the anaesthesia to overcome this. Catheterisation of the caudal thoracic air sac (Chapter 6) can allow anaesthesia to be maintained without tracheal intubation, leaving the head free to work on.

Metabolism

The avian metabolic rate is much faster than that of mammals. For this reason, lengthy fasting is potentially dangerous, as hypoglycaemia may develop. Conversely, regurgitation and aspiration of crop and proventricular contents are very real risks. It is generally recommended to fast avian patients for 3–4 hours before anaesthesia. If necessary, crop contents can be removed by lavage prior to anaesthesia. Keeping the patient's head elevated above the level of the crop may help to prevent passive regurgitation and aspiration during anaesthesia. Blood glucose can be determined using a glucometer and, if low, the administration of 5% glucose during anaesthesia can assist in maintaining the patient's metabolic rate.

Heat loss

Birds have a high surface to body volume ratio and will lose body heat quickly when not active or when their metabolic rate slows. The rate and amount of heat loss is directly proportional to the size of the patient (the smaller the bird, the faster and greater the heat loss). Under anaesthesia, profound hypothermia, resulting in the patient's death, can easily occur. Hypothermia has been shown to result in impaired cardiovascular function, hypoventilation, decreased metabolism and detoxification of anaesthetic drugs, weakness during recovery from anaesthesia, decreased resistance to infection, increased incidence of surgical wound infection, and increased postoperative protein catabolism. Birds begin to cool down within 20 minutes of induction.

Various techniques have been developed to maintain body temperature including warmed intravenous fluids, heating pads, warmed anaesthetic gases, and radiant heat. Undoubtedly, the best prevention is a well-thought-out surgical plan and rapid surgical techniques.

Chronically ill patients

Elective procedures are uncommon in avian medicine. (Endoscopic sex identification and the physical examination of large patients [1–5 kg] are examples of elective procedures.) Most patients requiring anaesthesia are chronically ill and require diagnostic investigation or surgery. On top of this, the masking phenomenon makes it likely that most avian patients requiring anaesthesia are often more compromised than at first realised.

However, the stress of physically restraining a bird to examine it may be far greater than the stress of an anaesthetic. The decision as to whether to anaesthetise a patient is ultimately the clinician's and will be determined by the patient's physical condition and the experience of the clinician.

Procedures should be planned, endeavouring to do as much as possible in as short a period as possible. For example, with planning, the patient can be anaesthetised, physically examined, radiographed, blood and culture swabs taken, fluids given, and initial treatment started – all in less than 10 minutes. This can be done with minimal stress to the patient and maximum return to the clinician.

The ideal requirements for avian anaesthetic agents and techniques are:

- Rapid induction and recovery to minimise stress and heat loss.
- Rapid metabolism and excretion, avoiding 'hangover' effects that may delay the bird's return to eating and its normal metabolic rate.
- Minimal cardiac and respiratory depressant effects combined with good ventilation.
- Operator health and safety must remain an important consideration.

A good anaesthetic plan should therefore include:

- Patient preparation
- A safe induction
- Good supportive measures
- Effective monitoring
- A safe recovery
- Planned emergency protocols

Anaesthetic drugs

Intravenous or intramuscular agents

Intravenous or intramuscular agents can be used in birds, but it must be recognised that induction and recovery are variable. Recovery is often prolonged, and there can be a significant hangover effect, with the bird not regaining its normal metabolic rate and appetite for several hours or longer. It is therefore recommended that these agents only be used as an induction agent where mask induction is not feasible (e.g., waterfowl, large ratites) or in the field where gas anaesthesia may be unavailable. Clinicians must be aware of the hazards of using intravenous or intramuscular anaesthesia and be prepared to deal with adverse effects and prolonged recovery times.

Ketamine

Ketamine alone is not recommended for anaesthesia. Although it does not appear to cause pulmonary depression or cardiovascular depression, ketamine (at anaesthetic doses) gives inadequate analgesia and muscle relaxation and is associated with spontaneous movements, muscular rigidity, and violent recoveries. It is excreted through the kidneys, and the recovery time is prolonged and variable. The dose is 20–50 mg/kg IM or IV.

Xylazine

When used alone, xylazine does not usually produce adequate surgical immobilisation. It may be associated with bradycardia, hypotension, arrhythmias, and respiratory depression. It frequently causes salivation, excitement, muscle tremors, and convulsions, sometimes aggravated by noise. Recovery is prolonged, but xylazine can be reversed with yohimbine hydrochloride (0.11–0.27 mg/kg). The dose of xylazine is 1 mg/kg IM.

Xylazine–ketamine

A combination of xylazine and ketamine is commonly used for restraint and anaesthesia, but waterfowl (especially ducks and Canada geese), owls, and accipiters do not respond well. When used together the combination is synergistic, giving a smooth induction and recovery with improved muscle relaxation and possibly enhanced analgesia. The combination can be associated with cardiac arrhythmias and hypotension. The dose is ketamine 5–20 mg/kg and xylazine 0.5–2.0 mg/kg IM or IV.

Diazepam

By itself, diazepam is used for sedation or seizure control. The dose is 0.05–0.15 mg/kg slowly IV, or 0.2–0.5 mg/kg IM.

Diazepam–ketamine

A combination of diazepam and ketamine can be used as an intravenous induction agent in large ratites, raptors and waterfowl, but it may not be a reliable maintenance agent. It can be used intramuscularly in parrots and pigeons (with a slow induction and recovery time). The dose is diazepam 0.2–2 mg/kg and ketamine 10–40 mg/kg.

Midazolam

Midazolam can be used as a sedative and anxiolytic, or as a premedication for anaesthesia. At sedative doses it has minimal effects on blood pressure, heart rate, and body temperature. The dose is 0.8–3 mg/kg IM or IV. It can be given intranasally for sedation at dose rates of 1–5 mg/kg.

Midazolam–ketamine

The dose for this combination is midazolam 0.2–4 mg/kg and ketamine 10–40 mg/kg IM.

Medetomidine–ketamine

The combination of medetomidine and ketamine is useful for intravenous induction of waterfowl, giving rapid induction and recovery. It is reversible with atipamezole. The dose is medetomidine 60–85 μg/kg and ketamine 1.5–2.0 mg/kg IV.

Zolazepam–tiletamine

Zolazepam–tiletamine is not recommended in birds due to violent recoveries.

Propofol

Propofol can be used as an intravenous induction agent following sedation with medetomidine–ketamine. Apnoea is common following intravenous induction. The dose is 1–5 mg/kg IV. Intubation and IPPV are strongly recommended if using this drug.

Alfaxalone

Alfaxalone is a synthetic neuroactive steroid that acts centrally and has been shown to have a high therapeutic index, minimally depress the cardiovascular system, and provide a rapid and smooth induction in the cat and dog. It is non-cumulative, non-irritant, and can be administered via multiple routes (SC, IM, IV, IO). It is best given following premedication to minimise excitation at 5–7mg/kg IV.

Inhalation anaesthetic agents

Gaseous anaesthetics such as isoflurane and sevoflurane are the most popular and convenient choice for induction and maintenance of anaesthesia in birds. However, to use gaseous anaesthetics a specialised vaporiser and anaesthetic machine is required. This makes the use of gaseous anaesthesia difficult outside the veterinary hospital setting. Furthermore, many gaseous anaesthetics have an unpleasant odour and are irritating to the respiratory airways, which impact the quality of induction due to increased patient stress. These issues are accentuated when inducing anaesthesia in large birds (ratites and geese) and breath-holding species (waterfowl, penguins).

Isoflurane

Isoflurane is poorly soluble in blood, fat, and other soft tissues; therefore induction and recovery times are rapid, and controlling the depth of anaesthesia is easier to achieve. It is excreted through the respiratory tract. It does cause some cardiac arrhythmias, but they are not as severe or life-threatening as those seen with halothane.

Isoflurane has a higher anaesthetic index than halothane, meaning that apnoea occurs much earlier in deep anaesthesia, preventing the spontaneous inhalation of more isoflurane. This lessens the probability of fatal cardiac arrhythmias developing, making it a safer anaesthetic agent.

Sevoflurane

Sevoflurane produces a more rapid induction and recovery than isoflurane but not so much as to be clinically significant. It must run at a higher concentration than isoflurane to maintain surgical anaesthesia. Cardiac arrhythmias are not noted. The anaesthetic index is higher than for isoflurane, but again this is not of clinical significance. The higher concentrations required mean that sevoflurane is significantly more expensive than isoflurane, but there are only small clinical advantages.

Anaesthetic technique

The anaesthetic technique should be tailored to each patient and situation. However, the technique can be broadly divided into:

- Patient preparation
- Induction
- Maintenance
- Recovery

Patient preparation

Evaluation

Patients about to undergo an anaesthetic should be assessed by a careful physical examination and, where indicated, by clinical pathology (or at least PCV, total serum solids, and blood glucose).

- Obtain an accurate weight of the patient. This allows the accurate calculation of dose rates and evaluation of body condition.
- Obtain a resting respiratory rate and evaluate respiratory effort. Look for mouth breathing, tail bobbing, bilateral wing drooping, and exaggerated sternal lift as evidence of dyspnoea. If present, this must be factored into the anaesthetic plan, or the anaesthetic postponed.

- Evaluate the patient's hydration status. Signs of dehydration include sunken eyes; thick, tenacious saliva; decreased urine output; delayed refill of the basilic vein (on the medial aspect of the elbow) after short occlusion; and wrinkling and tenting of skin.
- Auscultate the heart and lungs for abnormal noises, e.g., murmurs and excessive respiratory noises. The heart rate is difficult to obtain in most patients, but bradycardia can be detected.
- Palpate the patient's crop and abdomen. If the crop is full, it must be emptied by crop gavage. If there is abdominal distension, it must be determined if there is ascites present. If so, pre-anaesthetic coeliocentesis may be advantageous.
- If time and finances allow, a full haematology and biochemistry profile should be obtained. Measure the patient's PCV, TPP, and blood glucose as a minimal pre-anaesthetic baseline.

Any abnormalities detected should be addressed where possible prior to induction and, if possible, the procedure should be delayed until this has been done. This may require active warming, fluid therapy, assisted feeding, and analgesia. If necessary, stress can be addressed through the provision of privacy shelters and hides and, if needed, the judicious use of anxiolytic drugs such as benzodiazepines. In birds, this preparation may require 1–2 days; in some cases, it may take several weeks.

Fasting

Fasting is a compromise between (a) reducing the risk of regurgitation/vomiting (and possible aspiration of stomach/crop contents) and (b) inducing a hypoglycaemic crisis. Small birds may not require fasting at all, while larger birds can be fasted for 3–5 hours. Raptors, because of their unique gluconeogenesis abilities can be fasted for up to 12 hours if needed. As part of the pre-anaesthetic assessment, the crop should be palpated and, if indicated, emptied with a feeding tube prior to induction.

If painful procedures are anticipated, pre-emptive analgesia should be given to assist in preventing 'wind up' pain, i.e., the facilitation of nociception activation and recruitment, and the transmission and perception of neural signals.

Premedication

Premedicating a patient prior to anaesthesia has several advantages:

- It allows for a smoother induction and recovery, with reduced stress (and catecholamine release) to the patient.
- It reduces the dose rates and concentrations of anaesthetic drugs, lessening their potential cardiovascular effects and improving their safety.

The disadvantage of premedication is the sometimes prolonged recovery (especially in very short procedures) and the delayed return to voluntary movement and eating. For all but very short procedures, e.g., radiography, premedication should be included in the anaesthetic plan.

Benzodiazepines and opioids have similar sedating effects, and the two drugs can be used concurrently. As opioids will potentiate the sedation effect of the benzodiazepines, the dose of each should be reduced. Some clinicians prefer to use an opioid alone to obtain both analgesic and sedative effects.

Given that the immobility conferred on a bird when it is sedated, hypothermia can begin shortly after the premedication has been administered. Birds should be given supplemental warmth prior to anaesthesia and, if indicated, supplemental oxygen.

Induction

Induction of anaesthesia can be achieved using injectable drugs (in large birds) or inhalant gases via a mask. Premedication reduces the dose of the anaesthetic drug needed to induce anaesthesia and reduces stress. Preoxygenation via a face mask for a short time can help compensate for any apnoeic periods between induction and intubation. Placement of an IV catheter can be performed after the premedication has been given and facilitated using EMLA® cream. In many patients the placement of an IV or IO catheter may have to be delayed until anaesthesia has been induced.

Parrots and other small birds

Where appropriate or necessary, the patient should be wrapped loosely in a towel to prevent wing flapping and excessive struggling, and then preoxygenated. Mask induction is usually carried out with 3–5% isoflurane; this usually takes approximately 20–30 seconds, after which time the anaesthetic setting can be reduced to 2–3% (**Fig. 31.2a**).

Poultry, waterfowl, and other large birds

An intravenous catheter is placed in an accessible vein (e.g., the medial tarsal vein in waterfowl, the

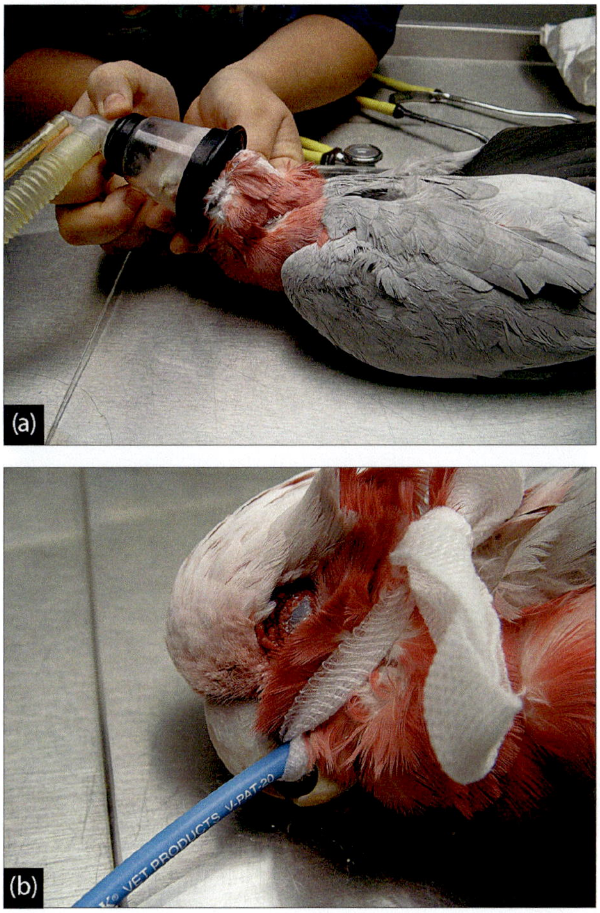

Figure 31.2 **(a) Galah undergoing anaesthesia using mask induction. Note this procedure is suitable for most companion birds. (b) Once anaesthetised, the galah was intubated. Note that for anything other than very short procedures this is the recommended practice.**

basilic vein in ratites). Intravenous induction with the drug(s) as described earlier is then followed by intubation and maintenance with a gaseous anaesthetic.

Intubation

For short procedures (e.g., radiography), maintenance of anaesthesia via a face mask may be all that is required. The mouth and nares should be encased in the mask, which should be sealed around the head to prevent the escape of anaesthetic gas and to assist in ventilation if required. For small patients this can be achieved by stretching a rubber glove over the mask and making a small hole for the patient's head. It is important to keep the patient's head elevated above the level of the crop throughout the procedure to prevent passive regurgitation and aspiration of crop or proventricular contents.

For procedures longer than 5 minutes, for compromised patients, or for rare or valuable birds, intubation is advisable (**Fig. 31.2b**). Most birds weighing more than 100 g can be intubated with minimal difficulty. Selection of an intubation tube should reflect the need to use the largest size possible to reduce air flow resistance and to effect a good seal to assist with ventilation, while avoiding iatrogenic trauma to the tracheal mucosa. Soft silicone or rubber endotracheal tubes are ideal as they are less likely to damage the tracheal mucosa. Only uncuffed tubes should be used; if cuffed tubes are used, they should not be inflated. (An inflated cuff can cause damage to the tracheal mucosa because of the complete cartilaginous rings.) Relatively shorter tubes than those used in mammals should be used and the diameter should be smaller than the width of the glottis would suggest (**Fig. 31.3**). Once the tube is placed, it should be secured in place with a tie to prevent accidental removal.

It is possible to intubate birds as small as 30 g in body weight, but they present a much greater challenge. In these smaller birds, an intravenous catheter (14 g–20 g) with the stylet removed can be used. These catheters are relatively inflexible, and care must be taken when using them to keep the head and neck extended so that the endotracheal tube is not kinked.

In some birds, laryngospasm can make intubation difficult. In this case, the application of a calculated dose of lignocaine to the glottis can assist. (Ensure the dose is precisely calculated to avoid any

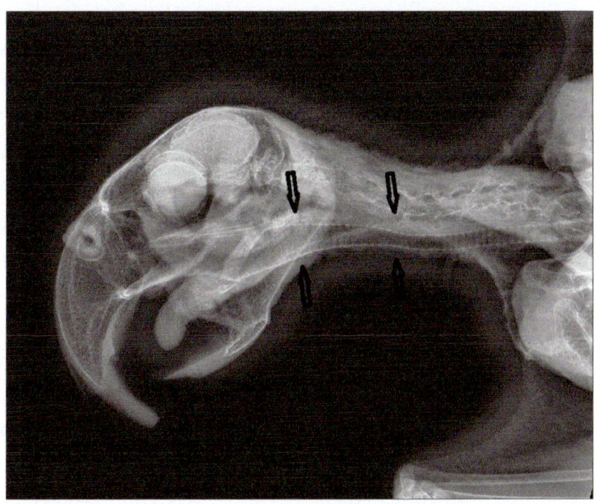

Figure 31.3 **Radiograph (lateral view) of a macaw shows how the trachea (highlighted) narrows as it progresses caudally. (Photo courtesy L Nemetz.)**

toxic effects from the lignocaine.) Laryngoscopes can assist in intubation but in small patients they will be too large. The use of smooth-tipped forceps, to grasp the tongue and bring it forward, can make intubation of even budgerigar-sized patients relatively straightforward.

Once the intubation is achieved, correct placement and patency of the airway can be assessed with a capnometer. If this is not available, airflow can be checked by holding a thread (or even a small feather) over the ET adapter – movement back and forth indicate the airway is patent. Once placement is confirmed, the endotracheal tube can be secured by tying or taping it in place. (Avoid taping the beak closed in waterfowl, as this may trigger a dive response with subsequent apnoea.)

If, after several attempts, intubation has not been possible, a face mask should be utilised, and the patient monitored for an obstructive breathing pattern suggesting oedema is obstructing the airway. (If this does occur the administration of an anti-inflammatory drug may be required. If necessary, the provision of alternative access to the respiratory tract may be required, e.g., an air sac catheter.)

Whether the patient is intubated or not, the patient's head should be elevated to prevent regurgitated stomach or crop contents reaching the pharynx and possible aspiration. This is easily achieved by placing a pad under the head to lift it above the level of the crop. Once intubated, ventroflexion of the neck must be avoided. This can bring the tip of the endotracheal tube into contact with the tracheal mucosa, causing pressure necrosis and possibly a tracheal stricture forming several days after intubation.

If the patient's condition precludes intubation (e.g., tracheal obstructions), an air sac catheter should be placed as described in Chapter 6.

Once a stable plane of anaesthesia has been reached, the isoflurane concentration can be reduced, usually to 1.5–3%. Oxygen flow rates are typically high, usually 1–3 litres/minute, even for small patients.

The patient should, wherever possible, be maintained in lateral or ventral recumbency. When placed in dorsal recumbency, the weight of the viscera can compromise the ability of the air sacs to function correctly. Sternal lift, necessary to expand the air sacs, also becomes more difficult for the patient in dorsal recumbency.

It should be noted that if a surgical procedure is being performed that opens an air sac to room air, spontaneous ventilation through the opened air sac will result in a lightening of the anaesthetic depth. This may require an increase in the oxygen flow rate and the level of anaesthesia, as well as manual ventilation, to maintain a surgical plane of anaesthesia.

Maintenance

Once the patient is intubated, the maintenance stage of anaesthesia begins; it concludes with extubation. Considerations during this stage include the depth of anaesthesia, the support provided to the patient, and how the patient is monitored.

Anaesthesia depth

As with other species, birds can pass through various depths of anaesthesia:

- Light – suitable for minor, non-painful procedures such as radiology. Respiration is smooth and regular, and muscle relaxation is good, but the patient may respond to stimulation.
- Medium – similar to a light plane of anaesthesia, but the patient is less responsive to stimuli.
- Deep – suitable for surgical procedures. There is increased abdominal or coelomic breathing due to a gradual loss of movement by the intercostal muscles and sternum. The respirations are shallow and very slow. The patient is non-responsive to stimuli.

Support

While under anaesthesia, the patient undergoes several changes: its body temperature will start to decrease, cardiac output and tissue perfusion will decrease, and respiratory compromise (hypoxia and hypercapnia) may develop. The respiratory acidaemia that may develop can lead to arrhythmias. These effects will worsen with blood loss, increasing anaesthetic depth, and positioning of the patient for the procedure.

Problems that have been identified in birds (and other animals) under anaesthesia include:

- Hypothermia. The patient's body temperature begins to decline when the premedication is administered. It can continue to drop until heart failure eventuates.
- Hypoglycaemia due to inappropriate fasting or prolonged anorexia can lead to hypoglycaemia.

- Circulatory collapse associated with low blood pressure or fluid overload can occur.

Patient support is therefore critical for minimising patient morbidity and mortality associated with anaesthesia.

Body temperature

Birds have a high surface to body volume ratio and will lose body heat quickly when not active or when their metabolic rate slows. The rate and amount of heat loss is directly proportional to the size of the patient (the smaller the bird, the faster and greater the heat loss). Numerous techniques have been devised to prevent hypothermia in anaesthetised patients and are listed next. It must be noted, though, that regardless of the thermal support provided, the patient's body temperature will start to decline after approximately 20 minutes of anaesthesia (although aggressive patient warming as described next will result in a slower rate of heat loss). One of the most important means of minimising this heat loss is to have a plan and preparations in place before inducing anaesthesia in order to minimise the anaesthetic time. Other techniques that have been employed include:

- Warmed intravenous fluids throughout the procedure
- Warmed and humidified anaesthetic gases
- Radiant heat from overhead heat lamps
- Heated air units (e.g., the Bair Hugger®)

Heat pads do not appear to be overly effective in providing heat support to avian patients and may lead to thermal burn injuries.

Circulatory support

The combined effects of anaesthetic drugs, patient positioning, lack of movement, and blood loss can lead to a decrease in cardiac output and systolic blood pressure. The resultant hypotension and decreased vascular perfusion can lead to decreased venous return to the heart or tissue damage. Orthostatic hypotension may play a role in the deaths of some patients at the end of procedure – care should be taken when moving an anaesthetised patient during or after the procedure, as rapid movements may transiently cause gravitational pooling of blood in the extremities leading to decreased cardiac output and potentially circulatory collapse.

Anaesthesia, especially when associated with surgical blood loss and/or pre-existing dehydration, can lead to hypotension and decreased vascular perfusion. Warmed fluids given subcutaneously, intravenously or intra-osseously will provide both circulatory and thermal support, but conversely the fluid rate must be calculated and controlled to prevent fluid overload. Isotonic electrolyte solutions such as lactated Ringer's solution, given at a rate of 10 ml/kg/hr and even up to 30–60 ml/kg/hr, appear to be safe and effective in most birds (**Fig. 31.4**).

Birds with pre-existing anaemia or coagulopathies, or those likely to experience excessive blood loss during surgery, will benefit from a fresh whole blood transfusion (see Chapter 5).

Respiratory support

Birds do not have a diaphragm and are reliant on the action of intercostal and sternal muscles for respiration. This effect is lost at surgical planes of anaesthesia. With prolonged anaesthesia (especially in patients that develop hypothermia, hypoglycaemia, or hypovolaemia), the patient's respiratory rate may slow and even stop. Cardiac arrest usually follows soon after. Adequate respiratory support, in the form of supplemental ventilation, is therefore essential for most avian anaesthetics. IPPV, performed either manually or by a ventilator, should be instituted early in the anaesthetic process rather than waiting for problems to develop (**Fig. 31.5**). Birds that are breathing by themselves while under anaesthesia will benefit from IPPV at a rate of 4–6 respirations per minute, with sufficient pressure to create normal inspiratory depth. (Excessive pressure should be

Figure 31.4 Cockatoo undergoing surgery receives intravenous fluids via the medial metatarsal vein.

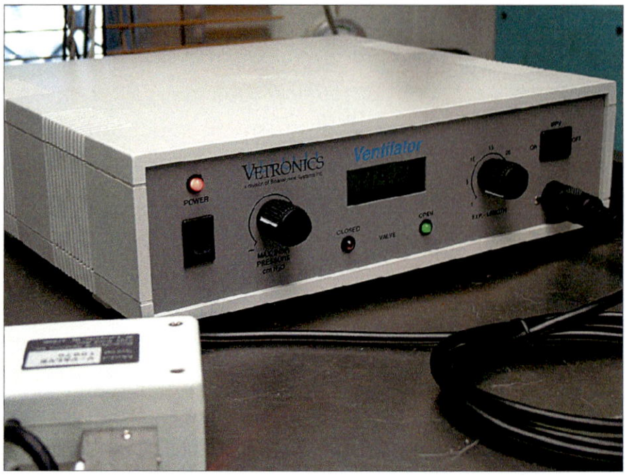

Figure 31.5 **Example of a commercially available respiratory ventilator. Note that these are suitable for small patients and can ease the anaesthetist's workload by ensuring that the patient is well ventilated without resorting to manual IPPV.**

avoided.) If the bird is apnoeic, IPPV should be given at a rate of 10–12 respirations per minute.

Monitoring

Monitoring of anaesthesia is vital in avian surgery (**Fig. 31.6**). For this reason, all anaesthetic procedures should be monitored by a staff member assigned to this task only. Safe and effective anaesthetic monitoring is a multimodal procedure, relying on visual assessment as well as monitoring equipment.

Manual assessment of the depth of anaesthesia requires monitoring of the following:

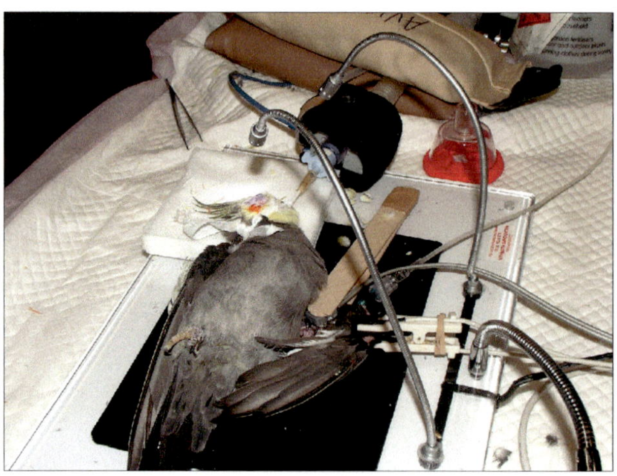

Figure 31.6 **Cockatiel about to undergo surgery is intubated, receiving intra-osseous fluids, and is been monitored with a Doppler, capnograph, and pulse oximetry.**

- The central nervous system.
  - The palpebral reflex (eyelid closure induced by stimulation of the ocular medial canthus or cere) disappears quickly as the anaesthetic plane deepens.
  - The pedal (toe withdrawal) reflex and the corneal reflex (third eyelid movement in response to corneal manipulation) slow, but can remain present even at surgical planes of anaesthesia.
  - Wing and jaw tone decrease with increasing anaesthesia depth.
- The respiratory system.
  - Respiratory rate and character are the most reliable indicators of anaesthesia depth in birds. The anaesthetised bird should breathe at least once every 2–7 seconds. At a light plane of anaesthesia, the respiratory rate and character can be rapid and deep. As the anaesthetic plane deepens, the respiration slows and eventually becomes shallower. As oxygen exchange only occurs when there is movement of air through the respiratory tract, periods of apnoea can result in cardiac arrest much sooner than that seen in dogs and cats.
- The cardiovascular system. Decreases or changes in heart rate and rhythm are often mirrored and preceded by changes in respiration.
  - Auscultation should be performed frequently, if not continuously (**Fig. 31.7**).
  - The use of a Doppler probe placed over the superficial ulnar artery (on the medial surface of the antebrachium or elbow) allows constant monitoring of the heart rate and rhythm. A pulse is usually easily detected over this artery with fingertip palpation.
  - In most patients an ECG can be used to monitor arrhythmias and changes in heart rate.
  - Indirect blood pressure can be monitored in larger patients.

Monitoring equipment

Capnography is becoming recognised as an essential means of monitoring avian anaesthesia. By measuring the concentration of expired CO_2 in each expiration, the capnometer produces a graph (capnogram) (**Fig. 31.8**) that characterises the elimination of CO_2 by the lungs. The capnogram shows the rate and depth of respiration, which often may change before changes in the heart rate and rhythm. Capnometers

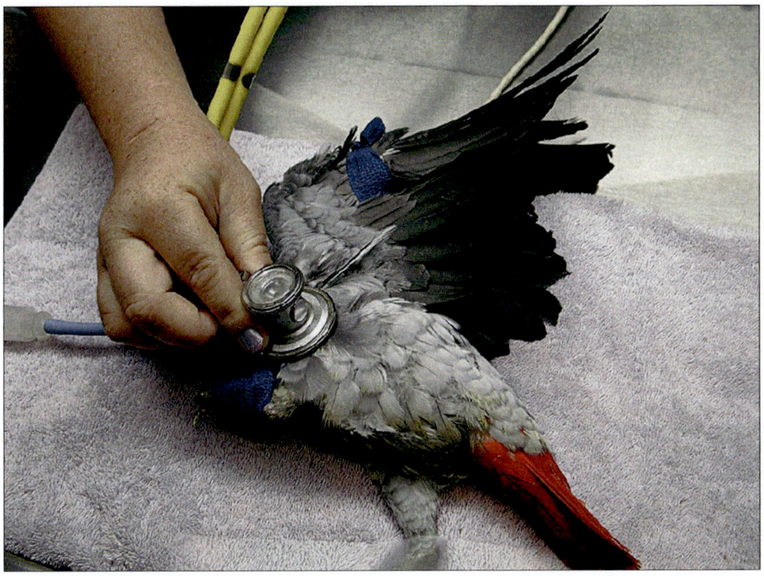

Figure 31.7 **An anaesthetised African grey undergoing surgery should be monitored via auscultation throughout. Note that this is an important monitoring tool.**

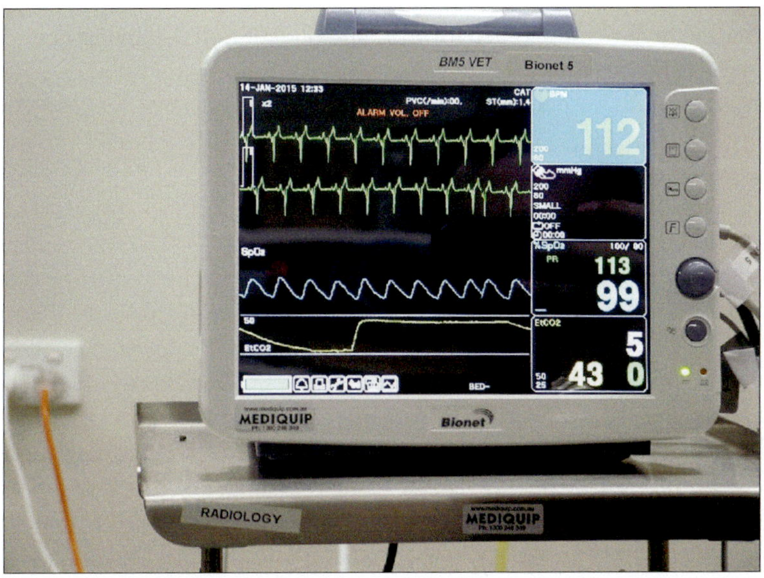

Figure 31.8 **A patient's rate and depth of respiration are monitored continuously during surgery along with heart rate and rhythm. The ECG trace is in green, at the top. The pulse oximetry is the blue trace in the middle, while the capnograph is in yellow at the bottom. Note that capnography is now a commonly used monitoring tool in avian surgery, and electrocardiography is occasionally employed, especially for larger birds.**

can be either mainstream (attached to the end of the endotracheal tube) or sidestream (drawing expired air from the endotracheal tube or adaptor). Mainstream capnometers may add to the dead space in the breathing circuit, potentially a major issue in small birds. However, sidestream capnometers that draw off a large amount of expired gas may exceed the respiratory volume of small birds. Microstream (50 mL/min) sidestream capnometers may be the most practical units to use in avian medicine. End-tidal CO_2 (ET CO_2) in anaesthetised birds should lie between 30–45 mmHg, which correlates to a $PaCO_2$ of 25–40 mmHg. Rising levels of ET CO_2 reflect hypoventilation and decreasing blood pH, resulting in respiratory acidosis and adverse cardiac effects.

Pulse oximetry is often considered to be unsuitable in birds because of the lack of calibration for avian blood. It may, however, provide some information on trends in blood oxygenation in avian patients and should not be discounted completely. By

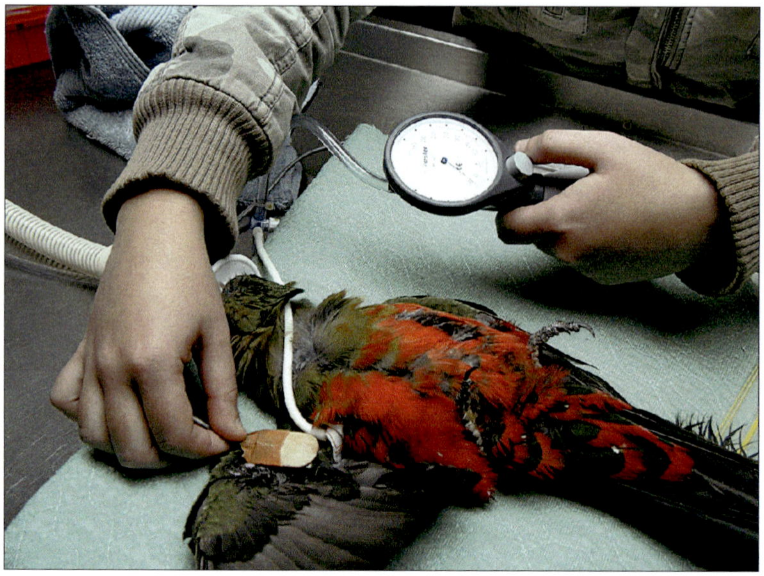

Figure 31.9 **King parrot being prepared for surgery has its indirect blood pressure monitored using Doppler ultrasound. Note that the pressure cuff has been wrapped around the humerus and the crystal over the ventral aspect of the radius and ulna.**

combining pulse oximetry (measuring oxygenation) and capnography (assessing ventilation), trends in the respiratory function of an anaesthetised patient can be closely monitored.

Blood pressure monitoring is becoming increasingly recognised as a valuable tool in anaesthetic monitoring. Although direct arterial pressure measurement is ideal, it is usually not practical in private practice. Indirect blood pressure measurement techniques, based on the detection of blood flow beneath an inflated cuff, are therefore recommended (**Fig. 31.9**). Systolic blood pressure determination via ultrasonic Doppler flow detection has been found to correlate well with direct blood pressure. The blood pressure of various avian species under gaseous anaesthesia lies between 90 and 140 mmHg systolic (see Chapter 6).

Electrocardiography (**Fig. 31.8**) is occasionally employed (especially in larger birds) and is a useful tool in monitoring the heart rate and rhythm, especially when auscultation or Doppler probe placement are difficult to perform without interfering with the surgery. Electrodes should be placed quickly after induction of anaesthesia and the ECG trace monitored (see Chapter 22 for details on recording and interpreting ECGs).

Recovery

Once the procedure for which anaesthesia was required is near completion, the anaesthetist can start to recover the patient. Depending on the type and duration of the anaesthetic, recovery to standing and perching usually takes 5–10 minutes (especially when the procedure was less than 20 minutes). Full recovery usually takes 30–60 minutes.

Given that over 25% of anaesthetic deaths occur in the first 3 hours after the anaesthetic has been discontinued, it is important that the patient be monitored carefully during this period. Care must be taken in repositioning a patient for recovery, as changes in body position can result in postural hypotension and cardiovascular collapse. Body temperature, heart and respiratory rate, and patient comfort should be assessed regularly and, if considered necessary, additional analgesia should be administered.

Birds are stimulated to spontaneously ventilate by rising CO_2 and, as they have a high oxygen requirement, they should be maintained on oxygen until spontaneous respiration and movement returns and the animal is attempting to self-extubate. Once the patient is extubated it should be maintained on oxygen by face mask until it is alert.

Once the patient has been extubated, it should be carefully observed to ensure it is not having breathing difficulties. Tracheal spasm or oedema, mucus plugs, or blood can obstruct the trachea once the ET tube is removed. An obstructive breathing pattern (slow, deep breathing with an exaggerated inspiratory effort) should alert the anaesthetist to a potential airway obstruction. In the event of an obstructed trachea, an alternative airway can be established using an air sac catheter.

During this period of recovery, monitoring the patient's body temperature and cardiac output should continue along with the supportive measures utilised during the anaesthetic. Once the patient has recovered sufficiently to be returned to its cage, fluid therapy can be stopped or reduced. The recovery cage should be warm and dark, and all perches, food, and water dishes should be removed. The bird should be lightly wrapped in a towel to prevent injury from wing flapping or excessive movement during recovery.

Birds should be kept warm, but care must be taken to avoid heat stress once the patient is normothermic; a bird in a small cage may overheat if unable to move away from a heat source. Once the patient can stand and move confidently, food and water should be returned to fuel the patient's energy requirements.

It may be advisable to hospitalise patients that have had lengthy anaesthetic procedures for 24 hours or more. Over 50% of anaesthetic deaths occur in this time and remaining in a controlled environment will help to reduce this mortality rate.

Anaesthetic emergencies

Careful monitoring of respiration, heart rate, and blood pressure will usually alert the avian anaesthetist to impending critical emergencies. Cardiac arrest, in many cases, is irreversible and fatal in the anaesthetised bird. However, prompt recognition and treatment of abnormalities can prevent this cardiac arrest in many cases.

Early signs of impending disaster include:

- Changes in respiratory rate and rhythm
- Changes in cardiac rate and rhythm
- Hypotension

Changes in respiratory rate and rhythm

The unique anatomy of the avian respiratory system contributes to their anaesthetic risk. Birds do not have substantial functional residual capacity (the quantity of air in the lung that is available for gas exchange at the end of expiration); 90% of the avian respiratory tract is composed of dead space (trachea, bronchi, and air sacs), and only 10% of the total intrapulmonary gas volume (located in the parabronchi and air capillaries) participates in gas exchange at any point in time. (In contrast, in many mammals only 4% of the total lung volume is dead space, and 96% of the intrapulmonary gas volume, located in alveoli, is available for gas exchange.) Birds compensate for the lack of functional residual capacity by moving air uni-directionally through their parabronchi during inspiration and expiration. Without proper airflow through the parabronchi, e.g., during apnoea or hypoventilation, efficient gas exchange cannot take place, and the physiologic effects of hypoventilation (hypoxaemia, hypercapnia) develop rapidly.

Tachypnoea (increased respiratory rate) can be due to:

- Lightening of the anaesthetic depth
- Pain perception in the lightly anaesthetised patient
- Pulmonary haemorrhage
- Hyperthermia

Dyspnoea (difficult or laboured breathing) can be due to:

- Airway obstruction (e.g., mucus plugs in the trachea or endotracheal tube)
- Poor circulatory perfusion due to heart disease, anaemia, or haemorrhage
- Restriction of body movement by internal organ enlargement, or improper restraint or positioning during the anaesthetic
- Pulmonary haemorrhage or oedema
- Pre-existing respiratory disease

Apnoea (complete cessation of breathing), usually preceded by a slowing of the respiratory rate, can be due to:

- Prolonged or very deep anaesthesia
- Hypothermia
- Inadequate ventilation
- Hypovolaemia

Management

Upon identifying an unexpected change in the patient's respiratory rate and rhythm, the anaesthetist should first check that the anaesthetic machine is functioning correctly and at what concentration the anaesthetic is being delivered. Connections to the patient should be checked to confirm that they are, in fact, connected and patent. The anaesthetist must alert the surgeon to the situation and ask if air sacs have been opened or if there is significant haemorrhage visible.

The patient should be assessed via auscultation (of the heart, lungs, and trachea) and monitoring devices such as capnography, Doppler, blood pressure, and ECG.

If tachypnoea is due to a lightening of the anaesthetic plane or perception of pain, the anaesthetic depth should be increased. If air sacs have been opened, the surgeon may have to temporarily 'close' the surgical site until a deeper plane of anaesthesia has been reached.

Tachypnoea or dyspnoea due to pulmonary haemorrhage or oedema does not respond to changes in anaesthetic depth or other treatments. Consideration should be given to terminating the procedure and recovering the patient. If dyspnoea is believed to be due to a mucus plug in the endotracheal tube, the tube should be removed and replaced. If the plug is in the trachea, placing an air sac catheter may be necessary.

Apnoea or a slowing of the respiratory rate should be treated by turning the anaesthetic down (or off), leaving the oxygen on and commencing IPPV at 10–12 breaths per minute. If the patient is not intubated, this should be performed where possible, and as quickly as possible. Gentle movement of the sternum ventrally and dorsally may assist ventilation if intubation is not possible, or until it can be performed.

If hypovolaemia is suspected, fluid rates should be increased.

Changes in cardiac rate and rhythm

Changes in cardiac rate and rhythm are usually preceded by changes in rate and depth of respiration. Tachycardia usually occurs as a result of decreased anaesthetic depth.

Bradycardia should be treated similarly to apnoea (i.e., the anaesthetic is turned down or off, IPPV is begun or increased, and fluid rates are increased). Sternal compression may assist circulation and ventilation. The sternum makes direct cardiac massage difficult, if not impossible. Atropine or glycopyrrolate can be given intravenously, as can epinephrine (1:1,000) at a rate of 0.5–1.0 ml/kg IM, IV, intraosseously, or intra-tracheally.

Once cardiac arrest has occurred, resuscitation is unlikely to be successful.

Hypotension

Hypotension occurs after prolonged or deep anaesthesia or severe blood loss. A slow bolus of IV fluids followed by increasing the fluid rate, lightening the anaesthetic depth, and commencing IPPV should be implemented once a significant drop in blood pressure is noted.

A quick guide to anaesthetic emergencies

PROBLEM	CAUSE	TREATMENT
Tachypnoea (increased respiratory rate)	Lightening of the anaesthetic depth Pain perception in the lightly anaesthetised patient	Increase anaesthetic depth. If air sacs have been opened, the surgeon may have to temporarily 'close' the surgical site until a deeper plane of anaesthesia has been reached.
	Pulmonary haemorrhage or congestion	Does not respond to changes in anaesthetic depth or other treatments. Consideration should be given to terminating the procedure and recovering the patient.
	Hyperthermia	Discontinue active heating. Consider terminating procedure.
Dyspnoea (difficult or laboured breathing)	Airway obstruction (e.g., mucus plugs in the trachea or endotracheal tube)	If dyspnoea is believed to be due to a mucus plug in the endotracheal tube, the tube should be removed and replaced. If the plug is in the trachea, placing an air sac catheter may be necessary.
	Poor circulatory perfusion due to heart disease, anaemia, or haemorrhage	Increase fluid rates. Consider blood transfusion.
	Restriction of body movement by internal organ enlargement, improper restraint, or positioning during the anaesthetic	Reposition the patient and begin IPPV.
	Pulmonary haemorrhage or oedema	Does not respond to changes in anaesthetic depth or other treatments. Consideration should be given to terminating the procedure and recovering the patient.
	Pre-existing respiratory disease	IPPV. Consider terminating the procedure.

PROBLEM	CAUSE	TREATMENT
Apnoea (complete cessation of breathing)	Prolonged or very deep anaesthesia	Turn the anaesthetic down or off, leave the oxygen on, and commence IPPV at 10–12 breaths per minute. Gentle movement of the sternum ventrally and dorsally may assist ventilation if no endotracheal tube or sealed face mask is being used.
	Hypothermia	Increase active heating.
	Inadequate ventilation	Commence IPPV at 10–12 breaths per minute.
	Hypovolaemia	Increase fluid rates.
Tachycardia (increased heart rate)	Decreased anaesthetic depth	Check that the anaesthetic machine is connected to patient. Increase anaesthetic rate.
	Pain	Additional analgesia.
Bradycardia (slowed heart rate)	Increased anaesthetic depth	Turn the anaesthetic down or off, and give a small fluid bolus IV (5–10 ml/kg).
	Hypoxia	Begin or increase IPPV. Check for airway obstruction.
	Cardiac disease	Sternal compression may assist circulation and ventilation. The sternum makes direct cardiac massage difficult, if not impossible.
		Adrenaline (1:1,000) can be used at a rate of 0.5–1.0 ml/kg IM, IV, IO, or intra-tracheally.
		Atropine @ 0.02 mg/kg IM or IV.
		Glycopyrrolate @ 0.01 mg/kg IV.
Hypotension	Prolonged or deep anaesthesia	Lighten the anaesthetic depth and commence IPPV.
	Severe blood loss	Increase the fluid rate. Blood transfusion.

CLASS	DRUG	DOSE AND ROUTE	COMMENTS
Anti-muscarinic agent used to treat bradycardia	Atropine	0.02 mg/kg IM or IV	May thicken respiratory secretions.
	Glycopyrrolate	0.01 mg/kg IV	May thicken respiratory secretions.
Alpha- and beta-adrenergic agonist agent used systemically for treating anaphylaxis and cardiac resuscitation	Adrenaline (epinephrine)	0.01–0.1 mg/kg IV	Care with hypovolaemic patients (not a substitute for adequate volume replacement).
CNS stimulant used to stimulate respiration	Doxapram	2mg/kg IM, IV	Not a substitute for ventilatory support. It will not increase respiratory frequency but will increase tidal volume and oxygen consumption, leading to decreased oxygen saturation. Consequently, its use is often not recommended.
Opiate antagonist	Naloxone	0.01–0.04 mg/kg IV q 2–3 minutes to effect	At reversal dosages, naloxone can negate opioid analgesic effects. The reversal effect may last for a shorter time than the opioid effect, so repeated dosing may be needed.
Benzodiazepine antagonist	Flumazenil	0.05–0.1 mg/kg IV or IM	Flumazenil is used for the reversal of benzodiazepine effects after overdoses or toxic effects.
Fluids	Crystalloid fluids (0.9% saline)	10 ml/kg IV bolus	Shock therapy.
	Colloid fluids	5 ml/kg IV bolus	Shock therapy.

FURTHER READING

Acierno MJ, da Cunha A, Smith J, Tully TN, Migallon-Guzman D, Serra V, Mitchell MA (2008) Agreement between direct and indirect blood pressure measurements obtained from anesthetized Hispaniolan Amazon parrots. *Journal of the American Veterinary Medical Association* 233(10): 1587–1990.

Acierno MJ, Smith J, Tully TN, Migallon-Guzman D, Mitchell MA (2007) Evaluation of indirect blood pressure measurement techniques (Doppler) and the comparing point of care blood gas analyzer (I-STAT) values. In: *Proceedings of the Annual Conference of the Association of Avian Veterinarians*, pp. 15–16.

Aliansyah E, Chng HT, Xie S (2021) A critical review of the pharmacokinetics and pharmacodynamics of opioid medications used in avian patients. *Birds* 3(1): 1–28.

Boedeker NC, Carpenter JW, Mason DE (2005) Comparison of body temperatures of pigeons (*Columba livia*) anesthetized by three different anesthetic delivery systems. *Journal of Avian Medicine and Surgery* 19(1): 1–6.

De Matos REC, Morrisey JK, Steffey M (2006) Postintubation tracheal stenosis in a Blue and Gold macaw (*Ara ararauna*) resolved with tracheal resection and anastomosis. *Journal of Avian Medicine and Surgery* 20(3): 167–174.

Dhondt L, Devreese M, Croubels S, De Baere S, Haesendonck R, Goessens T, Gehring R, De Backer P, Antonissen G (2017) Comparative population pharmacokinetics and absolute oral bioavailability of COX-2 selective inhibitors celecoxib, mavacoxib and meloxicam in cockatiels (Nymphicus hollandicus). *Scientific Reports* 7(1): 1–12.

Edling TM (2006) Updates in anesthesia and monitoring. In: *Clinical Avian Medicine, Vol 2*. GJ Harrison, TL Lightfoot (eds). Spix Publishing, Palm Beach, pp. 747–760.

Edling TM, Degernes LA, Flammer K, Horne WA (2001) Capnographic monitoring of anesthetized african grey parrots receiving intermittent positive pressure ventilation. *Journal of the American Veterinary Medical Association* 219(12): 1714–1718.

Heard DJ (1997) Anesthesia and analgesia. In: *Avian Medicine and Surgery*. RB Altman, SL Clubb, GM Dorrestein, K Quesenberry (eds). WB Saunders, Philadelphia, pp. 807–828.

Huckins GL, Carpenter JW, Dias S, KuKanich B (2020) Pharmacokinetics of oral mavacoxib in caribbean flamingos (phoenicopterus ruber ruber). *Journal of Zoo and Wildlife Medicine* 51(1): 53–58.

Klaphake E, Schumacher J, Greenacre C, Jones MP, Zagaya N (2006) Comparative anesthetic and cardiopulmonary effects of pre- versus postoperative butorphanol administration in Hispaniolan Amazon parrots (*Amazona ventralis*) anesthetised with sevoflurane. *Journal of Avian Medicine and Surgery* 20(1): 2–7.

Lichtenberger M (2005) Determination of indirect blood pressure in the companion bird. *Seminars in Avian Exotic Pet Medicine* 14(2): 149–152.

Lumeij JT, Deenik JW (2003) Medetomidine-ketamine and diazepam-ketamine anesthesia in racing pigeons (*Columba livia domestica*) — a comparative study. *Journal of Avian Medicine and Surgery* 17(4): 191–196.

Machin KL (2005) Avian pain: Physiology and evaluation. *Compendium on Continuing Education for the Practicing Veterinarian* 27(2): 98–109.

Mastakov H, Henning J, deGier R, Doneley R (2021) Induction of general anesthesia with alfaxalone in the domestic chicken. *Journal of Avian Medicine and Surgery* 35(3): 269–279.

Paul-Murphy J (2006) Pain management. In: *Clinical Avian Medicine, Vol 1*. GJ Harrison, TL Lightfoot (eds). Spix Publishing, Palm Beach, pp. 233–240.

Paul-Murphy J, Hess JC, Fialkowski JP (2004) Pharmacokinetic properties of a single intramuscular dose of buprenorphine in African Grey parrots (*Psittacus erithacus erithacus*). *Journal of Avian Medicine and Surgery* 18(4): 224–228.

CHAPTER 32

SOFT TISSUE SURGERY

INTRODUCTION

An understanding of surgical principles is necessary before operating on a bird. Even though there are many anatomical and physiological differences between birds and mammals, surgical principles and techniques remain similar. As discussed in Chapter 31, birds are prone to anaesthetic complications associated with pain perception, hypothermia, hypovolaemia, and hypoxia. These anaesthetic complications are closely linked with surgical principles.

To maximise surgical success, the following principles must be understood and applied:

- Minimise haemorrhage.
- Minimise tissue trauma.
- Minimise anaesthetic time.
- Minimise anaesthetic and metabolic complications.
- Provide postsurgical support and analgesia.

Complying with these principles requires the surgeon to:

- Ensure that the patient is in the best possible condition prior to surgery.
- Develop an analgesia and anaesthesia plan that maximises patient safety and comfort (see Chapter 31).
- Plan the procedure to minimise anaesthetic and surgical time.
- Ensure that surgical preparation and patient support procedures are in place to maximise patient safety.
- Use instruments, techniques, and suture materials that minimise tissue damage, blood loss, and inflammatory responses.

Presurgical assessment and conditioning

The presurgical assessment should take the form of a thorough physical examination and diagnostic testing.

Physical examination

The physical examination should include a comprehensive physical examination, weighing, and respiratory recovery time (a healthy patient should return to normal respirations within 3–5 minutes following capture).

Diagnostics

- Haematology and biochemistries. Ideally, preoperative PCV, total plasma protein, and blood glucose should be measured for most routine surgical procedures. If patients are obviously unwell, a more complete haematological and biochemical profile is indicated and can be performed while the patient is stabilised.
- Clotting time (crude estimate). A pinprick of the basilic vein should clot after 1 minute of direct pressure.
- Imaging such as radiology and ultrasound can be utilised to detect problems such as ascites and organomegaly.

Medical conditions that may necessitate postponement of surgery include respiratory distress, obesity, and concurrent disease problems unrelated to the surgical condition (e.g., cardiovascular disease, infectious conditions). Clinical pathology parameters that may necessitate postponing surgery are shown in **Table 32.1**.

DOI: 10.1201/9781003272519-32

Table 32.1 **Clinical pathology parameters that suggest surgery should be delayed until the patient is more stable**

PARAMETER	REFERENCE INTERVAL	RESULTS INDICATING PATIENT IS NOT SUITABLE FOR SURGERY
PCV	40–55%	<20% or >60%
Total plasma protein	20–50 g/l	<20 g/l
Glucose	12.6–24.4 mmol/l	<11 mmol/l
Uric acid	200–650 μmol/l	>700 μmol/l
AST	150–350 IU/l	>650 IU/l (in the presence of a normal CK)
Cholesterol	4.5–7.0 mmol/L	>18 mmol/l

Conditioning the patient

If any of the aforementioned abnormalities are detected, appropriate medical therapy should be instituted (e.g., fluid therapy, blood transfusion, tube feeding, diet conversion, or weight reduction). Vitamin supplementation may be of benefit in mildly malnourished birds, but it should be given several days in advance of the surgery to be of benefit. If a coagulopathy is suspected, vitamin K1 may be of benefit, but it must be administered at least 24 hours prior to the surgery.

If the patient's medical condition is critical and surgery cannot be postponed, the owner must be informed of the increased risks prior to commencing surgery.

Minimising anaesthetic time

A surgical plan should be developed. Although exploratory surgery is performed in birds, surgical time can be reduced if a plan has been developed by the surgeon. This should include:

- Revision of the appropriate anatomy if necessary.
- Determination of the extent of the lesion or disease process prior to the procedure can be achieved through careful physical examination, radiography, and endoscopy.
- Discussion with theatre and nursing staff about the procedure, so that all personnel involved know what is happening and what their roles are to be.

Prior to anaesthetising the patient, the surgeon should check that all equipment and materials likely to be used in the procedure are accessible and fully functional. The authors find that a verbal rehearsal with all involved personnel at this stage detects and corrects preparation errors that could have resulted in a delay during surgery.

Surgical preparation and patient support procedures

Preparation of the surgical site

The goal of site preparation is to create a sterile site while minimising the number of feathers removed to maximise heat retention after surgery (large areas of denuded skin will create problems for the bird's thermoregulation processes post-operatively). Feathers should be plucked, not cut, 2 to 3 cm around the surgical site. Feathers that are plucked regrow quickly (they are usually nearly regrown in 2 weeks at suture removal); however, cut feathers do not regrow until the next moult. Feathers should be plucked along the line of growth, rather than against it, to avoid tearing the skin. The shaft of the feather must be firmly gripped and pulled out cleanly so that the germinal layer of the feather papilla is not damaged, and a new feather will regrow. Soaking the feathers to be removed with the skin preparation solution or water-soluble lubrication gel prior to plucking will minimise the amount of loose feathers in the room that can potentially contaminate the surgical site. Retrices and remiges should be wrapped with bandage material if possible. The remaining feathers may be retracted from the surgical field with light adhesive tape such as masking tape.

The skin is prepared with routine cleansings using chlorhexidine alternating with sterile saline. The use of excessive amounts of water, alcohol, and chlorhexidine may result in hypothermia. Chlorhexidine is more commonly used because it has residual activity, has a broad spectrum of activity, causes fewer skin reactions, and is not inactivated by organic material. Saline is recommended instead of alcohol because it has proved to be as effective in preventing infection when alternated with the disinfectant solution and alcohol use results in greater heat loss.

Draping the surgical site

Transparent drapes allow visualisation of the patient and have heat-retaining characteristics. If adhesive drapes are not used, towel clamps may be placed around large feathers instead of through the skin.

Patient support

The provision of supportive measures, as discussed in Chapter 31, should be factored into the surgical plan. These measures include:

- Thermal support
- Circulatory support
- Respiratory support

Minimising tissue trauma and blood loss

Instrumentation

Magnification and illumination. Magnification is recommended to handle small and delicate tissues in small birds. The ideal surgical loupe system should:

- Be comfortable.
- Provide magnification of 3 to 5 times.
- Have lenses with a large depth of field so that more than one plane is in focus.
- Have a focal distance appropriate to the surgeon's working distance.
- Be ergonomic, minimising stress on the neck and back.
- Allow the surgeon to look around the lenses to accomplish tasks that do not need magnification.

Illumination with a focal light source is recommended. Head-mounted or lens-mounted lights direct the light where the surgeon needs it, keeping it in the field of view when the head is moved, but focal lights are also available separately from loupes.

Exposure. The small size of most avian patients means that surgical incisions are small, limiting exposure. This means that the use of retractors is a key element in improving surgical exposure. Three systems are currently in use by most avian veterinarians:

- Ophthalmic eyelid retractors
- Gelpi retractors
- Lonestar retractors

The first two systems can be somewhat limiting and often difficult to use on a small patient. The Lonestar retractor system offers greater flexibility and is lightweight compared to the stainless-steel options. However, all three retractors have a role to play.

Instrumentation. Due to the delicate nature of avian tissue and the small size of most patients, fine, well-constructed surgical instruments are often the best. Human and small animal ophthalmic and microsurgical instruments are often best suited for avian surgery.

Microsurgical instruments are long, counterbalanced instruments with round handles and miniaturised tips. Their proper use requires experience and training, but their length may make them more suitable for avian surgery than ophthalmic instruments.

Haemostasis. A healthy bird can generally tolerate greater blood loss than mammals, but many patients in need of surgery are often compromised; the amount of blood that a clinical avian patient can afford to lose during surgery may be considerably less than 1 ml/100 g. Careful consideration must be given to reducing haemorrhage to an absolute minimum. This can be achieved using:

- Gauze swabs and cotton-tipped applicators allow the surgeon to apply localised pressure to a small vessel and are less traumatic and more manageable in small, confined spaces.
- Vascular clips, such as Hemoclips (Weck), are a convenient and effective way to clamp vessels, serving the same purpose as ligatures. They are available in various sizes. Right-angled appliers are also available and are especially useful for applications deep in the body cavity of small patients.

Radiosurgery

Radiosurgery uses high-frequency radiowaves to cut through tissue by vaporising intracellular water. High-frequency radiowaves are modified by a waveform adaptor and then focused from the active electrode tip to the indifferent plate (antenna). These radiowaves produce alternating electromagnetic fields that generate heat due to resistance in the tissue. This localised heat volatilises intracellular fluids, causing cell disruption along the path of the electrode.

Electrosurgery, on the other hand, uses a lower frequency (2 MHz) electrical current passed through the patient to an antenna plate, heating the tissue without heating the electrode tip and destroying the tissue in contact with the probe. It often causes excessive lateral tissue damage compared to radiosurgery.

The advantages of radiosurgery are:

- Smooth, even cutting without pressure.
- A clear operating field with little or no pressure.
- Precise control of the incision.
- Minimal disuse destruction.
- Minimal scar tissue.
- Rapid and uncomplicated healing.

A frequency of 3.8–4.0 MHz has been found to be optimal for tissue incision, as it focuses the energy into a minimal area with little damage to surrounding tissue (lateral heat). Radiosurgery requires tissue contact, which provides tactile feedback but can result in tissue drag and increased tissue damage.

The radiosurgical waveform can be:

- Fully rectified, fully filtered – pure cutting with minimal coagulation, suitable for biopsies.
- Partially rectified – coagulation, no cutting, suitable for haemostasis.
- Fully rectified, unfiltered – 50% cutting, 50% coagulation, suitable for most surgeries.
- Fulguration – destruction of tissue.

Radiosurgery therefore offers both haemostasis and minimal tissue damage when used correctly. It has been shown to result in less collateral tissue damage than laser CO_2 in the skin of birds. Tissue damage (lateral heat) is determined by:

- Electrode size – the greater the size, the more lateral heat, and therefore the more tissue damage.
- The time the tissue is exposed to the electrode – the longer the time, the more lateral heat is created.

Radiosurgery uses different handpieces:

- Monopolar polar handpiece – a handheld pencil-like handpiece that must be used with an antenna plate under the patient. It comes with:
  - Straight-wire electrodes used for incision.
  - Loop electrodes used for tissue removal.
  - Ball electrodes used for coagulation and tissue destruction.
- Bipolar handpiece – a forcep-like handpiece that does not require an antenna plate. It is used for pinpoint haemostasis and limited incisions.

Some hints for using radiosurgery:

- Keep the electrodes clean during the procedure. Tissue and coagulated blood impede the radiosurgical current and cause tissue drag. The result is excessive lateral heat and tissue damage.
- Adjust the power setting as required. Too much power will cause sparking and smoke production as the tissue is cut. Too low a setting results in tissue drag and lateral heat. Power requirements will vary with the type and size of the electrode, the area of electrode surface in contact with the tissue, the nature of the tissue (e.g., normal skin vs. fibrotic tissue), the requirements of the procedure (cutting vs. coagulation), and the depth of the incision required.
- Use a good cutting technique.
  - The electrode should pass through the tissue without dragging or resistance.
  - There should be minimal/no tissue blanching along the incision.
  - No tissue shreds should adhere to the electrodes.
  - When cutting, always activate the electrode before contacting the tissue.
- When coagulating, the electrode should contact the tissue before being activated. Coagulation can be achieved by:
  - Using bipolar forceps.
  - Touching the ball electrode (monopolar handpiece) to a bleeding vessel or capillary bed.
  - Clamping a larger vessel with haemostats and then touching the ball electrode to the haemostat (the current will pass down the haemostats and seal the vessel).

Carbon dioxide and diode laser surgery

Carbon dioxide and diode laser surgery are slowly replacing radiosurgery. The light emitted from a laser has a wavelength that is absorbed by fluid in the cells, thus creating a thermal effect. CO_2 lasers are based on a carbon dioxide gas mixture, which is stimulated electrically to produce a wavelength of 10.6 μm Diode lasers generate a laser beam by means of the so-called seed laser and amplify it in specially designed glass fibres, which are supplied with energy via pump diodes. With a wavelength of 1.064 μm, diode lasers produce an extremely small focal diameter; as a result, their intensity is up to 100 times higher than that of CO_2 lasers with the same emitted average power.

The degree of penetration of the laser into tissue can be controlled, allowing for fine, controlled dissection of tissue. Advantages of the laser include:

- Haemostasis. The laser cuts and coagulates blood vessels up to about 0.5 mm in diameter. (Larger vessels may have to be cauterised or ligated.)
- Analgesia. As the laser cuts, it seals nerve endings and axons, reducing the sensation and transmission of pain.
- Decreased postoperative swelling. The laser seals lymphatic vessels, decreasing the extravasation of lymphatic fluid into the surgical site postoperatively.

Against these advantages are the following disadvantages:

- Laser units typically cost 3–10 times more than radiosurgery devices.
- The surgical plumes produced by CO_2 lasers have been shown to be hazardous to health and necessitate the use of masks and surgical-site smoke evacuation.
- Occupational health and safety regulations require:
    - Protective eyewear.
    - The prevention of unauthorised entry when a laser is in use.
    - The masking of all windows.

LigaSure™ vessel sealing system

The LigaSure™ vessel sealing system is a bipolar electrothermal device that seals blood vessels through an optimised combination of pressure and radio frequency.

After applying pressure on the tissues with the LigaSure forceps, the ForceTriad™ energy platform generates energy tailored to the tissue impedance, reducing fusion cycle time and tissue desiccation with consistently controlled tissue effect. The completion of coagulation is signaled by the feedback sensors, and the tissue is then cut along the line of coagulum.

LigaSure preserves the patient's collagen and uses it to form a permanent autologous seal that is strong enough to withstand up to three times the normal systolic (normal systolic: 120 mm/Hg) on vessels or tissue bundles. This result makes LigaSure comparable to the mechanical methods of vessel occlusion; it ensures complete coagulation of arteries and veins up to 5 mm in diameter with minimal surrounding thermal spread up to 2 mm in diameter and limited tissue charring.

Many of the handpieces currently available are not suitable for small birds but are useful in chickens.

Endosurgery

Endoscopy has traditionally been used as a diagnostic tool in avian medicine to examine and biopsy internal organs. Recent developments have seen it used as a surgical tool, including:

- Triangulation of various instruments using multiple ports or entry points.
- The use of radiosurgery and/or laser surgery to ablate lesions and provide haemostasis.
- The combination of these techniques provides haemostasis during surgical procedures.

Advantages include:

- The combination of both diagnosis and treatment at the same time.
- Reduced surgical stress, pain, and time.
- Decreased risk of infection.
- Increased anaesthetic safety.
- Faster postoperative recovery.

Suture choice

Desirable characteristics of suture material include high tensile strength while the tissues heal; good knot security; resistance to infection; and minimal inflammatory, immunogenic, or carcinogenic reactions. A wide selection of suture materials is available to the surgeon, classified on their absorption characteristics and whether they are monofilament or braided. It is incumbent on the surgeon to understand how long different tissues take to heal, how long different sutures maintain their tensile strength, and how long before the suture is absorbed. This then affects the selection of suture material for various procedures.

Table 32.2 **Suture characteristics**

SUTURE	TYPE	>50% TENSILE STRENGTH RETAINED	TIME FOR COMPLETE RESORPTION	COMMENTS
Chromic catgut	Biological monofilament	<7 days	60–90 days	• Poor knot security • Intense inflammatory reaction • Rapid loss of strength
Polyglactin 910 (Vicryl) Polyglycolic acid (Dexon)	Synthetic braided	14–21 days 7–14 days	90–110 days 60 days	• Increased tissue drag • Moderate inflammatory reaction • Can act as a wick for ascending infections
Polydioxanone (PDS)	Synthetic monofilament	21–28 days	180–210 days	• Stiff • Tendency to kink
Polyglyconate (Maxon)	Synthetic monofilament	21–28 days	180 days	• High degree of tensile memory (coil) when removed from package
Glycomer 631 (Biosyn) Poliglecaprone 25 (Monocryl)	Synthetic monofilament	14–21 days 7–14 days	90–110 days	• High degree of tensile memory (coil) when removed from package

Absorbable sutures of biological origin are broken down by phagocytosis, provoking a marked inflammatory response. Chromic catgut is the classic example of a biologically derived suture material. It rapidly loses its tensile strength and is quickly absorbed.

Synthetic polymers are absorbed by hydrolysis. Hydrolysis involves breaking down polymers into monomers by direct water cleavage; once broken down, the monomers are then absorbed and metabolised by the body. The speed at which this occurs, and the time to complete resorption, is determined by the composition of the monomers and polymers in the material.

The characteristics of various suture materials are summarised in **Table 32.2**. When these characteristics are taken into account, PDS emerges as the suture material of choice for many procedures, especially for the closure of muscle and other fascia. For gastrointestinal surgery (where tissue healing times are faster) Vicryl, Biosyn, and Monocryl are also suitable choices.

COMMON SOFT TISSUE SURGICAL PROCEDURES

Ingluviotomy

Indication

Ingluviotomy is indicated for the retrieval of foreign objects, endoscopic access to the proventriculus, placement of feeding tubes, and biopsy.

Technique

The bird is positioned in dorsal/lateral recumbency, with the head elevated. The bird should be intubated. The skin is incised over the left lateral wall of the crop, close to the thoracic inlet (**Fig. 32.1**). The skin edges are gently and bluntly dissected free of the underlying crop. A relatively avascular area is selected for an incision into the crop. The crop mucosa will often evert when the crop is opened. Care must be taken to prevent contamination of the subcutaneous tissues with any crop contents. Stay sutures may be required in some cases.

Closure

The crop is closed with two layers of 4–0 or 6–0 PDS in an inverting pattern. If unsure whether a good seal has been obtained, pass a feeding tube through the mouth into the crop and gently fill the crop with saline. If there is no leakage, the saline and the tube can be withdrawn prior to recovery. Flush the subcutaneous tissues thoroughly with sterile saline to remove any crop contents inadvertently spilt during the surgery, and then close the skin separately in either a continuous or interrupted pattern.

Postoperative care

Oral intake should be restricted for 6–12 hours, and then reintroduced in a controlled manner to prevent overstretching of the crop and placing tension on the suture line. After 3 days, if there have been no

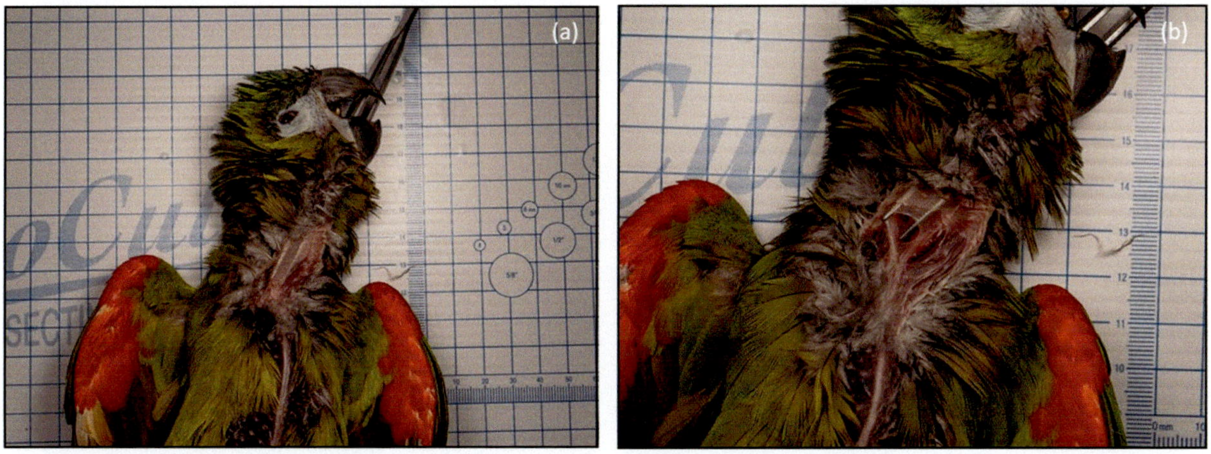

Figure 32.1 **(a) Localisation of the crop with the aid of forceps. (b) Skin incised over the lateral wall of the crop; avascular area selected for incision into the crop.**

complications, normal feeding and drinking can be resumed.

Crop fistula repair

Indications

Crop fistulae occur after crop burn due to microwaved or excessively heated and/or inadequately mixed hand-rearing food, or the ingestion of caustic substances. See Chapter 29.

Clinical presentation

Signs include slow crop emptying and erythema or blanching over the crop. In most cases, it will be 3–5 days before the delineation between healthy and devitalised tissues becomes apparent, and it may take as long as 7–14 days. At this stage, the fistula forms and food leaks out.

Technique

Analgesia, nutritional support, and antibiotics/antifungal medications should be provided as appropriate until the demarcation between live and dead tissue is apparent.

The patient is prepared and positioned as described earlier for an ingluviotomy. Once the fistula has formed, the crop mucosa will have adhered to the skin, forming a raised rim of granulation and fibrous tissue around the fistula. The skin is incised around the edges of this rim, and the incision is continued for a short distance both cranial and caudal to the fistula. The skin edges are bluntly dissected off the crop. The fistula is excised completely, leaving fresh crop edges (**Fig. 32.2a–d**). The crop and skin are then closed as for an ingluviotomy (**Fig. 32.2e**).

Postoperative care

Postoperative care is as described for an ingluviotomy.

Left lateral coeliotomy

Indications

A left lateral coeliotomy procedure provides access to the gonads, left kidney, oviduct, proventriculus, and ventriculus.

Technique

The bird is placed in right lateral recumbency, with the cranial end of the body elevated 30–40° to prevent fluid from entering the lungs. The wings are extended dorsally and secured into place. The left leg is abducted and drawn slightly forward. The inguinal skin web is incised between the coelomic wall and the left leg, and the leg is abducted further. This incision is continued from the sixth rib to the level of the left pubic bone (**Fig. 32.3**).

The superficial medial femoral artery and vein are cauterised where they traverse (in a dorsoventral direction) the lateral coelomic wall, medial to the coxofemoral joint. The muscles (external, internal coelomic oblique, and transverse coelomic muscles) are tented up, and a stab incision is made with pointed scissors while protecting the viscera. This incision is extended from the pubic bone to the eighth rib, which will require transecting the last two ribs. This is done by passing bipolar forceps around the ribs,

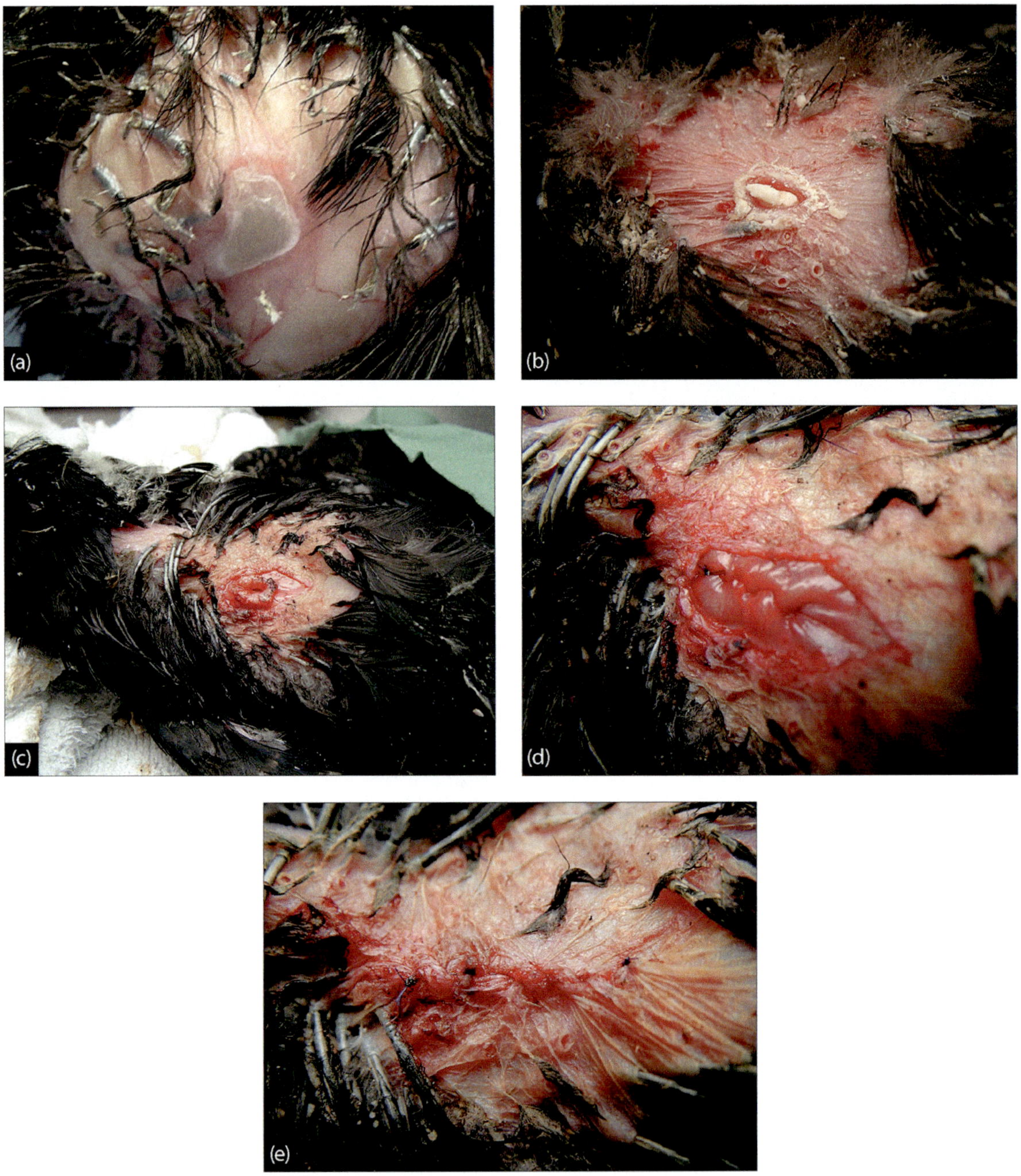

Figure 32.2 **Sequence of treatment and procedures to address a crop burn in a juvenile black cockatoo. (a) When presented, note the blanching of the now avascular skin and the surrounding erythema. (b) After a few days, note the limit of the burn. Crop necrosis has now become evident. (c) Surgical repair first involves dissecting the skin away from the crop, and the fistula is excised. (d) After excising the necrotic and granulating edges, the crop is repaired with a double-layer, continuous inverting suture pattern using a monofilament absorbable suture material (polydioxanone). (e) Finally, the skin is closed in a single layer with the same suture material.**

cauterising the intercostal blood vessels, and then cutting the ribs with scissors in turn.

A small retractor (e.g., a Lonestar) is placed to allow visualisation of the internal organs (**Figs. 32.4, 32.5**).

Closure

If air sacs have been transected, they are left to heal by second intention. The muscle and skin are closed in separate layers with absorbable sutures in a continuous or interrupted pattern. No attempt is made to rejoin the transected ribs.

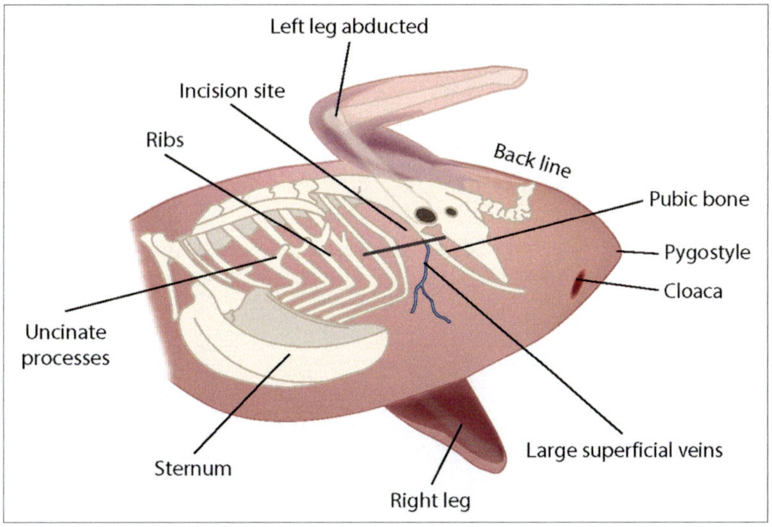

Figure 32.3 **Schematic diagram highlighting the anatomical landmarks for a left flank coeliotomy.**

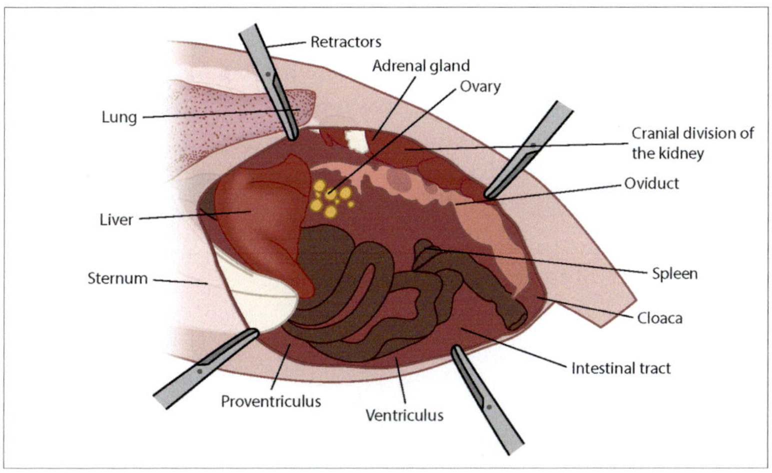

Figure 32.4 **Illustration of the typical view inside the coelom during a left flank coeliotomy.**

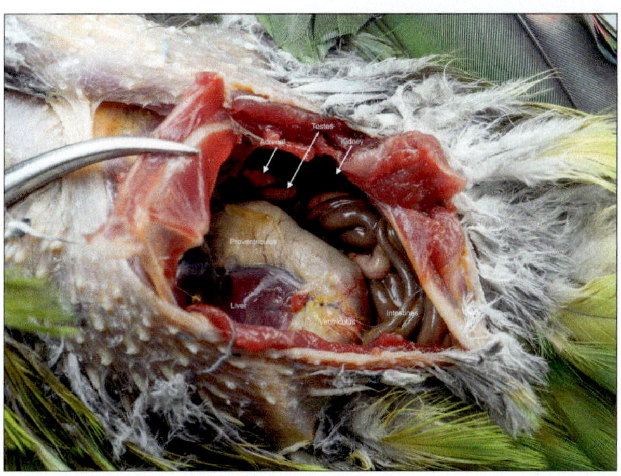

Figure 32.5 **Photo of the typical view inside the coelom during a left flank coeliotomy.**

Postoperative care

No special postoperative care is required other than analgesia and, if appropriate, antibiotics. An adhesive dressing should be placed over the incision for 3–5 days to minimise wound infection.

Ventral midline coeliotomy

Indications

A ventral midline coeliotomy procedure is used as an approach to the cloaca for cloacopexy, an approach to the oviduct or testes, or for a biopsy of the liver and pancreas.

Technique

The bird is placed in dorsal recumbency, and the legs are abducted caudally. The skin is tented and incised

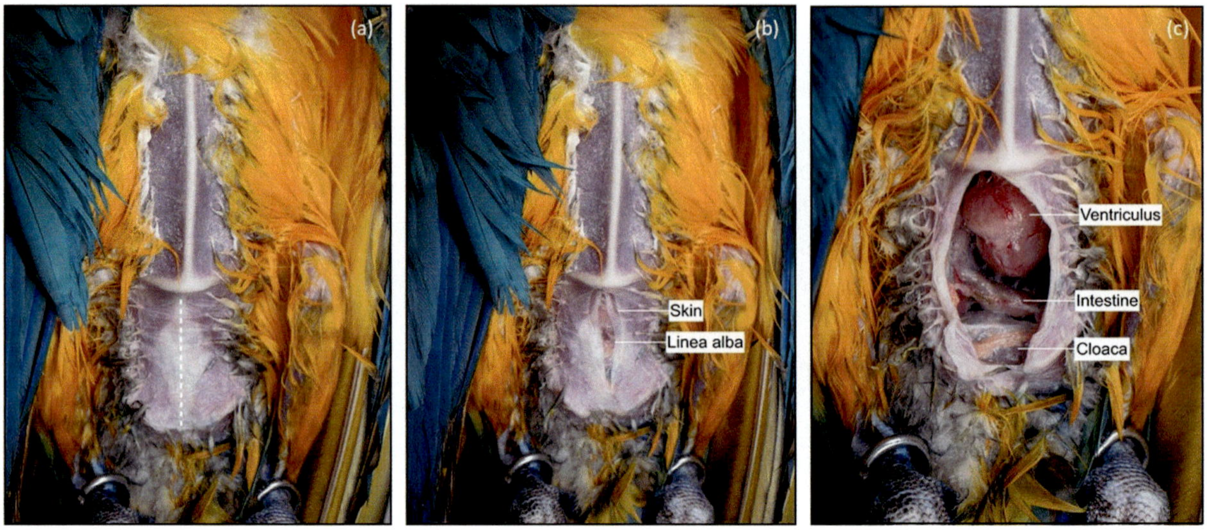

Figure 32.6 **(a) Landmarks of a simple ventral midline incision. (b) Exposure of linea alba. (c) Exposure of internal organs.**

in the ventral midline from the sternum to a point level with the pubic bone (**Figs. 32.6, 32.7**).

The linea alba is tented and incised in a craniodorsal direction, avoiding iatrogenic damage to the underlying viscera (**Figs. 32.8, 32.9**). To allow greater exposure, the incision can be extended laterally at the proximal and distal ends of the incision to create flaps. This will form a C-shaped incision with a unilateral flap or an I-shaped incision with bilateral flaps (**Fig. 32.10**). (Note that while lateral flaps increase exposure, they also allow increased anaesthetic gas and heat loss through the open coelomic cavity.) Again, a retractor can be used to increase exposure.

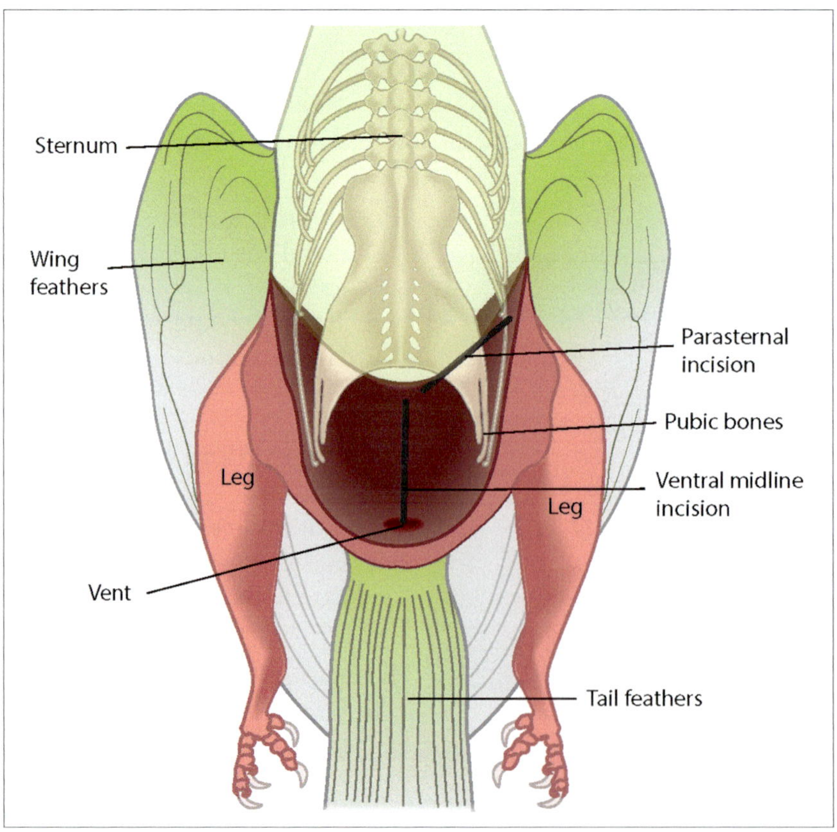

Figure 32.7 **Schematic diagram indicating appropriate incision sites for a ventral midline coeliotomy.**

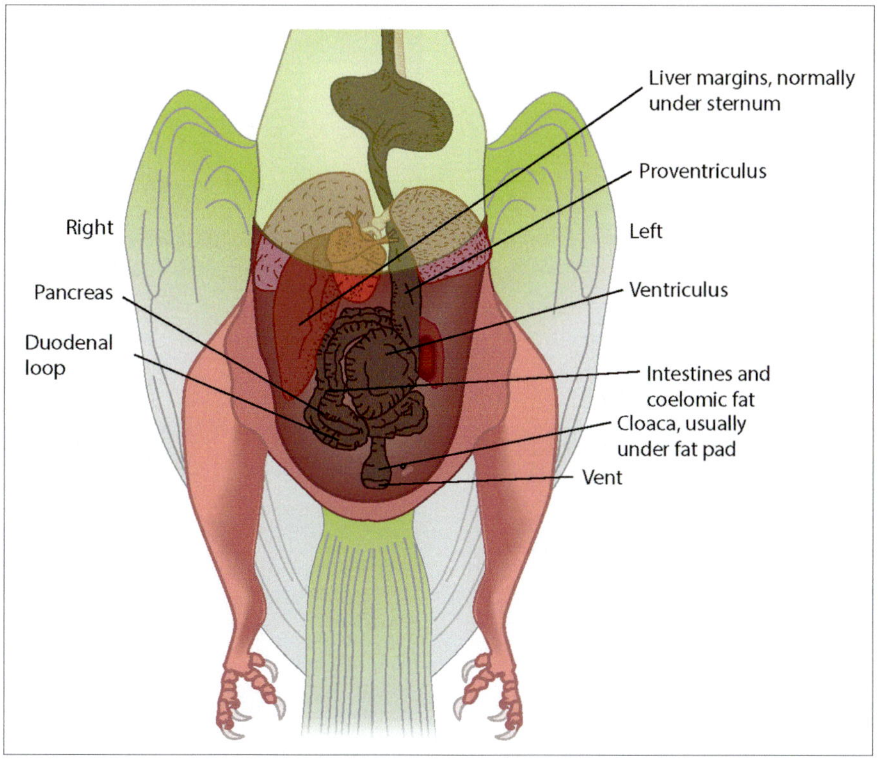

Figure 32.8 **Illustration of the typical view inside the abdomen following a ventral midline incision.**

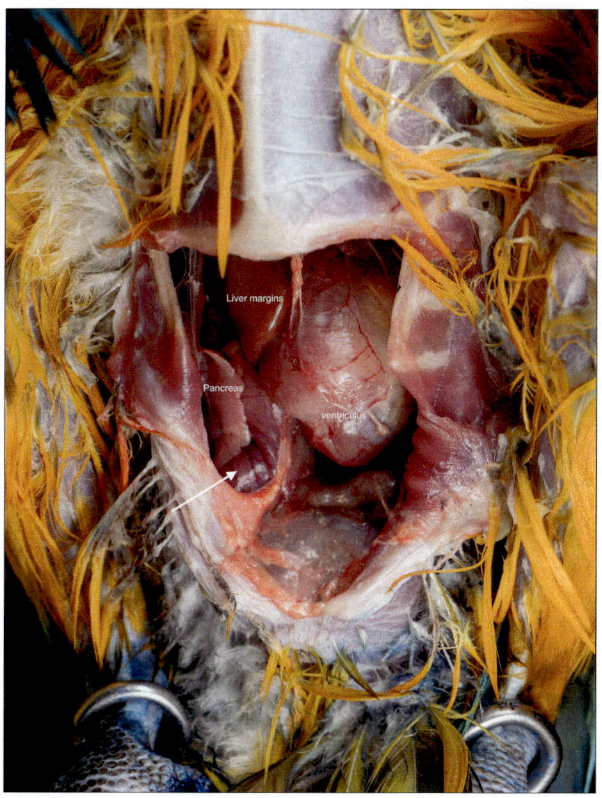

Figure 32.9 **Photo of the typical view inside the abdomen following a ventral midline incision.**

Closure

The muscle and skin are closed in separate layers with absorbable sutures in a continuous or interrupted pattern. No attempt is made to appose the peritoneum.

Postoperative care

Postoperative care is the same as for a left flank coeliotomy (see earlier).

Pancreatic biopsy

Indications

A pancreatic biopsy is used for the diagnosis of pancreatic disease.

Technique

With the bird in dorsal recumbency, a midline coeliotomy is performed as described earlier. The duodenal loop, found on the right side of the coelomic cavity, is gently exteriorised. Care must be taken not to pull too hard on the duodenal loop as this may damage the blood supply to both the duodenum and the splenic lobe of the pancreas.

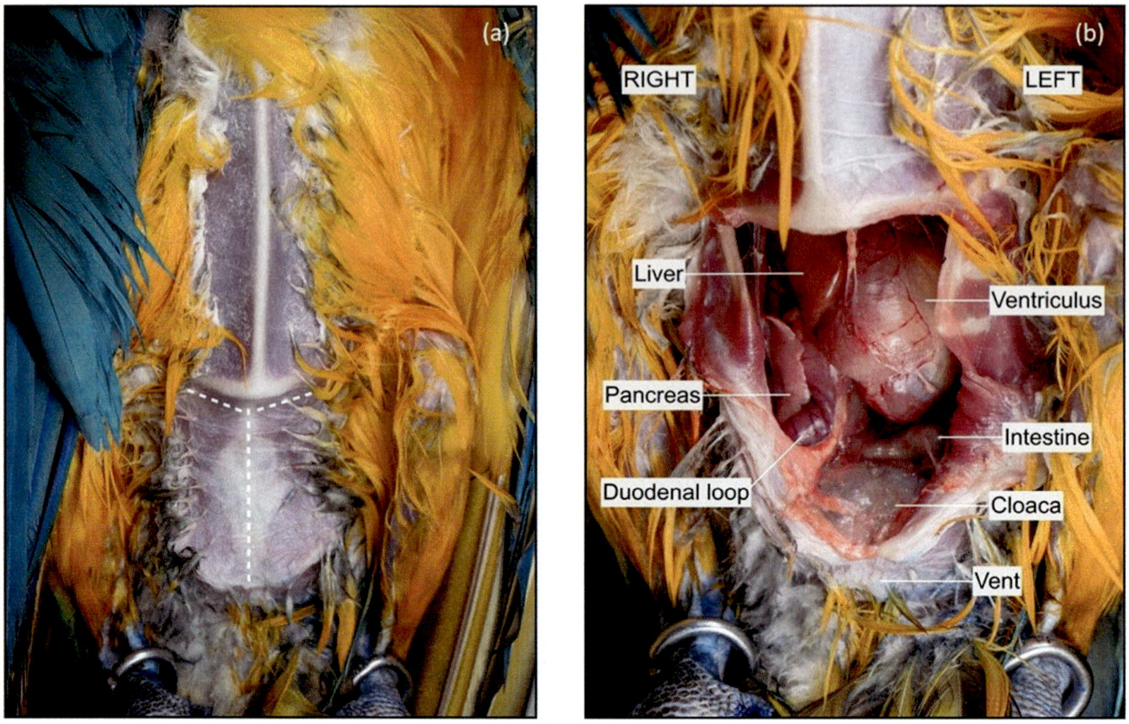

Figure 32.10 **(a) Landmarks of a Y-shaped incision to allow greater exposure. (b) Typical view inside the abdomen following a Y-shaped ventral midline incision.**

The pancreas is examined for gross focal lesions. If seen, they are carefully biopsied. If no focal lesions are visible, the distal edge of the ventral lobe of the pancreas, at the apex of the duodenal loop, is carefully reflected to reveal underlying vasculature. Once these blood vessels are located and avoided, the end of the ventral lobe is removed with iris scissors and fixed in formalin. Minimal bleeding usually results.

The duodenum is then replaced into the coelom, and the skin and muscle are closed as described earlier.

Postoperative care

Postoperative care is as described earlier. Iatrogenic pancreatitis does not appear to be a feature of avian pancreatic biopsies as it is with mammals.

Proventriculotomy

Indications

A proventriculotomy is used for the retrieval of foreign bodies from the proventriculus or ventriculus.

Approach

A left flank coeliotomy is performed as described earlier. The proventricular suspensory ligaments are broken down. Two stay sutures are placed in the tendinous part of the ventriculus, and it is brought up into the surgical field and attached to the skin. If possible, the rest of the coelom is packed off with saline-soaked gauze. The triangular lobe of the liver overlying the proventricular isthmus is identified and gently reflected with a sterile cotton swab.

A stab incision is made into the isthmus and extended with iris scissors. Suction is used to empty fluid from the proventriculus and ventriculus. If necessary, an endoscope can be placed in the incision to ensure all foreign objects have been removed.

Closure

The proventricular incision is closed with 6–0 or 8–0 synthetic monofilament absorbable suture in two layers: the first should be appositional, the second inverting. The liver is tacked down over the incision. The coelom is closed routinely.

Postoperative care

Postoperative care is as described earlier for a left flank approach. Pre-, peri-, and postoperative antibiotic coverage is advisable.

Ventriculotomy

Indications

A ventriculotomy is indicated for the removal of foreign bodies (e.g., metal or cotton fibres) in the ventriculus.

Approach

A ventral midline approach as described earlier is utilised. The ventriculus is brought up to the incision with careful traction, and stay sutures are placed in the muscle of the ventriculus. In birds with a well-muscled ventriculus, a stab incision is made through the muscle aponeurosis, seen as a white fibrous band. In birds with a poorly muscled ventriculus, e.g., lorikeets, a stab incision is made on the caudal wall of the ventriculus. Either incision is extended in a dorsoventral direction to gain access to the ventricular lumen.

Closure

The ventricular wall is closed with closely spaced simple interrupted PDS sutures. The surgical area is carefully flushed or cleaned with moistened Q-tips, and the coelom closed as described earlier.

Postoperative care

Postoperative care is as described for a proventriculotomy.

Coelomic hernia repair

Aetiology

Coelomic herniation is caused by a combination of increased intra-coelomic pressure (fat, ascites, and organomegaly) and weakened muscles due to hormonal influences, obesity, lack of exercise, and chronic malnutrition. It is commonly associated with females during the breeding season, as enlargement of the ovary and oviduct increases intra-coelomic pressure. Although most hernias occur on the ventral midline, they can also be seen on the lateral body wall and dorsal to the vent.

Precautions

If possible, the bird should be converted to a formulated diet before surgery to achieve significant weight loss. Reproductive activity can be reduced by hormonal and behavioural manipulation before surgery. These modifications and treatments may be sufficient to reduce the size of the hernia and avoid surgery.

The veterinarian should be aware that viscera may underlie and be attached to the skin. Replacing viscera back into the coelom may result in 'abdominal compartment syndrome', leading to decreased venous return and hypotension, and may also compress the air sacs, leading to dyspnoea. In this situation, cardiac arrest and death can occur suddenly.

Technique

The bird is positioned according to the location of the hernia. The skin is incised over the hernia, taking care not to cause iatrogenic trauma to the underlying viscera (which may have adhered to the skin). The skin over the hernia is gently and bluntly dissected until the borders of the body wall around the hernia are identified. Any adhesions to these borders are dissected away to free them from the hernia contents.

If possible, a salpingohysterectomy is performed, and as much intra-coelomic fat as possible is removed before closing. This lessens the risk of abdominal compartment syndrome and air sac compression.

Closure

The hernia deficit is closed with a monofilament absorbable suture in a simple interrupted pattern. The anaesthetist should watch closely at this stage for any changes in the bird's respiratory pattern (indicating excessive pressure on the air sacs) or slowing of the heart rate (indicating hypotension).

Very large hernias, or those that cannot be closed without cardiorespiratory compromise, may require

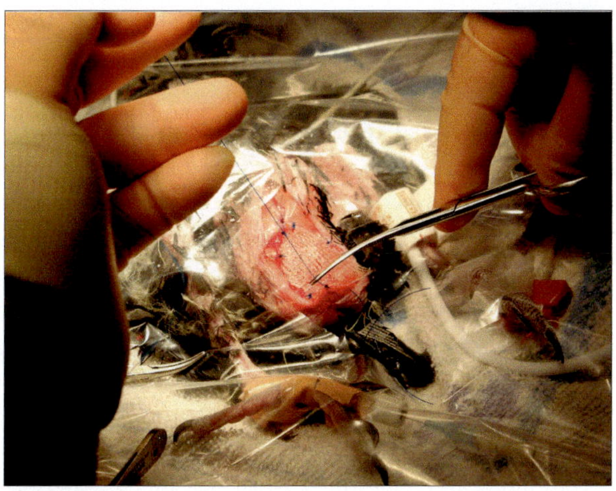

Figure 32.11 **Black cockatoo chick undergoing a mesh repair of a coelomic hernia.**

the use of non-absorbable mesh inside the body wall, attached to the muscles, last ribs, sternum, and pubis (**Fig. 32.11**).

The skin is closed routinely.

Postoperative care

Postoperative care is as for other coelomic surgeries. If mesh has been used, antibiotic coverage for 10–14 days should be considered. The owner should be advised that re-herniation, either at the surgery site or elsewhere, is possible if the bird's weight and reproductive activity are not addressed and controlled.

Cloacopexy

Indication

A cloacopexy is indicated for chronic cloacal prolapse, associated with hypersexuality in male cockatoos or excessive straining due to:

- Intestinal parasites
- Retained egg
- Adenomatous polyp
- Bacterial enteritis
- Neoplasia
- Coelomic mass
- Cloacal hyperplasia

Technique

A ventral midline incision is made as described earlier, although positioned more caudally. The prolapsed cloaca is replaced and an assistant places a gloved finger (large birds) or Q-tip (small birds) into the cloaca to define its extent and to lift it to the coelomic wall. Intra-coelomic fat ventral to the cloaca is removed.

Using monofilament non-absorbable material, sutures are placed through the full thickness of the cloacal wall and then around the last rib (at the junction of the sternal and vertebral portions) on each side or through the cartilaginous border of the sternum. All sutures are preplaced and tied once all are in position, anchoring the cloaca in a reduced position with the cloacal wall apposed to ribs, albeit more cranial than normal.

Following the placement of these sutures, the ventral cloacal serosa is incised to the level of the submucosa and tacking sutures are placed through the coelomic wall to suture the submucosa to the coelomic muscle. Some of these sutures are incorporated into the coelomic wall closure.

Postoperative care

Postoperative care is as for other coelomic surgeries.

Prognosis

A cloacopexy attaches the ventral aspect of the cloaca to the ribs and body wall. However, the dorsal cloaca is not dealt with. Therefore, if the bird continues to strain, this tissue can still prolapse. As time goes on, with continued straining, the sutures may break or cut through the attached tissues, and prolapse of all the cloacal tissue can occur.

Cloacal prolapse repair

Cloacal prolapse is not uncommon and is associated with hypersexuality in male cockatoos or excessive straining due to reasons outlined in the list in the Cloacopexy section.

Most commonly, the prolapse may only involve the cloaca itself, but prolapses of the oviduct or rectum can occur. See Chapter 19.

Surgical treatments for cloacal prolapse include:

- Cloacoplasty/ventoplasty – temporarily or permanently narrowing the vent.
- Cloacopexy – suturing the cloaca to the sternum and ribs.
- Colopexy – suturing the distal colon to the body wall.

Cloacoplasty/ventoplasty

Indications

A cloacoplasty/ventoplasty procedure provides temporary or permanent narrowing of the vent opening as treatment for cloacal prolapse (**Fig. 32.12**) or atony. A purse-string suture is contraindicated in birds as the vent is a lateral slit, not a circle.

Technique

If a permanent narrowing is required, the mucocutaneous border on the lateral third of the dorsal and ventral vent lips is trimmed on both sides, leaving the middle third untouched. If a temporary closure is required, this step is not indicated. The lateral third of the vent is closed on both sides with one to two vertical mattress sutures (see **Fig. 19.16** in

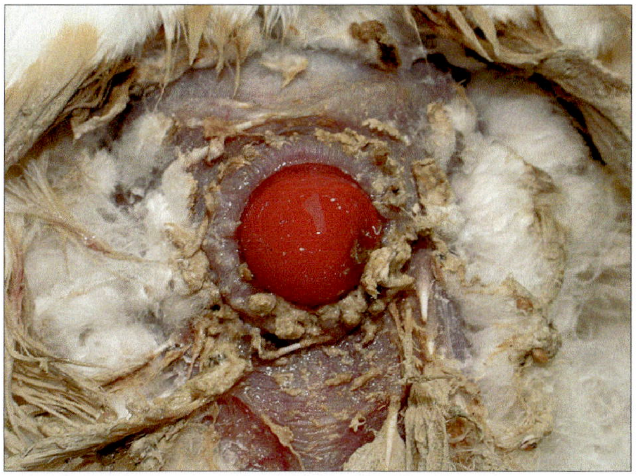

Figure 32.12 **Cockatoo with a cloacal prolapse.**

Chapter 19), leaving the middle third open for faeces and urine to be expelled.

Recurrence of cloacal prolapse may occur in traditional symmetrical cloacoplasty techniques (temporary or permanent) as the causative straining pressure remains. An alternative procedure in these cases is to perform a permanent asymmetrical cloacoplasty, creating a midline physical barrier (unlike symmetrical cloacoplasties described earlier). A semicircular excision at the mucocutaneous junction on the right side of the vent is made. (The right is closed as the rectum enters the left side of the cloaca.) This is then apposed with a two-layer closure (mucosal and skin edges) to reduce the cloacal opening by 50–75%. This two-layer closure prevents the inward curling of the skin edges, which can result in incomplete fusion of the vent lips. A tension-relieving suture is then applied by placing a mattress suture across the vent (but not through the cloacal sphincter) at the medial edge of the incision. This suture is removed after 2 weeks.

Postoperative care

The bird should be carefully observed to ensure it is able to defecate normally. If a temporary procedure has been performed, the sutures can be removed after 2–4 weeks. Care must be taken with egg-laying hens to ensure that either their reproductive drive is switched off (behaviour modification and hormonal therapy; see Chapter 27) or that the bird is monitored carefully for dystocia.

Colopexy

Indication

A colopexy is indicated for reducing cloacal prolapse when other techniques have failed.

Technique

A U-shaped incision, starting proximally at the level of the caudal ribs on both sides and extending to the level of the pubic bones just proximal to the vent, is used to gain access to the coelomic cavity. A moistened cotton swab or gloved finger is inserted into the cloaca to reduce the prolapse and restore the normal anatomy of the colon and cloaca.

An incisional colopexy is then performed by incising the seromuscular layer of the colon longitudinally at the antimesenteric side with a small scalpel blade, taking care not to enter the lumen of the colon. A corresponding incision is made in a craniocaudal direction in the opposing left body wall, incising the transverse abdominal muscle. This incision is made just cranial to the one in the colon to maintain slight traction on the colon and cloaca after fixation.

The dorsal and ventral edges of the seromuscular incision are sutured to the corresponding edges of the abdominal wall incision using 5-0 polypropylene in two simple continuous sutures. The coelomic incision is closed in two layers, and a temporary ventoplasty is performed (see earlier).

Postoperative care

Analgesia and antibiotic coverage are provided for 5–7 days. A faecal softener such as lactulose is given for 4 weeks. The temporary ventoplasty sutures are removed after 5 days.

Cloacotomy

Indications

A cloacotomy is used for debriding cloacal papillomas or removing cloacoliths.

Technique

The bird is placed in dorsal recumbency. A moistened cotton bud is inserted into the cloaca to outline the structure. An incision is made through the skin, cloacal sphincter, and cloacal mucosa from the vent to the cranial end of the cotton bud. This opens the ventral wall of the cloaca, exposing the coprodeum, urodeum, and proctodeum.

Closure

The mucosa is closed with a continuous suture pattern, the sphincter is closed with a single mattress suture, and the skin is closed routinely.

Postoperative care

A topical antibiotic ointment such as silver sulphadiazine can be introduced into the cloaca twice daily to soothe the cloacal wall and provide local antibiotic coverage.

Salpingohysterectomy

Indications

A salpingohysterectomy is indicated for chronic egg laying and any oviductal disease that cannot be managed medically (e.g., pyometra, yolk peritonitis, or coelomic herniation).

Presurgical evaluation and conditioning

Yolk peritonitis and underlying liver, lung, and kidney disease all carry a higher risk of surgical complications. An enlarged oviduct (due to hormonal influences) fills the left coelom and makes surgery difficult. Therefore, if time permits, the patient's nutritional status should be improved and the reproductive cycle 'turned off' through behavioural, social, and environmental modification and hormonal therapy (see Chapter 27).

Technique

The left lateral approach gives the best exposure, but the ventral midline approach can also be used depending on the surgeon's preferences. The ovary is left alone; the blood supply is tightly adherent to major blood vessels (the cranial renal artery and the caudal vena cava), making removal without an operating microscope extremely dangerous.

The infundibulum is identified and gently retracted out through the incision. A large blood vessel running between the infundibulum and ovary must be identified and ligated, and the relatively avascular ventral suspensory ligament is broken down. The more vascular dorsal suspensory ligament is broken down by sharp dissection, and the blood vessels (the cranial, middle, and caudal oviductal arteries) are cauterised or ligated as needed.

The oviduct is retracted through the incision, and the junction with the cloaca is identified (if necessary, a cotton bud or gloved finger is inserted into the cloaca to delineate the structure). Two vascular clips or sutures are placed across the oviduct near this junction, and the oviduct is removed. Closure is as previously described.

Postoperative care

Postoperative haemorrhage from ovarian or oviductal vessels is a major and common complication with salpingohysterectomy. Often these vessels are not identified at the time of surgery but bleed once the patient starts to move around. Weakness or obvious blood loss should be investigated with this in mind. Re-entering the coelom, identifying, and ligating the bleeding vessels is the most effective way of remedying this problem. The surgeon should be prepared to give the patient a blood transfusion.

If there has been ascites or coelomic bleeding, it is not uncommon for the patient to be mildly dyspnoeic and/or coughing up blood/fluid for a few days postoperatively. (This is due to fluids entering the air sacs and then the lungs.) This needs to be monitored carefully and, if necessary, supportive measures provided (such as an oxygen cage). In severe cases, it may be necessary to re-explore the coelom surgically.

If there are no complications, postoperative care is like that used for a coeliotomy (see earlier).

Follow-up

Continued ovulation and yolk release with subsequent yolk-related peritonitis have been reported in several birds following a salpingohysterectomy. This may be due to a persistent right oviduct or failure to remove larger follicles during surgery. Follow-up surgery may be required, although in many cases hormonal manipulation and anti-inflammatory medications may control this issue.

Orchidectomy

Indications

An orchidectomy may be carried out to treat orchitis or testicular neoplasia, or for behavioural modification.

Technique

The approach is via either a left lateral or bilateral coeliotomy or a ventral midline coeliotomy with cranial flaps. The testis is gently retracted ventrally, and the mesorchium over the testicle is carefully incised to allow it to retract and expose the blood supply to the testicle. Vascular clips are placed across the dorsal blood supply, and the testicle is excised, leaving the vascular clip in place. A radiosurgery unit is useful for this part of the procedure, cauterising small blood vessels that may have been missed in the ligation. The coelomic incision is closed as previously described.

Postoperative care

Postoperative care is as for other coelomic surgeries.

Follow-up

If all testicular tissue is not removed, regrowth is common. Some male characteristics are retained, indicating that testosterone may be produced in other tissues.

Enucleation

Indications

Enucleation is indicated for severe, irreversible panophthalmitis, perforating corneal ulceration, or neoplasia.

Precautions

Enucleation is more difficult in birds than in mammals because of the relatively larger globe compared with the size of the orbit. The optic nerve is short; excessive traction can result in contralateral blindness. The area is very vascular, and the surgeon should anticipate haemorrhage. An alternative technique is evisceration of the globe.

Enucleation technique

The lids are sutured together in a simple continuous pattern. A circumferential incision is made through the skin (not the conjunctiva) 1–2 mm from the lid margins. Note that the ligamentous attachments at the medial canthus are firm. Haemorrhage can be expected in this area and at the lateral canthus.

Dissection is carried out between the palpebral conjunctiva and the bony orbit, as it is not feasible to identify and transect each individual muscle. The sutured eyelids are manipulated to provide traction on the globe.

In some birds with a large globe, it may be necessary to collapse the globe prior to enucleation. In some species with thin orbital bones, it is feasible to extend the skin incision from the lateral canthus to the auditory meatus and transect the orbital rim at the lateral canthus. This opens the orbit and allows easier retraction of the globe.

Where feasible, a vascular clip is applied blindly to the optic stalk (incorporating the nerve and blood vessels). Care must be taken to apply minimal traction

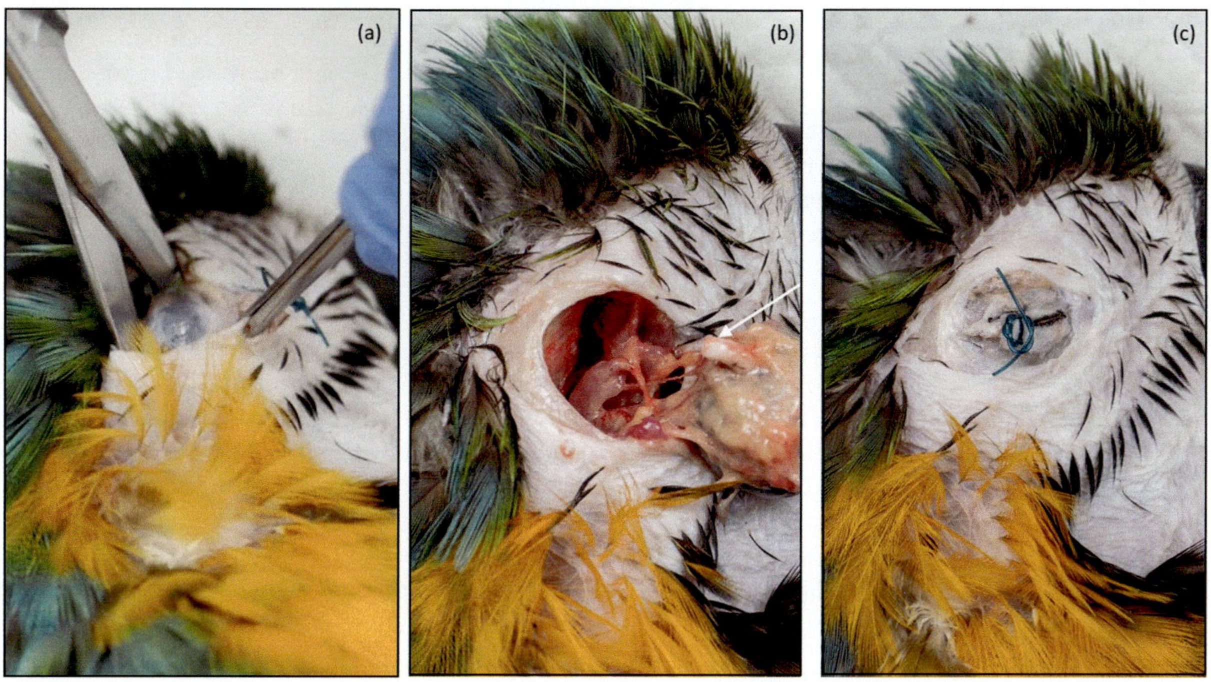

Figure 32.13 **(a) Dissection between the palpebral conjunctiva and the bony orbit. (b) The optic stalk (incorporating the nerve and blood vessels) as indicated by the arrow. This should be ligated where feasible. (c) The eyelid margins are excised, taking care not to remove or damage the superior and inferior lacrimal puncta located near the medial canthus, and then sutured in a simple interrupted pattern.**

to the globe at this point to avoid damage to the contralateral optic nerve. After the stalk is clipped, it is transected, and the eye removed. If it is not feasible to place a vascular clip, the globe is retracted with the use of sharp dissection. Haemorrhage can be expected at this stage and can usually be controlled by placing a vascular clip directly onto the now visible optic stalk. If this is not feasible, packing the orbit with absorbable gelatin sponges (e.g., Gelfoam®) is usually effective (**Fig. 32.13**).

Evisceration technique

A limbal stab incision is made in the cornea with a small scalpel blade. This incision is then extended, using tenotomy scissors, to completely remove the cornea. The iris, lens, vitreous body, retina, and uvea are then removed using gentle blunt dissection with a lens loop or sterile cotton-tipped applicators. Bleeding is usually not severe and can be controlled with gentle direct pressure by using eye spears (Eagle Vision Inc., Memphis, Tennessee, USA) or sterile cotton-tipped applicators. Complete removal of the choroidal tissue allows visualisation of the underlying, white-coloured sclera.

Once any bleeding has stopped, the intraocular space is lightly packed with absorbable gelatin sponge (e.g., Gelfoam®) to control any further haemorrhage. The third eyelid and conjunctival tissue are left in place.

Closure

The eyelid margins are excised, taking care not to remove or damage the superior and inferior lacrimal puncta located near the medial canthus, and then sutured in a simple interrupted pattern.

Postoperative care

Postoperative bleeding, epistaxis, and sneezing are common and usually self-limiting. Excessive bleeding (indicated by marked distension of the skin over the orbit) may indicate the need for surgical exploration to locate the source of the bleeding.

It is not unusual for glandular material to be inadvertently left in the orbit and for a mucoid discharge to persist after surgery. In these cases, surgical exploration of the orbit may be required to locate and remove this tissue.

Repairing skin lacerations

Skin lacerations are not uncommon in birds, with injuries inflicted by predator attacks, sharp objects, or even other birds. Some of these injuries are simple lacerations; others are more complex and associated with extensive soft tissue and even bone damage.

Avian skin is closely attached to underlying muscle and bone. Loose skin in most birds can be found on the neck and the inguinal area, leaving most of the skin on the torso, head, and limbs relatively immobile. This makes undermining and mobilising skin to close wounds more difficult than similar surgery in many mammalian species.

The skin is thin and often has a fatty subcutaneous layer. Subcutaneous suturing does not hold well and is therefore of limited value. Postoperative swelling is not as severe in birds as in mammals, and sutures can therefore be tied tighter and closer together.

Avian heterophils lack lysozymes, meaning that avian pus is usually caseated and therefore does not drain from a contaminated wound readily. The use of drains in these wounds may not be of significant benefit.

Skin laceration repairs in birds, therefore, require that the surgeon:

- Thoroughly debrides and flushes the wound. In most cases, the use of sterile saline is preferable as a flushing solution. Chlorhexidine and povidone iodine, even diluted, can be toxic to granulation tissue and should be avoided if possible.
- Mobilises available skin where possible to achieve primary closure without tension on the wound edges.
- Avoids the use of drains unless combined with a flushing or irrigation system. If necessary, dead space should be obliterated surgically.

Where primary closure cannot be achieved, either skin grafts and flaps, or healing by second intention is required. Using the same surgical techniques employed in mammals such as dogs and horses, skin grafting and flaps are feasible alternatives in birds. If skin grafting is not employed, good healing can often be achieved by secondary intention. In this situation, avoid the underlying subcutis and muscle drying out by the use of wet or hydrocolloid dressings (e.g., Duoderm®).

Aggressive postoperative antibiosis and analgesia will significantly improve morbidity and mortality rates in these patients.

FURTHER READING

Altman RB (1997) General surgical considerations. In: *Avian Medicine and Surgery.* RB Altman, SL Clubb, GM Dorrestein, K Quesenberry (eds). WB Saunders, Philadelphia, pp. 691–703.

Altman RB (1997) Soft tissue surgical procedures. In: *Avian Medicine and Surgery.* RB Altman, SL Clubb, GM Dorrestein, K Quesenberry (eds). WB Saunders, Philadelphia, pp. 704–732.

Altman RB (1997) Radiosurgery (Electrosurgery). In: *Avian Medicine and Surgery.* RB Altman, SL Clubb, GM Dorrestein, K Quesenberry (eds). WB Saunders, Philadelphia, pp. 767–772.

Bennett RA, Kuzma AB (1992) Fracture management in birds. *Journal of Zoo and Wildlife Medicine* 23(1): 5–38.

Bennett RA (1997) Orthopedic surgery. In: *Avian Medicine and Surgery.* RB Altman, SL Clubb, GM Dorrestein, K Quesenberry (eds). WB Saunders, Philadelphia, pp. 733–766.

Bennett RA, Harrison GJ (1994) Soft tissue surgery. In: *Avian Medicine: Principles and Application.* BW Ritchie, GJ Harrison, LR Harrison (eds). Wingers Publishing, Lake Worth, pp. 1096–1136.

Bowles HL, Odberg E, Harrison GJ, Kottwitz JJ (2006) Surgical resolution of soft tissue disorders. In: *Clinical Avian Medicine, Vol 2.* GJ Harrison, TL Lightfoot (eds). Spix Publishing Inc, Palm Beach, pp. 775–830.

Helmer P, Redig PT (2006) Surgical resolution of orthopedic disorders. In: *Clinical Avian Medicine, Vol 2.* GJ Harrison, TL Lightfoot (eds). Spix Publishing Inc, Palm Beach, pp. 761–774.

Hernandez-Divers SJ (2005) Minimally invasive endoscopic surgery of birds. *Journal of Avian Medicine and Surgery* 19(2): 107–120.

Martin HD, Ritchie BW (1994) Orthopedic surgical techniques. In: *Avian Medicine: Principles and Application.* BW Ritchie, GJ Harrison, LR Harrison (eds). Wingers Publishing, Lake Worth, pp. 1137–1170.

Murray M, Pizzirani S, Tseng F (2013) A technique for evisceration as an alternative to enucleation in birds of prey: 19 cases. *Journal of Avian Medicine and Surgery* 27(2): 120–127. https://doi.org/10.1647/2012-007.

Orosz SE, Ensley PK, Haynes CJ (1992) *Avian Surgical Anatomy: Thoracic and Pelvic Limbs.* WB Saunders, Philadelphia.

Van Zeeland YRA, Schoemaker NJ, van Sluijs FJ (2014) Incisional colopexy for treatment of chronic, recurrent colocloacal prolapse in a Sulphur-Crested Cockatoo (Cacatua galerita). *Veterinary Surgery* 43(7): 882–887. https://doi.org/10.1111/j.1532-950X.2014.12251.x.

Zaheer O, Beaufrère H, Jajou S, Gardhouse S (2020) Asymmetrical Cloacoplasty for the treatment of chronic cloacal prolapse in Psittaciformes: A case series. *Journal of Avian Medicine and Surgery* 34(2): 172–180.

CHAPTER 33

ORTHOPAEDIC SURGERY

FRACTURES are not uncommon in practice. Companion and aviary birds usually present soon after accidental injury (flying into aviary wire, window strikes, ceiling fan injuries, etc.). Wild birds, on the other hand, often present days (or even weeks) after the initial injury. Successful fracture management in the avian patient requires an understanding of orthopaedic principles, an appreciation of the unique characteristics of birds and their bones, and a clear picture of the desired long-term outcome.

FORCES ON FRACTURES

A fracture is a discontinuity in a bone resulting from mechanical forces that exceed the bone's ability to withstand them. This inability to withstand these forces can be due to:

- A normal bone subjected to acute overwhelming force, usually in the setting of trauma.
- Weakened bone from a focal lesion, e.g., neoplasia, osteomyelitis, or bone cyst (pathological fractures).
- Weakened bone due to metabolic abnormalities, such as nutritional secondary hyperparathyroidism or chronic/excessive egg laying.

Successful fracture repair requires a good understanding of bone anatomy and biomechanics, including the forces that not only contribute to the fracture but may also affect its healing.

This understanding leads to the following considerations:

- What forces are normally exerted on the bone under normal conditions/use?
- Can the fracture be anatomically reconstructed for load sharing?
- What fracture forces need to be countered?
- What implant(s) will be required to counter the forces?

> **MAJOR FRACTURE FORCES (FIG. 33.1)**
> Compression – the shortening of the bone.
> Tension – lengthening of the bone.
> Shear – as the fracture occurs, the bone shears. One end of the bone moves in one direction, while the other end moves in the opposite direction.
> Torsion – the bone twists around its long axis.
> Bending – creates a convex side to the bone (under tension) and a concave side (under compression).

BONE HEALING

Healing of bones is a combination of:

- Primary healing. Bone-to-bone healing through the Haversian system with minimal callus formation. This is only achieved with rigid fixation and perfect bone apposition.
- Secondary healing. This relies on the formation of a callus between bone fragments. This callus can be either endosteal (when the bones are well aligned) or periosteal callus (usually when the fracture is not aligned and there is movement at the fracture site).

Obviously, primary healing is the desired goal of fracture repair but is only occasionally achieved. Movement of the limb, combined with the strong pull of muscles (especially in the wing) and the thin cortices of avian bone, makes rigid fixation difficult to achieve. However, even if primary healing is not attained, endosteal healing makes for a solid fracture repair that can be very functional.

Immediately after the fracture occurs, a haematoma forms in and around the fracture site. Over a few days, multipotential stem cells invade, and a

DOI: 10.1201/9781003272519-33

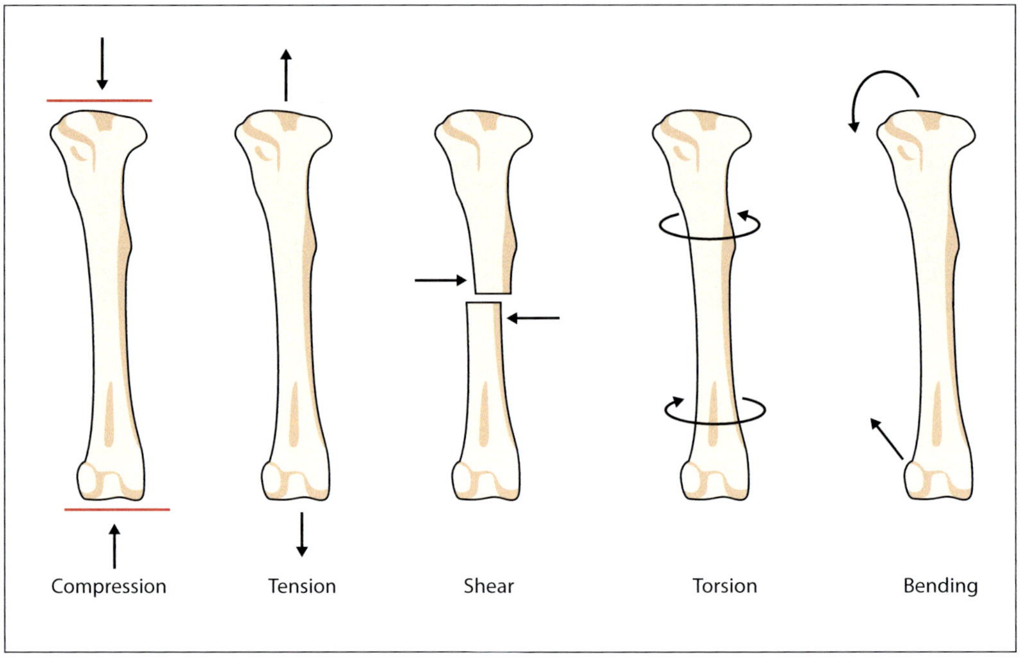

Figure 33.1 **Major fracture forces on bone.**

fibrous/fibrocartilaginous callus develops. If the fracture is relatively stable and has a good blood supply, osteoblasts lay down bone to form a bony callus. This is then remodeled over the next 12 months.

The rate of healing is dependent on:

- Displacement of bone fragments. Segmental fractures will heal well as long as the periosteal blood supply is intact. If devitalised, the fragment can be incorporated into the fracture site as a cortical bone graft. Healing is slower, as cancellous bone first bridges the gap, then the segment is demineralised and becomes cancellous itself. Healing may take 9–18 weeks.
- The blood supply to the bone. As mentioned earlier, the blood supply to bones arises from the periosteal, medullary, metaphyseal, and epiphyseal vessels. The periosteal blood supply is the predominant source of blood to the pneumatic bones (e.g., humerus and femur), while non-pneumatic bones (e.g., the radius, ulna, and tibiotarsus) may be more reliant on medullary or metaphyseal and epiphyseal blood vessels. The disruption of a bone's blood supply, either by trauma or surgical repair of a fracture, may result in delayed healing or even complete failure to heal (a non-union). The presence of medullary bone (osteopetrosis or hyperostotic polyostosis) in non-pneumatic bones means that their blood supply will be compromised even more by trauma, and, in the author's experience, fractures of these bones in hyperoestrogenic hens carry a guarded prognosis.
- Presence of infection. Osteomyelitis may result in a sequestrum of dead bone. Sequestra can add to the stability of a fracture and should not be removed until a bony callus has formed unless they are acting as a nidus of infection.
- Movement at the fracture site creates a large haematoma and a large cartilaginous bridge. This can delay or even prevent healing.

Stable, well-aligned fractures heal faster in birds than in mammals. Clinical stability of a fracture (often achievable through a firm fibrous callus at 2–3 weeks) may precede radiographic evidence that the bone is healed (4–8 weeks).

GENERAL CONSIDERATIONS FOR FRACTURE REPAIR

The aim of fracture repair is to aid healing by promoting the formation of an intact, strong bone, with no involvement of the surrounding structures. This aim is complicated by a number of factors:

- *Bones.* Avian bones are lightweight but possess great aerodynamic strength. They have thin, brittle cortices, which do not provide sufficient holding power for bone screws. Some, but not all, are pneumatic. Fractures are frequently open and comminuted due to minimal soft tissue coverage.

- *Blood supply to the bones*. The blood supply to bones arises from periosteal, medullary, metaphyseal, and epiphyseal blood vessels. The periosteal blood supply is especially important in callus formation, particularly in pneumatic bones such as the humerus and femur, and its importance may exceed that of the medullary blood supply in those bones. Exteriorised bone is often nonviable as it is frequently separated from its blood supply.
- *Joints*. The presence of significant joint involvement, such as luxations or intra-articular fractures, will influence prognosis and may be significant considerations in the decision to pursue treatment, especially for wild birds.
- If joints are immobilised for long periods of time, contracture of ligaments and tendons can result in a permanently reduced range of movement. Fracture calluses may impinge on joint range of motion, as may adhesions of ligaments and tendons.
- *Muscles*. Powerful flight muscles can cause rotational deformity of the humerus during the early healing phase.
- *Soft tissue cover*. The lack of deep soft tissue cover over most of the bones means that open fractures are more likely to occur more frequently, often with traumatic compromise of the nerve and blood supply.
- *Callus formation*. Callus formation in birds is rapid; a fibrous callus can form within a week. The amount of callus that forms during healing is proportional to the degree of instability at the fracture site – the more unstable a fracture is, the greater the size of the callus. Proliferative callus formation during healing can entrap nerves and tendons, leading to neuralgia and/or paresis below the fracture. Synostosis, a proliferative callus fusing two bones, is not uncommon in antebrachial fractures and can compromise the bird's flying abilities.

Assessment for fracture repair must take into consideration the future function of the bird. For instance, companion and aviary parrots do not need the capabilities of a wild falcon intended for release after treatment. When full restoration of function is not possible for whatever reason, it may be appropriate to set a goal of returning a patient to a comfortable and pain-free life. It must be possible to return wild birds to their normal habitat with a good chance of survival, and nothing less than perfection will give them that chance.

FRACTURE CLASSIFICATION

After careful examination of a fracture, including radiographic assessment, the fracture is classified to allow the surgeon to make decisions about the most appropriate treatment technique. The use of a standardised system will:

- Identify the mechanical forces acting on the fracture.
- Evaluate and compare techniques used to repair similar fractures.
- Determine a suitable fracture repair method.
- Predict patient prognosis.
- Enable effective communication of the case with colleagues.

There are a number of classification schemes in use, each with its own proponents. Regardless of the system used, the classification of fractures is performed in several steps or layers:

1. Location – which bone is fractured, what side of the body is affected, and the anatomical location (epiphyseal, metaphyseal, proximal, mid, or distal diaphyseal).
2. The cause of the injury – trauma, persistent abnormal strain on the bone, pathological causes (e.g., neoplasia), or malnutrition.
3. Presence or absence of an open wound.
4. Extent of the damage – fractures can be complete (through both cortices), incomplete (through only one cortex), comminuted (multiple fragments), segmental (two or more separate fractures), or fissured (a crack extending from a surface into, but not through, a long bone) (**Fig. 33.2**).
5. The direction, location, and number of fracture lines (**Fig. 33.3**).
    a. Transverse – perpendicular to the axis of the bone and is the diameter of the bone.
    b. Short oblique – the fracture is <2× the diameter of the bone.

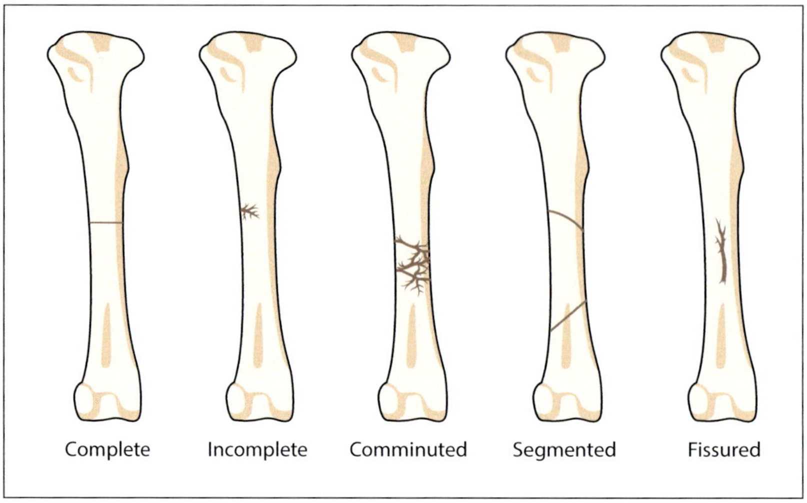

Figure 33.2 **Extent of damage.**

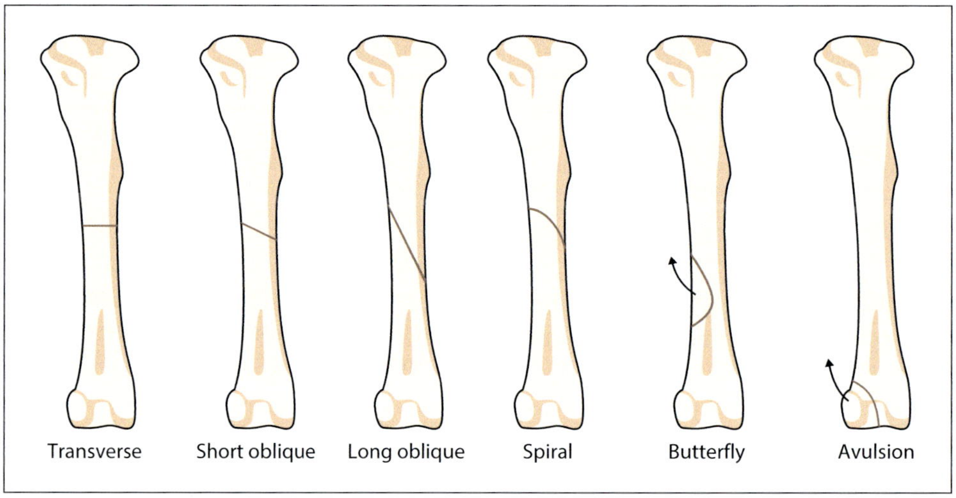

Figure 33.3 **Fracture lines.**

   c. Long oblique – fracture is >2× the diameter of the bone.
   d. Spiral – an oblique fracture with a twist.
   e. Butterfly – two oblique fracture lines meeting to create a large triangular or wedge-shaped fragment located between the proximal and distal fracture fragments, resembling a butterfly.
   f. Avulsion – fracture at the attachment of a joint capsule, or at the origin/insertion of a tendon or ligament.
6. Displacement of the distal fragment in relation to the proximal fragment.
   a. Minimal displacement – close relationship between the fragments.
   b. Displaced – distracted or overriding the ends of the fragment.
7. Involvement of articular surfaces.

PRINCIPLES OF FRACTURE REPAIR

Keeping in mind the issues discussed previously, the surgeon needs to consider:

1. The repair technique should promote a functional union of the fragments. This requires:
   a. Minimal soft tissue damage to preserve the periosteal blood supply.
   b. Accurate alignment of the bone to maintain the length of the bone, and rotational and angular orientation.
   c. Rigid stabilisation of the fracture while it heals.
2. Early return to normal function is needed to prevent 'fracture disease' (permanent contracture of muscles, tendons, ligaments, and joints). Load bearing must be shared equally during

healing and, in leg fractures, early return to function is needed to prevent pododermatitis in the opposite leg.

3. Hardware used should be biocompatible, lightweight, effective, adjustable, and associated with minimal complications.

DEALING WITH OPEN FRACTURES

Because of the thin sleeve of soft tissue overlying the bones, many fractures are open on initial presentation. These fractures may be:

- Grade 1 – The bone has come out from the skin but has gone back in.
- Grade 2 – The bone has come out from the skin and remains out.
- Grade 3 – The bone has come out from the skin, stayed out, and there is excessive soft tissue damage.

If the fracture is a grade 1 or 2, the protruding bone is usually desiccated and dead. Almost invariably, the fracture has become infected prior to the initial examination. The exposed medullary cavity is a portal of entry for bacteria and occasionally fungi or foreign bodies.

If the protruding bone end is still fresh, it must be protected from desiccation and contamination as soon as possible. This can be done by covering the wound with a sterile, water-based, water-soluble lubricant (e.g., K-Y® Jelly) or saline-moistened gauze as the first bandage layer (no direct contact between dry gauze and bone).

Necrotic or compromised bone must be removed with rongeurs. The medullary cavity must be picked clean if possible; do not flush the cavity, as lavage usually washes material deeper into the cavity where it is trapped by the internal struts or enters the respiratory tract via pneumatised bones.

Dead tissue surrounding the fracture site must be carefully debrided and irrigated with normal saline (taking precautions to prevent the flush from entering the medullary canal). Debridement of wounds should be performed without excessive manipulation to preserve soft tissues and avoid compromising circulation to bone fragments. Purulent material in birds is caseous and will not drain; the provision of drainage in cases where pus is present is therefore ineffective.

If the bone is known to be infected, the fracture should be repaired surgically; if some of the cortex must be removed, there is still support for the weakened bone whilst healing occurs. Multiple caseous masses are usually impossible to remove and the prognosis in these cases is very poor.

Repaired compound fractures should be radiographed weekly during the healing process, and purulent material should be removed surgically when it forms.

TYPES OF FRACTURE REPAIR

There are three options for the management of fractures: confinement/cage rest, external coaptation, and surgical fixation. The selection of the technique used is a function of:

- Patient factors – size, weight, species, activity level, and the requirement for full return of function.
- Client factors – finances, compliance, husbandry.
- Fracture factors – classification, location, forces, contamination.
- Veterinary factors – the skill and experience of the clinician and the equipment available.

The goal is to select the appropriate fixation method for each individual fracture based on these described factors. The decision whether to use a surgical or non-surgical fracture management technique in any given case is based on the nature of the fracture, the ability to meet orthopaedic objectives and goals, and the resources and expertise available. Overall outcomes will be improved if normal joint movement, range of motion, and bird mobility can be maintained during the healing process, goals often best achieved with surgical fixation.

Confinement or cage rest

Some fractures, usually of the wing, may best be treated without fixation. This reduces the potential for joint contractures.

- Young birds with greenstick fractures of wing or leg bones, single fractures of the radius or ulna, or fractures of the digits are often treatable in this manner. Young parrots can stay in their 'buckets'.
- Adult parrots can be confined in a wire mesh cage. If possible, the cage should have movable sides that are initially positioned sufficiently close so that the bird keeps its wings folded. As the fracture heals, the sides are moved apart so that the wing may be moved for preening, and

finally allowed to stretch to full length but not bear weight.

External coaptation

External coaptation (splints or bandages) may be used with or without surgical fixation. Although coaptation is most commonly used for initial or temporary presurgical stabilisation, it may be the definitive management for some fractures such as shoulder girdle fractures, non-displaced metacarpal fractures, and tibiotarsal or metatarsal fractures of small birds. Coaptation may also be used postsurgically to provide protection and/or additional support.

Although initially cheaper than surgical fixation and requiring less skill or experience, there are downsides to its use as primary stabilisation for many long-bone fractures in birds. In general, functional reduction and stabilisation are poorer with coaptation, resulting in poorer expected outcomes. Benefits of external coaptation include:

- Decreased chance of infection.
- Less damage to regional vascularisation.
- Minimal surgical and anaesthetic risk in high-risk (e.g., trauma) patients.
- Can be used in patients too small to be feasibly operated on.
- It is inexpensive and quick to apply.

However, the disadvantages of external coaptation include:

- 'Fracture disease' is common (restricted joint motion and soft tissue contraction resulting in joint ankylosis).
- Healed bone may be misaligned (shortened or rotated).
- The lack of rigid immobilisation may result in larger calluses, slower healing, usually by periosteal callus, and delayed or non-union is a major complication.
- It may allow the callus to incorporate other moving tissues, e.g., flexor tendons in the tarsometatarsal groove, or synostosis formation between the radius and ulna.
- Pressure necrosis can develop if the bandages are applied too tightly.
- Slippage of the bandage can result in the fracture becoming a fulcrum point, leading to movement at the fracture site and delayed or non-union.

The selection of the type of coaptation is dependent on these advantages and disadvantages, as well as the location of the fracture. Techniques include:

Figure-of-8. The figure-of-8 bandage (**Fig. 33.4**) is used to cover wounds on the antebrachium/manus and to stabilise fractures of the wing distal to the elbow. It should never be used for humeral fractures. A padded splint can be added over the fracture site to provide additional support.

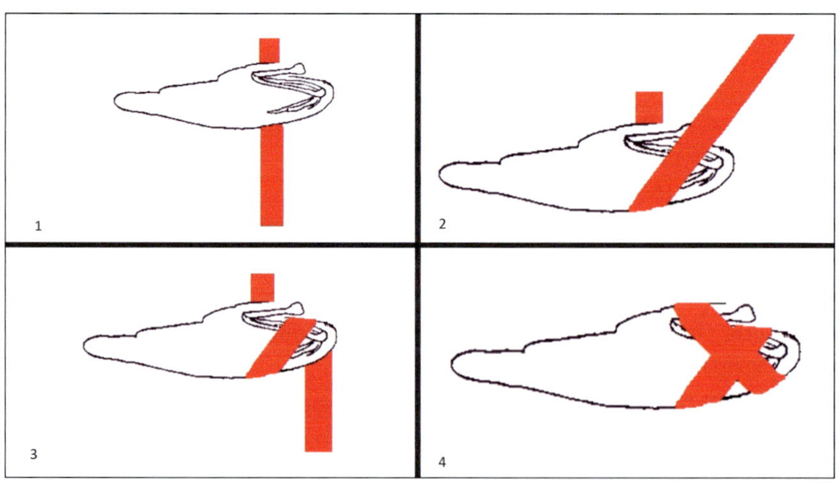

Figure 33.4 **Figure-of-8 bandage for antebrachial and metacarpal fractures. Lie the bird in lateral recumbency with the affected wing facing up and flexed. (1) Begin by placing the bandage UNDER the wing behind the humerus, with the short end at the top. The elbow must be outside of the bandage – the leading edge of the bandage must be caudal to the carpus. Be careful not to bunch the bandage up in the axilla – use narrow bandages rather than wide. (2) Bring the long end back over the top and fold it forwards over the antebrachium and then down under the wing. The cranial edge of the bandage must be caudal to the elbow joint. (3) Bring the long end of the bandage over the top of the wing, folding it backwards to meet the short end of the bandage. (4) If necessary, this can be repeated for additional strength, being careful not to add too much bulk.**

Short-term use (7–10 days) only is recommended as the figure-of-8 bandage is often associated with complications such as mid- to long-term reduced joint range of motion, hyperflexion of the carpus (recognisable by the primary feathers crossing over the secondaries), damage to the patagial membranes and ligaments, and poor stabilisation of fractures.

Body wrap. A body wrap may be used alone or in conjunction with a figure-of-8 bandage. Strap only the affected wing so that the bird can use the free wing for balance. When combined with physiotherapy, it is the recommended definitive treatment for shoulder girdle fractures (coracoid fractures and luxations, scapular fractures, or furcular fractures), even when highly displaced. It is also indicated for temporary stabilisation of wing fractures and may be combined with a figure-of-8 bandage for fractures of the radius or ulna (recognising the risk of synostosis) or with a metacarpal splint for treatment of distal wing fractures. Place a single wrap around the body, leaving the wings free, and then position the affected wing on this layer. It can then be taped or bandaged to the inner layer.

Curved edge splint. Using mouldable thermoplastic, a curved edge splint can be shaped to fit the carpometacarpal area to support and maintain alignment of metacarpal fractures. This may be a temporary technique prior to surgical stabilisation or definitive repair in cases with minimal displacement. The splint is secured in place with tape and a body wrap applied. It is important to monitor soft tissue swelling and adjust the fit as needed.

Splint. A tape splint is used for tibiotarsus and metatarsus fractures in small birds (pigeon size or smaller). A tape 'sandwich' is applied with the leg in a normal standing position and can be hardened by coating the completed splint in tissue glue (Vetbond®). In addition, a syringe case or other similar type of splint can be used in smaller birds (**Fig. 33.5**).

Ball bandage. This is used for fractured phalangeal bones and after repair of the flexor and extensor tendons of the digits. Complete immobilisation of the digit will result in a stiff digit.

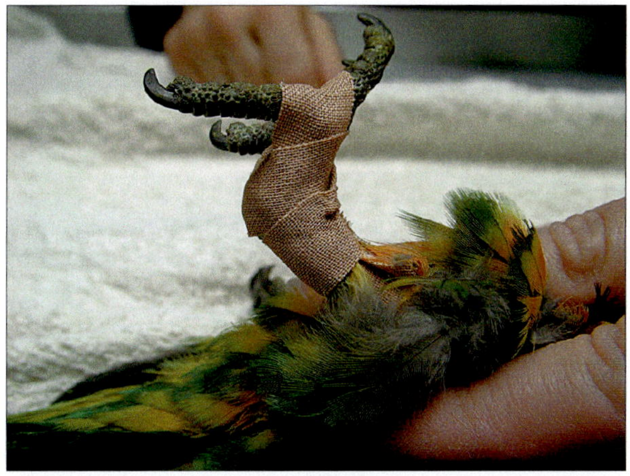

Figure 33.5 **External coaptation of a tarsometatarsal fracture.**

Surgical repair

Surgical fracture management often presents the best opportunity for a complete return to normal function in the avian patient. A well-constructed surgical fixator will meet all fracture management objectives and stabilise against the forces of tension, torsion, shearing, and bending. A variety of configurations using intramedullary (IM) pins, k-wires, external fixator pins, bars, and clamps provides the clinician with needed flexibility given the broad diversity in the size and shape of avian patients. Ideally, surgical fixation will be robust enough that prolonged coaptation is not required, thereby allowing the bird to have normal use of the limb during bone healing.

Intramedullary pins

Stainless steel pins or polymer rods can be used. (Polymer rods are introduced in a retrograde fashion and held in position with bone cement.) Intramedullary pins maintain alignment and length of the bone but may lack rotational stability (**Fig. 33.6**). This can be overcome by:

- Combining the pinning with semi- and full-cerclage wire, but great care must be taken not to disrupt the periosteal blood supply.
- Stack pinning with several small pins (most applicable to fractures of the humerus and femur).
- Combining with an external skeletal fixator to create a 'tie-in' fixation (see later).

Pin diameter should equal half to two-thirds of the medullary canal. Excessively large pins can interfere with the endosteal blood supply, which may cause avascular necrosis or iatrogenic fractures.

Intramedullary pins can be placed either:

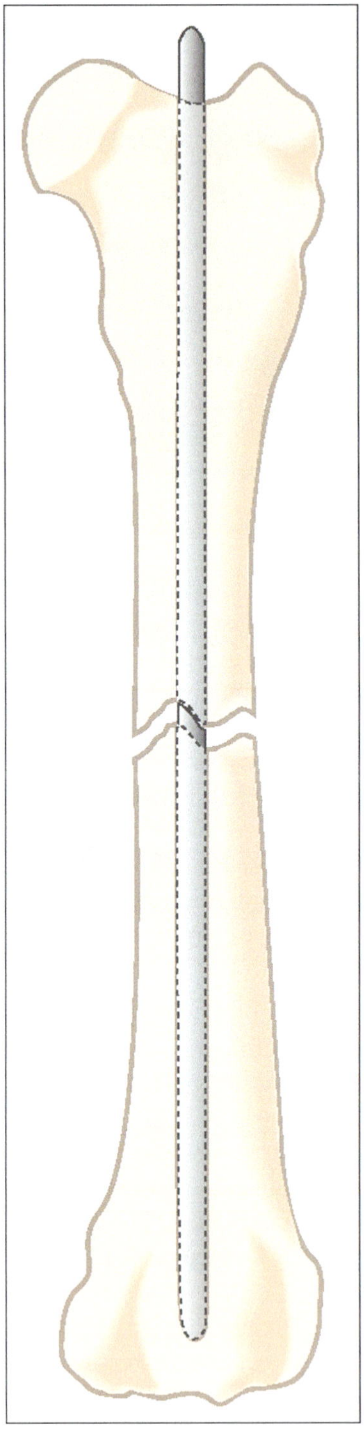

Figure 33.6 **Intramedullary pin placed in the femur.**

- Retrograde. The pin is inserted through the fracture site and advanced out either the proximal or distal end of the bone (avoiding the joint), and then the direction is reversed, and the pin is guided into the other fragment. This technique is usually used for femoral, humeral, and ulnar fractures, although some tibiotarsal fractures lend themselves to this technique.
- Normograde. Beginning at a natural end of the bone, the pin is advanced through the fracture site and into the other fragment. This is usually used for tibiotarsal, proximal humeral, or distal ulnar fractures.

IM pins can often be inserted in such a way as to avoid joint involvement by taking advantage of the curved anatomy of most avian long bones.

The pins should be removed when the bone has healed. Polymer rods and bone cement cannot be removed, nor are they reabsorbed. For this reason, their use should be carefully considered.

External skeletal fixation

External skeletal fixation (ESF) is lightweight, strong, inexpensive, and adaptable to many fractures (**Fig. 33.7**). In closed fractures, it can reduce the risk of osteomyelitis. It provides both rotational and lengthening stability, and moderate stability against bending forces. The stiffness and stability of an ESF are based on:

- Frame configuration: type I (unilateral), type II (bilateral), or type III (biplanar).
- Pin factors: type (non-threaded; threaded – positive or negative profile), size, number, location, length.
- Connecting bar material: stainless steel, acrylic, or epoxy-filled plastic tube/Penrose drain.
- Size and number of connecting bars.

The equipment needed includes:

- Transfixation pins. Positive-profile partially threaded pins, inserted through the predrilled holes, have been found to maintain solid bone-to-pin interfaces for prolonged periods (up to 3 months) in some birds.
- Connecting bars.
- Means of connecting pins to bars. Clamps are available, but dental acrylic and similar hardware materials have been used successfully.

Types commonly used in birds are the type I device, with half-pins connected on one side of the limb only, and the type II device, with full-pins running through the bone and connected on both sides of the limb.

The most distal and proximal pins are inserted first in a medial-to-lateral direction. Ideally, two to

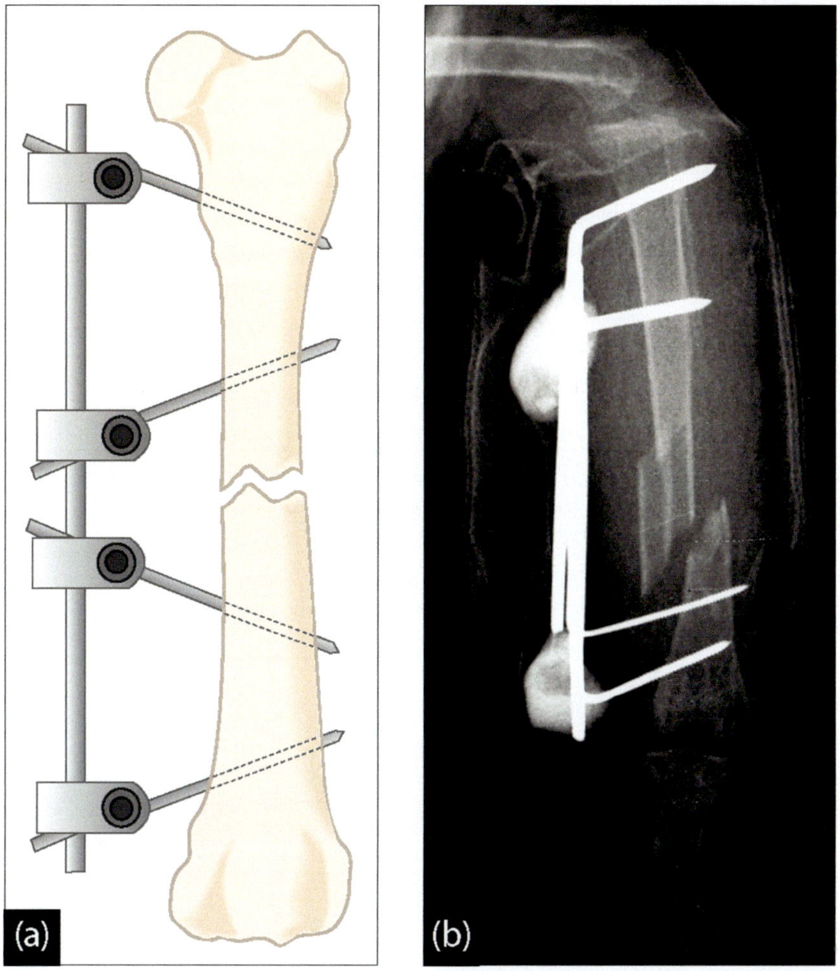

Figure 33.7 **(a) Type I external skeletal fixation. (b) Type I external skeletal fixation used to repair a tibiotarsal fracture in a cockatoo.**

three pins are placed through each side of the fracture at an angle of 30–90° to the bone, then the connecting bar is applied.

Tie-in fixator

A tie-in (or hybrid) fixator (TIF) is a combination of an intramedullary pin linked to external fixator pins (**Fig. 33.8**). It can be used to repair diaphyseal and periarticular fractures of all avian long bones except the tarsometatarsus (**Fig. 33.9**). When applying this system, every attempt should be made to reduce soft tissue disruption at the fracture site and preserve joint integrity. The intramedullary pin selected should fill 50–65% of the medullary cavity. It can be inserted using either a normograde or retrograde technique past the reduced fracture and emerges from the proximal end of the fracture. Positive-profile ESF pins are inserted perpendicular to the bone at locations where they will not be interfered with by the intramedullary pin. The protruding end of the intramedullary pin, where it emerges from the skin, is bent at 90° and rotated so that it is in the same plane as the ESF pins.

All pins are joined by a connecting bar. This connecting bar can be either a separate piece, or it can be formed from the intramedullary pin (if long enough) bent again at 90° to run down the length of the limb.

The degree of fixation achieved with this technique is such that primary bone healing often results (i.e., no periosteal callus) as long as the fracture is well aligned.

Plates

Although plate fixation has advantages over other fixation methods for certain indications, it is rarely used in avian surgery, especially in birds that weigh less than 1000 g.

Lightweight plates such as 'string of pearl' plates, cuttable plates, mini-plates, and finger plates may lend themselves to avian fracture repairs. Advantages of plates include:

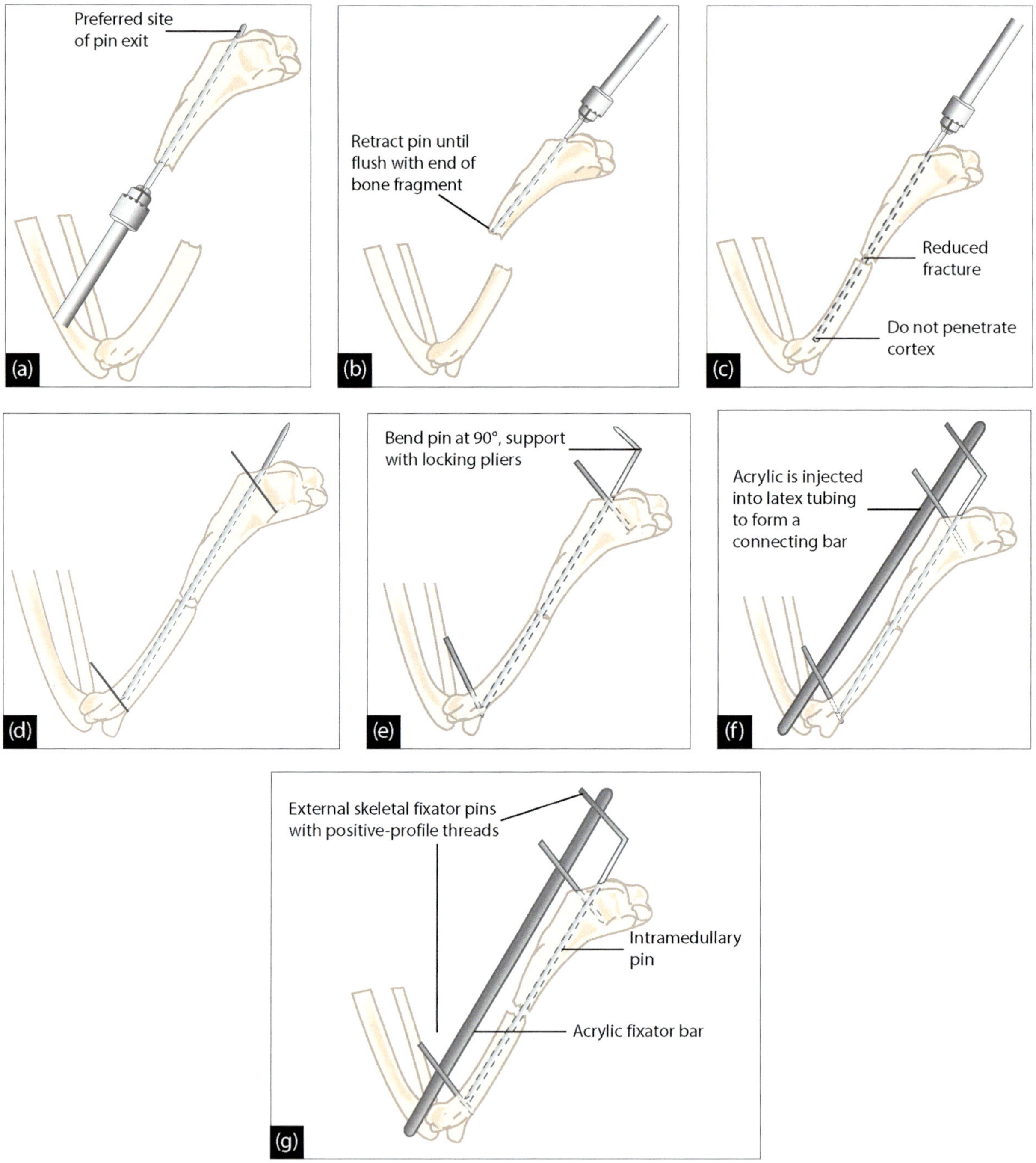

Figure 33.8 Schematic sequence of necessary steps required to address a diaphyseal fracture. (a) Retrograde placement of an intramedullary pin into the proximal fragment. (b) The pin is withdrawn proximally until the pin is flush with the fracture line. (c) The fracture is reduced and the pin advanced into the distal fragment. (d) Positive-profile threaded ESF pins or K-wires are placed at both ends of the bone, perpendicular to the bone. (e) The proximal end of the pin, protruding from the bone, is bent at 90° in the same plane as the ESF pins. (f) A connecting bar (pin or latex tube filled with acrylic) is used to join the ESF pins and the end of an intramedullary pin that has been bent over. (g) The completed tie-in fixator.

- Early return to function results in minimal fracture disease.
- Rigid immobilisation leads to primary bone healing.
- They provide rotational stability as well as maintain the length and alignment of the bone.

However, there are disadvantages to the use of bone plates, including:

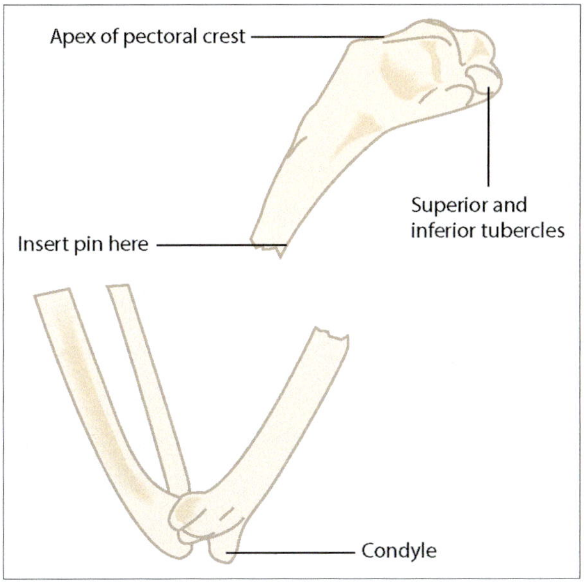

Figure 33.9 **Illustration of a typical diaphyseal fracture.**

- The cortices of most (but not all) avian bones are too thin to engage the thread of most bone screws. In some cases, this can be overcome by filling the medullary canal with bone cement before placing the screws, but the exothermic reaction as the cement cures can be detrimental to the bone. Titanium bone screws (which form a bond between the bone and the screw) can be used but present a major financial hurdle.
- Screws must be long enough to engage both bone cortices.
- Most plates are often too large and too heavy for avian bones. Titanium plates, especially if too short, may bend; stainless steel plates appear to resist bending pressure more effectively.
- There are technical difficulties involved in the placement and removal of plates and screws, involving a longer surgical time, and therefore increased risks of tissue compromise and prolonged anaesthesia.
- It may be difficult to obtain soft tissue coverage of the plate and screws.
- Bone plates and screws are expensive.

The use of a postoperative figure-of-8 bandage for up to 10 days may be advantageous and appears to reduce the risk of postoperative wound infection. As the plates become covered in callus, making them difficult to remove, they should be removed before day 28 after surgery.

SELECTION OF MEANS OF FIXATION

Wing

An inability to adequately immobilise the shoulder joint, as well as distortion produced by the powerful flight muscles, makes external coaptation a poor choice for humeral fractures if a return to flight is desired. If this is not a prerequisite, external coaptation may well be a suitable technique so long as the owner has realistic expectations of the bird's ability to fly afterwards.

Coracoid, scapula, clavicle

There are two schools of thought for the repair of fractures involving the pectoral girdle (**Fig. 33.10**). Conservative treatment, either strapping the wing to the body or simply using cage restraint for a month, may give good results if the coracoid is not severely displaced. However, surgical reduction and fixation of the coracoid with an IM pin may be indicated if it is displaced enough to press on the trachea or the heart.

Humeral fractures

Fractures in the humerus are often comminuted or spiral (**Fig. 33.11**). Both have the potential to shorten the bone by overriding. The strong pull of the pectoral and supracoracoideus flight muscles exerts both rotational and shortening forces on the bone. External fixation pins are therefore needed to counteract these forces, while an intramedullary pin keeps the line of the bone, as well as adding extra

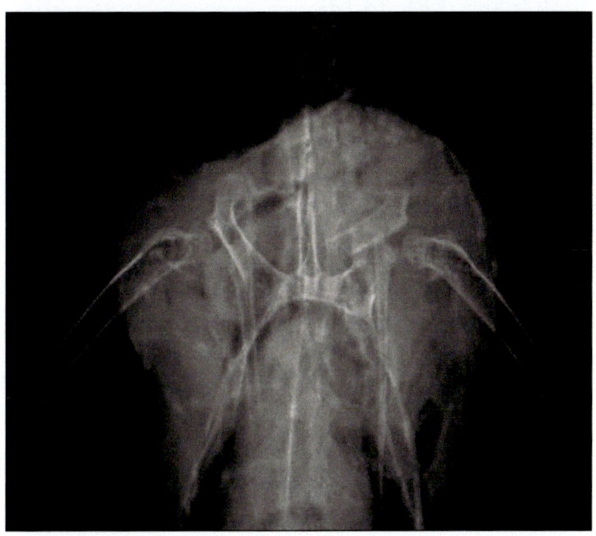

Figure 33.10 **Coracoid fracture (H-view).**

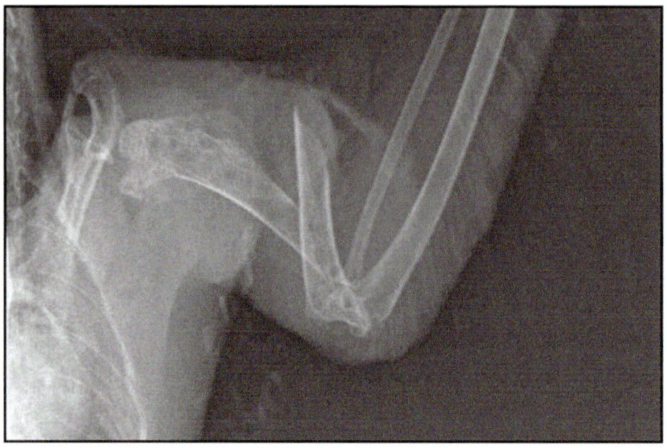

Figure 33.11 **Fractured humerus.**

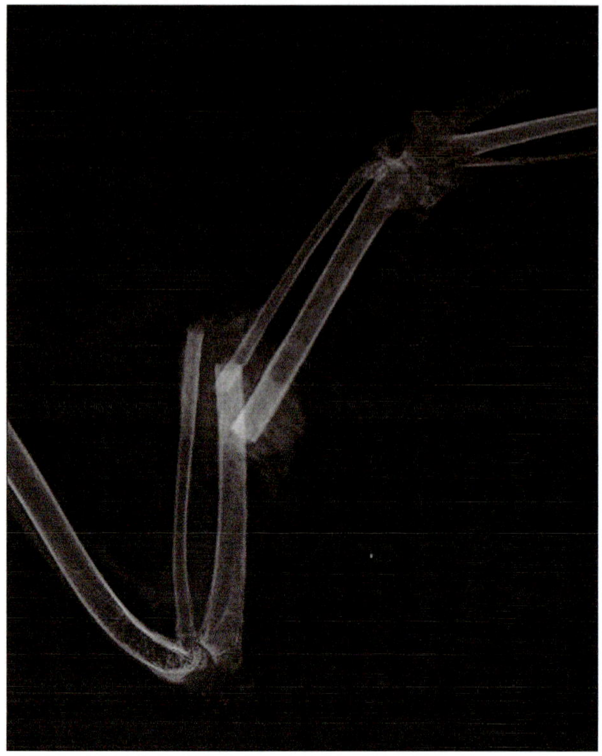

Figure 33.12 **Fractured antebrachium.**

stiffness to the repair. The humerus has a large pneumatised medullary cavity; fractures often lead to emphysema.

- *Proximal humerus.* ESF is difficult due to the lack of room in the proximal fragment. A tension band method using wire (in a figure-of-8 configuration) and an IM pin has worked, as has a TIF.
- *Mid-shaft humerus.* IM pinning or a TIF using a single ESF pin placed in the distal humerus are suitable repair methods for mid-shaft fractures of the humerus. Bone plating may be suitable for large birds (>3 kg).
- *Distal humerus.* Cross-pinning from the distal humerus, perhaps combined with a type I ESF, can be effective in repairing these fractures.

Antebrachium

Fractures can involve either the ulna, the radius, or both bones (**Fig. 33.12**).

- *Ulna.* If only the ulna is fractured (**Fig. 33.13**), it will usually heal with cage restraint, i.e., the bird is kept in a cage for 4 weeks with sufficient space to stretch its wings for grooming, etc., but not enough to flap them. Splinting is not necessary and increases the risk of the formation of a synostosis (**Fig. 33.14**).
- *Radius.* If the fracture is mid-shaft with little displacement, cage restraint will be sufficient for healing. If the fracture is close to the elbow, the biceps tendon distracts the proximal fragment and moves it sufficiently to prevent healing. Distal fractures can also displace significantly. This can result in a significant callus and subsequent synostosis. In both cases, the bone should be supported by an internal intramedullary pin. Care must be taken to avoid entering the elbow joint.
- *Both bones.* Fractures involving both bones will do better if surgically repaired. External coaptation often results in a synostosis joining the two bones, making pronation of the antebrachium difficult, and thereby restricting flight. Intramedullary pins, ESF, or TIF are suitable means of repair. A fracture of both the radius and ulna requires proper stabilisation by pinning the radius, which, combined with cage restraint, is usually sufficient. Pinning the ulna is more difficult as the bone has a large medullary cavity and a significant curve, making intramedullary pin support less satisfactory. The IM pin in the ulna should never be retrograded from the fracture site to exit at the elbow. Instead, the pin is inserted slightly distal to the olecranon (typically between the attachment points of the second and third secondary feathers) on the caudal aspect of the ulna and driven normograde to the fracture site. Alternatively, a type I ESF with small half-pins drilled in along the dorsal aspect will stabilise the bone and allow the wing to move, preventing a synostosis.

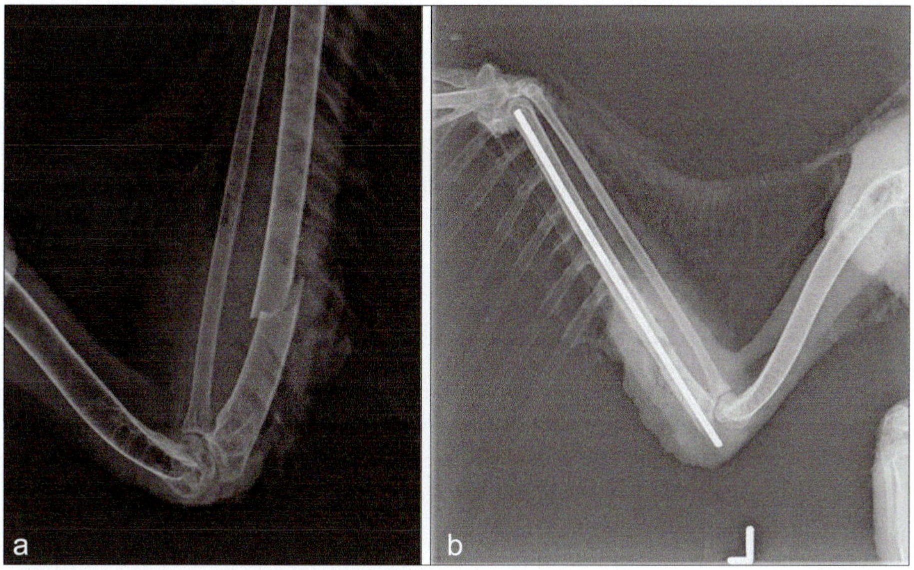

Figure 33.13 **(a) Fractured ulna. (b) IM pin placed in the ulna.**

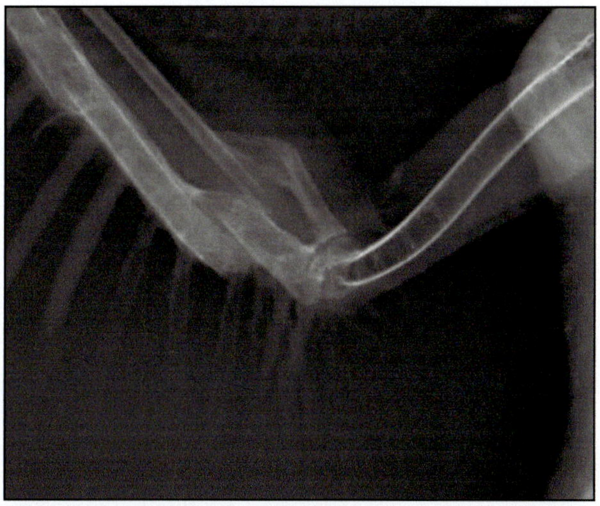

Figure 33.14 **Synostosis bridging the radius and ulna.**

Figure 33.15 **Metacarpal fracture.**

Metacarpus

For a fractured metacarpus (**Fig. 33.15**), external coaptation, type I ESF, or TIF are indicated depending on the size of the bird. Figure-of-8 bandaging allows many fractures to heal. Other techniques have been described, such as intramedullary pins or suturing foam and Vetlite over the fracture site. However, many cases fail due to loss of blood supply or infection. Amputation is usually the salvage technique.

Leg fractures

Femur

An inability to immobilise the coxofemoral joints, as well as distortion produced by the powerful leg muscles, makes external coaptation unsuitable for femoral fractures. The femur is not pneumatised in parrots but is in other species, e.g., raptors. It has a very large medullary cavity.

- *Proximal femur.* A TIF is usually successful, but tension band wiring using an IM pin and cerclage wire can also be used.
- *Mid-shaft femur.* IM pinning (especially stack pinning with two or three small pins), or a TIF with either a single ESF pin in the distal femur or a second pin in the proximal femur near the hip joint, is indicated. Plating can be used in large birds (>3 kg). The bulk of the leg muscles makes ESF pinning unsuitable for fractures in this position.

- *Distal femur*. Cross-pinning is used in combination with an IM pin.

Tibiotarsus

Fixation may be achieved with an IM pin, type I or II ESF, a TIF, or external coaptation. A TIF is the easiest and most reliable means of repair. External coaptation should only be used for fractures through the distal half of the bone. Fractures higher than this cannot be adequately stabilised by coaptation, leading to a high incidence of non-union.

Tarsometatarsus

The lack of a medulla in the tibiotarsus in most birds makes surgical repair difficult in most small birds. In large parrots, it is possible to reduce the fracture using a type II full-pin fixation, as the tarsometatarsus has a medullary cavity. The bird must be able to use the foot as freely as possible, or the callus will incorporate the flexor and extensor tendons, causing permanent disability. External coaptation is suitable in small birds, ensuring some (or all) of the toes are not included in the bandage. This allows the bird to grip a perch and, by flexing and extending the toes, it prevents adhesions between the callus and digital tendons.

JOINT SURGERY

The term joint luxation refers to the complete displacement of a bone from the joint, whereas in a subluxation, the bone is only partially displaced and some contact between the two articular surfaces still occurs. In most avian species, luxations occur infrequently compared with other orthopaedic conditions.

Fractures and luxations of joints have a guarded prognosis for a full return to normal function.

Shoulder joint

The coracoid, the scapula, and the clavicle comprise the avian thoracic girdle. As an ellipsoid joint (or biaxial joint), the shoulder allows movements in all angular motions – back and front, and side to side. In the resting position with the humerus adducted, it lies along the trunk, whereas during gliding flight, it is abducted by up to 90°.

It is not a very stable joint, and luxation is not uncommon. It is often accompanied by an avulsion fracture of the ventral tubercle of the proximal humerus, sometimes coracoid fractures, and sometimes by brachial plexus avulsion; careful radiographic assessment is warranted. If the ventral tubercle has been avulsed, surgical reattachment of the tubercle with a K-wire often results in a stable joint. If the tubercle is not avulsed, closed reduction is often successful. Regardless of the technique used, the wing should be strapped to the body for 10–14 days.

The prognosis is considered poor for severe luxations (e.g., those in which the tendon is unable to recover its function after luxation reduction) in individuals requiring complete flight restoration. Reluxation after treatment is common.

Elbow joint

The avian elbow is a shallow synovial joint consisting of three joints enclosed by a single capsule: between the humerus and the radius, the humerus and the ulna, and between the radius and ulna. Movement of the joint is mainly flexion and extension. The extension of the elbow results in the longitudinal displacement of the radius with respect to the ulna, which provokes the concordant extension of the carpus and, if fully extended, the rotation of the bones of the antebrachium about the longitudinal axis.

Luxation of the elbow may be dorsal, caudal, or caudodorsal (ventral luxations are usually associated with fractures of the radius). The presence of open wounds and fractures has been associated with a poor prognosis for return to normal function.

Reduction is accomplished by flexing the elbow while internally rotating the radius and ulna, and applying pressure to the dorsal (lateral) aspect of the radial head to force it into apposition with the dorsal (lateral) humeral condyle. The elbow is then gently extended; if the ligaments are not severely damaged, the joint should click into place. If the joint is stable following reduction, it may be supported with a figure-of-8 bandage for 7–12 days.

If closed reduction is not possible or the luxation readily recurs, open surgical reduction is recommended. Through a medial approach to the elbow joint, the common digital extensor is sutured to the scapulotriceps tendon distal to the elbow. A figure-of-8 bandage is then used to immobilise the joint for 2 weeks while it heals.

If further stabilisation is required, a trans-articular external skeletal fixation device may be applied to maintain reduction. At least two pins are placed in the humerus and two in the ulna. The elbow is flexed into a normal folded position, and the pins connected (proximal humeral pin to the distal ulnar pin and the distal humerus pin to the proximal ulnar pin).

Coxofemoral joint

The coxofemoral joint is diarthrodial, i.e., a freely movable joint, enclosed in a synovial capsule that produces synovial fluid. Its movement is basically flexion and extension; abduction and adduction movements are virtually non-existent, although a small degree of medial and lateral rotation of the joint is present.

Coxofemoral luxation is relatively uncommon but can be repaired surgically. The bird is placed in lateral recumbency, and feathers are plucked over the dorsolateral femur and pelvis.

A skin incision is made over the dorsolateral crest of the ilium, extending over the femoral trochanter. The iliotibialis lateralis muscle (cranially) and iliofibularis muscle (caudally) are separated from distal to proximal to the iliac crest, cutting the common tendinous insertion of these muscles on the ilium. The musculotendinous insertion of the iliofemoralis externus and the iliotrochantericus caudalis is transected, leaving enough tissue for reattachment. These two muscles lie dorsal to the acetabulum.

The bird is turned 180° to view the transparent membrane covering the joint, being aware of the branch of the femoral artery crossing this membrane and the femoral vein and nerve deep to it. The luxations can now be reduced, and the joint capsule sutured in two to three locations. Transected muscles and tendons are reattached, and stabilisation sutures are placed from the trochanter to the dorsolateral iliac crest (caudal to the central axis of the femur) and from the trochanter to the cranial rim of the acetabulum. The sutures are placed through the bone, and with the stifle maintained in a normal standing position, the sutures are tightened. These sutures prevent external rotation of the femur. The skin is closed routinely. Postoperatively, the limb should be supported in a spica splint or a sling for 7–10 days. It is best to maintain the bird in a cage with smooth walls and a perch near the floor to discourage attempts to climb.

An alternative to open reduction is to perform a femoral head and neck ostectomy (FHO). The approach to the hip described earlier is used, and the femoral head and neck are removed with an appropriately sized osteotome, scissors, or oscillating saw, and the bone is smoothed with a rongeur or file. Stabilising sutures (as described earlier) are used to prevent external rotation of the femur.

Following FHO, early use of the leg is encouraged to promote the formation of a fibrous pseudoarthrosis. Passive range of motion exercises can be started the day after surgery.

Femoral–tibiotarsal joint

The femorotibial (knee) joint consists of four intercommunicating synovial joints enclosed by a joint cavity. The main supporting structures of the stifle joint are the joint capsule, the medial and lateral collateral ligaments, the cranial and caudal cruciate ligaments, two menisci (medial and lateral), surrounding muscles, and tendons. The two cruciate ligaments limit flexion and extension, and the collateral ligaments restrict movement in the lateral–medial plane. The proximal fibula and tibiotarsus can move independently, allowing for the rotation of the joint. This rotational movement may predispose the joint to luxations; this is not uncommon and can be congenital or acquired (traumatic). Luxation can occur craniolaterally, craniomedially, caudolaterally, and caudomedially with concomitant damage to the collateral and cruciate ligaments, whereas meniscal damage is rarely diagnosed.

Clinical signs include:

- Non-weight-bearing lameness, often accompanied by swelling in the stifle region
- Joint hyperextension
- The 'cranial drawer' sign
- Medial and/or lateral instability
- Lack of crepitus or palpable fractures

The diagnosis can be confirmed radiologically.

The objective of the repair is to ankylose or arthrodese the joint in a normal perching position. The normal perching angle of the stifle should be researched or approximated prior to surgery. If the joint is not flexed sufficiently, the leg will be excessively straight; flexed too much, and the leg will be effectively shortened. Using the other leg as a comparator is not recommended, as the change in weight-bearing resulting from unilateral lameness may change the joint angle.

External coaptation in parrots is complicated by the presence of the inguinal skin web extending to the level of the stifle, preventing bandages/splints from extending proximally over the stifle, thus failing to immobilise it.

Three surgical techniques for repair have been described:

- In young birds with a developmental luxation, IM pins are introduced via a closed cranial approach to the joint, into the femur and the tibiotarsus such that the ends protrude through the cranial aspect of the joint. The joint luxation is reduced, the joint flexed into a normal perching/standing position, and the pins are locked together with acrylic (the excess lengths of the pin are removed). This technique does damage articular cartilage, and infection may track along the pins, resulting in synovitis or osteomyelitis.
- A type I or II ESF is constructed across the joint, with two pins in the femur and two in the tibiotarsus. Again, the joint is flexed into a normal perching/standing position. The pin in the proximal femur is connected to the pin in the distal tibiotarsus, and the distal femoral pin is connected to the proximal pin in the tibiotarsus.
- A lateral approach to the joint is used to expose the distal femur and the proximal tibiotarsus and reduce the luxation. A transverse hole is then drilled through the distal femur and the proximal tibiotarsus. A strong suture is placed through these holes (from lateral to medial in the distal femur and medial to lateral in the proximal tibiotarsus), creating a mattress suture that will provide collateral support (but not craniocaudal stability) to the joint. If needed, a type I ESF, as described earlier, can be placed to provide additional support.

After 2–4 weeks, the pins can be removed; the joint is usually ankylosed and stabilised.

Tibiotarsal–tarsometatarsal joint

The tibiotarsal–tarsometatarsal joint consists of two joints: the intertarsal joint and the metatarsal joint.

- The intertarsal joint is the synovial roll-and-glide articulation of the tibiotarsus with the tarsometatarsus. A lateral and usually also a medial C-shaped meniscus are interposed between the articulating bones and attached by ligaments to the tarsometatarsus and tibiotarsus.
- The metatarsal joint consists of the distal row of tarsal bones and the metatarsal bones of digits II, III, and IV.

The fibrocartilaginous tibial cartilage is attached to the caudal aspect of the intertarsal joint and presents canals for the deep flexors, the gastrocnemius muscle, and the superficial digital flexors.

Support for the joint is provided by the medial and lateral collateral ligaments (restricting the movement of the joint to flexion and extension), and the intercondylar tibiometatarsal ligament limits the cranial–caudal sliding movement.

This is often a developmental luxation seen in young birds. The tarsometatarsus may be rotated through 90° such the cranial aspect of the bone is now facing laterally or medially, with the toes reflecting this misalignment.

A lateral approach is made over the joint and the joint is opened with a stab incision using a scalpel. The articular surface of the joint can be debrided with a bone curette, and the tarsometatarsus is gently but firmly rotated back into a normal position. The joint is flexed into a normal perching position, and one or two K-wires are then introduced through the distal tibiotarsus into the tarsometatarsus to stabilise it. The joint and the skin are then closed, and the foot bandaged. After 3 weeks, the joint should have arthrodesed, and the pins can be removed.

ORTHOPAEDIC POSTOPERATIVE MANAGEMENT

Prevention of new trauma is a key concern immediately post-op. Patients should be adequately restrained during anaesthetic recovery and not returned to a cage until fully able to balance and control their activity. The recovery cage should be prepared by removing or lowering any perching and providing a padded cage bottom. Many birds are less anxious when able to perch yet are too unsteady to safely climb and perch immediately following the surgical procedure. A rolled-up towel on the bottom of the cage floor can be used to allow safe perching. Climbing should be prevented by using a solid-sided cage, which should be placed in a low-stress environment away from dogs, cats, humans, and any other predators.

Where possible, external coaptation should be used to prevent iatrogenic damage to a fracture site.

Excessive wing flapping or premature weight-bearing on a repaired limb can, and frequently does, undo the best efforts of any surgeon. This external coaptation should be placed immediately postoperatively before the patient is recovered from anaesthesia.

The patient must be encouraged to eat once capable of standing/perching normally. In addition to the type of food being offered, it is important to note the presentation of the food; consideration must be given to the patient's ability to access food, grasp and manipulate it, and eat.

Analgesia should be provided (see Chapter 31) for 1–2 weeks postoperatively.

Antibiotic coverage is provided using:

- Cephalosporins or enrofloxacin for 5 days postoperatively for uncomplicated fracture repairs
- Clindamycin or amoxicillin–clavulanic acid can be used for open, comminuted, and infected fractures.
- Antibiotic-impregnated methyl methacrylate beads can be used in some fractures if infection is likely.

Patagial (wing web) contraction, joint contracture, and ankylosis are common complications of orthopaedic repair. Physical therapy (passive range of motion exercises) can help to prevent or overcome these, and is usually done under anaesthesia, starting at 2 days (for humeral fractures) or 10 days (for other fractures). Five-minute sessions should be given twice weekly for 3 weeks. The limb is extended and held in a gentle stretch several times over a 5-minute period. Each joint is evaluated for range of motion relative to the normal limb. The fixation apparatus should be monitored for loosening of pins or exudate around pin tracts/surgical sites, as well as any signs of causing iatrogenic trauma.

As soon as possible, coaptation such as body and wing wraps should be removed, allowing the limb to return to normal function. These can normally be removed a few days after surgery (except in the case of metacarpal fractures, where additional support is required due to the weight of fixation at the end of the wing). If fixation allows for normal use of the limb with no coaptation, the requirement for physiotherapy is greatly reduced.

The fracture should be radiographed at 10–14 days and again at 3 weeks. These images should be evaluated for normal fracture healing, as well as potential complications such as reaction to the fixation device, ESF pin loosening, or development of osteomyelitis or sequestrum. If realignment is necessary, it must be done before 10 days. Dynamic destabilisation (partial dismantling of fixation), to begin gradual load sharing to the bone where appropriate, can be started at 21 days if healing is well underway. Sequestra can be seen radiographically at 21 days and can be removed surgically when their extent is clear.

An integral part of the follow-up care is the careful prevention of comorbidity situations. One of the most common sequelae to fracture repair in a leg is the development of contralateral pododermatitis (bumblefoot). This may require padded perches, prophylactic bandaging on the primary weight-bearing foot, and close monitoring of the feet for lesions during the convalescent phase.

Healing should be complete by 6 weeks, and all fixation removed.

COMPLICATIONS FOLLOWING ORTHOPAEDIC SURGERY

A retrospective study of 60 pelvic limb fractures treated with surgery, external coaptation, or cage rest showed that, while 85.5% of cases resulted in satisfactory resolution, complications were identified in 41.7% of the cases, and that complications were more common in cases treated with external coaptation. The most common complication reported was patient interference with bandages, splints, or both (Hollwarth and Dutton, 2023). Other complications can include non-union, osteomyelitis, or osteoarthritis.

Non-union

A non-union is when the bone does not heal properly, leaving the limb with pain and instability. A malunion occurs when a fractured bone does not heal properly, leaving the bone shortened and deformed. Both are painful and commonly produce swelling around the area, possible immobilisation, and deterioration of the bone and tissue. The cause is invariably instability of the fracture during repair. Constant movement leads to the bone ends becoming reduced in diameter, rounded, and smooth. No periosteal proliferation is present on either the proximal or distal fragments.

Healing can be induced by the removal of fibrous tissue and cartilage, if present. The bone ends are debrided, and internal fixation, preferably with some degree of compression, is used to support the bone while it heals. Provided that the bones are immobilised, the fracture heals.

Early non-unions may not require surgery; more complete support of the limb and usually less movement of the bird itself are all that is required to allow the bone to heal. Long-standing non-unions are difficult to repair, and a bone graft should be considered. The bone may be obtained from the carina of the sternum or the tibiotarsal crest, using rongeurs. The carina may be used as a single large piece laid onto or into the site of the malunion.

Osteomyelitis

Many avian fractures, especially in wild birds, are open. Almost invariably, the fracture has become infected prior to the initial examination. If possible, the affected bone should be removed at the time of surgery, but often the infection has invaded the medullary cavity and it proves impossible to prevent pus from forming. It is an unfortunate fact that purulent material in avian species collects as a caseous mass that will not clear with drainage. There is rarely formation of a sinus as there is in mammals. In long bones, the cortical bone becomes distorted by the continuing growth of the relatively hard purulent mass in the medullary cavity: this may be seen radiographically. Drainage is ineffective, and flushing the medullary cavity can push the infection further into the cavity, especially in pneumatised bones. Single purulent masses may be cured by surgical removal of the lump of purulent material, followed by lavage with antibiotic and administration of appropriate antibiotic therapy.

If the bone is known to be infected, the fracture should be repaired using a full-pin or half-pin fixation technique; if some of the cortex has to be removed, there is still support for the weakened bone whilst healing occurs. Multiple caseous masses are usually impossible to remove and the prognosis in these cases is very poor.

Sometimes, in bones that were fractured and not infected, radiographic evidence of small areas of apparent osteomyelitis frequently occurs around the pin during the healing process. Pin removal after the bone has healed will bring a coating of apparently purulent material with it. Antibiotic therapy is indicated after pin removal, and the radiographic lesions usually disappear within a few weeks.

Osteoarthritis

Osteoarthritis is rarely seen, but it is usually a sequel to a fracture or some other trauma involving a joint. Long-term analgesia, sometimes using multidrug therapy, is indicated.

APPROACHES TO THE BONES OF THE WING

Coracoid and clavicle

A ventral approach is used. A skin incision is made along the caudal edge of the furcula, starting laterally and continuing medially along the lateral edge of the keel for the first one-fifth or one-sixth of the length of the keel. An incision is made through the superficial pectoral muscle along the caudal edge of the furcula. Cranial and caudal pectoral vessels and nerves are seen between the superficial and deep pectoral muscles. The deep pectoral muscle is incised and reflected along the clavicle and keel. Haemorrhage from the clavicular artery, which supplies part of the pectoral muscles (encountered at the caudal midpoint of the furcula), must be controlled.

Muscles of the coracoid are now visible: the supracoracoideus and coracobrachialis caudalis. These can be separated to get better exposure of the coracoid. Multiple small intramedullary pins are introduced at the fracture site and exteriorised through the point of the shoulder, then the pins are normograded back through the distal fragment taking care not to perforate the pericardium and heart.

The supracoracoideus and coracobrachialis caudalis are sutured if separated. The pectorals are reattached to the periosteum in separate layers. The skin is closed.

Proximal humerus

A dorsal approach is used in fractures of the proximal third. The feathers are plucked over the proximal, medial, and ventral humerus. In raptors, the scapular coverts insert into a fascia, which continues from the cutaneous costohumeralis muscle. This should be maintained, if possible, as they may be involved in critical aspects of flight. The skin is incised along the shaft of the humerus.

Caution should be exercised at this stage, as:

- The axillary nerve is deep to the propatagialis complex in the proximal third of the humerus.
- A branch of the subscapular artery is medial and cranial to the nerve.
- The radial nerve crosses the humerus two-thirds of the way down the shaft, near the site of insertion of the deltoideus major muscle.

The propatagialis muscle can be transected in the distal third, or one can bluntly dissect through it, avoiding the muscle branch of the axillary nerve. The deltoideus may be retracted proximally, but it needs to be reattached to the bone when the procedure is completed.

There are two techniques for repair: stack pinning in birds <300 g or tension banding in birds >300 g. In the latter technique, two K-wires are inserted retrograde dorsally and ventrally to the pectoral crest, then the fracture is reduced, and the pins are advanced into the distal fragment. Two holes are drilled: one in the humerus 1 cm distal to the fracture, the other through the pectoral crest proximal to the K-wires. Cerclage wire is passed through the holes, made into a figure-of-8, and tightened.

If the deltoideus was elevated, a hole is drilled through the bone to allow a suture to reattach the muscle to the bone. The propatagialis muscle is sutured if it was transected.

Distal humerus

There are two approaches, both of which can be used to reduce and fix internal fractures of the humerus in the mid-shaft to distal third. The dorsal approach is more commonly used.

Dorsal approach

In the dorsal approach (**Fig. 33.16**), the bird is placed in sternal recumbency with the affected wing extended. The feathers are plucked from the pectoral crest to the proximal radius/ulna.

The radial nerve should be palpated before making the skin incision. The skin is incised from the proximal third of the humerus to the olecranon fossa and ventral epicondyle, or even more caudally. The deltoideus and triceps insertions on the caudal aspect of the humerus are identified, with the radial nerve emerging between them. The biceps brachii and the tendon of the tensor propatagialis pars brevis are identified on the cranial side of the humerus. Distally, three tendons are identified originating from the antebrachium: cranially, the extensor metacarpi radialis; caudally, the extensor metacarpi ulnaris; and between them, the supinator and common digital extensor.

In distal fractures, two K-wires can be used to cross-pin through the dorsal and ventral epicondyles. These can then be tied into an ESF.

Ventral approach

The ventral approach is used for distracted fractures of the distal two-thirds of the humerus (**Fig. 33.17**).

The bird is placed in dorsal recumbency with the affected wing extended. The feathers are plucked from the ventral humerus and dorsally and cranially over the pectoral crest.

The biceps brachii muscle is palpated cranial to the shaft of humerus. The ulnar and radial vessels and medianoulnar nerve run deep to or on the caudal edge of this muscle, as does the superficial basilic vein. To avoid these structures, the skin incision is made either over the belly of the biceps muscle or over the humeral shaft caudal to the vessels and nerves. The excision is continued distally to the elbow. The biceps are elevated and retracted with the vessels and nerve. The triceps can be elevated caudally for better exposure.

Closure is accomplished by simply suturing the skin and superficial fascia as a single layer.

Proximal radius and ulna

A dorsal approach is used for proximal radius and ulna fractures and elbow dislocation (**Fig. 33.18**).

The bird is placed in sternal recumbency, and the affected wing is extended. The feathers are plucked from mid-humerus to the distal antebrachium.

A curvilinear skin incision is made from the distal humerus, between the radius and ulna, and extending as far as needed for exposure. Care should be taken to avoid branches of the radial nerve and the insertion of the tensor propatagialis pars brevis tendon. The supinator muscle is retracted cranially and the extensor digitorum muscle is retracted caudally. For better exposure of the radial head, the tendon of insertion of the tensor propatagialis pars brevis must be transected. If there is still not enough exposure, the supinator muscle is transected at its distal third, avoiding the deep radial nerve. The ulna can be exposed by incising between the retinaculum of

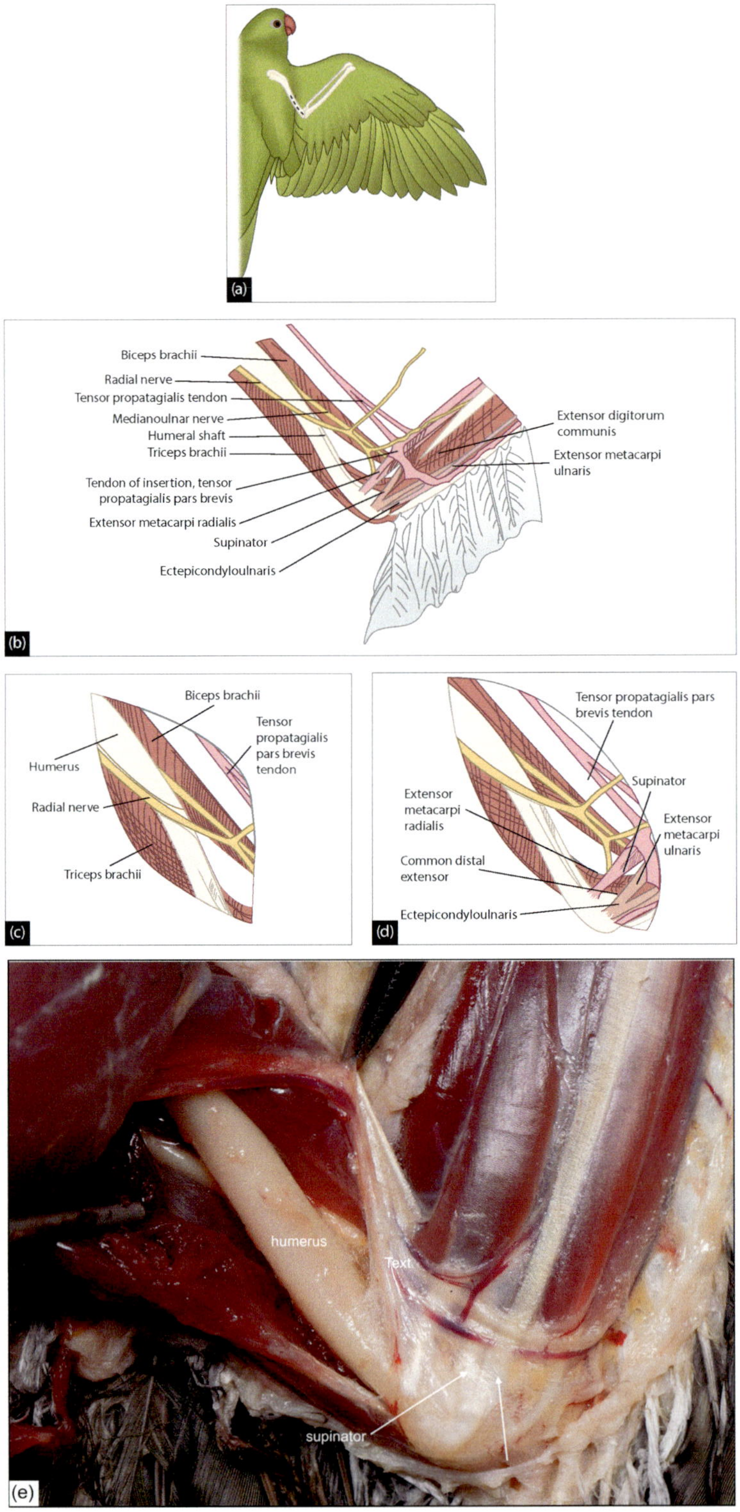

Figure 33.16 **The dorsal approach to the distal humerus. (a) Incision site. (b) Anatomy of the distal humerus (dorsal view). (c) Initial incision – superficial structures. (d) Extending the incision distally to reveal structures overlying the dorsal elbow. (e) Dissection view of dorsal approach to the distal humerus.**

the extensor metacarpi ulnaris and the muscle itself. The branch of the deep radial nerve and the interosseous dorsalis artery and vein that emerge here and run along the caudal aspect of the ulna should be avoided.

Distal radius and ulna

A dorsal approach is used for the open reduction of fractures of the radius and ulna and luxations of these bones (**Fig. 33.19**).

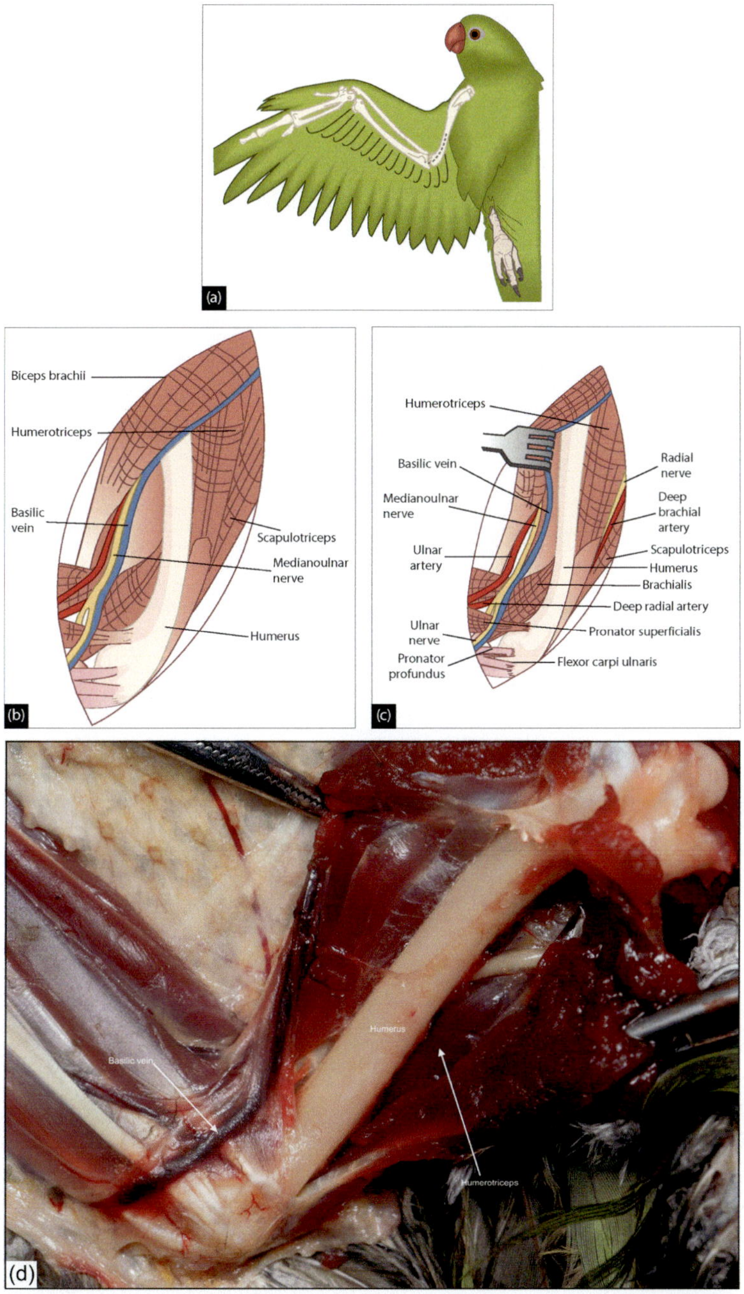

Figure 33.17 Ventral approach to the distal humerus. (a) Incision site. (b) Initial incision – superficial structures. (c) Reflection of the M. biceps brachii gains better exposure of the humeral shaft. If necessary, the M. humerotriceps can be dissected off the caudal humerus for more exposure. (d) Dissection view of ventral approach to the distal humerus.

The bird is placed in sternal recumbency, and the affected wing is extended. The feathers are plucked from the bone, but the secondary feathers are left intact.

The skin is incised between the radius and the ulna. If necessary, the extensor metacarpi radialis, on the cranial surface of the radius, and the extensor metacarpi ulnaris, on the dorsal surface of the ulna, can be retracted. If it is necessary to repair the ulna, the calamus of the secondary feathers between bone and skin can be cut, avoiding the follicle.

For ulnar fractures, intramedullary pins are introduced through the fracture site, retrograded out of the olecranon (avoiding the elbow) and normograded into the distal fragment. They can then be tied into a TIF.

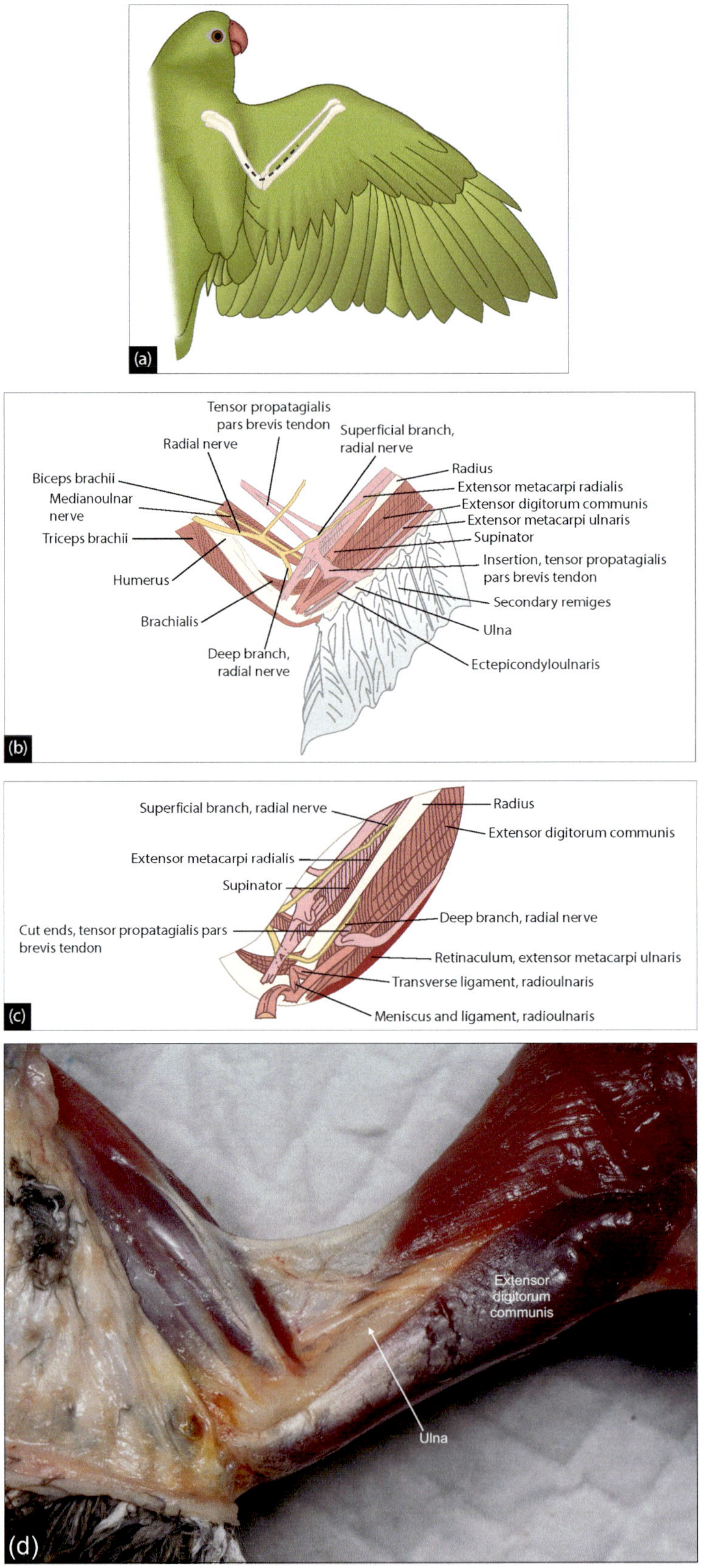

Figure 33.18 **Dorsal approach to the proximal radius and ulna. (a) Incision site. (b) Anatomy of the elbow and proximal radius and ulna (dorsal view). (c) Dorsal approach to the proximal ulna. (d) Dissection view of dorsal approach to the proximal ulna.**

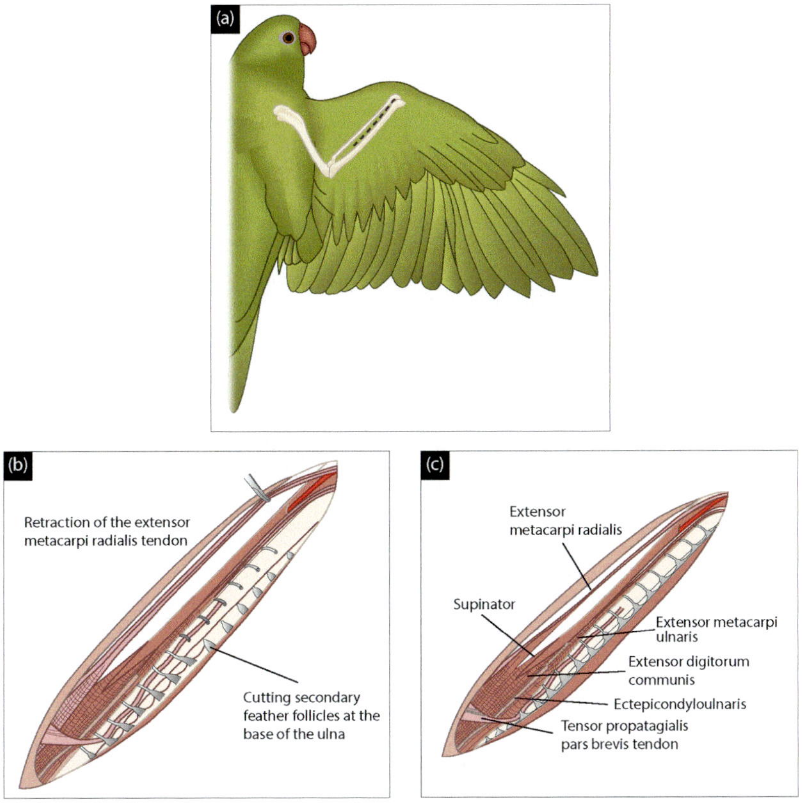

Figure 33.19 **Dorsal approach to the distal radius and ulna.**

Radius

A ventral approach is used for distal radial fractures only (**Fig. 33.20**). The dorsal approach is preferred for proximal fractures or ulnar fractures.

The bird is placed in dorsal recumbency, and the affected wing is extended. The feathers are plucked from the ventral antebrachium.

The superficial ulnar artery is palpated distal to the elbow joint. Starting distal to this artery, an incision is made over the caudal aspect of the radius between the extensor metacarpi radialis muscle anteriorly and the extensor digitorum communis over the intraosseous space. Numerous major arteries, veins, and nerves, as well as muscle bellies and tendons, are in this area. Care must be taken not to damage them. To expose the radius, the belly of the pronator superficialis must be reflected cranially away from the pronator profundus, taking care to avoid the arteries, veins, and nerves in this area.

If the fracture is displaced, intramedullary pins are inserted through the fracture site out toward the carpus (avoiding the joint) and then retrograded back through the proximal fragment, avoiding the elbow joint.

Ulna

A dorsal approach can be used for simple ulna fractures.

The bird is placed in sternal recumbency. Covert and down feathers are plucked from the caudal and dorsal aspects of antebrachium, leaving the secondary feathers in place.

An intramedullary pin is inserted at right angles to the skin on the caudal aspect of the ulna between the second and third last secondary feathers. As the trocar of the pin cuts into the cortex, the angle of the pin is gradually changed so that it becomes aligned with the ulna as it penetrates the medullary cavity. The pin can then be guided, by either a closed or open approach, into the distal fragment.

A slight variation of this technique is to use a larger pin to gain entrance to the medullary cavity, and then introduce a blunt-tipped smaller pin. This prevents accidental penetration of the carpal joint.

Metacarpus

A ventral approach is preferred due to the insertion of primary feathers on the dorsal surface (**Fig. 33.21**).

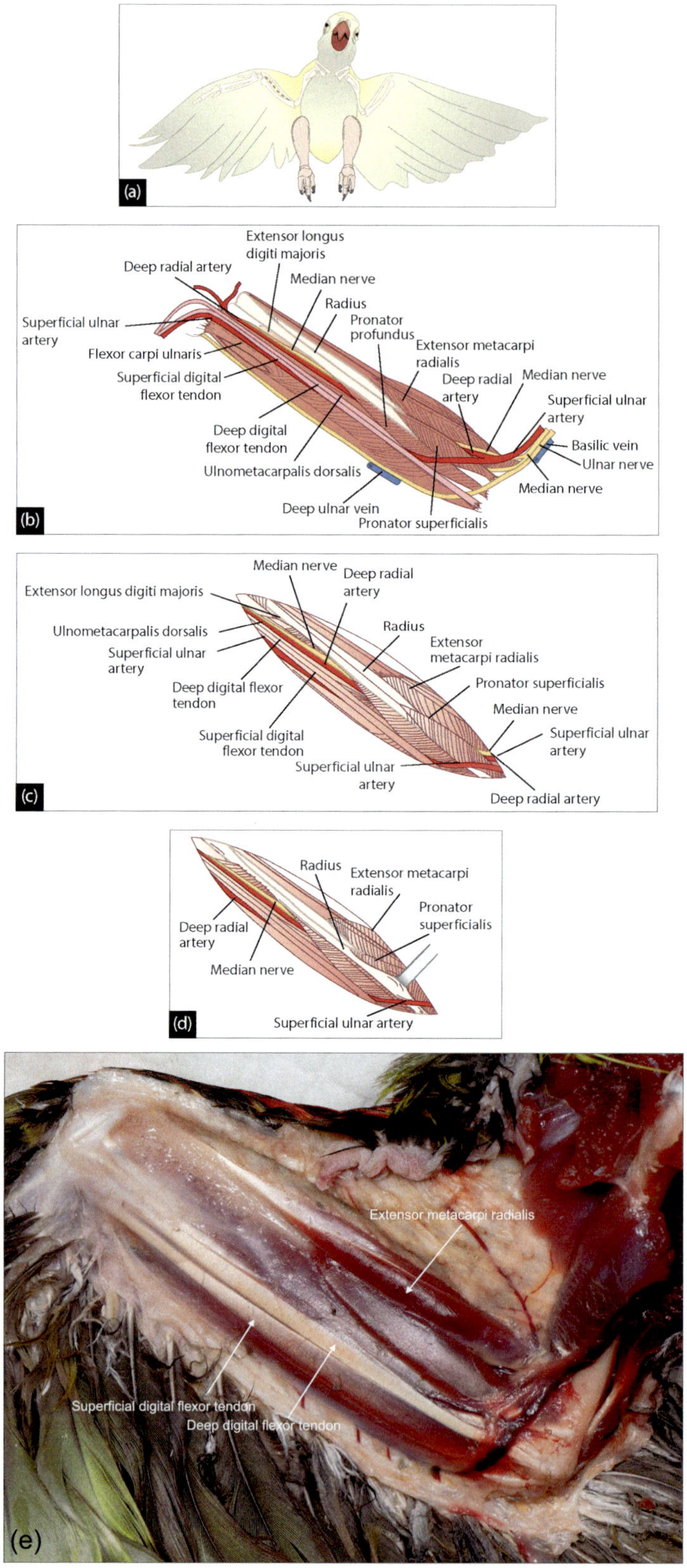

Figure 33.20 Ventral approach to the distal radius and ulna. (a) Incision site. (b) Anatomy of the distal radius and ulna (ventral view). (c) Initial incision, ventral approach to the distal radius and ulna. (d) Deeper dissection to expose the distal radius. (e) Dissection view of ventral approach to the distal radius and ulna.

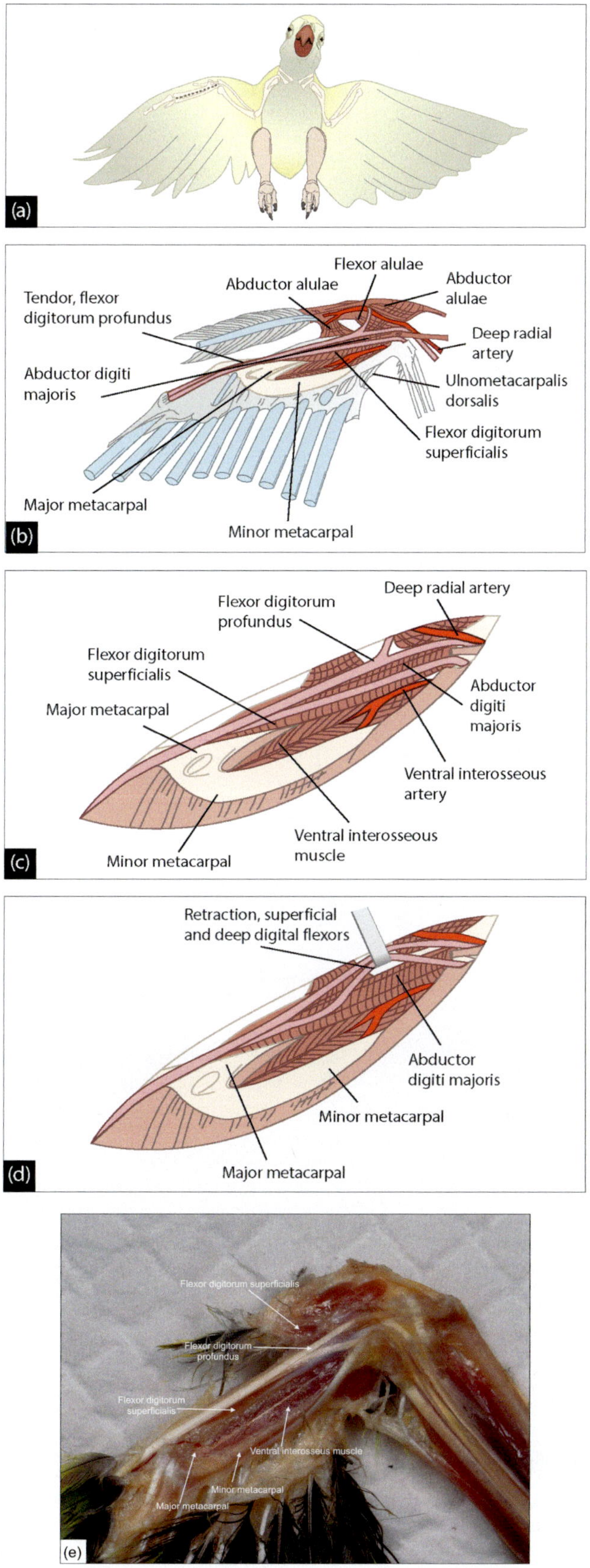

Figure 33.21 **Ventral approach to the metacarpus. (a) Incision site. (b) Anatomy of the metacarpal region, ventral view. (c) Superficial structures of the ventral metacarpal region. (d) Retraction of the deep and superficial digital flexors to expose the major metacarpal bone. (e) Dissection view of ventral approach to the metacarpus.**

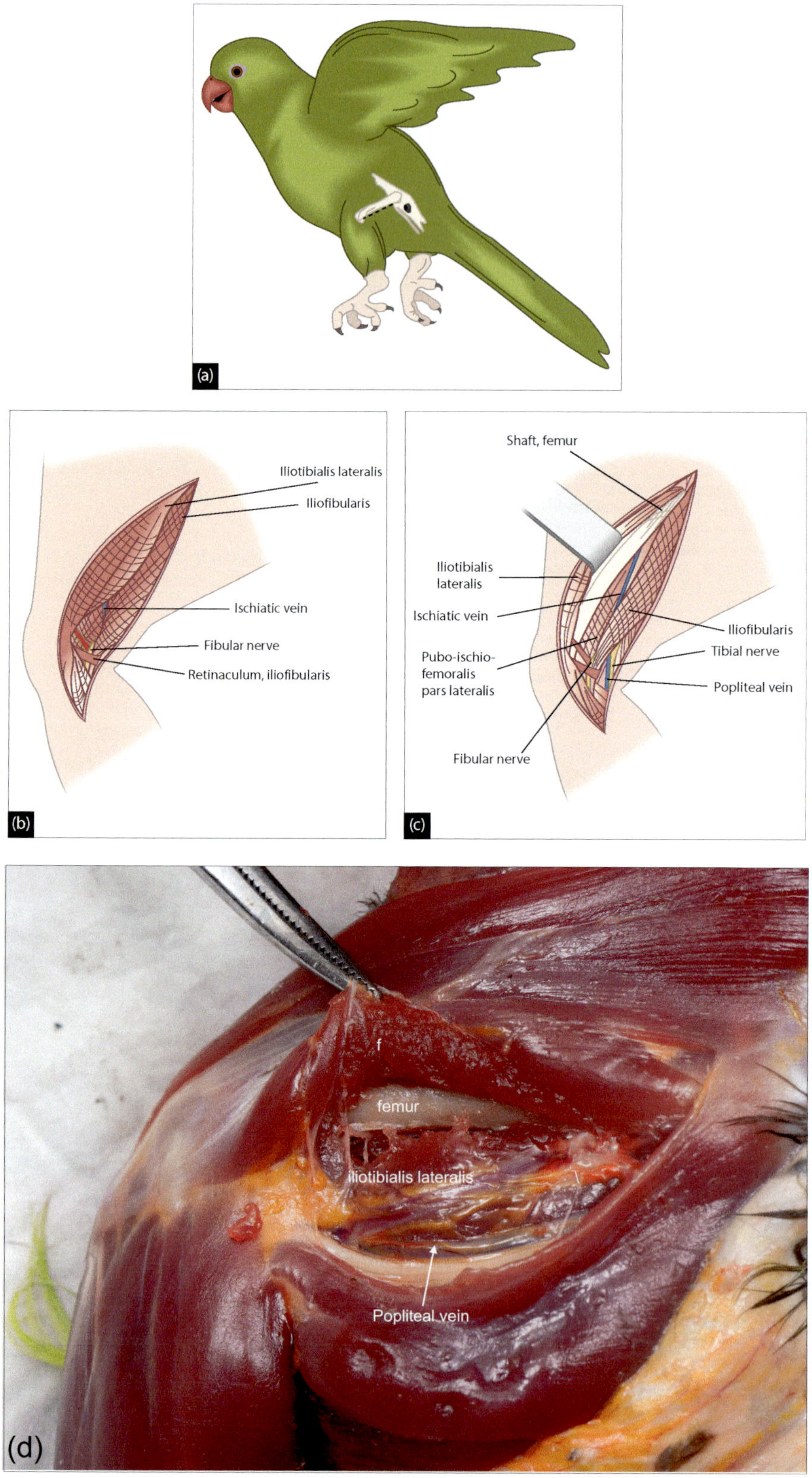

Figure 33.22 **Lateral approach to the femur. (a) Incision site. (b) Initial incision – superficial structures. (c) Exposure of the shaft of the femur. (d) Dissection view of lateral approach to the femur.**

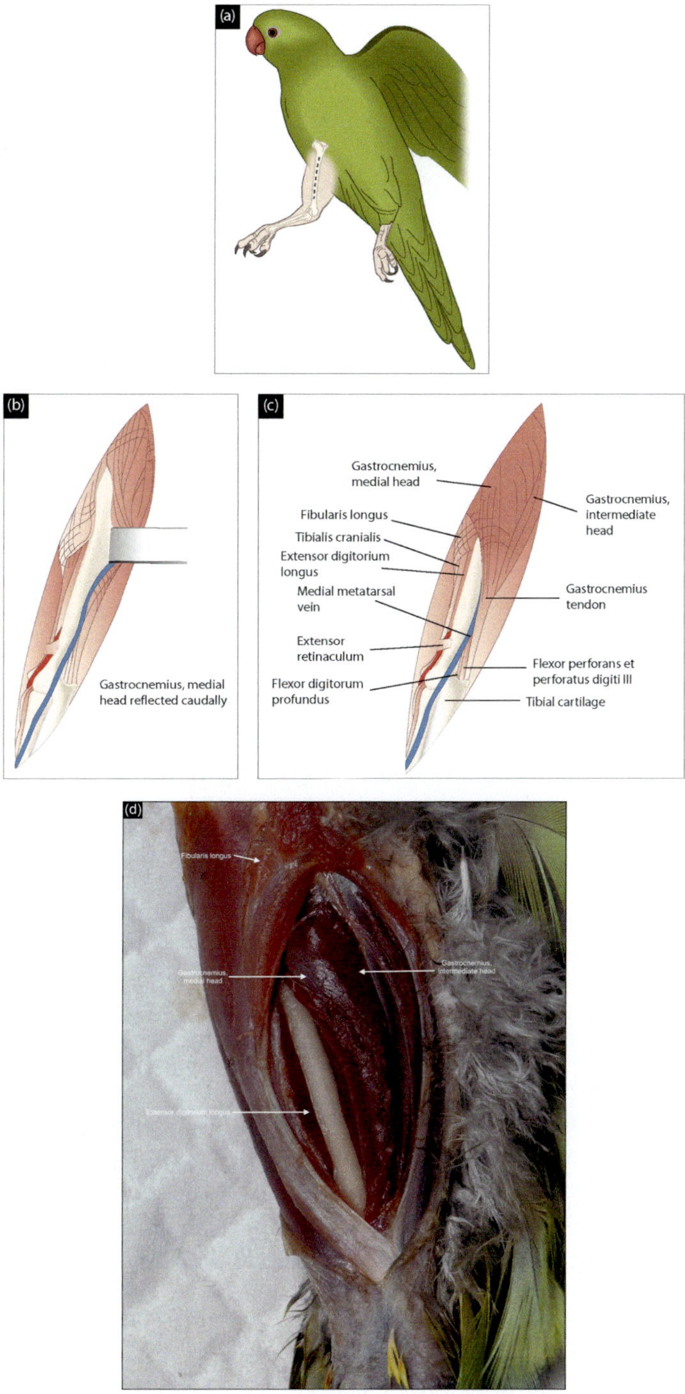

Figure 33.23 **Medial approach to the tibiotarsus. (a) Incision site. (b) Initial skin incision – superficial structures. (c) Exposure of the proximal shaft by retraction of the medial head of the gastrocnemius. (d) Dissection view of the medial approach to the tibiotarsus.**

The skin is incised between two metacarpals. The abductor digiti majoris muscle lies between two major tendons (deep and superficial digital flexors). These two flexor tendons are retracted cranially, and the muscle is transected and reflected. Soft tissues, tendons, and blood vessels are separated to approach the main, or primary, metacarpal bone.

APPROACHES TO THE BONES OF THE LEG

Femur

Two approaches are described. Vultures and gallinaceous birds have a well-developed iliotibialis lateralis muscle that covers the underlying structures; this muscle is not as well-developed in parrots, raptors, and owls.

The bird is placed in partial sternal or lateral recumbency. The femoral shaft is palpated, and an incision is made over it (**Fig. 33.22**) from the femoral trochanter to the lateral condyle. In vultures and chickens, the iliotibialis lateralis is bluntly separated, being careful not to go too far caudally and damage the sciatic nerve. In parrots, the iliotibialis lateralis (cranially, overlying the femorotibialis externus) and the iliofibularis (caudally, overlying the sciatic nerve) are separated. Retraction of the iliotibialis lateralis reveals the femorotibialis externus (cranial) overlying the femoral shaft, and the iliofibularis lying caudal to it. The femorotibialis externus can be retracted cranially to expose the femur. The ischiatic vein lies caudal to the femur in the middle third.

Tibiotarsus

There are four repair methods:

- Intramedullary pins retrograded out of the hock, then back up towards the stifle.
- Intramedullary pins are introduced from the tibial crest and normograded down through the proximal fragment and then into the distal fragment.
- ESF.
- TIF incorporates the latter two techniques.

A medial approach is used because of the muscle bulk on the lateral side (**Fig. 33.23**). Care must be taken, as the medial metatarsal vein crosses the hock and then runs caudally behind the tibiotarsal. Incise over the shaft of the tibiotarsus from just cranial to the hock to the medial femoral condyle.

The cranial complex of muscles (fibularis longus and tibialis cranialis) is separated from the medial head of the gastrocnemius. This division is easier to see in raptors than in parrots. To close the surgery site, this attachment is sutured before closing the skin.

Tarsometatarsus

Fractures of the tarsometatarsus tend to be open fractures and are best repaired with ESF. Closed fractures, especially in parrots, lend themselves to external coaptation.

If intramedullary pins are used, they should be introduced through the fracture and exteriorised retrograde laterally or medially to the joint, then passed normograde back into the distal fragment. The bone is U-shaped in cross-section, with a groove up the back. Flexor tendons run in this groove; extensor tendons, arteries, and the nerve supply run on the cranial aspect of the shaft; veins are on the medial and lateral sides. A lateral approach is used, with an incision made along the shaft.

FURTHER READING

Forbes NA. Fracture Management in Birds 2000 AAVAC Annual Conference Session:73-8.

Ponder JB, Redig PT. Avian Orthopedics: A Game Plan for Success. 2016 AAV Conference Proceedings Session #310.

Redig PT. Orthopedic Principles and Methods for the Humerus, Antebrachium, Femur, and Tibiotarsus of Birds 2003 AAV Conference Proceedings Session #0380b.

Redig PT. Orthopedic Fixation for Long Bone Fractures of Raptors and Other Large Birds: Pelvic Limb Raptors and Other Large Birds: Pelvic Limb, 2002 AAV Conference Proceedings Session #530b.

González MS. Avian Articular Orthopedics. *Veterinary Clinics of North America Exotic Animal Practice*, 2019: 22 (2): 239–251.

Hollwarth AJ, Dutton TAG. Retrospective Analysis of Pelvic Limb Fracture Management in Companion Psittacine Birds (60 Cases). *Journal of Avian Medicine and Surgery*, 2023: 37(2): 165–174.

CHAPTER 34

ONCOLOGY

INTRODUCTION

Advances in the prevention, diagnosis, and treatment of infectious diseases, malnutrition, and other non-infectious diseases of birds have meant longer and healthier lives for companion birds. With this rise in longevity comes a rise in degenerative diseases including neoplasia. The incidence of neoplasia in birds is unclear; while uncommon, it is considered that neoplastic diseases account for between 2% and 6% of cases presented to diagnostic laboratories.

It is now accepted that certain altered genes (oncogenes), whether inherited or caused by DNA damage, are a primary cause of neoplastic transformation in normal cells. In tumour cells, they are often found at high levels. While most normal cells undergo a programmed form of rapid cell death (apoptosis) when critical functions are altered, activated oncogenes can cause those cells to survive and proliferate instead. Most oncogenes require an additional step, such as mutations in another gene or environmental factors such as carcinogens (chemicals, irradiation), viral infections, or dietary influences, to cause cancer.

There also appears to be a hereditary predisposition to some cancers, with some avian species proportionally over-represented with certain neoplastic diseases. Among companion birds, Psittaciformes are most frequently diagnosed with neoplasia, with budgerigars being the most commonly affected species (17–24% overall incidence). Passeriformes have the lowest reported incidence of neoplasia among companion birds (**Table 34.1**).

TUMOUR CLASSIFICATION

Tumours can be classified by the cell types within the tumour:

- Carcinomas – malignancies of the epithelial cells.
- Sarcomas – malignancies of mesenchymal cells.
- Discrete or round cell neoplasms – the only common neoplasm of this type is lymphoid neoplasia.

(Note that poorly differentiated neoplasms produce cells having features of malignancy, but the cells are difficult to classify as carcinomas or sarcomas.)

They can also be classified by their tissue of origin:

- Ectodermal – skin, skin glands, mucous membranes.
- Endodermal – epithelium of the pharynx, digestive tract, respiratory tract, urinary tract.
- Mesodermal – connective tissue, bone, cartilage, muscle, blood, blood vessels, lymphoid organs, notochord, pleura, pericardium, kidneys, gonads.
- Teratomas (neoplasms of mixed tissue types, none of which originate from the site where the tumour is found) and dermoid cysts (fibrous wall lined with stratified epithelium, containing feather follicles, glands, and nerve elements, usually found in the skin, ovary, and eye) are generally considered rare.

DIAGNOSIS

History

A thorough history, including information on diet, past diseases and trauma, current medications, and reproductive history, should be obtained and a thorough physical examination performed. Because many tumours tend to affect animals of a particular age, gender, or species, this information is vital.

DOI: 10.1201/9781003272519-34

Table 34.1 **Commonly seen tumours, listed by species**

SPECIES	TUMOUR TYPES
African grey parrot (*Psittacus erithacus*)	Cutaneous papilloma Squamous cell carcinoma
Amazons (*Amazona* spp.)	GI papillomas, adenocarcinoma Bile duct carcinoma Lipomas
Budgerigar (*Melopsittacus undulatus*)	Renal adenocarcinoma Thyroid neoplasia Ovarian/oviductal adenocarcinoma Sertoli cell tumour (**Fig. 34.1**) Lipomas Xanthomas Fibroma/fibrosarcoma Uropygial adenoma/adenocarcinoma (**Fig. 34.2**)
Cockatiel (*Nymphicus hollandicus*)	Ovarian/oviductal adenocarcinoma Squamous cell carcinoma Fibroma/fibrosarcoma
Cockatoos (*Cacatua* spp.)	Soft tissue sarcoma Cloacal polyps Lipomas
Lovebirds (*Agapornis* spp.)	Soft tissue sarcoma Fibrosarcoma Lymphosarcoma
Macaws (*Ara* spp.)	GI papillomas Adenocarcinoma
Galah or rose-breasted cockatoo (*Eolophus roseicapilla*)	Lipomas

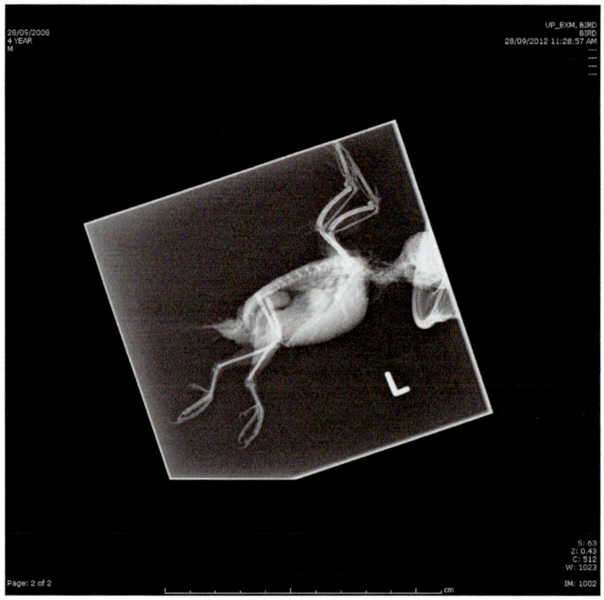

Figure 34.1 **Polyostotic hyperostosis in an adult male budgerigar with an oestrogen-secreting Sertoli cell tumour. The tumour is visible cranial to the kidney.**

Figure 34.2 **Preen gland carcinoma in a budgerigar.**

Physical examination

Many external neoplasms (e.g., lipomas, fibromas) are discrete masses easily detected on physical examination. In other cases, where the neoplasia is more diffuse (e.g., multicentric lymphosarcoma; **Fig. 34.3**), physical examination may reveal nothing more obvious than weight loss. Other tumours may exert their effect by compressing nearby structures (**Fig. 34.4**). As birds do not have lymph nodes, the spread of metastases relies on the overall evaluation of the patient to best determine the extent of the disease and the best treatment options.

Clinical pathology

Haematology and biochemistry analyses should be performed to assess overall health as well as to look for evidence of neoplasia.

a. Haematology may often be normal, but some abnormalities can include:
   - Anaemia (regenerative and non-regenerative) is often present with chronic neoplasia, an actively bleeding tumour, or neoplastic infiltration of the bone marrow.
   - Cytopaenia, which may result from the presence of haematopoietic malignancies in the bone marrow (myelophthisis).
   - Leucocytosis, either a mature lymphocytosis (chronic lymphocytic leukaemia), or a population of large immature blasts (acute

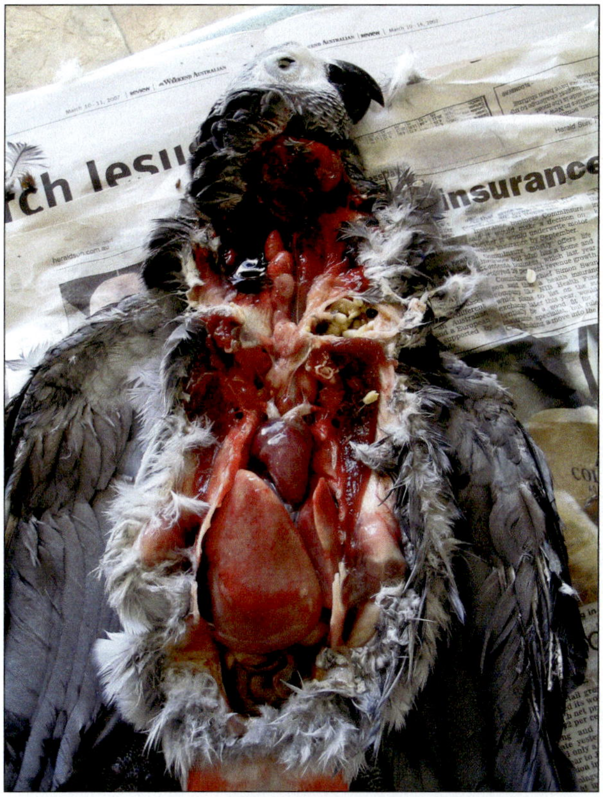

Figure 34.3 **A grey parrot with disseminated lymphosarcoma including the thymus and the liver.**

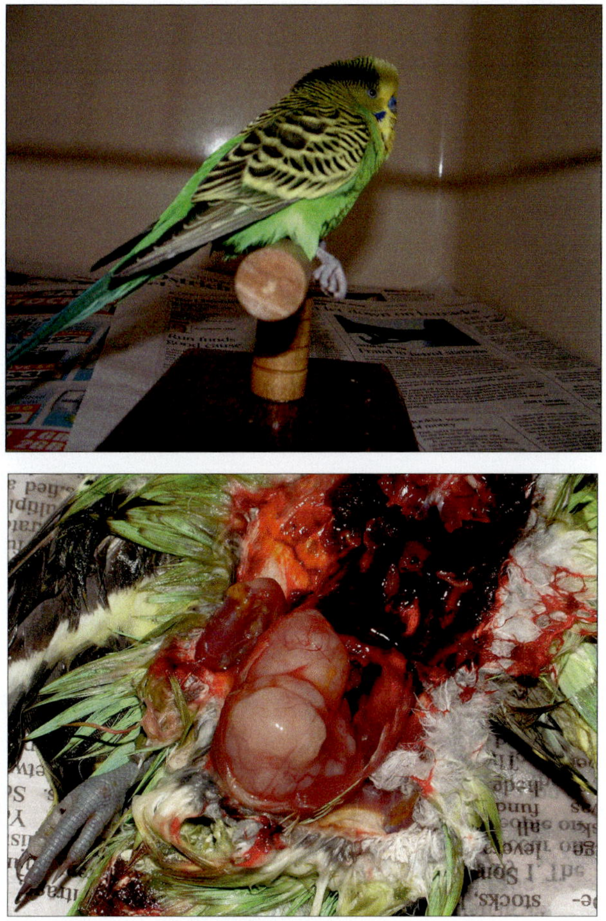

Figure 34.4 **(a) Unilateral paresis of a leg in a budgerigar is suspicious of a renal mass compressing the sciatic nerve. (b) On necropsy, the coelom was filled with a renal adenocarcinoma compressing the sciatic nerve as suspected.**

leukaemia), may be seen in patients with lymphoid neoplasia.

b. Biochemistry analysis. An abnormal biochemistry profile may be the sign of organ involvement by the tumour, or organ dysfunction related to age or other diseases, and may preclude the use of certain therapies.

Diagnostic imaging

Radiology, ultrasound, CT, and MRI are all useful adjuncts in the diagnosis of neoplastic disease. Although radiology may reveal skeletal neoplasia (**Fig. 34.5**) and some soft tissue masses (**Fig. 34.6**), CT (preferably with contrast) is becoming the preferred imaging modality in avian medicine (**Fig. 34.7**).

Molecular diagnostics

Given the known involvement of certain viruses (e.g., herpesvirus) in the development of some neoplastic diseases, the use of molecular diagnostic testing such as PCR can assist in determining the presence or aetiology of some neoplastic diseases.

Biopsy and cytology

Once the presence of a neoplastic mass has been confirmed or suspected, the use of cytology, incisional biopsy, or excisional biopsy can assist in further defining the disease process. Cytology of a sample collected by either fine needle aspirate biopsy or impression smears of an exfoliative mass can often be used to differentiate quickly between an inflammatory process (e.g., a granuloma) and neoplasia. If neoplastic, it may also assist in differentiating between a benign and malignant tumour.

Cytology can help differentiate neoplastic from non-neoplastic lesions and may broadly indicate tumour type, but accurate histogenesis and tumour grading require histology, with the exception of purely haematopoietic tumours, such as leukaemias. Special stains, including immunohistochemistry, may be needed to further determine various cellular antigens and cells of origin. However, the lack of

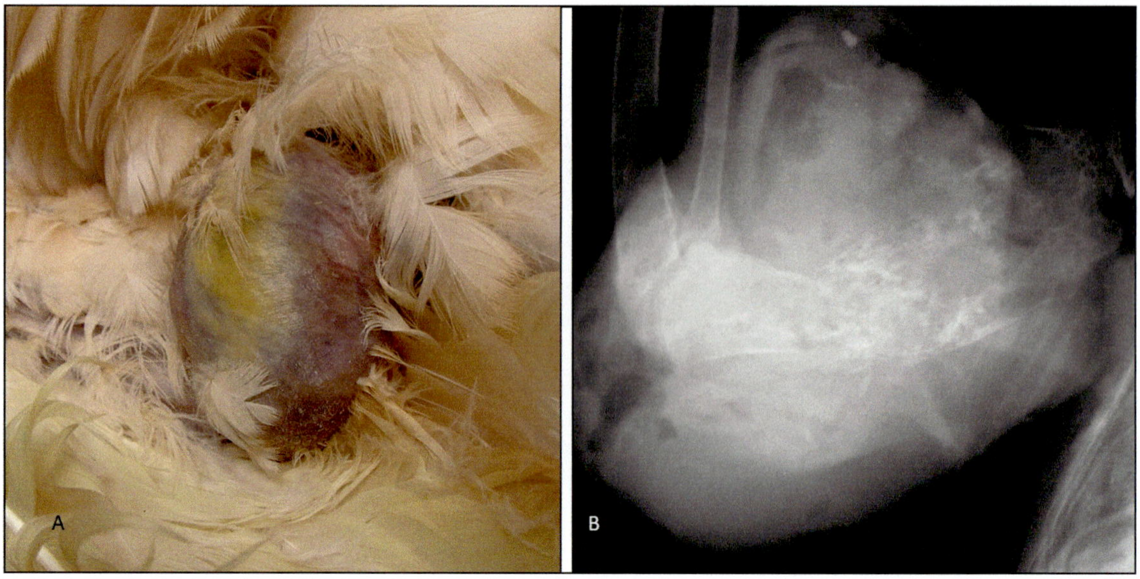

Figure 34.5 **This cockatoo presented with a drooping wing. A. Examination revealed a mass involving the left brachium. (The elbow is at the bottom of the image.) B. Radiology demonstrated extensive osteolysis and bone loss, consistent with an osteosarcoma.**

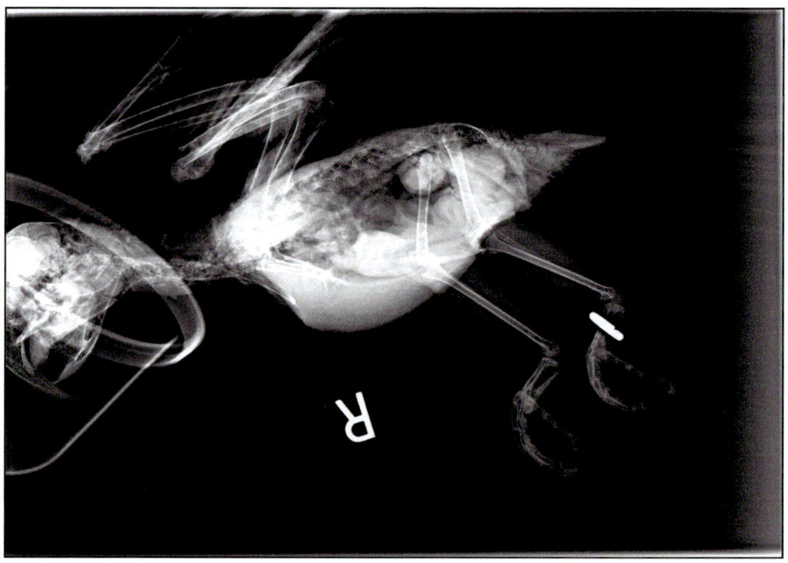

Figure 34.6 **This macaw presented with polydipsia and polyuria. Radiographs demonstrated renomegaly. Renal biopsy determined this was a renal adenocarcinoma.**

established immunohistochemistry positive controls in birds may hinder the interpretation of a sample.

TREATMENT OPTIONS

Considerations

Practical limitations, as well as anatomical or physiological differences, can complicate the application of certain cancer treatment modalities. Some of these considerations include:

- Metabolic rate. Birds have a higher metabolic rate compared with most mammals, and PK/PD data is available for only a few drugs and only in a few species. What we do know is that the extrapolation of dosing regimens from mammals to avian species is a good starting point.
- Vascular support. Repeated intravenous administration might be difficult in many avian patients because of both the limited continuous vascular access and the high vesicant and irritant properties of many chemotherapeutics. When

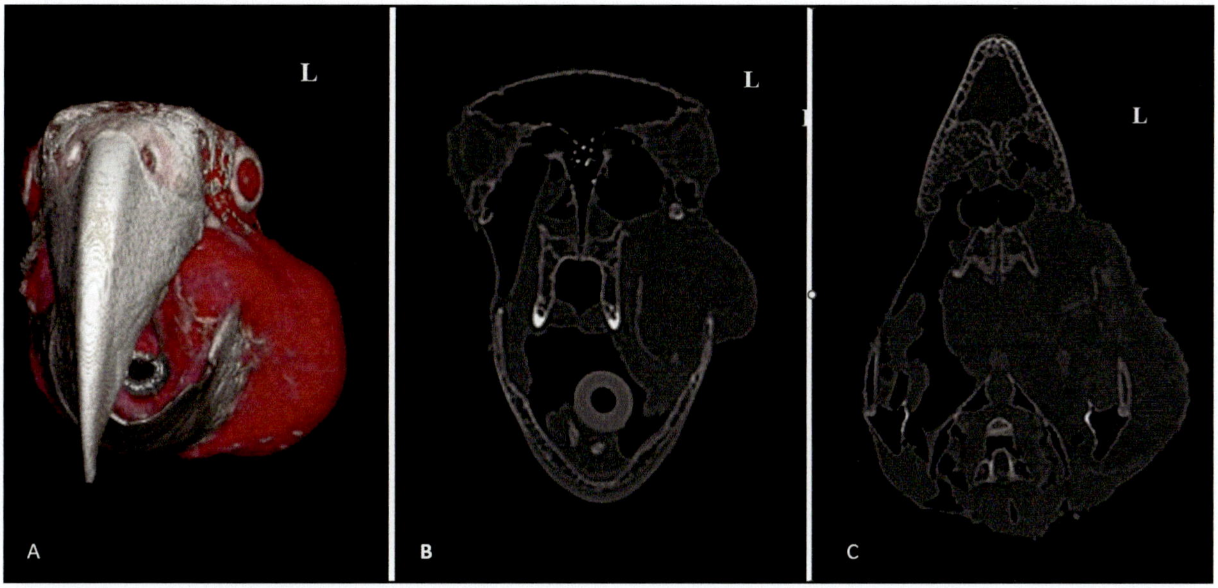

Figure 34.7 **An adult hyacinth macaw presented with facial swelling. The bird was examined by CT. (A) 3D reconstruction showing the extent of the mass. (B and C) Transverse and coronal images revealed the extent of the mass, which later proved to be an adenocarcinoma. CT allowed for a rapid determination of a poor prognosis, given the extensive state of the mass.**

long-term intravenous administration of tissue irritant drugs is contemplated, the use of surgically implanted vascular access ports should be considered.

- Renal excretion. Birds have a renal portal system and, in general, injection of chemotherapeutic drugs into leg vasculature should be avoided to prevent excretion of these agents before they enter the full circulatory system. Additionally, caution should be taken with the injection of any nephrotoxic agents into the leg vasculature for the same reason.
- Risks of corticosteroids and immunosuppressive agents. Any immunosuppressive therapy should be used with caution in most birds. Prophylactic antibiotic and antifungal therapy may be prudent whenever immunosuppressive drugs are used.
- Analgesia. Pain is a complex multifactorial phenomenon in patients with cancer. Extrapolating from medicine, it is now understood that the pain experienced by an oncological patient may not specifically be related to the tumour itself. There are numerous factors causing pain in cancer patients:
  - Pain from the tumour – compression on muscles, bones, nerves, spinal cord, and internal organs, or by stretching of skin.
  - Pain indirectly related to cancer, such as infections, metabolic imbalances, and myofascial pain.
  - Pain from diagnostic tests, such as biopsies and bone marrow collection.
  - Pain from treatment – surgery, chemotherapy, radiation, etc.

Analgesic plans should be incorporated into any treatment plan for a patient with cancer.

Staging

In most cancer patients, regardless of species, staging should be performed to determine if the neoplastic disease has spread to other parts of the body and, if so, how extensively, before embarking on a treatment course. In dogs and cats, there is a large body of information on this process, usually based on (a) the tumour type and likely behaviour, (b) spread to regional lymph nodes, and (c) the presence of metastatic disease. Unfortunately, much of this information is unavailable for birds with cancer. Staging should be performed by identifying the tumour type and its level of malignancy (based on cytology and/or biopsy), and then evaluating for metastatic disease (through physical examination and diagnostic imaging).

Goals and options

The goal of treating neoplasia is to either completely remove or destroy the neoplastic cells or, if this is not possible, to palliatively debulk the tumour to achieve a better quality of life for the patient. Treatment options are comparatively limited in many companion birds due to their relatively small size, the difficulty in gaining repeated intravascular access, and the lack of information regarding the prognosis and efficacy of anti-neoplastic treatments. Generally, it can be said that treatment options for avian neoplasia currently include:

- Surgical excision
- Chemotherapy
- Radiation
- Photodynamic therapy
- Cryosurgery

Each of these modalities can be used as a sole treatment or combined with any of the others.

Surgery

Surgery is usually reserved for localised and accessible skin or limb neoplasia. It is important to ensure that adequate margins are taken; however, this is not always feasible without amputation. Therefore, adjuvant chemotherapy or radiation is often used. Knowing the tumour type and behaviour is essential for predicting the likelihood of metastatic disease.

Adequate margins are needed to avoid (or minimise) recurrence of the tumour. These are harder to manage the second time because tissues have been disrupted and the resulting surgical field is much larger, making a second excision difficult. Recurrent tumours are often very aggressive and therefore carry a poorer prognosis. For this reason, the use of adjuvant chemotherapy or radiation may be used to destroy any cancer cells beyond the margins.

Chemotherapy

The basic concept of chemotherapy is that the drugs used target rapidly dividing cells, making cancer cells most susceptible to chemotherapy when actively dividing. However, not only do these drugs target the tumour cells, but they may also affect the bone marrow, gastrointestinal mucosa, and the feather follicle cells, causing unwanted side effects. There is no established standard of care for the treatment of avian cancer; dosages are mostly unknown and side effects difficult to predict. This should be discussed with clients so that expectations are realistic.

Most chemotherapy agents are dosed based on body surface area; however, this may not adequately correlate with the physiologic/pharmacologic factors that influence drug exposure. Another commonly used formula is based on weight (milligrams per kilogram).

When cancer cells are not dividing, they are partially/completely resistant to chemotherapy, making it often wise to combine drugs that have differing actions to maximise the chances of successful treatment with little/no resistance developing.

> **SAFETY MEASURES WHILE USING CHEMOTHERAPEUTICS**
>
> Chemotherapy is known to be mutagenic, carcinogenic, and teratogenic. Exposure may occur through preparation and administration of drugs, handling patient waste, transport, and waste disposal.
>
> Veterinarians and their staff administering chemotherapy need to be familiar with the chemotherapeutic agents used and the safety issues for themselves, owners, and staff.
>
> Luer-Lok syringes should be used to avoid drug leakage, and all personnel handling drugs or patients receiving chemotherapeutic drugs should wear protective clothing, including gloves, eye protection, closed-toe shoes, and long sleeves.
>
> Disposal of animal waste, unused/expired drugs, and all equipment used in treating a patient should follow workplace health and safety guidelines.

Chemotherapeutic drugs may be administered orally or parenterally (IV, IM, SC). An alternative route is delivering the drug directly into a solid tumour (intralesional chemotherapy). The goal of intralesional chemotherapy is to achieve a high concentration of chemotherapy at the tumour site, using a lower dose rate and thereby minimising systemic side effects. This route can be enhanced by the use of electrochemotherapy (ECT) (**Fig. 34.8**). An electrical current, when passed through the tumour, temporarily increases the permeability of the cells (electroporation) and enhances the uptake of the drug into the cells. This allows the use of relatively impermeable drugs (e.g., cisplatin and bleomycin) at doses much lower than when used systemically.

Close monitoring of patients for treatment-related side effects is important. These side effects include depression, vomiting, and heteropaenia.

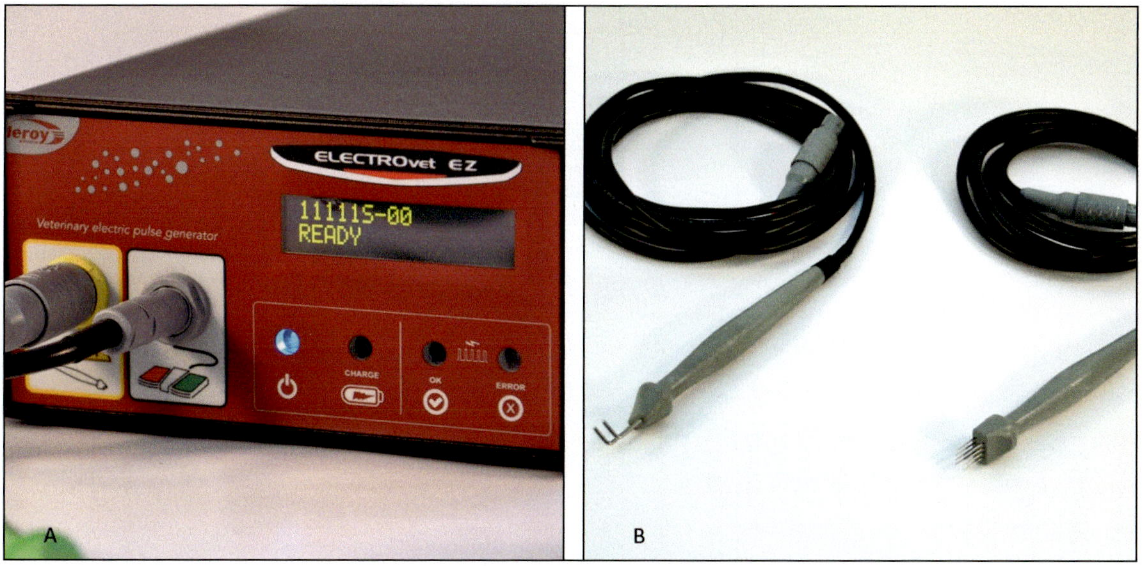

Figure 34.8 **(A) Electrochemotherapy unit (ElectroVET EZ, Leroy Biotech, Saint-Orens-de-Gameville, France). An electrical pulse transmitted between the two electrodes (B), while in contact with a tumour, increases the permeability of the neoplastic cells to chemotherapeutic drugs.**

Gastrointestinal side effects usually occur 3 to 5 days post-therapy and are readily managed with supportive medications. Bone marrow suppression and resulting heteropaenia happen most commonly at approximately 1 week after chemotherapy, and close monitoring of blood cell counts is, therefore, advisable.

Standard precautions for the use of chemotherapy apply, including appropriate personal protective equipment, suitable waste disposal, and separate caging. If treatment is given on an outpatient basis, owners must be cautioned to wear gloves when cleaning the bird's cage for 2–3 days post-therapy, especially as exact excretion times are unknown.

There are several classes of chemotherapeutic drugs (see **Table 34.2**).

Radiation therapy

Radiation therapy uses an external beam of radiation to destroy the proliferative capacity of the tumour without causing irreparable damage to surrounding normal tissues. The tumour, as well as a margin of normal surrounding tissue, is irradiated. However, the radiation response of the surrounding normal tissue often dictates the maximum tolerable dose that may be administered to a patient. Several reports of radiation therapy in birds exist and show mixed results in terms of success and side effects.

There are two forms of external beam radiation therapy, classified based on the energy of the photon emitted. They are:

- Orthovoltage radiation, a lower energy form of radiation, is more beneficial for treating surface tumours, but acute radiation side effects can be severe and quite uncomfortable for the patient.
- Megavoltage radiation, a higher energy radiation, is obtained from linear accelerators and cobalt machines. The penetrating ability of megavoltage radiation allows radiation therapy to be applied to deeper tumours for which orthovoltage would not be an option, as well as superficial tumours. Megavoltage radiation evenly distributes the radiation dose throughout the tissues in the radiation treatment field with a predictable interaction in tissues.

Another form of external radiation is plesiotherapy, i.e., beta (electron) irradiation with a Strontium-90 applicator via direct contact. Electrons produced have minimal penetration (<1 mm) and as such can only be used for very early, small superficial lesions such as squamous cell carcinoma. Strontium therapy can be repeated since there is minimal normal tissue damage. This form of therapy has been used to successfully treat uropygial squamous cell carcinomas in budgerigars.

Major problems in the use of radiation therapy include the lack of information on adequate fractionation and scheduling in avian patients, knowledge of the radiosensitivity of avian tumours, and the tolerance of normal tissues. Some of this information can be extrapolated from other species, in which haematopoietic tumours are exquisitely sensitive to

Table 34.2 **Chemotherapy drugs**

CLASS	EXAMPLES	ACTION	USES	COMMENTS
Vinca alkaloids/anti-mitotics	Vincristine	Arrests cell division in metaphase	• Leukaemic neoplasia myeloma • Soft tissue sarcomas • Mast cell tumours	• Given IV; perivascular irritation can be severe • Excreted in bile

Anti-tumour antibiotics	Doxorubicin	Binds to DNA and inhibits the synthesis of nucleic acids and cell division	• Carcinoma
			• Sarcoma
			• Round cell tumours
				• Given IV; very irritant if extravasated
				• Adverse effects:
				  • Bone marrow suppression
				  • GI tract effects
				  • Cardiotoxic
				  • Nephrotoxic
				  • Hypersensitivity reactions
				• Excreted in bile

Alkylating agents	Cyclophosphamide Chlorambucil Lomustine	Combines with DNA, damaging it	• Lymphomas • Carcinomas • Sarcomas	• Can be given orally or IV • Myelosuppressive • Inactivated in liver and excreted in urine
Platinum drugs	Cisplatin Carboplatin	Inhibits DNA synthesis and may have some alkylating effects	• Osteosarcoma • Carcinoma	• Given IV or intralesionally • Can be nephrotoxic, myelosuppressive, and neurotoxic • Carboplatin is not as toxic as cisplatin • Renal excretion
Other	L-Asparaginase	Asparaginase catalyses the conversion of L-asparagine to aspartic acid and ammonia, depriving the leukaemic cell of circulating asparagine, which leads to cell death	Leukaemia and lymphosarcoma	Can be given IM or SC

radiation therapy, whereas carcinomas are moderately sensitive and sarcomas much less so.

The key elements to consider when planning external beam radiation include:

- *Total radiation dose.* This is generally based on the tumour being treated and what normal tissues may be affected. Recommendations are usually based on treatment regimens for dogs and cats.
- *Fractionation.* Dividing the total dose into increments, which reduces long-term radiation side effects.
- *Treatment time.* Treatment over longer periods of time may help spare acutely affected tissues but may allow for tumour repopulation.

Photodynamic therapy

Photodynamic therapy utilises a plant extract or synthetic drug which, after injection, is selectively taken up by neoplastic cells. When irradiated by a laser, there is a reaction that results in cellular death. However, this technique requires a dedicated laser and direct access to the tumour, making it unsuitable for deep tumours. Patients may be photosensitive for 1–2 weeks after and need to be kept out of direct light. There are only limited reports of its use in avian medicine, but it has been used to successfully treat a sarcoma in an Indian ringneck (*Psittacula krameri*). It may be a useful tool to treat solid tumours in birds, but further studies are necessary to determine optimal technique and indications.

Cryotherapy

Cryotherapy uses liquid nitrogen to destroy targeted tissues. There are many different techniques described (open-spray, closed-spray, and cryoprobe), using single or multiple freeze–thaw cycles. The advantages of cryotherapy are the ease of treatment and low morbidity, as well as the possibility of performing repeat treatments if necessary. The disadvantage is that it is only suitable for superficial lesions of small diameter.

Hormonal therapy

Hormonal therapy using long-acting GnRH agonists has been used to treat reproductive neoplasia in both male and female birds. Several case reports exist on the use of leuprolide and deslorelin in birds with malignant ovarian, oviductal, and gonadal neoplasia.

FURTHER READING

Racnik J. et al. (2019) Electrochemotherapy with bleomycin in the treatment of squamous cell carcinoma of the uropygial gland in a cockatiel (*Nymphicus hollandicus*). *Journal of Exotic Pet Medicine* 29: 217–221.

Racnik J. et al. (2019) Electrochemotherapy with cisplatin for the treatment of a non-operable cutaneous fibroma in a cockatiel (*Nymphicus hollandicus*). *New Zealand Veterinary Journal* 67(3): 155–158.

Reavill DR (2004) Tumors of pet birds. *Veterinary Clinics of North America: Exotic Animal Practice* 7: 537–560.

Robat CS, Ammersbach M, Mans C (2017) Avian oncology: Diseases, diagnostics, and therapeutics. *Veterinary Clinics of North America: Exotic Animal Practice* 20(1): 57–86. https://doi.org/10.1016/j.cvex.2016.07.009. PMID: 27890293.

Zehnder A, Hawkins MG, Koski M, Kent MS (2010) Therapeutic considerations for squamous cell carcinoma: An avian case series. *Proceedings of the Annual Conference of the Association of Avian Veterinarians* 71.

Zehnder A, Graham J, Antonissen G (2018) Update on cancer treatment in exotics. The Veterinary Clinics of North America. *Exotic Animal Practice* 21(2): 465–509. https://doi.org/10.1016/j.cvex.2018.01.012.

APPENDIX 1: FORMULARY

EMERGENCY DRUGS

CLASS	DRUG	DOSE AND ROUTE	COMMENTS
Anti-muscarinic agent used to treat bradycardia	Atropine	0.02 mg/kg IM or IV	May thicken respiratory secretions.
	Glycopyrrolate	0.01 mg/kg IV	May thicken respiratory secretions.
Alpha- and beta-adrenergic agonist agent used systemically for treating anaphylaxis and cardiac resuscitation	Adrenaline (epinephrine)	0.01–0.1 mg/kg IV	Care with hypovolaemic patients (not a substitute for adequate volume replacement).
CNS stimulant used to stimulate respiration	Doxapram	2 mg/kg IM, IV	Not a substitute for ventilatory support. It will not increase respiratory frequency, but will increase tidal volume and oxygen consumption leading to decreased oxygen saturation. Consequently its use is often not recommended.
Opiate antagonist	Naloxone	0.01–0.04 mg/kg IV q 2–3 minutes to effect	At reversal dosages, naloxone can negate opioid analgesic effects. The reversal effect may last for a shorter time than the opioid effect, so repeated dosing may be needed.
Benzodiazepine antagonist	Flumazenil	0.05–0.1mg/kg IV or IM	Flumazenil is used for the reversal of benzodiazepine effects after overdoses or toxic effects.
Fluids	Crystalloid fluids (0.9% saline)	10 ml/kg IV bolus	Shock therapy.
	Colloid fluids	5 ml/kg IV bolus	Shock therapy.

ANTIBIOTICS

CLASS	DRUG	DOSE AND ROUTE	COMMENTS
Penicillins • Bactericidal. • Well distributed in extracellular spaces especially in inflamed tissues. • Do not readily penetrate the eye and CNS. • Excreted by the kidneys, largely unchanged, therefore high concentrations can be found in the urine. • Allergic reactions (anaphylaxis) have been reported, possibly from the procaine component in procaine penicillin. • Potentially synergistic in combination with aminoglycosides.	Amoxicillin	100 mg/kg q8h 200–800 mg/L water	Effective against gram-positive bacteria, especially *Staphylococcus*; most (all) gram-negative bacteria are resistant.
	Ampicillin	100 mg/kg IM q6h	
	Amoxicillin/ clavulanic acid	125 mg/kg PO q8–12h 60–120 mg/kg IM q12h	Clavulanic acid inhibits beta-lactamase (bacterial enzyme that inactivates many penicillins). Does not reach therapeutic plasma concentration in chickens.
	Carbenicillin	100–200 mg/kg IM q8h	Improved spectrum against *Pseudomonas* and other gram-negative bacteria. Oral form has poor bioavailability.
	Ticarcillin	200 mg/kg IM q8h	Similar to carbenicillin; much more active against *Pseudomonas.* Also found in combination with clavulanic acid and used at the same dose. Parenteral administration only.
	Piperacillin	100–200 mg/Kg IM q8–12h	Good activity against most gram-negative bacteria including *Pseudomonas*, *Klebsiella*, and *Enterobacter.* Piperacillin with tazobactam is commonly used to overcome beta-lactamase-containing bacteria. Parenteral administration only.
	Piperacillin/ tazobactam		
Cephalosporins • Three generations effective against both gram positives and gram negatives. • Well distributed in extracellular spaces. • Do not readily penetrate eye and CNS, except for cefotaxime and ceftazidime. • Synergistic with aminoglycosides.	Cephalothin	100 mg/Kg IM q6h	First-generation cephalosporin. Effective against most gram-positive cocci, many gram negatives and some anaerobes.
	Cephalexin	50–100 mg/Kg IM or PO q6–12h	
	Cefoxitin	75–100mg/Kg IM q8–12h	Second generation. Increased gram-negative activity.
	Ceftazidime	50–100 mg/Kg IM q6–8h	Third generation. Penetrates CSF.
	Ceftiofur crystalline-free acid (CCFA)	20 mg/kg IM at 80 (1) and 160 (2) hours Hawks: (gram-negative organisms) 10 mg/kg every 36 hours or 20 mg/kg every 96 hours (gram-positive organisms) 10 mg/kg every 45 hours or 20 mg/kg every 120 hours Ducks: 10 mg/kg q72h Egrets: 20 mg/kg q72h Flamingos: 10 mg/kg q72h	Third generation with activity against *Pasteurella*, *E. coli*, *Streptococcus*, *Staphylococcus*, and *Salmonella* spp. • Shows potential as a long lasting, injectable, bactericidal antibiotic. • Reported adverse effects have been largely limited to inflammation localised to injection sites. • Relatively long duration of plasma concentrations of active metabolites greater than the MIC of many bacterial.
	Ceftiofur sodium	10–20 mg/kg IM q12h	Third generation with activity against *Pasteurella*, *E. coli*, *Streptococcus*, *Staphylococcus*, and *Salmonella* spp. • In general shorter acting compared to CCFA. • Not to be confused with dose rates with CCFA.
	Cefotaxime	75–100 mg/kg IM q8–12h	Third generation. Expanded gram-negative spectrum. Penetrates CSF.

Continued

CLASS	DRUG	DOSE AND ROUTE	COMMENTS
Chloramphenicol • Bacteriostatic. • Highly lipid soluble, therefore good tissue penetration, including CNS and eye; tissue concentrations often exceed serum levels. • Reversible dose-related anaemia; CNS depression; loss of appetite has been reported in chickens, turkeys, and ducks. • Caution with people handling the drug. • Not to be used in food-producing animals.	Chloramphenicol palmitate (oral); Chloramphenicol succinate (injectable)	50–75 mg/kg IM or PO q6–12h	Broad spectrum against *Chlamydophila*, *Mycoplasma*, gram-positive and gram-negative bacteria, and some protozoa; many avian gram-negative isolates are resistant. Oral form (palmitate ester) leads to erratic blood concentrations. Injectable forms (succinate, propylene glycol based) provide more predictable serum concentrations.
Aminoglycosides • Bactericidal. • Synergistic with penicillins and cephalosporins. • Excellent spectrum against gram positives and gram negatives. • Ineffective against anaerobic bacteria and in proteinaceous environments such as abscesses and exudates. • Poor penetration into CNS and eye. • Nephrotoxic; in dehydrated birds and birds with compromised renal function the dose should be reduced or a less toxic drug selected.	Amikacin	10–20 mg/kg IM q12h 20 mg/kg q24h for 3 doses via regional limb perfusion (chickens)	• Greater activity against gram negatives including some resistant to gentamycin and tobramycin. • Achieves higher serum concentrations than gentamycin but is less toxic so levels are better tolerated. • Less toxic side effects and is the aminoglycoside of choice for use in birds.
	Gentamycin	5–10 mg/kg q12h	More toxic than amikacin. Not generally recommended due to narrow margin of safety.
	Tobramycin	5 mg/kg q12h	Pharmacology is similar to gentamycin but has greater activity against *Pseudomonas*. Neurotoxicity and nephrotoxicity may develop.
Quinolones • Bactericidal • Most avian gram-negative pathogens, some gram-positive pathogens, most *Mycoplasma* and possibly *Chlamydophila* are sensitive. • Not effective against anaerobes. • Achieves high levels everywhere especially in the liver and urinary tract; tissue concentrations may exceed serum concentrations. • May cause permanent articular defects in juveniles; use with extreme caution in growing birds. • Scattered anecdotal reports of aggressive, irritable behaviour in Amazon parrots treated with quinolones. • IM injection causes pain and necrosis at the site of injections. • Toxic reactions may be species-specific.	Enrofloxacin	10-30 mg/kg IM or PO q12h	• Excellent activity against *Mycoplasma*, some gram-positive, and most gram-negative bacteria. However, *Pseudomonas* resistance is common. Not suitable for anaerobic infections. • May have anti-chlamydial activity, but while treatment eliminates clinical signs it may not clear the carrier state. • Can cause vomiting in raptors when given orally. • If given in water, enrofloxacin may not achieve therapeutic levels except for highly susceptible bacteria. • Only a single IM injection should be given, as repeated injections cause significant bruising and muscle necrosis.
	Ciprofloxacin	15–40 mg/kg PO q12h	Antibacterial spectrum similar to enrofloxacin.
	Danofloxacin	15 mg/kg IM in brown pelicans	For *Mycoplasma* spp. and gram-positive bacteria.
	Marbofloxacin	2.5–15 mg/kg PO or IM q24h	Used in raptors, as it does not appear to cause nausea.
	Orbifloxacin	15–20 mg/kg PO q24h	Antibacterial spectrum similar to enrofloxacin.

Continued

CLASS	DRUG	DOSE AND ROUTE	COMMENTS
Trimethoprim–sulpha derivatives • Bacteriostatic. • Excellent gram-positive and gram-negative spectrum. • *Pseudomonas* and some strains of Enterobacteriaceae are resistant. • May be effective against some coccidia. • Has been used in combination therapy for sarcocytis infection. • Wide extracellular distribution. • Some birds (especially macaws) suffer GI upset and will regurgitate 1–3 hours after an oral dose. Incidence can be reduced if the drug is added to a small amount of food. • Sulphas are excreted via the same pathway as uric acid. In dehydrated birds and those with compromised renal function sulphas may form crystals and damage renal glomeruli. Therefore if serum uric acid is elevated, select another drug.	Trimethoprim–sulphadiazine (veterinary formulations)	50-100 mg/kg (of combined product) PO q12h or 8–20mg/kg IM q12h or 475–950 mg/L water for 5–7 days	Sterile abscesses and irritation have been reported following the use of the injectable veterinary products.
	Trimethoprim–sulphamethoxazole (medical formulation)	50–100 mg/kg (of combined product) PO q12h	

Tetracyclines
• Bacteriostatic
• Spectrum includes many gram-positive organisms; poor efficacy against most avian gram-negative isolates.
• Used for *Chlamydophila* and *Mycoplasma*.
• Wide volume of distribution.
• Side effects:
  • Anorexia, vomiting, diarrhoea.
  • Immunosuppression.
  • Hepatotoxicity (rare).
  • Localised tissue reactions to doxycycline injections formulations.
• Alteration of gut flora especially yeast overgrowth and secondary bacterial infections.
• Chelates calcium in gut and bone; dietary calcium interferes with the oral uptake of tetracyclines. Use cautiously in baby birds for extended periods of time.
	Oxytetracycline	10–50 mg/kg IM once every 3–5 days	• Poorly absorbed when given orally, therefore oral formulations are not recommended.
			• IM well absorbed and widely distributed, but it is irritating and will cause necrosis at the injection site.

	Chlortetracycline	40–50 mg/kg PO q12h or 1500 ppm in food or water	• Poor acceptance. • Immunosuppressive. • Can be used for flock treatment of chlamydiosis, but doxycycline is preferred.
	Doxycycline	25-50 mg/kg PO q24h or 100–500 mg/L water (African greys may require doses up to 800 mg/L) or 60–100 mg/kg IM once weekly	• More lipophilic than other tetracyclines; absorbed better and faster through the GI tract and greater bioavailability. • May offer significant advantages over other tetracyclines for treating chlamydiosis.

Continued

CLASS	DRUG	DOSE AND ROUTE	COMMENTS
Macrolides and lincosamides • Bacteriostatic. • Indicated for *Pasteurella*, *Bordetella*, some *Mycoplasma*, *Campylobacter* spp. and *Clostridia* spp. and obligate anaerobic bacteria. • Often used for susceptible URTI and osteomyelitis. • Active against gram-positive organisms and anaerobes, but virtually all aerobic gram-negative bacteria are resistant.	Erythromycin	60 mg/kg PO q12h	Active against chlamydiosis in people but not effective at the dose levels used in birds. Injectable form may cause severe tissue irritation.
	Clindamycin	25–150 mg/kg q12h	Used occasionally to treat osteomyelitis caused by susceptible gram-positive pathogens. Only oral forms are used in avian medicine because of injection site necrosis with injectable formulations.
	Lincomycin	75–100 mg/kg q12h or 500–750 mg/L water	Usually combined with spectinomycin. It has been used to treat respiratory and GI infections caused by gram-positive bacteria and *Mycoplasma*.
	Azithromycin	50–80 mg/kg PO q24h	New generation macrolide; it appears to be active against intracellular infections including *Chlamydophila*, *Toxoplasma*, *Plasmodium*, and *Cryptosporidium*.
	Tylosin	10–40 mg/kg IM q8–12h or 200–500 mg/L water	Used predominantly for suspected *Mycoplasma* infections and *Pasteurella*. Intramuscular injections can be very irritating.
Metronidazole • Bactericidal. • Effective against many gram-positive and most gram-negative obligate anaerobes. • Ineffective against aerobic bacteria. • Highly effective against many motile protozoa. • Well absorbed from GI tract. • highly lipophilic and penetrates bones, CNS, and abscesses.	Metronidazole	10–50 mg/kg PO q12–24h	Frequently used for anaerobic infections, (including clostridial infections) and motile protozoa (*Trichomonas*, *Giardia*, *and Cochlosoma*).

ANTIFUNGAL DRUGS

CLASS	DRUG	DOSE AND ROUTE	COMMENTS
Azoles • Inhibits ergosterol synthesis, increasing cellular membrane permeability; also causes secondary metabolic effects and growth inhibition. • Fungistatic, so months of therapy are often required. • Takes several days to reach steady-state concentrations.	Ketoconazole	20–30 mg/kg PO q12h for 10–30 days; should be administered with food	• For *Candida* infections. • Probably not as effective with *Aspergillus* infections as other azoles. • Distributes widely to tissues, but is highly protein-bound and does not penetrate CNS or eye fluids. • Water insoluble unless dissolved in acid first. • Toxic side effects seen in mammals are rarely seen in birds, but may be associated with hepatotoxicity.
	Enilconazole	Nebulisation solution; enilconazole 100 mg; DMSO 9 ml; 0.7% NaCl 90 ml. For topical use, dilute 1:10. Treat twice daily.	Administered topically or nebulised.
	Fluconazole	1.5–10 mg/kg PO q12h or 10–20 mg/kg PO q24–48h	• Effective against *Candida* (especially mycelial form), *Aspergillus*, and *Cryptococcus*. • Penetrates eye, CNS, and CSF. • Best safety margin of the azoles.
	Itraconazole	5–10 mg/kg PO q24h given with food for at least one month after signs have resolved	• Used for gastrointestinal and cutaneous candidiasis, dermatophyte infections. • Maybe more effective than other azoles and amphotericin B in *Aspergillus* infections. • Poorly distributed to CSF, ocular fluids, and plasma (tissue concentrations are higher than plasma). • Although less toxic than amphotericin B, it appears that African greys may be sensitive to this drug, causing anorexia and depression. It is not recommended to use it in this species, or to use it at 2.5–5 mg/KG PO q24h.
	Miconazole	5 mg/kg intratracheally q12h 5 days or Apply topical gel q12h	• Can be used topically, intratracheally, or nebulised. • Many toxic side effects – cardiac arrhythmias, cardiac arrest – if given IV too quickly.
	Clotrimazole	2 mg/kg intratracheally q24h or 10 mg/ml NaCl as a nasal flush q12h	• Poorly absorbed, therefore topical use or nebulised only. • Effective against *Aspergillus*. • It may cause local irritation.
	Voriconazole	12–18 mg/kg PO q12h. Higher doses may be needed to maintain plasma concentrations during long-term treatment 12.5 mg/kg IM q12h for 7 days (IM formulation: for pigeons)	• High oral bioavailability. • Can cross into the CNS. • IV formulation is safe given IM; no adverse effects seen despite biochemistry changes.

Continued

CLASS	DRUG	DOSE AND ROUTE	COMMENTS
Terbinafine • An allylamine; inhibits squalene epoxidase, an enzyme that is critical for fungal sterol and cell wall synthesis.	Terbinafine	10–15 mg/Kg PO q12–24h or Nebulisation (1 mg/ml; 500 mg terbinafine + 1 ml acetylcysteine as a 2% dilution, used for 30 minutes q8–12h	• *Aspergillus* and dermatophytes. • Good oral absorption; distributed well to fat and skin.
Polyenes • Acts by binding to sterols (primarily ergosterol) in the cell membrane and alters the permeability of the membrane allowing the loss of intracellular potassium and other cellular constituents.	Amphotericin B	1.5 mg/kg IV q8h for 3–5 days or Intratracheally 1 mg/kg q12h or Nebulise at 0.3–1 mg/ml for 15 minutes q612h or 100 mg/kg PO q12h 10–30 days for the treatment of *Macrorhabdus ornithogaster*	• Fungicidal. • Active against both yeast and hyphal fungi. • Used orally for the treatment of *Macrorhabdus ornithogaster.* • Not absorbed orally; for aspergillosis it must be administered IV, topically, or by nebulisation. • Widely distributed when given IV, but only minor systemic absorption with aerosol or topical administration. • Nephrotoxicity is common in mammals but uncommonly reported in birds. • Topical solutions must be diluted before flushing into closed spaces (e.g., sinuses) to prevent irritation. • Combine with itraconazole for best response.
	Nystatin	200,00–300,000 units/kg PO q812h	• Not systemically absorbed. • Must come into contact with yeast to be effective, therefore tube feeding may bypass oral lesions. • Do not mix with hand rearing formula to ensure concentration and contact time is maximised.
Fluorinated pyrimidines • Inhibit macromolecule synthesis.	5-Fluorocystine (flucytosine)	60 mg/kg PO q12h for birds >500 g 150 mg/kg PO q12h for birds <500 g	• Rarely used. • Can be used with amphotericin B to treat aspergillosis. • Resistance develops rapidly when used alone.

MYCOBACTERIAL TREATMENT PROTOCOLS

- Protocols involve combinations of antituberculous drugs, as resistance to a single drug can develop rapidly.
- Because these organisms can only be killed during replication, which occurs once every 16 to 20 hours, treatment often lasts 9 months or longer. Organisms can also persist for months in caseous lesions and macrophages where replication seldom occurs.
- Therapy is aimed at achieving the highest possible blood levels of drug rather than maintaining constant blood levels. Therefore, the highest tolerable doses of drugs are given once daily.
- Treatment protocols include the following combination of drugs, most given orally once daily:
    - Amikacin 20 mg/kg q24h IM, Enrofloxacin 30 mg/kg
    - Enrofloxacin 30 mg/kg, Ethambutol 30 mg/kg, Rifabutin 15 mg/kg
    - Clarithromycin 55 mg/kg, Rifabutin 6 mg/kg, Ethambutol 30 mg/kg, Enrofloxacin 30 mg/kg
    - Clarithromycin 55 mg/kg, Rifabutin 15 mg/kg, Ethambutol 30 mg/kg, Enrofloxacin 30 mg/kg
    - Clarithromycin 55 mg/kg, Rifabutin 15 mg/kg, Ethambutol 30 mg/kg

ANTIVIRAL DRUGS

CLASS	DRUG	DOSE AND ROUTE	COMMENTS
Acyclovir • Used for treating outbreaks of herpesvirus infection	Acyclovir	80–330 mg/kg PO q8h for a minimum of 7 days	• Works better on in-contact birds rather than affected birds. • Injectable form is very irritant.
Interferon • Group of small protein and glycoprotein cytokines naturally produced by the immune system following natural infection or vaccination. • Interferons protect the bird by suppressing cell proliferation, inhibiting viral replication, and augmenting the activity of macrophages and T lymphocytes. • They are species-specific and a mammalian interferon would not be expected to have a significant action in birds. However, cross-species reactivity has been reported in birds.	Avian interferon	1,000,000 units IM q24h 90 days or 1,000,000 units IM every 2–7 days for 3 treatments	• Initially the use of interferon was limited due to the difficulty in manufacturing the protein in large enough quantities but the recent development of recombinant DNA technologies has made interferon economic and easy to produce. • Not commercially available at time of writing.

ANTIPROTOZOAL DRUGS

CLASS	DRUG	DOSE AND ROUTE	COMMENTS
Nitroimidazoles • Used against motile protozoa (*Trichomonas, Giardia, Cochlosoma, Spironucleus, Histomonas*). • Candida overgrowth after prolonged therapy can occur, especially in cockatiels.	Carnidazole	20-30 mg/kg PO once	• Safe with good efficacy, but must be given on an empty stomach as vomiting may occur otherwise.
	Ronidazole	6-10 mg/kg PO or 60–600 mg/L water for 7 days	• Treatment of choice for *Trichomonas*; dose can be increased to 600 mg/L water if resistance develops.
	Dimetridazole	200–400 mg/L water for 5 days	• Toxic if overdosed: depression, anorexia, ataxia, seizures (incorrect water preparation or excessive water intake, e.g., hot weather, feeding chicks) • Do not use in pekin robins and finches. Use with care in lorikeets and mynahs (use lower dose).
	Metronidazole	50 mg/Kkg PO q12h–q24h or 40–80 mg/L water for 3 days	Frequently used for motile protozoa (*Trichomonas, Giardia and Cochlosoma,*).

Continued

CLASS	DRUG	DOSE AND ROUTE	COMMENTS

Sulphonamides
• Used to treat coccidia.
	Sulphachlorpyridazine	100–400 mg/L water for 3–5 days	• Contra-indicated with:
			  • Dehydration
			  • Liver disease
			  • Renal disease
			• Treatment periods greater than 2 weeks may require supplementation with folic acid.
			• All treatments should be repeated after 5 days to allow for the prepatent period of coccidia.

	Sulphadimethoxine	20–50 mg/kg PO q12h for 3–5 days or 250–500 mg/L water for 5–7 days	
	Sulphadimidine	50–150 mg/kg PO q12h or 3330–6660 mg/L water for 5 days	
	Sulphamethazine	75–185 mg/kg PO q24h for 3 days or 125 mg/L water for 3 days	
	Sulphaquinoxaline	100 mg/kg PO q24h for 3 days or 250–500 mg/L water for 5–7 days	
Benzeneacetonitrile derivatives • Used to treat coccidia. • All treatments should be repeated after 5 days to allow for the prepatent period of coccidia.	Clazuril	2.5 mg tablet per pigeon or 7 mg/kg PO q24h for 2–3 days	Single dose may suppress oocyst excretion for 2 weeks.
	Diclazuril	10 mg/kg PO q24h on days 0, 1, 2, 4,6, 8, and 10 or 5 mg/L water	Can be used for treating *Toxoplasma.*
	Toltrazuril	25–75mg/L for 5 days or 7–15 mg/Kg q24h for 3 days or 20–35 mg/kg PO once	Has also been used to treat *Atoxoplasma* in canaries.
Amprolium and amprolium–ethopabate • Used to treat coccidia. • Inhibits the active transport of thiamine into the cell. Coccidia are 50 times as sensitive to this inhibition as the host.	Amprolium Amprolium–ethopabate	15–30 mg/kg q24h for 1–5 days or 50–100 mg/L water for 5–7 days	• Resistance is common. • Treatment periods greater than 2 weeks may require supplementation with folic acid. • All treatments should be repeated after 5 days to allow for the prepatent period of coccidia.

Pyrimethamine
• Used for:
  • Toxoplasma
  • Atoxoplasma
  • Sarcocystis
  • Leucozytozoonosis
	Pyrimethamine	0.5 mg/kg PO q12h 30 days	

Continued

CLASS	DRUG	DOSE AND ROUTE	COMMENTS

Quinacrine HCl
• Used for:
  • Atoxoplasma
  • Plasmodium
	Quinacrine HCl	5–10 mg/kg PO q24h 10 days	Overdosage may cause hepatotoxicity.

Chloroquine
• Used for:
  • Plasmodium
  • Other blood parasites
	Chloroquine	10–25 mg/kg PO as a first dose, then 5–15 mg/kg at 6, 18, and 24 hours	Use in conjunction with primaquine.

Primaquine
• Used for:
  • Plasmodium, in combination with chloroquine
  • Sarcocystis
	Primaquine	0.3–1 mg/kg PO q24h for 3–10 days	Use in conjunction with chloroquine.

Mepacrine HCl • Used for plasmodium in canaries	Mepacrine HCl	0.24 mg/kg PO q12h	

INTERNAL PARASITICIDES

CLASS	DRUG	DOSE AND ROUTE	COMMENTS
Fumarate reductase inhibitors • Benzimidazole derivatives. • Used mainly for treating nematodes. • Interferes with energy metabolism; prevents the parasite from using sugars. • Usually requires several days of treatment, although single high doses can be effective. • Low toxicity, but can affect haematopoietic cells and intestinal epithelium. • Do not use during breeding, as embryotoxic effects have been observed. • Toxicity varies with species and drug.	Albendazole	10–50 mg/kg PO once	Can be used for treating Microsporidia, as well as some nematodes. Can be toxic in raptors at doses over 25 mg/kg.
	Thiabendazole	40–100 mg/kg q24h 7 days or 100–500 mg/kg once	Less effective than fenbendazole.
	Cambendazole	60–100 mg/kg q24h 3–7 days	
	Fenbendazole	25–50 mg/kg once or 8–10 mg/kg daily for 3–4 days or 50 mg/kg/day × 3 days (may treat *Giardia*)	• May also be effective against some cestodes, trematodes and *Giardia*. • Can interfere with feather growth during moulting. • Can be toxic to bone marrow, causing leucopaenia.
	Mebendazole	10–25 mg/kg q12h for 5 days or 10–20 mg/L water for 3–5 days	Effective against some cestodes and trematodes.
	Oxfendazole	10–40 mg/kg once	
	Febantel	30 mg/kg once	

Continued

CLASS	DRUG	DOSE AND ROUTE	COMMENTS
Imidithiazoles Used against nematodes. Stimulates cholinergic receptors, causing paralysis of parasite. Regurgitation, anorexia, diarrhoea, and neurological signs (ataxia, head tilt, and torticollis) are occasionally seen.	Levamisole	20–40 mg/kg PO once or 100–200 mg/L water for 3 days, repeat in 2 weeks	Not recommended in finches, lories, or debilitated birds. Care should be taken during hot weather or occasions where water intake is likely to be high, as this can lead to overdosage.

Gamma amino butyric acid (GABA) interfering drugs
• Macrocyclic lactones
• Used against:
  • Ascarids and other nematodes.
  • Blood-sucking external parasites
  • *Cnemidocoptes*
  • Microfilaria
• Toxicity is low, but overdosing and idiosyncratic reactions include severe depression, inactivity, excessive sleeping, and neurological signs.
• Budgerigars, African finches, and European finches may be more sensitive.
	Ivermectin
	Doramectin
	Moxidectin
		200–400 μg/kg IM, PO, or topically, once	Toxicity may be higher if the drug is injected (particularly via the IM route).

Adenosine triphosphate synthesis blockers • Block synthesis of ATP and produce paralysis by interfering with energy metabolism.	Niclosamide	50–100 mg/Kg PO once, repeat in 10–14 days	• Used for treating cestodes and trematodes. • Praziquantel is more effective.
	Rafoxanide	10 mg/kg PO	• Used for treating trematodes and cestodes.
Miscellaneous	Praziquantel	10–20 mg/kg PO or IM once	• Used for treating cestodes and trematodes. • Caution with IM in finches; sudden death has been reported. • Very unpalatable.
	Piperazine	100–250 mg/kg PO once or 1000–2000 mg/L water for 3 days	Used for ascarids only. Resistance is common.
	Pyrantel tartrate Pyrantel pamoate Pyrantel embonate	7–25 mg/kg PO once	• Used for nematodes. • Poorly absorbed from the GI tract. • Good safety margin.
Paromomycin	Paromomycin	100 mg/Kg PO q12h 7 days	• Used for treating Cryptosporidia.

EXTERNAL PARASITICIDES

CLASS	DRUG	DOSE AND ROUTE	COMMENTS
Pyrethrins and synthetic pyrethroids	Pyrethrin Permethrin	Spray or wash	Often combined with insect growth regulators.
Piperonyl butoxide	Piperonyl butoxide, usually combined with a pyrethrin or permethrin	Spray or wash	
Organophosphates	Malathion Maldison	Spray or wash	Toxic; use is not recommended because of hazards to birds and owners.
Fipronil	Fipronil	3 mg/kg spray or spot-on application	• Not registered for use in birds. • Spray works better than spot-on application. • Beware of inhalation or dermal absorption of the alcohol base. • Beware of the drying effect the alcohol base may have on feathers.
Carbaryl 5%	Carbaryl	Dust lightly or Add 1–2 teaspoons to nesting material	• Carbamate flea powder. • Use is illegal in many countries.

Gamma amino butyric acid (GABA) interfering drugs
• Macrocyclic lactones
• Used against:
  • Blood-sucking external parasites
  • *Cnemidocoptes*
• Toxicity is low, but overdosing and idiosyncratic reactions include severe depression, inactivity, excessive sleeping, and neurological signs.
• Budgerigars, African finches, and European finches may be more sensitive.
	Ivermectin
	Moxidectin
		200–400 ug/kg topically, IM, or PO once	Toxicity may be higher if the drug is injected (particularly via the IM route).

HORMONAL THERAPIES

CLASS	DRUG	DOSE AND ROUTE	COMMENTS
Reproductive hormones	Cabergoline	10-20 μg/kg PO q24hrs 30 μg/kg PO q24h × 14 days in quails	Inhibits prolactin release by direct stimulation of dopamine receptors in prolactin-releasing cells in the anterior pituitary. NIL effect on egg laying and prolactin concentration.
	Chorionic gonadotropin (HCG)	500–1,000 units IM on days 1, 3, and 7, and then every 2 weeks as required	• LH agonist. • Continued use is limited by formation of anti-HCG antibodies.
	Deslorelin	4.7 mg implant SC	GnRH agonist, will last for 6–18 months if environmental, behavioural, and dietary changes are made.
	Leuprolide acetate	100–700 μg/Kg every 2 weeks for 3 treatments, then as required	• GNRH agonist. • Must be combined with environmental, behavioural, and dietary changes for effect.
	Medroxyprogesterone acetate	5–25 mg/kg IM every 4–6 weeks	• Historically used for feather-picking behaviour, but side effects (PU/PD, polyphagia, diabetes mellitus, hepatopathy, weight gain, sudden death) have led to its use being discontinued. • At very low doses it may be useful in suppressing ovulation.
	Megestrol acetate	2.5 mg/kg PO q24h 7 days, then weekly as required	• Seldom used because of severe side effects, similar to medroxyprogesterone.
	Oxytocin	0.5–1.0 IU/kg repeated every 30–60 minutes	• Used to stimulate oviductal contractions, but is not a naturally occurring hormone in birds and it may not be effective. • Use is contra-indicated if uterovaginal sphincter is not dilated. • Should be used in conjunction with calcium gluconate injections.
	Prostaglandin E2 (dinoprost)	0.02–0.1 mg/kg applied topically into the cloaca	Used to induce egg laying in egg-bound birds; simultaneously relaxes utero-vaginal sphincter while contracting oviduct.
Thyroxine	Levothyroxine	5–200 μg/Kg PO q12h	• May induce moult. • Monitor for cardiotoxicity.
Insulin	Short-acting insulin NPH insulin Ultralente insulin	The immediate use of short-acting insulin (0.1–0.2 U/kg) can be used to initially stabilise the patient, but long-term control at home usually required longer-acting insulin (NPH or ultralente). Dose rates vary considerably and should be based on the observed effects. They range from 0.067 to 3.3 U/kg q12–24h.	For insulin-dependent diabetes.

Continued

CLASS	DRUG	DOSE AND ROUTE	COMMENTS
Sulfonylureas	Glipizide	0.5 mg/kg PO q12h	• Stimulates the beta cells in the pancreas to secrete additional endogenous insulin. • Extra-pancreatic effects include enhanced tissue sensitivity of circulating insulin.

Glucocorticoids
• Previously recommended for the treatment of shock and as an anti-inflammatory drug.
• Side effects are severe:
  • PU/PD
  • Catabolism
  • Liver and kidney disease
  • Immunosuppression
• Topical application seems to be as severe as, or worse than, parenteral administration.
	Dexamethasone	2–4 mg/kg SC, IM, IV once	• With the possible exception of prednisolone sodium succinate, the use of glucocorticoids in birds is dangerous and cannot be recommended.
			• Only indication may be immune-mediated disorders such as IMHA.

	Hydrocortisone	10 mg/kg IM once	
	Methylprednisolone acetate	0.5–1 mg/kg IM once	
	Prednisolone	0.5–1 mg/kg IM once	
	Prednisolone sodium succinate	2–4 mg/kg IV once	
	Triamcinolone	0.1–0.5 mg/kg IM once	

DRUGS USED TO TREAT LIVER DISEASE

CLASS	DRUG	DOSE AND ROUTE	COMMENTS
Anti-inflammatory • Anti-fibrotic. • Anti-inflammatory.	Colchicine	0.04–0.2 mg/kg PO q24h	May cause nausea and vomiting in some birds.
Chelating agent • Chelates iron.	Deferoxamine	100 mg/kg PO, IM or SC q24h	• Preferred iron chelator for iron storage disease. • Avoid in birds with renal disease.
Diuretic • Reduce ascites.	Furosemide	0.15–2 mg/kg IM, SC, PO q12–24h	Overdose can lead to dehydration and electrolyte abnormalities. Lorikeets are very sensitive to this drug – use with caution.
Lactulose • Does not treat liver disease. • Reduces absorption of ammonia from intestine by altering the pH of the intestinal lumen and promoting an osmotic catharsis. This reduces the ammonia levels presented to the liver. • Best results are seen in carnivorous birds, e.g., raptors, as vegetable protein lacks many encephalopathic precursors.	Lactulose	150–650 mg/kg PO q8–12h	Can cause diarrhoea.
Antioxidant • Unproven remedy that offers good promise in avian medicine. • Antioxidant. • Enhances protein synthesis and hepatocellular regeneration. • Protective effect against hepatotoxins. • Suppresses fibrogenesis. • Promotes fibrolysis.	Silibinin (silymarin, milk thistle)	50–75 mg/kg PO q12h 35 mg/kg PO q12h	Use a low alcohol or alcohol-free base.

Continued

CLASS	DRUG	DOSE AND ROUTE	COMMENTS
Hepatoprotective drugs	Ursodeoxycholic acid	10–15 mg/kg PO q24h	• Bile acid. • Cytoprotective. • Reduces involvement of hepatocytes and biliary epithelium in inflammatory process. • Changes mix of bile acids to eliminate toxic bile acids from liver.

DRUGS USED TO TREAT KIDNEY DISEASE

CLASS	DRUG	DOSE AND ROUTE	COMMENTS
Decrease production of uric acid	Allopurinol	10 to 15 mg/kg PO q12–24h	• May worsen renal disease. • Maintain good hydration.
Reduce inflammation associated with articular gout	Colchicine	0.04–0.2 mg/kg PO q24h	May cause nausea and vomiting in some birds.
Anti-inflammatory	Aspirin	1 mg/kg PO q24h	Used together for the treatment of membranous glomerulonephritis.
	Omega-3 and omega-6 fatty acid supplementation, mixed in a ratio of omega-6:omega-3 of 6:1	0.1 ml/kg PO q12h	

DRUGS USED TO TREAT CARDIOVASCULAR DISEASE

CLASS	DRUG	DOSE AND ROUTE	COMMENTS
Angiotensin converting enzyme (ACE) inhibitors • Block the formation of angiotensin II, thereby blocking the renin–angiotensin–aldosterone system. • They also have a diuretic effect.	Enalapril	0.5–1.25 mg/kg PO q12h	Side effects could include hypotension, reflex tachycardia, dehydration, gastrointestinal disorders, renal dysfunction, and hyperkalaemia.
	Benazapril	0.5 mg/kg PO q12h	
Calcium channel blockers • Indicated for hypertension. • Reduce the strength of cardiac contractions by hindering extracellular calcium influx in cardiomyocytes and vascular smooth muscle cells. • Net effect is peripheral vasodilation and reduced peripheral resistance.	Amlodipine	0.4 mg/kg PO q24h (0.1–0.4 mg/kg)	Clinical improvement seen in 2 months (average). Dose and frequency increased in response to clinical efficacy.
Statins (HMG-CoA reductase inhibitors) • Indicated for hypercholesterolemia. • Anti-atherosclerotic effects. • Beneficial effects in hepatic lipidosis.	Atorvastatin	10mg/kg PO q24h	Did not result in significant changes to plasma lipid profiles.
Cardiac glycoside • Indicated for myocardial dysfunction, chronic mitral insufficiency, and chronic volume overloads. • Contra-indicated for hypertrophic cardiomyopathy, ventricular tachycardia, and sinus or atrioventricular node disease.	Digoxin	0.02–0.05 mg/kg PO q24h	Adverse effects relate to myocardial toxicity, therefore patients should be monitored for clinical improvement and via ECG for prolonged PR time. Serum concentrations should be measured after 1 week of therapy or 1 week after the dose is changed. The dose can be increased if the serum concentration is less than 0.8 ng/L 8–10 hours after dosing.

Continued

CLASS	DRUG	DOSE AND ROUTE	COMMENTS
Positive Ionotrope • Similar in effect to digoxin but with less risk of adverse effects.	Pimobedan	0.15 mg/kg PO q12h	
Beta blocker • Used to treat ventricular and supraventricular tachycardia.	Propranolol	0.2 mg/kg q24h	
Diuretics • Reduce ascites associated with liver disease or congestive heart failure.	Furosemide	0.15–2 mg/kg IM, SC, PO q12–24h	Overdose can lead to dehydration and electrolyte abnormalities. Lorikeets are very sensitive to this drug – use with caution.
	Spironolactone	1–2 mg/kg PO q12h	Used in patients with congestive heart failure who do not adequately respond to furosemide and ACE inhibitors, or who develop hypokalaemia on other diuretics.
Vascular drugs	Pentoxifylline	10–25 mg/kg PO q6–12h	Increases erythrocyte flexibility and reduces inflammation.
	Isoxsuprine	5–10 mg/kg PO q24h	Peripheral vasodilation.

GASTROINTESTINAL DRUGS

CLASS	DRUG	DOSE AND ROUTE	COMMENTS
Anti-emetics • Indicated for the treatment and prevention of acute vomiting.	Maropitant	1–2 mg/kg q12–24h (chickens)	This dose appears to maintain above target plasma concentration similar to dogs (90 ng/ml).
Intestinal motility modifiers • Used for gastrointestinal motility disorders including slow emptying crops, crop stasis, and regurgitation or vomiting.	Metoclopramide	0.5–2 mg/kg PO or IM q6–12h	• Increases force and frequency of gastric contractions, relaxes pyloric sphincter, and promotes peristalsis in the duodenum and jejunum. • Usually only effective if given with IV fluids as a constant rate infusion.
	Cisapride	0.5–1.5 mg/kg PO q8–12h	• Stimulates gastrointestinal motility.
Intestinal protectants • Used when proventriculitis or gastric ulceration is present.	Cimetidine	5 mg/kg IM q12h	• Reduces gastric secretion of HCl and pepsin. • May impair the metabolism of concurrently administered drugs. • Potential side effects: depression, diarrhoea, tachycardia, respiratory failure.
	Ranitidine	1 mg/kg PO q8–12h	
	Sucralfate	25 mg/kg PO q8h	A complex disaccharide that reacts with stomach acid to form a complex that binds to proteins associated with an ulcer and produces a protective layer that protects the ulcerated mucosa from gastric acids and microbial pathogens.

Continued

CLASS	DRUG	DOSE AND ROUTE	COMMENTS
Laxatives and cathartics • Used to remove foreign bodies from the gastrointestinal tract.	Mineral oil (paraffin oil)	5 ml/kg PO	Acts unchanged as an emollient laxative. Care must be taken to prevent aspiration – give by crop gavage. It can be mixed with peanut butter in a ratio of 1 part mineral oil to 2 parts peanut butter.
	Methylcellulose	5 ml/kg PO q12h	Absorbs water and swells. Care must be taken not to mix too thickly or give too much, as it can absorb enough water to block the gastrointestinal tract.
	Magnesium sulphate (epsom salts)	0.25–1.0 g/kg PO q24h for 1–2 days	Retain or attract water via osmotic forces. Must be used with caution.
	Lactulose	150–650 mg/kg PO q8–12h	Has a laxative effect through osmotic catharsis. In birds with caecae, lactulose passes to the caecae where saccharolytic microflora ferment lactulose to produce acetic, lactic, and other organic acids, which lower the pH of the caecal content.
	Vegetable oils (e.g., castor oil, raw linseed oil, olive oil)	15 ml/kg	Act as irritant purgatives. Form sodium and potassium salts of released fatty acids after hydrolysis by pancreatic lipase, making them irritant soaps.

OPHTHALMIC DRUGS

CLASS	DRUG	DOSE AND ROUTE	COMMENTS
Aminosteroid non-depolarising neuromuscular blocker (muscle relaxant) • Used for mydriasis.	Rocuronium bromide	20 μL/eye topically, spaced 15 minutes apart before repeated dosing (2–3 times)	Bilateral topical administrations of two and three repeated rocuronium bromide doses in *Amazona aestiva* and *Amazona amazonica* parrots produced adequate mydriasis, with only transient eyelid paresis. Effect lasted for 24 hours.

ANTI-CONVULSANT THERAPY

CLASS	DRUG	DOSE AND ROUTE	COMMENTS
Barbiturate	Phenobarbital	1–7 mg/kg PO q12h	• Mild sedative effect used for long-term seizure control. • May cause deep sedation and ataxia. • Dose is best adjusted by response to treatment.
Others	Potassium bromide	25–80 mg/kg PO q24h	• Used for long-term seizure control alone or in conjunction with phenobarbital. • May take up to 90 days to establish steady state.
	Levetiracetam	50 mg/kg PO q8h or 100 mg/kg q12h	• Well tolerated. • Dose and frequencies may need escalation based on differences in species and individuals, and drug levels should be monitored.
	Zonisamide	10 mg/kg PO q12h	Used in conjunction with other anti-convulsant drugs such as levetiracetam.

ANAESTHETICS AND SEDATIVES

CLASS	DRUG	DOSE AND ROUTE	COMMENTS
Sedatives • Used for sedation or as a premedication for anaesthesia. • Often used in conjunction with an opioid, e.g., butorphanol.	Diazepam	0.05–0.15 mg/kg IV or 0.2–0.5 mg/kg IM	• Can be used as a premedication prior to anaesthesia if given 10–20 minutes before induction. • Onset of action 10–20 minutes when given IM. • Onset of action when given intranasally is 3 minutes. • At sedative doses there is minimal effects on blood pressure, heart rate, and temperature.
	Midazolam	0.5–1.5 mg/kg IV or 0.8-3 mg/kg IM or 2-5 mg/kg intranasally	
	Midazolam–butorphanol	4 mg/kg M + 2 mg/kg B (pigeons)	The combination in this study resulted in a mild–moderate level of sedation in pigeons and provides an additional option for immobilisation.
Local anaesthesia	Lignocaine	2–8 mg/kg SC	• The shorter acting local anaesthetics (e.g., lignocaine) have a faster onset, shorter duration of effect, lower potency, and higher therapeutic index than the long-acting local anaesthetics (e.g., bupivacaine). Lignocaine causes CNS effects (i.e., seizures) before cardiac effects are seen. However, bupivacaine induces adverse cardiac effects at lower doses than its CNS effect. • Note that topical application can reach systemic concentrations in birds greater than those achieved in mammals.
	Bupivacaine	1–2 mg/kg SC	
	EMLA (eutectic mixture of local anaesthetics) 5% cream • Lignocaine 2.5% • Prilocaine 2.5%	2–4 mg/kg topically	

Continued

CLASS	DRUG	DOSE AND ROUTE	COMMENTS
Injectable anaesthetics • See Chapter 31. • Injectable anaesthesia is a not recommended in companion birds; regimes and doses are included here to assist veterinarians without access to inhalation anaesthesia.	Ketamine	20–50 mg/kg IM, IV	• Ketamine alone is not recommended for anaesthesia. • Inadequate analgesia and muscle relaxation. • Associated with spontaneous movements and muscular rigidity. • Not associated with pulmonary depression or cardiovascular depression. • Violent recoveries. • Lack of coordination. • Excitement. • Head shaking. • Wing flapping.
	Xylazine	1 mg/kg IM, IV	• Does not usually produce adequate surgical immobilisation. • May lower the arrhythmogenic threshold to endogenous catecholamines. • Respiratory depression. • Hypotension. • Bradycardia. • Excitement and convulsions. • Prolonged recovery. • Muscle tremors, salivation, and movement in response to noise. • Reversible with yohimbine 0.11–0.27 mg/kg IM.
	Xylazine–ketamine	K 4.4 mg/kg + X 2.2 mg/kg) IV or K 20–30 mg/kg + X 2.5–4.0 mg/kg IM	• Used for restraint and anaesthesia. However, waterfowl (especially ducks and Canada geese), owls, and acciptres, do not respond well to this combination. • Combination is synergistic, giving a smooth induction and recovery with improved muscle relaxation and enhanced analgesia.
	Ketamine–diazepam	K 10–40 mg/kg + D 0.2–2 mg/kg IV, IM	• Can be used as an intravenous induction agent in large ratites, raptors, and waterfowl, but may not be a reliable maintenance agent. • Can be used intramuscularly in parrots and pigeons, with a slow induction and recovery time.
	Ketamine–medetomidine	K 1.5–2.0 mg/kg + M 60–85 µg/kg IV	• Useful for inducing waterfowl. • Can be reversed with atipamezole.
	Ketamine–midazolam	K 10–40 mg/kg + M 0.2–2 mg/kg IM	
	Propofol	1–5 mg/kg IV	• Can be used as an intravenous induction agent following sedation with medetomidine–ketamine. • Apnoea is common following IV induction. Intubation and IPPV are recommended.
	Alfaxalone	5–10 mg/kg IV 10–25 mg/kg IM 12.6 mg/kg IM (black-faced lovebirds)	• Brief apnoea on induction if not given slowly. • Dose to effect. • Short duration. • Resulted in muscle tremors and hyperexcitation during induction when used as a sole agent, which was attenuated by premedication with midazolam. • Greater recovery time with higher dosage. • Dosage for black-faced lovebirds provided nearly 1 hour of stable, non-analgesic sedation appropriate for non-invasive procedures like radiography.

Continued

CLASS	DRUG	DOSE AND ROUTE	COMMENTS
	Alfaxalone + ketamine	15 mg/kg A + 5 mg/kg K IM	• Induced loss of righting reflex in 78% (5/9) chickens.
	Alfaxalone + midazolam	10 mg–15 mg/kg A + 1 mg/kg M 15 mg/kg A + 2.5 mg/kg M IM	• 15 mg/kg A + 1 mg/kg M IM induced loss of righting reflex in 56% (5/9) of chickens. • 15 mg/kg A + 2.5 mg/kg M induces good sedation for non-painful procedures. • IM alfaxalone–midazolam resulting in significantly faster onset of sedative effects, significantly longer duration of recumbency, and higher peak sedation compared with IM butorphanol–alfaxalone combination.

ANALGESICS

CLASS	DRUG	DOSE AND ROUTE	COMMENTS
Opioids Some bird species appear to have more kappa opioid receptors in the forebrain than mu opioid receptors, explaining why some birds do not respond to mu agonists, such as morphine, buprenorphine, and fentanyl, in the same manner as mammals. Therefore kappa opioids, such as butorphanol, may be the more efficacious analgesics in birds.	Butorphanol	1-4 mg/kg IM, PO q6–8h	• Higher doses may be hyperalgesic. • Side effects are uncommon, but may include respiratory depression, nausea and vomiting, bradycardia, and constipation.
	Morphine	1 mg/kg IM q12h	
	Tapentadol	30 mg/kg PO (variable plasma concentration and short half-life in Hispaniolan amazon parrots)	
	Tramadol	10–30 mg/kg IM, PO q6–12h	
		2–5 mg/kg IV	
Non-steroidal anti-inflammatory drugs (NSAIDs) • NSAIDs act on peripheral tissues, and therefore are indicated when tissue damage and inflammation are the source of the pain. • They may be most effective when acting synergistically with opioids. • All have the potential to cause renal damage and gastrointestinal ulceration.	Carprofen	2–4 mg/kg PO q12h	
	Flunixin	2–4 mg/kg IM, PO q12h	Renal side effects and gastrointestinal ulceration are common; use with caution in all species.
	Ketoprofen	2 mg/kg IM q8–12h	Not commonly used.
	Meloxicam	1.0–1.5 mg/kg IM, PO q12–24h <0.2 mg/kg PO q72h (brown pelicans)	• Available as a palatable oral suspension or an injectable form. • Oral suspension can be diluted (1:5) with sterile water or methylcellulose to give a stable oral solution for small birds.
Others	Gabapentin	3–11 mg/kg PO q12–24h	Used to treat neuralgic pain.
	Cannabidiol	60–120 mg/kg PO single dose in Amazon parrots	• The serum half-life could not be determined in the 60 mg/kg treatment, but was 1.28 hours at 120 mg/kg. Adverse effects were not observed in any bird. The highly variable results and short half-life of the drug in Hispaniolan Amazon parrots, even at high doses, suggests that this drug formulation was inconsistent in achieving targeted concentrations as reported in other animal species.

ANTIDOTES AND OTHER DRUGS USED IN POISON/TOXIN CASES

CLASS	DRUG	DOSE AND ROUTE	COMMENTS
Antidotes	Atropine	0.01–0.1 mg/kg IM, repeat every 3–4 hours as required	• Organophosphate and carbamate toxicosis. • Does not cause pupillary dilation.
	Calcium EDTA (edentate calcium disodium)	10–50 mg/kg IM q12h 3–5 days	• Treatment of choice for lead and zinc toxicosis. • Often recommended as an oral treatment, but efficacy is uncertain. • Ongoing therapy may be required for chronic lead toxicosis, e.g., twice weekly for 6–8 weeks after initial stabilisation.
	Deferoxamine	20–100 mg/kg IM, PO q24h 3–5 months	• Preferred chelator for iron storage disease. • May take 3 months to achieve effect. • May cause reddish discolouration of urine.
	Dimercaprol (BAL)	25–35 mg/kg q12h PO 3–5 weeks	• Heavy metals other than lead or zinc. • Rarely used.
	Dimercaptosuccinic acid (DMSA)	25–35 mg/kg PO q12–24h for 3 weeks	• Oral chelation of lead. • Can be used with CaEDTA.
	D-Penicillamine	30–50 mg/kg PO q12h 1–3 weeks	• Preferred chelator for copper toxicosis but can be used for other heavy metal toxicosis. • Can cause nausea and vomiting.
	Pralidoxamine	10–100 mg/kg IM, repeat after 6 hours	• Organophosphate toxicosis; contraindicated for carbamate toxicosis. • Must be used with 24–36 hours of exposure.
	Vitamin K	0.2–2.2 mg/kg IM q6-8h until stable, then q24h orally for 2–4 weeks, depending on toxin	• Warfarin toxicosis.
Adsorbents	Activated charcoal	2,000–8,000 mg/kg PO	• May be mixed with hemicellulose to form a bulk laxative to remove toxins from the gastrointestinal tract.
	Phospholipid stabilised soybean oil (intralipid emulsion 20%)	0.25 ml/kg/min (all species) Loading dose of 1.5 ml/kg over 20 minutes followed by 23 ml/kg/h IV for 60 minutes (macaws) Loading dose of 1.5 ml/kg over 5 minutes followed by 15 ml/kg/h IV for 60 minutes (bald eagles)	• For toxicosis with oleander and avocado. • For toxicosis of bromethalin. • For toxicosis of pentobarbital relay in bald eagles from eating euthanised animal cadavers.
Antivenom	Crotalid polyvalent F(ab)2 antivenom	5 ml of antivenon (VenomVet, MRTvenom LLC, Canoga Park California, USA) diluted in 50 ml of 0.9% sodium chloride crystalloid solution (Baxter Healthcare Corporation, Deerfield, Illinois, USA) over 60 minutes.	

PSYCHOTROPIC DRUGS

CLASS	DRUG	DOSE AND ROUTE	COMMENTS
Benzodiazepines • Inhibit dopamine and potentiate GABA. • Muscle relaxants. • Anxiolytic. • Can interfere with learning.	Diazepam	0.5 mg/kg PO q8–12h or 0.25–0.5 mg/kg IM, IV	
	Lorazepam	0.1 mg/kg PO q12h	
Butyrophenones • Inhibit dopamine. • Used for self-mutilation and feather picking.	Haloperidol	0.1–0.4 mg/kg PO q12–24h or 1–2 mg/kg IM every 2–3 weeks	• May cause anorexia or depression.
Antihistamines • Inhibit histamine receptors, producing sedation.	Diphenhydramine	2–4 mg/kg PO q12h	• May work best if feather picking is allergic in origin.
	Hydroxyzine	2 mg/kg PO q8h	
Progestins • Potentiate GABA. • Have a calming effect and anti-inflammatory effect. • Multiple side effects.	Medroxyprogesterone acetate	5–25 mg/kg IM every 4–6 weeks	• Often used for feather-picking behaviour. • Side effects include PU/PD, polyphagia, diabetes mellitus, hepatopathy, weight gain, and sudden death.
	Megestrol acetate	2.5 mg/kg PO q24h 7 days, then weekly as required	• Seldom used because of severe side effects, similar to medroxyprogesterone.
Tricyclic antidepressants • Potentiate serotonin, given a sedative and anticholinergic activity. • Alleviate anxiety and depression. • Use with care because of side effects including constipation and arrhythmias.	Amitriptyline	1–5 mg/Kg PO q12–24hr	• Use for at least 30 days to assess effect.

	Clomipramine	0.5-2.0 mg/kg PO q12–24h	• Side effects may include:
			  • Regurgitation
			  • Drowsiness
			  • Death (possibly with pre-existing arrhythmias)
			• Adjust dose after 2–3 weeks.

	Doxepin	0.5–1.0 mg/kg PO q12h	• Adjust dose after 14 days.
Serotonin specific reuptake inhibitors • Have serotonergic effects. • Some effect in birds.	Fluoxetine	0.4–3.0 mg/kg PO q12–24h	• May cause sedation.
Narcotic antagonists and agonist/antagonists • Act at the opiate centres in the brain, blocking endorphin response to self-injurious behaviour. • May be useful in some birds.	Naltrexone	1.5 mg/kg PO q8–12h for 1–18 months	• Contra-indicated if liver disease is present. • May need to increase dose to effect.

NUTRITIONAL SUPPORT

CLASS	DRUG	DOSE AND ROUTE	COMMENTS
Calcium • Used to treat hypocalcaemic diseases, including hypocalcaemic tetany, egg binding, and metabolic/nutritional bone disease.	Calcium borogluconate	50–100 mg/kg IM or slowly IV	
	Calcium carbonate	3–7 mg/kg feed	Calcium requirements vary considerably between species and according to reproductive status, e.g., budgerigars have very low calcium requirements compared to chickens; egg-laying hens require more calcium than inactive hens or cocks.
	Calcium glubionate	20–25 mg/kg PO q24h or 750 mg/L water	
	Calcium gluconate	5–10 mg/kg slowly IV or 10–100 mg/kg IM	• For hypocalcaemic tetany or egg binding. • Dilute with saline before injecting.
Iodine	Lugol's iodine	0.2 ml/L water	Used to treat thyroid hyperplasia (goitre) in budgerigars.
Iron	Iron dextran	10 mg/kg IM repeated weekly	• Used for treating anaemia, although not used as frequently as it once was. • Use with caution in species prone to iron storage disease, e.g., mynahs, toucans, and lorikeets.

Vitamins	Vitamin A	There is considerable variation in recommended doses for vitamin A. Species susceptible to vitamin A toxicosis (e.g., cockatiels and budgerigars) should be dosed at 2,000–5,000 IU/kg IM weekly; those less susceptible may benefit from doses of 20,000–30,000 IU/kg IM weekly.	• Dietary sources:
			  • Fruit (caretenoids – provitamin A)
			  • Spinach
			  • Fish
			  • Rodents
			• Seed and insects are low in Vitamin A.
			• Beware of oversupplementing vitamin A.

	Vitamin B1(thiamine)	1-2 mg/kg IM, PO q24h	• Dietary sources:
			  • Grains
			  • Leafy green plant material

	Vitamin B7 (biotin)	0.05 mg/kg PO q24h 1–2 months	
	Vitamin B12	0.25–0.5 mg/kg IM weekly	May cause orange/pink urine and urates.
	Vitamin D3	3,000–6,000 IU/kg IM weekly	• Requires 11–45 minutes of unfiltered sunlight (or UVB at 290–320 nm) per day to activate vitamin D3 synthesis in the skin. • Beware of oversupplementation.
	Vitamin E	0.06 mg/kg IM weekly	• Dietary sources include green plants. • Sunflower oil, hazelnuts, almonds have insufficient levels of vitamin E, and the effect of the vitamin E is negated by polyunsaturated fatty acids. • Low in whole prey (rodents and fish).

	Vitamin K1	0.025–2.5 mg/kg IM q12h
		or
		0.2–2.2 mg/kg IM q6–8h until stable, then q24h orally for 2–4 weeks, depending on toxin
			• Dietary sources:
			  • Apple peel
			  • Spinach leaf
			  • Termites
			• Large fig parrots (*Psittaculirostris* spp.) require vitamin K supplementation in their diet.

CHEMOTHERAPY DRUGS

CLASS	DRUG	DOSE AND ROUTE	COMMENTS
Vinca alkaloids/antimitotics • Arrests cell division in metaphase.	Vincristine	0.75 mg/m^2 IV q7days × 3 or 0.1 mg/kg IV q7–14days or 0.5 mg/m^2 IV, then 0.75 mg/ m^2 q7days × 3	• Given IV – perivascular irritation can be severe. • Excreted in bile.

Anti-tumour antibiotics
• Binds to DNA and inhibits synthesis of nucleic acids and cell division.
	Doxorubicin	50–60 mg/m^2 IV
		or
		2 mg/kg IV
		or
		30 mg/m^2 IV q2days
			• Given IV; very irritant if extravasated.
			• Adverse effects:
			  • Bone marrow suppression
			  • GI tract effects
			  • Cardiotoxic
			  • Nephrotoxic
			  • Hypersensitivity reactions
			  • Excreted in bile

	Bleomycin with electochemotherapy (ECT)	1 mg/kg into the tumour once	Used to treat recurrence of a preen gland SCC after apparently successful ECT with cisplatin.
Alkylating agents • Combines with DNA, damaging it.	Cyclophosphamide	200 mg/m^2 IO q7days or 300 mg/m^2 PO once	• Can be given orally or IV. • Myelosuppressive. • Inactivated in liver and excreted in urine.
	Chlorambucil	1 mg/kg PO q3days	
	Lomustine		
Platinum drugs • Inhibits DNA synthesis and may have some alkylating effects.	Cisplatin	1 mg of cisplatin/cm^3 of tissue when given intralesionally (monthly for 3 treatments) with electrochemotherapy (ECT) or 15–27 mg/kg IV	• Given IV or intralesionally. • Can be nephrotoxic, myelosuppressive, and neurotoxic. • Carboplatin is not as toxic as cisplatin. • Renal excretion.
	Carboplatin	125 mg/m^2 IV slow bolus q2–3w. Dilute to a comfortable volume with 5% dextrose or 15 mg/kg IO or 5 mg/kg intralesional 1% carboplatin in poloxamer 407 (30%) gel dosed at 1.5 mg/cm^3 of carboplatin in poloxamer 407 (for a total volume of 0.2 mL per treatment) was administered intralesionally and perilesionally every 3 weeks for 3 applications in grey parrots after CO_2 surgical laser of the lesion.	
Other • Leukaemic cells and some other suspected tumour cells are unable to synthesise the amino acid asparagine (normal cells are able to make their own asparagine). Asparaginase catalyses the conversion of L-asparagine to aspartic acid and ammonia, depriving the leukaemic cell of circulating asparagine, which leads to cell death.	L-Asparaginase	400 IU/kg IM q7days or 1650 IU/kg SC once	• Used in the treatment of leukaemia and lymphosarcoma. • Can be given IM or SC.

APPENDIX 2: REFERENCE INTERVALS FOR COMMONLY KEPT COMPANION BIRDS (SI UNITS)

HAEMATOLOGY

SPECIES	PCV (%)	WCC ($\times 10^9$/L)	H (%)	L (%)	E (%)	M (%)	B (%)
Amazon	41–53	5–17	31–71	20–67	0	0–2	0–2
Budgerigar	44–58	3–10	40–68	22–60	0	0–2	0–2
Caique	47–55	8–15	39–72	20–61	0	0–2	0–1
Canary	45–55	3–10	21–60	20–65	0–1	0–1	0–1
Cockatiel	43–57	5–11	46–72	26–60	0–2	0–1	0–1
Cockatoo	40–55	5–13	45–72	20–50	0–2	0–2	0–1
Conure, *Aratinga* spp.	42–55	6–11	44–72	20–49	0	0–1	0–2
Conure, *Pyrrhura* spp.	42–55	5–10	46–71	22–53	0–1	0–2	0
Eclectus, red sided	45–55	9–15	46–70	23–57	0–1	0–1	0–1
Grey parrot	45–53	6–13	45–73	19–50	0–1	0–2	0–1
Lory	47–55	8–13	39–60	22–70	0–1	0–2	0–1
Lovebird	44–55	7–16	40–56	20–53	0–2	0–1	0–1
Macaw	42–56	10–20	50–75	23–53	0	0–1	0–1
Mynah	38–50	8–12	45–64	21–55	0–4	0–1	0–1
Pionus	45–54	5–13	55–74	19–70	0–1	0–1	0–1
Quaker (monk parakeet)	45–58	8–17	47–70	20–63	0–1	0–4	0–3
Ring-neck parrot (Indian)	45–54	8–14	40–70	30–55	0–1	0–1	0–1
Senegal	45–60	6–14	44–73	22–70	0–2	0–1	0–1

H = Heterophils; L = Lymphocytes; E= Eosinophils; M = Monocytes; B = Basophils.

BIOCHEMISTRIES

SPECIES	AST (IU/L)	CK (IU/L)	BILE ACIDS (MMOL/L)	GLDH (IU/L)	CHOLESTEROL (MMOL/L)	TRIGLYCERIDES (MMOL/L)	URIC ACID (MMOL/L)	GLUCOSE (MMOL/L)	TOTAL PROTEIN (G/L)	CALCIUM (MMOL/L)
Amazon	150–350	120–425	30–150	<8.0	4.3–10.9	1.1–2.1	130–620	14–21	26–45	2–3.4
Budgerigar	150–375	120–370	30–120	0.04–5.25	—	—	285–765	12–25	9–34	2–2.8
Caique	120–365	120–390	10–110	—	—	—	150–630	9.4–21	25–35	2.1–2.8
Canary	130–350	55–350	20–90	—	—	—	245–750	8.9–20	30–45	1.5–3.4
Cockatiel	120–400	160–420	40–110	0.0–8.0	4.7–6.7	1.1–1.7	200–650	12.6–24.4	20–50	2–2.7
Cockatoo	140–360	150–420	30–110	0.7–4.2	3.5–7.3	0.5–2.3	225–655	11.4–23.2	25–49	2–2.9
Conure, *Aratinga* spp.	140–350	150–370	10–90	—	3.4–10.1	0.6–3.4	130–690	9.3–21	24–45	2–2.8
Conure, *Pyrrhura* spp.	140–270	140–280	10–90	—	—	—	190–740	11.9–22	35–50	2–3.1
Eclectus, red sided	140–340	130–410	30–110	—	3.4–9.0	0.8–4.6	120–650	12–22	30–45	2–3
Grey parrot	110–340	140–410	10–100	<8.0	5.8–8.2	0.4–2.0	120–650	14.2–20	25–45	2–3.5
Lory	140–370	180–400	20–100	—	2.5–7.6	0.7–1.6	120–700	10.5–21.5	19–40	2–2.9
Lovebird	130–360	160–400	10–90	—	2.5–8.7	0.5–2.3	195–635	12.2–22	18–35	2.1–2.9
Macaw	65–170	90–360	7–100	0–2.0	3.1–6.7	0.4–2.5	110–700	11.7–20	24–44	2.1–3
Mynah	200–350	250–420	30–100	—	—	—	135–710	12.5–20.5	30–45	2–2.7
Pionus	130–360	120–410	15–90	—	3.4–8.6	0.7–2.6	120–685	12–20.5	20–43	2–2.6
Quaker (monk parakeet)	130–380	190–400	20–90	—	4.9–8.1	0.8–2.1	130–710	11.5–22	25–35	2.1–2.7
Ring-neck parrot (Indian)	150–390	—	—	1–2	8.0–9.5	—	195–700	12.2–20		2.1–2.6
Senegal	120–330	150–400	20–100	—	3.4–8.8	0.5–1.7	135–710	14.2–20	30–42	2.1–2.7

Adapted from Fudge A., Laboratory Medicine: Avian and Exotic Pets, W.B. Saunders, Philadelphia, 2000.

APPENDIX 3: BIOLOGICAL VALUES FOR SOME COMMON COMPANION BIRD SPECIES

SPECIES	WEANING AGE (DAYS)	SEXUAL MATURITY	ADULT BODY WEIGHT (G)	LIFE EXPECTANCY (YEARS)
Alexandrine parrot	85–100	2–3 years	200–300	25–35
Amazon, blue crowned	90–120	4–6 years	620–1000	50–70
Amazon, blue fronted	90–120	4–6 years	360–490	50–70
Amazon, double yellow headed	90–120	4–6 years	460–700	50–70
Amazon, yellow naped	90–120	4–6 years	470–800	50–70
Budgerigar	30–40	6–9 months	35–45	7–12
Caique	95–105	2–3 years	120–150	30–40
Canary	21	10 months	15–25	6–12
Cockatiel	40–50	8–12 months	80–100	10–15
Cockatoo, bare eyed (little corella)	110–120	3–4 years	450–600	30–40
Cockatoo, greater sulphur crested	120–150	3–4 years	850–950	50–70
Cockatoo, lesser sulphur crested	80–90	2–3 years	310–380	40–60
Cockatoo, Major Mitchell	80–90	2 years	350–450	40–60
Cockatoo, medium sulphur crested	100–120	3 years	275–350	40–60
Cockatoo, rose breasted (galah)	85–95	2–3 years	250–320	20–40
Cockatoo, citron crested	85–95	2–3 years	300–400	50–60
Cockatoo, Moluccan	100–120	3–5 years	700–1000	50–60
Cockatoo, umbrella	95–105	5–6 years	550–650	50–60
Conure, *Aratinga* spp.	55–75	2–4 years	110–130	15–25
Conure, *Pyrrhura* spp.	55–75	9–18 months	55–85	15–25
Eclectus, red sided	80–90	3 years	380–410	25–30
Grey parrot, Congo	100–120	4–6 years	390–500	50–60
Grey parrot, Timneh	95–105	3–5 years	280–360	25–35
Jardine parrot	95–105	3–4 years	200–250	20–25
Lorikeet, rainbow	60–70	1–2 years	130–150	15–20
Lorikeet				
Lory, chattering	85–100	2–3 years	160–180	30–35
Lory, red	85–100	2–3 years	100–130	25–30
Lovebird	45–55	6–12 months	40–50	7–15

Continued

SPECIES	WEANING AGE (DAYS)	SEXUAL MATURITY	ADULT BODY WEIGHT (G)	LIFE EXPECTANCY (YEARS)
Macaw, blue and gold	90–110	5–7 years	1000–1200	50–80
Macaw, green wing	110–140	3–5 years	1050–1320	60–90
Macaw, Hahn's	70–85	2–3 years	130–150	25–30
Macaw, hyacinth	180–270	5–7 years	1200–1700	60–90
Macaw, military	110–120	2–3 years	860–1100	50–60
Macaw, scarlet	90–110	5–7 years	950–1150	50–80
Meyer's parrot	70–90	2 years	80–110	25–35
Mynah	60	2–3 years	180–260	12
Pionus, blue headed	70	2–3 years	250	20–25
Pionus, white capped	70–100	2 years	130–160	25–30
Quaker (monk parakeet)	50–60	1 year	120–140	20–30
Ring–neck parrot (Indian)	70–90	2 years	110–120	25–30
Senegal	70	2–3 years	125	15–25

INDEX